Atomic Masses of the Elements

Name	Symbol	Atomic Number	Atomic Mass[a]	Name	Symbol	Atomic Number	Atomic Mass[a]
Actinium	Ac	89	(227)	Molybdenum	Mo	42	95.94
Aluminum	Al	13	26.98	Neodymium	Nd	60	144.2
Americium	Am	95	(243)	Neon	Ne	10	20.18
Antimony	Sb	51	121.8	Neptunium	Np	93	(237)
Argon	Ar	18	39.95	Nickel	Ni	28	58.69
Arsenic	As	33	74.92	Niobium	Nb	41	92.91
Astatine	At	85	(210)	Nitrogen	N	7	14.01
Barium	Ba	56	137.3	Nobelium	No	102	(259)
Berkelium	Bk	97	(247)	Osmium	Os	76	190.2
Beryllium	Be	4	9.012	Oxygen	O	8	16.00
Bismuth	Bi	83	209.0	Palladium	Pd	46	106.4
Bohrium	Bh	107	(264)	Phosphorus	P	15	30.97
Boron	B	5	10.81	Platinum	Pt	78	195.1
Bromine	Br	35	79.90	Plutonium	Pu	94	(244)
Cadmium	Cd	48	112.4	Polonium	Po	84	(209)
Calcium	Ca	20	40.08	Potassium	K	19	39.10
Californium	Cf	98	(251)	Praseodymium	Pr	59	140.9
Carbon	C	6	12.01	Promethium	Pm	61	(145)
Cerium	Ce	58	140.1	Protactinium	Pa	91	231.0
Cesium	Cs	55	132.9	Radium	Ra	88	(226)
Chlorine	Cl	17	35.45	Radon	Rn	86	(222)
Chromium	Cr	24	52.00	Rhenium	Re	75	186.2
Cobalt	Co	27	58.93	Rhodium	Rh	45	102.9
Copper	Cu	29	63.55	Roentgenium	Rg	111	(272)
Curium	Cm	96	(247)	Rubidium	Rb	37	85.47
Darmstadtium	Ds	110	(271)	Ruthenium	Ru	44	101.1
Dubnium	Db	105	(262)	Rutherfordium	Rf	104	(261)
Dysprosium	Dy	66	162.5	Samarium	Sm	62	150.4
Einsteinium	Es	99	(252)	Scandium	Sc	21	44.96
Erbium	Er	68	167.3	Seaborgium	Sg	106	(266)
Europium	Eu	63	152.0	Selenium	Se	34	78.96
Fermium	Fm	100	(257)	Silicon	Si	14	28.09
Fluorine	F	9	19.00	Silver	Ag	47	107.9
Francium	Fr	87	(223)	Sodium	Na	11	22.99
Gadolinium	Gd	64	157.3	Strontium	Sr	38	87.62
Gallium	Ga	31	69.72	Sulfur	S	16	32.07
Germanium	Ge	32	72.64	Tantalum	Ta	73	180.9
Gold	Au	79	197.0	Technetium	Tc	43	(98)
Hafnium	Hf	72	178.5	Tellurium	Te	52	127.6
Hassium	Hs	108	(269)	Terbium	Tb	65	158.9
Helium	He	2	4.003	Thallium	Tl	81	204.4
Holmium	Ho	67	164.9	Thorium	Th	90	232.0
Hydrogen	H	1	1.008	Thulium	Tm	69	168.9
Indium	In	49	114.8	Tin	Sn	50	118.7
Iodine	I	53	126.9	Titanium	Ti	22	47.87
Iridium	Ir	77	192.2	Tungsten	W	74	183.8
Iron	Fe	26	55.85	Uranium	U	92	238.0
Krypton	Kr	36	83.80	Vanadium	V	23	50.94
Lanthanum	La	57	138.9	Xenon	Xe	54	131.3
Lawrencium	Lr	103	(260)	Ytterbium	Yb	70	173.0
Lead	Pb	82	207.2	Yttrium	Y	39	88.91
Lithium	Li	3	6.941	Zinc	Zn	30	65.41
Lutetium	Lu	71	175.0	Zirconium	Zr	40	91.22
Magnesium	Mg	12	24.31	—	—	112	(285)
Manganese	Mn	25	54.94	—	—	113	(284)
Meitnerium	Mt	109	(268)	—	—	114	(289)
Mendelevium	Md	101	(258)	—	—	115	(288)
Mercury	Hg	80	200.6				

[a] Values in parentheses are the mass number of the most stable isotope.

Log on.

Tune in.

Succeed.

Your steps to success.

STEP 1: Register

All you need to get started is a valid email address and the access code below. To register, simply:

1. Go to www.aw-bc.com/chemplace
2. Click the appropriate book cover.
   Cover must match the textbook edition being used for your class.
3. Click **"Register"** under **"First-Time User?"**
4. Leave **"No, I Am a New User"** selected.
5. Using a coin, scratch off the silver coating below to reveal your access code.
   Do not use a knife or other sharp object, which can damage the code.
6. Enter your access code in lowercase or uppercase, without the dashes.
7. Follow the on-screen instructions to complete registration.
   During registration, you will establish a personal login name and password to use for logging into the website. You will also be sent a registration confirmation email that contains your login name and password.

Your Access Code is:

*

Note: If there is no silver foil covering the access code, it may already have been redeemed, and therefore may no longer be valid. In that case, you can purchase access online using a major credit card. To do so, go to www.aw-bc.com/chemplace, click the cover of your textbook, click **"Buy Now"**, and follow the on-screen instructions.

STEP 2: Log in

1. Go to www.aw-bc.com/chemplace and click the appropriate book cover.
2. Under **"Established User?"** enter the login name and password that you created during registration. *If unsure of this information, refer to your registration confirmation email.*
3. Click **"Log In"**.

STEP 3: (Optional) Join a class

Instructors have the option of creating an online class for you to use with this website. If your instructor decides to do this, you'll need to complete the following steps using the Class ID your instructor provides you. By "joining a class," you enable your instructor to view the scored results of your work on the website in his or her online gradebook.

To join a class:

1. Log into the website. For instructions, see "STEP 2: Log in."
2. Click **"Join a Class"** near the top right.
3. Enter your instructor's **"Class ID"** and then click **"Next"**.
4. At the Confirm Class page you will see your instructor's name and class information. If this information is correct, click **"Next"**.
5. Click **"Enter Class Now"** from the Class Confirmation page.
- *To confirm your enrollment in the class, check for your instructor and class name at the top right of the page. You will be sent a class enrollment confirmation email.*
- *As you complete activities on the website from now through the class end date, your results will post to your instructor's gradebook, in addition to appearing in your personal view of the Results Reporter.*

To log into the class later, follow the instructions under "STEP 2: Log in."

Got technical questions?

Visit http://247.aw.com. Email technical support is available 24/7.

Register and log in

Join a Class

Important: Please read the Subscription and End-User License agreement, accessible from the book website's login page, before using the Chemistry Place website. By using the website, you indicate that you have read, understood, and accepted the terms of this agreement.

Basic Chemistry

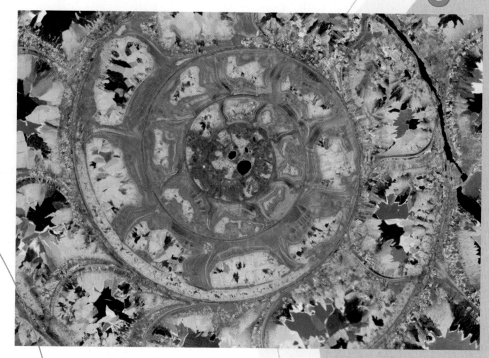

SECOND EDITION

Karen C. Timberlake
Los Angeles Valley College

William Timberlake
Los Angeles Harbor College

PEARSON

Prentice
Hall

Upper Saddle River, NJ 07458

Library of Congress Cataloging-in-Publication Data

Timberlake, Karen.
   Basic chemistry.—2nd ed. / Karen C. Timberlake, William Timberlake.
      p. cm.
   ISBN 0-8053-4469-1
   1. Chemistry—Textbooks.   I. Timberlake, William E.   II. Title
QD31.3.T54 2007
540—dc22   2006052308

Vice President and Editorial Director: Adam Black

Publisher: Jim Smith

Project Editor: Katherine Brayton

Editorial Assistant: Grace Joo

Managing Editor: Corinne Benson

Production Supervisor: Caroline Ayres

Production Service: Progressive Publishing Alternatives; Jean Lake

Compositor: Progressive Information Technologies

Interior Designer: tani hasegawa

Cover Designer: Jeanne Calabrese

Illustrator: Precision Graphics

Photo Researcher: Clare Maxwell

Director, Image Resource Center: Melinda Patelli

Image Rights and Permissions Manager: Zina Arabia

Manufacturing Buyer: Pam Augspurger

Executive Marketing Manager: Scott Dustan

Text printer: QuebecorWorld, Dubuque

Cover printer: Phoenix Color

Cover Photo Credit: Ammonite fossil with crystal inclusions by © Dirk Wiersma/Photo Researchers, Inc.

© 2008 Pearson Education, Inc.
Pearson Prentice Hall
Pearson Education, Inc.
Upper Saddle River, NJ 07458

Printed in the United States of America
2   3   4   5   6   7   8   9   10

ISBN 10: 0-8053-4469-1 (Student Edition)

ISBN 13: 978-08053-4469-1 (Student Edition)

ISBN 10: 0-321-46905-4 (P-copy)

ISBN 13: 978-03214-6905-2 (P-copy)

ISBN 10: 0-13-134685-7 (NASTA Edition)

ISBN 13: 978-01313-4685-7 (NASTA Edition)

Pearson Education LTD., *London*
Pearson Education Australia PTY, Limited, *Sydney*
Pearson Education Singapore, Pte. Ltd.
Pearson Education North Asia Ltd., *Hong Kong*
Pearson Education Canada, Ltd., *Toronto*
Pearson Educación de Mexico, S.A. de C.V.
Pearson Education—Japan, *Tokyo*
Pearson Education Malaysia, Pte. Ltd.

Brief Contents

Contents

4 Atoms and Elements   93

5 Names and Formulas of Compounds   127

6 Chemical Quantities   163

18   Biochemistry   645

Applications and Activities

About the Authors

Karen Timberlake is professor emeritus of chemistry at Los Angeles Valley College, where she taught chemistry for allied health and preparatory chemistry for 36 years. She received her bachelor's degree in chemistry from the University of Washington and her Master's degree in biochemistry from the University of California at Los Angeles.

Professor Timberlake has been writing chemistry textbooks for 30 years. During that time, her name has become associated with the strategic use of pedagogical tools that promote student success in chemistry and the application of chemistry to real-life situations. More than one million students have learned chemistry using texts, laboratory manuals, and study guides written by Karen Timberlake. In addition to *Basic Chemistry,* she is also the author of *General, Organic, and Biological Chemistry, Structures of Life, Second Edition* and *Chemistry: An Introduction to General, Organic, and Biological Chemistry, Ninth Edition* with the accompanying *Study Guide with Solutions for Selected Problems, Laboratory Manual,* and *Essentials Laboratory Manual.*

Professor Timberlake belongs to numerous science and educational organizations including the American Chemical Society (ACS) and the National Science Teachers Association (NSTA). In 1987, she was the Western Regional Winner of Excellence in College Chemistry Teaching Award given by the Chemical Manufacturers Association. In 2004, she received the McGuffey Award in Physical Sciences by the Textbook Author Association, awarded for textbooks whose excellence has been demonstrated over time. In 2006, she also received the Texty "Textbook" Excellence Award from the Textbook Authors Association for the first edition of *Basic Chemistry*. She has participated in education grants for science teaching including the Los Angeles Collaborative for Teaching Excellence (LACTE) and a Title III grant at her college. She often speaks at conferences and educational meetings on the use of student-centered teaching methods in chemistry to promote the learning success of students.

Bill Timberlake is also a professor emeritus of chemistry and has contributed to writing this text. He taught preparatory and organic chemistry at Los Angeles Harbor College for 36 years. He received his bachelor's degree in chemistry from Carnegie Mellon University and his Master's degree in organic chemistry from the University of California at Los Angeles. When the Professors Timberlake are not writing textbooks, they relax by hiking, traveling to Mexico and Europe, trying new restaurants, and playing lots of tennis. The Timberlakes' son, John, and daughter-in-law, Cindy, are involved in preparing materials for English language learners. Their grandson, Daniel, and granddaughter, Emily, don't know what they want to be yet.

To the Student

We hope that this textbook helps you discover exciting new ideas and gives you a rewarding experience as you develop an understanding and appreciation of the role of chemistry in your life. If you would like to share your experience with chemistry, or have questions and comments about this text, we would appreciate hearing from you.

Karen and Bill Timberlake
Email: khemist@aol.com

Preface

Welcome to the second edition of *Basic Chemistry*. Our main objective in writing this text is to prepare students with little or no chemistry experience for future chemistry courses or careers in nursing, dietetics, respiratory therapy, or as laboratory technicians. Thus, this text includes topics that are essential to chemistry, have high value to students' future science classes and careers, have real-life applications, and can be learned in the timeframe of a one-quarter or one-semester class.

Our goal in this text is to provide a learning environment that makes the study of chemistry an engaging and positive experience. It is also our goal to help every student become a critical thinker by understanding scientific concepts that will form a basis for making important decisions about issues concerning health and the environment. Thus, we have utilized materials that

- Help students to develop problem-solving skills that lead to success in chemistry
- Motivate students to learn and enjoy chemistry
- Relate chemistry to careers in science that interest students
- Provide pedagogy that promotes learning

Features of the Text

Students are often challenged by the study of chemistry and have difficulty seeing the relevance of chemistry to their career paths. A common view of chemistry is that it is just a lot of facts to be memorized. To change this perception, we have included many features to give students an appreciation of chemistry in their lives and to help students learn chemistry successfully. These pedagogical features include connections to real life, a visual guide to problem solving, and in-chapter problem sets that immediately reinforce the learning of a small group of new concepts. A successful learning program in each chapter provides many learning tools, which are discussed below.

Professional interviews with scientists Chapter openers begin with interviews with scientists, engineers, and medical personnel who discuss the importance of chemistry in their careers.

Learning Goals directed learning Learning Goals in every section preview the concepts the student is to learn.

Chem Notes connections to real life Real-life applications throughout the text relate chemistry chapters to real-life topics in science and medicine that are interesting and motivating to students and support the role of chemistry in the real world.

Guide to Problem Solving visualizations As part of a comprehensive learning program, an abundant number of Sample Problems and Study Checks in every chapter model successful problem-solving techniques for the student. A unique Guide to Problem Solving strategy illustrates solutions to problems with color blocks that visually guide students through the solution pathway.

Questions and Problems in every section Every section includes a comprehensive set of Questions and Problems to encourage students to apply critical thinking and problem solving to a small set of concepts. Within each problem set, the questions are in order of simple to more complex to enable students to build a knowledge base for problem solving. By solving a set of problems after studying each section, students immediately reinforce newly learned concepts rather than waiting until they get to the end of the chapter. Students are encouraged to be active learners by working problems frequently as they progress through each chapter. Answers for all the Study Checks are located at the end of each chapter.

Matched pairs of problems All the questions and problems in the text are written as pairs of matched problems. Thus, each question or problem is followed with a similar type of question or problem. Answers for the odd-numbered problems are located at the end of each chapter rather than at the end of the text to give immediate feedback to problem-solving efforts by students. The even-numbered problems do not have answers, allowing instructors to use them for homework and/or quiz questions.

End-of-chapter questions Understanding the Concepts encourages students to think about the concepts they have learned. Additional Questions and Problems and Challenge Questions integrate the topics from the entire chapter to promote further study and critical thinking.

Macro-to-micro art illustrations of atomic organization Throughout the text, a robust and vibrant art program visually connects the real-life world of materials familiar to students with their atomic-level structures. Macro-to-micro illustrations show students that everyday things have an atomic level of organization and structure that determines their behavior and functions. Every figure in the text also contains a question that encourages the student to study that figure and relate the visual representation to the content in the text. The plentiful use of three-dimensional structures takes the descriptions of molecules to a visual level, which stimulates the imagination and aids the understanding of structures by the student.

End-of-chapter aids Concept Maps connect and guide students through the topics and concepts in the chapter. Chapter Reviews summarize the important concepts in each section of the chapter. Key Terms remind students of the new vocabulary presented in each chapter.

New for the Second Edition

New features that have been added to every chapter of this second edition include the following:

- More Guides to Problem Solving (GPS) that illustrate problem-solving strategies.
- New ChemNotes including "Early Chemist: the Alchemists" and "Toxicology and Risk-Benefit Assessment."
- New Career Focus interviews, including Geologist and Toxicologist.
- Measurement of parts per million (ppm) and parts per billion (ppb).
- Molecular models that increase visual understanding of a formula.
- Concept Maps that visually connect and summarize chapter topics.
- Understanding the Concepts that correlate visual examples to conceptual learning.
- Challenge Questions that integrate concepts from the entire chapter.
- Combining Ideas interchapter problem sets that provide problems with greater depth, combining topics from several chapters.

Chapter Organization of the Second Edition

In each textbook we write, we consider it essential to relate every chemical concept to real-life issues of health and environment. Because a course of chemistry may be taught in different time frames, it may be difficult to cover all the chapters in this text. However, each chapter is a complete package, which allows some chapters to be skipped or the order of presentation to be changed.

Chapters 1–3 Chemistry in Our Lives, Measurements, and Matter and Energy

Chapter 1 Chemistry in Our Lives introduces the concepts of chemicals and chemistry, discusses the scientific method, and asks students to develop a study plan for learning chemistry. **Chapter 2 Measurements** looks at measurement and the need to understand numerical structures of the metric system in the sciences. An explanation of scientific notation and working with a calculator has been incorporated into the chapter. **Chapter 3 Matter and Energy** classifies matter, describes temperature measurement, and discusses energy and its measurement. *Combining Ideas from Chapters 1–3* follows as an interchapter problem set.

Chapters 4–6 Atoms and Elements, Names and Formulas of Compounds, and Chemical Quantities.

In **Chapter 4 Atoms and Elements**, we look at elements and their atoms. The element name and symbol Roentgenium (Rg) 111 was added along with elements up to 115. The new section *Electron Energy Levels* discusses electron arrangement and was added to prepare students for compound formation. **Chapter 5 Names and Formulas of Compounds** describes how atoms form ionic and covalent compounds. Students learn to write formulas and to name ionic compounds, including those with polyatomic ions and covalent compounds. A new *Career Focus* about a geologist was added. **Chapter 6 Chemical Quantities** introduces the mole and molar masses of compounds, which are used in calculations to determine the mass or number of particles in a quantity as well as the percent composition and empirical and molecular formulas of compounds. *Combining Ideas from Chapters 4–6* follows as an interchapter problem set.

Chapters 7–8 Chemical Reactions and Chemical Quantities in Reactions

Chapter 7 Chemical Reactions looks at the interaction of atoms and molecules in chemical reactions. Chemical equations are balanced and organized into combination, decomposition, replacement, and combustion reactions. A section on *Energy in Chemical Reactions* and endothermic and exothermic reactions completes this chapter. **Chapter 8 Chemical Quantities in Reactions** describes the mole and mass relationships among the reactants and products, and provides calculations of percent yields and limiting reactants. *Combining Ideas from Chapters 7 and 8* follows as an interchapter problem set.

Chapters 9–10 Atomic Structure and Periodic Trends and Molecular Structure: Solids and Liquids

Chapter 9 Atomic Structure and Periodic Trends uses the electromagnetic spectrum to explain atomic spectra and develop the concept of energy levels and sublevels. Electrons in sublevels and orbitals are represented using orbital diagrams and electron configurations. Periodic properties of elements including atomic radii are related to their valence electrons. **Chapter 10 Molecular Structure: Solids and Liquids** introduces electron formulas for multiple bonds and resonance. Electronegativity leads to a discussion of the polarity of bonds and molecules. Electron-dot formulas and VSEPR theory illustrate covalent bonding and the three-dimensional shapes of molecules and ions. The attractive forces between particles and their impact on states of matter and changes of state are described. *Combining Ideas from Chapters 9 and 10* follows as an interchapter problem set.

Chapters 11–14 Gases, Solutions, Chemical Equilibrium, and Acids and Bases

Chapter 11 Gases discusses the properties of a gas and calculates changes in gases using the gas laws and ideal gas law. The amounts of gases required or produced in chemical reactions are calculated. **Chapter 12 Solutions** describes solutions, saturation and solubility, and concentrations. The volumes and molarities of solutions are used in calculations of reactants and products in chemical reactions, as well as dilutions and titrations. **Chapter 13 Chemical Equilibrium** looks at the rates of reactions and the equilibrium condition when forward and reverse rates for a reaction become equal. Equilibrium expressions for reactions are written and equilibrium constants are calculated. Using equilibrium constants, reactions are evaluated to determine if the equilibrium favors the reactants or the products and the concentrations of components are calculated. LeChâtelier's Principle is used to evaluate the impact on concentrations when a stress is placed on the equilibrium system. The equilibrium of dissolving and crystallizing in saturated solutions is evaluated using solubility product constants. **Chapter 14 Acids and Bases** discusses acids and bases and their strengths, conjugate acid–base pairs, the dissociation of weak acids and bases and water, pH and pOH, and buffers. Acid–base titration uses the neutralization reactions between acids and bases to calculate quantities of acid in a sample. *Combining Ideas from Chapters 11–14* follows as an interchapter problem set.

Chapters 15–16 Oxidation–Reduction: Transfer of Electrons and Nuclear Radiation

In **Chapter 15 Oxidation–Reduction: Transfer of Electrons** looks at the characteristics of oxidation and reduction reactions. Oxidation numbers are assigned to the atoms in elements, molecules, and ions to determine the components that lose electrons during oxidation and gain electrons during reduction. Changes in oxidation numbers and the half-reaction method are both utilized to balance oxidation–reduction reactions. The production of electrical energy in voltaic cells and the requirement of electrical energy in electrolytic cells are diagrammed using half-cells. The activity series is used to determine the spontaneous direction of an oxidation–reduction reaction. In **Chapter 16 Nuclear Radiation**, we look at the type of radioactive particles that are emitted from the nuclei of radioactive atoms. Equations are written and balanced for both naturally occurring radioactivity and artificially produced radioactivity. The half-lives of radioisotopes are discussed and the amount of time for a sample to decay is calculated. Radioisotopes important in the field of nuclear medicine are described. *Combining Ideas from Chapters 15 and 16* follows as an interchapter problem set.

Chapters 17–18 Organic Chemistry and Biochemistry

Chapters 17 and 18 describe the chemistry of organic compounds and biochemical compounds. In **Chapter 17 Organic Chemistry**, the physical and chemical properties of organic compounds are related to the structure and functional groups for each family of organic compounds, which forms a basis for understanding the biomolecules of living systems. In **Chapter 18 Biochemistry**, we look at the functional groups found in the structures of carbohydrates, lipids, proteins, and nucleic acids. The shape of proteins is related to the activity and regulation of enzyme activity. Finally, we look at the metabolic pathways of glycolysis and the citric acid cycle to the production of energy in the form of ATP. *Combining Ideas from Chapters 17–18* follows as an interchapter problem set.

Instructional Package

Basic Chemistry, Second Edition is the nucleus of an integrated teaching and learning pack of support material for both students and professors.

For Students

Study Guide for *Basic Chemistry, Second Edition* by Karen Timberlake, is keyed to the learning goals in the text and designed to promote active learning through a variety of exercises with answers as well as mastery exams. The *Study Guide* also contains complete solutions to odd-numbered problems. (ISBN 0-321-49635-3)

The Chemistry Place™ for *Basic Chemistry, Second Edition*
www.aw-bc.com/chemplace
This website, tailored specifically to the second edition of *Basic Chemistry,* offers comprehensive and interactive teaching and learning tools for chemistry. Features include **interactive tutorials, review questions,** chapter **quizzes, career resources,** a **glossary, flashcards,** an **interactive periodic table,** and **case studies** that show how chemical concepts apply to familiar situations. **Interact Math** provides the student with math practice and review specific to each chapter in the text. Over 2000 **PowerPoint® slides**—complete with art, photos, and tables from the text—teach key concepts from every section of the book. Instructors can also view student quiz results through the Gradebook feature. Access to the Chemistry Place is packaged with every new copy of the book.

The Chemistry Tutor Center
www.aw-bc.com/tutorcenter
The Tutor Center provides one-to-one tutoring by phone, fax, email, and/or the Internet. Qualified instructors answer questions to help students understand concepts and solve problems in the textbook.

Media Grid, Walkthrough, and Media Icons
Located in the front of the book, the Media Grid correlates all of the media available on the Chemistry Place website with the content in *Basic Chemistry, Second Edition.* The Walkthrough in the front of the book helps students and instructors find the book's features easily. Media Icons in the margins throughout the text direct students to tutorials and case studies on the Chemistry Place website for *Basic Chemistry, Second Edition.*

For Instructors

Instructor Solutions Manual, by Karen Timberlake, includes answers and solutions for all questions and problems in the text. (ISBN 0-321-49634-5)

Media Manager for *Basic Chemistry, Second Edition*
This CD-ROM includes all the art and tables from the book in high-resolution (150 dpi) format for use in classroom projection or when creating study materials and tests. In addition, the instructor can access over 2000 PowerPoint lecture outlines, written by Karen Timberlake, or create their own easily with the art provided in PPT format. Also available on the Media Manager are downloadable files of the *Instructor Solutions Manual* and *Test Bank,* as well as a set of "clicker questions" designed for use with classroom-response systems. (ISBN 0-321-49688-4)

Transparency Acetates A set of 125 color transparencies is available. (ISBN 0-321-46154-1)

Printed Test Bank, by William Timberlake, includes more than 1600 multiple-choice, matching, and True/False questions. (ISBN 0-321-46153-3)

Computerized Test Bank includes more than 1600 multiple-choice, matching, and True/False questions on a multiformat CD-ROM. (ISBN 0-321-46917-8)

Blackboard, CourseCompass, and WebCT provide powerful course management capability.

Visit the Prentice Hall catalog page for Timberlake's *Basic Chemistry, Second Edition,* at **www.prenhall.com** to download available instructor supplements.

Acknowledgments

The preparation of a new text is a continuous effort by many people over a two-year period. As in our work on other text books, we are thankful for the support, encouragement, and dedication of many people who put in hours of tireless effort to produce a high quality book that provides an outstanding learning package. Once again the editorial team at Benjamin Cummings has done an exceptional job. We appreciate the work of our publisher, Jim Smith, who supported our vision of this new edition with a new topic sequence, new conceptual problems in Understanding the Concepts, and more Guided Problem Solving. Our project editors, Katherine Brayton and Grace Joo, were like angels who continually encouraged us at each step during the development of this new text, while skillfully coordinating reviews, art, website materials, and all the things it takes to make a book come together. Kate Brayton and Grace Joo also worked carefully and swiftly to complete all the supplements and media materials that accompany the text. Caroline Ayres, production supervisor, and Jean Lake, production editor, brilliantly coordinated all phases of the manuscript to the final pages of a beautiful book.

We are especially proud of the art program in this text, which lends beauty and understanding to chemistry. We would like to thank tani hasegawa, designer, whose creative ideas provided an outstanding design for the pages of the book. Clare Maxwell, photo researcher, was invaluable in researching and selecting vivid photos for the text so that students can see the beauty of chemistry. The macro-to-micro illustrations designed by J.B. Woolsey Associates give students visual impressions of the atomic and molecular organization of everyday things and are a fantastic learning tool. We want to thank Progressive Publishing Alternatives for their precise edit of the manuscript, and the hours they spent proofreading the pages. It has been especially helpful to have accuracy reviewers for this new text. Without them, much would have gone unnoticed and uncorrected. We also appreciate all the hard work in the field put in by the marketing team: Stacy Treco, director of marketing, and Scott Dustan, chemistry marketing manager. Without them, no one would know about this text.

This text also reflects the contributions of many professors who took the time to review and edit the manuscript, provide outstanding comments, help, and suggestions. We are extremely grateful to an incredible group of peers for their careful assessment of all the new ideas for the text, suggested additions, corrections, changes, and deletions, and for providing an incredible amount of feedback about the best direction for the text. In addition, we appreciate the time scientists took to let us take photos and discuss their work with them. We admire and appreciate every one of you.

Reviewers

Michelle Driessen, *University of Minnesota*

Wesley Fritz, *College of DuPage*

Amy Waldman Grant, *El Camino College*

Richard Lavallee, *Santa Monica College*

MaryKay Orgill, *University of Nevada, Las Vegas*

Cyriacus Chris Uzomba, *Austin Community College*

Previous Edition Reviewers

Bal Barot, *Lake Michigan College*

Sharmistha Basu-Dutt, *State University of West Georgia*

Les Battles, *Arkansas State University*

Ernest Baughman, *University of La Verne*

John L. Bonte, *Clinton Community College*

Kim Browning, *Catawba Valley Community College*

Angela Carraway, *Meridian Community College*

Jeanne Cassara, *California State University, Northridge*

Juan Pablo Claude, *University of Alabama-Birmingham*

Cristen Colantoni-Painter, *Community College of Allegheny County*

Joanne Cox, *Cleveland Community College*

C. Irvin Drew, *El Camino College*

Michelle Driessen, *University of Minnesota*

Jeffrey Emig, *Pennsylvania State University-York*

Mildred Hall, *Clark State Community College*

Helen Hauer, *Delaware Technical and Community College*

Richard Jarman, *College of DuPage*

Allen Johnson, *University of Nevada-Las Vegas*

Matthew Johnston, *Lewis-Clark State College*

Carolyn Sweeney Judd, *Houston Community College*

Matt Koutroulis, *Rio Honda College*

Richard Lavallee, *Santa Monica College*

Larry Manno, *Triton College*

Carol Martinez, *Albuquerque Technical Vocational Institute*

C. Michael McCallum, *University of the Pacific*

Douglas McLemore, *Front Range Community College*

Carl Minnier, *The Community College of Baltimore County-Essex*

Divina T. Miranda, *Southern University at New Orleans*

Aaron Monte, *University of Wisconsin-La Crosse*

Wyatt Murphy, *Seton Hall University*

Dan Nogales, *Northwest Nazarene University*

Becky Osmond, *Oklahoma State University-Oklahoma City*

Richard Pendarvis, *Central Florida Community College*

Cortlandt Pierpont, *University of Colorado*

Miriam Rossi, *Vassar College*

Gerald C. Swanson, *Daytona Beach Community College*

Vernon Thielmann, *Southwest Missouri State University*

Dave Thomasson, *Fontbonne University*

Amy Waldman, *El Camino College*

Gary L. Wood, *Valdosta State University*

Linda Zarzana, *American River College*

Student-Friendly Approach
Keeping Students Engaged Is the Ultimate Goal

Learning Goals

At the beginning of each section, a **Learning Goal** clearly identifies the key concept of the section, providing a roadmap for studying. All information contained in that section relates back to the Learning Goal.

Learning Goal

Given the formula of a covalent compound, write its correct name; given the name of a covalent compound, write its formula.

WEB TUTORIAL
Covalent Bonds

5.5 Covalent Compounds and Their Names

When ionic compounds form, metals lose their valence electrons and nonmetals gain electrons. However, when two nonmetals combine to form **covalent compounds,** the valence electrons are not transferred from one atom to another. Atoms in covalent compounds achieve stability by sharing their valence electrons. When atoms share electrons, the resulting bond is a **covalent bond.**

Formation of a Hydrogen Molecule

The simplest covalent molecule is hydrogen gas, H_2. When two hydrogen atoms are far apart, they are not attracted to each other. As the atoms move closer, the positive charge of each nucleus attracts the electron of the other atom. This attraction pulls the atoms closer until they share a pair of valence electrons and form a covalent bond. (See Figure 5.5.) In the covalent bond in H_2, the shared electrons give the noble gas configuration of He to each of the H atoms. Thus the atoms bonded in H_2 are more stable than two individual H atoms.

Macro-to-Micro Art

Photographs and drawings illustrate the atomic structure of recognizable objects, putting chemistry in context and connecting the atomic world to the macroscopic world.

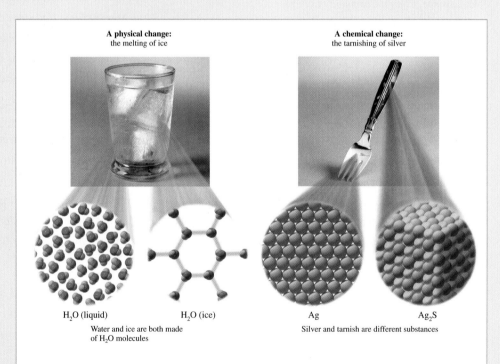

A physical change: the melting of ice

A chemical change: the tarnishing of silver

H_2O (liquid) H_2O (ice)

Water and ice are both made of H_2O molecules

Ag Ag_2S

Silver and tarnish are different substances

Figure 7.1 A chemical change produces new substances; a physical change does not.

Q Why is the formation of tarnish considered a chemical change?

Questions paired with figures challenge students to think critically about photos and illustrations.

Chapter Opener

Each chapter begins with an interview with a professional in science, engineering, or medicine. The interviews illustrate how these professionals interact with chemistry in their careers.

Career Focus

Within the chapters are additional examples of professionals using chemistry in their careers. In addition, the Chemistry Place website features in-depth resources for each of the professions in the book.

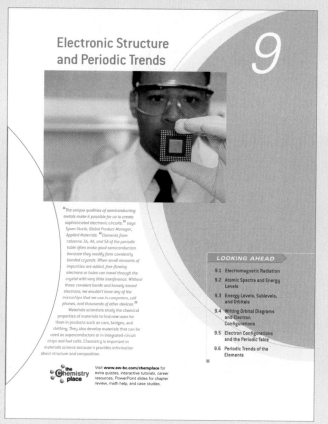

Electronic Structure and Periodic Trends

9

"The unique qualities of semiconducting metals make it possible for us to create sophisticated electronic circuits," says Syeen Strelis, Global Product Manager, Applied Materials. "Elements from columns 3A, 4A, and 5A of the periodic table often make good semiconductors because they readily form covalently bonded crystals. When small amounts of impurities are added, free-flowing electrons or holes can travel through the crystal with very little interference. Without these covalent bonds and loosely bound electrons, we wouldn't have any of the microchips that we use in computers, cell phones, and thousands of other devices."
Materials scientists study the chemical properties of materials to find new uses for them in products such as cars, bridges, and clothing. They also develop materials that can be used as superconductors or in integrated-circuit chips and fuel cells. Chemistry is important in materials science because it provides information about structure and composition.

LOOKING AHEAD

9.1 Electromagnetic Radiation
9.2 Atomic Spectra and Energy Levels
9.3 Energy Levels, Sublevels, and Orbitals
9.4 Writing Orbital Diagrams and Electron Configurations
9.5 Electron Configurations and the Periodic Table
9.6 Periodic Trends of the Elements

the Chemistry place

Visit www.aw-bc.com/chemplace for extra quizzes, interactive tutorials, career resources, PowerPoint slides for chapter review, math help, and case studies.

Writing Style

Karen Timberlake is known for her accessible writing style, based on a carefully paced and simple development of chemical ideas, suited to the background of preparatory students. She precisely defines terms and sets clear goals for each section of the text. Her clear analogies help students to visualize and understand key chemical concepts.

Chem Note

A rich array of **Chem Notes** in each chapter apply chemical concepts to relevant topics in science and medicine. These real-life, high interest topics include energy and nutrition, artificial fats, alcohol, global warming, acid rain, and pheromones.

CHEM NOTE

GLOBAL WARMING

The amount of carbon dioxide (CO_2) gas in our atmosphere is on the increase as we burn more gasoline, coal, and natural gas. The algae in the oceans and the plants and trees in the forests normally absorb carbon dioxide, but they cannot keep up with the continued increase. The cutting of trees in the rain forests (deforestation) reduces the amount of carbon dioxide removed from the atmosphere. Many of the trees are also burned as land is cleared. It has been estimated that deforestation may account for 15–30% of the carbon dioxide that remains in the atmosphere each year.

The carbon dioxide in the atmosphere acts like the glass in a greenhouse. When sunlight warms the Earth's surface, some of the heat is trapped by carbon dioxide. As CO_2 levels rise, more heat is trapped. It is not yet clear how severe the effects of global warming might be. Some scientists estimate that by around the year 2030, the atmospheric level of carbon dioxide could double and cause the temperature of Earth's atmosphere to rise by 2–5°C. If that should happen, it would have a profound impact on Earth's climate. For example, an increase in the melting of snow and ice could raise the ocean levels by as much as 2 m, which is enough to flood many cities located on the ocean shorelines.

Worldwide efforts are being made to reduce fossil fuel use and to slow or stop deforestation. It will require cooperation throughout the world to avoid the bleak future that some scientists predict should global warming continue unchecked.

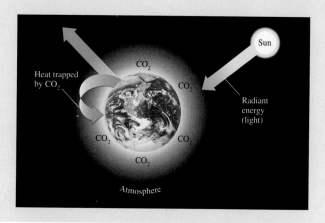

Problem Solving
Many Tools Show Students How to Solve Problems

A Visual Guide to Problem Solving

The authors understand the learning challenges facing students in this course, so they walk students through the problem-solving process step by step. For each type of problem, they use a unique, color-coded flow chart that is coordinated with parallel worked examples to visually guide students through each problem-solving strategy.

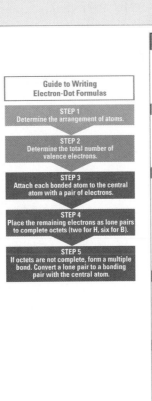

Guide to Writing Electron-Dot Formulas

STEP 1
Determine the arrangement of atoms.

STEP 2
Determine the total number of valence electrons.

STEP 3
Attach each bonded atom to the central atom with a pair of electrons.

STEP 4
Place the remaining electrons as lone pairs to complete octets (two for H, six for B).

STEP 5
If octets are not complete, form a multiple bond. Convert a lone pair to a bonding pair with the central atom.

Sample Problem 10.2 ▷ **Writing Electron-Dot Formulas with Lone Pairs**

Write the electron-dot formulas for phosphorus trichloride, PCl_3.

Solution

STEP 1 **Determine the arrangement of atoms.** In PCl_3, the central atom is P.

Cl P Cl
Cl

STEP 2 **Determine the total number of valence electrons.** We can use the group numbers to determine the valence electrons for each of the atoms in the molecule.

Element	Group	Atoms	Valence Electrons	=	Total
P	5A (15)	1 P	$\times 5\,e^-$	=	$5\,e^-$
Cl	7A (17)	3 Cl	$\times 7\,e^-$	=	$21\,e^-$
			Total valence electrons for PCl_3	=	$26\,e^-$

STEP 3 **Attach the central atom to each bonded atom by a pair of electrons.**

Cl:P:Cl   or   Cl—P—Cl
  Cl   Cl

STEP 4 **Arrange the remaining electrons as lone pairs to complete octets.** A total of 6 electrons ($3 \times 2e^-$) are needed to bond the central P atom to three Cl atoms. There are 20 valence electrons left.

$$26 \text{ valence } e^- - 6 \text{ bonding } e^- = 20\,e^- \text{ remaining}$$

The remaining electrons are placed as lone pairs around the outer Cl atoms first, which uses 18 more electrons.

:Cl:P:Cl: or   :Cl—P—Cl:
   :Cl: :Cl:

The remaining 2 electrons are used to complete the octet for the P atom. Octets are written for all the atoms using 26 valence electrons.

P has an octet

:Cl:P:Cl: or   :Cl—P—Cl:
   :Cl: :Cl:

Study Check

Write the electron-dot formula for Cl_2O.

Sample Problems with Study Checks

Numerous **Sample Problems** appear throughout the text to immediately demonstrate the application of each new concept. The worked-out solution gives step-by-step explanations, provides a problem-solving model, and illustrates required calculations. Each Sample Problem is followed by a **Study Check** question that allows students to test their understanding of the problem-solving strategy.

Integrated Questions and Problems

Questions and Problems at the end of each section encourage students to apply concepts and begin problem solving after each section. Answers to odd-numbered problems are given at the end of each chapter.

10.5 Determine the total number of valence electrons in each of the following:
a. H_2S
b. I_2
c. CCl_4
d. OH^-

10.6 Determine the total number of valence electrons in each of the following:
a. SBr_2
b. NBr_3
c. CH_3OH
d. NH_4^+

10.7 Write the electron-dot formula for each of the following molecules or ions:
a. HF
b. SF_2
c. NBr_3
d. BH_4^-
e. CH_3OH (methyl alcohol)

f. N_2H_4 (hydrazine)   H N N H

End-of-Chapter Questions and Problems

Understanding the Concepts questions encourage students to think about the concepts they have learned. **Additional Questions and Problems** integrate the topics from the entire chapter to promote further study and critical thinking. **Challenge Questions** are designed for group work in cooperative learning environments.

▶ Understanding the Concepts

6.55 A dandruff shampoo contains pyrithion, $C_{10}H_8N_2O_2S_2$, which acts as antibacterial and antifungal agent.

a. What is the empirical formula of pyrithion?
b. What is the molar mass of pyrithion?
c. What is the percent composition of pyrithion?
d. How many C atoms are in 25.0 g pyrithion?
e. How many moles of pyrithion contain 8.2×10^{24} nitrogen atoms?

6.56 Ibuprofen, the anti-inflammatory ingredient, has the formula $C_{13}H_{18}O_2$.

a. What is the percent by mass of oxygen in ibuprofen?
b. How many atoms of carbon are in 0.425 g ibuprofen?
c. How many grams of hydrogen are in 3.75×10^{22} molecules of ibuprofen?
d. How many atoms of hydrogen are in 0.245 g ibuprofen?

6.57 Using the following models of the molecules, determine (black = C, light blue = H, yellow = S, green = Cl)

1.   2.

a. molecular formula
b. empirical formula
c. molar mass
d. percent composition

▶ Additional Questions and Problems

6.59 Calculate the formula mass of each of the following:
a. $FeSO_4$, ferrous sulfate, iron supplement
b. $Ca(IO_3)_2$, calcium iodate, iodine source in table salt
c. $C_5H_8NNaO_4$, monosodium glutamate, flavor enhancer
d. $C_6H_{12}O_2$, isoamyl formate used to make artificial fruit syrups

6.60 Calculate the formula mass of each of the following:
a. $Mg(HCO_3)_2$, magnesium hydrogen carbonate

b. $Au(OH)_3$, gold(III) hydroxide, used in gold plating
c. $C_{18}H_{34}O_2$, oleic acid from olive oil
d. $C_{21}H_{26}O_5$, prednisone, anti-inflammatory

6.61 Calculate the percent composition for each of the following compounds:
a. K_2CrO_4
b. $Al(HCO_3)_3$
c. $C_6H_{12}O_6$

▶ Challenge Questions

9.117 Compare the speed, wavelengths, and frequencies of ultraviolet light and microwaves.

9.118 Radio waves with a frequency of $5.0 \times 10^5 \text{ s}^{-1}$ are used to communicate with satellites in space.
a. What is the wavelength in meters and nanometers of the radio waves?
b. How many minutes will it take for a message from flight control to reach a satellite that is 5.0×10^7 km from Earth?

9.119 How do scientists explain the colored lines observed in the spectra of heated atoms?

9.120 Even though H has only one electron, there are many lines in the spectrum of H. Explain.

9.121 What is meant by a principal energy level, a sublevel, and an orbital?

9.122 In some periodic charts, H is placed in Group 1A (1). In other periodic charts, H is placed in Group 7A (17). Why?

Combining Ideas

This new feature appears after every 2–3 chapters and is a set of integrated problems designed to test students' understanding of the previous chapters.

Combining Ideas from Chapters 4–6

CI 6 The active ingredient in Tums has the percent composition Ca 40.0%, C 12.0%, and O 48.0%.

a. If the empirical and molecular formulas are the same, what is its molecular formula?
b. What is the name of the ingredient?
c. If one Tums tablet contains 500. mg of this ingredient and a person takes two tablets a day, how many calcium ions does this person obtain from the Tums tablets?

CI 7 Oxalic acid found in plants and vegetables such as rhubarb has the percent composition C 26.7%, H 2.24%, and O 71.1%. The oxalic acid present in rhubarb can be toxic when large quantities of raw or cooked leaves are ingested because oxalic acid can interfere with respiration. Oxalic acid can cause kidney or bladder stones. Rhubarb leaves contain about 0.5% oxalic acid. The lethal dose (LD_{50}) in rats for oxalic acid is 375 mg/kg.

a. What is the empirical formula of oxalic acid?
b. If oxalic acid has a molar mass of about 90. g/mol, what is the molecular formula?

c. Using the LD_{50}, how many grams of oxalic acid would be toxic for a 160-lb person?
d. How many kilograms of rhubarb leaves would the person in part c need to eat to reach the toxic level of oxalic acid?

CI 8 The compound butyric acid gives rancid butter its characteristic odor.

butyric acid

a. If black (grey) spheres are carbon atoms, light blue spheres are hydrogen atoms, and red spheres are oxygen atoms, what is the molecular formula?
b. What is the empirical formula of butyric acid?
c. What is the percent composition of butyric acid?
d. How many grams of carbon are in 0.850 g butyric acid?
e. How many grams of butyric acid contain 3.28×10^{23} oxygen atoms?
f. Butyric acid has a density of 0.959 g/mL at 20°C. How many moles of butyric acid are contained in 0.565 mL butyric acid?

CI 9 Tamiflu (Oseltamivir), $C_{16}H_{28}N_2O_4$, is an antiviral drug that is used to treat influenza. The preparation of Tamiflu begins with the extraction of shikimic acid from the seedpods of the Chinese spice star anise. However, the star anise has no antiviral activity itself.

shikimic acid

Media and Additional Supplements

Valuable Ancillaries for Both Students and Instructors

The Chemistry Place

Study and Practice

On the Chemistry Place website, **Review Questions** give students more interactive practice for each section of the book. **Study Goals** help them lay out a study plan. The website includes chapter **Quizzes,** with questions unique to the site for extra learning practice. Hints and feedback are provided and results can be emailed to instructors.

Review

The Chemistry Place website also features a comprehensive set of math review documents as well as **InterAct Math,** a powerful tool that allows students to practice math problems that pertain directly to the content in each chapter.

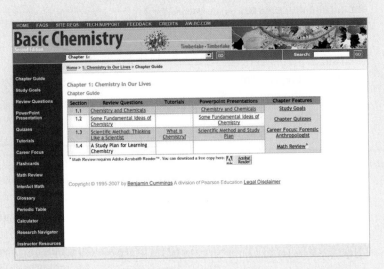

Interactivity

The Chemistry Place website is packed with comprehensive sets of interactive **Tutorials** that bring chemical reactions and concepts to life. The website has a multitude of **Quizzes** specifically tailored to the text. A **Flashcards** feature enables students to study vocabulary in an electronic format. Students also have access to over 2000 **PowerPoint slides** with learning checks created by Karen Timberlake, covering topics in every section of the text.

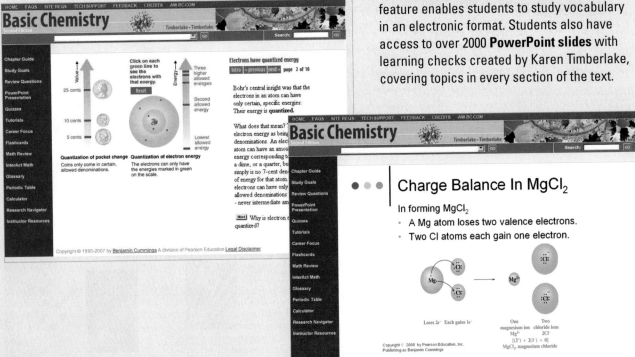

Supplements for the Student

Study Guide

by Karen Timberlake

(0-321-49635-3)

The Study Guide is keyed to the learning goals in the text and designed to promote active learning. It includes practice tests and complete solutions to odd-numbered problems.

Tutor Center

www.aw-bc.com/tutorcenter

This center provides one-to-one tutoring by phone, fax, e-mail, and the Internet. Qualified instructors answer questions and provide instructions regarding examples, exercises, and other content found in the text.

Additional Instructor Supplements

Instructor Solutions Manual

by Karen Timberlake

(0-321-49634-5)

Highlights chapter topics and includes suggestions for the laboratory. Contains answers and solutions to all problems in the text.

Printed Test Bank

by William Timberlake and G. Lynn Carlson

(0-321-46153-3)

Includes over 1600 multiple choice, matching, and short-answer questions.

Computerized Test Bank

(0-321-46917-8)

Over 1600 questions in multiple-choice, matching, true/false, and short-answer format on a dual-platform CD-ROM.

Media Manager

(0-321-49688-4)

This CD-ROM puts the outstanding art of *Basic Chemistry, Second Edition* in the palm of your hand. It includes all the art and tables from the book in high-resolution (150 dpi) format for use in classroom projection or when creating study materials and tests. In addition, customizable PowerPoint® lecture outlines are available on the CD, as well as downloadable files of the *Instructor Solutions Manual* and *Test Bank.* "Clicker Questions" suitable for classroom response systems are also available.

Transparency Acetates

(0-321-46154-1)

Includes 125 full-color transparency acetates.

Chemistry in Our Lives

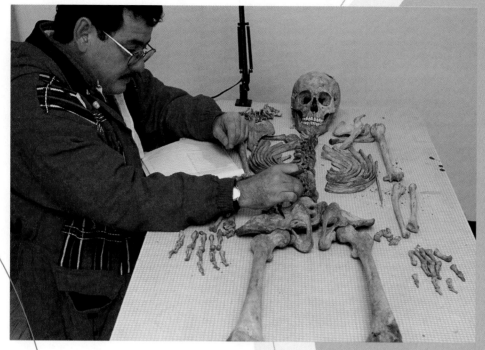

"*Chemistry plays an integral part in all aspects of medical legal death investigation,*" *says Charles L. Cecil, forensic anthropologist and medical legal death investigator, San Francisco Medical Examiner's Office.* "*Crime-scene analysis of blood droplets determines whether they are human or nonhuman, while the analyses of toxicological samples of blood and/or other fluids help determine the cause and time of death. Specialists in forensic anthropology can analyze trace-element ratios in bones to identify the number of individuals in mixed human bone situations. These conditions are quite often found during investigations of massive human-rights violations, such as the site of El Mozote in El Salvador.*"

Forensic anthropologists help police identify skeletal remains by determining gender, approximate age, height, and possible cause of death. Bone analysis can also provide information about illnesses or trauma a person may have experienced.

LOOKING AHEAD

1.1 Chemistry and Chemicals

1.2 Scientific Method: Thinking like a Scientist

1.3 A Study Plan for Learning Chemistry

the Chemistry place

Visit **www.aw-bc.com/chemplace** for extra quizzes, interactive tutorials, career resources, PowerPoint slides for chapter review, math help, and case studies.

Now that you are in a chemistry class, you may be wondering what you will be learning. Chemistry is actually a part of many sciences as well as much of everything around you. Perhaps you have been curious about something that happens in our environment. How does car exhaust produce smog? Chemistry can help you understand the process of forming smog. In a car engine where the temperature is high, nitrogen (N_2) and oxygen (O_2) from the air are converted to nitrogen oxide written as NO by chemists. The NO in the car exhaust reacts with oxygen in the air to produce nitrogen dioxide (NO_2), which gives the reddish-brown color to smog.

Or perhaps you have wondered about the depletion of ozone in the upper atmosphere. In the 1970s, scientists determined that substances called chloro-fluorocarbons are broken down by sunlight into smaller substances that destroy the ozone layer. As a result, the use of chlorofluorocarbons has been discontinued.

Perhaps you have been curious about how aspirin helps reduce muscle pain. Chemistry can help explain this too. When a part of the body is injured, sub-stances called prostaglandins are released, causing inflammation and pain. Aspirin acts by blocking the production of prostaglandin, thus reducing inflam-mation and pain. Chemists and other scientists around the world use chemistry to understand medical and environmental conditions so they can design new

Molecules of NO_2

treatments and processes that reduce environmental impact. For the chemist in the laboratory, the physician in the dialysis unit, and the agricultural scientist, chemistry plays a central role in providing understanding, assessing solutions, and making important decisions.

1.1 Chemistry and Chemicals

Chemistry is the study of the composition, structure, properties, and reactions of matter. *Matter* is another word for all the substances that make up our world. Perhaps you imagine that chemistry is done only in a laboratory by a chemist wearing a lab coat and goggles. Actually, chemistry happens all around you every day and has a big impact on everything you use and do. You are doing chemistry when you cook food, add chlorine to a swimming pool, or start your car. A chemical reaction has taken place when a nail rusts or an antacid tablet is dropped into water. Plants grow because chemical reactions convert carbon dioxide, water, and energy to carbohydrates. Chemical reactions take place when you digest food and break it down into substances that you need for energy and health.

All the things you see around you right now are composed of one or more chemicals. A **chemical** is any material used in or produced by a chemical process. Chemical processes take place every day in nature and in the body, as well as in chemistry laboratories, manufacturing plants, and pharmaceutical labs. A **substance** is a chemical that consists of one type of matter and always has the same composition and properties wherever it is found. Often the terms *chemical* and *substance* are used interchangeably to describe a specific type of matter.

Each day you use products containing substances that were developed and prepared by chemists. When you shower in the morning, chemicals in soaps and shampoos combine with oils on your skin and scalp and are removed by rinsing with water. When you brush your teeth, the chemicals in toothpaste clean your teeth and prevent plaque and tooth decay. Toothpaste contains chemicals such as abrasives, antibacterial agents, enamel strengtheners, colorings, and flavorings. Some of the substances used to make toothpaste are listed in Table 1.1. In cosmetics and lotions, chemicals are used to moisturize, prevent deterioration of the product, fight bacteria, and thicken the product. Your clothes may be made of natural fibers such as cotton or synthetics such as nylon or polyester. Perhaps you wear a ring or watch made of gold, silver, or platinum. Your breakfast cereal is probably fortified with iron, calcium,

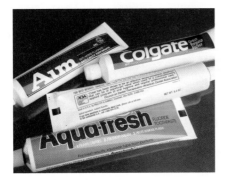

Table 1.1   **Chemicals Commonly Used in Toothpaste**

Chemical	Function
Calcium carbonate	An abrasive used to remove plaque
Sorbitol	Prevents loss of water and hardening of toothpaste
Carrageenan	Keeps the toothpaste from hardening or separating
Glycerin	Makes the toothpaste foam in the mouth
Sodium lauryl sulfate	A detergent used to loosen plaque
Titanium dioxide	Makes the toothpaste base white and opaque
Triclosan	An antibacterial agent used to inhibit bacteria that cause plaque and gum disease
Sodium monofluorophosphate	Prevents formation of cavities by strengthening tooth enamel with fluoride
Methyl salicylate	Gives a pleasant flavor of wintergreen

and phosphorus, while the milk you drink is enriched with vitamins A and D. Antioxidants are chemicals added to your cereal to prevent it from spoiling. Some of the chemicals you may encounter when you cook in the kitchen are shown in Figure 1.1.

Sample Problem 1.1   **Everyday Chemicals**

Identify the chemical described by each of the following statements:
a. Aluminum is used to make soda cans.
b. Throughout history salt (sodium chloride) has been used to preserve meat and fish.
c. Some people sweeten coffee or tea with sugar (sucrose).

Solution
a. aluminum   **b.** salt (sodium chloride)   **c.** sugar (sucrose)

Study Check

Which of the following are chemicals?
a. iron   **b.** tin   **c.** a low temperature   **d.** water

Answers to all of the Study Checks can be found at the end of this chapter in the Answer section. Checking your answers will let you know if you understand the material in this section.

Figure 1.1 Many of the items found in a kitchen are obtained using chemical processes.

Q *What are some other chemicals found in a kitchen?*

Silicon dioxide (glass)

Caffeine (coffee)

Chemically treated water

Synthetic material

Metal alloy

Baking carbohydrates, fats, and proteins

Natural gas

Fruits grown with fertilizers and pesticides

Questions and Problems   **Chemistry and Chemicals**

In every chapter, each red, odd-numbered exercise is paired with the next even-numbered exercise. The answers for all the Study Checks and the red, odd-numbered exercises are given at the end of this chapter. The complete solutions to the odd-numbered exercises are in the *Study Guide*.

1.1   Using a reference book or dictionary, write a one-sentence definition of the following:
     a. chemistry   **b.** chemist   **c.** chemical

1.2   Ask two of your friends (not in this class) to define the terms in problem 1.1. Do their answers agree with the definitions in problem 1.1?

1.3   Obtain a vitamin bottle or other packaged item and observe the list of ingredients. List four. Which ones are chemicals?

1.4   Obtain a box of breakfast cereal and observe the list of ingredients. List four. Which ones are chemicals?

1.5   Read the labels on some items found in your medicine cabinet. What are the names of some chemicals contained in those items?

1.6   Read the labels on products used to wash and clean your car. What are the names of some chemicals contained in those products?

1.7   A "chemical-free" shampoo includes the ingredients: water, cocomide, glycerin, and citric acid. Is the shampoo "chemical-free"?

1.8   A "chemical-free" sunscreen includes the ingredients: titanium dioxide, vitamin E, and vitamin C. Is the sunscreen "chemical-free"?

1.9   Pesticides are chemicals. Give one advantage and one disadvantage of using pesticides.

1.10 Sugar is a chemical. Give one advantage and one disadvantage of eating sugar.

1.2   Scientific Method: Thinking like a Scientist

Learning Goal

Describe the activities that are part of the scientific method.

When you were very young, you explored the things around you by touching and tasting. When you got a little older, you asked questions about the world you live in. What is lightning? Where does a rainbow come from? Why is water blue? As an adult you may have wondered how antibiotics work. Each day you ask questions and seek answers as you organize and make sense of the world you live in.

When Nobel Laureate Linus Pauling described his student life in Oregon, he recalled that he read many books on chemistry, mineralogy, and physics. "I mulled over the properties of materials: why are some substances colored and others not, why are some minerals or inorganic compounds hard and others soft." He said, "I was building up this tremendous background of empirical knowledge and at the same time asking a great number of questions." Linus Pauling won two Nobel Prizes: the first, in 1954, was in chemistry for his work on the structure of proteins, and the second, in 1962, was the Peace Prize.

Scientific Method

Although the process of trying to understand nature is unique to each scientist, there is a set of general principles, called the **scientific method,** that describes the thinking of a scientist.

1. **Observations** The first step in the scientific method is to observe, describe, and measure some event in nature. Observations based on measurements are called *data.*

2. **Hypothesis** After sufficient data are collected, a *hypothesis* is proposed that states a possible interpretation of the observations. The hypothesis must be stated in a way that it can be tested by experiments.

3. **Experiments** Experiments are tests that determine the validity of the hypothesis. Often many experiments are performed and a large amount of data collected. If the results of experiments provide different results than predicted by the hypothesis, a new or modified hypothesis is proposed and a new group of experiments are performed.

4. **Theory** When experiments can be repeated by many scientists with consistent results that confirm the hypothesis, the hypothesis becomes a *theory*. Each theory, however, continues to be tested and, based on new data, sometimes needs to be modified or even replaced. Then a new hypothesis is proposed, and the process of experimentation takes place once again.

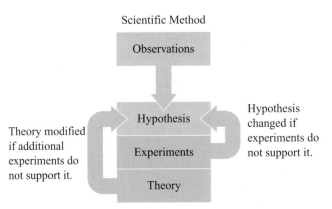

Using the Scientific Method in Everyday Life

You may be surprised to realize that you use the scientific method in your everyday life. Let's suppose that you visit a friend in her home. Soon after you arrive, your eyes start to itch and you begin to sneeze. Then you observe that your friend has a new cat. Perhaps you ask yourself why you are sneezing and form the hypothesis that you are allergic to cats. To test your hypothesis, you leave your friend's home. If the sneezing stops, perhaps your hypothesis is correct. You test your hypothesis further by visiting another friend who also has a cat. If you start to sneeze again, your experimental results indicate that you are allergic to cats. However, if you continue sneezing after you leave your friend's home, your hypothesis is not supported. Now you need to form a new hypothesis, which could be that you have a cold.

Sample Problem 1.2   **Scientific Method**

Identify each of the following statements as an observation or a hypothesis:
a. A silver tray turns a dull gray color when left uncovered.
b. Water freezes at 0°C.
c. Ice cubes have a greater volume than the liquid from which they were formed because the water molecules are further apart in the ice structure than in the liquid form.

Solution
a. observation   **b.** observation   **c.** hypothesis

Study Check

The following statements are found in a student's notebook. Identify each of the following as (**1**) observation, (**2**) hypothesis, or (**3**) experiment.

a. "Today, I planted two tomato seedlings in the garden. Two more tomato seedlings are placed in a closet. I will give all the plants the same amount of water and fertilizer."

b. "After 50 days, the tomato plants in the garden are 3 feet high with green leaves. The plants in the closet are 8 inches tall and yellow."

c. "Tomato plants need sunlight to grow."

CHEM NOTE

EARLY CHEMISTS: THE ALCHEMISTS

For many centuries, chemistry has been the study of changes in matter. From the time of the Greeks to about the sixteenth century, alchemists described matter in terms of four components of nature: earth, air, fire, and water, with the qualities of hot, cold, damp, or dry. By the eighth century, alchemists believed that they could rearrange these qualities in such a way as to change metals such as copper and lead into gold and silver. They searched for an unknown substance called a philosopher's stone that they thought would turn metals into gold as well as retain youth and postpone death. Although these efforts failed, the alchemists did provide information on the processes and chemical reactions involved in the extraction of metals from ores. The alchemists also designed some of the first laboratory equipment and developed early laboratory procedures. These early efforts were some of the first observations and experiments using the scientific method.

One alchemist, who used the name Paracelsus (1493–1541), thought that alchemy was not about producing gold, but preparing new medicines. Using observation and experiment, he viewed the body as a series of chemical processes that could be unbalanced by certain chemical compounds and rebalanced by using minerals and medicines. For example, he determined that the lung diseases of miners were caused by inhalation of dust, not underground spirits. He also thought that goiter was a problem with drinking water and treated syphilis with compounds of mercury. His opinion of medicines was that the right dose makes the difference between a poison and a cure. Today this idea is part of the risk analysis of medicines. Paracelsus changed alchemy in ways that helped to establish modern medicine and chemistry.

Science and Technology

When scientific information is applied to industrial and commercial uses, it is called technology. Such uses have made the chemical industry one of the largest industries in the United States. Every year, technology provides new materials or procedures that produce more energy, cure diseases, improve crops, and produce new kinds of synthetic materials. Table 1.2 lists some of the important scientific discoveries, laws, theories, and technological innovations that were made over the past 300 years.

However, there have also been unintended consequences of scientific research. The production of some substances has contributed to the development of hazardous conditions in our environment. We have become concerned about the energy requirements of new products and how some materials may

Table 1.2   Some Important Scientific and Technological Discoveries or Theories

Discovery/Theory	Date	Name	Country
Law of gravity	1687	Isaac Newton	England
Oxygen	1774	Joseph Priestley	England
Electric battery	1800	Alessandro Volta	Italy
Atomic theory	1803	John Dalton	England
Anesthesia, ether	1842	Crawford Long	U.S.
Nitroglycerine	1847	Ascanio Sobrero	Italy
Germ theory	1865	Louis Pasteur	France
Antiseptic surgery	1865	Joseph Lister	England
Discovery of nucleic acids	1869	Friedrich Miescher	Switzerland
Radioactivity	1896	Henri Becquerel	France
Discovery of radium	1898	Marie and Pierre Curie	France
Quantum theory	1900	Max Planck	Germany
Theory of relativity	1905	Albert Einstein	Germany
Identification of components of RNA and DNA	1909	Phoebus Theodore Levene	U.S.
Insulin	1922	Frederick Banting, Charles Best, John Macleod	Canada
Penicillin	1928	Alexander Fleming	England
Nylon	1937	Wallace Carothers	U.S.
Discovery of DNA as genetic material	1944	Oswald Avery	U.S.
Synthetic production of transuranium elements	1944	Glenn Seaborg, Arthur Wahl, Joseph Kennedy, Albert Ghiorso	U.S
Determination of DNA structure	1953	Francis Crick, Rosalind Franklin, James Watson	England U.S.
Polio vaccine	1954 1957	Jonas Salk Albert Sabin	U.S. U.S.
Laser	1958 1960	Charles Townes Theodore Maiman	U.S. U.S.
MRI	1980	Paul Lauterbur	U.S
Cellular phones	1973	Martin Cooper	U.S.
Prozac	1988	Ray Fuller	U.S.
HIV protease inhibitor	1995	Joseph Martin, Sally Redshaw	U.S

CHEM NOTE
DDT—GOOD PESTICIDE, BAD PESTICIDE

DDT (Dichlorodiphenyltrichloroethane) was once one of the most commonly used pesticides. Although it was first made in 1874, it was not used as an insecticide until 1939. Before DDT was widely used, insect-borne diseases such as malaria and typhus were rampant in many parts of the world. Paul Mueller, who discovered that DDT was an effective pesticide, was recognized for saving many lives and received the Nobel Prize for Medicine and Physiology in 1948. DDT was considered the ideal pesticide because it was toxic to many insects, had a low toxicity to humans and animals, and was inexpensive to prepare.

In the United States, DDT was extensively used on home gardens as well as farm crops, particularly cotton and soybeans. Because of its stable chemical structure, DDT did not break down quickly in the environment, which meant that it did not have to be applied as often. At first, everyone was pleased with DDT as crop yields increased and diseases such as malaria and typhus were under control.

However, in the early 1950s, problems attributed to DDT began to arise. Insects were becoming more resistant to the pesticide. At the same time, there was increasing public concern about the long-term impact of a substance that could remain in the environment for many years. The metabolic systems of humans and animals cannot break down DDT, which is soluble in fats but not in water and stored in the fatty tissues of the body. Although the concentrations of DDT applied to crops was very low, the concentrations of DDT found in fish and the birds that ate fish were as much as 10 million times greater. DDT was

found to reduce the calcium in eggshells causing the incubating eggs to crack open early. Due to this difficulty with reproduction, the populations of birds such as the bald eagle and brown pelican dropped significantly.

By 1972, DDT was banned in the United States. The EPA (Environmental Protection Agency) reported that the DDT levels were reduced by 90% in fish in Lake Michigan by 1978. Today new types of pesticides that are more water soluble and do not persist in the environment have replaced the long-lasting pesticides such as DDT. However, these new pesticides are much more toxic to humans.

cause changes in our oceans and atmosphere. We want to know if the new materials can be recycled, how they are broken down, and whether there are processes that are safer. The ways in which we continue to utilize scientific research will impact our planet in the future. These decisions can best be made if every citizen has an understanding of science.

Questions and Problems
Scientific Method: Thinking like a Scientist

1.11 Define each of the following terms of the scientific method:
- **a.** hypothesis
- **b.** experiment
- **c.** theory
- **d.** observation

1.12 Identify each of the following activities in the scientific method as
- (1) observation
- (2) hypothesis
- (3) experiment
- (4) theory
- **a.** Formulate a possible explanation for your experimental results.
- **b.** Collect data.
- **c.** Design an experimental plan that will give new information about a problem.
- **d.** State a generalized summary of your experimental results.

1.13 At a popular restaurant, where Chang is the head chef, the following occur:
- (1) Chang determines that sales of the chef's salad have dropped.
- (2) Chang decides that the chef's salad needs a new dressing.
- (3) In a taste test, four bowls of lettuce are prepared with four new dressings: sesame seed, oil and vinegar, blue cheese, and anchovies.
- (4) The tasters rate the dressing with sesame seeds the best.
- (5) After two weeks, Chang notes that the orders for the chef's salad with the new sesame dressing have doubled.

(6) Chang decides that the sesame dressing improved the sales of the chef's salad because the sesame dressing improved the taste of the salad.

Identify each activity as an

a. observation   **b.** hypothesis
c. experiment   **d.** theory

1.14 Lucia wants to develop a process for dyeing shirts so that the color will not fade when the shirt is washed. She proceeds with the following activities:

(1) Lucia notices that the dye in a design on T-shirts fades when the shirt is washed.

(2) Lucia decides that the dye needs something to help it set in the T-shirt fabric.

(3) She places a spot of dye on each of four T-shirts, and then places each one separately in water, salt water, vinegar, and baking soda and water.

(4) After 1 hour, all the T-shirts are removed and washed with a detergent.

(5) Lucia notices that the dye has faded on the tee shirts in water, salt water, and baking soda, while the dye held up in the T-shirts soaked in vinegar.

(6) Lucia thinks that the vinegar binds with the dye so it does not fade when the shirt is washed.

Identify each activity as an

a. observation   **b.** hypothesis
c. experiment   **d.** theory

Learning **Goal**

Develop a study plan for learning chemistry.

1.3 A Study Plan for Learning Chemistry

Here you are taking chemistry, perhaps for the first time. Whatever your reasons for choosing to study chemistry, you can look forward to learning many new and exciting ideas.

Features in This Text to Help You Study Chemistry

This text has been designed with a variety of study aids to complement different learning styles. On the inside of the front cover is a periodic table of the elements that provides information about the elements. On the inside of the back cover are tables that summarize useful information needed throughout the study of chemistry. Each chapter begins with *Looking Ahead,* which outlines the topics in the chapter. A *Learning Goal* at the beginning of each section previews the concepts you are to learn.

Before you begin reading the text, obtain an overview of the chapter by reviewing the *Looking Ahead* list of topics. As you prepare to read a section of the chapter, look at the section title and turn it into a question. For example, for Section 1.1, titled "Chemistry and Chemicals," you could write a question that asks "What is chemistry?" or "What are chemicals?" When you are ready to read through that section, review the *Learning Goal.* A *Learning Goal* tells you what to expect in that section and what you need to accomplish. As you read, try to answer the questions you wrote for the section. The *Sample Problems* illustrate a step-by-step approach, enhanced, where appropriate, by a visual *Guide to Problem Solving (GPS).* When you come to a sample problem, take the time to work it through, try each *Study Check,* and check your answer. If your answer matches, you most likely understand the topic. If your answer does not match, you need to study the section again. At the end of each section, you will find a set of *Questions and Problems* that allow you to apply problem solving immediately to the new concepts.

Items appear throughout each chapter to help you connect chemistry to real-life events. The figures and diagrams use illustrations to depict the atomic level of organization of ordinary objects. These visual models illustrate the concepts described in the text and allow you to "see" the world in a microscopic way. Boxes called *Chem Notes* connect the chemical concepts you are studying to real-life situations.

At the end of each chapter, you will find several study aids that complete the chapter. The *Key Terms* are bold-faced in the text and listed again at the end of

---

Table 1.3   Steps in Active Learning

1. Read the set of *Looking Ahead* topics for an overview of the material.

2. Form a question from the title of the section you are going to read.

3. Read the section looking for answers to your question.

4. Self-test by working *Sample Problems* and *Study Checks* within each section.

5. Complete the *Questions and Problems* that follow each section and check the odd-numbered answers.

6. Proceed to the next section and repeat the steps.

---

the chapter. *Chapter Reviews* and *Concept Maps* at the end of each chapter give a summary and show the connections between important concepts. *Additional Questions and Problems* at the end of each chapter provide more problems to test your understanding of the topics in the chapter. All the problems are paired, and the odd-numbered problems have answers at the end of the chapter.

Using Active Learning to Learn Chemistry

A student who is an active learner thinks about the new chemical concepts while reading the text and attending lecture. Let's see how this is done.

As you read and practice problem solving, you remain actively involved in studying, which enhances the learning process. In this way, you learn small bits of information at a time and establish the necessary foundation for understanding the next section. You may also note questions you have about the reading to discuss with your professor and laboratory instructor. Table 1.3 summarizes these steps for active learning. The time you spend in lecture can also be useful as a learning time. By keeping track of the class schedule and reading the assigned material before lecture, you become aware of the new terms and concepts you need to learn. Some questions that occur during your reading may be answered during the lecture. If not, you can ask for further clarification from your professor.

Many studies now indicate that studying with a group can be beneficial to learning for many students. In a group, students motivate each other to study, fill in gaps, and correct misunderstandings by teaching and learning together. Studying alone does not allow the process of peer correction that takes place when you work with a group of students in your class. In a group, you can cover the ideas more thoroughly as you discuss the reading and problem solving with other students. Waiting to study until the night before an exam does not give you time to understand concepts and practice problem solving. Ideas may be ignored or avoided that turn out to be important on test day.

Thinking Scientifically About Your Study Plan

As you embark on your journey into the world of chemistry, think about your approach to studying and learning chemistry. You might consider some of the ideas in the following list. Check those ideas that will help you learn chemistry successfully. Commit to them now. Your success depends on you.

My study of chemistry will include the following:

_____ Reviewing the *Learning Goals.*

_____ Keeping a problem notebook.

_____ Reading the text as an active learner.

_____ Self-testing by working the chapter problems and checking solutions in the text.

_____ Reading the chapter before lecture.

_____ Being an active learner in lecture.

_____ Going to lecture.

_____ Organizing a study group.

_____ Seeing the professor during office hours.

_____ Attending review sessions.

_____ Organizing my own review sessions.

_____ Studying a little bit as often as I can.

Sample Problem 1.3   A Study Plan for Learning Chemistry

Which of the following activities would you include in a study plan for learning chemistry successfully?
a. skipping lecture
b. forming a study group
c. keeping a problem notebook
d. waiting to study the night before the exam
e. becoming an active learner

Solution

Your success in chemistry can be helped if you
b. form a study group
c. keep a problem notebook
e. become an active learner

Study Check

Which of the following would help you learn chemistry?
a. skipping review sessions
b. working assigned problems
c. attending the professor's office hours
d. staying up all night before an exam
e. reading the assignment before lecture

Questions and Problems   A Study Plan for Learning Chemistry

1.15 What are four things you can do to help your success in chemistry?

1.16 What are four things that would make it difficult for you to learn chemistry?

1.17 A student in your class asks you for advice on learning chemistry. Which of the following might you suggest?
    a. Form a study group.
    b. Skip lecture.
    c. Visit the professor during office hours.
    d. Wait until the night before an exam to study.
    e. Become an active learner.

1.18 A student in your class asks you for advice on learning chemistry. Which of the following might you suggest?
    a. Do the assigned problems.
    b. Don't read the book; it's never on the test.
    c. Attend review sessions.
    d. Read the assignment before lecture.
    e. Keep a problem notebook.

Concept Map   **Chemistry in Our Lives**

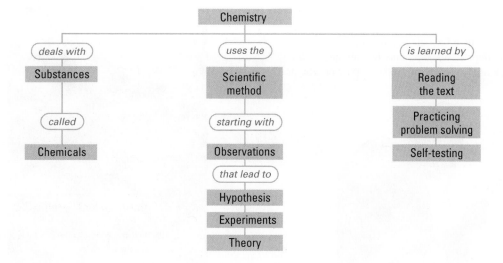

Chapter Review

1.1 Chemistry and Chemicals

Chemistry is the study of the composition of substances and the way in which they interact with other substances. A chemical is any substance used in or produced by a chemical process.

1.2 Scientific Method: Thinking like a Scientist

The scientific method is a process of explaining natural phenomena beginning with observations, hypothesis, and experiments, which may lead to a theory when experimental results support the hypothesis. Technology involves the application of scientific information to industrial and commercial uses.

1.3 A Study Plan for Learning Chemistry

A study plan for learning chemistry utilizes the visual features in the text and develops an active learning approach to study. By using the *Learning Goals* in the chapter and working the *Sample Problems* and *Study Checks* and the problems at the end of each section, the student can successfully learn the concepts of chemistry.

Key Terms

chemical A substance used in or produced by a chemical process.

chemistry A science that studies the composition of substances and the way they interact with other substances.

experiment A procedure that tests the validity of a hypothesis.

hypothesis An unverified explanation of a natural phenomenon.

observation Information determined by noting and recording a natural phenomenon.

scientific method The process of making observations, proposing a hypothesis, testing the hypothesis, and developing a theory that explains a natural event.

substance A particular kind of matter that has the same composition and properties wherever it is found.

theory An explanation of an observation that has been validated by experiments that support a hypothesis.

Understanding the Concepts

1.19 According to Sherlock Holmes, "One must follow the rules of scientific inquiry, gathering, observing, and testing data, then formulating, modifying, and rejecting hypotheses, until only one remains." Did Sherlock use the scientific method? Why or why not?

1.20 In "A Scandal in Bohemia," Sherlock Holmes receives a mysterious note. He states, "I have no data yet. It is a capital mistake to theorize before one has data. Insensibly one begins to twist facts to suit theories instead of theories to suit facts." What do you think Sherlock meant?

1.21 Classify each of the following statements as an observation, hypothesis, or theory:
   a. Aluminum melts at 660°C.
   b. Dinosaurs became extinct when a large meteorite struck Earth and caused a huge dust cloud that severely decreased the amount of light reaching Earth.
   c. The 100-yard dash was run in 9.8 seconds.

1.22 Classify each of the following statements as an observation, hypothesis, or theory:
   a. Analysis of ten ceramic dishes showed that four dishes contained lead levels that exceeded federal safety standards.
   b. Marble statues undergo corrosion in acid rain.
   c. Statues corrode in acid rain because the acidity is sufficient to dissolve calcium carbonate, the major substance of marble.

 ## Additional Questions and Problems

1.23 Why does the scientific method include a hypothesis?

1.24 Why is experimentation an important part of the scientific method?

1.25 Select the correct phrase(s) to complete the following statement: If experimental results do not support your hypothesis, you should
   a. pretend that the experimental results do support your hypothesis.
   b. write another hypothesis.
   c. do more experiments.

1.26 Select the correct phrase(s) to complete the following statement: A hypothesis becomes a theory when

   a. one experiment proves the hypothesis.
   b. many experiments by many scientists validate the hypothesis.
   c. you decide to call it a theory.

1.27 Which of the following will help you develop a successful study plan?
   a. Skip lecture and just read the book.
   b. Work the sample problems as you go through a chapter.
   c. Go to your professor's office hours.
   d. Read through the chapter, but work the problems later.

1.28 Which of the following will help you develop a successful study plan?
  a. Study all night before the exam.
  b. Form a study group and discuss the problems together.
  c. Work problems in a notebook for easy reference.
  d. Copy the answers to homework from a friend.

Challenge Questions

1.29 Classify each of the following as (1) an observation, (2) an hypothesis, or (3) an experiment:
  a. The bicycle tire is flat.
  b. If I add air to the bicycle tire, it will expand to the proper size.
  c. When I added air to the bicycle tire, it was still flat.
  d. The bicycle tire must have a leak in it.

1.30 Classify each of the following as (1) an observation, (2) an hypothesis, or (3) an experiment:
  a. A big log in the fire does not burn well.
  b. If I chop the log into smaller wood pieces, it will burn better.
  c. The smaller pieces of wood burn brighter and make a hotter fire.
  d. The small wood pieces are used up faster than burning the big log.

Answers

Answers to Study Checks

1.1 a, b, and d

1.2 a. experiment (3)   **b.** observation (1)
  c. hypothesis (2)

1.3 b, c, and e

Answers to Selected Questions and Problems

1.1 a. Chemistry is the science of the composition and properties of matter.
  b. A chemist is a scientist who studies the composition and changes of matter.
  c. A chemical is a substance that is used in or produced by a chemical process.

1.3 Many chemicals are listed on a vitamin bottle such as vitamin A, vitamin B_3, vitamin B_{12}, vitamin C, folic acid, etc.

1.5 Typical items found in a medicine cabinet and chemicals they contain:
  Antacid tablets: calcium carbonate, cellulose, starch, stearic acid, silicon dioxide
  Mouthwash: water, alcohol, glycerol, sodium benzoate, benzoic acid
  Cough suppressant: menthol, beta-carotene, sucrose, glucose

1.7 No. All of the ingredients are chemicals.

1.9 An advantage of a pesticide is that it gets rid of insects that bite or damage crops. A disadvantage is that a pesticide can destroy beneficial insects or be retained in a crop that is eventually eaten by animals or humans.

1.11 a. A hypothesis proposes a possible explanation for a natural phenomenon.
  b. An experiment is a procedure that tests the validity of a hypothesis.
  c. A theory is a hypothesis that has been validated many times by many scientists.
  d. An observation is a description or measurement of a natural phenomenon.

1.13 a. 1, 4, and 5 are observations.
  b. 2 is a hypothesis.
  c. 3 is an experiment.
  d. 6 is a theory.

1.15 There are several, including forming a study group, going to lecture, working sample problems and study checks, working problems and checking answers, reading the assignment ahead of class, keeping a problem notebook, etc.

1.17 a, c, and e

1.19 Yes. Sherlock's investigation includes observations (gathering data), formulating a hypothesis, testing the hypothesis and modifying it until one of the hypotheses is validated.

1.21 a. observation   **b.** hypothesis or theory
  c. observation

1.23 A hypothesis, which is a possible explanation for an observation, can be tested with experiments.

1.25 b and c

1.27 b and c

1.29 a. (1) observation   **b.** (2) hypothesis
  c. (3) experiment   **d.** (2) hypothesis

Measurements

"*I use measurement in just about every part of my nursing practice,*" says registered nurse Vicki Miller. "*When I receive a doctor's order for a medication, I have to verify that order. Then I draw a carefully measured volume from an IV or a vial to create that particular dose. Some dosage orders are specific to the size of the patient. I measure the patient's weight and calculate the dosage required for the weight of that patient.*"

Nurses use measurement each time they take a patient's temperature, height, weight, or blood pressure. Measurement is used to obtain the correct amounts for injections and medications and to determine the volumes of fluid intake and output. For each measurement, the amounts and units are recorded in the patient's records.

the Chemistry place

Visit **www.aw-bc.com/chemplace** for extra quizzes, interactive tutorials, career resources, PowerPoint slides for chapter review, math help, and case studies.

Measurements are important in our everyday life. Think about your day; you probably made some measurements. Perhaps you checked your weight by stepping on a scale. If you made some rice, you added 2 cups of water to 1 cup of rice. If you did not feel well, you may have taken your temperature. Whenever you made a measurement, you used some measuring device such as a scale, a measuring cup, or a thermometer. Over the years, you have learned to read the markings on each device to make a correct measurement.

In science, we use measurement to understand the world around us. Scientists measure the amounts of the materials that make up everything in our universe. An engineer determines the amount of metal in an alloy or the volume of seawater flowing through a desalination plant. A physician orders laboratory tests to measure substances in the blood such as glucose or cholesterol. An environmental scientist measures the levels of pollutants in our soil and atmosphere such as lead and carbon monoxide.

By learning about measurement, we develop skills for solving problems and learn how to work with numbers in chemistry. An understanding of measurement is essential to evaluate our health and surroundings.

2.1 Units of Measurement

Learning Goal

Write the names and abbreviations for the metric or SI units used in measurements of length, volume, mass, temperature, and time.

Suppose today you walked 2.1 kilometers (km) to campus carrying a backpack that weighed 12 kilograms (kg). The temperature at 8:30 A.M. was 7 degrees Celsius (°C). If these measurements seem unusual to you, it is because the *metric system* was used to measure each one. Perhaps you are more familiar with these measurements stated in the U.S. system; then you walked 1.3 miles (mi) to school carrying a backpack that weighed 26 pounds (lb). The temperature at 8:30 A.M. was 45 degrees Fahrenheit (°F).

The **metric system** is used by scientists and health professionals throughout the world. It is also the common measuring system in all but a few countries of the world. In 1960, a modification of the metric system called the *International System of Units,* Système International **(SI),** was adopted by scientists to provide additional uniformity of units throughout the world. For many types of measurements, the metric units used by chemists and SI units are the same. For other measurements, smaller metric units are used.

Some Metric (SI) Units in Chemistry

Before we get to details of measurement and calculations, we will look at the units typically used in chemistry. They include units for length, mass, volume, temperature, and time.

Length

The metric and SI unit of length is the **meter (m).** In comparison to the U.S. system, 1 meter is slightly longer than a yard (1.094 yd) or equivalent to 39.37 inches, as seen in Figure 2.1. A smaller unit of length, the **centimeter (cm),** is more commonly used in chemistry and is about as wide as your little finger.

Figure 2.1 Length in the metric (SI) system is based on the meter, which is slightly longer than a yard.

Q How many centimeters are in a length of 1 inch?

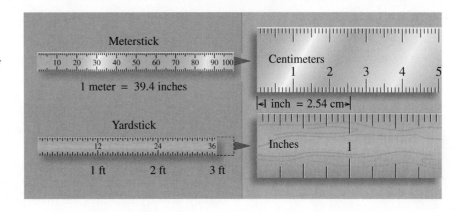

1 meter = 39.4 inches

1 inch = 2.54 cm

1 meter = 1.094 yd.
1 meter = 39.37 in.
2.54 cm = 1 in.

Volume

Volume (V) is the amount of space a substance occupies. The SI unit of volume, **cubic meter (m^3)**, is the volume of a cube that has sides that measure 1 meter in length. In a chemistry laboratory, the cubic meter is too large for practical use. Instead, chemists work with metric units of volume that are smaller and more convenient, such as the **liter (L)** and **milliliter (mL).** A liter is slightly larger than a quart (1 L = 1.057 qt) and contains 1000 mL, as shown in Figure 2.2. A cubic meter is the same volume as 1000 L.

$1 \text{ m}^3 = 1000 \text{ L}$
$1 \text{ L} = 1.057 \text{ qt}$
$1 \text{ L} = 1000 \text{ mL}$

Figure 2.2 Volume is the space occupied by a substance. In the metric system, volume is based on the liter, which is slightly larger than a quart.

Q How many quarts are in 1 liter?

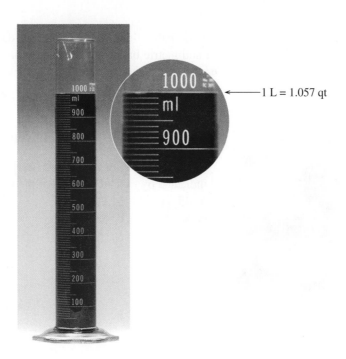

1 L = 1.057 qt

Figure 2.3 On an electronic balance, mass is shown in grams as a digital readout.

Q How many grams are in 1 pound of candy?

Figure 2.4 A thermometer is used to determine the temperature of a substance.

Q What kinds of temperature readings have you made today?

Mass

The **mass** of an object is the quantity of material it contains. You may be more familiar with the term *weight* than with mass. The weight of an object depends on its mass and the pull on it by gravity. Therefore, the weight of an object changes as the gravitational pull changes. On the moon an object weighs much less than on Earth because the gravitational pull on the moon is much less. However, the mass is the same because the amount of material in that object is constant.

The SI unit of mass is the **kilogram (kg).** For smaller masses, the metric unit **gram (g)** is used. There are 1000 g in 1 kg. A balance in a chemistry laboratory actually measures the mass in grams of a substance, not its weight. (See Figure 2.3.) In comparison to the U.S. system, the mass of 1 kilogram is equivalent to 2.205 lb and 1 pound (lb) is equivalent to 453.6 g.

$$1 \text{ kg} = 1000 \text{ g}$$
$$1 \text{ kg} = 2.205 \text{ lb}$$
$$453.6 \text{ g} = 1 \text{ lb}$$

Temperature

You probably use a thermometer to see how hot something is, or how cold it is outside, or perhaps to determine if you have a fever. (See Figure 2.4.) The **temperature** of an object tells us how hot or cold that object is. A typical laboratory thermometer consists of a glass bulb with a liquid in it that expands as the temperature increases. On a **Celsius (°C) scale,** water freezes at 0°C and boils at 100°C, while on a Fahrenheit (°F) scale, water freezes at 32°C and boils at 212°C. In the SI system, temperature is measured using the **Kelvin (K) scale** on which the lowest temperature possible is assigned a value of 0 K. Note that the units on the Kelvin scale are called kelvins (**K**) and do not have a degree sign.

Time

You probably think of time as years, days, minutes, or seconds. Of these, the SI and metric basic unit of time is the **second (s).** The standard now used to determine a second is an atomic clock. A comparison of metric and SI units for measurement is shown in Table 2.1.

Sample Problem 2.1 ▶ **SI Units of Measurement**

State the type of measurement (mass, length, volume, temperature, or time) indicated by the unit in each of the following:
a. 45.6 kg   **b.** 1.85 m³
c. 45 s   **d.** 315 K

Solution

a. mass   **b.** volume
c. time   **d.** temperature

Table 2.1 Units of Measurement

Measurement	Metric	SI
Length	Meter (m)	Meter (m)
Volume	Liter (L)	Cubic meter (m^3)
Mass	Gram (g)	Kilogram (kg)
Time	Second (s)	Second (s)
Temperature	Celsius (°C)	Kelvin (K)

Study Check

Give the SI unit and abbreviation that would be used to express the following measurements:

a. the length of a football field   **b.** the daytime temperature on Mars
c. the mass of an electric car

Answers to all of the Study Checks can be found at the end of this chapter in the Answer section. Checking your answers will let you know if you understand the material in this section.

Questions and Problems   **Units of Measurement**

In every chapter, each red, odd-numbered exercise is paired with the next even-numbered exercise. The answers for all the Study Checks and the red, odd-numbered exercises are given at the end of each chapter. The complete solutions to the odd-numbered exercises are in the *Study Guide*.

2.1 State the name of the unit and the type of measurement indicated for each of the following quantities:
a. 4.8 m   **b.** 325 g   **c.** 1.5 L   **d.** 480 s
e. 28°C

2.2 State the name of the unit and the type of measurement indicated for each of the following quantities:
a. 0.8 L   **b.** 3.6 m   **c.** 14 kg   **d.** 35 g
e. 373 K

2.3 State the name of the unit and identify that unit as an SI unit, a metric unit, both, or neither.

a. 5.5 m   **b.** 45 kg   **c.** 5 ft   **d.** 25 s
e. 22°C

2.4 State the name of the unit and identify that unit as an SI unit, a metric unit, both, or neither.
a. 8 m^3 **b.** 245 K   **c.** 45°F   **d.** 125 L
e. 125 g

2.5 State the name of the unit and identify that unit as an SI unit, a metric unit, both, or neither.
a. 25.2 g   **b.** 1.5 L   **c.** 15°C   **d.** 5.5 m
e. 15 s

2.6 State the name of the unit and identify that unit as an SI unit, a metric unit, both, or neither.
a. 245 K   **b.** 45.8 kg   **c.** 0.48 L   **d.** 28.6 m
e. 4.2 m^3

2.2 Scientific Notation

Learning Goal

Write a number in scientific notation.

In chemistry and science in general, measurements involve numbers that may be very small and sometimes very large. For example, the width of a human hair is about 0.000 008 m, and there are typically 100 000 hairs on the average human scalp. (See Figure 2.5.) Sometimes spaces are used between sets of three digits to make the places easier to count. For both measurements, it is convenient to use *scientific notation,* an efficient way to write very large and very small numbers.

Item Measured	Measurement	Scientific Notation
Width of a human hair	0.000 008 m	8×10^{-6} m
Hairs on a human scalp	100 000 hairs	1×10^5 hairs

Writing a Number in Scientific Notation

When a number is written in **scientific notation,** there are two parts: a coefficient and a power of 10. For example, the number 2400 in scientific notation is 2.4×10^3. The coefficient is 2.4 and 10^3 shows the power of 10. The coefficient is determined by moving the decimal point three places to the left

Figure 2.5 Humans have an average of 1×10^5 hairs on their scalps. Each hair is about 8×10^{-6} m wide.

Q Why are large and small numbers written in scientific notation?

8×10^{-6} m

to give a number between 1 and 10. Because we moved the decimal point three places to the left, the power of 10 is a positive 3, which is written as 10^3. For a number greater than 1, the power of 10 is positive.

$$\underset{\underset{\text{3 places}}{\longleftarrow}}{2400.} = 2.4 \times 1000 = \underset{\text{Coefficient}}{2.4} \times \underset{\substack{\text{Power} \\ \text{of 10}}}{10^3}$$

When a number less than 1 is written in scientific notation, the power of 10 is negative. For example, to write the number 0.000 86 in scientific notation the decimal point is moved to the right four places to give a coefficient of 8.6, which is greater than 1 but less than 10. By moving the decimal point four places to the right, the power of 10 becomes a negative 4, or 10^{-4}.

$$\underset{\underset{\text{4 places}}{\longrightarrow}}{0.00086} = \frac{8.6}{10\,000} = \frac{8.6}{10 \times 10 \times 10 \times 10} = \underset{\text{Coefficient}}{8.6} \times \underset{\substack{\text{Power} \\ \text{of 10}}}{10^{-4}}$$

Table 2.2 gives some examples of numbers written as positive and negative powers of 10. The powers of 10s are really a way of keeping track of the decimal point in the decimal number. Table 2.3 gives several examples of writing standard numbers in scientific notation.

Table 2.2   Some Numbers Written as Powers of 10

Standard Number	Multiples of 10		Scientific Notation	
10 000	$10 \times 10 \times 10 \times 10$		1×10^4	
1 000	$10 \times 10 \times 10$		1×10^3	
100	10×10		1×10^2	Some positive
10	10		1×10^1	powers of 10
1	0		1×10^0	
0.1	$\dfrac{1}{10}$		1×10^{-1}	
0.01	$\dfrac{1}{10} \times \dfrac{1}{10}$	$= \dfrac{1}{100}$	1×10^{-2}	Some negative
0.001	$\dfrac{1}{10} \times \dfrac{1}{10} \times \dfrac{1}{10}$	$= \dfrac{1}{1\,000}$	1×10^{-3}	powers of 10
0.000 1	$\dfrac{1}{10} \times \dfrac{1}{10} \times \dfrac{1}{10} \times \dfrac{1}{10}$	$= \dfrac{1}{10\,000}$	1×10^{-4}	

Table 2.3 Writing Numbers in Scientific Notation

Measured Quantity	Standard Number	Scientific Notation
Diameter of Earth	12 800 000 m	1.28×10^7 m
Depth of Lake Baikal	1740 m	1.74×10^3 m
Mass of a typical human	68 kg	6.8×10^1 kg
Mass of a hummingbird	0.002 kg	2×10^{-3} kg
Length of a pox virus	0.000 000 3 m	3×10^{-7} m
Mass of bacterium (mycoplasma)	0.000 000 000 000 000 000 1kg	1×10^{-19} kg

Scientific Notation and Calculators

You can enter numbers in scientific notation on many calculators using the EE or EXP key. After you enter the coefficient, push the EXP (or EE) key and enter only the power of 10, because the EXP function key includes the $\times 10$ value. To enter a negative power of 10, push the plus/minus key $(+/-)$, *not* the minus $(-)$ key. Some calculators require entering the sign before the power.

Number to Enter	Method	Display Reads	
4×10^6	4 EXP (EE) 6	4 06 or	4 ^{06}
2.5×10^{-4}	2.5 EXP (EE) 4 +/-	2.5 −04 or	2.5^{-04}

When a calculator answer appears in scientific notation, it is usually shown in the display as a number between 1 and 10 followed by a space and the power of 10. To express this display in scientific notation, write the number, insert "$\times 10$," and use the power of 10 as an exponent.

Calculator Display			Expressed in Scientific Notation
7.52 04 or	7.52^{04}		7.52×10^4
5.8 −02 or	5.8^{-02}		5.8×10^{-2}

On many scientific calculators, a number can be converted into scientific notation using the appropriate keys. For example, the number 0.000 52 can be entered followed by hitting the 2nd or 3rd function key and the SCI key. The scientific notation appears in the calculator display as a coefficient and the power of 10.

0.000 52 [2nd or 3rd function key] [SCI] = 5.2^{-4} OR 5.2−04 = 5.2×10^{-4}
                      Key   Key   Display

Converting a Number in Scientific Notation to a Standard Number

When a number in scientific notation has a positive power of 10, the standard number is written by multiplying the coefficient by the value of the power of 10. Zeros are used to give the correct place to the nonzero digits.

$$4.3 \times 10^2 = 4.3 \times 10 \times 10 = 430$$

For a number with a negative power of 10, the standard number is written by multiplying the coefficient by the value of the power of 10. Zeros are added in front of the nonzero digits to locate the decimal point.

$$2.5 \times 10^{-5} = 2.5 \times \frac{1}{10^5} = 2.5 \times 0.00001 = 0.000025$$

Sample Problem 2.2   **Scientific Notation**

1. Write the following measurements using scientific notation:
    a. 350 g   **b.** 0.000 16 L   **c.** 5 220 000 m
2. Write the following measurements as a standard number:
    a. 2.85×10^2 L   **b.** 7.2×10^{-3} m   **c.** 2.4×10^5 g

Solution

1. **a.** 3.5×10^2 g   **b.** 1.6×10^{-4} L   **c.** 5.22×10^6 m
2. **a.** 285 L   **b.** 0.007 2 m   **c.** 240 000 g

Study Check

Write the following measurements using scientific notation:
a. 425 000 m   **b.** 0.000 000 8 g

Questions and Problems   **Scientific Notation**

Answers for odd-numbered Questions and Problems are given at the end of this chapter. The complete solutions to the odd-numbered Questions and Problems are in the *Study Guide*.

2.7 Write the following measurements in scientific notation:
   a. 55 000 m   **b.** 480 g   **c.** 0.000 005 cm
   d. 0.000 14 s   **e.** 0.007 85 L   **f.** 670 000 kg
2.8 Write the following measurements in scientific notation:
   a. 180 000 000 g   **b.** 0.000 06 m   **c.** 750 000 g
   d. 0.15 m   **e.** 0.024 s   **f.** 1500 m^3
2.9 Which number in each pair is larger?
   a. 7.2×10^3 or 8.2×10^2
   b. 4.5×10^{-4} or 3.2×10^{-2}

   c. 1×10^4 or 1×10^{-4}
   d. 0.000 52 or 6.8×10^{-2}
2.10 Which number in each pair is smaller?
   a. 4.9×10^{-3} or 5.5×10^{-9}
   b. 1250 or 3.4×10^2
   c. 0.000 000 4 or 5×10^{-8}
   d. 4×10^8 or 4×10^{-10}
2.11 Write the following as standard numbers:
   a. 1.2×10^4 **b.** 8.25×10^{-2}
   c. 4×10^6 **d.** 5×10^{-3}
2.12 Write the following as standard numbers:
   a. 3.6×10^{-5} **b.** 8.75×10^4
   c. 3×10^{-2} **d.** 2.12×10^5

Learning Goal

Identify a number as measured or exact; determine the number of significant figures in a measured number.

2.3 Measured Numbers and Significant Figures

Whenever you make a measurement, you use some type of measuring device. For example, you may use a meterstick to measure your height, a scale to check your weight, and a thermometer to take your temperature. **Measured numbers** are the numbers you obtain when you measure a quantity such as your height, weight, or temperature.

Measured Numbers

Suppose you are going to measure the lengths of the objects in Figure 2.6. You would select a ruler with a scale marked on it. By observing the lines on

(a)

(b)

(c)

Figure 2.6 The lengths of the rectangular objects are measured as **(a)** 4.5 cm and **(b)** 4.55 cm.

Q What is the length of the object in (c)?

WEB TUTORIAL
Significant Figures

the scale, you determine the measurement for each object. Perhaps the divisions on the scale are marked as 1 cm. Another ruler might be marked in divisions of 0.1 cm. To report the length, you would first read the numerical value of the marked line. Finally, you *estimate* between the smallest marked lines. This estimated number is the final digit in a measured number.

An estimation is made by visually dividing the space between the smallest marked lines. For example, in Figure 2.6a, the end of the object falls between the lines marked 4 cm and 5 cm. That means that the length is 4 cm plus an estimated digit. If you estimate that the end is halfway between 4 cm and 5 cm, you would report its length as 4.5 cm. However, someone else might report the length as 4.4 cm. The last digit in a measured number can differ because people do not estimate in the same way. The ruler shown in Figure 2.6b is marked with lines spaced 0.1 cm apart. With this ruler, you can estimate the value of the hundredths place (0.01 cm). Perhaps you would report the length of the object as 4.55 cm, while someone else may report its length as 4.56 cm. Both results are acceptable.

There is always *uncertainty* in any measurement. When a measurement ends right on a marked line, a zero is written as the estimated digit. For example, in Figure 2.6c, the measurement for length is written as 3.0 cm, not 3. This means that the uncertainty of the measurement is in the tenths place. In any measurement, the last digit is the uncertain digit.

Significant Figures

In a measured number, the **significant figures** are all the digits including the estimated digit. All *nonzero* numbers are counted as significant. Zeros may or may not be significant depending on their position in a number. Table 2.4 gives the rules and examples for counting significant figures.

Table 2.4 Significant Figures in Measured Numbers

Rule	Examples of Measured Numbers	Number of Significant Figures
1. A number is a *significant figure* if it is		
a. not a zero	4.5 g	2
	122.35 m	5
b. a zero between nonzero digits	205 m	3
	5.082 kg	4
c. a zero at the end of a decimal number	50. L	2
	25.0°C	3
	16.00 g	4
d. any digit in the coefficient of a number written in scientific notation	4.0×10^5 m	2
	5.70×10^{-3} g	3
2. A zero is *not significant* if it is		
a. at the beginning of a decimal number	0.000 4 lb	1
	0.075 m	2
b. used as a placeholder in a large number without a decimal point	850 000 m	2
	1 250 000 g	3

When one or more zeros in a large number are significant digits, they can be shown by writing the number using scientific notation. For example, if the first zero in the measurement 500 m is significant, it can be shown by writing 5.0×10^2 m. In this text, we will place a decimal point after a significant zero at the end of a number. For example, a measurement written as 250. g has three significant figures, which includes the zero. It could also be written as 2.50×10^2 g. Unless noted otherwise, we will assume that zeros at the end of large standard numbers are not significant; that is, we would interpret 400 000 g as 4×10^5 g, with one significant figure.

Exact Numbers

Exact numbers are numbers obtained by counting items or from a definition that compares two units in the same measuring system. Suppose a friend asks you about the number of bicycles you have or the number of classes you are taking in school. You would answer by counting the items. It is not necessary for you to use any type of measuring tool. Suppose someone asks you to state the number of seconds in 1 minute. Without using any measuring device, you would give the definition: 60 seconds in 1 minute. Exact numbers are not measured, do not have a limited number of significant figures, and do not affect the number of significant figures in a calculated answer. For more examples of exact numbers, see Table 2.5.

Table 2.5 Examples of Some Exact Numbers

	Defined Equalities	
Counted Numbers	U.S. System	Metric System
Eight doughnuts	1 ft = 12 in.	1 L = 1000 mL
Two baseballs	1 qt = 4 cups	1 m = 100 cm
Five caps	1 lb = 16 ounces	1 kg = 1000 g

Sample Problem 2.3   Significant Figures

Identify each of the following numbers as measured or exact and give the number of significant figures in each measured number.

a. 42.2 g   **b.** 3 eggs   **c.** 0.000 5 cm
d. 450 000 km   **e.** 9 planets   **f.** 3.500×10^5 s

Solution

a. measured, three   **b.** exact   **c.** measured, one
d. measured, two   **e.** exact   **f.** measured, four

Study Check

State the number of significant figures in each of the following measured numbers:

a. 0.000 35 g   **b.** 2000 m   **c.** 2.0045 L

Questions and Problems ▷ **Measured Numbers and Significant Figures**

2.13 What is the estimated digit in each of the following measured numbers?
a. 8.6 m   **b.** 45.25 g   **c.** 25.0°C

2.14 What is the estimated digit in each of the following measured numbers?
a. 125.04 g   **b.** 5.057 m   **c.** 525.8°C

2.15 Identify the numbers in each of the following statements as measured or exact numbers:
a. A person weighs 155 lb.
b. The basket holds 8 apples.
c. In the metric system, 1 kg is equal to 1000 g.
d. The distance from Denver, Colorado, to Houston, Texas, is 1720 km.

2.16 Identify the numbers in each of the following statements as measured or exact numbers:
a. There are 31 students in the laboratory.
b. The oldest known flower lived 120 000 000 years ago.
c. The largest gem ever found, an aquamarine, has a mass of 104 kg.
d. A laboratory test shows a blood cholesterol level of 184 mg/dL.

2.17 In each set of numbers, identify the measured number(s), if any.
a. 3 hamburgers and 6 oz of meat
b. 1 table and 4 chairs
c. 0.75 lb of grapes and 350 g of butter
d. 60 s equals 1 min

2.18 In each set of numbers, identify the exact number(s), if any.
a. 5 pizzas and 50.0 g of cheese
b. 6 nickels and 16 g of nickel
c. 3 onions and 3 lb of potatoes
d. 5 miles and 5 cars

2.19 For each measurement, indicate if the zeros are significant figures.
a. 0.0038 m   **b.** 5.04 cm   **c.** 800. L
d. 3.0×10^{-3} kg   **e.** 85 000 g

2.20 For each measurement, indicate if the zeros are significant figures.
a. 20.05 g   **b.** 5.00 m   **c.** 0.000 02 L
d. 120 000 years   **e.** 8.05×10^2 g

2.21 How many significant figures are in each of the following measured quantities?
a. 11.005 g   **b.** 0.000 32 m   **c.** 36 000 000 m
d. 1.80×10^4 g   **e.** 0.8250 L   **f.** 30.0°C

2.22 How many significant figures are in each of the following measured quantities?
a. 20.60 L   **b.** 1036.48 g   **c.** 4.00 m
d. 20.8°C   **e.** 60 800 000 g   **f.** 5.0×10^{-3} L

2.23 In which of the following pairs do both numbers contain the same number of significant figures?
a. 11.0 m and 11.00 m
b. 600.0 K and 60 K
c. 0.000 75 s and 75 000 s
d. 250.0 L and 0.02500 L

2.24 In which of the following pairs do both numbers contain the same number of significant figures?
a. 0.00575 g and 5.75×10^{-3} g
b. 0.0250 and 0.205 m
c. 150 000 s and 1.50×10^4 s
d. 3.8×10^{-2} L and 7.5×10^5 L

2.25 Write each of the following in scientific notation with two significant figures:
a. 5000 L   **b.** 30 000 g   **c.** 100 000 m
d. 0.000 25 cm

2.26 Write each of the following in scientific notation with two significant figures:
a. 5 100 000 g   **b.** 26 000 s   **c.** 40 000 m
d. 0.000 820 kg

2.4 Significant Figures in Calculations

Learning Goal

Adjust calculated answers to give the correct number of significant figures.

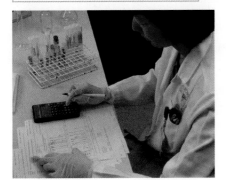

In the sciences, we measure many things: the length of a bacterium, the volume of a gas sample, the temperature of a reaction mixture, or the mass of iron in a sample. The numbers obtained from these types of measurements are often used in calculations. The number of significant figures in the measured numbers limits the number of significant figures in the calculated answer.

Using a calculator will help do calculations faster. However, calculators cannot think for you. It is up to you to enter the numbers correctly, press the right function keys, and give an answer with the correct number of significant figures.

Rounding Off

To calculate the area of a carpet that measures 5.5 m by 3.5 m, multiply 5.5 times 3.5 to obtain the number 19.25 as the area in square meters. However, all four digits cannot be used in the answer because they are not all significant figures. The measurements of length and width each have two significant figures. This means that the calculated result must be *rounded off* to give an answer that also has two significant figures, which would be $19\,m^2$. When you obtain a calculator result, determine the number of significant figures needed for the answer and round off using the following rules.

Rules for Rounding Off

1. If the first digit to be dropped is *4 or less,* it and all following digits are dropped from the number.

2. If the first digit to be dropped is *5 or greater,* the last retained digit of the number is increased by 1.

Note: The value of a large number is retained by using zeros to replace dropped digits.

	Three Significant Figures	Two Significant Figures
Example 1: 8.4234 rounds off to	8.42	8.4
Example 2: 14.780 rounds off to	14.8	15
Example 3: 3256	3260	3300

Sample Problem 2.4 ▶ **Rounding Off**

Round off each of the following numbers to three significant figures:
a. 35.7823 m   **b.** 0.002627 L   **c.** 3826.8 g   **d.** 1.2836 kg

Solution
a. 35.8 m   **b.** 0.00263 L   **c.** 3830 g or $3.83 \times 10^3\,g$
d. 1.28 kg

Study Check

Round off each of the numbers in Sample Problem 2.4 to two significant figures.

Multiplication and Division

In multiplication and division, the final answer can only have the same number of digits as the measurement with the *fewest* significant figures (SFs).

Example 1

Multiply the following measured numbers: 24.65×0.67

24.65 ⨉ 0.67 =   *16.5155* ⟶ 17
Four SFs   Two   Calculator   Final answer,
            SFs   display   rounded to two SFs

The answer in the calculator display has more digits than the data allow. The measurement 0.67 has the least number of significant figures, two. Therefore, the calculator answer is rounded off to two significant figures.

Example 2

Solve the following:

$$\frac{2.85 \times 67.4}{4.39}$$

To do this problem on a calculator, enter the number and then press the operation key. In this case, we might press the keys in the following order:

2.85 ⊠ 67.4 ÷ 4.39 = 43.756264 ⟶ 43.8

Three SFs   Three SFs   Three SFs   Calculator display   Final answer, rounded to three SFs

All of the measurements in this problem have three significant figures. Therefore, the calculator result is rounded off to give an answer, 43.8, that has three significant figures.

Adding Significant Zeros

Sometimes a calculator displays a small whole number. To give an answer with the correct number of significant figures, significant zeros are written after the calculator result. For example, suppose the calculator display is 4, but you used measurements that have three significant numbers. The answer 4.00 is obtained by placing two significant zeros after the 4.

$$\frac{8.00}{2.00} = \quad 4 \quad \longrightarrow \quad 4.00$$

Three SFs   Calculator display   Final answer, two zeros added to give three SFs

---

Sample Problem 2.5   **Significant Figures in Multiplication and Division**

Perform the following calculations of measured numbers. Give the answers with the correct number of significant figures.

a. 56.8×0.37 **b.** $\frac{71.4}{11.0}$ **c.** $\frac{(2.075)(0.585)}{(8.42)(0.0045)}$ **d.** $\frac{25.0}{5.00}$

Solution
a. 21   **b.** 6.49   **c.** 32   **d.** 5.00 (Must add significant zeros)

Study Check

Perform the following calculations of measured numbers. Give the answers with the correct number of significant figures.

a. 45.26×0.01088 **b.** $2.60 \div 324$ **c.** $\frac{4.0 \times 8.00}{16}$

Addition and Subtraction

In addition or subtraction, the answer is written so it has the same number of places as the measurement having the largest place.

Example 3

Add:

2.045	Three decimal places
+ 34.1	One decimal place
36.145	Calculator display
36.1	Answer, rounded to one decimal place

Example 4

Subtract:

255	Ones place
− 175.65	Two decimal places
79.35	Calculator display
79	Answer, rounded to ones place

When numbers are added or subtracted to give answers ending in zero, the zero does not appear after the decimal point in the calculator display. For example, 14.5 g − 2.5 g = 12.0 g. However, if you do the subtraction on your calculator, the display shows 12. To give the correct answer, a significant zero is written after the decimal point.

Example 5

14.5 g	One decimal place
− 2.5 g	One decimal place
12.	Calculator display
12.0 g	Answer, zero written after the decimal point

Sample Problem 2.6   **Significant Figures in Addition and Subtraction**

Perform the following calculations and give the answers with the correct number of decimal places:

a. 27.8 cm + 0.235 cm   **b.** 104.45 mL + 0.838 mL + 46 mL
c. 153.247 g − 14.82 g

Solution

a. 28.0 cm   **b.** 151 mL   **c.** 138.43 g

Study Check

Perform the following calculations and give the answers with the correct number of decimal places:

a. 82.45 mg + 1.245 mg + 0.000 56 mg   **b.** 4.259 L − 3.8 L

2.27 Why do we usually need to round off calculations that use measured numbers?

2.28 Why do we sometimes add a zero to a number in a calculator display?

2.29 Round off each of the following numbers to three significant figures:
a. 1.854   **b.** 184.203 8   **c.** 0.004 738 265
d. 8807   **e.** 1.832 149

2.30 Round off each of the numbers in problem 2.29 to two significant figures.

2.31 Round off or add zeros to the following calculator answers to give a final answer with three significant figures:
a. 56.855 m   **b.** 0.002 282 5 g
c. 11 527 s   **d.** 8.1 L

2.32 Round off or add zeros to each of the calculated answers to give a final answer with two significant figures:
a. 3.2805 m   **b.** 1.855×10^2 g
c. 0.002 341 m   **d.** 2 L

2.33 For the following problems, give answers with the correct number of significant figures:
a. 45.7×0.034 **b.** 0.00278×5
c. $\dfrac{34.56}{1.25}$ **d.** $\dfrac{(0.2465)(25)}{1.78}$
e. $(2.8 \times 10^4)(5.05 \times 10^{-6})$
f. $\dfrac{(3.45 \times 10^{-2})(1.8 \times 10^5)}{(8 \times 10^3)}$

2.34 For the following problems, give answers with the correct number of significant figures:
a. 400×185 **b.** $\dfrac{2.40}{(4)(125)}$
c. $0.825 \times 3.6 \times 5.1$ **d.** $\dfrac{(3.5)(0.261)}{(8.24)(20.0)}$
e. $\dfrac{(5 \times 10^{-5})(1.05 \times 10^4)}{(8.24 \times 10^{-8})}$
f. $\dfrac{(4.25 \times 10^2)(2.56 \times 10^{-3})}{(2.245 \times 10^{-3})(56.5)}$

2.35 For the following problems, give answers with the correct number of decimal places:
a. 45.48 cm + 8.057 cm
b. 23.45 g + 104.1 g + 0.025 g
c. 145.675 mL − 24.2 mL
d. 1.08 L − 0.585 L

2.36 For the following problems, give answers with the correct number of decimal places:
a. 5.08 g + 25.1 g
b. 85.66 cm + 104.10 cm + 0.025 cm
c. 24.568 mL − 14.25 mL
d. 0.2654 L − 0.2585 L

2.5 Prefixes and Equalities

The special feature of the metric system of units is that a **prefix** can be attached to any unit to increase or decrease its size by some factor of 10. For example, the prefixes *milli* and *micro* are used to make the smaller units milligram (mg) and microgram (μg). Table 2.6 lists some of the metric prefixes, their symbols, and their decimal values.

Table 2.6   Metric and SI Prefixes

Prefix[a]	Symbol	Meaning	Numerical Value	Scientific Notation
Prefixes That Increase the Size of the Unit				
tera	T	trillion	1 000 000 000 000	10^{12}
giga	G	billion	1 000 000 000	10^9
mega	M	million	1 000 000	10^6
kilo	k	thousand	1 000	10^3
Prefixes That Decrease the Size of the Unit				
deci	d	tenth	0.1	10^{-1}
centi	c	hundredth	0.01	10^{-2}
milli	m	thousandth	0.001	10^{-3}
micro	μ	millionth	0.000 001	10^{-6}
nano	n	billionth	0.000 000 001	10^{-9}
pico	p	trillionth	0.000 000 000 001	10^{-12}
femto	f	quadrillionth	0.000 000 000 000 001	10^{-15}

[a] The prefixes used by chemists most often are in boldface type.

The prefix *centi* is like cents in a dollar. One cent would be a centidollar, or $\frac{1}{100}$ of a dollar. That also means that one dollar is the same as 100 cents. The prefix *deci* is like dimes in a dollar. One dime would be a decidollar, or $\frac{1}{10}$ of a dollar. That also means that one dollar is the same as 10 dimes.

The relationship of a unit to its base unit can be expressed by replacing the prefix with its numerical value. For example, when the prefix *kilo* in *kilometer* is replaced with its value of 1000, we find that a kilometer is equal to 1000 meters. Other examples follow:

1 **kilo**meter (1 km) = **1000** meters (1000 m)

1 **kilo**liter (1 kL) = **1000** liters (1000 L)

1 **kilo**gram (1 kg) = **1000** grams (1000 g)

Sample Problem 2.7 Prefixes

Fill in the blanks with the correct numerical value:
a. kilogram = ____ grams
b. millisecond = ____ second
c. deciliter = ____ liter

Solution
a. The numerical value of *kilo* is 1000; 1000 grams.
b. The numerical value of *milli* is 0.001; 0.001 second.
c. The numerical value of *deci* is 0.1; 0.1 liter.

Study Check

Write the correct prefix in the blanks:
a. 1 000 000 seconds = ____ seconds
b. 0.01 meter = ____ meter

Measuring Length

An ophthalmologist may measure the diameter of the retina of the eye in centimeters (cm), whereas a surgeon may need to know the length of a nerve in millimeters (mm). When the prefix *centi* is used with the unit *meter*, it indicates the unit *centimeter*, a length that is one-hundredth of a meter (0.01 m). A *millimeter* measures a length of 0.001 m. There are 100 cm and 1000 mm in a meter.

If we compare the lengths of a millimeter and a centimeter, we find that 1 mm is 0.1 cm; there are 10 mm in 1 cm. These comparisons are examples of **equalities,** which show the relationship between two units that measure the same quantity. For example, in the equality 1 m = 100 cm, both quantities describe the same length but in different units. In every equality expression, each quantity always has both a number and a unit.

First Quantity   Second Quantity

1 m = 100 cm

Number + unit   Number + unit

Some Length Equalities

1 m = 100 cm

1 m = 1000 mm

1 cm = 10 mm

Some metric units for length are compared in Figure 2.7.

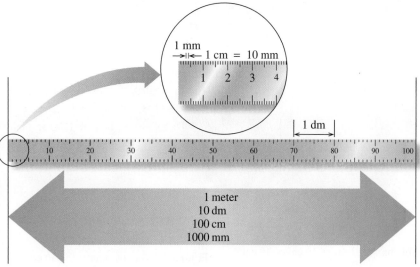

Figure 2.7 The metric length of 1 meter is the same length as 10 dm, 100 cm, and 1000 mm.

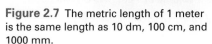 *How many millimeters (mm) are in 1 centimeter (cm)?*

Measuring Volume

Volumes of 1 L or smaller are common in the sciences. When a liter is divided into 10 equal portions, each portion is a deciliter (dL). There are 10 dL in 1 L. When a liter is divided into a thousand parts, each of the smaller volumes is called a *milliliter.*

Some Volume Equalities

1 L = 10 dL

1 L = 1000 mL

1 dL = 100 mL

The **cubic centimeter (cm³ or cc)** is the volume of a cube whose dimensions are 1 cm on each side. A cubic centimeter has the same volume as a milliliter, and the units are often used interchangeably.

$$1 \text{ cm}^3 = 1 \text{ cc} = 1 \text{ mL}$$

When you see *1 cm*, you are reading about length; when you see *1 cc* or *1 cm³* or *1 mL,* you are reading about volume. A comparison of units of volume is illustrated in Figure 2.8.

Measuring Mass

When you get a physical examination, your mass is recorded in kilograms, whereas the results of your laboratory tests are reported in grams, milligrams (mg), or micrograms (μg). A kilogram is equal to 1000 g. One gram represents the same mass as 1000 mg, and 1 mg equals 1000 μg.

Some Mass Equalities

1 kg = 1000 g

1 g   = 1000 mg

1 mg = 1000 μg

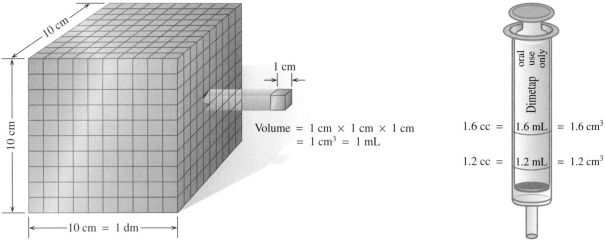

$$\text{Volume} = 1 \text{ cm} \times 1 \text{ cm} \times 1 \text{ cm}$$
$$= 1 \text{ cm}^3 = 1 \text{ mL}$$

$$\text{Volume} = 10 \text{ cm} \times 10 \text{ cm} \times 10 \text{ cm}$$
$$= 1000 \text{ cm}^3$$
$$= 1000 \text{ mL}$$
$$= 1 \text{ L}$$

1.6 cc = 1.6 mL = 1.6 cm³

1.2 cc = 1.2 mL = 1.2 cm³

Figure 2.8 A cube measuring 10 cm on each side has a volume of 1000 cm^3, or 1 L; a cube measuring 1 cm on each side has a volume of 1 cm^3 (cc) or 1 mL.

Q What is the relationship between a milliliter (mL) and a cubic centimeter (cm^3)?

CHEM NOTE

TOXICOLOGY AND RISK-BENEFIT ASSESSMENT

Each day we make choices about what we do or what we eat, often without thinking about the risks associated with these choices. We are aware of the risks of cancer from smoking or lead paints, and we know there is a greater risk of crossing a street where there is no light or crosswalk.

A basic concept of toxicology is the statement of Paracelsus that the right dose is the difference between a poison and a cure. To evaluate the level of danger from various substances, natural or synthetic, a risk assessment is made by exposing laboratory animals to the substances and monitoring the health effects.

Often, doses very much greater than humans might encounter are given to the test animals. Many hazardous chemicals or substances have been identified by these tests. One measure of toxicity is the LD_{50} or "lethal dose," which is the concentration of the substance that causes death in 50% of the test animals. A dosage is typically measured in milligrams per kilogram of body mass or micrograms per kilogram.

Dosage	Units
parts per million (ppm) =	milligrams per kilogram (mg/kg)
parts per billion (ppb) =	micrograms per kilogram (μg/kg)

There are other evaluations that need to be made, but it is easy to compare LD_{50}s. Parathion, a pesticide, with an LD_{50} of 3 mg/kg would be highly toxic. That means that half the test animals given 3 mg/kg body mass would be expected to die. But salt (sodium chloride) with an LD_{50} of 3000 mg/kg would have a much lower toxicity. By comparison to parathion, a huge amount of salt would need to be ingested before any toxic effect would be observed. Although risk to animals based on dose can be evaluated in the laboratory, it is more difficult to determine the impact in the environment since there is also a difference between continued exposure and a single, large dose of the substance.

Table 2.7 lists some LD_{50}s comparing pesticides and common substances in our everyday life in order of increasing toxicity.

Table 2.7 Some LD_{50} Values for Pesticides and Common Materials Tested in Rats

Substance	LD_{50} (mg/kg)
Table sugar	29 700
Baking soda	4220
Table salt	3000
Ethanol	2080
Aspirin	1100
Caffeine	192
Sodium cyanide	6
Parathion	3

> **Sample Problem 2.8** ▶ **Writing Metric Relationships**
>
> **1.** Identify the larger unit in each of the following pairs:
>   **a.** centimeter or kilometer   **b.** L or dL   **c.** mg or μg
> **2.** Complete the following list of metric equalities:
>   **a.** 1 L = _____ dL   **b.** 1 km = _____ m
>   **c.** 1 m = _____ cm   **d.** 1 cm^3 = _____ mL
>
> **Solution**
>
> **1. a.** kilometer   **b.** L   **c.** mg
> **1. a** 10 dL   **b.** 1000 m   **c.** 100 cm   **d.** 1 mL
>
> **Study Check**
>
> Complete the following equalities:
> **a.** 1 kg = _____ g   **b.** 1 mL = _____ L

Questions and Problems ▶ **Prefixes and Equalities**

2.37 The speedometer shown above is marked in both km/hr and mi/hr or mph. What is the meaning of each abbreviation?

2.38 In Canada, a highway sign gives a speed limit as 80 km/hr. According to the speedometer in problem 2.37, would you be exceeding the speed limit of 55 mph if you were in the United States?

2.39 How does the prefix *kilo* affect the gram unit in *kilogram*?

2.40 How does the prefix *centi* affect the meter unit in *centimeter*?

2.41 Write the abbreviation for each of the following units:
   a. milligram   **b.** deciliter   **c.** kilometer
   d. kilogram   **e.** microliter   **f.** nanosecond

2.42 Write the complete name for each of the following units:
   a. cm   **b.** kg   **c.** dL
   d. Gm   **e.** μg   **f.** pg

2.43 Write the numerical values for each of the following prefixes:
   a. centi   **b.** kilo   **c.** milli
   d. deci   **e.** mega   **f.** nano

2.44 Write the complete name (prefix + unit) for each of the following numerical values:
   a. 0.10 g   **b.** 0.000 001 g
   c. 1000 g   **d.** $\frac{1}{100}$ g
   e. 0.001 g   **f.** 0.000 000 000 001 g

2.45 Complete the following relationships:
   a. 1 m = ____ cm   **b.** 1 km = ____ m
   c. 1 mm = ____ m   **d.** 1 L = ____ mL

2.46 Complete the following relationships:
   a. 1 kg = ____ g   **b.** 1 mL = ____ L
   c. 1 g = ____ kg   **d.** 1 g = ____ mg

2.47 For each of the following pairs, which is the larger unit?
   a. milligram or kilogram   **b.** milliliter or microliter
   c. cm or km   **d.** kL or dL

2.48 For each of the following pairs, which is the smaller unit?
   a. mg or g   **b.** centimeter or millimeter
   c. mm or μm   **d.** mL or dL

2.6 Writing Conversion Factors

Many problems in chemistry require a change of units. You make changes in units every day. For example, suppose you spent 2.0 hours (hr) on your homework, and someone asked you how many minutes that was. You would answer 120 minutes (min). You knew how to change from hours to minutes because you knew an equality (1 hr = 60 min) that related the two units. To do the problem, the equality is written in the form of a fraction called a **conversion factor.** One of the quantities is the numerator, and the other is the denominator. Be sure to include the units when you write the conversion factors. Two factors are always possible from any equality.

Two Conversion Factors for the Equality 1 hr = 60 min

$$\frac{\text{Numerator}}{\text{Denominator}} \longrightarrow \frac{60 \text{ min}}{1 \text{ hr}} \quad \text{and} \quad \frac{1 \text{ hr}}{60 \text{ min}}$$

Table 2.8 Some Common Equalities

Quantity	U.S.	Metric (SI)	Metric–U.S.
Length	1 ft = 12 in.	1 km = 1000 m	2.54 cm = 1 in. (exact)
	1 yard = 3 ft	1 m = 1000 mm	1 m = 39.37 in.
	1 mile = 5280 ft	1 cm = 10 mm	1 km = 0.6214 mi
Volume	1 qt = 4 cups	1 L = 1000 mL	1 L = 1.057 qt
	1 qt = 2 pints	1 dL = 100 mL	
	1 gallon = 4 qt	1 mL = 1 cm^3	
Mass	1 lb = 16 oz	1 kg = 1000 g	1 kg = 2.205 lb
		1 g = 1000 mg	453.6 g = 1 lb
Time		1 hr = 60 min	
		1 min = 60 s	

These factors are read as "60 minutes per 1 hour," and "1 hour per 60 minutes." The term *per* means "divide." Some common relationships are given in Table 2.8. It is important that the equality you select to construct a conversion factor is a true relationship.

When an equality shows the relationship for two units from the same system, it is considered a definition and exact. It is not used to determine significant figures. When an equality shows the relationship of units from two different systems, the number is measured and counts toward the significant figures in a calculation. For example, in the equality 1 lb = 453.6 g, the measured number 453.6 has four significant figures. The number 1 in 1 lb is considered exact. An exception is the relationship of 1 in. = 2.54 cm; the value 2.54 has been defined as exact.

Metric Conversion Factors

We can write metric conversion factors for the metric relationships we have studied. For example, from the equality for meters and centimeters, we can write the following factors:

Metric Equality	**Conversion Factors**
1 m = 100 cm	$\dfrac{100 \text{ cm}}{1 \text{ m}}$ and $\dfrac{1 \text{ m}}{100 \text{ cm}}$

Both are proper conversion factors for the relationship; one is just the inverse of the other. The usefulness of conversion factors is enhanced by the fact that we can turn a conversion factor over and use its inverse.

Metric–U.S. System Conversion Factors

Suppose you need to convert from pounds, a unit in the U.S. system, to kilograms in the metric (or SI) system. A relationship you could use is

1 kg = 2.205 lb

The corresponding conversion factors would be

$$\frac{2.205 \text{ lb}}{1 \text{ kg}} \quad \text{and} \quad \frac{1 \text{ kg}}{2.205 \text{ lb}}$$

Figure 2.9 In the United States, the contents of many packaged foods are listed in both U.S. and metric units.

Q What are some advantages of using the metric system?

Figure 2.9 illustrates the contents of some packaged foods in both metric and U.S. units.

Conversion Factors with Powers

Sometimes we need to use a factor that is squared or cubed. This is often the case for areas and volume.

Distance = length

Area = length × length = length^2

Volume = length × length × length = length^3

To obtain the necessary conversion factor, we can use a known equality and raise both sides of the equality to the same power.

Measurement	Equality	Conversion Factors	
Length	1 in. = 2.54 cm	$\dfrac{1 \text{ in.}}{2.54 \text{ cm}}$ and	$\dfrac{2.54 \text{ cm}}{1 \text{ in.}}$
Area	$(1 \text{ in.})^2 = (2.54 \text{ cm})^2$	$\dfrac{(1 \text{ in.})^2}{(2.54 \text{ cm})^2}$ and	$\dfrac{(2.54 \text{ cm})^2}{(1 \text{ in.})^2}$
Volume	$(1 \text{ in.})^3 = (2.54 \text{ cm})^3$	$\dfrac{(1 \text{ in.})^3}{(2.54 \text{ cm})^3}$ and	$\dfrac{(2.54 \text{ cm})^3}{(1 \text{ in.})^3}$

Suppose you want to write an equality and conversion factors for square centimeters and square meters.

Equality: 1 m = 100 cm

Area = length × length = $(1 \text{ m})^2 = (100 \text{ cm})^2$

Conversion factors: $\dfrac{(1 \text{ m})^2}{(100 \text{ cm})^2}$ and $\dfrac{(100 \text{ cm})^2}{(1 \text{ m})^2}$

Sample Problem 2.9 Writing Conversion Factors

Write conversion factors for the relationship between the following pairs of units:

a. milligrams and grams

b. minutes and hours

c. quarts and liters

d. square inches and square feet

Solution

Equality	Conversion Factors
a. $1 \text{ g} = 1000 \text{ mg}$	$\dfrac{1 \text{ g}}{1000 \text{ mg}}$ and $\dfrac{1000 \text{ mg}}{1 \text{ g}}$
b. $1 \text{ hr} = 60 \text{ min}$	$\dfrac{1 \text{ hr}}{60 \text{ min}}$ and $\dfrac{60 \text{ min}}{1 \text{ hr}}$
c. $1.057 \text{ qt} = 1 \text{ L}$	$\dfrac{1.057 \text{ qt}}{1 \text{ L}}$ and $\dfrac{1 \text{ L}}{1.057 \text{ qt}}$
d. $(12 \text{ in.})^2 = (1 \text{ ft})^2$	$\dfrac{(1 \text{ ft})^2}{(12 \text{ in.})^2}$ and $\dfrac{(12 \text{ in.})^2}{(1 \text{ ft})^2}$

Study Check

Write the equality and conversion factors for the relationship between inches and centimeters.

Conversion Factors Stated Within a Problem

Many times, an equality is specified within a problem that is true only for that problem. It might be the cost of 1 kilogram of oranges or the speed of a car in kilometers per hour. Such equalities are easy to miss when you first read a problem. Let's see how conversion factors are written from statements made within a problem.

1. The motorcycle was traveling at a speed of 85 kilometers per hour.

> **Equality:** $1 \text{ hr} = 85 \text{ km}$
>
> **Conversion Factors:** $\dfrac{85 \text{ km}}{1 \text{ hr}}$ and $\dfrac{1 \text{ hr}}{85 \text{ km}}$

2. One tablet contains 500 mg of vitamin C.

> **Equality:** $1 \text{ tablet} = 500 \text{ mg vitamin C}$
>
> **Conversion Factors:** $\dfrac{500 \text{ mg vitamin C}}{1 \text{ tablet}}$ and $\dfrac{1 \text{ tablet}}{500 \text{ mg vitamin C}}$

Conversion Factors from a Percentage, ppm, and ppb

Sometimes a percentage is given in a problem. The term percent (%) means parts per 100 parts. To write a percentage as a conversion factor, we choose a unit and express the numerical relationship of the parts to 100 parts of the whole. For example, an athlete might have 18% (percent) body fat by mass. (See Figure 2.10.) The percent quantity can be written as 18 mass units of body fat in every 100 mass units of body mass. Different mass units, such as grams, kilograms (kg), or pounds (lb) can be used, but both units in the factor must be the same.

Percent quantity:	18% body fat by mass
Equality:	18 kg body fat $=$ 100 kg body mass
Conversion factors:	$\dfrac{100 \text{ kg body mass}}{18 \text{ kg body fat}}$ and $\dfrac{18 \text{ kg body fat}}{100 \text{ kg body mass}}$

Figure 2.10 The thickness of the skin fold at the waist measured in millimeters (mm) is used to determine the percent of body fat.

Q *What is the percent body fat of an athlete with a body mass of 120 kg and 18 kg body fat?*

or

Equality: 18 lb body fat = 100 lb body mass

Conversion factors: $\dfrac{100 \text{ lb body mass}}{18 \text{ lb body fat}}$ and $\dfrac{18 \text{ lb body fat}}{100 \text{ lb body mass}}$

When scientists want to indicate ratios with very small percent values, they use parts per million (ppm) or parts per billion (ppb). The ratio of parts per million indicates the milligrams of a substance per kilogram (mg/kg). The ratio of parts per billion gives the micrograms per kilogram (μg/kg).

Sample Problem 2.10 **Writing Conversion Factors Stated in a Problem**

Write possible conversion factors for each of the following statements:
a. There are 325 mg of aspirin in 1 tablet.
b. One kilogram of bananas costs $1.25.
c. The permissible level of arsenic in water is 10 ppb.

Solution

a. $\dfrac{325 \text{ mg aspirin}}{1 \text{ tablet}}$ and $\dfrac{1 \text{ tablet}}{325 \text{ mg aspirin}}$

b. $\dfrac{\$1.25}{1 \text{ kg bananas}}$ and $\dfrac{1 \text{ kg bananas}}{\$1.25}$

c. $\dfrac{10 \ \mu g}{1 \text{ kg}}$ and $\dfrac{1 \text{ kg}}{10 \ \mu g}$

Study Check

What conversion factors can be written for the following statements?
a. A cyclist in the Tour de France bicycle race rides at the average speed of 62.2 km/hr.
b. A 86.7 g sample of sterling silver has a volume of 8.5 cm^3.

Questions and Problems **Writing Conversion Factors**

2.49 Why can two conversion factors be written for an equality such as 1 m = 100 cm?
2.50 How can you check that you have written the correct conversion factors for an equality?
2.51 What equality is expressed by the conversion factor $\dfrac{1000 \text{ g}}{1 \text{ kg}}$?
2.52 What equality is expressed by the conversion factor $\dfrac{1 \text{ m}}{100 \text{ cm}}$?
2.53 Write the equality and conversion factors for each of the following:
a. One yard is 3 ft.   b. One liter is 1000 mL.
c. One minute is 60 s.   d. One deciliter is 100 mL.

2.54 Write the equality and conversion factors for each of the following:
a. One gallon is 4 quarts.
b. One meter is 1000 millimeters.
c. There are 7 days in 1 week.
d. One dollar has four quarters.
2.55 Write the equality and conversion factors for the following pairs of units:
a. centimeters and meters
b. milligrams and grams
c. liters and milliliters
d. kilograms and milligrams
e. cubic meters and cubic centimeters

2.56 Write the equality and conversion factors for the following pairs of units:
   a. centimeters and inches
   b. pounds and kilograms
   c. pounds and grams
   d. quarts and liters
   e. square centimeters and square inches
2.57 Write the conversion factors for each of the following statements:
   a. A bee flies at an average speed of 3.5 meters per second.
   b. One milliliter of gasoline has a mass of 0.65 g.
   c. An automobile traveled 46.0 km on 1.0 gallon of gasoline.
   d. Sterling silver is 93% by mass silver.
   e. The pesticide level in plums was 29 ppb.
2.58 Write the conversion factors for each of the following statements:
   a. The highway gas mileage was 32 miles per gallon.
   b. There are 20 drops in 1 milliliter of water.
   c. The nitrate level in well water was 32 ppm.
   d. Gold jewelry contains 58% by mass gold.
   e. The price of a gallon of gas is $3.19.

2.7 Problem Solving

Learning Goal

Use conversion factors to change from one unit to another.

The process of problem solving in chemistry often requires the conversion of an initial quantity given in one unit to the same quantity but in different units. By using one or more of the conversion factors we discussed in the previous section, the initial unit can be converted to the final unit.

Given quantity $\times$ one or more conversion factors = desired quantity
(Initial unit) $\longrightarrow$ (Final unit)

You may use a sequence similar to the steps in the following guide for problem solving (GPS):

Guide to Problem Solving (GPS) with Conversion Factors

Guide to Problem Solving Using Conversion Factors

STEP 1
State the given and needed units.

STEP 2
Write a unit plan to convert the given unit to the final unit.

STEP 3
State the equalities and conversion factors to cancel units.

STEP 4
Set up problem to cancel units and calculate answer.

STEP 1 **Given/Need** State the initial unit given in the problem and the final unit needed.

STEP 2 **Plan** Write out a sequence of units that starts with the initial unit and progresses to the final unit for the answer. Be sure you can supply the equality for each unit conversion.

STEP 3 **Equalities/Conversion Factors** For each change of unit in your plan, state the equality and corresponding conversion factors. Recall that equalities are derived from the metric (SI) system, the U.S. system, and statements within a problem.

STEP 4 **Set Up Problem** Write the initial quantity and unit and set up conversion factors that connect the units. Be sure to arrange the units in each factor so the unit in the denominator cancels the preceding unit in the numerator. Check that the units cancel properly to give the final unit. Carry out the calculations, count the significant figures in each measured number, and give a final answer with the correct number of significant figures.

Suppose a problem requires the conversion of 165 lb to kilograms. One part of this statement (165 lb) is the given quantity (initial unit), while another part (kilograms) is the final unit needed for the answer. Once you identify these units, you can determine which equalities you need to convert the initial unit to the final unit.

STEP 1 **Given** 165 lb **Need** kg

STEP 2 **Plan** It is helpful to decide on a plan of units. When we look at the initial units given and the final units needed, we see that one is a metric unit and the other is a unit in the U.S. system of measurement. Therefore, the connecting conversion factor must be one that includes a metric and a U.S. unit.

lb [Metric–U.S. factor] kg

STEP 3 **Equalities/Conversion Factors** From the discussion on U.S. and metric equalities, we can write the following equality and conversion factors:

$$1 \text{ kg} = 2.205 \text{ lb}$$

$$\frac{2.205 \text{ lb}}{1 \text{ kg}} \quad \text{and} \quad \frac{1 \text{ kg}}{2.205 \text{ lb}}$$

STEP 4 **Set Up Problem** Now we can write the setup to solve the problem using the unit plan and a conversion factor. First, write down the initial unit, 165 lb. Then multiply by the conversion factor that has the unit lb in the denominator to cancel out the initial unit. The unit kg in the numerator (top number) gives the final unit for the answer.

Unit for answer goes here

$$165 \text{ lb} \quad \times \quad \frac{1 \text{ kg}}{2.205 \text{ lb}} \quad = \quad 74.8 \text{ kg}$$

Given   Conversion factor   Answer
(initial unit)   (cancels initial unit)   (desired unit)

Take a look at how the units cancel. The unit that you want in the answer is the one that remains after all the other units have cancelled out. This is a way to check that a problem is set up properly.

$$\text{lb} \times \frac{\text{kg}}{\text{lb}} = \text{kg} \quad \textit{Unit needed for answer}$$

The calculation done on a calculator gives the numerical part of the answer. The calculator answer is adjusted to give a final answer with the proper number of significant figures (SFs).

$$165 \times \frac{1}{2.205} = 165 \; \boxed{\div} \; 2.205 = \boxed{74.829932} = 74.8$$

3 SF   4 SF   Calculator   3 SF
                                              answer   (rounded)

The value of 74.8 combined with the final unit, kg, gives the final answer of 74.8 kg. With few exceptions, answers to numerical problems contain a number and a unit.

Sample Problem 2.11   **Problem Solving Using Metric Factors**

The recommended amount of sodium in the diet per day is 2400 mg. How many grams of sodium is that?

Solution

STEP 1   **Given**   2400 mg   **Need**   g

STEP 2 **Plan** When we look at the initial units given and the final units needed, we see that both are metric units. Therefore, the connecting conversion factor must relate two metric units.

mg   Metric factor   g

STEP 3 **Equalities/Conversion Factors** From the discussion on prefixes and metric equalities, we can write the following equality and conversion factors:

$$1 \text{ g} = 1000 \text{ mg}$$

$$\frac{1 \text{ g}}{1000 \text{ mg}} \quad \text{and} \quad \frac{1000 \text{ mg}}{1 \text{ g}}$$

STEP 4 **Set Up Problem** We write the setup using the unit plan and a conversion factor starting with the initial unit, 2400 mg. The final answer (g) is obtained by using the conversion factor that cancels the unit mg.

Unit for answer goes here

$$2400 \text{ mg} \times \frac{1 \text{ g}}{1000 \text{ mg}} = 2.4 \text{ g}$$

Given   Metric factor   Answer

Study Check

If 1890 mL of orange juice is prepared from orange juice concentrate, how many liters of orange juice is that?

Using Two or More Conversion Factors

In many problems, two or more conversion factors are needed to complete the change of units. In setting up these problems, one factor follows the other. Each factor is arranged to cancel the preceding unit until the final unit is obtained. Up to this point, we have used the conversion factors one at a time and calculated an answer. If you work the problem in single steps, you may keep one or two extra digits in the intermediate answers and round off the final answer to the correct number of significant figures. A more efficient way to do this problem is to use a series of two or more conversion factors set up so that the unit in the denominator of each factor cancels the unit in the preceding numerator. Both these approaches are illustrated in the following sample problem.

Sample Problem 2.12 **Problem Solving Using Two Factors**

A recipe for salsa requires 3 cups of tomato sauce, which can be measured accurately as 3.0 cups. If only metric measures are available, how many liters of tomato sauce are needed? (There are 4 cups in 1 quart.)

Solution

STEP 1 **Given** 3.0 cups tomato sauce **Need** liters (L)

STEP 2   **Plan**   We see that the initial unit, cups, needs to be changed to a final unit of liters, but we do not know an equality for cups and liters. However, we do know the U.S. equality for cups and quarts and the metric–U.S. equality for quarts and liters, as shown in the following plan:

cups [U.S. factor] quarts [Metric–U.S. factor] liters

STEP 3   **Equalities/Conversion Factors**   From the discussion on U.S. and metric equalities, we can write the following equalities and conversion factors:

$$1 \text{ qt} = 4 \text{ cups}$$
$$\frac{1 \text{ qt}}{4 \text{ cups}} \quad \text{and} \quad \frac{4 \text{ cups}}{1 \text{ qt}}$$

$$1 \text{ L} = 1.057 \text{ qt}$$
$$\frac{1 \text{ L}}{1.057 \text{ qt}} \quad \text{and} \quad \frac{1.057 \text{ qt}}{1 \text{ L}}$$

STEP 4   **Set Up Problem**   Working in single steps, we can use the U.S factor to convert from cups to quarts:

$$3.0 \text{ cups} \times \frac{1 \text{ qt}}{4 \text{ cups}} = 0.75 \text{ qt}$$

Then we use the metric–U.S. factor to cancel quarts and give liters as the unit.

$$0.75 \text{ qt} \times \frac{1 \text{ L}}{1.057 \text{ qt}} = 0.71 \text{ L}$$

When set up as a series, the first factor cancels cups, and the second factor cancels quarts, which gives liters as the final unit for the answer.

$$\text{cups} \times \frac{\text{qt}}{\text{cups}} \times \frac{\text{L}}{\text{qt}} = \text{L}$$

$$3.0 \text{ cups} \times \frac{1 \text{ qt}}{4 \text{ cups}} \times \frac{1 \text{ L}}{1.057 \text{ qt}} = 0.71 \text{ L}$$

Given (initial unit)	U.S. factor	Metric–U.S. factor	Answer (desired unit)

The calculation is done in a sequence on a calculator to give the numerical part of the answer. The calculator answer is adjusted to give a final answer with the proper number of significant figures (SFs).

3.0 [÷] 4 [÷] 1.057 =   *0.709555*   = 0.71

2 SFs	Exact	Calculator answer	2 SFs (rounded)

Study Check

One medium bran muffin contains 4.2 g of fiber. How many ounces (oz) of fiber are obtained by eating three medium bran muffins if 1 lb = 16 oz? (*Hint*: number of muffins ⟶ g of fiber ⟶ lb ⟶ oz)

Using a sequence of two or more conversion factors is a very efficient way to set up and solve problems, especially if you are using a calculator. Once you have the problem set up, the calculations can be done without writing out the intermediate values. This process is worth practicing until you understand unit cancellation and the mathematical calculations.

Sample Problem 2.13 **Using a Factor from a Word Problem**

During a volcanic eruption on Mauna Loa, Hawaii, the lava flowed at a rate of 33 meters per minute. At this rate, how far in kilometers can the lava travel in 45 minutes?

Solution

STEP 1 **Given** 45 minutes **Need** kilometers

STEP 2 **Plan**

min → Rate factor → m → Metric factor → km

STEP 3 **Equalities/Conversion Factors** In the problem, the information for the rate of lava flow is given as 33 m/min. We will use this rate as one of the equalities as well as the metric equality for meters and kilometers and write conversion factors for each.

$$1 \text{ min} = 33 \text{ m} \qquad\qquad 1 \text{ km} = 1000 \text{ m}$$
$$\frac{1 \text{ min}}{33 \text{ m}} \text{ and } \frac{33 \text{ m}}{1 \text{ min}} \qquad \frac{1 \text{ km}}{1000 \text{ m}} \text{ and } \frac{1000 \text{ m}}{1 \text{ km}}$$

STEP 4 **Set Up Problem** The problem can be set up using the rate as a conversion factor to cancel minutes, and then the metric factor to obtain kilometers in the final factor.

$$45 \text{ min} \times \frac{33 \text{ m}}{1 \text{ min}} \times \frac{1 \text{ km}}{1000 \text{ m}} =$$

The calculation is done as follows:

$$45 \boxed{\times} 33 \boxed{\div} 1000 = 1.485 \qquad \text{(Calculator answer)}$$

By counting the significant figures in the measured quantities, we write the final answer with two significant figures.

$$45 \text{ min} \times \frac{33 \text{ m}}{1 \text{ min}} \times \frac{1 \text{ km}}{1000 \text{ m}} = 1.5 \text{ km}$$
$$\text{2 SF} \qquad\qquad \text{2 SF} \qquad\qquad \text{Exact} \qquad\qquad \text{2 SF}$$

Study Check

How many hours are required for the lava to flow a distance of 5.0 kilometers?

Sample Problem 2.14 **Using a Percent as a Conversion Factor**

Bronze is 80.0% by mass copper and 20.0% by mass tin. A sculptor is preparing to cast a figure that requires 1.75 lb bronze. How many kilograms of copper are needed for the bronze figure?

Solution

STEP 1 **Given** 1.75 lb bronze   **Need** kg copper

STEP 2 **Plan**

lb bronze → Metric–U.S. factor → kg bronze → Percent factor → kg copper

STEP 3 **Equalities/Conversion Factors** Now we can write the equalities and conversion factors. One is the U.S.–metric factor for kg and lb. The second is the percent factor derived from the information given in the problem.

1 kg bronze = 2.205 lb

$$\frac{1 \text{ kg bronze}}{2.205 \text{ lb}} \quad \text{and} \quad \frac{2.205 \text{ lb}}{1 \text{ kg bronze}}$$

100 kg bronze = 80.0 kg copper

$$\frac{80.0 \text{ kg copper}}{100 \text{ kg bronze}} \quad \text{and} \quad \frac{100 \text{ kg}}{80.0 \text{ kg copper}}$$

STEP 4 **Set Up Problem** We can set up the problem using conversion factors to cancel each unit, starting with lb bronze, until we obtain the final factor, kg copper, in the numerator. After we count the significant figures in the measured quantities, we write the final answer with three significant figures.

$$1.75 \text{ lb bronze} \times \frac{1 \text{ kg bronze}}{2.205 \text{ lb bronze}} \times \frac{80.0 \text{ kg copper}}{100 \text{ kg bronze}} = 0.635 \text{ kg copper}$$

3 SF   4 SF   3 SF   3 SF

You could set up this problem in a different way by using the following relationships and conversion factors:

1 lb = 453.6 g

$$\frac{1 \text{ lb}}{453.6 \text{ g}} \quad \text{and} \quad \frac{453.6 \text{ g}}{1 \text{ lb}}$$

100 g bronze = 80.0 g copper

$$\frac{80.0 \text{ g copper}}{100 \text{ g bronze}} \quad \text{and} \quad \frac{100 \text{ g bronze}}{80.0 \text{ g copper}}$$

1 kg = 1000 g

$$\frac{1 \text{ kg}}{1000 \text{ g}} \quad \text{and} \quad \frac{1000 \text{ g}}{1 \text{ kg}}$$

Then the setup would appear as follows:

$$1.75 \text{ lb bronze} \times \frac{453.6 \text{ g bronze}}{1 \text{ lb bronze}} \times \frac{80.0 \text{ g copper}}{100.0 \text{ g bronze}} \times \frac{1 \text{ kg copper}}{1000 \text{ g copper}} = 0.635 \text{ kg copper}$$

3 SF   4 SF   3 SF   Exact   3 SF

Study Check

A lean hamburger is 22% fat by weight. How many grams of fat are in 0.25 lb of the hamburger?

Questions and Problems   **Problem Solving**

2.59 When you convert one unit to another, how do you know which unit of the conversion factor to place in the denominator?

2.60 When you convert one unit to another, how do you know which unit of the conversion factor to place in the numerator?

2.61 Use metric conversion factors to solve the following problems:
  a. The height of a student is 175 cm. How tall is the student in meters?
  b. A cooler has a volume of 5500 mL. What is the capacity of the cooler in liters?
  c. A hummingbird has a mass of 0.0055 kg. What is the mass of the hummingbird in grams?
  d. A balloon has a volume of 350 cm^3. What is the volume in m^3?

2.62 Use metric conversion factors to solve the following problems:
  a. The daily requirement of phosphorus is 800 mg. How many grams of phosphorus are recommended?
  b. A glass of orange juice contains 0.85 dL of juice. How many milliliters of orange juice is that?
  c. A package of chocolate instant pudding contains 2840 mg of sodium. How many grams of sodium is that?
  d. A park has an area of 150 000 m^2. What is the area in km^2?

2.63 Solve the following problems using one or more conversion factors:
  a. A container holds 0.750 qt of liquid. How many milliliters of lemonade will it hold?
  b. In England, a person is weighed in *stones*. If one stone is 14.0 lb, what is the mass in kilograms of a person who weighs 11.8 stones?
  c. The femur, or thighbone, is the longest bone in the body. In a 6-ft-tall person, the femur is 19.5 in. long. What is the length of that femur in millimeters?
  d. How many inches thick is an arterial wall that measures 0.50 μm?

2.64 Solve the following problems using one or more conversion factors:
  a. You need 4.0 oz of a steroid ointment. If there are 16 oz in 1 lb, how many grams of ointment does the pharmacist need to prepare?
  b. During surgery, a person receives 5.0 pints of plasma. How many milliliters of plasma were given? (1 quart = 2 pints)
  c. Solar flares containing hot gases can rise to 120 000 miles above the surface of the sun. What is that distance in kilometers?
  d. A filled gas tank contains 18.5 gal of unleaded fuel. If a car uses 46 L, how many gallons of fuel remain in the tank?

2.65 The singles portion of a tennis court is 27.0 ft wide and 78.0 ft long.
  a. What is the length of the court in meters?
  b. What is the area of the court in square meters (m^2)?

c. If a serve is measured at 185 km/h, how many seconds does it take for the tennis ball to travel the length of the court?

d. How many liters of paint are needed to paint the court if 1 gal of paint covers 150 ft^2?

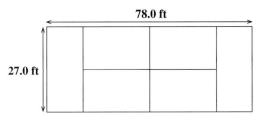

2.66 A football field is 160. ft wide and 300. ft long between goal lines.

  a. How many meters does a player run if he catches the ball on his own goal line and scores a touchdown?
  b. If a player catches the football and runs 45 yd, how many meters did he gain?
  c. How many square meters of Astroturf are required to completely cover the playing field?
  d. If a player runs at a speed of 36 km/h, how many seconds does it take to run from the 50-yd line to the 20-yd line?

2.67 **a.** Oxygen makes up 46.7% by mass of Earth's crust. How many grams of oxygen are present if a sample of the crust has a mass of 325 g?
  b. Magnesium makes up 2.1% by mass of Earth's crust. How many grams of magnesium are present if a sample of the crust has a mass of 1.25 g?
  c. A plant fertilizer contains 15% by mass nitrogen (N). In a container of soluble plant food, there are 10.0 oz of fertilizer. How many grams of nitrogen are in the container?
  d. In a candy factory, the nutty chocolate bars contain 22.0% by mass pecans. If 5.0 kg of pecans were used for candy last Tuesday, how many pounds of nutty chocolate bars were made?

2.68 **a.** Water is 11.2% by mass hydrogen. How many kilograms of water would contain 5.0 g of hydrogen?
  b. Water is 88.8% by mass oxygen. How many grams of water would contain 2.25 kg of oxygen?
  c. Blueberry fiber cakes contain 51% dietary fiber. If a package with a net weight of 12 oz contains six cakes, how many grams of fiber are in each cake?
  d. A jar of crunchy peanut butter contains 1.43 kg of peanut butter. If you use 8.0% of the peanut butter for a sandwich, how many ounces of peanut butter did you take out of the container?

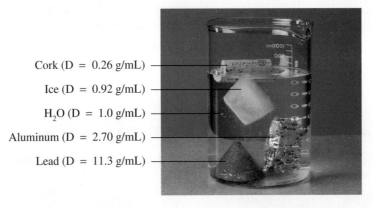

Cork (D = 0.26 g/mL)

Ice (D = 0.92 g/mL)

H_2O (D = 1.0 g/mL)

Aluminum (D = 2.70 g/mL)

Lead (D = 11.3 g/mL)

Figure 2.11 Objects that sink in water are more dense than water; objects float if they are less dense.

Q *Why does a cork float and a piece of lead sink?*

2.8 Density

Differences in density determine whether an object will sink or float. In Figure 2.11, the density of lead is greater than the density of water, and the lead object sinks. The cork floats because cork is less dense than water.

The mass and volume of any substance can be measured. However, the separate measurements do not tell us how tightly packed the substance might be. If we compare the mass of the object to its volume, we obtain a relationship called **density.**

$$\text{Density} = \frac{\text{mass of substance}}{\text{volume of substance}}$$

In the metric system, the densities of solids and liquids are usually expressed as grams per cubic centimeter (g/cm^3) or grams per milliliter (g/mL). The density of a gas is usually stated as grams per liter (g/L). Table 2.9 gives the densities of some common substances.

Table 2.9   Densities of Some Common Substances

Solids (at 25°C)	Density (g/cm^3 or g/mL)	Liquids (at 25°C)	Density (g/mL)	Gases (at 0°C)	Density (g/L)
cork	0.26	gasoline	0.66	hydrogen	0.090
ice	0.92	ethyl alcohol	0.785	helium	0.179
sugar	1.59	olive oil	0.92	methane	0.714
salt (NaCl)	2.16	water (at 4°C)	1.000	neon	0.90
aluminum	2.70	milk	1.04	nitrogen	1.25
diamond	3.52	mercury	13.6	air (dry)	1.29
copper	8.92			oxygen	1.43
silver	10.5			carbon dioxide	1.96
lead	11.3				
gold	19.3				

Guide to Calculating Density
STEP 1 State the given and needed quantities.
STEP 2 Write the density expression.
STEP 3 Express mass in grams and volume in milliliters (mL) or cm³.
STEP 4 Substitute mass and volume into density expression and solve.

Sample Problem 2.15   **Calculating Density**

A 44.65-g copper sample has a volume of 5.0 cm^3. What is the density of copper?

Solution

STEP 1 **Given**   mass of copper sample = 44.65 g; volume = 5.0 cm^3
Need   density (g/cm^3)

STEP 2 **Write the density expression.**

$$\text{Density} = \frac{\text{mass of substance}}{\text{volume of substance}}$$

STEP 3 **Express mass in grams and volume in milliliters (mL) or cm^3.**

Mass of copper sample = 44.65 g

Volume of copper sample = 5.0 cm^3

STEP 4 **Substitute mass and volume into density expression and solve.**

$$\text{Density} = \frac{\overset{\text{4 SF}}{44.65 \text{ g}}}{\underset{\text{2 SF}}{5.0 \text{ cm}^3}} = \frac{\overset{\text{2 SF}}{8.9 \text{ g}}}{1 \text{ cm}^3} = 8.9 \text{ g/cm}^3$$

Study Check

What is the density (g/cm^3) of a silver bar that has a volume of 28.0 cm^3 and a mass of 294 g?

Density of Solids

The density of a solid is calculated using its mass and volume. When a solid is completely submerged, it displaces a volume of water that is *equal to its own volume.* In Figure 2.12, the water level rises from 35.5 mL to 45.0 mL. This means that 9.5 mL of water is displaced and that the volume of the object is 9.5 mL.

Figure 2.12 The density of a solid can be determined by volume displacement because a submerged object displaces a volume of water equal to its own volume.

Q *What is the density of the zinc object?*

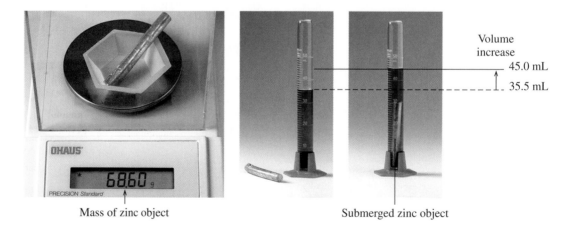

Mass of zinc object   Submerged zinc object

| Sample Problem 2.16 | **Using Volume Displacement to Calculate Density** |

A lead weight used in the belt of a scuba diver has a mass of 226 g. When the weight is carefully placed in a graduated cylinder containing 200.0 mL of water, the water level rises to 220.0 mL. What is the density of the lead weight (g/mL)?

Solution

STEP 1   **Given**   mass of lead = 226 g; water level before object submerged = 200.0 mL; water level after object submerged = 220.0 mL
   Need   density (g/cm^3)

STEP 2   **Write the density expression.**

$$\text{Density} = \frac{\text{mass of substance}}{\text{volume of substance}}$$

STEP 3   **Express mass in grams and volume in milliliters (mL) or cm^3.**

Mass of lead sample = 226 g

Water level after object submerged	= 220.0 mL
−Water level before object submerged	= 200.0 mL
Water displaced (volume of lead)	= 20.0 mL

STEP 4   **Substitute mass and volume into density expression and solve.**

$$\text{Density} = \frac{\overset{3\,\text{SF}}{226\ \text{g}}}{\underset{3\,\text{SF}}{20.0\ \text{mL}}} = \frac{11.3\ \text{g}}{1\ \text{mL}} = \overset{3\,\text{SF}}{11.3}\ \text{g/mL}$$

Study Check

A total of 0.50 lb of glass marbles is added to 425 mL of water. The water level rises to a volume of 528 mL. What is the density (g/mL) of the glass marbles?

Problem Solving Using Density

Density can be used as a conversion factor. For example, if the volume and the density of a sample are known, the mass in grams of the sample can be calculated.

| Sample Problem 2.17 | **Problem Solving Using Density** |

If the density of milk is 1.04 g/mL, how many grams of milk are in 0.50 qt of milk?

Solution

STEP 1   **Given**   0.50 qt   **Need**   g

STEP 2   **Unit Plan**

qt → U.S.–metric factor → L → Metric factor → mL → Density factor → g

Guide to Using Density

STEP 1
State the given and needed quantities.

STEP 2
Write a plan to calculate the needed quantity.

STEP 3
Write equalities and their conversion factors including density.

STEP 4
Set up problem to solve for the needed quantity.

STEP 3   Equalities/Conversion Factors

$$1 \text{ L} = 1.057 \text{ qt}$$
$$\frac{1 \text{ L}}{1.057 \text{ qt}} \text{ and } \frac{1.057 \text{ qt}}{1 \text{ L}}$$

$$1 \text{ L} = 1000 \text{ mL}$$
$$\frac{1 \text{ L}}{1000 \text{ mL}} \text{ and } \frac{1000 \text{ mL}}{1 \text{ L}}$$

$$1 \text{ mL} = 1.04 \text{ g}$$
$$\frac{1 \text{ mL}}{1.04 \text{ g}} \text{ and } \frac{1.04 \text{ g}}{1 \text{ mL}}$$

STEP 4   Set Up Problem

$$0.50 \text{ qt} \times \frac{1 \text{ L}}{1.057 \text{ qt}} \times \frac{1000 \text{ mL}}{1 \text{ L}} \times \frac{1.04 \text{ g}}{1 \text{ mL}} = 490 \text{ g} (4.9 \times 10^2 \text{ g})$$

| 2 SF | 4 SF | Exact | 3 SF | 2 SF |

Study Check

How many milliliters of mercury are in a thermometer that contains 20.4 g of mercury? (See Table 2.9 for the density of mercury.)

Questions and Problems   Density

2.69 In an old trunk, you find a piece of metal that you think may be aluminum, silver, or lead. In lab, you find it has a mass of 217 g and a volume of 19.2 cm³. Using Table 2.9, what is the metal you found?

2.70 Suppose you have two 100-mL graduated cylinders. In each cylinder there is 40.0 mL of water. You also have two cubes: one is lead and the other is aluminum. Each cube measures 2.0 cm on each side. After you carefully lower each cube into the water of its own cylinder, what will the new water level be in each of the cylinders?

2.71 Determine the density (g/mL) for each of the following:
  a. A 20.0-mL sample of a salt solution that has a mass of 24.0 g.
  b. A cube of butter that weighs 0.250 lb and has a volume of 130.3 mL.
  c. A gem has a mass of 45.0 g. When the gem is placed in a graduated cylinder containing 20.0 mL of water, the water level rises to 34.5 mL.
  d. A solid with a volume of 114 cm³ that has a mass of 485.6 g.
  e. A syrup added to an empty container with a mass of 115.25 g. When 0.100 pint of syrup is added, the total mass of the container and syrup is 182.48 g. (1 qt = 2 pt)

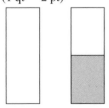

  115.25 g   182.48 g

2.72 Determine the density (g/mL) for each of the following:
  a. a plastic material that weighs 2.68 lb and has a volume of 3.5 L
  b. the fluid in a car battery, if it has a volume of 125 mL and a mass of 155 g

  c. a 5.00-mL urine sample from a patient suffering from diabetes mellitus that has a mass of 5.025 g
  d. an ebony carving that has a mass of 275 g and a volume of 207 cm³
  e. a 10.00-L sample of oxygen gas that has a mass of 0.014 kg

2.73 Use the density values in Table 2.9 to solve the following problems:
  a. How many liters of ethyl alcohol contain 1.50 kg of alcohol?
  b. How many grams of mercury are present in a barometer that holds 6.5 mL of mercury?
  c. A sculptor has prepared a mold for casting a bronze figure. The figure has a volume of 225 mL. If bronze has a density of 7.8 g/mL, how many ounces of bronze are needed in the preparation of the bronze figure?
  d. What is the mass in grams of a cube of copper that has a volume of 74.1 cm³?
  e. How many kilograms of gasoline fill a 12.0-gal gas tank? (1 gal = 4 qt)

2.74 Use the density values in Table 2.9 to solve the following problems:
  a. A graduated cylinder contains 28.0 mL of water. What is the new water level after 35.6 g of silver metal is submerged in the water?
  b. A thermometer containing 8.3 g of mercury has broken. What volume of mercury spilled?
  c. A fish tank holds 35 gal of water. How many pounds (lb) of water are in the fish tank?
  d. The mass of an empty container is 88.25 g. The mass of the container and a liquid with a density of 0.758 g/mL is 150.50 g. What is the volume (mL) of the liquid in the container?
  e. A cannonball made of iron has a volume of 115 cm³. If iron has a density of 7.86 g/cm³, what is the mass in kilograms of the cannonball?

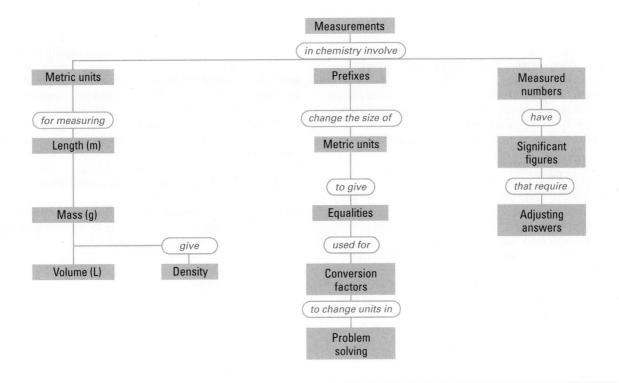

Chapter Review

2.1 Units of Measurement

In science, physical quantities are described in units of the metric or International System (SI). Some important units are meter (m) for length, liter (L) for volume, gram (g) and kilogram (kg) for mass, second (s) for time, and Celsius (°C) for temperature.

2.2 Scientific Notation

Large and small numbers can be written using scientific notation in which the decimal point is moved to give a coefficient between 1 and 10 and the number of decimal places moved shown as a power of 10. A number greater than 1 will have a positive power of 10, while a number less than 1 will have a negative power of 10.

2.3 Measured Numbers and Significant Figures

A measured number is any number derived from using a measuring device to determine a quantity. An exact number is obtained by counting items or from a definition; no measuring device is used. Significant figures are the numbers reported in a measurement including the estimated digit.

2.4 Significant Figures in Calculations

In multiplication or division, the measured number in the calculation that has the smallest number of significant figures determines the number of significant figures in an answer. In addition or subtraction, the final answer reflects the last place that all the measured numbers have in common.

2.5 Prefixes and Equalities

Prefixes placed in front of a unit change the size of the unit by factors of 10. Prefixes such as *centi, milli,* and *micro* provide smaller units; prefixes such as *kilo* provide larger units. An equality relates two metric units that measure the same quantity of length, volume, or mass. Examples of metric equalities are 1 m = 100 cm; 1 L = 1000 mL; 1 kg = 1000 g.

2.6 Writing Conversion Factors

Conversion factors are used to express a relationship in the form of a fraction. Two factors can be written for any relationship in the metric or U.S. system.

2.7 Problem Solving

Conversion factors are useful when changing a quantity expressed in one unit to a quantity expressed in another unit. In the process, a given unit is multiplied by one or more conversion factors that cancel units until the desired answer is obtained.

2.8 Density

The density of a substance is a ratio of its mass to its volume, usually g/mL or g/cm^3. The units of density can be used as a factor to convert between the mass and volume of a substance.

Key Terms

Celsius (°C) temperature scale A temperature scale on which water has a freezing point of 0°C and a boiling point of 100°C.

centimeter (cm) A unit of length in the metric system; there are 2.54 cm in 1 in.

conversion factor A ratio in which the numerator and denominator are quantities from an equality or given relationship. For example, the conversion factors for the relationship 1 kg = 2.205 lb are written as the following:

$$\frac{2.205 \text{ lb}}{1 \text{ kg}} \quad \text{and} \quad \frac{1 \text{ kg}}{2.205 \text{ lb}}$$

cubic centimeter (cm³, cc) The volume of a cube that has 1-cm sides; 1 cubic centimeter is equal to 1 mL.

cubic meter (m³) The SI unit of volume; the volume of a cube with sides that measure 1 m.

density The relationship of the mass of an object to its volume expressed as grams per cubic centimeter (g/cm³), grams per milliliter (g/mL), or grams per liter (g/L).

equality A relationship between two units that measure the same quantity.

exact number A number obtained by counting or definition.

gram (g) The metric unit used in measurements of mass.

Kelvin (K) temperature scale A temperature scale on which the lowest possible temperature is 0 K, which makes the freezing point of water 273 K and the boiling point of water 373 K.

kilogram (kg) A metric mass of 1000 g, equal to 2.205 lb. The kilogram is the SI standard unit of mass.

liter (L) The metric unit for volume that is slightly larger than a quart.

mass A measure of the quantity of material in an object.

measured number A number obtained when a quantity is determined by using a measuring device.

meter (m) The metric unit for length that is slightly longer than a yard. The meter is the SI standard unit of length.

metric system A system of measurement used by scientists and in most countries of the world.

milliliter (mL) A metric unit of volume equal to one-thousandth of a L (0.001 L).

prefix The part of the name of a metric unit that precedes the base unit and specifies the size of the measurement. All prefixes are related on a decimal scale.

scientific notation A form of writing large and small numbers using a coefficient between 1 and 10, followed by a power of 10.

second(s) The unit of time in both the SI and metric systems.

SI An International System of units that modifies the metric system.

significant figures The numbers recorded in a measurement.

temperature An indicator of the hotness or coldness of an object.

volume (V) The amount of space occupied by a substance.

▶ Understanding the Concepts

2.75 Indicate if each of the following is answered with an exact number or a measured number:

a. number of legs
b. height of table
c. number of chairs at the table
d. area of tabletop

2.76 Read the temperature on each of the Celsius thermometers.

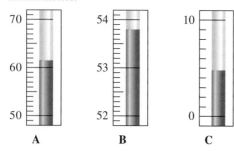

2.77 Measure the length and width of the rectangle using a metric rule.

a. What is the length and width of this rectangle measured in centimeters?
b. What is the length and width of this rectangle measured in millimeters?
c. How many significant figures are in the length measurement?
d. How many significant figures are in the width measurement?
e. What is the area of the rectangle in cm²?
f. How many significant figures are in the calculated answer for area?

2.78 Measure the length of each of the objects in figure (**a**), (**b**), and (**c**) using the metric rule in the figure. Indicate the number of significant figures for each and the estimated digit for each.

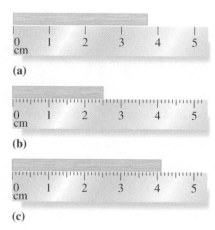

(a)

(b)

(c)

2.79 What is the density of the solid object that is weighed and submerged in water?

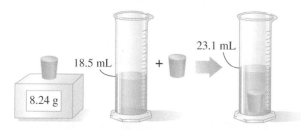

2.80 Each of the following diagrams represents a container of water and a cube. Some cubes float while others sink. Match diagrams 1, 2, 3, or 4 with one of the following descriptions and explain your choices:

a. The cube has a greater density than water.
b. The cube has a density that is 0.60–0.80 g/mL.
c. The cube has a density that is one-half the density of water.
d. The cube has the same density of water.

Solid Water

    1   2   3   4

2.81 A graduated cylinder contains three liquids A, B, and C, which have different densities and are not soluble in each other: mercury (D = 13.5 g/mL), vegetable oil (D = 0.92 g/mL), and water (D = 1.0 g/mL). Identify the liquids A, B, and C in the cylinder.

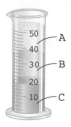

2.82 The solids A, B, and C represent gold, silver, and aluminum. If each has a mass of 10.0 g, what is the identity of each solid?

Density aluminum = 2.70 g/mL

Density gold = 19.3 g/mL

Density silver = 10.5 g/mL

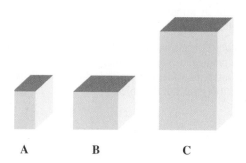

A   B   C

▶ Additional Questions and Problems

2.83 Round off or add zeros to the following calculated answers to give a final answer with three significant figures:
a. 0.000 012 58 L
b. 3.528×10^2 kg
c. 125 111 m^3
d. 58.703 m
e. 3×10^{-3} s
f. 0.010 826 g

2.84 What is the total mass in grams of a dessert containing 137.25 g vanilla ice cream, 84 g fudge sauce, and 43.7 g nuts?

2.85 During a workout at the gym, you set the treadmill at a pace of 55.0 m per minute. How many minutes will you walk if you cover a distance of 7500 ft?

2.86 A fish company delivers 22 kg of salmon, 5.5 kg of crab, and 3.48 kg of oysters to your seafood restaurant.
a. What is the total mass, in kilograms, of the seafood?
b. What is the total number of pounds?

2.87 Bill's recipe for onion soup calls for 4.0 lb of thinly sliced onions. If an onion has an average mass of 115 g, how many onions does Bill need?

2.88 The price of 1 pound (lb) of potatoes is $1.75. If all the potatoes sold today at the store bring in $1420, how many kilograms (kg) of potatoes did grocery shoppers buy?

2.89 The following nutrition information is listed on a box of crackers:
Serving size 0.50 oz (6 crackers)
Fat 4 g per serving **Sodium** 140 mg per serving
a. If the box has a net weight (contents only) of 8.0 oz, about how many crackers are in the box?
b. If you ate 10 crackers, how many ounces of fat are you consuming?
c. How many grams of sodium are used to prepare 50 boxes of crackers?

2.90 An aquarium store unit requires 75 000 mL of water. How many gallons of water are needed? (1 gal = 4 qt)

2.91 In Mexico, avocados are 48 pesos per kilogram. What is the cost in cents of an avocado that weighs 0.45 lb if the exchange rate is 10.8 pesos to the dollar?

2.92 Celeste's diet restricts her intake of protein to 24 g per day. If she eats an 8.0-oz burger that is 15.0% protein, has she exceeded her protein limit for the day? How many ounces of a burger would be allowed for Celeste?

2.93 A sunscreen preparation contains 2.50% by mass benzyl salicylate. If a tube contains 4.0 oz of sunscreen, how many kilograms of benzyl salicylate are needed to manufacture 325 tubes of sunscreen?

2.94 An object has a mass of 3.15 oz. When it is submerged in a graduated cylinder initially containing 325.2 mL of water, the water level rises to 442.5 mL. What is the density (g/mL) of the object?

2.95 What is a cholesterol level of 1.85 g/L in units of mg/dL?

2.96 If a recycling center collects 1254 aluminum cans and there are 22 aluminum cans in 1 lb, what volume in liters of aluminum was collected? (See Table 2.9.)

2.97 The water level in a graduated cylinder initially at 215 mL rises to 285 mL after a piece of lead is submerged. What is the mass in grams of the lead? (See Table 2.9.)

2.98 A graduated cylinder contains 155 mL of water. A 15.0-g piece of iron (density = 7.86 g/cm^3) and a 20.0-g piece of lead are added. What is the new water level in the cylinder? (See Table 2.9.)

2.99 How many cubic centimeters (cm^3) of olive oil have the same mass as 1.00 L of gasoline? (See Table 2.9.)

2.100 What is the volume, in quarts, of 1.50 kg of ethyl alcohol? (See Table 2.9.)

2.101 a. Some athletes have as little as 3.0% body fat. If such a person has a body mass of 45 kg, how many pounds of body fat does that person have?
b. In a process called liposuction, a doctor removes fat deposits from a person's body. If body fat has a density of 0.94 g/mL and 3.0 liters of fat is removed, how many pounds of fat were removed from the patient?

2.102 A mouthwash is 21.6% alcohol by mass. If each bottle contains 0.358 pint of mouthwash with a density of 0.876 g/mL, how many kilograms of alcohol are in 180 bottles of the mouthwash?

2.103 Sterling silver is 92.5% silver by mass with a density of 10.3 g/cm^3. If a cube of sterling silver has a volume of 27.0 cm^3, how many ounces of pure silver are present?

2.104 A typical adult body contains 55% water. If a person has a mass of 65 kg, how many pounds of water does she have in her body?

Challenge Questions

2.105 A balance measures mass to 0.001 g. If you determine the mass of an object that weighs about 30 g, would you record the mass as 30 g, 32.5 g, 31.25 g, 34.075 g, or 3000 g? Explain your choice by writing 2–3 complete sentences that describe your thinking.

2.106 When three students use the same meterstick to measure the length of a paperclip, they obtain results of 5.8 cm, 5.75 cm, and 5.76 cm. If the meterstick has millimeter markings, what are some reasons for the different values?

2.107 A car travels at 55 miles per hour and gets 11 kilometers per liter of gasoline. How many gallons of gasoline are needed for a 3.0 hr trip?

2.108 For a 180-lb person, calculate the quantities of each of the following that must be ingested to provide the LD50 for caffeine given in Table 2.7:
a. cups of coffee if one cup is 12 fluid ounces and there is 100 mg caffeine per 6-fl oz of drip-brewed coffee
b. cans of cola if one can contains 50 mg caffeine
c. tablets of No-Doz if one tablet contains 100 mg caffeine

2.109 A package of aluminum foil is $66\frac{2}{3}$ yd long, 12 in. wide, and 0.000 30 in. thick. If aluminum has a density of 2.7 g/cm^3, what is the mass in grams of the foil?

2.110 A circular pool with a diameter of 27 ft is filled to a depth of 50. in. Assume the pool is a cylinder ($V_{cylinder} = \pi r^2 h$).
   a. What is the volume of water in the pool in cubic meters?
   b. The density of water is 1.0 g/cm^3. What is the mass in kilograms of the water in the pool?

2.111 An 18-karat gold necklace is 75% gold by mass, 16% silver, and 9.0% copper.

a. What is the mass in grams of the necklace if it contains 0.24 oz silver?
b. How many grams of copper are in the necklace?
c. If 18-karat gold has a density of 15.5 g/cm^3, what is the volume in cubic centimeters?

2.112 In the manufacturing of computer chips, cylinders of silicon are cut into thin wafers that are 3.00 in. in diameter and have a mass of 1.50 g of silicon. How thick (mm) is each wafer if silicon has a density of 2.33 g/cm^3? (The volume of a cylinder is $V = \pi r^2 h$.)

2.113 A 50.0-g silver object and a 50.0-g gold object are both added to 75.5 mL of water contained in a graduated cylinder. What is the new water level in the cylinder?

Answers

Answers to Study Checks

2.1 a. meters (m)   **b.** kelvins (K)
     c. kilograms (kg)

2.2 a. 4.25×10^5 m   **b.** 8×10^{-7} g

2.3 a. two   **b.** one   **c.** five

2.4 a. 36 m   **b.** 0.0026 L
     c. 3800 g   **d.** 1.3 kg

2.5 a. 0.4924   **b.** 0.00802 or 8.02×10^{-3} **c.** 2.0

2.6 a. 83.70 mg   **b.** 0.5 L

2.7 a. mega   **b.** centi

2.8 a. 1000   **b.** 0.001 or $\dfrac{1}{1000}$

2.9 Equality: 1 in. = 2.54 cm

   Conversion Factors: $\dfrac{1 \text{ in.}}{2.54 \text{ cm}}$ and $\dfrac{2.54 \text{ cm}}{1 \text{ in.}}$

2.10 a. 62.2 km/1 hr and 1 hr/62.2 km
      b. 8.5 cm^3/86.7 g and 86.7 g/8.5 cm^3

2.11 1.89 L

2.12 0.44 oz

2.13 2.5 hr

2.14 25 g fat

2.15 10.5 g/cm^3

2.16 2.2 g/mL

2.17 1.50 mL mercury

Answers to Selected Questions and Problems

2.1 a. meter, length   **b.** gram, mass
     c. liter, volume   **d.** second, time
     e. Celsius, temperature

2.3 a. meter, both   **b.** kilogram, both
     c. foot, neither   **d.** second, both
     e. Celsius, metric

2.5 a. gram, metric   **b.** liter, metric
     c. Celsius, metric   **d.** meter, both
     e. second, both

2.7 a. 5.5×10^4 m   **b.** 4.8×10^2 g
     c. 5×10^{-6} cm   **d.** 1.4×10^{-4} s
     e. 7.85×10^{-3} L **f.** 6.7×10^5 kg

2.9 a. 7.2×10^3 **b.** 3.2×10^{-2}
     c. 1×10^4 **d.** 6.8×10^{-2}

2.11 a. 12 000   **b.** 0.0825
      c. 4 000 000   **d.** 0.005

2.13 a. first decimal place (0.6)
      b. second decimal place (0.05)
      c. first decimal place (0.0)

2.15 a. measured   **b.** exact
      c. exact   **d.** measured

2.17 a. 6 oz of meat   **b.** none
      c. 0.75 lb, 350 g
      d. none (definitions are exact)

2.19 a. not significant   **b.** significant
      c. significant   **d.** significant
      e. not significant

2.21 a. 5   **b.** 2   **c.** 2   **d.** 3   **e.** 4   **f.** 3

2.23 Both measurements in part c have two significant figures and both measurements in part d have four significant figures.

2.25 a. 5.0×10^3 L   **b.** 3.0×10^4 g
      c. 1.0×10^5 m   **d.** 2.5×10^{-4} cm

2.27 The number of figures in the answer is limited by the measurements used in the calculation.

2.29 a. 1.85   **b.** 184   **c.** 0.004 74
      d. 8810   **e.** 1.83

2.31 a. 56.9 m   **b.** 0.002 28 g
      c. 11 500 s (1.15×10^4 s)   **d.** 8.10 L

2.33 a. 1.6   **b.** 0.01
      c. 27.6   **d.** 3.5
      e. 1.4×10^{-1} (0.14)   **f.** 8×10^{-1} (0.8)

2.35 **a.** 53.54 cm   **b.** 127.6 g
    c. 121.5 mL   **d.** 0.50 L

2.37 km/hr is kilometers per hour; mi/hr is miles per hour

2.39 The prefix *kilo* means to multiply by 1000; 1 kg is the same mass as 1000 g.

2.41 **a.** mg   **b.** dL   **c.** km
    d. kg   **e.** μL   **f.** ns

2.43 **a.** 0.01   **b.** 1000   **c.** 0.001
    d. 0.1   **e.** 1 000 000   **f.** 10^{-9}

2.45 **a.** 100 cm   **b.** 1000 m   **c.** 0.001 m
    d. 1000 mL

2.47 **a.** kilogram   **b.** milliliter   **c.** km   **d.** kL

2.49 A conversion factor can be inverted to give a second conversion factor.

2.51 1 kg = 1000 g

2.53 **a.** 3 ft = 1 yd, $\dfrac{3\ \text{ft}}{1\ \text{yd}}$ and $\dfrac{1\ \text{yd}}{3\ \text{ft}}$

   b. 1 L = 1000 mL, $\dfrac{1000\ \text{mL}}{1\ \text{L}}$ and $\dfrac{1\ \text{L}}{1000\ \text{mL}}$

   c. 1 min = 60 s, $\dfrac{60\ \text{s}}{1\ \text{min}}$ and $\dfrac{1\ \text{min}}{60\ \text{s}}$

   d. 1 dL = 100 mL, $\dfrac{100\ \text{mL}}{1\ \text{dL}}$ and $\dfrac{1\ \text{dL}}{100\ \text{mL}}$

2.55 **a.** 100 cm = 1 m, $\dfrac{100\ \text{cm}}{1\ \text{m}}$ and $\dfrac{1\ \text{m}}{100\ \text{cm}}$

   b. 1000 mg = 1 g, $\dfrac{1000\ \text{mg}}{1\ \text{g}}$ and $\dfrac{1\ \text{g}}{1000\ \text{mg}}$

   c. 1 L = 1000 mL, $\dfrac{1000\ \text{mL}}{1\ \text{L}}$ and $\dfrac{1\ \text{L}}{1000\ \text{mL}}$

   d. 1 kg = 10^6 mg, $\dfrac{10^6\ \text{mg}}{1\ \text{kg}}$ and $\dfrac{1\ \text{kg}}{10^6\ \text{mg}}$

   e. $(1\ \text{m})^3 = (100\ \text{cm})^3$, $\dfrac{(100\ \text{cm})^3}{(1\ \text{m})^3}$ and $\dfrac{(1\ \text{m})^3}{(100\ \text{cm})^3}$

2.57 **a.** $\dfrac{3.5\ \text{m}}{1\ \text{s}}$ and $\dfrac{1\ \text{s}}{3.5\ \text{m}}$

   b. $\dfrac{0.65\ \text{g}}{1\ \text{mL}}$ and $\dfrac{1\ \text{mL}}{0.65\ \text{g}}$

   c. $\dfrac{46.0\ \text{km}}{1.0\ \text{gal}}$ and $\dfrac{1.0\ \text{gal}}{46.0\ \text{km}}$

   d. $\dfrac{93\ \text{g silver}}{100\ \text{g sterling}}$ and $\dfrac{100\ \text{g sterling}}{93\ \text{g silver}}$

   e. $\dfrac{29\ \mu\text{g}}{1\ \text{kg}}$ and $\dfrac{1\ \text{kg}}{29\ \mu\text{g}}$

2.59 The unit in the denominator must cancel with the preceding unit in the numerator.

2.61 **a.** 1.75 m   **b.** 5.5 L   **c.** 5.5 g
    d. 3.5×10^{-4} m^3

2.63 **a.** 710. mL   **b.** 74.9 kg   **c.** 495 mm
    d. 2.0×10^{-5} in.

2.65 **a.** 23.8 m   **b.** 196 m^2 **c.** 0.463 s   **d.** 53.2 L

2.67 **a.** 152 g oxygen   **b.** 0.026 g magnesium
    c. 43 g N   **d.** 50. lb chocolate bars

2.69 lead, 11.3 g/cm^3

2.71 **a.** 1.20 g/mL   **b.** 0.870 g/mL   **c.** 3.10 g/mL
    d. 4.26 g/cm^3 **e.** 1.42 g/mL

2.73 **a.** 1.91 L   **b.** 88 g   **c.** 62 oz
    d. 661 g   **e.** 30. kg

2.75 **a.** exact   **b.** measured
    c. exact   **d.** measured

2.77 **a.** length = 6.96 cm, width = 4.75 cm
    b. length = 69.6 mm, width = 47.5 mm
    c. 3 significant figures
    d. 3 significant figures
    e. 33.1 cm^2
    f. 3 significant figures

2.79 1.8 g/mL

2.81 A is vegetable oil, B is water, and C is mercury.

2.83 **a.** 0.000 012 6 L; 1.26×10^{-5} L
    b. 3.53×10^2 kg
    c. 125 000 m^3 or 1.25×10^5 m^3
    d. 58.7 m
    e. 3.00×10^{-3} s
    f. 0.010 8 g

2.85 42 min

2.87 16 onions

2.89 **a.** 96 crackers   **b.** 0.2 oz fat
    c. 110 g sodium

2.91 91 cents

2.93 0.92 kg

2.95 185 mg/dL

2.97 790 g

2.99 720 cm^3

2.101 **a.** 3.0 lb body fat   **b.** 6.2 lb

2.103 9.07 oz pure silver

2.105 Because the balance can measure mass to 0.001 g, the mass should be given to 0.001 g. You should record the mass of the object as 34.075 g.

2.107 6.4 gal

2.109 3.8×10^2 g aluminum

2.111 **a.** 43 g 18-karat gold   **b.** 3.9 g copper   **c.** 2.8 cm^3

2.113 82.9 mL

Matter and Energy

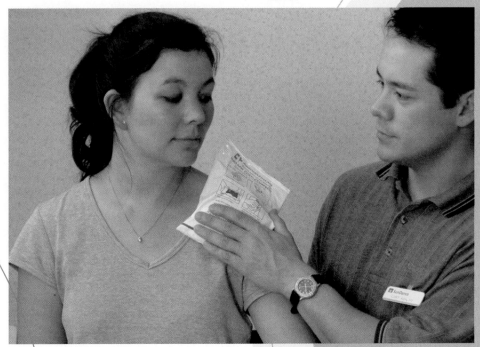

"If you've had first aid for a sports injury," says Cort Kim, physical therapist at the Sunrise Sports Medicine Clinic, *"you've likely been treated with a cold pack or hot pack. We use them for several kinds of injury. Here, I'm showing how I can use a cold pack to reduce swelling in my patient's shoulder."*

When you hit or open a hot or cold pack to activate it, your action mixes chemicals together. In a cold pack, the reaction is one that absorbs heat energy, chills the pack, and draws heat from the injury. Hot packs use reactions that release energy, thus warming the pack. In both cases, the reaction proceeds at a moderate pace, so the pack stays active for a long time and doesn't get too cold or hot.

the **Chemistry place**

Visit **www.aw-bc.com/chemplace** for extra quizzes, interactive tutorials, career resources, PowerPoint slides for chapter review, math help, and case studies.

Every day, we see around us a variety of materials with many shapes and forms. To a scientist, all of this material is *matter*. There is matter everywhere around you: the orange juice you had for breakfast, the water you put in the coffee maker, the aluminum foil you wrap your sandwich in, your toothbrush and toothpaste, the oxygen you inhale, and the carbon dioxide you exhale are all forms of matter.

When we look around us, we see that matter takes the physical form of a solid, liquid, or gas. A familiar example is water, which is a compound that exists in all three states. In an ice cube or an ice rink, water is in the solid state. Water is a liquid when it comes out of a faucet or fills a pool. Water forms a gas when it evaporates from wet clothes or boils in a pan. Substances change states by losing or gaining energy. For example, energy is needed to melt ice cubes and to boil water in a teakettle. In contrast, energy is released when water vapor (gas) condenses to liquid and liquid water in an ice cube tray freezes.

Most of everything we do involves energy. We use energy when we walk, play tennis, study, and breathe. We also use energy when we warm water, cook food, turn on lights, use computers, use a washing machine, or drive our cars. Of course, that energy has to come from something. In our bodies, the food we eat provides us with energy. If we don't eat for a while, we run out of energy. In our homes, schools, and automobiles, burning fossil fuels such as oil, propane, or gasoline provides energy.

3.1 Classification of Matter

Matter is any material that has mass and occupies space. The materials we use such as water, wood, plates, plastic bags, clothes, and shoes are all matter. Because there are so many kinds, we categorize matter by the types of components it contains. A pure substance has a definite composition, while a mixture is made up of two or more substances in varying amounts. (See Figure 3.1.)

Learning **Goal**

Classify matter as pure substances or mixtures.

Figure 3.1 Matter is organized by its components: elements, compounds, and mixtures. An element or a compound has a fixed composition, whereas a mixture has a variable composition. A homogeneous mixture has a uniform composition, but a heterogeneous mixture does not.

Q Why are copper and water pure substances, but brass is a mixture?

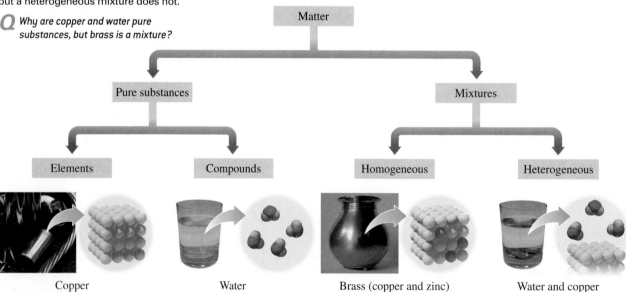

Copper   Water   Brass (copper and zinc)   Water and copper

Pure Substances

A **pure substance** is a type of matter that has a fixed or definite composition. There are two kinds of pure substances, elements and compounds. **Elements** are the simplest type because they are composed of only one type of substance. Perhaps you are aware of the elements silver, iron, and aluminum, which contain matter of one type. **Compounds** are pure substances too, but they consist of a combination of two or more elements bonded together in the same ratio. For example, in all samples of water, H_2O, there is the same proportion of the elements hydrogen and oxygen. In another compound, hydrogen peroxide, H_2O_2, hydrogen and oxygen are also combined, but in a different ratio. Water, H_2O, and hydrogen peroxide, H_2O_2, are different compounds with different properties.

An important difference between elements and compounds is that chemical processes can break down compounds into simpler substances such as elements. You may know that ordinary table salt consists of the compound NaCl, which can be broken down into sodium and chlorine, as seen in Figure 3.2. Compounds are not broken down through physical methods such as boiling or sifting. Elements cannot be broken down further by chemical or physical processes.

Mixtures

Much of the matter in our everyday lives consists of mixtures. In a **mixture,** two or more substances are physically mixed, but not chemically combined. The air we breathe is a mixture of mostly oxygen and nitrogen gases. The steel in buildings and railroad tracks is a mixture of iron, nickel, carbon, and chromium. The brass in knobs and fixtures is a mixture of zinc and copper. Solutions such as tea, coffee, and ocean water are mixtures too. In any mixture, the composition can vary. For example, two sugar–water mixtures would appear the same, but the one that has a higher ratio of sugar to water

Figure 3.2 The decomposition of salt, NaCl, produces the elements sodium and chlorine.

Q How do elements and compounds differ?

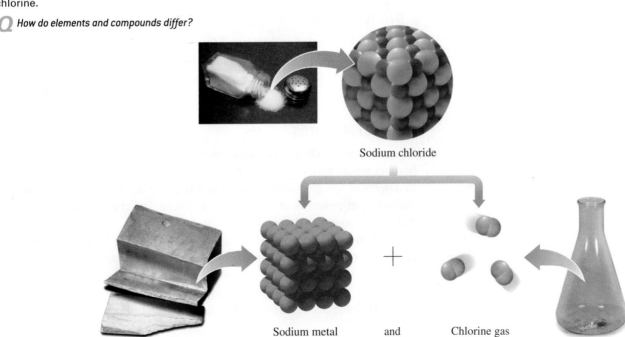

Sodium chloride

Sodium metal   and   Chlorine gas

Figure 3.3 A mixture of spaghetti and water is separated using a strainer, a physical method of separation.

Q *Why can physical methods be used to separate mixtures but not compounds?*

Physical method of separation

would taste sweeter. Different types of brass have different properties depending on the ratio of copper to zinc.

Physical processes can be used to separate most mixtures between the components. For example, different coins such as nickels, dimes, and quarters can be separated by size; iron particles mixed with sand can be picked up with a magnet; and water is separated from cooked spaghetti using a strainer. (See Figure 3.3.)

Sample Problem 3.1 **Pure Substances and Mixtures**

Classify each of the following as a pure substance or a mixture.
a. sugar (sucrose)   **b.** nickels and dimes in a box
c. coffee with milk and sugar   **d.** aluminum

Solution
a. Pure substance; one type of matter.
b. Mixture; the ratio of nickels and dimes can vary.
c. Mixture; the amount of coffee, milk, and sugar can vary.
d. Pure substance; one type of matter.

Study Check

Pure gold is labeled as 24-carat. Many jewelry items are made of 14-carat gold, which is 24-carat gold combined with another metal such as copper. Is 14-carat gold a pure substance or a mixture?

Types of Mixtures

Mixtures can be classified further as homogeneous or heterogeneous. In a **homogeneous mixture,** the composition is uniform throughout the sample. Examples of familiar homogeneous mixtures are air, which contains oxygen and nitrogen gases; bronze, which is a mixture of copper and tin; and salt water, a solution of salt and water.

In a **heterogeneous mixture,** the components do not have a uniform composition throughout the sample. For example, a mixture of oil and water is heterogeneous because the oil floats on the surface of the water. Other examples of heterogeneous mixtures include the raisins in a cookie and the bubbles in a soda. Table 3.1 summarizes the classification of matter.

Table 3.1   Classification of Matter

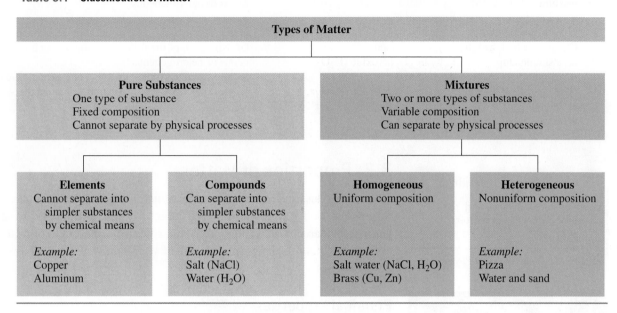

Sample Problem 3.2   **Classifying Mixtures**

Classify each of the following mixtures as heterogeneous or homogeneous.
a. ice-cream float   **b.** chocolate chip cookie
c. air in a scuba tank   **d.** vinegar (acetic acid and water)

Solution

a. Heterogeneous; ice cream floats on soda, which is a nonuniform composition.
b. Heterogeneous; the chocolate chips are not uniformly distributed in a cookie.
c. Homogeneous; the gases oxygen and nitrogen have a uniform composition in air.
d. Homogeneous; acetic acid and water in vinegar have a uniform composition.

Study Check

A salad dressing is prepared with oil, vinegar, and chunks of blue cheese. Is this a homogeneous or heterogeneous mixture?

Questions and Problems   **Classification of Matter**

In every chapter, each red, odd-numbered exercise is paired with the next even-numbered exercise. The answers for all the Study Checks and the red, odd-numbered exercises are given at the end of each chapter. The complete solutions to the odd-numbered exercises are in the *Study Guide*.

3.1 Classify each of the following as a pure substance or a mixture.
a. baking soda
b. a blueberry muffin
c. ice
d. aluminum foil

3.2 Classify each of the following as a pure substance or a mixture.
   a. a soft drink   **b.** vitamin C
   c. a cheese sandwich   **d.** an iron nail

3.3 Classify each of the following as a compound or element.
   a. a silicon chip   **b.** hydrogen peroxide (H_2O_2)
   c. oxygen   **d.** vitamin A

3.4 Classify each of the following as a compound or element.
   a. helium gas   **b.** mercury in a thermometer
   c. sugar   **d.** sulfur

3.5 Classify each of the following mixtures as homogeneous or heterogeneous.
   a. vegetable soup   **b.** sea water
   c. tea   **d.** tea with ice and lemon slices

3.6 Classify each of the following mixtures as homogeneous or heterogeneous.
   a. homogenized milk   **b.** chocolate chip ice cream
   c. gasoline   **d.** bubbly champagne

3.2 Properties of Matter

Learning Goal

Describe some physical and chemical properties of matter.

One way to describe matter is to observe its properties. For example, if you were asked to describe yourself, you might list your **properties,** which are your particular characteristics. You could describe the color of your hair, eyes, and skin, or the length and texture of your hair. Perhaps you have freckles or dimples.

Physical Properties

In chemistry, these types of descriptions are called physical properties and include the shape, physical state, color, melting, and boiling point of a substance. **Physical properties** are those properties that can be observed or measured without affecting the identity of a substance. For example, you might describe a copper penny as orange-red, solid, and shiny. Table 3.2 gives more examples of physical properties of copper found in pennies, electrical wiring, and copper pans.

States of Matter

Matter exists in one of three forms called the **states of matter:** solid, liquid, and gas. Each state has a different set of physical properties. A **solid** such as a pebble or a baseball has a definite shape and volume. You can probably recognize several solids within your reach right now such as books, pencils, or a computer mouse. A **liquid** has a definite volume, but not a definite shape. Thus a liquid such as water takes the shape of a bottle and then the shape of the glass you pour it into. A **gas** does not have a definite shape or volume. When you inflate a bicycle tire, the air, which is a gas, fills the entire shape and volume of the tire.

Water is a substance that is commonly found in all three states: solid, liquid, and gas. When matter undergoes a **physical change,** its state or its appearance will change, but its identity or composition remains the same. The solid form of water, snow or ice, has a different appearance than its liquid or gaseous form, but all three forms are water. (See Figure 3.4.)

Some Physical Changes of Water

$$\text{Solid water (ice)} \underset{\text{freezing}}{\overset{\text{melting}}{\rightleftharpoons}} \text{liquid water} \underset{\text{condensing}}{\overset{\text{boiling}}{\rightleftharpoons}} \text{water vapor (gas)}$$

The physical appearance of a substance can change in other ways too. Suppose that you dissolved some salt in water. The appearance of the salt changes, but you could re-form the salt crystals by heating the mixture and

Table 3.2 Some Physical Properties of Copper

Color	Reddish-orange
Odor	Odorless
Melting point	1083°C
Boiling point	2567°C
State at 25°C	Solid
Luster	Very shiny
Conduction of electricity	Excellent
Conduction of heat	Excellent

Figure 3.4 Water exists in **(a)** an ice cube as a solid, **(b)** water as a liquid, **(c)** water vapor as a gas.

Q *In what state of matter does water have a definite volume, but not a definite shape?*

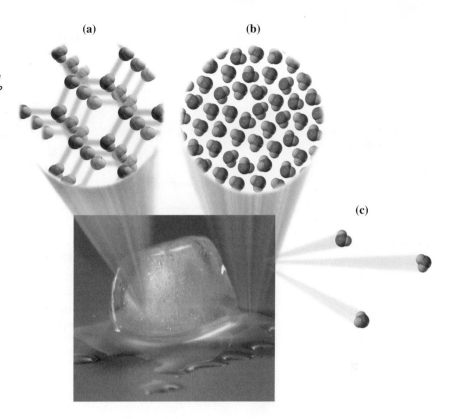

(a)   (b)

(c)

evaporating the water. Thus in a physical change there are no new substances produced. Table 3.3 gives more examples of physical changes.

Matter and Chemical Properties

The **chemical properties** are those that describe the ability of a substance to change into a new substance. During a **chemical change,** the original substance is converted into one or more new substances with different chemical and physical properties. For example, wood can burn because it has the chemical property of being flammable. When wood burns, it is converted to ashes and smoke, which have different chemical and physical properties. Rusting or corrosion is a chemical property of iron. In the rain, an iron nail undergoes a chemical change when it reacts with oxygen in the air to form rust, a new substance. Table 3.4 gives some examples of chemical changes. Table 3.5 summarizes physical and chemical properties and changes.

Table 3.3 Examples of Some Physical Changes

Type of Physical Change	Example
Change of state	Water boiling Water freezing
Change of appearance	Dissolving sugar in water
Change of shape	Hammering a gold ingot into shiny gold leaf Drawing copper into thin copper wire
Change of size	Cutting paper into tiny pieces for confetti Grinding pepper into smaller particles

Table 3.4   Examples of Some Chemical Changes

Type of Chemical Change	Changes in Chemical Properties
Tarnishing of silver	Shiny, silver metal reacts in air to give a black, grainy coating.
Burning wood	A piece of pine burns with a bright flame producing heat, ashes, carbon dioxide, and water vapor.
Caramelizing sugar	At high temperatures, white, granular sugar changes to a smooth, caramel-colored substance.
Formation of rust	Iron, which is gray and shiny, combines with oxygen to form orange-red rust

Table 3.5   Summary of Physical and Chemical Properties and Changes

	Physical	Chemical
Property	A characteristic of the substance such as color, shape, odor, luster, size, melting point, and density.	A characteristic that indicates the ability of a substance to form another substance: paper can burn, iron can rust, and silver can tarnish.
Change	A change in a physical property that retains the identity of the substance: a change of state, a change in size, or a change in shape.	A change in which the original substance is converted to one or more new substances: burning paper, rusting iron, tarnishing silver.

Sample Problem 3.3   **Physical and Chemical Properties**

Classify each of the following as a physical or chemical property.
a. Water is a liquid at room temperature.
b. Helium does not react with other elements.
c. Gasoline is flammable.
d. Aluminum foil has a shiny appearance.

Solution
a. Physical property; liquid is a state of matter.
b. Chemical property; describes a chemical characteristic of helium.
c. Chemical property; describes the ability of gasoline to change.
d. Physical property; shininess of a substance is a physical characteristic.

Study Check

Identify each of the following as a physical or chemical property.
a. Silver is shiny.
b. Silver can tarnish.

Sample Problem 3.4 **Physical and Chemical Changes**

Classify each of the following as a physical or chemical change.
a. An ice cube melts to form liquid water.
b. Bleach removes a stain.
c. An enzyme breaks down the lactose in milk.
d. Peppercorns are ground into flakes.

Solution

a. Physical change; the ice cube changes state.
b. Chemical change; a change occurs in the composition of the stain.
c. Chemical change; a change occurs in the composition of lactose.
d. Physical change; a change of size does not change composition.

Study Check

Which of the following are chemical changes?
a. Water freezes on a pond.
b. Gas bubbles form when baking powder is placed in vinegar.
c. A log is chopped for firewood.
d. A log is burned in a fireplace.

Questions and Problems **Properties of Matter**

3.7 Describe each of the following as physical or chemical properties.
   a. Chromium is a steel-gray solid.
   b. Hydrogen reacts readily with oxygen.
   c. Nitrogen freezes at −210°C.
   d. Milk will sour when left in a warm room.
3.8 Describe each of the following as physical or chemical properties.
   a. Neon is a colorless gas at room temperature.
   b. Apple slices turn brown when they are exposed to air.
   c. Phosphorus will ignite when exposed to air.
   d. At room temperature, mercury is a liquid.
3.9 What type of change, physical or chemical, takes place in each of the following?
   a. Water vapor condenses to form rain.
   b. Cesium metal reacts explosively with water.
   c. Gold melts at 1064°C.
   d. A puzzle is cut into 1000 pieces.
   e. Sugar dissolves in water.
3.10 What type of change, physical or chemical, takes place in each of the following?

   a. Gold is hammered into thin sheets.
   b. A silver pin tarnishes in the air.
   c. A tree is cut into boards at a sawmill.
   d. Food is digested.
   e. A chocolate bar melts.
3.11 Describe each property of the element fluorine as physical or chemical.
   a. is highly reactive
   b. is a gas at room temperature
   c. has a pale, yellow color
   d. will explode in the presence of hydrogen
   e. has a melting point of −220°C
3.12 Describe each property of the element zirconium as physical or chemical.
   a. melts at 1852°C
   b. is resistant to corrosion
   c. has a grayish-white color
   d. ignites spontaneously in air when finely divided
   e. is a shiny metal

3.3 Temperature

Temperature is a measure of how hot or cold a substance is compared to another substance. Heat always flows from a substance with a higher temperature to a substance with a lower temperature. When you drink hot coffee or touch a hot burner, heat flows to your mouth or hand, which is at a lower temperature. When you touch an ice cube, it feels cold because heat flows from your hand to the colder ice cube.

Celsius and Fahrenheit Temperatures

Temperatures in science, and in most of the world, are measured and reported in *Celsius (°C)* units. In the United States, everyday temperatures are commonly reported in *Fahrenheit (°F)* units. A typical room temperature of 22°C would be the same as 72°F. A normal body temperature of 37.0°C is 98.6°F.

On the Celsius and Fahrenheit scales, the temperatures of melting ice and boiling water are used as reference points. On the Celsius scale, the freezing point of water is defined as 0°C and the boiling point as 100°C. On a Fahrenheit scale, water freezes at 32°F and boils at 212°F. On each scale, the temperature difference between freezing and boiling is divided into smaller units, or degrees. On the Celsius scale, there are 100 degree units between the freezing and boiling temperatures of water compared with 180 degree units on the Fahrenheit temperature scale. That makes a Celsius degree almost twice the size of a Fahrenheit degree: 1°C = 1.8°F. (See Figure 3.5.)

Figure 3.5 A comparison of the Fahrenheit, Celsius, and Kelvin temperature scales between the freezing and boiling points of water.

Q What is the difference in the values for freezing on the Celsius and Fahrenheit temperature scales?

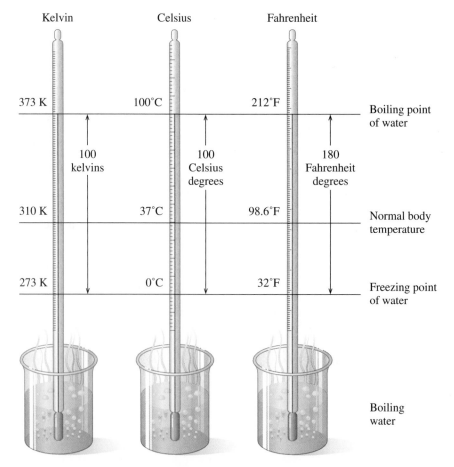

$$180 \text{ Fahrenheit degrees} = 100 \text{ Celsius degrees}$$

$$\frac{180 \text{ Fahrenheit degrees}}{100 \text{ Celsius degrees}} = \frac{1.8°F}{1°C}$$

In a chemistry laboratory, temperatures are measured in Celsius degrees. To convert to a Fahrenheit temperature, the Celsius temperature is multiplied by 1.8, and then 32 degrees is added. The 32 degrees adjusts the freezing point of 0°C on the Celsius scale to 32°F on the Fahrenheit scale. Both values, 1.8 and 32, are exact numbers. The equation for this conversion follows:

$$T_F = \underbrace{\frac{1.8°F(T_C)}{1°C}}_{\substack{\text{Changes} \\ °C \text{ to } °F}} + \underbrace{32°}_{\substack{\text{Adjusts freezing} \\ \text{point}}} \quad \text{or} \quad T_F = 1.8(T_C) + 32°$$

CHEM NOTE

GLOBAL WARMING

The amount of carbon dioxide (CO_2) gas in our atmosphere is on the increase as we burn more gasoline, coal, and natural gas. The algae in the oceans and the plants and trees in the forests normally absorb carbon dioxide, but they cannot keep up with the continued increase. The cutting of trees in the rain forests (deforestation) reduces the amount of carbon dioxide removed from the atmosphere. Many of the trees are also burned as land is cleared. It has been estimated that deforestation may account for 15–30% of the carbon dioxide that remains in the atmosphere each year.

The carbon dioxide in the atmosphere acts like the glass in a greenhouse. When sunlight warms the Earth's surface, some of the heat is trapped by carbon dioxide. As CO_2 levels rise, more heat is trapped. It is not yet clear how severe the effects of global warming might be. Some scientists estimate that by around the year 2030, the atmospheric level of carbon dioxide could double and cause the temperature of Earth's atmosphere to rise by 2–5°C. If that should happen, it would have a profound impact on Earth's climate. For example, an increase in the melting of snow and ice could raise the ocean levels by as much as 2 m, which is enough to flood many cities located on the ocean shorelines.

Worldwide efforts are being made to reduce fossil fuel use and to slow or stop deforestation. It will require cooperation throughout the world to avoid the bleak future that some scientists predict should global warming continue unchecked.

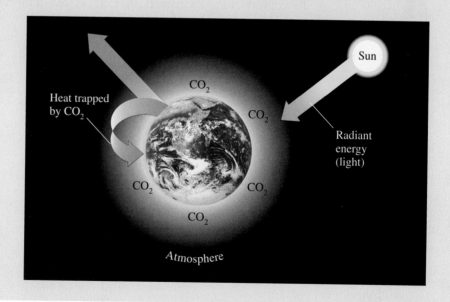

| Sample Problem 3.5 | Converting Celsius to Fahrenheit |

While traveling in China, you discover that your temperature is 38.2°C. What is your temperature in Fahrenheit degrees?

Solution

STEP 1 **Given** 38.2°C   **Need** T_F

STEP 2 **Plan**

$$T_C \quad \boxed{\text{Temperature equation}} \quad T_F$$

STEP 3 **Equality/Conversion Factor**

$$T_F = 1.8(T_C) + 32°$$

STEP 4 **Set Up Problem** Substitute the Celsius temperature into the equation and solve.

$$T_F = 1.8(38.2) + 32°$$

$$T_F = 68.8° + 32° \qquad \text{1.8 is exact; 32 is exact}$$
$$\quad = 100.8°F \qquad \text{Answer to the first decimal place}$$

In the equation, *the values of 1.8 and 32 are exact numbers.*

Study Check

When making ice cream, rock salt is used to chill the mixture. If the temperature drops to $-11°C$, what is it in °F?

To convert from Fahrenheit to Celsius, the temperature equation is rearranged for °C. Start with

$$T_F = 1.8(T_C) + 32°$$

Then subtract 32 from both sides.

$$T_F - 32° = 1.8(T_C) + 32° - 32°$$
$$T_F - 32° = 1.8(T_C)$$

Solve the equation for T_C by dividing both sides by 1.8.

$$\frac{T_F - 32°}{1.8} = \frac{\cancel{1.8}(T_C)}{\cancel{1.8}}$$

$$\frac{T_F - 32°}{1.8} = T_C$$

| Sample Problem 3.6 | Converting Fahrenheit to Celsius |

You are going to cook a pizza at 425°F. What is that temperature in Celsius degrees?

Solution

STEP 1   **Given**   425°F   **Need** T_C

STEP 2   **Plan**

$$T_F \quad \boxed{\begin{array}{c}\text{Temperature}\\ \text{equation}\end{array}} \quad T_C$$

STEP 3   **Equality/Conversion Factor**

$$\frac{T_F - 32°}{1.8} = T_C$$

STEP 4   **Set Up Problem**   To solve for T_C, substitute the Fahrenheit temperature into the equation and solve.

$$T_C = \frac{T_F - 32°}{1.8}$$

$$T_C = \frac{(425°C - 32°)}{1.8} \quad \text{32 is exact; 1.8 is exact}$$

$$= \frac{393°C}{1.8} = 218°C \quad \text{Answer to the 1's place}$$

Study Check

A child has a temperature of 103.6°F. What is this temperature on a Celsius thermometer?

Kelvin Temperature Scale

Scientists tell us that the coldest temperature possible is −273°C (more precisely −273.15°C). On the Kelvin scale, this temperature, called *absolute zero,* has the value of 0 kelvins (0 K). Because there are no lower temperatures possible, the Kelvin scale has no negative numbers. Between the freezing and boiling points of water, there are 100 kelvins, which makes a kelvin equal to a Celsius degree.

$$1 \text{ K} = 1°C$$

To convert a temperature in Celsius to kelvins, add 273 to the Celsius temperature:

$$T_K = T_C + 273$$

Table 3.6   A Comparison of Temperatures

Example	Fahrenheit (°F)	Celsius (°C)	Kelvin (K)
Sun	9937	5503	5776
A hot oven	450	232	505
A desert	120	49	322
A high fever	104	40	313
Room temperature	72	22	295
Water freezes	32	0	273
A northern winter	−76	−60	213
Helium boils	−452	−269	4
Absolute zero	−459	−273	0

To convert kelvins to a Celsius temperature, 273 is subtracted.

$$T_K - 273 = T_C + 273 - 273$$
$$T_K - 273 = T_C$$
$$T_C = T_K - 273$$

Table 3.6 gives a comparison of some temperatures on the three scales.

Sample Problem 3.7   **Converting Degrees Celsius to Kelvins**

What is the temperature of a laboratory experiment run at 185°C on the Kelvin scale?

Solution

STEP 1   **Given**   185°C   **Need** T_K

STEP 2   **Plan**

$$T_C \quad \boxed{\text{Kelvin equation}} \quad T_K$$

STEP 3   **Equality/Conversion Factor**

$$T_K = T_C + 273$$

STEP 4   **Set Up Problem**   Substitute the Celsius temperature into the equation and solve.

$$T_K = T_C + 273$$
$$T_K = 185 + 273$$
$$= 458 \text{ K} \qquad \text{Answer to the 1's place}$$

Study Check

On the planet Mercury, the average night temperature is 13 K and the average day temperature is 683 K. What are these temperatures on the Celsius scale?

CHEM NOTE

VARIATION IN BODY TEMPERATURE

Normal body temperature is considered to be 37.0°C, although it varies throughout the day and from person to person. Oral temperatures of 36.1°C are common in the morning and climb to a high of 37.2°C between 6 P.M. and 10 P.M. Temperatures above 37.2°C for a person at rest are usually an indication of illness. Individuals involved in prolonged exercise may also experience elevated temperatures. Body temperatures of marathon runners can range from 39°C to 41°C as heat production during exercise exceeds the body's ability to lose heat.

Changes of more than 3.5°C from the normal body temperature begin to interfere with bodily functions. Temperatures above 41°C can lead to convulsions, particularly in children, which may cause permanent brain damage. Heatstroke occurs above 41.1°C. Sweat production stops and the skin becomes hot and dry. The pulse rate is elevated, and respiration becomes weak and rapid. The person can become lethargic and lapse into a coma. Damage to internal organs is a major concern, and treatment, which must be immediate, may include immersing the person in an ice-water bath.

In hypothermia, body temperature can drop as low as 28.5°C. The person may appear cold and pale and have an irregular heartbeat. Unconsciousness can occur if the body temperature drops below 26.7°C. Respiration becomes slow and shallow, and oxygenation of the tissues decreases. Treatment involves providing oxygen and increasing blood volume with glucose and saline fluids. Internal temperature may be restored by injecting warm fluids (37.0°C) into the abdominal cavity.

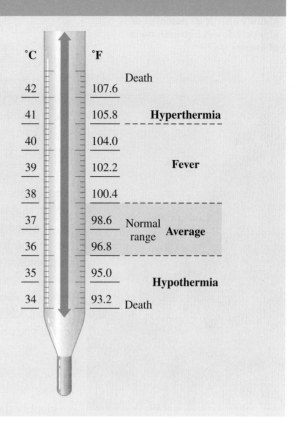

Questions and Problems   Temperature

3.13 Your friend who is visiting from France just took her temperature. When she reads 99.8, she becomes concerned that she is quite ill. How would you explain the temperature to your friend?

3.14 You have a friend who is using a recipe for flan from a Mexican cookbook. You notice that he set your oven temperature at 175°F. What would you advise him to do?

3.15 Solve the following temperature conversions:
  a. 37.0°C = _____ °F
  b. 65.3°F = _____ °C
  c. −27°C = _____ K
  d. 62°C = _____ K
  e. 114°F = _____ °C
  f. 72°F = _____ K

3.16 Solve the following temperature conversions:
  a. 25°C = _____ °F
  b. 155°C = _____ °F

  c. −25°F = _____ °C
  d. 224 K = _____ °C
  e. 545 K = _____ °C
  f. 875 K = _____ °C

3.17 a. A person with heatstroke has a temperature of 106°F. What does this read on a Celsius thermometer?
  b. Because high fevers can cause convulsions in children, the doctor wants to be called if the child's temperature goes over 40.0°C. Should the doctor be called if a child has a temperature of 103°F?

3.18 a. Water is heated to 145°F. What is the temperature of the hot water in °C?
  b. During extreme hypothermia, a young woman's temperature dropped to 20.6°C. What was her temperature on the Fahrenheit scale?

Identify energy as potential or kinetic; convert between units of energy.

3.4 Energy

When you are running, walking, dancing, or thinking, you are using energy to do work. In fact, **energy** is defined as the ability to do work or to supply heat. Suppose you are climbing a steep hill. Perhaps you become too tired to go on. We could say that you do not have sufficient energy to do any more work. Now suppose you sit down and have lunch. In a while you will have obtained some energy from the food, and you will be able to do more work and complete the climb. (See Figure 3.6.)

Potential and Kinetic Energy

All energy can be classified as potential energy or kinetic energy. **Potential energy** is stored energy, whereas **kinetic energy** is the energy of motion. Any object that is moving has kinetic energy. The potential energy of an object is determined by its position or by the chemical bonds it contains. A boulder resting on top of a mountain has potential energy because of its location. If the boulder rolls down the mountain, the potential energy becomes kinetic energy. Water stored in a reservoir has potential energy. When the water goes over the dam, the potential energy is converted to kinetic energy. The food you eat has potential energy stored in its chemical bonds. When you digest food, you convert its potential energy to kinetic energy to do biological work.

Figure 3.6 Work is done as the rock climber moves up the cliff. At the top, the climber has more potential energy than when she started the climb.

Q *What happens to the potential energy of the climber when she descends?*

Sample Problem 3.8 ▸ **Forms of Energy**

Identify the energy in each of the following as potential or kinetic.
a. gasoline   **b.** skating   **c.** a candy bar

Solution
a. potential energy (stored)   **b.** kinetic energy (motion)
c. potential energy (stored)

Study Check

Would the energy in a stretched rubber band be kinetic or potential?

Forms of Energy

There are many forms of energy that include electrical energy, radiant energy such as light, mechanical energy, chemical energy, and nuclear energy. In an electric power plant, energy from the burning of fossil fuels such as natural gas or coal is used to produce steam, which turns the turbines and generates electrical energy. In your home, electrical energy is converted to light energy when you switch on a light bulb, to mechanical energy when you use a mixer or a washing machine, and to heat when you use a hair dryer or a toaster (see Figure 3.7). In your body, chemical energy from the food you eat is converted to mechanical energy when you use muscles to pedal a bicycle, run a marathon, or mow the lawn.

Heat is the energy that flows from a warmer object to a cooler one. A frozen pizza feels cold because heat flows from your hand into the pizza. A cup of hot

Figure 3.7 Examples of electrical energy converted into light, heat, and mechanical energy.

Q *Is heat a form of kinetic or potential energy? Why?*

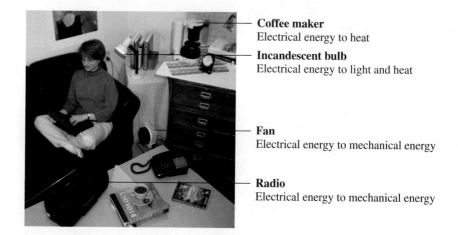

Coffee maker
Electrical energy to heat

Incandescent bulb
Electrical energy to light and heat

Fan
Electrical energy to mechanical energy

Radio
Electrical energy to mechanical energy

coffee feels hot because heat flows from the coffee into the mouth. Heat is associated with the motion of particles. The faster particles move, the greater the heat or thermal energy of the substance. In the frozen pizza, the particles are moving very slowly. As the pizza cooks, heat is transferred, the motions of the particles increase, and the pizza becomes warm. (See Figure 3.8.) Eventually, the particles have enough energy to make the pizza hot and ready to eat.

The SI unit of energy and work is the **joule (J),** pronounced, "jewel." The joule is a small amount of energy, so scientists often use the kilojoule (kJ), 1000 joules. When you heat water for one cup of tea, you use about 75 000 J or 75 kJ of heat.

Figure 3.8 As heat is transferred to the particles in a frozen pizza **(a)**, their motions increase and the pizza warms up **(b)**. When the pizza is eaten, the potential energy in its chemical bonds is converted to kinetic energy as the body does work.

Q *Why does the cooked pizza feel hot?*

(a)

(b)

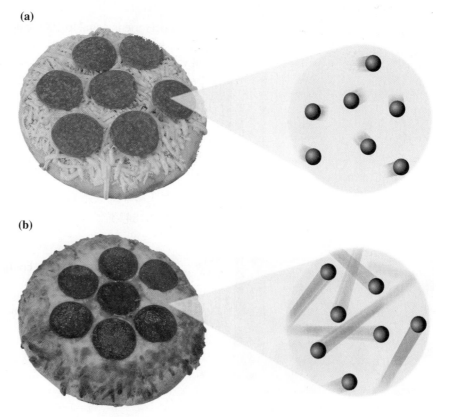

Energy in joules

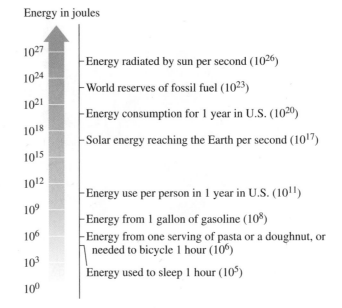

10^{27}	
10^{24}	Energy radiated by sun per second (10^{26})
10^{21}	World reserves of fossil fuel (10^{23})
10^{18}	Energy consumption for 1 year in U.S. (10^{20})
10^{15}	Solar energy reaching the Earth per second (10^{17})
10^{12}	
10^{9}	Energy use per person in 1 year in U.S. (10^{11})
10^{6}	Energy from 1 gallon of gasoline (10^{8})
10^{3}	Energy from one serving of pasta or a doughnut, or needed to bicycle 1 hour (10^{6})
10^{0}	Energy used to sleep 1 hour (10^{5})

You may be more familiar with the older unit **calorie (cal),** from the Latin *caloric,* meaning "heat." The calorie was originally defined as the amount of energy (heat) needed to raise the temperature of 1 gram of water by 1°C (from 14.5°C to 15.5°C). Now, 1 calorie is defined as exactly 4.184 J. This equality can also be written as a conversion factor.

1 cal = 4.184 J (exact)

$$\frac{4.184 \text{ J}}{1 \text{ cal}} \quad \text{and} \quad \frac{1 \text{ cal}}{4.184 \text{ J}}$$

One kilocalorie is equal to 1000 calories (small *c*). The international unit of nutritional energy is the kilojoule (kJ). For example, a baked potato has a nutritional value of 120 Calories, which is the same as 120 kcal or 500 kJ of energy.

1 kcal = 1000 cal

1 kJ = 1000 J

When we talk about food and nutrition, we often use Calorie (Cal) with a capital *C*. This kind of "Calorie" is actually a kilocalorie, which is indicated by writing Calorie (Cal) with a capital *C.*

1 Cal = 1 kcal

1 Cal = 4.184 kJ

Sample Problem 3.9   **Energy Units**

When 10. g octane in gasoline burns in an automobile engine, 8500 J is released. Convert this quantity of energy to each of the following units.
a. kilojoules   **b.** calories

Solution
a. kilojoules

STEP 1   **Given**   8500 J   **Need**   kilojoules

STEP 2   **Plan**   J   Energy factor   kJ

STEP 3   **Equalities/Conversion Factors**

$$1 \text{ kJ} = 1000 \text{ J}$$

$$\frac{1000 \text{ J}}{1 \text{ kJ}} \quad \text{and} \quad \frac{1 \text{ kJ}}{1000 \text{ J}}$$

STEP 4   **Set Up Problem** $8500 \cancel{J} \times \dfrac{1 \text{ kJ}}{1000 \cancel{J}} = 8.5 \text{ kJ}$

b. calories

STEP 1   **Given**   8500 J   **Need**   calories (cal)

STEP 2   **Plan**   J   Energy factor   cal

STEP 3   **Equalities/Conversion Factors**

$$1 \text{ cal} = 4.184 \text{ J}$$

$$\frac{1 \text{ cal}}{4.184 \text{ J}} \quad \text{and} \quad \frac{4.184 \text{ J}}{1 \text{ cal}}$$

STEP 4   **Set Up Problem**

$$8500 \cancel{J} \times \frac{1 \text{ cal}}{4.184 \cancel{J}} = 2000, \text{ or } 2.0 \times 10^3 \text{ cal}$$

Study Check

How many kcal are in a jelly doughnut if it provides 110 kJ of energy?

Questions and Problems   **Energy**

3.19 Discuss the changes in the potential and kinetic energy of a roller-coaster ride as the roller coaster climbs up a ramp and goes down the other side.

3.20 Discuss the changes in the potential and kinetic energy of a ski jumper taking the elevator to the top of the jump and going down the ramp.

3.21 Indicate whether each item describes potential or kinetic energy.
  a. water at the top of a waterfall
  b. kicking a ball
  c. the energy in a lump of coal
  d. a skier at the top of a hill

3.22 Indicate whether each item describes potential or kinetic energy.
  a. the energy in your food
  b. a tightly wound spring

  c. an earthquake
  d. a car speeding down the freeway

3.23 What are the forms of energy involved in the following examples? *Example:* Striking a match converts chemical energy into heat and light.
  a. using a hair dryer
  b. using an electric fan
  c. burning gasoline in a car engine
  d. sunlight falling on a solar water heater

3.24 What are the forms of energy involved in the following examples?
  a. turning on a light switch
  b. cooking food on a gas stove
  c. using a microwave oven
  d. burning a log in the fireplace

3.25 Match the device with the following changes in energy:
(1) heater   (2) light bulb
(3) hydroelectric power plant
a. electrical energy to radiant energy
b. mechanical energy to electrical energy
c. electrical energy to heat (thermal energy)

3.26 Match the device with the following changes in energy:
(1) 6-volt battery   (2) automobile engine
(3) ceiling fan

a. electrical energy to mechanical energy
b. chemical energy to mechanical energy
c. chemical energy to electrical energy

3.27 Convert each of the following energy units.
a. 3500 cal to kcal   **b.** 415 J to cal
c. 28 cal to J   **d.** 4.5 kJ to cal

3.28 Convert each of the following energy units.
a. 8.1 kcal to cal   **b.** 325 J to kJ
c. 2550 cal to kJ   **d.** 2.50 kcal to J

3.5 Specific Heat

Use specific heat to calculate heat loss or gain, temperature change, or mass of a sample.

Every substance can absorb heat. When you want to bake a potato, you place it in a hot oven. If you are cooking pasta, you add the pasta to boiling water. Every substance has its own characteristic ability to absorb heat. Some substances must absorb more heat than others to reach a certain temperature. These energy requirements for different substances are described in terms of a physical property called specific heat. **Specific heat** (*SH*) is the amount of heat (*q*) needed to raise the temperature of exactly 1 g of a substance by exactly 1°C. This temperature change is written as ΔT (*delta T*).

$$\text{Specific heat } (SH) = \frac{\text{heat } (q)}{\text{grams} \times \Delta T} = \frac{\text{J (or cal)}}{\text{g} \times °C}$$

Now we can write the specific heat for water using our definition of the calorie and joule.

$$\text{Specific heat } (SH) \text{ of } H_2O(l) = 4.184 \frac{J}{g \, °C} = 1.00 \frac{cal}{g \, °C}$$

If we look at Table 3.7, we see that the specific heat of water is much larger than the specific heat of aluminum or copper. We see that 1 g of water requires 4.184 J to increase its temperature by 1°C. However, adding the same amount of heat (4.814 J) will raise the temperature of 1 g of aluminum by about 5°C and 1 g of copper by 10°C. Because of its high specific heat, the water in the body can absorb or release large amounts of heat to maintain an almost constant temperature.

Table 3.7   Specific Heats of Some Substances

	Substance	Specific Heat (J/g °C)
Elements	Aluminum, Al	0.897
	Copper, Cu	0.385
	Gold, Au	0.129
	Iron, Fe	0.450
	Silver, Ag	0.235
	Titanium, Ti	0.523
Compounds	Ammonia, $NH_3(g)$	2.04
	Ethanol, $C_2H_5OH(l)$	2.46
	Sodium chloride, $NaCl(s)$	0.864
	Water, $H_2O(l)$	4.184
	Water, $H_2O(s)$	2.03

Sample Problem 3.10   **Calculating Specific Heat**

What is the specific heat of lead if 57.0 J are needed to raise the temperature of 35.6 g Pb by 12.5°C?

Solution

STEP 1   **Given**

heat 57.0 J mass 35.6 g Pb temperature change 12.5°C

Need specific heat (J/g °C)

STEP 2   **Plan** The specific heat (*SH*) is calculated by dividing the heat by the mass (g) and by the temperature difference (ΔT).

$$SH = \frac{\text{heat}}{\text{mass} \;\; \Delta T}$$

STEP 3   **Write the specific heat equation.**

STEP 4   **Substitute the given values into the equation.**

$$\text{Specific heat } (SH) = \frac{57.0 \text{ J}}{35.6 \text{ g} \;\; 12.5°C} = 0.128 \frac{\text{J}}{\text{g °C}}$$

Study Check

What is the specific heat of sodium if 123 J are needed to raise the temperature of 4.00 g of Na by 25.0°C?

When we know the specific heat of a substance, we can calculate the heat lost or gained by measuring the mass of the substance and the initial and final temperature. We can substitute these measurements into the specific heat expression that is rearranged to solve for heat, which we call the *heat equation.*

Heat	=	mass	×	temperature change	×	specific heat
Heat (q)	=	mass (m)	×	ΔT	×	SH
J	=	grams	×	°C	×	$\dfrac{\text{J}}{\text{g °C}}$

In heat calculations, the temperature change is always the difference between the final temperature and the initial temperature.

$$\Delta T = T_{\text{final}} - T_{\text{initial}}$$

When a substance absorbs energy, the temperature rises. Then the sign of ΔT is positive (+), which gives a positive sign (+) for the heat (q) that is calculated. However, if a substance loses energy, the temperature drops. Then the sign of ΔT is negative (−), which gives a negative sign (−) for the heat (q). Thus, a positive (+) sign for heat (q) means that heat flows into the substance, whereas a negative (−) sign for heat (q) means that heat flows out of the substance. In the following sample problems, we will see how the heat equation is used to calculate heat and how it can be rearranged to solve for mass.

Guide to Calculations Using Specific Heat
STEP 1 List given and needed data.
STEP 2 Calculate temperature change (ΔT).
STEP 3 Write the equation for heat $q = m \times \Delta T \times SH$ and rearrange for unknown.
STEP 4 Substitute given values and solve, making sure units cancel.

Sample Problem 3.11 ▶ **Calculating Heat with Temperature Increase**

The element aluminum has a specific heat of 0.897 J/g °C. How many joules are absorbed by 45.2 g aluminum if its temperature rises from 12.5°C to 76.8°C?

Solution

STEP 1 **List given and needed data.**

Given Mass (m) = 45.2 g

SH for aluminum = 0.897 J/g°C

Initial temperature = 12.5°C

Final temperature = 76.8°C

Need heat (q) in joules (J)

STEP 2 **Calculate the temperature change.** The temperature change ΔT is the difference between the two temperatures.

$$\Delta T = T_{final} - T_{initial} = 76.8°C - 12.5°C = 64.3°C$$

STEP 3 **Write the heat equation.**

$$\text{Heat }(q) = \text{mass }(m) \times \Delta T \times SH$$

STEP 4 **Substitute the given values into the equation and solve, making sure units cancel.**

$$\text{Heat }(q) = 45.2 \, \cancel{g} \times 64.3°\cancel{C} \times \frac{0.897 \text{ J}}{\cancel{g}°\cancel{C}} = 2610 \text{ J} \, (2.61 \times 10^3 \text{ J})$$

Study Check

Some cooking pans have a layer of copper on the bottom. How many kilojoules are needed to raise the temperature of 125 g copper from 22°C to 325°C if the specific heat of copper is 0.385 J/g °C?

Sample Problem 3.12 ▶ **Calculating Heat Loss**

A 225-g sample of hot tea cools from 74.6°C to 22.4°C. How much heat in kilojoules is lost, assuming that tea has the same specific heat as water?

Solution

STEP 1 **List given and needed data.**

Given Mass (m) = 225 g   SH for tea = 4.184 J/g°C

Initial temperature = 74.6°C   Final temperature = 22.4°C

Heat factor: 1 kJ = 1000 J

Need heat (q) lost in kilojoules (kJ)

STEP 2 **Calculate the temperature change.**   The temperature change ΔT is the difference between the two temperatures.

$$\Delta T = T_f - T_i = 22.4°C - 74.6°C = -52.2°C$$

STEP 3 **Write the heat equation.**

$$q = m \times \Delta T \times SH$$

STEP 4 **Substitute the given values into the equation and solve, making sure units cancel.**

$$q = 225\ \cancel{g} \times (-52.2°\cancel{C}) \times \frac{4.184\ \cancel{J}}{\cancel{g}°\cancel{C}} \times \frac{1\ kJ}{1000\ \cancel{J}} = -49.1\ kJ$$

Study Check

How much heat in joules is lost when 15.5 g gold cools from 215°C to 35°C? The specific heat of gold is 0.129 J/g °C.

Sample Problem 3.13   **Calculating Mass Using Specific Heat**

Ethanol has a specific heat of 2.46 J/g °C. When 655 J are added to a sample of ethanol, its temperature rises from 18.2°C to 32.8°C. What is the mass of the ethanol sample?

Solution

STEP 1 **List given and needed data.**

Given   Heat = 655 J SH for ethanol = 2.46 J/g°C

   Initial temperature = 18.2°C   Final temperature = 32.8°C

Need   mass (m) of ethanol sample

STEP 2 **Calculate the temperature change.**   The temperature change ΔT is the difference between the two temperatures.

$$\Delta T = T_f - T_i = 32.8°C - 18.2°C = 14.6°C$$

STEP 3 **Write the heat equation.**

$$q = m \times \Delta T \times SH$$

The heat equation must be rearranged to solve for mass (m), which is the heat divided by the temperature change and the specific heat.

$$\text{Mass}\ (m) = \frac{\text{heat}\ (q)}{\Delta T\ SH}$$

STEP 4   **Substitute the given values into the equation and solve, making sure units cancel.**

$$\text{Mass } (m) = \frac{655\ \cancel{J}}{14.6\ \cancel{°C}\ \dfrac{2.46\ \cancel{J}}{g\ \cancel{°C}}}$$

$$\text{Mass } (m) = 18.2\ g$$

Study Check

When 8.81 kJ are absorbed by a piece of iron, its temperature rises from 15°C to 122°C. What is the mass, in grams, if iron has a specific heat of 0.450 J/g °C?

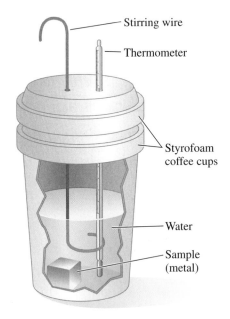

Stirring wire

Thermometer

Styrofoam coffee cups

Water

Sample (metal)

Measuring Heat Changes

In chemistry and nutrition laboratories, a type of equipment called a **calorimeter** is used to measure the temperature when a sample gains or loses heat. A simple type of calorimeter can be made from a Styrofoam coffee cup, a measured amount of water, and a thermometer. First the initial temperature of the water is measured. Then an object of known mass is heated to a higher temperature and added to the water. While submerged in the water, the object loses heat (cools down) and the water gains heat (warms up). The final temperature is attained when both the object and the water reach the same temperature. The amount of heat transferred from the object is equal to the heat that was gained by the water.

Heat $(-q)$ lost by object $=$ heat (q) gained by water

Sample Problem 3.14   **Using a Calorimeter**

A 35.20-g sample of a metal heated to 100.0°C is placed in a calorimeter containing 42.5 g water at an initial temperature of 19.2°C. If the final temperature of the metal and the water is 29.5°C, what is the specific heat of the solid, assuming all the heat is transferred to the water?

Solution

STEP 1   **List the given and needed data.**

Unknown Object	Water
Mass $(m) = 35.20$ g	Mass $(m) = 42.5$ g
Initial temperature $= 100.0$°C	Initial temperature $= 19.2$°C
Final temperature $= 29.5$°C	Final temperature $= 29.5$°C
	SH of water $= 4.184$ J/g°C

Need SH of metal (J/g °C)

STEP 2 **Determine the temperature changes.**   Determine the temperature change for the solid and the water. The final temperature of the solid is the same as the final temperature of the water, which is 29.5°C.

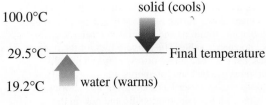

$$\Delta T_{\text{water}} = T_{\text{f}} - T_{\text{i}} = 29.5°C - 19.2°C = 10.3°C$$
$$\Delta T_{\text{solid}} = T_{\text{f}} - T_{\text{i}} = 29.5°C - 100°C = -70.5°C$$

STEP 3 **Write the heat equation.**

$$q = m \times \Delta T \times SH$$

STEP 4 **Substitute the given values into the equations and solve.**   We first calculate the heat gained by the water because we have all the data.

$$q_{\text{water}} = m \times \Delta T \times SH$$

$$q_{\text{water}} = 42.5 \text{ g} \times (10.3°C) \times 4.184 \, \frac{J}{g°C} = 1830 \text{ J}$$

The heat gained by the water is equal to the heat lost by the object. If water gained 1830 J, then the object lost 1830 J.

Heat gained by water (q = 1830 J) = heat lost by object ($-q = -1830$ J)

Now we rearrange the heat equation to solve for the specific heat of the metal.

$$q_{\text{metal}} = m \times \Delta T \times SH$$

$$SH_{\text{metal}} = \frac{q}{m \times \Delta T}$$

$$SH_{\text{metal}} = \frac{-1830 \text{ J}}{(35.20 \text{ g})(-70.5°C)} = 0.737 \text{ J/g °C}$$

Study Check

A piece of granite weighing 250.0 g is heated in boiling water to 100.0°C. When the granite is placed in a calorimeter containing 400.0 g water, the temperature of the water increases from 20.0°C to 28.5°C. What is the specific heat of the granite, assuming all the heat is transferred to the water?

CHEM NOTE

ENERGY AND NUTRITION

The food we eat provides energy to do work in the body, which includes the growth and repair of cells. The components of foods that provide us with energy are carbohydrates, fats, and proteins, which break down during digestion to glucose, fatty acids, and amino acids. Vitamins and minerals are necessary for health, but have no energy value. The primary fuel for the body is carbohydrates, but if carbohydrate reserves are exhausted, fats and then proteins can be used for energy. For many years, the energy in foods was expressed as Calories or kilocalories, but kilojoules is the unit used internationally and now in the United States. Thus, a typical diet of 2000 Cal (kcal) is the same as an 8400 kJ diet.

In the nutrition laboratory, foods undergo combustion in a calorimeter to determine their energy value. The energy released when the food sample is burned is determined by measuring the mass of water and the increase in the temperature of the water. For example, suppose 2.3 g butter is placed in a calorimeter containing 1900 g water at an initial temperature of 17°C. After combustion of the butter, the water has a temperature of 28°C. We can calculate the heat from the combustion using the mass, temperature change, and specific heat of water ($\Delta T = 11$°C).

$$1900 \; \cancel{g} \; \times \; 11\cancel{°C} \times 4.184 \; \frac{\cancel{J}}{\cancel{g}°\cancel{C}} \times \frac{1 \text{ kJ}}{1000 \; \cancel{J}} = 87 \text{ kJ}$$

The energy used to heat the water was obtained from the combustion of the butter. Thus we can determine the energy value of butter as kilojoules per gram of butter.

$$\text{Energy value of butter} = \frac{87 \text{ kJ}}{2.3 \text{ g}} = 38 \text{ kJ/g}$$

Because the energy values for the metabolism of carbohydrates, fats, and proteins vary somewhat, the energy values determined using a calorimeter represent an average. Typical values for carbohydrates, fats, and proteins are listed in Table 3.8 along with the older kcal values, which are still prevalent on food labels in the United States.

Table 3.8 Typical Energy Values for the Three Food Types

Food Type	kJ/g	kcal/g
Carbohydrate	17	4
Fat	38	9
Protein	17	4

If the composition of a food is known in grams, the energy value for each type of food can be calculated as

$$\text{Kilojoules} = \cancel{g} \times \frac{\text{kJ}}{\cancel{g}}$$

$$\text{Kilocalories} = \cancel{g} \times \frac{\text{kcal}}{\cancel{g}}$$

The energy content of a packaged food is listed in the nutritional information, usually in terms of the number of Calories in a serving. Suppose that a piece of chocolate cake (1 serving) contains 34 g of carbohydrate, 10 g of fat, and 5 g of protein. We can calculate the total energy content as follows.

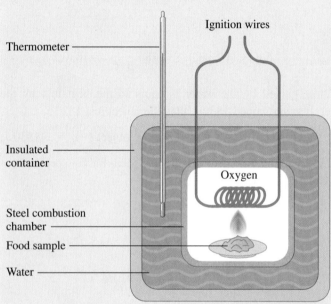

| | Thermometer |
| Ignition wires |
| Insulated container |
| Oxygen |
| Steel combustion chamber |
| Food sample |
| Water |

Food Type	Mass	Energy Value		Energy
Carbohydrate	$34 \; \cancel{g}$	$\times \; 17 \frac{\text{kJ}}{\cancel{g}}$	$\frac{(4 \text{ kcal})}{\cancel{g}}$ =	580 kJ (140 kcal)
Fat	$10 \; \cancel{g}$	$\times \; 38 \frac{\text{kJ}}{\cancel{g}}$	$\frac{(9 \text{ kcal})}{\cancel{g}}$ =	380 kJ (90 kcal)
Protein	$5 \; \cancel{g}$	$\times \; 17 \frac{\text{kJ}}{\cancel{g}}$	$\frac{(4 \text{ kcal})}{\cancel{g}}$ =	90 kJ (20 kcal)
	Total energy content		=	1050 kJ (250 kcal)

Table 3.9 gives the composition and energy content of some foods.

Table 3.9 General Composition and Caloric Content of Some Foods

Food	Carbohydrate	Fat	Protein	Energy*
Banana, 1 medium	26	0	1	460 kJ (110 kcal)
Carrots, raw, 1 cup	11	0	1	205 kJ (50 kcal)
Beef, ground, 3 oz	0	14	22	910 kJ (220 kcal)
Chicken, no skin, 3 oz	0	3	20	460 kJ (110 kcal)
Egg, 1 large	0	6	6	330 kJ (80 kcal)
Milk, 4% fat, 1 cup	12	9	9	700 kJ (170 kcal)
Milk, nonfat, 1 cup	12	0	9	360 kJ (90 kcal)
Potato, baked	23	0	3	440 kJ (100 kcal)
Salmon, 3 oz	0	5	16	460 kJ (110 kcal)
Steak, 3 oz	0	27	19	1350 kJ (320 kcal)

*Energy values are rounded to the nearest ten.

The number of kilojoules or kilocalories you need in your daily diet depends on your age, sex, and physical activity. (See Table 3.10.) When food intake exceeds energy output, a person's body weight increases. Food intake is usually regulated by the hunger center in the hypothalamus, located in the brain. The regulation of food intake is normally proportional to the nutrient stores in the body. If these nutrient stores are low, you feel hungry; if they are high, you do not feel like eating.

Weight loss occurs when food intake is less than energy output. Many diet products contain cellulose, which has no nutritive value but provides bulk and makes you feel full. Diet drugs, which depress the hunger center, can also excite the nervous system and elevate blood pressure. Exercise is an important way to expend energy and increase weight loss. Table 3.11 lists some types of exercise and the amount of energy they require.

Snack Crackers

Nutrition Facts
Serving Size 14 crackers (31g)
Servings Per Container About 7

Amount Per Serving

Calories 120 Calories from Fat 35
Kilojoules 500 kJ from Fat 150

	% Daily Value*
Total Fat 4g	6%
Saturated Fat 0.5g	3%
Trans Fat 0g	
Polyunsaturated Fat 0.5%	
Monounsaturated Fat 1.5g	
Cholesterol 0mg	0%
Sodium 310mg	13%
Total Carbohydrate 19g	6%
Dietary Fiber Less than 1g	4%
Sugars 2g	
Proteins 2g	

Vitamin A 0% • Vitamin C 0%
Calcium 4% • Iron 6%

*Percent Daily Values are based on a 2,000 calorie diet. Your daily values may be higher or lower depending on your calorie needs.

	Calories:	2,000	2,500
Total Fat	Less than	65g	80g
Sat Fat	Less than	20g	25g
Cholesterol	Less than	300mg	300mg
Sodium	Less than	2,400mg	2,400mg
Total Carbohydrate		300g	375g
Dietary Fiber		25g	30g

Calories per gram:
Fat 9 • Carbohydrate 4 • Protein 4

Table 3.10 Typical Energy Requirements for 70-kg Adult

Sex	Energy (kJ)	Energy (kcal)
Female	10 000	2200
Male	12 500	3000

Table 3.11 Energy Expended by a 70-kg Person

Activity	Energy (kJ/hr)	Energy (kcal/hr)
Sleeping	250	60
Sitting	420	100
Walking	840	200
Swimming	2100	500
Running	3100	750

Questions and Problems   **Specific Heat**

3.29 If the same amount of heat is supplied to samples of 10.0 g each of aluminum, iron, and copper all at 15.0°C, which sample would reach the highest temperature?

3.30 Substances A and B are the same mass and at the same initial temperature. When the same amount of heat is added to both, the final temperature of A is 75°C and B is 35°C. What does this tell you about the specific heats of A and B?

3.31 Calculate the following specific heats.
  a. the specific heat of the zinc if a 13.5-g sample of zinc heated from 24.2°C to 83.6°C absorbs 312 J of heat
  b. the specific heat of a metal if 48.2 g of the metal absorbs 345 J and changes its temperature from 35.0°C to 57.9°C

3.32 Calculate the following specific heats.
  a. the specific heat of tin if the temperature of 18.5 g tin rises from 35.0°C to 78.6°C when the sample absorbs 183 J of heat
  b. the specific heat of a metal if 22.5 grams of the metal absorbs 645 J and the temperature changes from 36.2°C to 92.0°C

3.33 Calculate the energy in joules and calories
  a. required to heat 25.0 g water from 12.5°C to 25.7°C
  b. required to heat 38.0 g copper (Cu) from 122°C to 246°C
  c. lost when 15.0 g ethanol, C_2H_5OH, cools from 60.5°C to −42.0°C
  d. lost when 125 g iron, Fe, cools from 118°C to 55°C

3.34 Calculate the energy in joules and calories
  a. required to heat 5.25 g water, H_2O, from 5.5°C to 64.8°C
  b. lost when 75.0 g water, H_2O, cools from 86.4°C to 2.1°C
  c. required to heat 10.0 g silver (Ag) from 112.°C to 275°C
  d. lost when 18.0 g gold (Au) cools from 224°C to 118°C

3.35 Calculate the mass in each of the following.
  a. a gold (Au) sample that absorbs 225 J to change its temperature from 15.0°C to 47.0°C
  b. an iron (Fe) object that loses 8.40 kJ when its temperature drops from 168.0°C to 82.0°C
  c. a sample of aluminum (Al) that absorbs 8.80 kJ when heated from 12.5°C to 26.8°C
  d. a sample of titanium (Ti) that loses 14 200 J when it cools from 185°C to 42°C

3.36 Calculate the mass in each of the following.
  a. water that absorbs 8250 J when its temperature rises from 18.4°C to 92.6°C
  b. a pure silver (Ag) sample that loses 3.22 kJ when its temperature drops from 145°C to 24°C

  c. a sample of aluminum (Al) that absorbs 1.65 kJ when its temperature rises from 65°C to 187°C
  d. an iron (Fe) bar that loses 2.52 kJ when its temperature drops from 252°C to 75°C

3.37 Calculate the rise in temperature for each of the following.
  a. 20.0 g iron (Fe) that absorbs 1580 J
  b. 150.0 g water that absorbs 7.10 kJ
  c. 85.0 g gold (Au) that absorbs 7680 J
  d. 50.0 g copper (Cu) that absorbs 6.75 kJ

3.38 Calculate the decrease in temperature for each of the following.
  a. 115 g copper (Cu) that loses 2.45 kJ
  b. 22.0 g silver (Ag) that loses 625 J
  c. 0.650 kg water that loses 5.48 kJ
  d. 35.0 g silver (Ag) that loses 472 J

3.39 Using the following data, determine the kilojoules and kilocalories for each food burned in a calorimeter.
  a. one stalk of celery that heats 505 g water from 25.2°C to 35.7°C
  b. a waffle that heats 4980 g water from 20.6°C to 62.4°C
  c. one cup of popcorn that changes the temperature of 1250 g water from 25.5°C to 50.8°C

3.40 When a 0.500-g sample of octane is burned in a calorimeter, the heat released increases the temperature of 357 g water from 22.7°C to 38.8°C. What is the heat in kilojoules and kilocalories produced when 1.00 g octane burns?

3.41 Using the energy values for foods, determine the kilojoules and kilocalories for each of the following (round the answers to the nearest 10).
  a. one cup of orange juice that has 26 g carbohydrate, 2 g protein, and no fat
  b. one apple that has 18 g carbohydrate, no protein, and no fat
  c. one tablespoon of vegetable oil that contains 14 g fat
  d. a diet that contains 68 g carbohydrate, 150 g protein, and 5.0 g fat

3.42 Using the energy values for foods, determine the kilojoules and kilocalories for each of the following (round the answers to the nearest 10).
  a. two tablespoons of crunchy peanut butter that contain 6 g carbohydrate, 16 g fat, and 7 g protein
  b. one cup of soup that has 7 g fat, 9 g carbohydrate, and 2 g protein
  c. one can of cola that has 35 g carbohydrate, no fat, and no protein
  d. one cup of clam chowder that contains 16 g carbohydrate, 9 g protein, and 12 g fat

Concept Map ▶ **Matter and Energy**

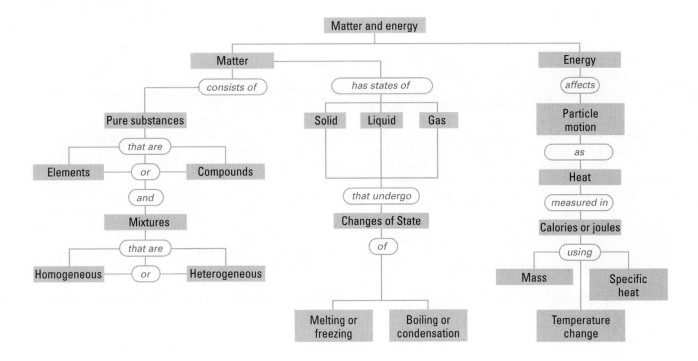

Chapter Review

3.1 Classification of Matter

Matter is everything that occupies space and has mass. Matter is classified as pure substances or mixtures. Pure substances, which are elements or compounds, have fixed compositions. Mixtures have variable compositions and can be homogeneous or heterogeneous. The substances in mixtures can be separated using physical methods.

3.2 Properties of Matter

Chemistry is the study of the properties of matter. A physical property is a characteristic unique to a substance. A physical change occurs when physical properties change but not the identity of the substance. A chemical property describes the properties of a substance interacting with another substance during a chemical change.

3.3 Temperature

In science, temperature is measured in Celsius units, °C, or kelvins, K. In the United States, the Fahrenheit scale, °F, is still in use.

3.4 Energy

Energy is the ability to do work. Potential energy is stored energy; kinetic energy is the energy of motion. Common units of energy are the calorie (cal), kilocalorie (kcal), joule (J), and kilojoule (kJ).

3.5 Specific Heat

Specific heat is the amount of energy required to raise the temperature of exactly 1 g of a substance by exactly 1°C. The heat lost or gained by a substance is determined by multiplying its mass (m), the temperature change (ΔT), and its specific heat (J/g °C). The nutritional Calorie is the same as 1 kcal or 1000 calories. The energy content of a food is the sum of the joules from carbohydrate, fat, and protein.

Key Terms

calorie (cal) The amount of heat energy that raises the temperature of exactly 1 g water exactly 1°C.

chemical change A change during which the original substance is converted into a new substance that has a different composition and new chemical and physical properties.

chemical properties The properties that characterize the ability of a substance to change into a new substance.

compound A pure substance consisting of two or more elements, with a definite composition, that can be broken down into simpler substances by chemical methods.

element A pure substance containing only one type of matter, which cannot be broken down by chemical methods.

energy the ability to do work.

gas A state of matter that fills the entire shape and volume of its container.

heat The energy that flows from a hotter object to a cooler one.

heterogeneous mixture A mixture of two or more substances that are not mixed uniformly.

homogeneous mixture A mixture of two or more substances that are mixed uniformly.

joule (J) The SI unit of heat energy; 4.184 J = 1 cal.

kinetic energy A type of energy that is required for actively doing work; energy of motion.

liquid A state of matter that has it own volume, but takes the shape of the container.

matter The material that makes up a substance and has mass and occupies space.

mixture The physical combination of two or more substances that does not change the identities of the mixed substances.

physical change A change that occurs in a substance without any change in its identity.

physical properties The properties that can be observed or measured without affecting the identity of a substance.

potential energy An inactive type of energy that is stored for use in the future.

properties Characteristics that are unique to a substance.

pure substance A type of matter with a fixed composition: elements and compounds.

solid A state of matter that has its own shape and volume.

specific heat A quantity of heat that changes the temperature of exactly 1 g of a substance by exactly 1°C.

states of matter Three forms of matter: solid, liquid, and gas.

 ## Understanding the Concepts

3.43 Identify the following as an element, compound, or mixture.

a.
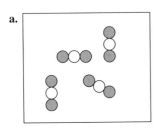

b.
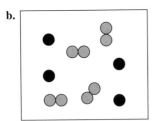

c.
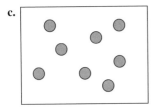

For Problems 3.44 and 3.45, consider the mixtures in the following diagrams.

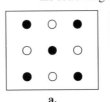   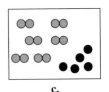

a.   b.   c.

3.44 Which diagram(s) illustrates a homogeneous mixture? Explain your choice.

3.45 Which diagram(s) illustrates a heterogeneous mixture? Explain your choice.

3.46 Classify each of the following as a homogeneous mixture or heterogeneous mixture.
  a. lemon-flavored water
  b. stuffed mushrooms
  c. chicken noodle soup
  d. ketchup
  e. hard-boiled egg
  f. eye drops

3.47 Determine the energy to heat three cubes (gold, aluminum, and silver) each with a volume of 10.0 cm³ from 15°C to 25°C. Refer to Tables 2.9 and 3.7. What do you notice about the energy needed for each?

3.48 A 70.0-kg person has just eaten a quarter-pound cheeseburger, French fries, and a chocolate shake. According to Table 3.11, determine each of the following.

a. the number of hours of sleep needed to "burn off" the kilocalories in this meal

b. the number of hours of running needed to "burn off" the kilocalories in the meal

Item	Protein (g)	Fat (g)	Carbohydrate (g)
Cheeseburger	31	29	34
French fries	3	11	26
Chocolate shake	11	9	60

Additional Questions and Problems

3.49 Classify each of the following as an element, compound, or mixture.
a. carbon in pencils
b. carbon dioxide (CO_2) we exhale
c. orange juice
d. neon gas in lights
e. a salad dressing of oil and vinegar

3.50 Classify each of the following as a homogenous or heterogeneous mixture.
a. hot fudge sundae
b. herbal tea
c. vegetable oil
d. water and sand
e. mustard

3.51 Identify each of the following as a physical or chemical property.
a. Gold is shiny.
b. Gold melts at 1064°C.
c. Gold is a good conductor of electricity.
d. When gold reacts with yellow sulfur, a black sulfide compound forms.

3.52 Identify each of the following as a physical or chemical property.
a. A candle is 10 inches high and 2 inches in diameter.
b. A candle burns.

c. The wax of a candle softens on a hot day.
d. A candle is blue.

3.53 Identify each of the following as a physical or chemical change.
a. A plant grows a new leaf.
b. Frost forms on the lawn.
c. Wood is chopped for the fireplace.
d. Wood burns in a wood stove.

3.54 Identify each of the following as a physical or chemical change.
a. Short hair grows until it is long.
b. Carrots are grated for use in a salad.
c. Malt undergoes fermentation to make beer.
d. A copper pipe reacts with air and turns green.

3.55 Identify each of the following as solid, liquid, or gas.
a. vitamin tablets in a bottle
b. helium in a balloon
c. milk in a bottle
d. the air you breathe
e. charcoal briquettes on a barbecue

3.56 Identify each of the following as solid, liquid, or gas.
a. popcorn in a bag
b. water in a garden hose
c. a computer mouse
d. air in a tire
e. hot tea in a teacup

3.57 Calculate the following temperatures in degrees Celsius and kelvins.
   a. The highest recorded temperature in the continental United States was 134°F in Death Valley, California, July 10, 1913.
   b. The lowest recorded temperature in the continental United States was −69.7°F in Rodgers Pass, Montana, January 20, 1954.

3.58 Calculate the following temperatures in kelvins and degrees Fahrenheit.
   a. The highest recorded temperature in the world was 57.8°C in Death Valley, California, July 10, 1913.
   b. The lowest recorded temperature in the world was −89.4°C in Vostok, Antarctica, July 21, 1983.

3.59 What is −15°F in degrees Celsius and in kelvins?

3.60 The highest recorded body temperature that a person has survived is 46.5°C. Calculate that temperature in °F and in kelvins.

3.61 If you want to lose 1 pound of "fat," which is 15% water, how many kilocalories do you need to expend?

3.62 Calculate the Cal (kcal) in 1 cup of whole milk: 12 g of carbohydrate, 9 g of fat, and 9 g of protein.

3.63 On a hot day, the beach sand gets hot, but the water stays cool. Compare the specific heat of sand to that of water.

3.64 A large bottle of water (883 g) at 4°C is removed from the refrigerator. How many kilojoules (kJ) are absorbed to warm the water to a room temperature of 27°C?

3.65 A hot-water bottle contains 725 g water at 65°C. If the water cools to body temperature (37°C), how many kilojoules of heat could be transferred to sore muscles?

3.66 Copper is sometimes coated on the bottom surface of cooking pans. Copper has a specific heat of 0.385 J/g °C. A 25.0-g sample of copper heated to 85.0°C is dropped into water at an initial temperature of 14.0°C. If the final temperature reached by the copper–water mixture is 36.0°C, how many grams of water are present?

3.67 A 25.0-g sample of an alloy at 98.0°C is placed in 50.0 g water at 18.0°C. If the final temperature of the alloy and water is 27.4°C, what is the specific heat of the alloy?

3.68 A 0.50-g sample of vegetable oil is placed in a calorimeter. When the sample is burned, 18.9 kJ is given off. What is the caloric value (kcal/g) of the oil?

3.69 If you used the 8400 kJ you expend in energy in one day to heat 50 000 g water at 20°C, what would be its new temperature?

Challenge Questions

3.70 When 1.0 g gasoline burns, 11 500 calories of energy are given off. If the density of gasoline is 0.66 g/mL, how many kilocalories of energy are obtained from 1.5 gallons of gasoline?

3.71 A piece of copper metal (specific heat 0.385 J/g °C) at 86.0°C is placed in 50.0 g water at 16.0°C. The metal and water come to the same temperature of 24.0°C. What was the mass of the piece of copper?

3.72 Your friend has just eaten a pizza, cola soft drink, and ice cream. What is the total kilocalories your friend obtained from this meal? How many hours will your friend need to swim to "burn off" the kilocalories in this meal if your friend has a mass of 70.0 kg? (See Table 3.11.)

Item	Protein (g)	Fat (g)	Carbohydrate (g)
Pizza	13	10	29
Cola	0	0	51
Ice cream	8	28	44

3.73 A typical diet in the United States provides 15% of the calories from protein, 45% from carbohydrates, and the remainder from fats. Calculate the grams of protein, carbohydrate, and fat to be included each day in diets having the following calorie requirements:
   a. 1200 kcal   **b.** 1900 kcal
   c. 2600 kcal

3.74 A 125-g piece of metal is heated to 288°C and dropped into 85.0 g water at 26°C. If the final temperature of the water and metal is 58.0°C, what is the specific heat of the metal (J/g °C)?

3.75 Rearrange the heat equation to solve for each of the following.
   a. the mass in grams of water that absorbs 8250 J when its temperature rises from 18.3°C to 92.6°C
   b. the mass in grams of a gold sample that absorbs 225 J when the temperature rises from 15.0°C to 47.0°C
   c. the rise in temperature (ΔT) when a 20.0-g sample of iron absorbs 1580 J
   d. the specific heat of a metal when 8.50 g of the metal absorbs 28 cal and the temperature rises from 12°C to 24°C

Answers

Answers to Study Checks

3.1 24-carat gold is a pure substance; 14-carat gold is a mixture.

3.2 This is a heterogeneous mixture because it does not have a uniform composition.

3.3 **a.** physical property   **b.** chemical property

3.4 b and d are chemical changes.

3.5 12°F

3.6 39.8°C

3.7 night, −260.°C; day, 410.°C

3.8 potential energy

3.9 26 kcal

3.10 $SH = 1.23$ J/g °C

3.11 14.6 kJ

3.12 −360. J

3.13 183 g Fe

3.14 $SH = 0.79$ J/g °C

Answers to Selected Questions and Problems

3.1 **a.** pure substance   **b.** mixture
c. pure substance   **d.** pure substance

3.3 **a.** element   **b.** compound
c. element   **d.** compound

3.5 **a.** heterogeneous   **b.** homogeneous
c. homogeneous   **d.** heterogeneous

3.7 **a.** physical   **b.** chemical
c. physical   **d.** chemical

3.9 **a.** physical   **b.** chemical
c. physical   **d.** physical
e. physical

3.11 **a.** chemical   **b.** physical
c. physical   **d.** chemical
e. physical

3.13 In the United States, we still use the Fahrenheit temperature scale. In °F, normal body temperature is 98.6. On the Celsius scale, her temperature would be 37.7°C, a mild fever.

3.15 **a.** 98.6°F   **b.** 18.5°C
c. 246 K   **d.** 335 K
e. 46°C   **f.** 295 K

3.17 **a.** 41°C
b. No. The temperature is equivalent to 39°C.

3.19 When the car is at the top of the ramp, it has its maximum potential energy. As it descends, potential energy changes to kinetic energy. At the bottom, all the energy is kinetic.

3.21 **a.** potential   **b.** kinetic
c. potential   **d.** potential

3.23 **a.** Electrical energy changes to heat and mechanical energy.
b. Electrical energy changes to mechanical energy.
c. Chemical energy changes to mechanical energy and heat.
d. Radiant energy changes to heat.

3.25 **a.** (2) light bulb
b. (3) hydroelectric power plant
c. (1) heater

3.27 **a.** 3.5 kcal   **b.** 99.2 cal
c. 120 J   **d.** 1100 cal

3.29 Copper, which has the lowest specific heat, would reach the highest temperature.

3.31 **a.** 0.389 J/g °C   **b.** 0.313 J/g °C

3.33 **a.** 1380 J; 330. cal   **b.** 1810 J; 434 cal
c. −3780 J; −904 cal   **d.** −3500 J; −850 cal

3.35 **a.** 54.5 g   **b.** 217 g
c. 686 g   **d.** 190. g

3.37 **a.** 176°C   **b.** 11.3°C
c. 700.°C   **d.** 351°C

3.39 **a.** 22.2 kJ; 5.30 kcal   **b.** 871 kJ; 208 kcal
c. 132 kJ; 31.6 kcal

3.41 **a.** 480 kJ; 110 kcal   **b.** 310 kJ; 70 kcal
c. 530 kJ; 130 kcal   **d.** 3900 kJ; 930 kcal

3.43 **a.** compound   **b.** mixture
c. element

3.45 b and c are not the same throughout the mixture and b and c are heterogeneous mixtures.

3.47 gold 250 J or 59 cal; aluminum 240 J or 58 cal; silver 250 J or 59 cal
The heat needed for 10.0-cm^3 samples of the metals is almost the same.

3.49 **a.** element   **b.** compound
c. mixture   **d.** element
e. mixture

3.51 **a.** Appearance is a physical property.
b. The melting point of gold is a physical property.
c. The ability of gold to conduct electricity is a physical property.
d. The ability of gold to form a new substance with sulfur is a chemical property.

3.53 **a.** Plant growth is a chemical change.
b. A change of state from liquid to solid is a physical change.
c. Chopping wood into smaller pieces is physical change.
d. Burning wood, which forms new substances, is a chemical change.

3.55 **a.** Tablets are solid.
b. Helium in a balloon is a gas.
c. Milk is a liquid.
d. Air is mixture of gases.
e. Charcoal is a solid.

3.57 **a.** 56.7°C, 330. K   **b.** −56.5°C, 216 K

3.59 −26°C, 247 K

3.61 3500 kcal

3.63 Water has a higher specific heat than sand, which means that a large amount of energy is required to cause a significant temperature change. Even a small amount of energy will cause a significant temperature change in the sand.

3.65 85 kJ

3.67 1.1 J/g °C

3.69 60°C

3.71 70. g copper

3.73 **a.** 45 g protein, 140 g carbohydrate, 53 g fat
b. 71 g protein, 210 g carbohydrate, 84 g fat
c. 98 g protein, 290 g carbohydrate, 120 g fat

3.75 **a.** 26.5 g
b. 54.5 g
c. 176°C
d. 0.27 cal/g °C

Combining Ideas from Chapters 1–3

CI 1 Gold, one of the most sought-after metals in the world, has a density of 19.3 g/cm³, melts at 1064°C, and has a specific heat of 0.129 J/g °C. In 1998, a gold nugget was found in Alaska that weighed 294.10 troy ounces.

a. How many significant figures are in the measurement of the mass of the nugget?

b. If 1 troy ounce is 31.1035 g, what is the mass of the nugget in grams? In kilograms?

c. If the nugget were pure gold, what is its volume in cm³?

d. What is the area in m² if the gold nugget is hammered to a foil with a thickness of 0.0035 in.?

e. What is the melting point of gold in degrees Fahrenheit and kelvins?

f. How many kilojoules are required to raise the temperature of the nugget from 72°F to 85°F?

CI 2 The mileage of a motorcycle with a fuel tank that holds 22 L of gasoline is 35 miles per gallon.

a. How long a trip in kilometers can be made on one full tank of gasoline?

b. If the price of gasoline is $2.67 per gallon, what would be the cost of fuel for the trip?

c. If the average speed during the trip is 44 mi/hr, how long will it take to reach the destination?

d. If the density of gasoline is 0.71 g/mL, what is the mass in grams of the fuel in the tank?

e. When 1.00 g gasoline burns, 42 kJ of energy are released. How many kilojoules are produced when the fuel in one full tank is burned?

CI 3 A box of iron nails weighing 0.25 lb contains 75 nails. The density of iron is 7.86 g/cm³. The specific heat of iron is 0.450 J/g °C.

a. What is the volume of the iron nails in the box?

b. If 30 nails are added to a graduated cylinder containing 17.6 mL water, what is the new level of water in the cylinder?

c. How many joules must be added to the nails in the box to raise the temperature from 15.6°F to 125.2°F?

d. If all the iron nails at 55.0°C are added to 325 g water at 4.0°C, what is the final temperature (°C) of the water?

CI 4 The "nutritional facts" on the label of a lemon power bar says that one serving size (1 bar of 48 g) contains 5 g fat, 25 g carbohydrate, and 10 g protein.

a. Using the caloric values of carbohydrate 4 kcal/g, fat 9 kcal/g, and protein 4 kcal/g, what are the total kilocalories (Calories) listed for the lemon bar?
b. What is the energy value of the bar in kilojoules?
c. If the bar has a mass of 48 g, how many kilojoules are obtained from eating 10. g of the bar?
d. If you are walking (840 kJ/hr), how many minutes of walking will you need to do if you eat two power bars?

CI 5 A hot tub with a surface area of 25 ft² is filled with water to a depth of 28 in.

a. What is the volume of water in the tub in liters?
b. What is the mass in kilograms of water it contains (D_{H_2O} 1.0 g/mL)?
c. How many kilojoules are needed to heat the water from 62°F to 105°F?
d. If the hot tub heater provides 6000 kJ per minute, how long in hours will it take to heat the water in the hot tub from part c?

Answers CI

CI 1   a. 5 significant figures   b. 9147.5 g, 9.1475 kg
          c. 474 cm³   d. 5.3 m²
          e. 1947°F, 1337 K   f. 8 kJ

CI 3   a. 14 cm³   b. 23.4 mL
          c. 3100 J   d. 5.8°C

CI 5   a. 1.7×10^3 L   b. 1.7×10^3 kg
          c. 1.7×10^5 kJ   d. 0.47 hour

Atoms and Elements

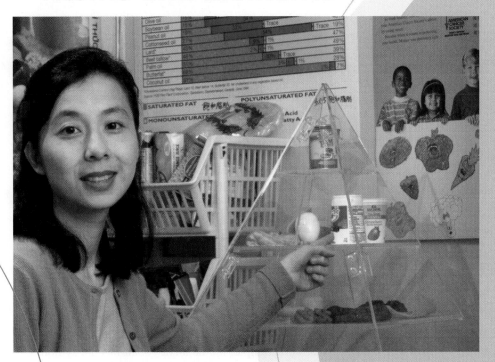

"*Many of my patients have diabetes, ulcers, hypertension, and cardiovascular problems,*" *says Sylvia Lau, registered dietitian.* "*If a patient has diabetes, I discuss foods that raise blood sugar such as fruit, milk, and starches. I talk about how dietary fat contributes to weight gain and complications from diabetes. For stroke patients, I suggest diets low in fat and cholesterol because high blood pressure increases the risk of another stroke.*"

If a lab test shows low levels of iron, zinc, iodine, magnesium, or calcium, a dietitian discusses foods that provide those essential elements. For instance, she may recommend more beef for an iron deficiency, whole grain for zinc, leafy green vegetables for magnesium, dairy products for calcium, and iodized table salt and seafood for iodine.

the Chemistry place

Visit **www.aw-bc.com/chemplace** for extra quizzes, interactive tutorials, career resources, PowerPoint slides for chapter review, math help, and case studies.

All matter is composed of *elements,* of which there are about 110 different kinds. Of the 88 elements that occur in nature, one or more is used to make all the substances in the universe. Many elements are already familiar to you. Perhaps you use aluminum in the form of foil or drink a soft drink from an aluminum can. You may have a ring or necklace made of gold, or silver, or perhaps platinum. If you play tennis or golf, your racket or clubs may be made from the elements titanium or carbon. In the body, calcium and phosphorus form the structure of bones and teeth, iron and copper are needed in the formation of red blood cells, and iodine is required for the proper functioning of the thyroid.

4.1 Elements and Symbols

Learning Goal

Given the name of an element, write its correct symbol; from the symbol, write the correct name.

Earlier we learned that elements are the substances from which all matter is made. Many of the elements were named for planets, mythological figures, minerals, colors, geographic locations, and famous people. Some sources of names of elements are listed in Table 4.1.

Chemical Symbols

Chemical symbols are one- or two-letter abbreviations for the names of the elements. Only the first letter of an element's symbol is capitalized; a second letter, if there is one, is lowercase. That way, we know when a different element is indicated. If both letters are capitalized, it represents the symbols of two different elements. For example, the element cobalt has the symbol Co. However, the two capital letters CO specify two elements, carbon (C) and oxygen (O).

Table 4.1 Some Elements and Their Names

Element	Source of Name
Uranium	The planet Uranus
Titanium	Titans (mythology)
Chlorine	*Chloros,* "greenish yellow" (Greek)
Iodine	*Ioeides,* "violet" (Greek)
Magnesium	Magnesia, a mineral
Californium	California
Curium	Marie and Pierre Curie

One-Letter Symbols		Two-Letter Symbols	
C	carbon	Co	cobalt
S	sulfur	Si	silicon
N	nitrogen	Ne	neon
I	iodine	Ni	nickel

Although most of the symbols use letters from the current names, some are derived from their ancient Latin or Greek names. For example, Na, the symbol for sodium, comes from the Latin word *natrium.* The symbol for iron, Fe, is derived from the Latin name *ferrum.* Table 4.2 lists the names and symbols of some common elements. Learning their names and symbols will greatly help your learning of chemistry. A complete list of all the elements and their symbols appears on the inside front cover of this text.

Sample Problem 4.1 ▶ **Writing Chemical Symbols**

What are the chemical symbols for the following elements?

a. carbon   **b.** nitrogen   **c.** chlorine   **d.** copper

Solution

a. C   **b.** N   **c.** Cl   **d.** Cu

Study Check

What are the chemical symbols for the elements silicon, sulfur, and silver?

Table 4.2   Names and Symbols of Some Common Elements

Aluminum

Carbon

Copper

Gold

Name[a]	Symbol
Aluminum	Al
Argon	Ar
Arsenic	As
Barium	Ba
Boron	B
Bromine	Br
Cadmium	Cd
Calcium	Ca
Carbon	C
Chlorine	Cl
Chromium	Cr
Cobalt	Co
Copper (*cuprum*)	Cu
Fluorine	F
Gold (*aurum*)	Au
Helium	He
Hydrogen	H
Iodine	I
Iron (*ferrum*)	Fe
Lead (*plumbum*)	Pb
Lithium	Li
Magnesium	Mg
Manganese	Mn
Mercury (*hydrargyrum*)	Hg
Neon	Ne
Nickel	Ni
Nitrogen	N
Oxygen	O
Phosphorus	P
Platinum	Pt
Potassium (*kalium*)	K
Radium	Ra
Silicon	Si
Silver (*argentum*)	Ag
Sodium (*natrium*)	Na
Strontium	Sr
Sulfur	S
Tin (*stannum*)	Sn
Titanium	Ti
Uranium	U
Zinc	Zn

Iron

Silver

Sulfur

[a]Names given in parentheses are ancient Latin or Greek words from which the symbols are derived.

Sample Problem 4.2   **Naming Chemical Elements**

Give the name of the element that corresponds to each of the following chemical symbols:

a. Zn   **b.** K   **c.** H   **d.** Fe

Solution

a. zinc   **b.** potassium   **c.** hydrogen   **d.** iron

Study Check

What are the names of the elements with the chemical symbols Mg, Al, and F?

Questions and Problems   **Elements and Symbols**

4.1 Write the symbols for the following elements:
   a. copper   **b.** silicon
   c. potassium   **d.** nitrogen
   e. iron   **f.** barium
   g. lead   **h.** strontium

4.2 Write the symbols for the following elements:
   a. oxygen   **b.** lithium
   c. sulfur   **d.** aluminum
   e. hydrogen   **f.** neon
   g. tin   **h.** gold

4.3 Write the name of the element for each symbol.
   a. C   **b.** Cl
   c. I   **d.** Hg
   e. F   **f.** Ar
   g. Zn   **h.** Ni

4.4 Write the name of the element for each symbol.
   a. He   **b.** P
   c. Na   **d.** Mg
   e. Ca   **f.** Br
   g. Cd   **h.** Si

4.5 What elements are in the following substances?
   a. table salt, $NaCl$
   b. plaster casts, $CaSO_4$
   c. Demerol, $C_{15}H_{22}ClNO_2$
   d. antacid, $CaCO_3$

4.6 What elements are in the following substances?
   a. water, H_2O
   b. baking soda, $NaHCO_3$
   c. lye, $NaOH$
   d. sugar, $C_{12}H_{22}O_{11}$

4.2 The Periodic Table

Learning Goal

Use the periodic table to identify the group and the period of an element and decide whether it is a metal or a nonmetal.

As more and more elements were discovered, it became necessary to organize them with some type of classification system. By the late 1800s, scientists recognized that certain elements looked alike and behaved much the same way. In 1872, a Russian chemist, Dmitri Mendeleev, arranged the 60 elements known at that time into groups with similar properties and placed them in order of increasing atomic masses. Today, this arrangement of over 110 elements is known as the **periodic table.** (See Figure 4.1.)

Periods and Groups

Each horizontal row in the table is called a **period.** Each row is counted from the top of the table as Period 1 to Period 7. The first period contains only the elements hydrogen (H) and helium (He). The second period contains 8 elements: lithium (Li), beryllium (Be), boron (B), carbon (C), nitrogen (N), oxygen (O), fluorine (F), and neon (Ne). The third period also contains 8 elements, beginning with sodium (Na) and ending with argon (Ar). The

Periodic Table of Elements

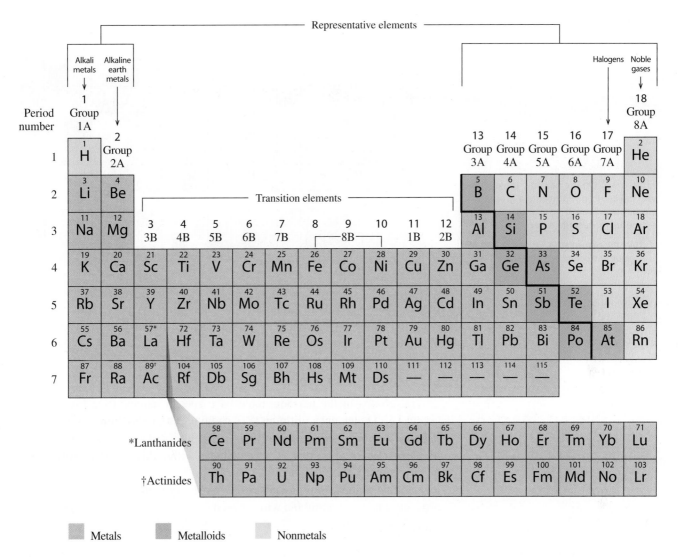

Figure 4.1 Groups and periods in the periodic table.

Q What is the symbol of the alkali metal in Period 3?

fourth period, which begins with potassium (K), and the fifth period, which begins with rubidium (Rb), have 18 elements each. The sixth period, which begins with cesium (Cs), has 32 elements. The seventh period as of today contains the 29 remaining elements, although it could go up to 32. (See Figure 4.2.)

Each vertical column on the periodic table contains a **group** (or family) of elements that have similar properties. At the top of each column is a number that is assigned to each group. The elements in the first two columns on the left of the periodic table and the last six columns on the right are called the **representative elements** or the *main group elements.* For many years, they have been given group numbers 1A–8A. On some periodic tables, the group numbers may be written with Roman numerals: IA–VIIIA. In the center of the periodic table is a block of elements known as the **transition elements** or *transition metals,* which are designated with the letter "B." A newer numbering system assigns group numbers of 1–18 going across the periodic table.

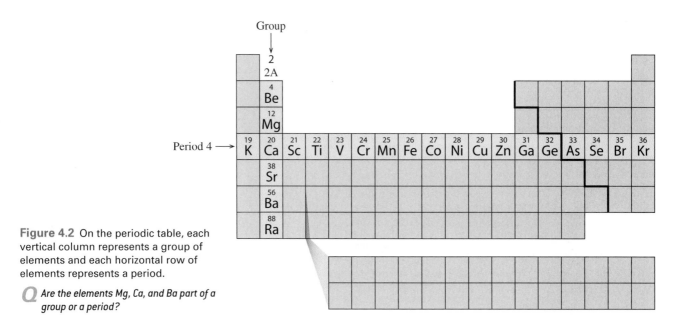

Figure 4.2 On the periodic table, each vertical column represents a group of elements and each horizontal row of elements represents a period.

Q *Are the elements Mg, Ca, and Ba part of a group or a period?*

Because both systems of group numbers are currently in use, they are both indicated on the periodic table in this text and are included in our discussions of elements and group numbers.

Classification of Groups

Several groups in the periodic table have special names. (See Figure 4.3.) Group 1A (1) elements—lithium (Li), sodium (Na), potassium (K), rubidium (Rb), cesium (Cs), and francium (Fr)—are a family of elements known as the **alkali metals** (see Figure 4.4). The elements within this group are soft, shiny metals that are good conductors of heat and electricity and have relatively low melting points. Alkali metals react vigorously with water and form white products when they combine with oxygen.

Figure 4.3 Certain groups on the periodic table have common names.

Q *What is the common name for the group of elements that includes helium and argon?*

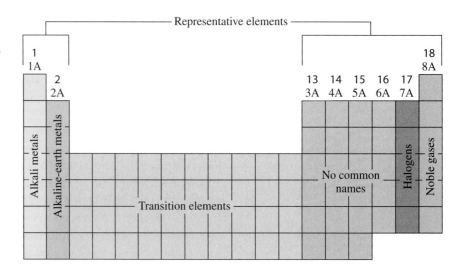

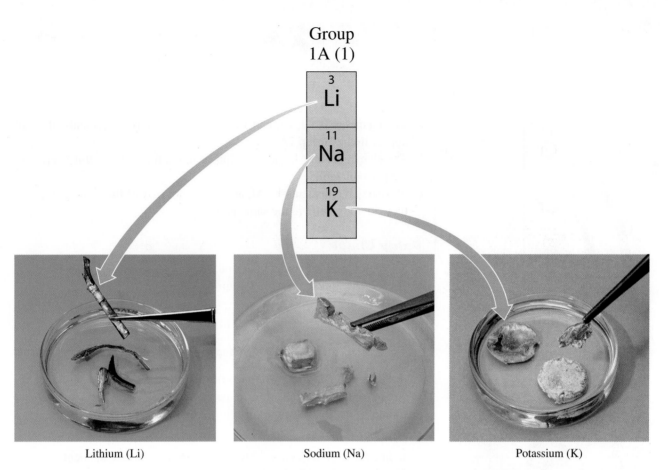

Figure 4.4 Lithium (Li), sodium (Na), and potassium (K) are some alkali metals from Group 1A (1).

Q What physical properties do these alkali metals have in common?

Although hydrogen (H) is at the top of Group 1A (1), hydrogen is not an alkali metal and has very different properties from the rest of the elements in this group. Thus hydrogen is not included in the classification of alkali metals. In some periodic tables, H is sometimes placed at the top of Group 7A (17).

Group 2A (2) elements—beryllium (Be), magnesium (Mg), calcium (Ca), strontium (Sr), barium (Ba), and radium (Ra)—are called the **alkaline earth metals.** They are also shiny metals like those in Group 1A, but they are not as reactive.

The **halogens** are found on the right side of the periodic table in Group 7A (17). They include the elements fluorine (F), chlorine (Cl), bromine (Br), and iodine (I), as shown in Figure 4.5. The halogens, especially fluorine and chlorine, are strongly reactive and form compounds with most of the elements.

Group 8A (18) contains the **noble gases**—helium (He), neon (Ne), argon (Ar), krypton (Kr), xenon (Xe), and radon (Rn). They are quite unreactive and are seldom found in combination with other elements.

Group
7A (17)

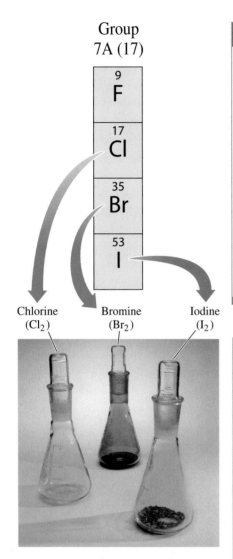

Figure 4.5 Chlorine (Cl₂), bromine (Br₂), and iodine (I₂) are samples of halogens from Group 7A (17).

Q *What elements are in the halogen group?*

Sample Problem 4.3 **Groups and Periods in the Periodic Table**

State whether each set represents elements in a group, a period, or neither.
a. F, Cl, Br, I   **b.** Na, Al, P   **c.** K, Al, O

Solution

a. The elements F, Cl, Br, and I are part of a group of elements; they all appear in the Group 7A (17).
b. The elements Na, Al, and P all appear in the third row or third period in the periodic table.
c. Neither. The elements K, Al, and O are not part of the same group and they do not belong to the same period.

Study Check

a. What elements are found in Period 2?
b. What elements are found in Group 2A (2)?

Sample Problem 4.4 **Group and Period Numbers of Some Elements**

Give the period and group for each of the following elements and identify as a representative or transition element:
a. iodine   **b.** manganese   **c.** barium   **d.** gold

Solution

a. Iodine (I), Period 5, Group 7A (17), is a representative element.
b. Manganese (Mn), Period 4, Group 7B (7), is a transition element.
c. Barium (Ba), Period 6, Group 2A (2), is a representative element.
d. Gold (Au), Period 6, Group 1B (11), is a transition element.

Study Check

Strontium is an element that gives a brilliant red color to fireworks.
a. In what group is strontium found?
b. What is the name of this chemical family?
c. What is the period of strontium?
d. For the same group, what element is in Period 3?
e. What alkali metal, halogen, and noble gas are in the same period as strontium?

Metals, Nonmetals, and Metalloids

Another feature of the periodic table is the heavy zigzag line that separates the elements into the *metals* and the *nonmetals*. The metals are those elements on the left of the line *except for hydrogen,* and the nonmetals are the elements on the right.

In general, most **metals** are shiny solids. They can be shaped into wires (ductile) or hammered into a flat sheet (malleable). Metals are good conductors of heat and electricity. They usually melt at higher temperatures than nonmetals. All of the metals are solids at room temperature, except for mer-

CHEM NOTE

TOXICITY OF MERCURY

Mercury is a silvery, shiny element that is a liquid at room temperature. Mercury can enter the body through inhaled mercury vapor, contact with the skin, or foods or water contaminated with mercury. In the body, mercury destroys proteins and disrupts cell function. Long-term exposure to mercury can damage the brain and kidneys, cause mental retardation, and decrease physical development. Blood, urine, and hair samples are used to test for mercury.

In fresh and salt water, bacteria convert mercury into toxic methylmercury, which primarily attacks the central nervous system (CNS). Because fish absorb methylmercury, we are exposed to mercury by consuming mercury-contaminated fish. As levels of mercury ingested from fish became a concern, the Food and Drug Administration (FDA) set a maximum level of

one part mercury per million parts seafood (1 ppm), which is the same as 1 μg mercury in every gram of seafood. Fish higher in the food chain, such as swordfish and shark, can have such high levels of mercury that the Environmental Protection Agency (EPA) recommends they be consumed no more than once a week.

One of the worst incidents of mercury poisoning occurred in Minamata and Niigata, Japan, in 1950. At that time, the ocean was polluted with high levels of mercury from industrial wastes. Because fish were a major food in the diet, more than 2000 people were affected with mercury poisoning and died or developed neural damage. In the United States, industry decreased the use of mercury between 1988 and 1997 by 75% by banning mercury in paint and pesticides, reducing mercury in batteries, and regulating mercury in other products.

This mercury fountain, housed in glass, was designed by Alexander Calder for the 1937 World's Fair in Paris.

cury (Hg), which is a liquid. Some typical metals are sodium (Na), magnesium (Mg), copper (Cu), gold (Au), silver (Ag), iron (Fe), and tin (Sn).

Nonmetals are not very shiny, malleable, or ductile, and they are often poor conductors of heat and electricity. They typically have low melting points and low densities. You may have heard of nonmetals such as hydrogen (H), carbon (C), nitrogen (N), oxygen (O), chlorine (Cl), and sulfur (S). Several nonmetals such as H_2, O_2, N_2, and Cl_2 are gases at room temperature.

Metalloids are elements that exhibit properties that are typical both of metals and nonmetals. They are better conductors of heat and electricity than nonmetals, but not as good as metals. Metalloids are used as semiconductors because they can function as conductors or insulators. On the periodic chart, the metalloids—B, Si, Ge, As, Sb, Te, Po, and At—are located along the heavy black line that separates the metals from the nonmetals. (See Figure 4.6.) Table 4.3 compares characteristics of silver, a metal, with those of antimony, a metalloid, and sulfur, a nonmetal.

Figure 4.6 Along the heavy zigzag line on the periodic table that separates the metals and nonmetals are metalloids that exhibit characteristics of both metals and nonmetals.

Q *On which side of the heavy zigzag line are the nonmetals located?*

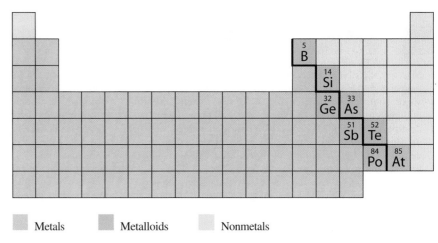

Metals   Metalloids   Nonmetals

Table 4.3 Some Characteristics of a Metal, a Metalloid, and a Nonmetal

Silver (Ag)	Antimony (Sb)	Sulfur (S)
Metal	Metalloid	Nonmetal
Shiny	Blue-grey, shiny	Dull, yellow
Extremely ductile	Brittle	Brittle
Can be hammered into sheets (malleable)	Shatters when hammered	Shatters when hammered
Good conductor of heat and electricity	Poor conductor of heat and electricity	Poor conductor, good insulator
Used in coins, jewelry, tableware	Used to harden lead, color glass and plastics	Used in gunpowder, rubber, fungicides
Density 10.5 g/mL	Density 6.7 g/mL	Density 2.1 g/mL
Melting point 962°C	Melting point 630°C	Melting point 113°C

Sample Problem 4.5 **Metals, Nonmetals, and Metalloids**

Use a periodic table to classify each of the following elements as a metal, nonmetal, or metalloid:
a. Na   **b.** Si   **c.** I   **d.** Sn

Solution

a. Sodium, which lies to the left of the heavy zigzag line, is a metal.
b. Silicon, which lies on the heavy zigzag line, is a metalloid.
c. Iodine, to the right of the zigzag line, is a nonmetal.
d. Tin, which lies on the left of the heavy line, is a metal.

Study Check

Identify each of the following elements as a metal, nonmetal, or metalloid.
a. A shiny, silver-colored element that melts at 962°C.
b. An element with a bright, lustrous reddish color that melts at 2570°C and is malleable, ductile, and an excellent conductor of heat and electricity.
c. A black, elemental substance that breaks into smaller chunks when hit with a hammer.

4.7 Identify the group or period number described by each of the following statements:
  a. contains the elements C, N, and O
  b. begins with helium
  c. the alkali metals
  d. ends with neon

4.8 Identify the group or period number described by each of the following statements:
  a. contains Na, K, and Rb
  b. the row that begins with Li
  c. the noble gases
  d. contains F, Cl, Br, and I

4.9 Classify the following as an alkali metal, alkaline earth metal, transition element, halogen, or noble gas:
  a. Ca   **b.** Fe
  c. Xe   **d.** Na
  e. Cl

4.10 Classify the following as an alkali metal, alkaline earth metal, transition element, halogen, or noble gas:
  a. Ne   **b.** Mg
  c. Cu   **d.** Br
  e. Ba

4.11 Give the symbol of the element described by the following:
  a. Group 4A, Period 2
  b. a noble gas in Period 1

  c. an alkali metal in Period 3
  d. Group 2, Period 4
  e. Group 13, Period 3

4.12 Give the symbol of the element described by the following:
  a. an alkaline earth metal in Period 2
  b. Group 15, Period 3
  c. a noble gas in Period 4
  d. a halogen in Period 5
  e. Group 4A, Period 4

4.13 Is each of the following elements a metal, nonmetal, or metalloid?
  a. calcium   **b.** sulfur
  c. an element that is shiny   **d.** does not conduct heat
  e. located in Group 7A   **f.** phosphorus
  g. boron   **h.** silver

4.14 Is each of the following elements a metal, nonmetal, or metalloid?
  a. located in Group 2A
  b. a good conductor of electricity
  c. chlorine
  d. arsenic
  e. an element that is not shiny
  f. oxygen
  g. nitrogen
  h. aluminum

4.3 The Atom

Learning Goal

Describe the electrical charge and location in an atom of a proton, a neutron, and an electron.

the Chemistry place

WEB TUTORIAL
Atoms and Isotopes

All the elements listed on the periodic table are made up of small particles called atoms. An **atom** is the smallest particle of an element that retains the characteristics of that element. You have probably seen the element aluminum. Imagine that you are tearing a piece of aluminum foil into smaller and smaller pieces. Now imagine that you tear the foil until you have a piece so small that you can no longer break it down further. Then you would have an atom of aluminum, the smallest particle of an element that still retains the characteristics of that element.

Billions of atoms are packed together to build you and everything around you. The paper in this book contains atoms of carbon, hydrogen, and oxygen. The ink on this paper, even the dot over the letter *i*, contains huge numbers of atoms. There are as many atoms in that dot as there are seconds in 10 billion years.

The concept of the atom is relatively recent. Although the Greek philosophers in 500 B.C.E. reasoned that everything must contain minute particles they called *atomos,* the idea of atoms did not become a scientific theory until 1808. Then John Dalton (1766–1844) developed an atomic theory that proposed that atoms were responsible for the combinations of elements found in compounds.

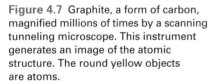

CHEM NOTE
ELEMENTS ESSENTIAL
TO HEALTH

Many elements are essential for the well-being and survival of the human body. The four elements oxygen, carbon, hydrogen, and nitrogen are the elements that make up carbohydrates, fats, proteins, and DNA. Most of the hydrogen and oxygen is found in water, which makes up 55–60% of our body mass. Some examples and the amounts present in a 60-kg person are listed in Table 4.4.

Table 4.4   Elements Essential to Health

Element	Symbol	Amount in a 60-kg Person
Oxygen	O	39 kg
Carbon	C	11 kg
Hydrogen	H	6 kg
Nitrogen	N	1.5 kg
Calcium	Ca	1 kg
Phosphorus	P	600 g
Potassium	K	120 g
Sulfur	S	120 g
Sodium	Na	86 g
Chlorine	Cl	81 g
Magnesium	Mg	16 g
Iron	Fe	3.6 g
Fluorine	F	2.2 g
Zinc	Zn	2.0 g
Copper	Cu	60 mg
Iodine	I	20 mg

Dalton's Atomic Theory

1. Every element is made up of tiny particles called atoms.
2. All atoms of a given element are identical to one another and different from atoms of other elements.
3. Atoms of two or more different elements combine to form compounds. A particular compound is always made up of the same kinds of atoms and the same number of each kind of atom.
4. A chemical reaction involves the rearrangement, separation, or combination of atoms. Atoms are never created or destroyed during a chemical reaction.

Dalton's atomic theory formed the basis of current atomic theory, although we have modified some of Dalton's statements. We now know that atoms of the same element are not completely identical to each other and consist of even smaller particles. However, an atom is still the smallest particle that retains the properties of an element.

Atoms are the building blocks of everything we see around us; yet we cannot see an atom or even a billion atoms with the naked eye. However, when billions and billions of atoms are packed together, the characteristics of each atom are added to those of the next until we can see the characteristics we associate with the element. For example, a small piece of the shiny, copper-colored element we call copper consists of many, many copper atoms. Through a special kind of microscope called a scanning tunneling microscope, we can now "see" images of individual atoms, such as the atoms of carbon in graphite shown in Figure 4.7.

Parts of an Atom

By the early part of the twentieth century, growing evidence indicated that atoms were not solid spheres, as Dalton had imagined. As more was discovered about atoms, they were found to contain smaller particles called **subatomic particles.** Although there are many subatomic particles, we will be concerned only with protons, neutrons, and electrons. It is the number and arrangement of these subatomic particles that determine the type of atom found in an element.

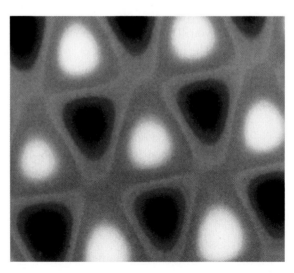

Figure 4.7 Graphite, a form of carbon, magnified millions of times by a scanning tunneling microscope. This instrument generates an image of the atomic structure. The round yellow objects are atoms.

Q Why is a microscope with extremely high magnification needed to see these atoms?

Positive charges
repel

Negative charges
repel

Unlike charges
attract

Figure 4.8 Like charges repel and unlike charges attract.

Q *Why are the electrons attracted to the protons in the nucleus of an atom?*

Of the three subatomic particles of interest to us, protons, neutrons, and electrons, two of these carry electrical charges. The **proton** has a positive charge (+), and the **electron** carries a negative charge (−). The **neutron** has no electrical charge; it is neutral.

Like charges repel; they push away from each other. Opposite or unlike charges attract. When you brush your hair on a dry day, electrical charges build up on the brush and in your hair; as a result, your hair is attracted to the brush. The crackle of clothes taken from the clothes dryer indicates the presence of electrical charges. The clinginess of the clothing is due to the attraction of opposite, unlike charges, as shown in Figure 4.8.

Structure of the Atom

By the early 1900s, scientists knew that atoms contained subatomic particles, but they did not know how those particles were arranged. In 1911, in an experiment by Ernest Rutherford, positively charged particles were aimed at a very thin sheet of gold foil, as illustrated in Figure 4.9. Unexpectedly, a few particles changed direction as they passed through the gold foil. Some were deflected so much that they seemed to go back in the direction of the source of the particles. According to Rutherford, it was as though he had shot a cannonball at a piece of tissue paper, and it bounced back at him.

Nucleus of the Atom

From the gold-foil experiment, Rutherford concluded that the positively charged particles would change direction only if they came close to some type of dense, positively charged region in the atom. This dense core of an atom, called the **nucleus,** where the protons and neutrons are located, has a positive charge and contains most of the mass of an atom. Most of an atom is empty space, which is occupied only by fast-moving electrons. (See Figure 4.10.) If an atom were the size of a football stadium, the nucleus would be about the size of a golf ball placed in the center of the field.

Mass of Subatomic Particles

In a neutral atom, there are the same numbers of protons as electrons, which means that an atom has a net electrical charge of zero (0). A proton has a mass of 1.673×10^{-24} g and the mass of a neutron is almost the same. The

Figure 4.9 **(a)** Positive particles are aimed at a piece of gold foil. **(b)** Particles that come close to the atomic nuclei are deflected from their straight path.

Q *Why are some particles deflected while most pass through the gold foil undeflected?*

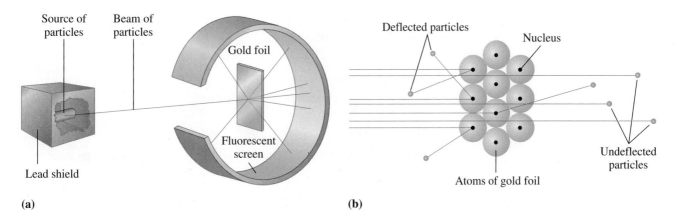

(a)   (b)

Figure 4.10 In an atom, the protons (positive charge) and neutrons (neutral) that make up almost all the mass of the atom are packed into the tiny volume of the nucleus. The rapidly moving electrons (negative charge) surround the nucleus and account for the large volume of the atom.

Q *Why do scientists consider the atom to be mostly empty space?*

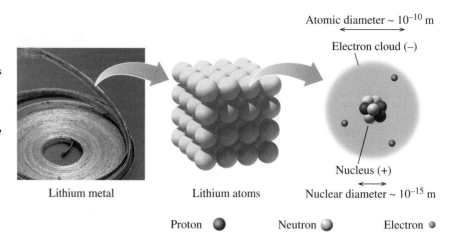

Atomic diameter ~ 10^{-10} m

Electron cloud (–)

Nucleus (+)

Nuclear diameter ~ 10^{-15} m

Lithium metal   Lithium atoms

Proton   Neutron   Electron

mass of the electron, which is 9.110×10^{-28} g, is much less. The mass of atoms is so small that chemists use the atomic mass scale, in which a carbon with 6 protons and 6 neutrons has a mass of 12 amu. This makes one **atomic mass unit (amu)** equal to one-twelfth (1/12) of the mass of this carbon atom. Using the amu scale, a proton and a neutron each have a mass of about 1 amu. By comparison, the electron has a much smaller mass (0.000 549 amu), which means that a proton or neutron is about 2000 times more massive than an electron. Because the mass of the electron is so small, it is usually ignored in atomic mass calculations. Table 4.5 summarizes information about the particles in an atom.

Table 4.5   Subatomic Particles in the Atom

Particle	Symbol	Relative Charge	Mass (g)	Mass (amu)	Location in Atom
Proton	p or p^+	1+	1.673×10^{-24}	1.007	Nucleus
Neutron	n or n^0	0	1.675×10^{-24}	1.008	Nucleus
Electron	e^-	1–	9.110×10^{-28}	0.000 55	Outside nucleus

Sample Problem 4.6   **Identifying Subatomic Particles**

Identify the subatomic particle that has the following characteristics:
a. has no charge
b. is located outside nucleus
c. has a mass of 0.000 55 amu.
d. has a mass about the same as a neutron

Solution
a. neutron   **b.** electron   **c.** electron   **d.** proton

Study Check
Give the symbol, electrical charge, and location of a proton in an atom.

4.15 Is a proton, neutron, or electron described by each of the following?
  a. has the smallest mass
  b. carries a positive charge
  c. is found outside the nucleus
  d. is electrically neutral

4.16 Is a proton, neutron, or electron described by each of the following?
  a. has a mass about the same as a proton's
  b. is found in the nucleus
  c. is found in the larger part of the atom
  d. carries a negative charge

4.17 What did Rutherford's gold-foil experiment tell us about the organization of the subatomic particles in an atom?

4.18 Why does the nucleus in every atom have a positive charge?

4.19 Which of the following particles have opposite charges?
  a. two protons
  b. a proton and an electron
  c. two electrons
  d. a proton and a neutron

4.20 Which of the following pairs of particles have the same charges?
  a. two protons
  b. a proton and an electron
  c. two electrons
  d. an electron and a neutron

4.21 On a dry day, your hair is attracted to your brush. How would you explain this?

4.22 Sometimes clothes removed from the dryer cling together. What kinds of charges are on the clothes?

4.4 Atomic Number and Mass Number

Learning Goal

Given the atomic number and the mass number of an atom, state the number of protons, neutrons, and electrons.

All of the atoms of the same element always have the same number of protons. This feature distinguishes atoms of one element from atoms of all the other elements.

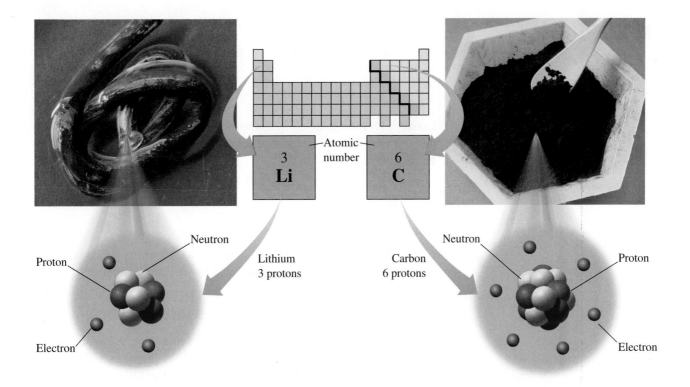

Atomic number

Lithium
3 protons

Carbon
6 protons

WEB TUTORIAL
Atoms and Isotopes

Atomic Number

An **atomic number,** which is equal to the number of protons in the nucleus of an atom, is used to identify each element.

Atomic number = number of protons in an atom

On the inside front cover of this text is a periodic table, which gives all of the elements in order of increasing atomic number. The atomic number is the whole number that appears above the symbol for each element. For example, a hydrogen atom, with atomic number 1, has 1 proton; a lithium atom, with atomic number 3, has 3 protons; an atom of carbon, with atomic number 6, has 6 protons; and gold, with atomic number 79, has 79 protons.

Atoms Are Neutral

An atom is electrically neutral. That means that the number of protons in an atom is equal to the number of electrons. This electrical balance gives an atom an overall charge of zero. Thus, in every atom, the atomic number also gives the number of electrons.

Sample Problem 4.7 **Using the Atomic Number to Find the Number of Protons and Electrons**

Using the periodic table, state the atomic number, number of protons, and number of electrons for an atom of each of the following elements:
a. nitrogen   **b.** magnesium   **c.** bromine

Solution

a. atomic number 7; 7 protons and 7 electrons
b. atomic number 12; 12 protons and 12 electrons
c. atomic number 35; 35 protons and 35 electrons

Study Check

Consider an atom that has 26 electrons.
a. How many protons are in its nucleus?
b. What is its atomic number?
c. What is its name, and what is its symbol?

Mass Number

We now know that the protons and neutrons determine the mass of the nucleus. For any atom, the **mass number** is the sum of the number of protons and neutrons in the nucleus.

Mass number = number of protons + number of neutrons

For example, an atom of oxygen that contains 8 protons and 8 neutrons has a mass number of 16. An atom of iron that contains 26 protons and 30 neutrons has a mass number of 56. Table 4.6 illustrates the relationship between atomic number, mass number, and the number of protons, neutrons, and electrons in some atoms of different elements.

Table 4.6   Composition of Some Atoms of Different Elements

Element	Symbol	Atomic Number	Mass Number	Number of Protons	Number of Neutrons	Number of Electrons
Hydrogen	H	1	1	1	0	1
Nitrogen	N	7	14	7	7	7
Chlorine	Cl	17	37	17	20	17
Iron	Fe	26	56	26	30	26
Gold	Au	79	197	79	118	79

Sample Problem 4.8   Calculating Mass Number

Calculate the mass number of an atom by using the information given.
a. 5 protons and 6 neutrons   **b.** 18 protons and 22 neutrons
c. atomic number 48 and 64 neutrons

Solution
a. mass number $= 5 + 6 = 11$ **b.** mass number $= 18 + 22 = 40$
c. mass number $= 48 + 64 = 112$

Study Check

What is the mass number of a silver atom that has 60 neutrons?

Sample Problem 4.9   Calculating Numbers of Protons and Neutrons

For an atom of phosphorus that has a mass number of 31, determine the following:
a. the number of protons   **b.** the number of neutrons
c. the number of electrons

Solution
a. On the periodic table, the atomic number of phosphorus is 15. A phosphorus atom has 15 protons.
b. The number of neutrons in this atom is found by subtracting the atomic number from the mass number. The number of neutrons is 16.

Mass number − atomic number = number of neutrons
        31   −   15   =   16

c. Because an atom is neutral, there is an electrical balance of protons and electrons. Because the number of electrons is equal to the number of protons, the phosphorus atom has 15 electrons.

Study Check

How many neutrons are in the nucleus of a bromine atom that has a mass number of 80?

4.23 Would you use atomic number, mass number, or both to obtain the following?
  a. number of protons in an atom
  b. number of neutrons in an atom
  c. number of particles in the nucleus
  d. number of electrons in a neutral atom

4.24 What do you know about the subatomic particles from the following?
  a. atomic number
  b. mass number
  c. mass number − atomic number
  d. mass number + atomic number

4.25 Write the names and symbols of the elements with the following atomic numbers:
  a. 3 **b.** 9 **c.** 20 **d.** 30
  e. 10 **f.** 14 **g.** 53 **h.** 8

4.26 Write the names and symbols of the elements with the following atomic numbers:
  a. 1 **b.** 11 **c.** 19 **d.** 26
  e. 35 **f.** 47 **g.** 15 **h.** 2

4.27 How many protons and electrons are there in a neutral atom of the following?
  a. magnesium **b.** zinc
  c. iodine **d.** potassium

4.28 How many protons and electrons are there in a neutral atom of the following?
  a. carbon **b.** fluorine
  c. calcium **d.** sulfur

4.29 Complete the following table for neutral atoms:

Name of Element	Symbol	Atomic Number	Mass Number	Number of Protons	Number of Neutrons	Number of Electrons
	Al		27			
		12			12	
Potassium					20	
				16	15	
			56			26

4.30 Complete the following table for neutral atoms:

Name of Element	Symbol	Atomic Number	Mass Number	Number of Protons	Number of Neutrons	Number of Electrons
	N		15			
Calcium			42			
				38	50	
		14			16	
		56	138			

4.5 Isotopes and Atomic Mass

WEB TUTORIAL
Atoms and Isotopes

We have seen that all atoms of the same element have the same number of protons and electrons. However, the atoms of any one element are not completely identical because they can have different numbers of neutrons. **Isotopes** are atoms of the same element that have different numbers of neutrons. For example, all atoms of the element magnesium (Mg) have 12 protons. However, some magnesium atoms have 12 neutrons, others have 13 neutrons, and still others have 14 neutrons. The differences in numbers of neutrons for these magnesium atoms cause their mass numbers to be different but not their chemical behavior. The three isotopes of magnesium have the same atomic number but different mass numbers. Often, the term *isotope* is used to describe radioactive atoms. However, most elements, such as magnesium, have one or more nonradioactive isotopes.

To specify a particular isotope, we write a **nuclear symbol,** which gives the number of protons and neutrons in the nucleus. The nuclear symbol shows the symbol of the element with the mass number as a superscript and the atomic number as a subscript.

Nuclear Symbol for an Isotope of Magnesium

Mass number ⟶ 24
Symbol of element ⟶ Mg
Atomic number ⟶ 12

Table 4.7	**Isotopes of Magnesium**		
Atomic symbol	$^{24}_{12}Mg$	$^{25}_{12}Mg$	$^{26}_{12}Mg$
Number of protons	12	12	12
Number of electrons	12	12	12
Number of neutrons	**12**	**13**	**14**
Mass number	**24**	**25**	**26**
% abundance	78.9%	10.0%	11.1%

Atomic structure of Mg

Isotopes of Mg

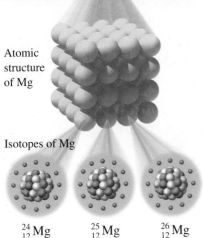

$^{24}_{12}Mg$ $^{25}_{12}Mg$ $^{26}_{12}Mg$

Table 4.8	**The Atomic Mass of Some Elements**	
Element	Isotopes	Atomic Mass (Weighted Average)
Lithium	^6Li, ^7Li	6.941 amu
Carbon	^{12}C, ^{13}C, ^{14}C	12.01 amu
Fluorine	^{19}F	19.00 amu
Oxygen	^{16}O, ^{17}O, ^{18}O	16.00 amu
Sulfur	^{32}S, ^{33}S, ^{34}S, ^{36}S	32.07 amu
Copper	^{63}Cu, ^{65}Cu	63.55 amu

An isotope may be referred to by its name or symbol followed by the mass number, such as magnesium-24 or Mg-24. Magnesium has three naturally occurring isotopes, as shown in Table 4.7.

Identifying Protons and Neutrons in Isotopes

State the number of protons and neutrons in the following isotopes of neon (Ne):
a. $^{20}_{10}Ne$ **b.** $^{21}_{10}Ne$ **c.** $^{22}_{10}Ne$

Solution

The atomic number of Ne is 10; each isotope has 10 protons. The number of neutrons in each isotope is found by subtracting the atomic number (10) from each mass number.
a. 10 protons; 10 neutrons $(20 - 10)$
b. 10 protons; 11 neutrons $(21 - 10)$
c. 10 protons; 12 neutrons $(22 - 10)$

Study Check

Write a symbol for the following isotopes:
a. a nitrogen atom with 8 neutrons
b. an atom with 20 protons and 22 neutrons
c. an atom with mass number 27 and 14 neutrons

Atomic Mass

Anytime we work with elements in the laboratory, the samples contain all the naturally occurring isotopes of that element. Although the mass of each isotope is known, we actually use the mass of an "average atom" of that element. The **atomic mass** of an element is the weighted average mass of all the naturally occurring isotopes of that element based on the abundance and mass of each isotope. This is the number shown under the symbol of each element listed on the periodic table. Because most elements consist of several isotopes, the atomic masses on the periodic table are not usually whole numbers. Table 4.8 lists the isotopes of some selected elements and their atomic masses.

Calculating Atomic Mass

To calculate an atomic mass, the contribution of each isotope is determined by multiplying the mass of each isotope by its percent abundance and adding the results. For example, in a sample of chlorine (Cl), 75.78% of the Cl atoms have a mass of 34.969 amu and 24.22% of the Cl atoms have a mass of 36.966.

Isotope	Mass (amu)	×	Abundance (%)	=	Contribution to Average Cl Atom
^{35}Cl	34.969	×	$\dfrac{75.78}{100}$	=	26.50 amu
^{37}Cl	36.966	×	$\dfrac{24.22}{100}$	=	8.953 amu
			Atomic mass of Cl	=	35.45 amu

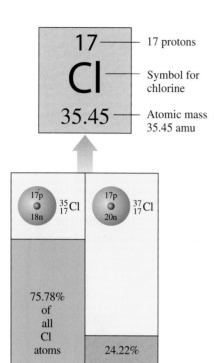

The atomic mass 35.45 amu is the weighted average mass of a sample of Cl atoms, although no individual Cl atom actually has this mass.

Sample Problem 4.11   **Calculating Atomic Mass**

Calculate the atomic mass for magnesium with the following isotopes and abundance:

Isotope	Mass	Abundance (%)
^{24}Mg	23.985	78.70
^{25}Mg	24.986	10.13
^{26}Mg	25.983	11.17

Solution

$$^{24}Mg \quad 23.985 \quad \times \quad \frac{78.70}{100} \quad = \quad 18.88 \text{ amu}$$

$$^{25}Mg \quad 24.986 \quad \times \quad \frac{10.13}{100} \quad = \quad 2.531 \text{ amu}$$

$$^{26}Mg \quad 25.983 \quad \times \quad \frac{11.17}{100} \quad = \quad 2.902 \text{ amu}$$

$$\text{Atomic mass of Mg} \quad = \quad 24.31 \text{ amu}$$

Study Check

There are two naturally occurring isotopes of boron. ^{10}B has a mass of 10.013 amu with an abundance of 19.80% and ^{11}B has a mass of 11.009 amu with an abundance of 80.20%. What is the atomic mass of boron?

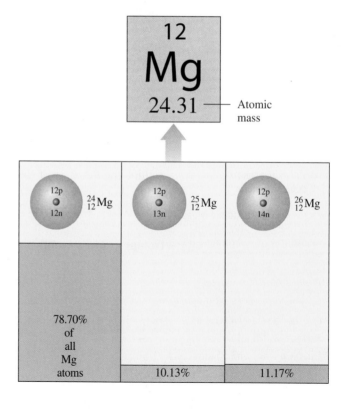

4.31 What are the number of protons, neutrons, and electrons in the following isotopes?
   a. $^{27}_{13}$Al   **b.** $^{52}_{24}$Cr   **c.** $^{34}_{16}$S   **d.** $^{56}_{26}$Fe

4.32 What are the number of protons, neutrons, and electrons in the following isotopes?
   a. $^{2}_{1}$H   **b.** $^{14}_{7}$N   **c.** $^{26}_{14}$Si   **d.** $^{70}_{30}$Zn

4.33 Write the atomic symbols for isotopes with the following:
   a. 15 protons and 16 neutrons
   b. 35 protons and 45 neutrons
   c. 13 electrons and 14 neutrons
   d. a chlorine atom with 18 neutrons

4.34 Write the atomic symbols for isotopes with the following:
   a. an oxygen atom with 10 neutrons
   b. 4 protons and 5 neutrons
   c. 26 electrons and 30 neutrons
   d. a mass number of 24 and 13 neutrons

4.35 There are four isotopes of sulfur, with mass numbers 32, 33, 34, and 36.
   a. Write the atomic symbol for each of these atoms.
   b. How are these isotopes alike?
   c. How are they different?
   d. Why is the atomic mass of sulfur listed on the periodic table not a whole number?

4.36 There are four isotopes of strontium, with mass numbers 84, 86, 87, 88.
   a. Write the atomic symbol for each of these atoms.
   b. How are these isotopes alike?

   c. How are they different?
   d. Why is the atomic mass of strontium listed on the periodic table not a whole number?

4.37 What is the difference between the mass of an isotope and the atomic mass of an element?

4.38 What is the difference between the mass number and atomic mass of the element?

4.39 Copper consists of two isotopes, ^{63}Cu and ^{65}Cu. If the atomic mass for copper on the periodic table is 63.55, are there more atoms of ^{63}Cu or ^{65}Cu in a sample of copper?

4.40 A fluorine sample consists of only one type of atom, ^{19}F, which has a mass of 19.00 amu. How would the mass of ^{19}F atom compare to the atomic mass listed on the periodic table?

4.41 There are four naturally occurring isotopes of iron: ^{54}Fe, ^{56}Fe, ^{57}Fe, and ^{58}Fe. Use the atomic mass of iron listed on the periodic table to identify the most abundant isotope.

4.42 Zinc consists of five naturally occurring isotopes: ^{64}Zn, ^{66}Zn, ^{67}Zn, ^{68}Zn, and ^{70}Zn. None of these isotopes has the atomic mass of 65.38 listed for zinc on the periodic table. Explain.

4.43 Two isotopes of gallium are naturally occurring, with ^{69}Ga at 60.10% (68.926 amu) and ^{71}Ga at 39.90% (70.925 amu). What is the atomic mass of gallium?

4.44 Two isotopes of copper are naturally occurring, with ^{63}Cu at 69.15% (62.930 amu) and ^{65}Cu at 30.85% (64.928 amu). What is the atomic mass of copper?

4.6 Electron Energy Levels

Learning **Goal**

Given the name or symbol of one of the first 18 elements in the periodic table, write the electron arrangement.

Electrons are constantly moving within the large space of an atom, which means they possess energy. However, they do not all have the same energy. Electrons of similar energy are grouped in **energy levels.** We may think of the energy levels of an atom as similar to the rungs on a ladder. The lowest energy level would be the first rung of the ladder; the second energy level would be the second rung. As you climb up or down the ladder, you must step from one rung to the next. You cannot stop at a level between the rungs. In atoms, electrons exist only in the available energy levels. Generally, the energy levels closest to the nucleus contain electrons with the lowest energies, whereas energy levels farther away contain electrons with higher energies. Unlike a ladder, the lower energy levels are far apart compared to the higher energy levels.

The maximum number of electrons allowed in each energy level is given by the formula $2n^2$, where n is the number of the main energy level. As shown in Table 4,9, Level 1, the lowest energy level, can hold up to 2 electrons; level 2 can hold up to 8 electrons; level 3 can take 18 electrons; and level 4 has room for 32 electrons. In the atoms of the elements known today, electrons occupy seven energy levels.

Table 4.9   Capacity of Energy Levels 1–4

Energy Level (n)	Maximum Number of Electrons ($2n^2$)
1	$2(1)^2 = 2$
2	$2(2)^2 = 8$
3	$2(3)^2 = 18$
4	$2(4)^2 = 32$

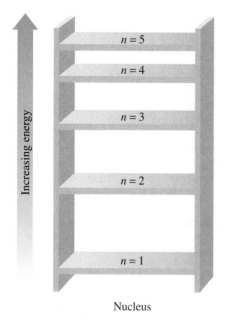

Nucleus

Electron Arrangements for the First 18 Elements

The electron arrangement of an atom gives the number of electrons in each energy level. The electron arrangements for the first 18 elements can be written by placing electrons in energy levels beginning with the lowest. The single electron of hydrogen and the 2 electrons of helium are placed in energy level 1.

As shown in Table 4.10, the elements of the second period (lithium, Li, to neon, Ne) have enough electrons to fill the first energy level and part or all of the second energy level. For example, lithium has 3 electrons. Two of those electrons complete energy level 1. The remaining electron goes into the second energy level. As we go across Period 2, more and more electrons enter the second energy level. For example, an atom of carbon, with a total of 6 electrons, fills energy level 1 with 2 electrons, and 4 remaining electrons enter the second energy level. The last element in Period 2 is neon. The 10 electrons in an atom of neon completely fill the first and second energy levels.

In an atom of sodium, atomic number 11, the first and second energy levels are filled and the last electron enters the third energy level. The rest of the elements in the third period continue to add to the third level. For example, a sulfur atom with 16 electrons has 2 electrons in the first level, 8 electrons in the second level, and 6 electrons in the third level. At the end of Period 3, we find that argon has 8 electrons in the third level.

WEB TUTORIAL
Bohr's Shell Model of the Atom

Table 4.10   Electron Level Arrangements for the First 18 Elements

Element	Symbol	Atomic Number	Number of Electrons in Energy Level		
			1	2	3
hydrogen	H	1	1		
helium	He	2	2		
lithium	Li	3	2	1	
beryllium	Be	4	2	2	
boron	B	5	2	3	
carbon	C	6	2	4	
nitrogen	N	7	2	5	
oxygen	O	8	2	6	
fluorine	F	9	2	7	
neon	Ne	10	2	8	
sodium	Na	11	2	8	1
magnesium	Mg	12	2	8	2
aluminum	Al	13	2	8	3
silicon	Si	14	2	8	4
phosphorus	P	15	2	8	5
sulfur	S	16	2	8	6
chlorine	Cl	17	2	8	7
argon	Ar	18	2	8	8

Sample Problem 4.12   **Writing Electron Arrangements**

Write the electron arrangement for each of the following:
a. oxygen   **b.** chlorine

Solution

a. Oxygen has an atomic number of 8. There are 8 electrons arranged as $2e^-, 6e^-$.
b. An atom of chlorine has atomic number of 17. The 17 electrons are arranged as $2e^-, 8e^-, 7e^-$.

Study Check

What element has an electron arrangement of $2e^-, 8e^-, 2e^-$?

Group Number and Valence Electrons

The chemical properties of representative elements are mostly due to the **valence electrons,** which are the electrons in the outermost energy levels. The **group numbers** indicate the number of valence (outer) electrons for the elements in each vertical column. For example, the elements in Group 1A (1), such as lithium, sodium, and potassium, all have one electron in the outer energy level. Elements in Group 2A (2), the alkaline earth metals, have two (2) valence electrons. The halogens in Group 7A (17) have seven (7) valence electrons. (See Table 4.11.)

Table 4.11   Comparison of Electron Level Arrangements, by Group, for Some Representative Elements

Group Number	Element	Number of Electrons in Energy Level			
		1	2	3	4
2A (2)	beryllium	2	2		
	magnesium	2	8	2	
	calcium	2	8	8	2
7A (17)	fluorine	2	7		
	chlorine	2	8	7	
	bromine	2	8	18	7

Sample Problem 4.13   **Using Group Numbers**

Using the periodic table, write the group number and the number of electrons in the outer electron level of the following elements:
a. sodium   **b.** sulfur   **c.** aluminum

Solution

a. Sodium is in Group 1A (1); sodium has 1 electron in the outer energy level.
b. Sulfur is in Group 6A (16); sulfur has 6 electrons in the outer energy level.
c. Aluminum is in Group 3A (13); aluminum has 3 electrons in the outer energy level.

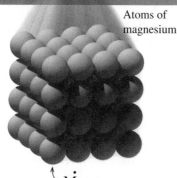

Atoms of magnesium

Ṁg·

Electron-dot symbol
Electron arrangement Mg 2, 8, ☐2

Study Check

What is the group number and name of the element with atoms that have 5 electrons in the third energy level?

Electron-Dot Symbols

An **electron-dot symbol** is a convenient way to represent the valence electron. Valence electrons are shown as dots placed on the sides, top, or bottom of the symbol for the element. It does not matter on which of the four sides you place the dots. However, 1 to 4 valence electrons are arranged as single dots. When there are more than 4 electrons, the electrons begin to pair up. Any of the following would be an acceptable electron-dot symbol for magnesium, which has 2 valence electrons:

Possible Electron-Dot Symbols for the 2 Valence Electrons in Magnesium

Ṁg·   Ṁg   ·Ṁg   ·Mg·   Mg·   ·Mg

Electron-dot symbols for selected elements are given in Table 4.12.

Sample Problem 4.14   **Writing Electron-Dot Symbols**

Write the electron-dot symbol for each of the following elements:
a. bromine   **b.** aluminum

Solution

a. Because the group number for bromine is 7A (17), bromine has 7 valence electrons.

·Br:

b. Aluminum, in Group 3A (3), has 3 valence electrons.

·Al·

Study Check

What is the electron-dot symbol for phosphorus?

Table 4.12   **Electron-Dot Symbols for Selected Elements in Periods 1–4**

Group Number	1A (1)	2A (2)	3A (13)	4A (14)	5A (15)	6A (16)	7A (17)	8A (18)
Valence Electrons	1	2	3	4	5	6	7	8
Electron-Dot Symbols	H·							He:
	Li·	Be·	·B·	·C·	·N:	·O:	·F:	:Ne:
	Na·	Mg·	·Al·	·Si·	·P·	·S:	·Cl:	:Ar:
	K·	Ca·	·Ga·	·Ge·	·As·	·Se:	·Br:	:Kr:

Li atom

Na atom

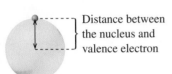

Distance between the nucleus and valence electron

K atom

Ionization Energy Decreases

Ionization Energy

Electrons are held in atoms by their attraction to the nucleus. Therefore, energy is required to remove an electron from an atom. The **ionization energy** is the energy needed to remove the least tightly bound electron from an atom in the gaseous (g) state. When an electron is removed from a neutral atom, a particle called a cation (pronounced cat·ion), with a 1+ charge, is formed.

$$Na(g) + \text{energy (ionization)} \longrightarrow Na^+(g) + e^-$$

The ionization energy generally decreases going down a group. Less energy is needed to remove an electron as nuclear attraction for electrons decreases farther from the nucleus. *Going across a period from left to right, the ionization energy generally increases.* As the number of protons in the nucleus increases across a period, more energy is required to remove an electron. In general, the ionization energy is low for the metals and high for the nonmetals.

In Period 1, the valence electrons are close to the nucleus and strongly held. H and He have high ionization energies because a large amount of energy is required to remove an electron. The ionization energy for He is the highest of any element because He has a full, stable, energy level which is disrupted by removing an electron. The high ionization energies of the noble gases indicate that their electron arrangements are especially stable.

Sample Problem 4.15   Ionization Energy

Indicate the element in each that has the higher ionization energy and explain your choice.
a. K or Na   **b.** Mg or Cl   **c.** F or N

Ionization energy →

Groups

	1A	2A		3A	4A	5A	6A	7A	8A
	1	2		13	14	15	16	17	18
1	H								He
2	Li	Be		B	C	N	O	F	Ne
3	Na	Mg		Al	Si	P	S	Cl	Ar
4	K	Ca		Ga	Ge	As	Se	Br	Kr
5	Rb	Sr		In	Sn	Sb	Te	I	Xe
6	Cs	Ba		Tl	Pb	Bi	Po	At	Rn

Ionization energy ↑

Solution

a. Na. In Na, the valence electron is closer to the nucleus.

b. Cl. Attraction for the valence electrons increases across a period, going left to right.

c. F. With more protons, the F nucleus has a stronger attraction for valence electrons in the second energy level than N.

Study Check

Arrange Cl, Br, and I in order of increasing ionization energy.

Questions and Problems ▸ **Electron Energy Levels**

4.45 Electrons exist in specific energy levels. Explain.

4.46 In what order do electrons fill the energy levels 1–3 for the first 18 elements on the periodic table?

4.47 How many electrons are in energy level 2 of the following elements?
a. sodium **b.** nitrogen **c.** sulfur
d. helium **e.** chlorine

4.48 How many electrons are in energy level 3 of the following elements?
a. oxygen **b.** sulfur **c.** phosphorus
d. argon **e.** fluorine

4.49 Write the electron arrangement for each of the following elements.
Example: sodium 2, 8, 1
a. carbon **b.** argon
c. sulfur **d.** silicon
e. an atom with 13 protons and 14 neutrons
f. nitrogen

4.50 Write the electron arrangement for each of the following atoms.
Example: sodium 2, 8, 1
a. phosphorus **b.** neon
c. oxygen **d.** an atom with atomic number 18
e. aluminum **f.** silicon

4.51 Identify the elements that have the following electron arrangements:

Energy Level:	1	2	3
a.	$2e^-$	$1e^-$	
b.	$2e^-$	$8e^-$	$2e^-$
c.	$1e^-$		
d.	$2e^-$	$8e^-$	$7e^-$
e.	$2e^-$	$6e^-$	

4.52 Identify the elements that have the following electron arrangements:

Energy Level:	1	2	3
a.	$2e^-$	$5e^-$	
b.	$2e^-$	$8e^-$	$6e^-$
c.	$2e^-$	$4e^-$	
d.	$2e^-$	$8e^-$	$8e^-$
e.	$2e^-$	$8e^-$	$3e^-$

4.53 The elements boron and aluminum are in the same group on the periodic table.
a. Write the electron arrangements for B and Al.
b. How many valence electrons are in the outer energy level of each atom?
c. What is their group number?

4.54 The elements fluorine and chlorine are in the same group on the periodic table.
a. Write the electron arrangements for F and Cl.
b. How many valence electrons are in each of their outer energy levels?
c. What is their group number?

4.55 What is the number of electrons in the outer energy level and the group number for each of the following elements?
Example: fluorine $7e^-$; Group 7A (17)
a. magnesium **b.** chlorine **c.** oxygen
d. nitrogen **e.** barium **f.** bromine

4.56 What is the number of electrons in the outer energy level and the group number for each of the following elements?
Example: fluorine $7e^-$; Group 7A (17)
a. lithium **b.** silicon **c.** neon
d. argon **e.** tin **f.** cesium

4.57 Write the group number and electron-dot symbol for each element:
a. sulfur **b.** nitrogen **c.** calcium
d. sodium **e.** potassium

4.58 Write the group number and electron-dot symbol for each element:
a. carbon **b.** oxygen **c.** fluorine
d. lithium **e.** chlorine

4.59 Using the symbol M for a metal atom, draw the electron-dot symbol for an atom of a metal in the following groups:
a. Group 1A (1) **b.** Group 2A (2)

4.60 Using the symbol Nm for a nonmetal atom, draw the electron-dot symbol for an atom of a nonmetal in the following groups:
a. Group 5A (15) **b.** Group 7A (17)

4.61 The alkali metals are in the same family on the periodic table. What is their group number and how many valence electrons does each have?

4.62 The halogens are in the same family on the periodic table. What is their group number and how many valence electrons does each have?

4.63 Arrange each set of elements in order of increasing ionization energy.
a. F, Cl, Br   **b.** Na, Cl, Al   **c.** Na, K, Cs

4.64 Arrange each set of elements in order of increasing ionization energy.
a. C, N, O   **b.** P, S, Cl   **c.** As, P, N

4.65 Select the element in each pair with the higher ionization energy.
a. Br or I   **b.** K or Al   **c.** Si or P

4.66 Select the element in each pair with the higher ionization energy.
a. O or Ne   **b.** K or Br   **c.** Ca or Ba

Concept Map ▸ Atoms and Elements

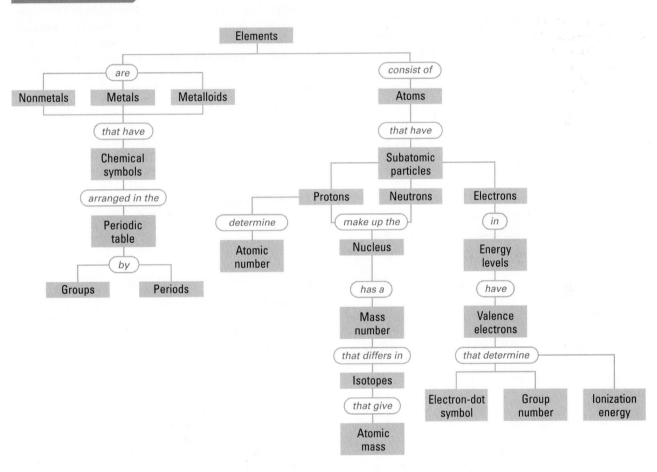

Chapter Review

4.1 Elements and Symbols

Elements are the primary substances of matter. Chemical symbols are one- or two-letter abbreviations of the names of the elements.

4.2 The Periodic Table

The periodic table is an arrangement of the elements by increasing atomic number. A vertical column on the periodic table containing elements with similar properties is called a group. A horizontal row is called a period. Elements in Group 1A (1) are called the alkali metals; Group 2A (2), alkaline earth metals; Group 7A (17), the halogens; and Group 8A (18), the noble gases. On the periodic table, metals are located on the left of the heavy zigzag line, and nonmetals are to the right of the heavy zigzag line. Elements located on the heavy line are called metalloids.

4.3 The Atom

An atom is the smallest particle that retains the characteristics of an element. Atoms are composed of three subatomic particles. Protons have a positive charge (+), electrons carry a negative charge (−), and neutrons are electrically neutral. The protons and neutrons are found in the tiny, dense nucleus. Electrons are located outside the nucleus.

4.4 Atomic Number and Mass Number

The atomic number gives the number of protons in all the atoms of the same element. In a neutral atom, there is an equal number of protons and electrons. The mass number is the total number of protons and neutrons in an atom.

4.5 Isotopes and Atomic Mass

Atoms that have the same number of protons but different numbers of neutrons are called isotopes. The atomic mass of an element is the weighted average mass of all the isotopes in a naturally occurring sample of that element.

4.6 Electron Energy Levels

Every electron has a specific amount of energy. In an atom, the electrons of similar energy are grouped in specific energy levels. The first level nearest the nucleus can hold 2 electrons, the second level can hold 8 electrons, the third level will take up to 18 electrons. The properties of elements are related to the valence electrons of the atoms. The electron arrangement is written by placing the number of electrons in that atom in order from the lowest energy levels and filling to higher levels. The similarity of behavior for the elements in a group is related to having the same number of electrons in their outermost, valence, level. The group number for an element gives the number of electrons in its outermost energy level. The energy required to remove a valence electron is the ionization energy, which generally decreases going down a group and generally increases going across a period.

Key Terms

alkali metals Elements of Group 1A (1) except hydrogen; these are soft, shiny metals.

alkaline earth metals Group 2A (2) elements.

atom The smallest particle of an element that retains the characteristics of the element.

atomic mass The weighted average mass of all the naturally occurring isotopes of an element.

atomic mass unit (amu) A small mass unit used to describe the mass of very small particles such as atoms and subatomic particles; 1 amu is equal to one-twelfth the mass of a carbon-12 atom.

atomic number A number that is equal to the number of protons in an atom.

chemical symbol An abbreviation that represents the name of an element.

electron A negatively charged subatomic particle having a very small mass that is usually ignored in calculations; its symbol is e^-.

electron-dot symbol The representation of an atom that shows valence electrons as dots around the symbol of the element.

energy level Level within the atom that contains electrons of similar energy.

group A vertical column in the periodic table that contains elements having similar physical and chemical properties.

group number A number that appears at the top of each vertical column (group) in the periodic table and indicates the number of valence electrons.

halogen Group 7A (17) elements of fluorine, chlorine, bromine, iodine, and astatine.

ionization energy The energy needed to remove the least tightly bound electron from the outermost energy level of an atom.

isotope An atom that differs only in mass number from another atom of the same element. Isotopes have the same atomic number (number of protons) but different numbers of neutrons.

mass number The total number of neutrons and protons in the nucleus of an atom.

metal An element that is shiny, malleable, ductile, and a good conductor of heat and electricity. The metals are located to the left of the zigzag line in the periodic table.

metalloid Elements with properties of both metals and nonmetals located along the heavy zigzag line on the periodic table.

neutron A neutral subatomic particle having a mass of 1 amu and found in the nucleus of an atom; its symbol is n or n^0.

noble gas An element in Group 8A (18) of the periodic table, generally unreactive and seldom found in combination with other elements.

nonmetal An element with little or no luster that is a poor conductor of heat and electricity. The nonmetals are located to the right of the zigzag line in the periodic table.

nuclear symbol An abbreviation used to indicate the mass number and atomic number of an isotope.

nucleus The compact, very dense center of an atom, containing the protons and neutrons of the atom.

period A horizontal row of elements in the periodic table.

periodic table An arrangement of elements by increasing atomic number such that elements having similar chemical behavior are grouped in vertical columns.

proton A positively charged subatomic particle having a mass of 1 amu and found in the nucleus of an atom; its symbol is p or p^+.

representative element An element found in Groups 1A (1) through 8A (18) excluding B groups (3–12) of the periodic table.

subatomic particle A particle within an atom; protons, neutrons, and electrons are subatomic particles.

transition element An element located between Groups 2A (2) and 3A (13) on the periodic table.

valence electrons Electrons in the outermost energy level of an atom.

Understanding the Concepts

4.67 According to Dalton's atomic theory, which of the following are true?
 a. Atoms of an element are identical to atoms of other elements.
 b. Every element is made of atoms.
 c. Atoms of two different elements combine to form compounds.
 d. In a chemical reaction, some atoms disappear and new atoms appear.

4.68 For each of the following, write the symbol and name for X and the number of protons and neutrons. Which are isotopes of each other?
 a. $^{37}_{17}X$
 b. $^{56}_{26}X$
 c. $^{116}_{50}X$
 d. $^{124}_{50}X$
 e. $^{116}_{48}X$

4.69 Indicate if the atoms in each pair have the same number of protons? Neutrons? Electrons?
 a. ^{37}Cl, ^{36}S
 b. ^{36}S, ^{35}S
 c. ^{79}Se, ^{81}Br
 d. ^{40}Ar, ^{39}Cl

4.70 Complete the following table for the three naturally occurring isotopes of silicon, the major component in computer chips.

	Isotope		
	^{28}Si	^{29}Si	^{30}Si
Number of protons			
Number of neutrons			
Number of electrons			
Atomic number			
Mass number			

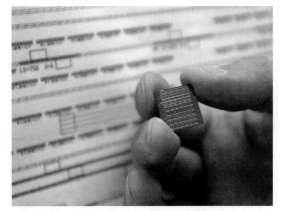

4.71 For each nucleus, write the nuclear symbol, and determine which are isotopes.

proton ●   neutron ◐

a   b   c   d   e

4.72 Identify the element represented by each nucleus in problem 4.71 as a metal, nonmetal, or metalloid.

4.73 Provide the following information:
  a. the atomic number and symbol of the lightest alkali metal

  b. the atomic number and symbol of the heaviest noble gas

  c. the atomic mass and symbol of the alkaline earth metal in Period 3

  d. the atomic mass and symbol of the halogen with the fewest electrons

4.74 Provide the following information:
  a. the atomic number and symbol of the heaviest metalloid in Group 4A (14)

  b. the atomic number and symbol of the only metal in Group 5A (15)

  c. the atomic mass and symbol of the alkali metal in Period 4

  d. the atomic mass and symbol of the heaviest halogen

Additional Questions and Problems

4.75 Why is Co the symbol for cobalt, not CO?

4.76 Which of the following is correct? Write the correct symbol if needed.
  a. copper, Co   **b.** silicon, SI
  c. iron, Fe   **d.** fluorine, Fl
  e. potassium, P   **f.** sodium, Na
  g. gold, Au   **h.** lead, PB

4.77 Give the symbol and name of the element found in the following group and period on the periodic table:
  a. Group 2A, Period 3   **b.** Group 7A, Period 4
  c. Group 13, Period 3   **d.** Group 16, Period 2

4.78 Give the group and period number for the following elements:
  a. potassium   **b.** phosphorus   **c.** carbon   **d.** neon

4.79 Write the names of two elements that are in the following groups:
  a. halogens   **b.** noble gases
  c. alkali metals   **d.** alkaline earth metals

4.80 The following are trace elements that have been found to be crucial to the biochemical and physiological processes in the body. Indicate whether each is a metal or nonmetal.
  a. zinc   **b.** cobalt
  c. manganese (Mn)   **d.** iodine
  e. copper   **f.** selenium (Se)

4.81 Indicate if each of the following statements is *true* or *false:*
  a. The proton is a negatively charged particle.
  b. The neutron is 2000 times as heavy as a proton.
  c. The atomic mass unit is based on a carbon atom with 6 protons and 6 neutrons.
  d. The nucleus is the largest part of the atom.
  e. The electrons are located outside the nucleus.

4.82 Indicate if each of the following statements is *true* or *false:*
  a. The neutron is electrically neutral.
  b. Most of the mass of an atom is due to the protons and neutrons.
  c. The charge of an electron is equal, but opposite, to the charge of a neutron.
  d. The proton and the electron have about the same mass.
  e. The mass number is the number of protons.

4.83 Complete the following statements:
  a. The atomic number gives the number of _____ in the nucleus.
  b. In an atom, the number of electrons is equal to the number of _____.
  c. Sodium and potassium are examples of elements called _____.

4.84 Complete the following statements:
  a. The number of protons and neutrons in an atom is also the _____ number.
  b. The elements in Group 7A (17) are called the _____.
  c. Elements that are shiny and conduct heat are called _____.

4.85 Write the names and symbols of the elements with the following atomic numbers:
  a. 3   **b.** 9   **c.** 20
  d. 33   **e.** 50   **f.** 55
  g. 79   **h.** 8

4.86 Write the names and symbols of the elements with the following atomic numbers:
  a. 1   **b.** 11   **c.** 20
  d. 26   **e.** 35   **f.** 47
  g. 83   **h.** 92

4.87 Give the number of protons and electrons in atoms of the following:
  a. Mn   **b.** zinc   **c.** iodine
  d. Mg   **e.** potassium

4.88 Give the number of protons and electrons in atoms of the following:
  a. carbon   **b.** Ca   **c.** copper
  d. chlorine   **e.** Cd

4.89 For the following atoms, determine the number of protons, neutrons, and electrons:
  a. $^{27}_{13}\text{Al}$ **b.** $^{52}_{24}\text{Cr}$ **c.** $^{34}_{16}\text{S}$
  d. $^{56}_{26}\text{Fe}$ **e.** $^{136}_{54}\text{Xe}$

4.90 For the following atoms, give the number of protons, neutrons, and electrons:
  a. $^{22}_{10}\text{Ne}$ **b.** $^{127}_{53}\text{I}$ **c.** $^{75}_{35}\text{Br}$
  d. $^{133}_{55}\text{Cs}$ **e.** $^{195}_{78}\text{Pt}$

4.91 Write the symbol and mass number for each of the following:
  a. an atom with 4 protons and 5 neutrons
  b. an atom with 12 protons and 14 neutrons
  c. a calcium atom with a mass number of 46
  d. an atom with 30 electrons and 40 neutrons
  e. a copper atom with 34 neutrons

4.92 Write a symbol and mass number for each of the following:
  a. an aluminum atom with 14 neutrons
  b. an atom with atomic number 26 and 32 neutrons
  c. a strontium atom with 50 neutrons
  d. an atom with a mass number of 72 and atomic number 33

4.93 Complete the following table:

Name	Nuclear Symbol	Number of Protons	Number of Neutrons	Number of Electrons
	$^{34}_{16}\text{S}$			
		30	40	
magnesium			14	
	$^{220}_{86}\text{Rn}$			

4.94 Complete the following table:

Name	Nuclear Symbol	Number of Protons	Number of Neutrons	Number of Electrons
potassium			22	
	$^{51}_{23}\text{V}$			
		48	64	
barium			82	

4.95 The most abundant isotope of iron is Fe-56.
  a. How many protons, neutrons, and electrons are in this isotope?
  b. What is the symbol of another isotope of iron with 25 neutrons?
  c. What is the symbol of an atom with mass number 51 and 27 neutrons?

4.96 Cadmium, atomic number 48, consists of eight naturally occurring isotopes. Do you expect any of the isotopes to have the atomic mass listed on the periodic table for cadmium? Explain.

4.97 Consider the following atoms in which the chemical symbol of the element is represented by X:

$$^{16}_{8}\text{X} \qquad ^{16}_{9}\text{X} \qquad ^{18}_{10}\text{X} \qquad ^{17}_{8}\text{X} \qquad ^{18}_{8}\text{X}$$

  a. What atoms have the same number of protons?
  b. Which atoms are isotopes? Of what element?
  c. Which atoms have the same mass number?
  d. What atoms have the same number of neutrons?

4.98 Five isotopes of zinc are zinc-64, zinc-66, zinc-67, zinc-68, and zinc-70.
  a. Write the atomic symbols, including atomic number and mass number, for each of these atoms.
  b. Give the number of protons, electrons, and neutrons for each of the zinc isotopes.

4.99 The most abundant isotope of lead is ^{208}Pb.
  a. How many protons, neutrons, and electrons are in the atom?
  b. What is the symbol of an isotope of lead with 132 neutrons?
  c. What is the name and symbol of an isotope with the same mass number as in part b, and 131 neutrons?

4.100 The most abundant isotope of silver is ^{107}Ag.
  a. How many protons, neutrons, and electrons are in the atom?
  b. What is the symbol of an isotope of silver with 62 neutrons?
  c. What is the name and symbol of an isotope with the same mass number as in part b, and 61 neutrons?

4.101 Write an electron arrangement, the group number, and electron-dot symbol for an atom of each of the following:
  a. nitrogen   **b.** sodium
  c. sulfur   **d.** boron

4.102 Write an electron arrangement, the group number, and electron-dot symbol for an atom of each of the following:
  a. carbon   **b.** silicon
  c. phosphorus   **d.** argon

4.103 Why is the ionization energy of Ca higher than K, but lower than Mg?

4.104 Why is the ionization energy of Cl lower than F, but higher than S?

4.105 Of the elements Na, P, Cl, and F, which
  a. is a metal?
  b. has the highest ionization energy?
  c. loses an electron most easily?
  d. is found in Group 7A (17), Period 3?

4.106 Of the elements: Mg, Ca, Br, Kr, which
  a. is a noble gas?
  b. has the lowest ionization energy?
  c. requires the most energy to remove an electron?
  d. is found in Group 2A (2), Period 4?

 Challenge Questions

4.107 Lead consists of four naturally occurring isotopes. Calculate the atomic mass of lead.

Isotope	Mass (amu)	Abundance (%)
^{204}Pb	203.97	1.40
^{206}Pb	205.97	24.10
^{207}Pb	206.98	22.10
^{208}Pb	207.98	52.40

4.108 Indium (In), with an atomic mass of 114.8 amu, consists of two naturally occurring isotopes, ^{113}In and ^{115}In. If 4.30% of indium is ^{113}In, which has a mass of 112.90 amu, what is the mass of the ^{115}In?

4.109 Silicon has three isotopes that occur in nature. ^{28}Si (27.977 amu) has an abundance of 92.23%, ^{29}Si

(28.976 amu) has a 4.68% abundance, and ^{30}Si (29.974 amu) has a 3.09% abundance. What is the atomic mass of silicon?

4.110 Antimony (Sb), which has an atomic weight of 121.75 amu, has two naturally occurring isotopes: Sb-121 and Sb-123. If a sample of antimony is 42.70% Sb-123, which has a mass of 122.90 amu, what is the mass of Sb-121?

4.111 If the diameter of a sodium atom is 3.14×10^{-8} cm, how many sodium atoms would fit along a line exactly 1 inch long?

4.112 A lead atom has a mass of 3.4×10^{-22} g. How many lead atoms are in a cube of lead that has a volume of 2.00 cm^3 if the density of lead is 11.3 g/cm^3?

Answers

Answers to Study Checks

4.1 Si, S, and Ag

4.2 magnesium, aluminum, and fluorine

4.3 **a.** Period 2: Li, Be, B, C, N, O, F, Ne
  b. Group 2A (2): Be, Mg, Ca, Sr, Ba, Ra

4.4 **a.** Strontium is in Group 2A (2).
  b. This is the alkaline earth metals.
  c. Strontium is in Period 5.
  d. Magnesium, Mg
  e. Alkali metal-Rb, halogen-I, noble gas-Xe

4.5 **a.** metal   **b.** metal   **c.** nonmetal

4.6 A proton, symbol p or p^+ with a charge of 1+, is found in the nucleus of an atom.

4.7 **a.** 26   **b.** 26   **c.** iron, Fe

4.8 Because the atomic number of silver is 47, it has 47 protons. The mass number is 107, which is the sum of 47 protons and 60 neutrons.

4.9 45 neutrons

4.10 **a.** $^{15}_{7}$N   **b.** $^{42}_{20}$Ca   **c.** $^{27}_{13}$Al

4.11 10.81 amu

4.12 magnesium

4.13 5A (15); phosphorus

4.14 $\cdot\ddot{\text{P}}\cdot$

4.15 I < Br < Cl

Answers to Selected Questions and Problems

4.1 **a.** Cu   **b.** Si   **c.** K
  d. N   **e.** Fe   **f.** Ba
  g. Pb   **h.** Sr

4.3 **a.** carbon   **b.** chlorine   **c.** iodine
  d. mercury   **e.** fluorine   **f.** argon
  g. zinc   **h.** nickel

4.5 **a.** sodium, chlorine
  b. calcium, sulfur, oxygen
  c. carbon, hydrogen, chlorine, nitrogen, oxygen
  d. calcium, carbon, oxygen

4.7 **a.** Period 2   **b.** Group 8A (18)
  c. Group 1A (1)   **d.** Period 2

4.9 a. alkaline earth metal
   b. transition element
   c. noble gas
   d. alkali metal
   e. halogen

4.11 a. C   **b.** He   **c.** Na   **d.** Ca   **e.** Al

4.13 a. metal   **b.** nonmetal
   c. metal   **d.** nonmetal
   e. nonmetal   **f.** nonmetal
   g. metalloid   **h.** metal

4.15 a. electron   **b.** proton
   c. electron   **d.** neutron

4.17 Rutherford determined that an atom contains a small, compact nucleus that is positively charged.

4.19 b.

4.21 In the process of brushing your hair, unlike charges on the hair and brush attract each other.

4.23 a. atomic number   **b.** both
   c. mass number   **d.** atomic number

4.25 a. lithium, Li   **b.** fluorine, F
   c. calcium, Ca   **d.** zinc, Zn
   e. neon, Ne   **f.** silicon, Si
   g. iodine, I   **h.** oxygen, O

4.27 a. 12   **b.** 30   **c.** 53   **d.** 19

4.29 See Table 4.13.

4.31 a. 13 protons, 14 neutrons, 13 electrons
   b. 24 protons, 28 neutrons, 24 electrons
   c. 16 protons, 18 neutrons, 16 electrons
   d. 26 protons, 30 neutrons, 26 electrons

4.33 a. $^{31}_{15}P$ **b.** $^{80}_{35}Br$
   c. $^{27}_{13}Al$ **d.** $^{35}_{17}Cl$

4.35 a. $^{32}_{16}S$ $^{33}_{16}S$ $^{34}_{16}S$ $^{36}_{16}S$
   b. They all have the same number of protons and electrons.
   c. They have different numbers of neutrons, which gives them different mass numbers.
   d. The atomic mass of S listed on the periodic table is the average atomic mass of all the isotopes.

4.37 The mass of an isotope is the mass of an individual atom. The atomic mass is the weighted average of all the naturally occurring isotopes of that element.

4.39 Since the atomic mass of copper is closer to 63 amu, there are more atoms of ^{63}Cu.

4.41 Since the atomic mass of iron is 55.85 amu, the most abundant isotope is ^{56}Fe.

4.43 69.72 amu

4.45 The energy of electrons is of a specific quantity for each energy level. Electrons cannot have energies that are between the energy levels.

4.47 a. 8   **b.** 5   **c.** 8
   d. 0   **e.** 8

4.49 a. 2, 4   **b.** 2, 8, 8   **c.** 2, 8, 6
   d. 2, 8, 4   **e.** 2, 8, 3   **f.** 2, 5

4.51 a. Li   **b.** Mg   **c.** H
   d. Cl   **e.** O

4.53 a. B 2, 3 Al 2, 8, 3   **b.** 3
   c. Group 3A (13)

4.55 a. $2e^-$, Group 2A (2)
   b. $7e^-$, Group 7A (17)
   c. $6e^-$, Group 6A (16)
   d. $5e^-$, Group 5A (15)
   e. $2e^-$, Group 2A (2)
   f. $7e^-$, Group 7A (17)

4.57 a. Group 6A (16) $\cdot\ddot{S}\cdot$
   b. Group 5A (15) $\cdot\ddot{N}\cdot$
   c. Group 2A (2) $\cdot Ca\cdot$
   d. Group 1A (1) $Na\cdot$
   e. Group 1A (1) $K\cdot$

4.59 a. $M\cdot$ **b.** $\cdot M\cdot$

4.61 They are all in Group 1A (1); each has 1 valence electron.

4.63 a. Br, Cl, F   **b.** Na, Al, Cl   **c.** Cs, K, Na

4.65 a. Br   **b.** Al   **c.** P

4.67 a. false   **b.** true
   c. true   **d.** false

Table 4.13

Name of Element	Symbol	Atomic Number	Mass Number	Number of Protons	Number of Neutrons	Number of Electrons
aluminum	Al	13	27	13	14	13
magnesium	Mg	12	24	12	12	12
potassium	K	19	39	19	20	19
sulfur	S	16	31	16	15	16
iron	Fe	26	56	26	30	26

4.69 a. Both have 20 neutrons.
    b. Both have 16 protons.
    c. Not the same.
    d. Both have 22 neutrons.

4.71 a. $^{9}_{4}\text{Be}$ **b.** $^{11}_{5}\text{B}$ **c.** $^{13}_{6}\text{C}$
    d. $^{10}_{5}\text{B}$ **e.** $^{12}_{6}\text{C}$
    b and d are isotopes of boron; c and e are isotopes of carbon.

4.73 a. 3, Li   **b.** 86, Rn
    c. 24.31 amu, Mg   **d.** 19.00 amu, F

4.75 The first letter of a symbol is a capital, but a second letter is lowercase. The symbol Co is for cobalt, but the symbols in CO are for carbon and oxygen.

4.77 a. Mg, magnesium   **b.** Br, bromine
    c. Al, aluminum   **d.** O, oxygen

4.79 a. any two elements in Group 7A (17), such as fluorine, chlorine, bromine, or iodine
    b. any two elements in Group 8A (18), such as helium, neon, argon, krypton, xenon, or radon
    c. any two elements in Group 1A (1), such as lithium, sodium, and potassium, except hydrogen
    d. any two elements in Group 2A (2), such as magnesium, calcium, and barium

4.81 a. false   **b.** false   **c.** true
    d. false   **e.** true

4.83 a. protons   **b.** protons
    c. alkali metals

4.85 a. lithium, Li   **b.** fluorine, F
    c. calcium, Ca   **d.** arsenic, As
    e. tin, Sn   **f.** cesium, Cs
    g. gold, Au   **h.** oxygen, O

4.87 a. 25 protons, 25 electrons
    b. 30 protons, 30 electrons
    c. 53 protons, 53 electrons
    d. 12 protons, 12 electrons
    e. 19 protons, 19 electrons

4.89 a. 13 protons, 14 neutrons, 13 electrons
    b. 24 protons, 28 neutrons, 24 electrons
    c. 16 protons, 18 neutrons, 16 electrons
    d. 26 protons, 30 neutrons, 26 electrons
    e. 54 protons, 82 neutrons, 54 electrons

4.91 a. ^{9}Be **b.** ^{26}Mg **c.** ^{46}Ca
    d. ^{70}Zn **e.** ^{63}Cu

4.93

Name	Nuclear Symbol	Number of Protons	Number of Neutrons	Number of Electrons
sulfur	$^{34}_{16}\text{S}$	16	18	16
zinc	$^{70}_{30}\text{Zn}$	30	40	30
magnesium	$^{26}_{12}\text{Mg}$	12	14	12
radon	$^{220}_{86}\text{Rn}$	86	134	86

4.95 a. 26 protons, 30 neutrons, 26 electrons
    b. $^{51}_{26}\text{Fe}$ **c.** $^{51}_{24}\text{Cr}$

4.97 a. $^{16}_{8}\text{X}$, $^{17}_{8}\text{X}$, $^{18}_{8}\text{X}$ all have 8 protons.
    b. $^{16}_{8}\text{X}$, $^{17}_{8}\text{X}$, $^{18}_{8}\text{X}$ are all isotopes of O.
    c. $^{16}_{8}\text{X}$ and $^{16}_{9}\text{X}$ have mass number 16, and $^{18}_{10}\text{X}$ and $^{18}_{8}\text{X}$ have a mass number of 18.
    d. $^{16}_{8}\text{X}$ and $^{18}_{10}\text{X}$ both have 8 neutrons.

4.99 a. 82 protons, 126 neutrons, 82 electrons
    b. ^{214}Pb
    c. ^{214}Bi, bismuth

4.101 a. N Group 5A (15), 2, 5, $\cdot\ddot{\text{N}}\cdot$
    b. Na Group 1A (1), 2, 8, 1, Na$\cdot$
    c. S Group 6A (16), 2, 8, 6, $\cdot\ddot{\text{S}}\cdot$
    d. B Group 3A (13), 2, 3, $\cdot\text{B}\cdot$

4.103 Calcium has a greater number of protons than K. The least tightly bound electron in Ca is further from the nucleus than in Mg and needs less energy to remove.

4.105 a. Na   **b.** F
    c. Na   **d.** Cl

4.107 $0.0140(203.97) + 0.2410(205.97) + 0.2210(206.98) + 0.5240(207.98) = 207.22$ amu

4.109 28.09 amu

4.111 8.09×10^{7} sodium atoms

Names and Formulas of Compounds

"*Dentures replace natural teeth that are extracted due to cavities, bad gums, or trauma,*" *says Dr. Irene Hilton, dentist, La Clinica De La Raza.* "*I make an impression of teeth using alginate, which is a polysaccharide extracted from seaweed. I mix the compound with water and place the gel-like material in the patient's mouth, where it becomes a hard, cement-like substance. I fill this mold with gypsum ($CaSO_4$) and water, which form a solid to which I add teeth made of plastic or porcelain. When I get a good match to the patient's own teeth, I prepare a preliminary wax denture. This is placed in the patient's mouth to check the bite and adjust the position of the replacement teeth. Then a permanent denture is made using a hard plastic polymer (methyl methacrylate).*"

the Chemistry place

Visit **www.aw-bc.com/chemplace** for extra quizzes, interactive tutorials, career resources, PowerPoint slides for chapter review, math help, and case studies.

In nature, atoms of almost all the elements on the periodic table are found in combination with other atoms. Only the atoms of the noble gases—He, Ne, Ar, Kr, Xe, and Rn—are found as individual atoms. Some elements consist of molecules of two or more atoms of the same element. When atoms of one element combine with atoms of a different element, they form a compound. In an ionic compound, one or more electrons are transferred from the atoms of metals to atoms of nonmetals. When metal atoms lose electrons, they become positively charged ions, while atoms of nonmetals gain electrons and become negative ions. The attraction that results between the oppositely charged particles is called an ionic bond.

We use many ionic compounds every day. When we cook or bake we use ionic compounds such as salt, $NaCl$, and baking soda, $NaHCO_3$. Epsom salts, $MgSO_4$, may be used to soak sore feet. Milk of magnesia, $Mg(OH)_2$, or calcium carbonate, $CaCO_3$, may be taken to settle an upset stomach. In a mineral supplement, iron is present as iron(II) sulfate, $FeSO_4$. Certain sunscreens contain zinc oxide, ZnO, and the tin(II) fluoride, SnF_2, in toothpaste provides fluoride to help prevent tooth decay.

The structures of ionic crystals result in the beautiful facets seen in gems. Sapphires and rubies are made of aluminum oxide, Al_2O_3: impurities of chromium make rubies red, and iron and titanium make sapphires blue.

In compounds of nonmetals, covalent bonding occurs by atoms sharing one or more valence electrons. There are many more covalent compounds than there are ionic ones and many simple covalent compounds are present in our everyday lives. For example, water (H_2O), oxygen (O_2), and carbon dioxide (CO_2) are all covalent compounds.

Covalent compounds consist of molecules, which are discrete groups of atoms. A molecule of oxygen gas (O_2) consists of two oxygen atoms; a molecule of water (H_2O) consists of two atoms of hydrogen and one atom of oxygen. Your food contains much bigger molecules such as starch, which contains many covalent bonds between atoms of carbon, hydrogen, and oxygen. Carbohydrates break down in digestion to provide us with glucose $(C_6H_{12}O_6)$ for energy. When you have iced tea, perhaps you add molecules of sugar (sucrose), which is a covalent compound $(C_{12}H_{22}O_{11})$. Other covalent compounds include propane C_3H_8, alcohol C_2H_6O, the antibiotic amoxicillin $(C_{16}H_{19}N_3O_5S)$, and the antidepressant Prozac $(C_{17}H_{18}F_3NO)$.

5.1 Octet Rule and Ions

Learning Goal

Using the octet rule, write the symbols of the simple ions for the representative elements.

Most of the elements on the periodic table combine to form compounds. **Compounds** result from the formation of chemical bonds between two or more different elements. In ionic bonding, electrons are transferred from atoms of metals to atoms of nonmetals. In covalent bonds, which usually form between atoms of nonmetals, valence electrons are shared. In both ionic and covalent compounds, the atoms tend to acquire the electron arrangement of the nearest noble gas. This is known as the **octet rule** because atoms form compounds by losing, gaining, or sharing electrons to acquire an octet of 8 valence electrons. A few elements achieve the stability of helium with 2 valence electrons.

Noble gases may be used when it is necessary to have a substance that is unreactive. Scuba divers normally use a pressurized mixture of nitrogen and oxygen gases for breathing under water. However, when the air mixture is used at depths where pressure is high, the nitrogen gas is absorbed into the blood, where it can cause mental disorientation. To avoid this problem, a breathing mixture of oxygen and helium may be substituted. The diver still obtains the necessary oxygen, but the unreactive helium that dissolves in the blood does not cause mental disorientation. However, its lower density does change the vibrations of the vocal cords, and the diver will sound like Donald Duck.

Helium is also used to fill blimps and balloons. When dirigibles were first designed, they were filled with hydrogen, a very light gas. However, contact with a spark or heating source could cause a violent explosion because of the extreme reactivity of hydrogen gas with oxygen present in the air. Today blimps are

filled with unreactive helium gas, which presents no danger of explosion.

Lighting bulbs are generally filled with a noble gas such as argon. While the electrically heated filaments that produce the light get very hot, the surrounding noble gases do not react with the hot filament. If heated in air, the filament would soon burn up.

Positive Ions

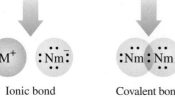

Loss and gain of electrons   Sharing electrons

Ionic bond   Covalent bond

M is a metal
Nm is a nonmetal

In ionic bonding, the valence electrons of a metal are transferred to a non-metal. Because the ionization energies of metals of Groups 1A (1), 2A (2), and 3A (13) are low, these metal atoms readily lose their valence electrons to nonmetals. In doing so, they acquire the electron arrangement of a noble gas (usually 8 valence electrons) and form **ions** with positive charges. For example, when a sodium atom loses its single valence electron, the remaining electrons have the noble gas arrangement of neon. By losing an electron, sodium has 10 electrons instead of 11. Because there are still 11 protons in its nucleus, the atom is no longer neutral. It has become a sodium ion and has an electrical charge, called an **ionic charge,** of 1+. In the symbol for the sodium ion, the ionic charge is written in the upper right-hand corner as Na^+.

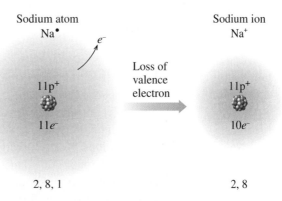

Positive ions are also called **cations** (pronounced *cat'-i-ons*). Magnesium, a metal in Group 2A (2), attains an octet by losing 2 valence electrons and forming an ion with a 2+ ionic charge, Mg^{2+}. A metal ion is named by its element name. Thus, Mg^{2+} is named the magnesium ion.

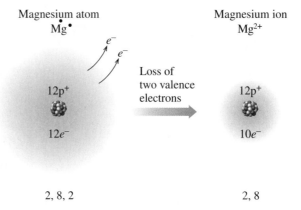

Magnesium atom
Mg·

Magnesium ion
Mg^{2+}

12p⁺

Loss of
two valence
electrons

12p⁺

12e⁻

10e⁻

2, 8, 2

2, 8

Negative Ions

Because the ionization energies of nonmetals of Groups 5A (15), 6A (16), and 7A (17) are high, nonmetal atoms gain electrons from metals. Most nonmetals gain 1 or more electrons to become ions with negative charges. For example, when an atom of chlorine in Group 7A (17) with 7 valence electrons gains an electron, it attains the arrangement of argon. The resulting chloride ion has a negative 1− charge (Cl^-). A nonmetal ion is named by changing the end of its name to *ide*. Negatively charged ions are also called **anions** (pronounced *an'-i-ons*).

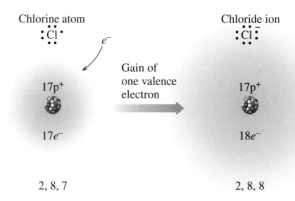

Chlorine atom
:Cl·

Chloride ion
:Cl:⁻

17p⁺

Gain of
one valence
electron

17p⁺

17e⁻

18e⁻

2, 8, 7

2, 8, 8

Ionic Charges from Group Numbers

Group numbers can be used to determine the ionic charges for ions of the representative elements. The elements in Groups 1A (1), 2A (2), and 3A (13) lose 1, 2, and 3 electrons, respectively, to form positive ions. Group 1A (1) metals form ions with 1+ charges, Group 2A (2) metals form ions with 2+ charges, and Group 3A (13) metals form ions with 3+ charges.

The nonmetals from Groups 5A (15), 6A (16), and 7A (17) gain 3, 2, or 1 electron, respectively, to form negative ions in ionic compounds. Group 5A (15) nonmetals form ions with 3− charges, Group 6A (16) nonmetals form ions with 2− charges, and Group 7A (17) nonmetals form ions with 1− charges. The elements of Group 4A (14) do not typically form ions. Table 5.1 lists the ionic charges for typical ions of these representative elements.

Table 5.1 Positive and Negative Ions Have the Same Electron Arrangement as the Nearest Noble Gases

| Noble Gases | Electron Arrangement | Metals Lose Valence Electrons | | | Nonmetals Gain Valence Electrons | | | Electron Arrangement | Noble Gases |
		1A (1)	2A (2)	3A (13)	5A (15)	6A (16)	7A (17)		
He	⇐	Li^+							
Ne	⇐	Na^+	Mg^{2+}	Al^{3+}	N^{3-}	O^{2-}	F^-	⇒	Ne
Ar	⇐	K^+	Ca^{2+}		P^{3-}	S^{2-}	Cl^-	⇒	Ar
Kr	⇐	Rb^+	Sr^{2+}				Br^-	⇒	Kr
Xe	⇐	Cs^+	Ba^{2+}				I^-	⇒	Xe

Sample Problem 5.1   Writing Ions

Consider the elements aluminum and oxygen.
a. Identify each as a metal or a nonmetal.
b. State the number of valence electrons for each.
c. State the number of electrons that must be lost or gained for each.
d. Write the symbol of each resulting ion, including its ionic charge.

Solution

Aluminum
a. metal
b. three
c. loses $3e^-$
d. Al^{3+}

Oxygen
nonmetal
six
gains $2e^-$
O^{2-}

Study Check

What are the symbols for the ions formed by potassium and sulfur?

Questions and Problems   Octet Rule and Ions

5.1 a. How does the octet rule explain the formation of a sodium ion?
b. Why do you think Group 1A (1) and Group 2A (2) elements are found in many compounds, but not Group 8A (18) elements?

5.2 a. How does the octet rule explain the formation of a chloride ion?

b. Why do you think Group 7A (17) elements are found in many compounds, but not Group 8A (18) elements?

5.3 State the number of electrons that must be lost by atoms of each of the following elements to acquire a noble gas electron arrangement:
a. Li   b. Mg   c. Al   d. Cs   e. Ba

5.4   State the number of electrons that must be gained by atoms of each of the following elements to acquire a noble gas electron arrangement:
  a. Cl   **b.** O   **c.** N   **d.** I   **e.** P

5.5   What noble gas has the same electron arrangement as each of the following ions?
  a. Na^+ **b.** Mg^{2+} **c.** K^+ **d.** O^{2-} **e.** F^-

5.6   What noble gas has the same electron arrangement as each of the following ions?
  a. Li^+ **b.** Sr^{2+} **c.** S^{2-} **d.** Al^{3+} **e.** Br^-

5.7   State the number of electrons lost or gained when the following elements form ions:
  a. Mg   **b.** P   **c.** Group 7A   **d.** Na   **e.** Al

5.8   State the number of electrons lost or gained when the following elements form ions:
  a. O   **b.** Group 2A   **c.** F   **d.** Li   **e.** N

5.9   Write the symbols of the ions with the following number of protons and electrons:
  a.   3 protons, 2 electrons
  b.   9 protons, 10 electrons
  c.   12 protons, 10 electrons
  d.   26 protons, 23 electrons
  e.   30 protons, 28 electrons

5.10   How many protons and electrons are in the following ions?
  a. O^{2-} **b.** K^+ **c.** Br^- **d.** S^{2-} **e.** Sr^{2+}

CHEM NOTE

SOME IMPORTANT IONS IN THE BODY

A number of ions in body fluids have important physiological and metabolic functions. Some of them are listed in Table 5.2.

Table 5.2   Ions in the Body

Ion	Occurrence	Function	Source	Result of Too Little	Result of Too Much
Na^+	Principal cation outside the cell	Regulation and control of body fluids	Salt	Hyponatremia, anxiety, diarrhea, circulatory failure, decrease in body fluid	Hypernatremia, little urine, thirst, edema
K^+	Principal cation inside the cell	Regulation of body fluids and cellular functions	Bananas, orange juice, milk, prunes, potatoes	Hypokalemia (hypopotassemia), lethargy, muscle weakness, failure of neurological impulses	Hyperkalemia (hyperpotassemia), irritability, nausea, little urine, cardiac arrest
Ca^{2+}	Cation outside the cell; 90% of calcium in the body in bone as $Ca_3(PO_4)_2$ or $CaCO_3$	Major cation of bone; muscle smoothant	Milk, yogurt, cheese, greens, and spinach	Hypocalcemia, tingling fingertips, muscle cramps, osteoporosis	Hypercalcemia, relaxed muscles, kidney stones, deep bone pain
Mg^{2+}	Cation outside the cell: 70% of magnesium in the body in bone structure	Essential for certain enzymes, muscles, and nerve control	Widely distributed (part of chlorophyll of all green plants), nuts, whole grains	Disorientation, hypertension, tremors, slow pulse	Drowsiness
Cl^-	Principal anion outside the cell	Gastric juice, regulation of body fluids	Salt	Same as for Na^+	Same as for Na^+

5.2 Ionic Compounds

Learning **Goal**

Using charge balance, write the correct formula for an ionic compound.

Ionic compounds consist of positive and negative ions. The ions are held together by strong electrical attractions between the opposite charges called **ionic bonds.**

Properties of Ionic Compounds

The physical and chemical properties of an ionic compound such as NaCl are very different from those of the original elements. For example, the original elements of NaCl were sodium, a soft, shiny metal, and chlorine, a yellow-green poisonous gas. Yet, as positive and negative ions, they form table salt, NaCl, a white, crystalline substance that is common in our diet. In ionic compounds, the attraction between the ions is very strong, which makes the melting points of ionic compounds high, often more than 300°C. For example, the melting point of NaCl is 800°C. At room temperature, ionic compounds are solids.

The structure of an ionic solid depends on the arrangement of the ions. In a crystal of NaCl, which has a cubic shape, the larger Cl^- ions are packed close together in a lattice structure, as shown in Figure 5.1. The smaller Na^+ ions occupy the holes between the Cl^- ions.

Charge Balance in Ionic Compounds

The **formula** of an ionic compound indicates the number and kinds of ions that make up the ionic compound. The sum of the ionic charges in the formula is always zero. For example, the NaCl formula indicates that there is one sodium ion, Na^+, for every chloride ion, Cl^-, in the compound. Note that the ionic charges do not appear in the formula of the compound.

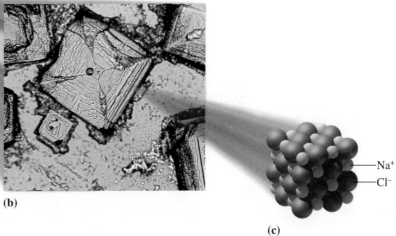

Figure 5.1 (a) The elements sodium and chlorine react to form the ionic compound sodium chloride, the compound that makes up table salt. **(b)** Crystals of NaCl under magnification. **(c)** A diagram of the arrangements of Na^+ and Cl^- packed together in a NaCl crystal.

Q What is the type of bonding between Na^+ and Cl^- ions in salt?

(a)

(b)

(c)

Na^+
Cl^-

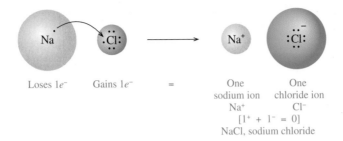

Subscripts in Formulas

The formula of any ionic compound has a zero overall charge. That means that the total amount of positive charge is equal to the total amount of negative charge. Consider a compound of magnesium and chlorine. To achieve an octet, a Mg atom loses its two valence electrons to form Mg^{2+}. Each Cl atom gains 1 electron to complete the octet and form Cl^-. In this example, two Cl^- ions are needed to balance the positive charge of Mg^{2+}. This gives the formula, $MgCl_2$, magnesium chloride, in which a subscript of 2 shows that two Cl^- were needed for charge balance.

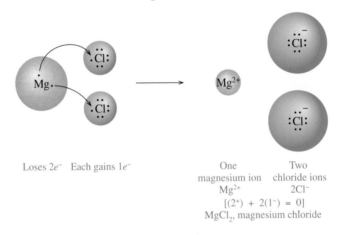

Sample Problem 5.2   **Diagramming an Ionic Compound**

Diagram the formation of the ionic compound aluminum fluoride, AlF_3.

Solution

In their electron-dot symbols, aluminum has 3 valence electrons and fluorine has 7. The aluminum loses its 3 valence electrons, and each fluorine atom gains an electron, to give ions with noble gas arrangements in the ionic compound AlF_3.

Study Check

Diagram the formation of lithium sulfide, Li_2S.

Writing Ionic Formulas from Ionic Charges

The subscripts in the formula of an ionic compound represent the number of positive and negative ions that give an overall charge of zero. Thus, we can now write a formula directly from the ionic charges of the positive and negative ions. Suppose we wish to write the formula of the ionic compound containing Na^+ and S^{2-} ions. To balance the ionic charge of the S^{2-} ion, we will need to place two Na^+ ions in the formula. This gives the formula Na_2S, which has an overall charge of zero. In an ionic formula, the positive ion is written first, followed by the negative ion.

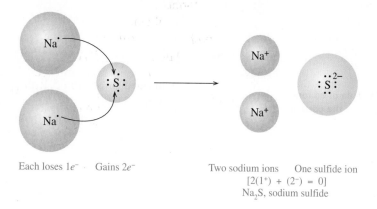

Each loses $1e^-$ Gains $2e^-$ Two sodium ions   One sulfide ion
$$[2(1^+) + (2^-) = 0]$$
Na_2S, sodium sulfide

Sample Problem 5.3   **Writing Formulas from Ionic Charges**

Use ionic charge balance to write the formula for the ionic compound containing K^+ and N^{3-}.

Solution

Determine the number of each ion needed for charge balance. The charge for nitrogen (3–) is balanced by three K^+ ions (3+). Writing the positive ion first gives the formula K_3N.

Study Check

Use ionic charges to determine the formula of the compound that would form when calcium and chlorine react.

Questions and Problems   **Ionic Compounds**

5.11 Which of the following pairs of elements are likely to form an ionic compound?
   a. lithium and chlorine
   b. oxygen and chlorine
   c. potassium and oxygen
   d. sodium and neon
   e. sodium and magnesium

5.12 Which of the following pairs of elements are likely to form an ionic compound?
   a. helium and oxygen
   b. magnesium and chlorine

   c. chlorine and bromine
   d. potassium and sulfur
   e. sodium and potassium

5.13 Using electron-dot symbols, diagram the formation of the following ionic compounds:
   a. KCl   **b.** $CaCl_2$ **c.** Na_3N

5.14 Using electron-dot symbols, diagram the formation of the following ionic compounds:
   a. MgS   **b.** $AlCl_3$ **c.** Li_2O

5.15 Write the correct ionic formula for compounds formed between the following ions:
  a. Na^+ and O^{2-}
  b. Al^{3+} and Br^-
  c. Ba^{2+} and O^{2-}
  d. Mg^{2+} and Cl^-
  e. Al^{3+} and S^{2-}

5.16 Write the correct ionic formula for compounds formed between the following ions:
  a. Al^{3+} and Cl^-
  b. Ca^{2+} and S^{2-}
  c. Li^+ and S^{2-}
  d. K^+ and N^{3-}
  e. K^+ and I^-

5.17 Write the correct formula for ionic compounds formed by the following:
  a. sodium and sulfur
  b. potassium and nitrogen
  c. aluminum and iodine
  d. lithium and oxygen

5.18 Write the correct formula for ionic compounds formed by the following:
  a. calcium and chlorine
  b. barium and bromine
  c. sodium and phosphorus
  d. magnesium and oxygen

5.3 Naming and Writing Ionic Formulas

Learning Goal

Given the formula of an ionic compound, write the correct name.

As we mentioned in Section 5.1, the name of a metal ion is the same as its elemental name. The name of a nonmetal ion is obtained by replacing the end of its elemental name with *ide*. Table 5.3 lists the names of some important metal and nonmetal ions.

Naming Ionic Compounds Containing Two Elements

In the name of an ionic compound made up of two elements, the metal ion is named followed by the name of the nonmetal ion. Subscripts are never mentioned; they are understood as a result of the charge balance of the ions in the compound.

Guide to Naming Ionic Compounds with Metals That Form a Single Ion

STEP 1 Identify the cation and anion.

STEP 2 Name the cation by its element name.

STEP 3 Name the anion by changing the last part of its element name to *ide*.

STEP 4 Write the name of the cation first and the name of the anion second.

Compound	Metal Ion	Nonmetal Ion	Name
NaF	Na^+ sodium	F^- fluoride	sodium fluoride
$MgBr_2$	Mg^{2+} magnesium	Br^- bromide	magnesium bromide
Al_2O_3	Al^{3+} aluminum	O^{2-} oxide	aluminum oxide

Table 5.3 Formulas and Names of Some Common Ions

Group Number	Formula of Ion	Name of Ion	Group Number	Formula of Ion	Name of Ion
	Metals			Nonmetals	
1A (1)	Li^+	lithium	5A (15)	N^{3-}	nitride
	Na^+	sodium		P^{3-}	phosphide
	K^+	potassium	6A (16)	O^{2-}	oxide
2A (2)	Mg^{2+}	magnesium		S^{2-}	sulfide
	Ca^{2+}	calcium	7A (17)	F^-	fluoride
	Ba^{2+}	barium		Cl^-	chloride
3A (13)	Al^{3+}	aluminum		Br^-	bromide
				I^-	iodide

> **Sample Problem 5.4**   **Naming Ionic Compounds**
>
> Write the name of each of the following ionic compounds:
> **a.** Na_2O **b.** Mg_3N_2
>
> *Solution*

Compound	Ions and Names		Name of Compound
a. Na_2O	Na^+	O^{2-}	
	sodium	oxide	sodium oxide
b. Mg_3N_2	Mg^{2+}	N^{3-}	
	magnesium	nitride	magnesium nitride

> *Study Check*
>
> Name the compound $CaCl_2$.

Metals That Form More Than One Positive Ion

The transition metals in Group B (3–12) and the metals in Group 4A (14) and Group 5A (15) also form positive ions. However, these metals typically form more than one positive ion. For example, in some ionic compounds, iron is in the Fe^{2+} form, but in different compounds, it takes the Fe^{3+} form. Copper also forms two different ions: Cu^+ is present in some compounds and Cu^{2+} in others. When a metal can form more than one ion, it is not possible to predict the ionic charge from the group number. We say that it has a variable charge.

When different ions are possible, a naming system is needed to identify the particular cation in a compound. To do this, a Roman numeral that matches the ionic charge is placed in parentheses after the elemental name of the metal. For example, Fe^{2+} is named iron(II), and Fe^{3+} is named iron(III). Table 5.4 lists the ions of some common metals that produce more than one ion.

Figure 5.2 shows some common ions and their location on the periodic table. Typically, the transition metals form more than one positive ion. However, zinc, cadmium, and silver form only one ion. The ionic charges of silver, cadmium, and zinc are fixed like Group 1A (1), 2A (2), and 3A (13) metals, so their elemental names are sufficient when naming their ionic compounds.

When a Roman numeral in parentheses must be included, the selection of the correct Roman numeral depends upon the calculation of the ionic charge of the metal in the formula of a specific ionic compound. For example, we know that in the formula $CuCl_2$ the positive charge of the copper ion must balance the negative charge of two chloride ions. Because we know that chloride ions each have a $1-$ charge, there must be a total negative charge of $2-$. Balancing the $2-$ by the positive charge gives a charge of $2+$ for Cu (a Cu^{2+} ion):

$CuCl_2$

Cu charge + Cl^- charge = 0

(?)   + 2(1−)   = 0

(2+)   + 2−   = 0

Because copper forms ions, Cu^+ and Cu^{2+}, we need to use the Roman numeral system and place (II) after *copper* when naming the compound:

copper(II) chloride

Table 5.5 lists names of some ionic compounds in which the metals form more than one positive ion.

Table 5.4   Some Metals That Form More Than One Positive Ion

Element	Possible Ions	Name of Ion
chromium	Cr^{2+}	chromium(II)
	Cr^{3+}	chromium(III)
cobalt	Co^{2+}	cobalt(II)
	Co^{3+}	cobalt(III)
copper	Cu^+	copper(I)
	Cu^{2+}	copper(II)
gold	Au^+	gold(I)
	Au^{3+}	gold(III)
iron	Fe^{2+}	iron(II)
	Fe^{3+}	iron(III)
lead	Pb^{2+}	lead(II)
	Pb^{4+}	lead(IV)
manganese	Mn^{2+}	manganese(II)
	Mn^{3+}	manganese(III)
mercury	Hg_2^{2+}	mercury(I)*
	Hg^{2+}	mercury(II)
nickel	Ni^{2+}	nickel(II)
	Ni^{3+}	nickel(III)
tin	Sn^{2+}	tin(II)
	Sn^{4+}	tin(IV)

*mercury(I) ions form pairs with $2+$ charge

Table 5.5   Some Ionic Compounds of Metals That Form Two Kinds of Positive Ions

Compound	Systematic Name
$FeCl_2$	iron(II) chloride
$FeCl_3$	iron(III) chloride
Cu_2S	copper(I) sulfide
$CuCl_2$	copper(II) chloride
$SnCl_2$	tin(II) chloride
$PbBr_4$	lead(IV) bromide

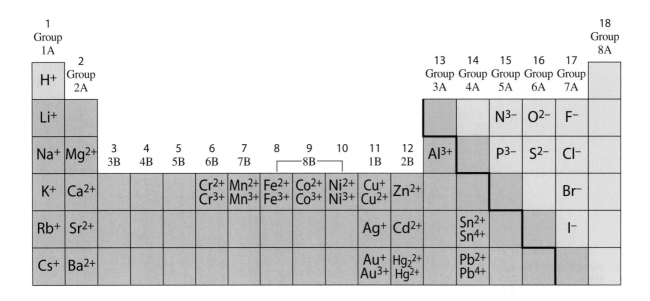

Metals   Metalloids   Nonmetals

Figure 5.2 On the periodic table, positive ions are produced from metals and negative ions are produced from nonmetals.

Q *What are the typical ions produced by calcium, copper, and oxygen?*

Guide to Naming Ionic Compounds with Variable Charge Metals
STEP 1 Determine the charge of the cation from the anion.
STEP 2 Name the cation by its element name and a Roman numeral in parentheses for the charge.
STEP 3 Name the anion by changing the last part of its element name to *ide*.
STEP 4 Write the name of the cation first and the name of the anion second.

Sample Problem 5.5   **Naming Ionic Compounds with Variable Charge Metal Ions**

Write the name for Cu_2S.

Solution

STEP 1 **Determine the charge of the cation from the anion.**   The nonmetal S in Group 6A (16) forms the S^{2-} ion. Because there are two Cu ions to balance the S^{2-}, the charge of each Cu ion is 1+.

	Cation	**Anion**
Ions	Cu?	S^{2-}
Groups	transition	6A (16)
Charge balance	2(+1) + (2−) = 0	
Symbols	Cu^+	S^{2-}

STEP 2 **Name the cation by its element name and use a Roman numeral in parentheses for the charge.**

copper(I)

STEP 3 **Name the anion by changing the last part of its element names to *ide*.**

sulfide

STEP 4 **Write the name of the cation first and the name of the anion second.**

copper(I) sulfide

Study Check

Write the name of the compound whose formula is $AuCl_3$.

Writing Formulas from the Name of an Ionic Compound

The formula of an ionic compound is written from the first part of the name that describes the metal ion and the second part that specifies the nonmetal. Subscripts are added as needed to balance the charge. The steps for writing a formula from the name of an ionic compound are shown in Sample Problem 5.6.

Sample Problem 5.6   **Writing Formulas of Ionic Compounds**

Write the formula of each of the following ionic compounds:
a. sodium nitride   **b.** aluminum sulfide

Solution

a. sodium nitride

Guide to Writing Formulas from the Name of an Ionic Compound
STEP 1 Identify the cation and anion.
STEP 2 Balance the charges.
STEP 3 Write the formula, cation first, using subscripts from charge balance.

STEP 1 **Identify the cation and anion.**   The metal Na in Group 1A (1) forms the Na^+ ion. The nonmetal N in Group 5A (15) forms the N^{3-} ion.

	Cation	**Anion**
Ions	sodium	nitride
Groups	1A (1)	5A (15)
Symbols	Na^+	N^{3-}

STEP 2 **Balance the charges.**

$$Na^+ \qquad N^{3-}$$
$$Na^+$$
$$Na^+$$
$$\overline{3(1+) \; + \; 1(3^-) = 0}$$

Becomes a subscript in the formula

STEP 3 **Write the formula, cation first, using the subscripts from the charge balance.**

Na_3N

b. aluminum sulfide

STEP 1 **Identify the cation and anion.**   The metal Al in Group 3A (13) forms the Al^{3+} ion. The nonmetal S in Group 6A (16) forms the S^{2-} ion.

	Cation	**Anion**
Ions	aluminum	sulfide
Groups	3A (13)	6A (16)
Symbols	Al^{3+}	S^{2-}

STEP 2 **Balance the charges.**
Two Al^{3+} ions (6+) are needed to balance the charges of three S^{2-} ions (6−).

$$Al^{3+} \qquad S^{2-}$$
$$Al^{3+} \qquad S^{2-}$$
$$\qquad\qquad S^{2-}$$
$$\overline{2(3+) \; + \; 3(2-) = 0}$$

Use as subscripts in formula

| STEP 3 | **Write the formula, cation first, using the subscripts from the charge balance.** |

Al_2S_3

Study Check

Write the ions and formulas for each of the following ionic compounds:
a. magnesium bromide   **b.** lithium oxide

Sample Problem 5.7   **Writing Formulas of Ionic Compounds**

Write the formula for iron(III) chloride.

Solution

| STEP 1 | **Identify the cation and anion.**   The Roman numeral (III) indicates that the charge of the iron ion is 3+, Fe^{3+}. |

	Cation	**Anion**
Ions	iron(III)	chloride
Groups	transition	7A (17)
Symbols	Fe^{3+}	Cl^-

| STEP 2 | **Balance the charges.** |

$$Fe^{3+} \qquad Cl^-$$
$$Cl^-$$
$$Cl^-$$
$$\overline{1(3+) \quad + \quad 3(1-) = 0}$$

Becomes a subscript in the formula

| STEP 3 | **Write the formula, cation first, using the subscripts from the charge balance.** |

$FeCl_3$

Study Check

Write the correct formula for chromium(III) oxide.

Questions and Problems   **Naming and Writing Ionic Formulas**

5.19 Write the symbol for the ion of each of the following:
   a. chloride   **b.** potassium
   c. oxide   **d.** aluminum

5.20 Write the symbol for the ion of each of the following:
   a. fluoride   **b.** calcium
   c. sodium   **d.** lithium

5.21 What is the name of each of the following ions?
   a. K^+ **b.** S^{2-} **c.** Ca^{2+} **d.** N^{3-}

5.22 What is the name of each of the following ions?
   a. Mg^{2+} **b.** Ba^{2+} **c.** I^- **d.** Cl^-

5.23 Write names for the following ionic compounds:
   a. Al_2O_3 **b.** $CaCl_2$ **c.** Na_2O **d.** Mg_3N_2
   e. KI **f.** BaF_2

5.24 Write names for the following ionic compounds:
   a. $MgCl_2$ **b** K_3P **c.** Li_2S **d.** $LiBr$
   e. MgO **f.** $SrBr_2$

5.25 Why is a Roman numeral placed after the name of most transition metal ions?

5.26 The compound $CaCl_2$ is named calcium chloride; the compound $CuCl_2$ is named copper(II) chloride. Explain why a Roman numeral is used in one name but not the other.

5.27 Write the names of the following Group 4A (14) and transition metal ions (include the Roman numeral when necessary):
a. Fe^{2+} **b.** Cu^{2+} **c.** Zn^{2+} **d.** Pb^{4+}
e. Cr^{3+} **f.** Mn^{2+}

5.28 Write the names of the following Group 4A (14) and transition metal ions (include the Roman numeral when necessary):
a. Ag^+ **b.** Cu^+ **c.** Fe^{3+} **d.** Sn^{2+}
e. Au^{3+} **f.** Ni^{2+}

5.29 Write names for the following ionic compounds:
a. $SnCl_2$ **b.** FeO **c.** Cu_2S **d.** CuS
e. $CdBr_2$ **f.** $HgCl_2$

5.30 Write names for the following ionic compounds:
a. Ag_3P **b.** PbS **c.** SnO_2 **d.** $AuCl_3$
e. Cr_2O_3 **f.** CoS

5.31 Indicate the charge on the metal ion in each of the following:
a. $AuCl_3$ **b.** Fe_2O_3 **c.** PbI_4 **d.** $SnCl_2$

5.32 Indicate the charge on the metal ion in each of the following:
a. $FeCl_2$ **b.** CuO **c.** Fe_2S_3 **d.** AlP

5.33 Write formulas for each of the following ionic compounds:
a. magnesium chloride   **b.** sodium sulfide
c. copper(I) oxide   **d.** zinc phosphide
e. gold(III) nitride

5.34 Write formulas for each of the following ionic compounds:
a. iron(III) oxide   **b.** barium fluoride
c. tin(IV) chloride   **d.** silver sulfide
e. copper(II) chloride

5.35 Write the formula of the following ionic compounds:
a. cobalt(III) chloride   **b.** lead(IV) oxide
c. silver chloride   **d.** calcium nitride
e. copper(I) phosphide   **f.** chromium(II) chloride

5.36 Write the formula of the following ionic compounds:
a. tin(IV) oxide   **b.** iron(III) sulfide
c. manganese(IV) oxide   **d.** chromium(III) iodide
e. lithium nitride   **f.** gold(I) oxide

5.4 Polyatomic Ions

Learning **Goal**

Write a formula of a compound containing a polyatomic ion.

In a **polyatomic ion,** a group of atoms has an electrical charge. The charge is shared among the atoms that form the polyatomic ion. Most polyatomic ions consist of a nonmetal such as phosphorus, sulfur, carbon, or nitrogen covalently bonded to oxygen atoms. These oxygen-containing polyatomic ions have an ionic charge of $1-$, $2-$, or $3-$. Only one of the common polyatomic ions, NH_4^+, is positively charged. Some polyatomic ions are shown in Figure 5.3.

Figure 5.3 Many products contain polyatomic ions, which are groups of atoms that carry an ionic charge.

Q What is the charge of a sulfate ion?

Plaster molding
$CaSO_4$

Fertilizer
$NaNO_3$

$2+$ $2-$ $+$ $-$

Ca^{2+} SO_4^{2-}
Sulfate ion

Na^+ NO_3^-
Nitrate ion

Naming Polyatomic Ions

The names of the most common polyatomic anions end in *ate*. The *ite* ending is used for the names of related ions that have one less oxygen atom. Recognizing these endings will help you identify polyatomic anions in the name of a compound. The hydroxide ion (OH^-) and cyanide ion (CN^-) are exceptions to this naming pattern. There is no easy way to learn polyatomic ions. You will need to memorize the number of oxygen atoms and the charge associated with each ion, as shown in Table 5.6. By memorizing the formulas and the names of the ions shown in the boxes, you can derive the related ions. For example, the sulfate ion is SO_4^{2-}. We write the formula of the sulfite ion, which has one less oxygen atom, as SO_3^{2-}. The formula of hydrogen carbonate, or *bicarbonate*, can be written by placing a hydrogen cation (H^+) in front of the formula for carbonate (CO_3^{2-}) and decreasing the charge from 2− to 1− to give HCO_3^-.

The elements in Group 7A (17) can form more than two types of polyatomic anions. Prefixes are added to the names and the ending is changed to distinguish among these ions. The prefix *per* is used for the polyatomic ion that has one more oxygen atom than the *ate* form of the polyatomic ion. The ending *ate* is changed to *ite* for the ion with one less oxygen. The prefix *hypo* is used for the polyatomic ion that has one less oxygen than in the *ite* form.

Table 5.6   Names and Formulas of Some Common Polyatomic Ions

Nonmetal	Formula of Ion[a]	Name of Ion
hydrogen	OH^-	hydroxide
nitrogen	NH_4^+	ammonium
	$\boxed{NO_3^-}$	**nitrate**
	NO_2^-	nitrite
chlorine	ClO_4^-	perchlorate
	$\boxed{ClO_3^-}$	**chlorate**
	ClO_2^-	chlorite
	ClO^-	hypochlorite
carbon	$\boxed{CO_3^{2-}}$	**carbonate**
	HCO_3^-	hydrogen carbonate (or bicarbonate)
	CN^-	cyanide
	$C_2H_3O_2^-$ (CH_3COO^-)	acetate
	SCN^-	thiocyanate
sulfur	$\boxed{SO_4^{2-}}$	**sulfate**
	HSO_4^-	hydrogen sulfate (or bisulfate)
	SO_3^{2-}	sulfite
	HSO_3^-	hydrogen sulfite (or bisulfite)
phosphorus	$\boxed{PO_4^{3-}}$	**phosphate**
	HPO_4^{2-}	hydrogen phosphate
	$H_2PO_4^-$	dihydrogen phosphate
	PO_3^{3-}	phosphite
chromium	$\boxed{CrO_4^{2-}}$	**chromate**
	$Cr_2O_7^{2-}$	dichromate
manganese	MnO_4^-	permanganate

[a]Boxed formulas are the most common polyatomic ion for that element.

We can see this in the polyatomic ions of chlorine combined with oxygen. Note that all the polyatomic ions of chlorine have the same ionic charge.

ClO_4^-	*per*chlorate ion	one more O than common (*ate*) ion
ClO_3^-	chlorate ion	most common ion
ClO_2^-	chlor*ite* ion	one less O than common (*ate*) ion
ClO^-	*hypo*chlorite ion	one less O than *ite* ion

Writing Formulas for Compounds Containing Polyatomic Ions

No polyatomic ion exists by itself. Like any ion, a polyatomic ion must be associated with ions of opposite charge. The bonding between polyatomic ions and other ions is one of electrical attraction. For example, the compound sodium sulfate consists of sodium ions (Na^+) and sulfate ions (SO_4^{2-}) held together by ionic bonds.

To write correct formulas for compounds containing polyatomic ions, we follow the same rules of charge balance that we used for writing the formulas of simple ionic compounds. The total negative and positive charges must equal zero. For example, consider the formula for a compound containing calcium ions and carbonate ions. The ions are written as

$$Ca^{2+} \qquad CO_3^{2-}$$
calcium ion　carbonate ion

Ionic charge: $(2+) \; + \; (2-) = 0$

Because one ion of each balances the charge, the formula is written as

$$CaCO_3$$
calcium carbonate

When more than one polyatomic ion is needed for charge balance, parentheses are used to enclose the formula of the ion. A subscript is written outside the closing parenthesis to indicate the number of polyatomic ions. Consider the formula for magnesium nitrate. The ions in this compound are the magnesium ion and the nitrate ion, a polyatomic ion.

$$Mg^{2+} \qquad NO_3^-$$
magnesium ion　nitrate ion

To balance the positive charge of 2+, two nitrate ions are needed. The formula, including the parentheses around the nitrate ion, is as follows:

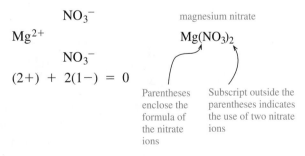

$$NO_3^-$$
$$Mg^{2+}$$
$$NO_3^-$$
$$(2+) \; + \; 2(1-) = 0$$

magnesium nitrate

$$Mg(NO_3)_2$$

Parentheses enclose the formula of the nitrate ions

Subscript outside the parentheses indicates the use of two nitrate ions

CHEM NOTE

IONS IN BONE AND TEETH

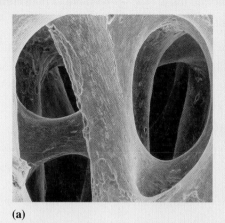

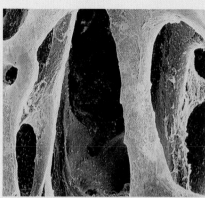

(a)   **(b)**

Bone structure consists of two parts: a solid mineral material and a second phase made up primarily of collagen protein. The mineral substance is a compound called hydroxyapatite, a solid formed from calcium ions, phosphate ions, and hydroxide ions. This material is deposited in the web of collagen to form a very durable bone material.

$$Ca_{10}(PO_4)_6(OH)_2$$

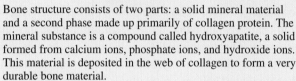

Calcium hydroxyapatite

In most individuals, bone material is continuously being absorbed and reformed. After age 40, more bone material may be lost than formed, a condition called osteoporosis. Bone mass reduction occurs at a faster rate in women than in men and at different rates in different parts of the skeleton. The reduction in bone mass can be as much as 50% over a period of 30 to 40 years. Scanning electron micrographs (SEMs) show **(a)** normal bone and **(b)** bone in osteoporosis due to calcium loss. It is recommended that persons over 35, especially women, include a daily calcium supplement in their diet.

Sample Problem 5.8   **Writing Formulas Having Polyatomic Ions**

Write the formula of aluminum bicarbonate.

Solution

STEP 1 **Identify the cation and anion.** The cation is from aluminum, Al^{3+}, and the cation is bicarbonate, which is a polyatomic anion, HCO_3^-.

	Cation	**Anion**
Ions	Al^{3+}	HCO_3^-

STEP 2 **Balance the charges.**

$$Al^{3+} \qquad HCO_3^-$$
$$HCO_3^-$$
$$HCO_3^-$$
$$\overline{1(3+) \quad + \quad \mathbf{3}(1-) = 0}$$

Becomes a subscript in the formula

STEP 3 **Write the formula, cation first, using the subscripts from the charge balance.** The formula for the compound is written by enclosing the formula of the bicarbonate ion, HCO_3^-, in parentheses and writing the subscript 3 outside the last parenthesis.

$$Al(HCO_3)_3$$

Study Check

Write the formula for a compound containing ammonium ions and phosphate ions.

Naming Compounds Containing Polyatomic Ions

When naming ionic compounds containing polyatomic ions, we write the positive ion, usually a metal, first, and then we write the name of the polyatomic ion. It is important that you learn to recognize the polyatomic ion in the formula and name it correctly. As with other ionic compounds, no prefixes are used.

Na_2SO_4	$FePO_4$	$Al_2(CO_3)_3$
Na₂ SO₄	Fe PO₄	Al₂(CO₃)₃
sodium sulfate	iron(III) phosphate	aluminum carbonate

Table 5.7 lists the formulas and names of some ionic compounds that include polyatomic ions and also gives their uses.

Table 5.7   Some Compounds That Contain Polyatomic Ions

Formula	Name	Use
$BaSO_4$	barium sulfate	x-ray contrast medium
$CaCO_3$	calcium carbonate	antacid, calcium supplement
$Ca_3(PO_4)_2$	calcium phosphate	calcium replenisher
$CaSO_3$	calcium sulfite	preservative in cider and fruit juices
$CaSO_4$	calcium sulfate	plaster casts
$AgNO_3$	silver nitrate	topical anti-infective
$NaHCO_3$	sodium bicarbonate	antacid
$Zn_3(PO_4)_2$	zinc phosphate	dental cements
$FePO_4$	iron(III) phosphate	food and bread enrichment
K_2CO_3	potassium carbonate	alkalizer, diuretic
$Al_2(SO_4)_3$	aluminum sulfate	antiperspirant, anti-infective
$AlPO_4$	aluminum phosphate	antacid
$MgSO_4$	magnesium sulfate	cathartic, epsom salts

Sample Problem 5.9   **Naming Compounds Containing Polyatomic Ions**

Name the following ionic compounds:
a. $CaSO_4$ **b.** $Cu(NO_2)_2$ **c.** $KClO_3$
d. $Mn(OH)_2$ **e.** Na_3PO_4

Solution

We can name compounds with polyatomic ions by separating the compound into a cation and anion, which is usually the polyatomic ion.

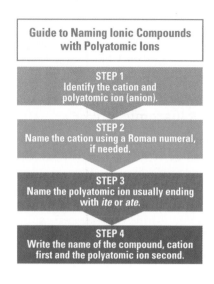

	STEP 1		STEP 2 Name of Cation	STEP 3 Name of Anion	STEP 4 Name of Compound
Formula	**Cation**	**Anion**			
a. $CaSO_4$	Ca^{2+}	SO_4^{2-}	calcium ion	sulfate ion	calcium sulfate
b. $Cu(NO_2)_2$	Cu^{2+}	NO_2^-	copper(II) ion	nitrite ion	copper(II) nitrite
c. $KClO_3$	K^+	ClO_3^-	potassium ion	chlorate ion	potassium chlorate
d. $Mn(OH)_2$	Mn^{2+}	OH^-	manganese (II) ion	hydroxide ion	manganese(II) hydroxide
e. Na_3PO_4	Na^+	PO_4^{3-}	sodium ion	phosphate ion	sodium phosphate

Study Check

What is the name of $Ca_3(PO_4)_2$?

Summary of Naming Ionic Compounds

Throughout this chapter we have examined strategies for naming ionic compounds. Now we can summarize the rules, as illustrated in Figure 5.4. In general, ionic compounds having two elements are named by stating the first element, followed by the second element with an *ide* ending. For ionic compounds, it is necessary to determine whether the metal can form more than one positive ion; if so, a Roman numeral following the name of the metal indicates the particular ionic charge. Ionic compounds having three or more elements include some type of polyatomic ion. They are named by ionic rules, but usually have an *ate* or *ite* ending when the polyatomic ion has a negative charge. Table 5.8 summarizes some naming rules.

Figure 5.4 A flowchart for naming ionic compounds.

 *Why are the names of some metal ions followed by a Roman numeral in the name of a compound?*

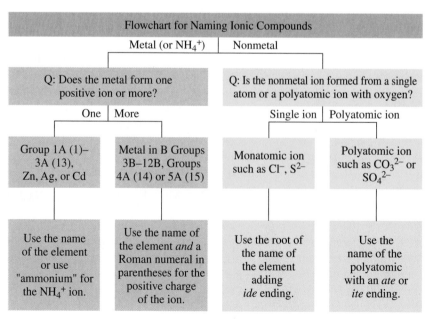

Table 5.8   Rules for Naming Ionic Compounds

Type	Formula Feature	Naming Procedure
Ionic compound (two elements)	Symbol of metal followed by symbol of nonmetal; subscripts used for charge balance.	Use element name for metal; Roman numeral required if more than one positive ion is possible. For nonmetal use element name with *ide* ending.
	Examples: Na_2O Fe_2S_3	*Examples:* sodium oxide iron(III) sulfide
Ionic compound (more than two elements)	Usually symbol of metal followed by a polyatomic ion composed of nonmetals; parentheses may enclose polyatomic ion for charge balance.	Use element name for metal, with Roman numeral if needed, followed by name of polyatomic ion.
	Examples: $Mg(NO_3)_2$ $CuSO_4$ $(NH_4)_2CO_3$	*Examples*: magnesium nitrate copper(II) sulfate ammonium carbonate

Sample Problem 5.10   **Naming Ionic Compounds**

Name the following compounds:

a. Na_3P **b.** $CuSO_4$ **c.** $Cr(ClO)_3$

Solution

Ionic Compound	Ions	Feature	Name
a. Na_3P	Na^+ sodium	forms one ion	sodium phosphide
	P^{3-} phosphide	ion from a single atom	
b. $CuSO_4$	Cu^{2+} copper(II)	two ions possible	copper(II) sulfate
	SO_4^{2-} sulfate	polyatomic ion	
c. $Cr(ClO)_3$	Cr^{3+} chromium(III)	two ions possible	chromium(III) hypochlorite
	ClO^- hypochlorite	polyatomic ion with one O less than chlorite	

Study Check

What is the name of $Fe(NO_3)_2$?

Questions and Problems   ▶   **Polyatomic Ions**

5.37 Write the formulas including the charge for the
following polyatomic ions:
   a. bicarbonate   **b.** ammonium
   c. phosphate   **d.** hydrogen sulfate
   e. hypochlorite
5.38 Write the formulas including the charge for the following polyatomic ions:
   a. nitrite   **b.** sulfite   **c.** hydroxide
   d. phosphite   **e.** acetate
5.39 Name the following polyatomic ions:
   a. SO_4^{2-} **b.** CO_3^{2-} **c.** PO_4^{3-}
   d. NO_3^{-} **e.** ClO_4^{-}
5.40 Name the following polyatomic ions:
   a. OH^{-} **b.** HSO_3^{-} **c.** CN^{-}
   d. NO_2^{-} **e.** CrO_4^{2-}
5.41 Complete the following table with the formula and
name of the compound:

	NO_2^{-}	CO_3^{2-}	HSO_4^{-}	PO_4^{3-}
Li^{+}				
Cu^{2+}				
Ba^{2+}				

5.42 Complete the following table with the formula and
name of the compound:

	NO_3^{-}	HCO_3^{-}	SO_3^{2-}	HPO_4^{2-}
NH_4^{+}				
Al^{3+}				
Pb^{4+}				

5.43 Write the formula for the polyatomic ion in each of the
following and name each compound:
   a. Na_2CO_3 **b.** NH_4Cl **c.** Li_3PO_4
   d. $Cu(NO_2)_2$ **e.** $FeSO_3$ **f.** $KC_2H_3O_2$
5.44 Write the formula for the polyatomic ion in each of the
following and name each compound:
   a. KOH **b.** $NaNO_3$ **c.** $CuCO_3$
   d. $NaHCO_3$ **e.** $BaSO_4$ **f.** $Ca(ClO)_2$
5.45 Write the correct formula for the following compounds:
   a. barium hydroxide   **b.** sodium sulfate
   c. iron(II) nitrate   **d.** zinc phosphate
   e. iron(III) carbonate
5.46 Write the correct formula for the following compounds:
   a. aluminum chlorate   **b.** ammonium oxide
   c. magnesium bicarbonate   **d.** sodium nitrite
   e. copper(I) sulfate
5.47 Name the compounds that are found in the following
sources:
   a. $Al_2(SO_4)_3$ antiperspirant
   b. $CaCO_3$ antacid
   c. Cr_2O_3 green pigment
   d. Na_3PO_4 laxative
   e. $(NH_4)_2SO_4$ fertilizer
   f. Fe_2O_3 pigment
5.48 Name the compounds that are found in the following
sources:
   a. $Co_3(PO_4)_2$ violet pigment
   b. $Mg_3(PO_4)_2$ antacid
   c. $FeSO_4$ iron supplement in vitamins
   d. $MgSO_4$ Epsom salts
   e. Cu_2O fungicide
   f. SnF_2 tooth decay preventative

5.5 Covalent Compounds and Their Names

Learning Goal

> Given the formula of a covalent
> compound, write its correct name;
> given the name of a covalent
> compound, write its formula.

the
Chemistry
place

WEB TUTORIAL
Covalent Bonds

When ionic compounds form, metals lose their valence electrons and nonmetals gain electrons. However, when two nonmetals combine to form
covalent compounds, the valence electrons are not transferred from one
atom to another. Atoms in covalent compounds achieve stability by sharing
their valence electrons. When atoms share electrons, the resulting bond is a
covalent bond.

Formation of a Hydrogen Molecule

The simplest covalent molecule is hydrogen gas, H_2. When two hydrogen
atoms are far apart, they are not attracted to each other. As the atoms move
closer, the positive charge of each nucleus attracts the electron of the other
atom. This attraction pulls the atoms closer until they share a pair of valence
electrons and form a covalent bond. (See Figure 5.5.) In the covalent bond in
H_2, the shared electrons give the noble gas configuration of He to each of the
H atoms. Thus the atoms bonded in H_2 are more stable than two individual H
atoms.

Figure 5.5 A covalent bond forms as H atoms move close together to share electrons.

Q *What determines the attraction between two H atoms?*

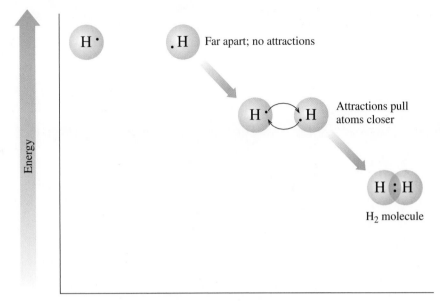

Distance between nuclei decreases ⟶

In covalent compounds, the valence electrons and the shared pairs of electrons can be shown using electron-dot formulas. In an electron-dot formula, a shared pair of electrons is written as two dots or a single line between the atomic symbols. This notation is shown in the formation of the covalent bond in the electron-dot formula for the H_2 molecule.

$$H\bullet \ + \ \bullet H \ \longrightarrow \ H\!:\!H \ \longrightarrow \ H\!-\!H \ = \ H_2$$

| Electrons to share | A shared pair of electrons | A covalent bond | A hydrogen molecule |

Formation of Octets in Covalent Molecules

In most covalent compounds, atoms share electrons to achieve octets for the valence electrons. For example, a fluorine molecule, F_2, consists of two fluorine atoms. Each fluorine atom has 7 valence electrons. By sharing 1 valence electron, each F atom achieves an octet. In the electron-dot formula, the shared electrons, or **bonding pair,** are written between atoms with the non-bonding pairs of electrons, or **lone pairs,** on the outside. This is shown in the formation of the covalent bond for the F_2 molecule.

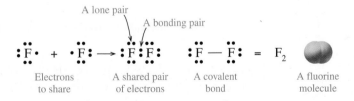

A lone pair

A bonding pair

Electrons to share   A shared pair of electrons   A covalent bond   A fluorine molecule

Hydrogen (H_2) and fluorine (F_2) are examples of nonmetals whose natural state is diatomic; that is, they contain two atoms. The elements listed in Table 5.9 exist naturally as diatomic molecules.

Table 5.9   Elements That Exist as Diatomic, Covalent Molecules

Element	Diatomic Molecule	Name
H	H_2	hydrogen
N	N_2	nitrogen
O	O_2	oxygen
F	F_2	fluorine
Cl	Cl_2	chlorine
Br	Br_2	bromine
I	I_2	iodine

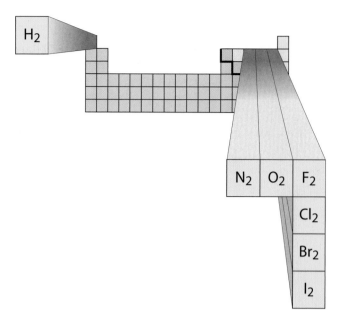

Sharing Electrons Between Atoms of Different Elements

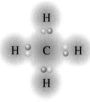

Methane, CH_4

In Period 2, the number of electrons that an atom shares and the number of covalent bonds it forms are usually equal to the number of electrons needed to acquire a noble gas arrangement. For example, carbon has 4 valence electrons. Because carbon needs to acquire 4 more electrons for an octet, it forms 4 covalent bonds by sharing its 4 valence electrons.

Methane, a component of natural gas, is a compound made of carbon and hydrogen. To attain an octet, each carbon shares 4 electrons and each hydrogen shares 1 electron. In this molecule, a carbon atom forms 4 covalent bonds with 4 hydrogen atoms. The electron-dot formula for the molecule is written with the carbon atom in the center and the hydrogen atoms on the sides. Table 5.10 gives the formulas of some covalent molecules for Period 2 elements.

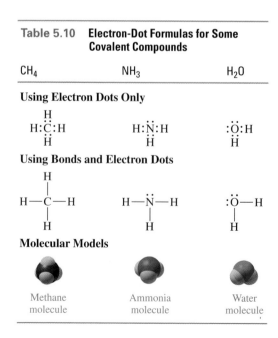

Table 5.10	Electron-Dot Formulas for Some Covalent Compounds	
CH_4	NH_3	H_2O

Using Electron Dots Only

Using Bonds and Electron Dots

Molecular Models

| Methane molecule | Ammonia molecule | Water molecule |

ize !.oks.

Names and Formulas of Covalent Compounds

Unlike ionic compounds, the names of nonmetals in covalent compounds need prefixes because several different compounds can be formed from the same two nonmetals. For example, carbon and oxygen can form two different compounds, carbon monoxide, CO, and carbon dioxide, CO_2. Nitrogen and oxygen also form several different covalent molecules. We could not distinguish between them by using the name *nitrogen oxide*. Therefore, prefixes are used with molecular compounds composed of two nonmetals.

In the name of a covalent compound, the first nonmetal in the formula is named by its elemental name; the second nonmetal is named by its elemental name with the ending changed to *ide*. Subscripts indicating two or more atoms of an element are expressed as prefixes placed in front of each name. Table 5.11 lists some prefixes used in naming covalent compounds.

Table 5.11 Prefixes Used in Naming Covalent Compounds

1 mono	6 hexa
2 di	7 hepta
3 tri	8 octa
4 tetra	9 nona
5 penta	10 deca

Some Covalent Compounds Formed by Nitrogen and Oxygen

NO	nitrogen oxide
N_2O	dinitrogen oxide
N_2O_3	dinitrogen trioxide
N_2O_4	dinitrogen tetroxide
N_2O_5	dinitrogen pentoxide

In the names of covalent compounds, the prefix *mono* is usually omitted. When the vowels *o* and *o* or *a* and *o* appear together, the first vowel is omitted. Table 5.12 lists the formulas, names, and commercial uses of some other covalent compounds.

Table 5.12 Some Common Covalent Compounds

Formula	Name	Commercial Uses
CS_2	carbon disulfide	manufacture of rayon
CO_2	carbon dioxide	carbonation of beverages, fire extinguishers, propellant in aerosols, dry ice
SiO_2	silicon dioxide	manufacture of glass, computer parts
NCl_3	nitrogen trichloride	bleaching of flour in some countries (prohibited in U.S.)
SO_2	sulfur dioxide	preserving fruits, vegetables; disinfectant in breweries; bleaching textiles
SO_3	sulfur trioxide	manufacture of explosives
SF_6	sulfur hexafluoride	electrical circuits (insulation)
ClO_2	chlorine dioxide	bleaching pulp (for making paper), flour, leather
ClF_3	chlorine trifluoride	rocket propellant

Guide to Naming Covalent Compounds
with Two Nonmetals

STEP 1
Name the first nonmetal
by its element name.

STEP 2
Name the second nonmetal by changing
the last part of its element name to *ide*.

STEP 3
Add prefixes to indicate the number
of atoms (subscripts).

Sample Problem 5.11 **Naming Covalent Compounds**

Name the covalent compound P_4O_6.

Solution

STEP 1 **Name the first nonmetal by its element name.** In P_4O_6, the first nonmetal (P) is phosphorus.

STEP 2 **Name the second nonmetal by changing the last part of its element name to *ide*.** The second nonmetal (O) is named oxide.

STEP 3 **Add prefixes to indicate the number of atoms of each nonmetal.** Because there are four P atoms, we use the prefix *tetra* to write *tetraphosphorus*. The six oxygen atoms use the prefix *hexa*, which gives the name *hexoxide*. When the vowels *a* and *o* appear together, as in *hexa + oxide,* the ending (*a*) of the prefix is dropped.

P_4O_6 *tetra*phosphorus *hex*oxide

Study Check

Write the name of each of the following compounds:
a. $SiBr_4$ **b.** Br_2O

Writing Formulas from the Names of Covalent Compounds

In the name of a covalent compound, the names of two nonmetals are given along with prefixes for the number of atoms of each. To obtain a formula, we write the symbol for each element and a subscript if a prefix indicates two or more atoms.

Guide to Writing Formulas for
Covalent Compounds

STEP 1
Write the symbols in the order of the
elements in the name.

STEP 2
Write any prefixes as subscripts.

Sample Problem 5.12 **Writing Formulas for Covalent Compounds**

Write the formula for each of the following covalent compounds:
a. sulfur dichloride   **b.** diboron trioxide

Solution
a. sulfur dichloride

STEP 1 **Write the symbols in order of the elements in the names.** The first nonmetal is sulfur and the second nonmetal is chlorine.

S   Cl

STEP 2 **Write any prefixes as subscripts.** Since there is no prefix for sulfur, we know that there is one atom of sulfur. The prefix *di* in *dichloride* indicates that there are two atoms of chlorine, shown as a subscript 2 in the formula.

SCl_2

b. diboron trioxide

STEP 1 **Write the symbols in order of the elements in the names.** The first nonmetal is boron and the second nonmetal is oxygen.

B   O

STEP 2   **Write any prefixes as subscripts.** The prefix *di* in *diboron* indicates that there are two atoms of boron, shown as a subscript 2 in the formula. The prefix *tri* in *trioxide* indicates that there are three atoms of oxygen, shown as a subscript 3 in the formula.

$$B_2O_3$$

Study Check

What is the formula of iodine pentafluoride?

Summary of Naming Compounds

Throughout this chapter we have examined strategies for naming ionic and covalent compounds. Now we can summarize the rules, as illustrated in Figure 5.6. In general, compounds having two elements are named by stating the first element, followed by the second element with an *ide* ending. If the first element is a metal, the compound is usually ionic; if the first element is a nonmetal, the compound is usually covalent. For ionic compounds, it is necessary to determine whether the metal can form more than one type of positive ion; if so, a Roman numeral following the name of the metal indicates the particular ionic charge. One exception is the ammonium ion, NH_4^+, which is also written first as a positively charged polyatomic ion. In naming covalent compounds having two elements, prefixes are necessary to indicate the number of atoms of each nonmetal as shown in that particular formula. Ionic compounds having three or more elements include some type of polyatomic ion. They are named by ionic rules but have an *ate* or *ite* ending when the polyatomic ion has a negative charge.

Figure 5.6 A flow chart of how ionic and covalent compounds are named.

Q Why does the name sulfur dichloride have a prefix but magnesium chloride does not?

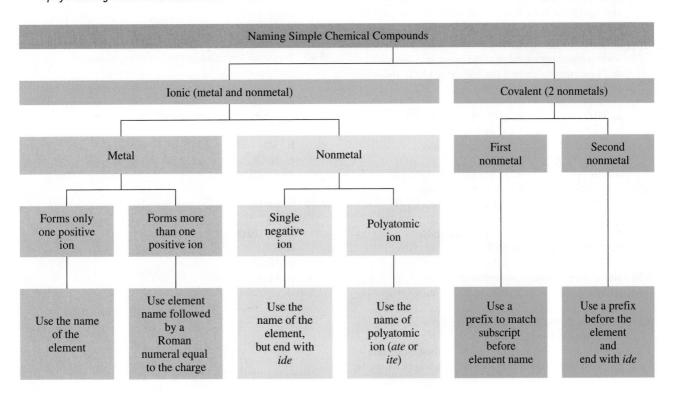

> **Sample Problem 5.13** **Naming Ionic and Covalent Compounds**
>
> Name the following compounds:
> **a.** Ca_3N_2 **b.** Cu_3PO_4 **c.** SO_3
>
> *Solution*
> **a.** Ca_3N_2 is an ionic compound. Ca is a metal that forms a single ion, Ca^{2+}, which is named calcium. The single negative ion, N^{3-}, is named nitride. The compound is named calcium nitride.
> **b.** Cu_3PO_4 is an ionic compound. Cu is a transition metal that forms more than one ion. Each of the positive ions are Cu^+, named copper(I) because 3(+1) balances the 3− charge on the phosphate ion PO_4^{3-}. The compound is named copper(I) phosphate.
> **c.** SO_3 is a covalent compound of two nonmetals. The first element, S, is named sulfur (no prefix is needed). The second element, O, named oxide, has a subscript 3, which requires a prefix *tri*. The compound is named sulfur trioxide.
>
> *Study Check*
> What is the name of $CoCO_3$?

Questions and Problems Covalent Compounds and Their Names

5.49 What elements on the periodic table are most likely to form covalent compounds?

5.50 How does the bond that forms between Na and Cl differ from a bond that forms between N and Cl?

5.51 State the number of valence electrons, bonding pairs, and lone pairs in each of the following electron-dot formulas:
a. H:H   **b.** H:$\ddot{\ddot{Br}}$:   **c.** :$\ddot{\ddot{Br}}$:$\ddot{\ddot{Br}}$:

5.52 State the number of valence electrons, bonding pairs, and lone pairs in each of the following electron-dot formulas:
a. H:$\ddot{\ddot{O}}$:   **b.** H:$\ddot{N}$:H (with H below)   **c.** :$\ddot{\ddot{Br}}$:$\ddot{\ddot{O}}$:$\ddot{\ddot{Br}}$:

5.53 Name the following covalent compounds:
a. PBr_3 **b.** CBr_4 **c.** SiO_2
d. HF   **e.** NI_3

5.54 Name the following covalent compounds:
a. CS_2 **b.** P_2O_5 **c.** Cl_2O
d. PCl_3 **e.** N_2O_4

5.55 Name the following covalent compounds:
a. N_2O_3 **b.** NCl_3 **c.** $SiBr_4$
d. PCl_5 **e.** SO_3

5.56 Name the following covalent compounds:
a. SiF_4 **b.** IBr_3 **c.** CO_2
d. SO_2 **e.** N_2O

5.57 Write the formulas of the following covalent compounds:
a. carbon tetrachloride   **b.** carbon monoxide
c. phosphorus trichloride   **d.** dinitrogen tetroxide

5.58 Write the formulas of the following covalent compounds:
a. sulfur dioxide   **b.** silicon tetrachloride
c. iodine pentafluoride   **d.** dinitrogen oxide

5.59 Write the formulas of the following covalent compounds:
a. oxygen difluoride   **b.** boron trifluoride
c. dinitrogen trioxide   **d.** sulfur hexafluoride

5.60 Write the formulas of the following covalent compounds:
a. sulfur dibromide   **b.** carbon disulfide
c. tetraphosphorus hexoxide   **d.** dinitrogen pentoxide

5.61 Name the following compounds:
a. $AlCl_3$
b. SO_3
c. N_2O
d. $Sn(NO_3)_2$
e. $Cu(ClO_2)_2$

5.62 Name the following compounds:
a. N_2
b. $Mg(BrO)_2$
c. SiF_4
d. $NiSO_4$
e. Fe_2S_3

Concept Map ▶ **Names and Formulas of Compounds**

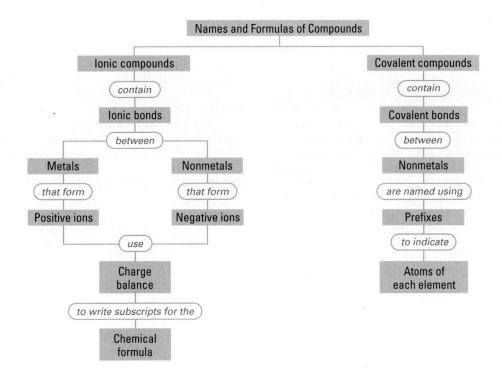

Chapter Review

5.1 Octet Rule and Ions

Metals with low ionization energies lose electrons easily compared to nonmetals. The stability of the noble gases is associated with 8 valence electrons (an octet). Helium is stable with a complete first energy level of 2 electrons. Metals in Groups 1A (1) to 3A (13) form octets by losing electrons and forming cations with 1+, 2+, and 3+ charges, respectively. When reacting with metals, nonmetals in Groups 5A (15), 6A (16), and 7A (17) form octets by gaining electrons and forming anions with 3−, 2−, and 1− charges.

5.2 Ionic Compounds

An ionic compound contains positive and negative ions. The formula of an ionic compound is neutral, which means that the total charge of the positive ions and the total charge of the negative ions is zero. Subscripts are written after the symbol of each ion to balance the sum of positive and negative charges.

5.3 Naming and Writing Ionic Formulas

In naming ionic compounds, the positive ion is given first, followed by the name of the negative ion. Ionic compounds containing two elements end with *ide*. When the metal can

form more than one positive ion, its ionic charge is determined from the total negative charge in the formula. Typically, transition metals form cations with two or more ionic charges. The charge is given as a Roman numeral in the name, such as iron(II) and iron(III) for the cations of iron with 2+ and 3+ ionic charges.

5.4 Polyatomic Ions

A polyatomic ion is a group of atoms that has an electrical charge. Most polyatomic ions contain a nonmetal and one or more oxygen atoms. The common polyatomic ions have charges of 1−, 2−, or 3−, indicating that 1, 2, or 3 electrons were added to complete octets. The ammonium ion, NH_4^+, is a positive polyatomic ion.

5.5 Covalent Compounds and Their Names

Two nonmetals can form two or more different covalent compounds. In a covalent bond, atoms of nonmetals share valence electrons. In most covalent compounds, the atoms achieve a noble gas configuration. In the names of covalent compounds, prefixes are used to indicate the subscripts in the formulas. The ending of the second nonmetal is changed to *ide*.

Key Terms

anion A negatively charged ion with a noble gas configuration, such as Cl^-, O^{2-}, or S^{2-}.

bonding pair A pair of electrons shared between two atoms.

cation A positively charged ion with a noble gas configuration, such as Na^+, Mg^{2+}, or Al^{3+}.

compound A combination of atoms in which noble gas arrangements are attained through electron transfer.

covalent bond A sharing of valence electrons by atoms.

covalent compound A combination of atoms in which noble gas arrangements are attained by electron sharing.

formula The group of symbols representing the elements in a compound with subscripts for the number of each.

ion An atom or group of atoms having an electrical charge because of a loss or gain of electrons.

ionic bond The attraction between oppositely charged ions.

ionic charge The difference between the number of protons (positive) and the number of electrons (negative), written in the upper right corner of the symbol for the ion.

ionic compound A compound of positive and negative ions held together by ionic bonds.

lone pair Electrons in a molecule that are not shared in a bond, but complete the octet for an element.

octet rule Representative elements react with other elements to produce a noble gas configuration with 8 valence electrons.

polyatomic ion A group of atoms that has an electrical charge.

Understanding the Concepts

5.63 Identify each of the following atoms or ions:

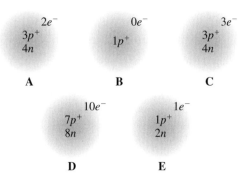

A   B   C   D

5.64 Consider the following electron-dot formulas for elements X and Y.

X·   ·Ÿ·

a. What are the group numbers of X and Y?

b. Will a compound of X and Y be ionic or covalent?

c. What ions would be formed by X and Y?

d. What would be the formula of a compound of X and Y?

e. What would be the formula of a compound of X and chlorine?

f. What would be the formula of a compound of Y and chlorine?

5.65 Identify the following as an atom or ion of
1. H   **2.** Li   **3.** Li^+ **4.** H^+ **5.** N^{3-}

$2e^-$
$3p^+$
$4n$

A

$0e^-$
$1p^+$

B

$3e^-$
$3p^+$
$4n$

C

$10e^-$
$7p^+$
$8n$

D

$1e^-$
$1p^+$
$2n$

E

5.66 As discussed in the Health Note "Polyatomic Ions in Bone and Teeth," the mineral component of bone and teeth is composed of calcium hydroxyapatite, $Ca_{10}(PO_4)_6(OH)_2$. Name the ions in the formula.

5.67 Write the formulas and names of the ionic compounds for the elements indicated by the period and electron-dot symbols in the following table:

Period	Electron-Dot Symbols	Formula of Compound	Name of Compound
2	X· and ·Ÿ·		
4	Ẋ· and :Ÿ·		
3	·Ẋ· and :Ÿ·		

5.68 Write the formulas and names of the ionic compounds for the elements indicated by the period and electron-dot symbols in the following table:

Period	Electron-Dot Symbols	Formula of Compound	Name of Compound
2	$\dot{X}\cdot$ and $\cdot\ddot{Y}\cdot$		
3	$\cdot\dot{X}\cdot$ and $:\ddot{Y}\cdot$		
5	$\dot{X}\cdot$ and $:\ddot{Y}\cdot$		

5.69 Using the electron arrangement for the elements, write the electron-dot symbols of the atoms, the cations and anions that form, and the formulas and names of their ionic compounds.

Electron Arrangements		Electron-Dot Symbols	Cations	Anions	Formula of Compound	Name of Compound
2, 8, 2	2, 5					
2, 8, 8, 1	2, 6					
2, 8, 3	2, 8, 18, 7					

5.70 Write the electron-dot symbols, formulas, and names of their ionic compounds using the electron arrangement for the elements.

Electron Arrangements		Electron-Dot Symbols	Formula of Compound	Name of Compound
2, 1	2, 8, 6			
2, 8, 8, 2	2, 8, 5			
2, 8, 1	2, 8, 7			

Additional Questions and Problems

5.71 What noble gas has the same electron arrangement as the following ions?
a. N^{3-} **b.** Mg^{2+} **c.** Cl^- **d.** O^{2-} **e.** Li^+

5.72 What noble gas has the same electron arrangement as the following ions?
a. Al^{3+} **b.** Br^- **c.** Ca^{2+} **d.** Na^+ **e.** S^{2-}

5.73 Consider an ion with the symbol X^{2+} formed from a representative element.
a. What is the group number of the element?
b. What is the electron-dot symbol of the element?
c. If X is in Period 2, what is the element?
d. What is the formula of the compound formed from X and the nitride ion?

5.74 Consider the following electron-dot symbols of representative elements X and Y:

$X\cdot$ $\cdot\ddot{Y}\cdot$

a. What are the group numbers of X and Y?
b. What ions would be formed by X and Y?
c. What would be the formula of a compound of X and Y?
d. What would be the formula of a compound of X and chlorine?

5.75 One of the ions of tin is tin(IV).
a. What is the symbol for this ion?
b. How many protons and electrons are in the ion?
c. What is the formula of tin(IV) oxide?
d. What is the formula of tin(IV) phosphate?

5.76 One of the ions of gold is gold(III).
a. What is the symbol for this ion?
b. How many protons and electrons are in the ion?
c. What is the formula of gold(III) sulfate?
d. What is the formula of gold(III) chloride?

5.77 Identify the group number in the periodic table of X, a representative element, in each of the following ionic compounds:
a. XCl_3 **b.** Al_2X_3 **c.** XCO_3

5.78 Identify the group number in the periodic table of X, a representative element, in each of the following ionic compounds:
a. X_2O_3 **b.** X_2SO_3 **c.** Na_3X

5.79 Name the following ionic compounds:
a. $FeCl_3$ **b.** $Ca_3(PO_4)_2$ **c.** $Al_2(CO_3)_3$
d. $PbCl_4$ **e.** $MgCO_3$ **f.** $SnSO_4$
g. CuS

5.80 Name the following ionic compounds:
a. $CaSO_4$ **b.** $Ba(NO_3)_2$ **c.** MnS
d. $LiClO_4$ **e.** $CrPO_3$ **f.** Na_2HPO_4
g. $CaCl_2$

5.81 Write the formula for the following ionic compounds:
a. copper(I) nitride
b. potassium hydrogen sulfite
c. lead(IV) sulfide
d. gold(III) carbonate
e. zinc perchlorate

5.82 Write the formula for the following ionic compounds:
  a. iron(III) nitrate
  b. copper(II) hydrogen carbonate
  c. tin(IV) sulfite
  d. barium dihydrogen phosphate
  e. cadmium hypochlorite

5.83 Write the formula of the following ionic compounds:
  a. gold(III) chloride
  b. lead(IV) oxide
  c. silver chloride
  d. calcium nitride
  e. copper(I) phosphide
  f. chromium(II) chloride

5.84 Write the formula of the following ionic compounds:
  a. tin(IV) oxide
  b. iron(III) sulfide
  c. lead(IV) sulfide
  d. chromium(III) iodide
  e. lithium nitride
  f. gold(I) oxide

5.85 Write the name for each of the following:
  a. MgO   **b.** $Cr(HCO_3)_3$ **c.** $Mn_2(CrO_4)_3$

5.86 Write the name for each of the following
  a. Cu_2S **b.** $Fe_3(PO_4)_2$ **c.** $Ca(ClO)_2$

5.87 Name each of the following covalent compounds:
  a. NCl_3 **b.** SCl_2 **c.** N_2O
  d. F_2 **e.** PCl_5 **f.** P_2O_5

5.88 Name each of the following covalent compounds:
  a. CBr_4 **b.** SF_6 **c.** Br_2
  d. N_2O_4 **e.** SO_2 **f.** CS_2

5.89 Give the formula for each of the following:
  a. carbon monoxide
  b. diphosphorus pentoxide
  c. dihydrogen sulfide
  d. sulfur dichloride

5.90 Give the formula for each of the following:
  a. silicon dioxide
  b. carbon tetrabromide
  c. sulfur trioxide
  d. dinitrogen oxide

5.91 Classify each of the following compounds as ionic or covalent, and give its name:
  a. $FeCl_3$ **b.** Na_2SO_4 **c.** N_2O
  d. F_2 **e.** PCl_5 **f.** CF_4

5.92 Classify each of the following compounds as ionic or covalent, and give its name:
  a. $Al_2(CO_3)_3$ **b.** SF_6 **c.** Br_2
  d. Mg_3N_2 **e.** SO_2 **f.** $CrPO_4$

5.93 Write the formulas for the following:
  a. tin(II) carbonate _____
  b. lithium phosphide _____
  c. silicon tetrachloride _____
  d. iron(III) sulfide _____
  e. carbon dioxide _____
  f. calcium bromide _____

5.94 Write the formulas for the following:
  a. sodium carbonate _____
  b. nitrogen dioxide _____
  c. aluminum nitrate _____
  d. copper(I) nitride _____
  e. potassium phosphate _____
  f. lead(IV) oxide _____

5.95 When sodium reacts with sulfur, an ionic compound forms. If a sample of this compound contains 4.8×10^{22} sodium ions, how many sulfide ions does it contain?

5.96 When magnesium reacts with nitrogen, an ionic compound forms. If a sample of this compound contains 4.8×10^{22} magnesium ions, how many nitride ions does it contain?

Challenge Questions

5.97 Why are only the valence electrons of the representative elements involved in the formation of positive and negative ions?

5.98 How does the octet rule determine the loss or gain of electrons by representative elements?

5.99 How are the ions of Group 2A (2) and Group 6A (16) different? How are they similar?

5.100 Identify the errors in the following formulas or names. Write a correct formula or name.
  a. $Ca(NO_3)_2$ is calcium dinitrate.
  b. Copper(II) oxide has the formula Cu_2O.
  c. Potassium carbonate has the formula $(K)_2CO_3$.
  d. Na_2S is sodium sulfate.
  e. Silver sulfate has the formula $Ag_3(SO_4)$.

5.101 Indicate the type of compound (ionic or covalent) and complete the table.

Formula of Compound	Type of Compound	Name of Compound
$FeSO_4$		
		silicon dioxide
		ammonium nitrate
$Al_2(SO_4)_3$		
		cobalt(III) sulfide

5.102 Give the symbol of the noble gas that has the same electron arrangement as
  a. Cl^- **b.** Sr^{2+} **c.** Se^{2-}

5.103 Classify the following compounds as ionic or covalent and name each.
  a. Li_2O **b.** N_2O **c.** CF_4
  d. Cl_2O **e.** MgF_2 **f.** CO
  g. $CaCl_2$ **h.** K_3PO_4

5.104 Name the following compounds:
  a. $FeCl_2$ **b.** Cl_2O_7 **c.** N_2
  d. $Ca_3(PO_4)_2$ **e.** PCl_3 **f.** $Al(NO_3)_2$
  g. $PbCl_4$ **h.** $MgCO_3$ **i.** NO_2
  j. $SnSO_4$ **k.** $Ba(NO_3)_2$ **l.** CuS

Answers

Answers to Study Checks

5.1 K^+ and S^{2-}

5.2

5.3 $CaCl_2$

5.4 calcium chloride

5.5 gold(III) chloride

5.6 a. Mg^{2+}, Br^- $MgBr_2$
  b. Li^+, O^{2-} Li_2O

5.7 Cr_2O_3

5.8 $(NH_4)_3PO_4$

5.9 calcium phosphate

5.10 iron(II) nitrate

5.11 a. silicon tetrabromide
  b. dibromine oxide

5.12 IF_5

5.13 cobalt(II) carbonate

Answers to Selected Questions and Problems

5.1 a. When a sodium atom loses its valence electron, its second energy level has a complete octet.
  b. Group 1A (1) and 2A (2) elements can lose one or two electrons to attain a noble gas arrangement. Group 8A (18) elements already have an octet of valence electrons, so they do not lose or gain electrons and are not normally found in compounds.

5.3 a. 1   **b.** 2   **c.** 3   **d.** 1   **e.** 2

5.5 a. neon   **b.** neon   **c.** argon
  d. neon   **e.** neon

5.7 a. lose $2e^-$ **b.** gain $3e^-$ **c.** gain $1e^-$
  d. lose $1e^-$ **e.** lose $3e^-$

5.9 a. Li^+ **b.** F^- **c.** Mg^{2+}
  d. Fe^{3+} **e.** Zn^{2+}

5.11 a, c

5.13 a.

  b.

  c.

5.15 a. Na_2O **b.** $AlBr_3$ **c.** BaO
  d. $MgCl_2$ **e.** Al_2S_3

5.17 a. Na_2S **b.** K_3N **c.** AlI_3 **d.** Li_2O

5.19 a. Cl^- **b.** K^+ **c.** O^{2-} **d.** Al^{3+}

5.21 a. potassium   **b.** sulfide
  c. calcium   **d.** nitride

5.23 a. aluminum oxide   **b.** calcium chloride
  c. sodium oxide   **d.** magnesium nitride
  e. potassium iodide   **f.** barium fluoride

5.25 Most of the transition metals form more than one positive ion. The specific ion is indicated in a name by writing a Roman numeral that is the same as the ionic charge. For example, iron forms Fe^{2+} and Fe^{3+} ions, which are named iron(II) and iron(III).

5.27 a. iron(II)   **b.** copper(II)
  c. zinc   **d.** lead(IV)
  e. chromium(III)   **f.** manganese(II)

5.29 a. tin(II) chloride   **b.** iron(II) oxide
  c. copper(I) sulfide   **d.** copper(II) sulfide
  e. cadmium bromide   **f.** mercury(II) chloride

5.31 a. Au^{3+} **b.** Fe^{3+}
  c. Pb^{4+} **d.** Sn^{2+}

5.33 a. $MgCl_2$ **b.** Na_2S **c.** Cu_2O
  d. Zn_3P_2 **e.** AuN

5.35 a. $CoCl_3$ **b.** PbO_2 **c.** $AgCl$
  d. Ca_3N_2 **e.** Cu_3P **f.** $CrCl_2$

5.37 a. HCO_3^- **b.** NH_4^+ **c.** PO_4^{3-}
  d. HSO_4^- **e.** ClO^-

5.39 a. sulfate   **b.** carbonate   **c.** phosphate
  d. nitrate   **e.** perchlorate

5.41

	NO_2^-	CO_3^{2-}	HSO_4^-	PO_4^{3-}
Li^+	$LiNO_2$ lithium nitrite	Li_2CO_3 lithium carbonate	$LiHSO_4$ lithium hydrogen sulfate	Li_3PO_4 lithium phosphate
Cu^{2+}	$Cu(NO_2)_2$ copper(II) nitrite	$CuCO_3$ copper(II) carbonate	$Cu(HSO_4)_2$ copper(II) hydrogen sulfate	$Cu_3(PO_4)_2$ copper(II) phosphate
Ba^{2+}	$Ba(NO_2)_2$ barium nitrite	$BaCO_3$ barium carbonate	$Ba(HSO_4)_2$ barium hydrogen sulfate	$Ba_3(PO_4)_2$ barium phosphate

5.43 **a.** CO_3^{2-}, sodium carbonate
b. NH_4^+, ammonium chloride
c. PO_4^{3-}, lithium phosphate
d. NO_2^-, copper(II) nitrite
e. SO_3^{2-}, iron(II) sulfite
f. $C_2H_3O_2^-$, potassium acetate

5.45 **a.** $Ba(OH)_2$ **b.** Na_2SO_4 **c.** $Fe(NO_3)_2$
d. $Zn_3(PO_4)_2$ **e.** $Fe_2(CO_3)_3$

5.47 **a.** aluminum sulfate
b. calcium carbonate
c. chromium(III) oxide
d. sodium phosphate
e. ammonium sulfate
f. iron(III) oxide

Answers to Selected Questions and Problems

5.49 The nonmetallic elements are most likely to form covalent bonds.

5.51 **a.** 2 valence electrons, 1 bonding pair and 0 lone pairs
b. 8 valence electrons, 1 bonding pair and 3 lone pairs
c. 14 valence electrons, 1 bonding pair and 6 lone pairs

5.53 **a.** phosphorus tribromide
b. carbon tetrabromide
c. silicon dioxide
d. hydrogen monofluoride; hydrogen fluoride
e. nitrogen triiodide

5.55 **a.** dinitrogen trioxide
b. nitrogen trichloride
c. silicon tetrabromide
d. phosphorus pentachloride
e. sulfur trioxide

5.57 **a.** CCl_4 **b.** CO **c.** PCl_3 **d.** N_2O_4

5.59 **a.** OF_2 **b.** BF_3 **c.** N_2O_3 **d.** SF_6

5.61 **a.** aluminum chloride
b. sulfur trioxide
c. dinitrogen oxide
d. tin(II) nitrate
e. copper(II) chlorite

5.63 **a.** P^{3-} ion **b.** O atom
c. Zn^{2+} ion **d.** Fe^{3+} ion

5.65 1. H (E) 2. Li (C) 3. Li^+ (A)
4. H^+ (B) 5. N^{3-} (D)

5.67

Period	Electron-Dot Symbols	Formula of Compound	Name of Compound
2	X· and ·Ÿ·	Li_3N	lithium nitride
4	Ẋ· and :Ÿ·	$CaBr_2$	calcium bromide
3	·Ẋ· and :Ÿ·	Al_2S_3	aluminum sulfide

5.69

Electron Arrangements		Electron-Dot Symbols		Cations	Anions	Formula of Compound	Name of Compound
2, 8, 2	2, 5	Mġ·	·N̈·	Mg^{2+}	N^{3-}	Mg_3N_2	magnesium nitride
2, 8, 8, 1	2, 6	K·	:Ö·	K^+	O^{2-}	K_2O	potassium oxide
2, 8, 3	2, 8, 18, 7	·Al·	:B̈r·	Al^{3+}	Br^-	$AlBr_3$	aluminum bromide

Answers to Additional Questions and Problems

5.71 a. Ne   **b.** Ne   **c.** Ar
     d. Ne   **e.** He

5.73 a. Group 2A   **b.** ·X·   **c.** Be   **d.** X_3N_2

5.75 a. Sn^{4+} **b.** 50 protons, 46 electrons
     c. SnO_2 **d.** $Sn_3(PO_4)_4$

5.77 a. 3A (13)   **b.** 6A (16)   **c.** 2A (2)

5.79 a. iron(III) chloride
     b. calcium phosphate
     c. aluminum carbonate
     d. lead(IV) chloride
     e. magnesium carbonate
     f. tin(II) sulfate
     g. copper(II) sulfide

5.81 a. Cu_3N **b.** $KHSO_3$ **c.** PbS_2
     d. $Au_2(CO_3)_3$ **e.** $Zn(ClO_4)_2$

5.83 a. $AuCl_3$ **b.** PbO_2 **c.** AgCl
     d. Ca_3N_2 **e.** Cu_3P **f.** $CrCl_2$

5.85 a. magnesium oxide
     b. $Cr(HCO_3)_3$ is chromium(III) hydrogen carbonate or chromium(III) bicarbonate.
     c. manganese(III) chromate.

5.87 a. nitrogen trichloride
     b. sulfur dichloride
     c. dinitrogen oxide
     d. fluorine
     e. phosphorus pentachloride
     f. diphosphorus pentoxide

5.89 a. CO   **b.** P_2O_5 **c.** H_2S
     d. SCl_2

5.91 a. ionic, iron(III) chloride
     b. ionic, sodium sulfate
     c. covalent, dinitrogen oxide
     d. covalent, fluorine
     e. covalent, phosphorus pentachloride
     f. covalent, carbon tetrafluoride

5.93 a. $SnCO_3$ **b.** Li_3P **c.** $SiCl_4$
     d. Fe_2S_3 **e.** CO_2 **f.** $CaBr_2$

5.95 Na_2S is the compound, which has two Na^+ to every one S^{2-}.

$$4.8 \times 10^{22} \, \cancel{Na^+ \, ions} \times \frac{1 \, S^{2-} \, ion}{2 \, \cancel{Na^+ \, ions}} = 2.4 \times 10^{22} \, S^{2-}$$

5.97 The valence electrons are the electrons in the highest energy levels lost or gained in the formation of ionic compounds.

5.99 Elements in Group 2A (2) will lose two electrons to attain an octet; elements in Group 6A (16) will gain two electrons to attain an octet. Both either gain or lose two electrons.

5.101

Formula of Compound	Type of Compound	Name of Compound
$FeSO_4$	ionic	iron(II) sulfate
SiO_2	covalent	silicon dioxide
NH_4NO_3	ionic	ammonium nitrate
$Al_2(SO_4)_3$	covalent	aluminum sulfate
Co_2S_3	ionic	cobalt(III) sulfide

5.103 Compounds with a metal and nonmetal are classified as ionic; two nonmetals as covalent.
     a. ionic, lithium oxide
     b. covalent, dinitrogen oxide
     c. covalent, carbon tetrafluoride
     d. covalent, dichlorine oxide
     e. ionic, magnesium fluoride
     f. covalent, carbon monoxide
     g. ionic, calcium chloride
     h. ionic, potassium phosphate

Chemical Quantities

6

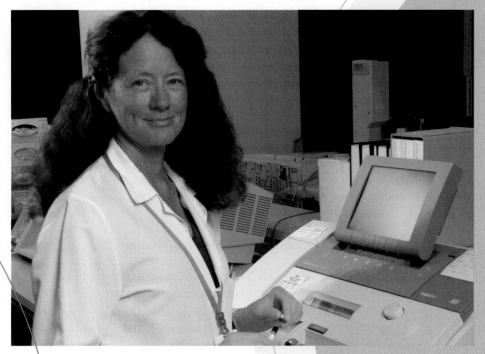

"*In a stat lab, we are sent blood samples of patients in emergency situations,*" *says Audrey Trautwein, clinical laboratory technician, Stat Lab, Santa Clara Valley Medical Center.* "*We may need to assess the status of a trauma patient in ER or a patient who is in surgery. For example, an acidic blood pH diminishes cardiac function, and affects the actions of certain drugs. In a stat situation, it is critical that we obtain our results fast. This is done using a blood gas analyzer. As I put a blood sample into the analyzer, a small probe draws out a measured volume, which is tested simultaneously for pH, P_{O_2} and P_{CO_2} as well as electrolytes, glucose, and hemoglobin. In about one minute we have our test results, which are sent to the doctor's computer.*"

the Chemistry place

Visit **www.aw-bc.com/chemplace** for extra quizzes, interactive tutorials, career resources, PowerPoint slides for chapter review, math help, and case studies.

In chemistry, we calculate and measure the amounts of substances to use in the lab. Actually, measuring the amount of a substance is something you do every day. When you cook, you measure out the proper amounts of ingredients so you don't have too much of one and too little of another. At the gas station, you measure out a certain amount of fuel in your gas tank. If you paint the walls of a room, you measure the area and purchase the amount of paint that will cover the walls. If you take a pain reliever, you read the label to see how much aspirin or ibuprofen is present. We read the nutrition labels to determine amounts of carbohydrate, fat, sodium, iron, or zinc. At a pharmacy, a pharmacist determines the amounts of medication to fill a prescription properly.

Each substance that we use in the chemistry lab is measured so it functions properly in an experiment. The formula of a substance tells us the number and kinds of atoms it has, which we then use to determine the mass of the substance to use in an experiment. Similarly, when we know the percentage and mass of the elements in a substance, we can determine its formula.

6.1 Atomic Mass and Formula Mass

In Chapter 4, we learned that each element on the periodic table has an atomic mass based on the weighted average of the masses of the isotopes. The atomic mass unit (amu) was defined as one-twelfth (1/12) of the mass of a carbon-12 atom (^{12}C), which is assigned a mass of exactly 12 amu. In most samples of elements, we treat every atom as if it has the same atomic mass. In this text, the atomic masses on the periodic table are given to four significant figures. Thus, a nitrogen atom has a mass of 14.01 amu, a zinc atom has a mass of 65.41 amu, and a tin atom has a mass of 118.7 amu.

Mass of a Sample of Atoms

The atomic mass written under the symbol on the periodic table is the mass of a single atom of that element. Because atomic mass is the mass in amu of one atom, two conversion factors can be written from atomic mass. For example, the atomic mass of sulfur can be written as

Atomic mass conversion factor

$$\frac{32.07 \text{ amu}}{1 \text{ S atom}} \quad \text{or} \quad \frac{1 \text{ S atom}}{32.07 \text{ amu}}$$

The atomic mass can be used to convert a known quantity of atoms to their mass (amu). For example, the masses of different quantities of S atoms are calculated as follows:

Number of S Atoms	Atomic Mass of S	Calculation	Mass of S Sample
10 S atoms	$\dfrac{32.07 \text{ amu}}{1 \text{ S atom}}$	$10 \text{ S atoms} \times \dfrac{32.07 \text{ amu}}{1 \text{ S atom}}$	320.7 amu
500 S atoms	$\dfrac{32.07 \text{ amu}}{1 \text{ S atom}}$	$500 \text{ S atoms} \times \dfrac{32.07 \text{ amu}}{1 \text{ S atom}}$	1.604×10^4 amu
10 000 S atoms	$\dfrac{32.07 \text{ amu}}{1 \text{ S atom}}$	$10\,000 \text{ S atoms} \times \dfrac{32.07 \text{ amu}}{1 \text{ S atom}}$	3.207×10^5 amu

Using Atomic Mass to Count Atoms

The atomic mass is also used to determine the number of atoms present in a specific mass of an element. For example, the atomic mass for sulfur can be used to calculate the number of sulfur atoms in the following sulfur samples.

Mass of S Sample	Atomic Mass of S	Calculation	Number of S Atoms
6420 amu	$\dfrac{1\ S\ atom}{32.07\ amu}$	$6420\ \cancel{amu} \times \dfrac{1\ S\ atom}{32.07\ \cancel{amu}}$	2.00×10^2 S atoms
3.21×10^4 amu	$\dfrac{1\ S\ atom}{32.07\ amu}$	$3.21 \times 10^4\ \cancel{amu} \times \dfrac{1\ S\ atom}{32.07\ \cancel{amu}}$	1.00×10^3 S atoms
9.63×10^{16} amu	$\dfrac{1\ S\ atom}{32.07\ amu}$	$9.63 \times 10^{16}\ \cancel{amu} \times \dfrac{1\ S\ atom}{32.07\ \cancel{amu}}$	3.00×10^{15} S atoms

---

Sample Problem 6.1   **Calculating the Mass of a Sample of Atoms**

Calculate the mass in amu of 100 atoms of each of the following:
a. Mg atoms   **b.** Fe atoms

Solution

a. 100 Mg atoms

STEP 1 **Given** 100 Mg atoms   **Need** mass in amu

STEP 2 **Plan** atoms Atomic mass factor amu

STEP 3 **Equalities/Conversion Factors** Obtain the atomic mass for Mg from the periodic table.

$$1\ Mg\ atom = 24.31\ amu$$

$$\dfrac{24.31\ amu}{1\ Mg\ atom} \quad and \quad \dfrac{1\ Mg\ atom}{24.31\ amu}$$

STEP 4 **Set Up Problem** The mass for the 100 Mg atoms is calculated using the atomic mass. The answer has four significant figures because 100 Mg atoms is a counted number and exact.

$$100\ \cancel{Mg\ atoms} \times \dfrac{24.31\ amu}{1\ \cancel{Mg\ atom}} = 2431\ amu$$

b. 100 Fe atoms

STEP 1 **Given** 100 Fe atoms   **Need** mass in amu

STEP 2 **Plan** atoms Atomic mass factor amu

STEP 3 **Equalities/Conversion Factors** Obtain the atomic mass for Fe from the periodic table.

$$1\ Fe\ atom = 55.85\ amu$$

$$\dfrac{55.85\ amu}{1\ Fe\ atom} \quad and \quad \dfrac{1\ Fe\ atom}{55.85\ amu}$$

STEP 4 **Set Up Problem** The mass for the 100 Fe atoms is calculated using the atomic mass. The answer has four significant figures because 100 Fe atoms is a counted number and therefore exact.

$$100 \text{ Fe atoms} \times \frac{55.85 \text{ amu}}{1 \text{ Fe atom}} = 5585 \text{ amu}$$

Study Check

What is the mass in amu of 750 Ba atoms?

Sample Problem 6.2 **Calculating the Number of Atoms in a Specified Mass**

Determine the number of atoms in a sample of copper that has a mass of 7.82×10^{10} amu.

Solution

STEP 1 **Given** 7.82×10^{10} amu Cu   **Need** atoms of Cu

STEP 2 **Plan** amu   Atomic mass factor   atoms

STEP 3 **Equalities/Conversion Factors** Obtain the atomic mass for Cu from the periodic table.

$$1 \text{ Cu atom} = 63.55 \text{ amu}$$
$$\frac{63.55 \text{ amu}}{1 \text{ Cu atom}} \text{ and } \frac{1 \text{ Cu atom}}{63.55 \text{ amu}}$$

STEP 4 **Set Up Problem** Calculate the number of Cu atoms:

$$7.82 \times 10^{10} \text{ amu} \times \frac{1 \text{ Cu atom}}{63.55 \text{ amu}} = 1.23 \times 10^9 \text{ Cu atoms}$$

Study Check

How many gold atoms are present in a gold sample that has a mass of 9.85×10^5 amu?

Formula Mass

For a compound, the **formula mass** is the sum of the atomic masses of all the atoms in its formula. To determine the formula mass, we multiply each element's subscript by its atomic mass and add the results. For example, the formula mass of carbon dioxide, CO_2, is the sum of the atomic masses of one carbon atom and two oxygen atoms. (See Figure 6.1.)

Figure 6.1 At −78.5°C, dry ice (solid CO_2) changes to a gas. Dry ice is used in ice cream stores to keep ice cream frozen, in fire extinguishers, and for making "fog" during stage performances.

Q *How would you determine the formula mass of CO_2?*

1 Molecule of CO_2

$$1 \text{ C } (12.01 \text{ amu/C}) + 2 \text{ O } (16.00 \text{ amu/O}) = 44.01 \text{ amu}$$

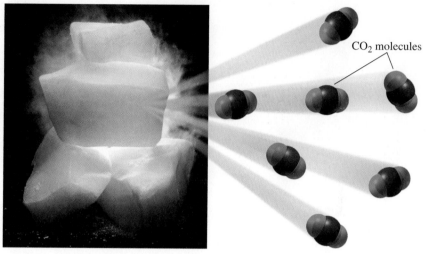

Dry ice (solid CO_2)

CO_2 molecules

Number of Atoms in the Formula		Atomic Mass		Total Mass for Each Element
1 C atom	×	$\dfrac{12.01 \text{ amu}}{1 \text{ C atom}}$	=	12.01 amu
2 O atoms	×	$\dfrac{16.00 \text{ amu}}{1 \text{ O atom}}$	=	32.00 amu
		Formula mass of CO_2	=	44.01 amu

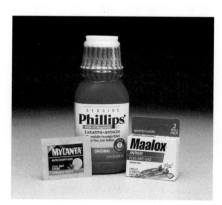

Sample Problem 6.3 **Calculating Formula Mass**

Calculate the formula mass of magnesium hydroxide, $Mg(OH)_2$, which is present in some antacids.

Solution

STEP 1 **Given** $Mg(OH)_2$ **Need** formula mass in amu

STEP 2 **Plan** The formula mass is the sum of the atomic masses of all the atoms in the formula.

1 [amu Mg] + 2 [amu O] + 2 [amu H] = formula mass $Mg(OH)_2$

STEP 3 **Equalities/Conversion Factors** Obtain the atomic mass for each element from the periodic table.

$$1 \text{ Mg atom} = 24.31 \text{ amu}$$
$$\frac{24.31 \text{ amu}}{1 \text{ Mg atom}} \quad \text{and} \quad \frac{1 \text{ Mg atom}}{24.31 \text{ amu}}$$

$$1 \text{ O atom} = 16.00 \text{ amu}$$
$$\frac{16.00 \text{ amu}}{1 \text{ O atom}} \quad \text{and} \quad \frac{1 \text{ O atom}}{16.00 \text{ amu}}$$

$$1 \text{ H atom} = 1.008 \text{ amu}$$
$$\frac{1.008 \text{ amu}}{1 \text{ H atom}} \quad \text{and} \quad \frac{1 \text{ H atom}}{1.008 \text{ amu}}$$

STEP 4 **Set Up Problem**

$$1 \text{ Mg atom} \times \frac{24.31 \text{ amu}}{1 \text{ Mg atom}} = 24.31 \text{ amu}$$

$$2 \text{ O atoms} \times \frac{16.00 \text{ amu}}{1 \text{ O atom}} = 32.00 \text{ amu}$$

$$2 \text{ H atoms} \times \frac{1.008 \text{ amu}}{1 \text{ H atom}} = 2.016 \text{ amu}$$

$$\text{Formula mass of Mg(OH)}_2 = 58.33 \text{ amu}$$

Study Check

Calcium hydrogen sulfite, $Ca(HSO_3)_2$, is used in papermaking and to disinfect brewery casks. What is the formula mass of calcium hydrogen sulfite?

Questions and Problems **Atomic Mass and Formula Mass**

6.1 Calculate the mass in amu of each of the following samples:
  a. 25 atoms O   **b.** 1.50×10^5 atoms Cr
  c. 6.24×10^{20} atoms F

6.2 Calculate the mass in amu of each of the following samples:
  a. 75 atoms Ba   **b.** 2.80×10^3 atoms P
  c. 9.15×10^{25} atoms Cl

6.3 Determine the number of atoms in each of the following samples:
  a. 540 amu Al   **b.** 2.95×10^4 amu Ag
  c. 35 000 amu Ne

6.4 Determine the number of atoms in each of the following samples:
  a. 1.60×10^4 amu Cu   **b.** 6400 amu Kr
  c. 128 000 amu Sr

6.5 Determine the formula mass of each of the following:
  a. SF_6 **b.** $Ca(NO_3)_2$
  c. C_2H_5OH, ethanol   **d.** $C_6H_{12}O_6$, glucose

6.6 Determine the formula mass of each of the following:
  a. Cl_2O_7 **b.** $(NH_4)_2CO_3$
  c. $C_3H_8O_3$, glycerol   **d.** $C_8H_{10}N_4O_2$, caffeine

6.7 Determine the formula mass of each of the following:
  a. Na_2SO_3, used in photography
  b. $Al(OH)_3$, used as an antacid
  c. $FeSO_4$, used as an iron supplement
  d. $Ca_3(PO_4)_2$, used as a calcium supplement

6.8 Determine the formula mass of each of the following:
  a. $CaCO_3$, used as a calcium supplement
  b. $(NH_4)_2SO_4$, a fertilizer
  c. $Zn_3(PO_4)_2$, a dental material
  d. $Pt(NH_3)_2Cl_2$, cisplatin, used to treat tumors

6.2 The Mole

At the store, you buy eggs by the dozen. In an office, pencils are ordered by the gross and paper by the ream. In a restaurant, soda is ordered by the case. In each of these examples, the terms *dozen, gross, ream,* and *case* count the number of items present. For example, when you buy a dozen eggs, you know you will get 12 eggs in the carton.

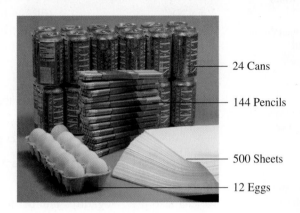

24 Cans

144 Pencils

500 Sheets

12 Eggs

In chemistry, particles such as atoms, molecules, and ions are counted by the **mole** (abbreviated *mol* in calculations), a unit that contains 6.022×10^{23} items. This very large number, called **Avogadro's number** after Amedeo Avogadro, an Italian physicist, looks like this when written with four significant figures:

Avogadro's Number

$$602\ 200\ 000\ 000\ 000\ 000\ 000\ 000 = 6.022 \times 10^{23}$$

One mole of an element always has Avogadro's number of atoms. For example, 1 mol of carbon contains 6.022×10^{23} carbon atoms; 1 mol of aluminum contains 6.022×10^{23} aluminum atoms; 1 mol of iron contains 6.022×10^{23} iron atoms.

$$1 \text{ mol of an element} = 6.022 \times 10^{23} \text{ atoms of that element}$$

One mole of a compound contains Avogadro's number of molecules or formula units. Molecules are the particles of covalent compounds; **formula units** are the group of ions given by the formula of an ionic compound. For example, 1 mol CO_2, a covalent compound, contains 6.022×10^{23} molecules CO_2. One mol NaCl, an ionic compound, contains 6.022×10^{23} formula units NaCl (Na^+, Cl^-). Table 6.1 gives examples of the number of particles in some 1-mol quantities.

Avogadro's number tells us that 1 mol of any element or compound contains 6.022×10^{23} of the particular type of particles that make up a substance. We can use Avogadro's number as a conversion factor to convert between the moles of a substance and number of particles it contains.

$$\frac{6.022 \times 10^{23} \text{ particles}}{1 \text{ mol}} \quad \text{and} \quad \frac{1 \text{ mol}}{6.022 \times 10^{23} \text{ particles}}$$

Table 6.1   Number of Particles in 1-Mol Samples

Substance	Number and Type of Particles
1 mol carbon	6.022×10^{23} carbon atoms
1 mol aluminum	6.022×10^{23} aluminum atoms
1 mol iron	6.022×10^{23} iron atoms
1 mol water (H_2O)	6.022×10^{23} H_2O molecules
1 mol NaCl	6.022×10^{23} NaCl formula units
1 mol sucrose ($C_{12}H_{22}O_{11}$)	6.022×10^{23} sucrose molecules
1 mol vitamin C ($C_6H_8O_6$)	6.022×10^{23} vitamin C molecules

Calculating Particles and Moles

Moles of element or compound

⬆

Avogadro's number

⬇

Particles: atoms, ions, molecules, or formula units

For example, we use Avogadro's number to convert 4.00 mol iron to atoms of iron.

$$4.00 \; \cancel{\text{mol Fe atoms}} \times \frac{6.022 \times 10^{23} \text{ Fe atoms}}{1 \; \cancel{\text{mol Fe atoms}}} = 2.41 \times 10^{24} \text{ Fe atoms}$$

Avogadro's number as a conversion factor

We can also use Avogadro's number to convert 3.01×10^{24} molecules CO_2 to moles of CO_2

$$3.01 \times 10^{24} \; \cancel{CO_2 \text{ molecules}} \times \frac{1 \text{ mol } CO_2 \text{ molecules}}{6.022 \times 10^{23} \; \cancel{CO_2 \text{ molecules}}} = 5.00 \text{ mol } CO_2 \text{ molecules}$$

Avogadro's number as a conversion factor

Sample Problem 6.4   **Calculating the Number of Molecules in a Mole**

How many molecules of ammonia, NH_3, are present in 1.75 mol ammonia?

Solution

STEP 1   **Given**   1.75 mol NH_3 **Need**   molecules of NH_3

STEP 2   **Plan**   Moles   Avogadro's number   molecules

STEP 3   **Equalities/Conversion Factors**

$$1 \text{ mol } NH_3 = 6.022 \times 10^{23} \text{ molecules } NH_3$$

$$\frac{6.022 \times 10^{23} \text{ molecules } NH_3}{1 \text{ mol } NH_3} \quad \text{and} \quad \frac{1 \text{ mol } NH_3}{6.022 \times 10^{23} \text{ molecules } NH_3}$$

STEP 4   **Set Up Problem**   Calculate the number of NH_3 molecules:

$$1.75 \; \cancel{\text{mol } NH_3} \times \frac{6.022 \times 10^{23} \text{ molecules } NH_3}{1 \; \cancel{\text{mol } NH_3}} = 1.05 \times 10^{24} \text{ molecules } NH_3$$

Study Check

How many moles of water, H_2O, contain 2.60×10^{23} molecules of water?

Subscripts State Moles of Elements

We have seen that the subscripts in a chemical formula indicate the number of atoms of each type of element in a compound. For example, one molecule of NH_3 consists of 1 N atom and 3 H atoms. If we have 6.022×10^{23} molecules (1 mol) of NH_3, it would contain 6.022×10^{23} atoms (1 mol) N and $3 \times 6.022 \times 10^{23}$ molecules (3 mol) H. Thus, each subscript in a formula also refers to the moles of each kind of atom in 1 mol of a compound. For example, the subscripts in the NH_3 formula specify that 1 mol NH_3 contains 1 mol N atoms and 3 mol H atoms.

The formula subscript specifies the

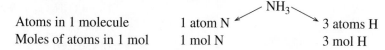

		NH_3
Atoms in 1 molecule	1 atom N	3 atoms H
Moles of atoms in 1 mol	1 mol N	3 mol H

In a different example, aspirin, which has a formula of $C_9H_8O_4$, there are 9 C atoms, 8 H atoms, and 4 O atoms. The subscripts also state the number of moles of each element in 1 mol of aspirin: 9 mol C, 8 mol H, and 4 mol O.

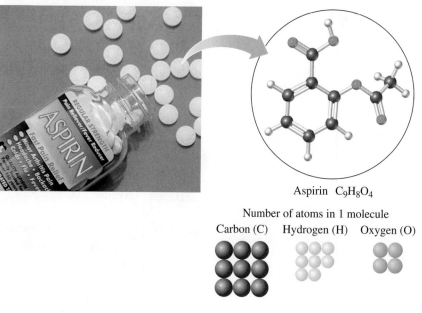

Aspirin $C_9H_8O_4$

Number of atoms in 1 molecule
Carbon (C)   Hydrogen (H)   Oxygen (O)

The formula subscript specifies the

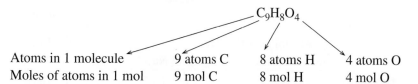

		$C_9H_8O_4$	
Atoms in 1 molecule	9 atoms C	8 atoms H	4 atoms O
Moles of atoms in 1 mol	9 mol C	8 mol H	4 mol O

The subscripts in a formula are helpful when we need to determine the amount of any of the elements in the formula. For the aspirin formula, $C_9H_8O_4$, we can write the following sets of conversion factors for each of the elements in 1 mol of aspirin.

$$\frac{9 \text{ mol C}}{1 \text{ mol } C_9H_8O_4} \qquad \frac{8 \text{ mol H}}{1 \text{ mol } C_9H_8O_4} \qquad \frac{4 \text{ mol O}}{1 \text{ mol } C_9H_8O_4}$$

$$\frac{1 \text{ mol } C_9H_8O_4}{9 \text{ mol C}} \qquad \frac{1 \text{ mol } C_9H_8O_4}{8 \text{ mol H}} \qquad \frac{1 \text{ mol } C_9H_8O_4}{4 \text{ mol O}}$$

Sample Problem 6.5 **Calculating the Moles of an Element in a Compound**

How many moles of carbon atoms are present in 1.50 mol aspirin, $C_9H_8O_4$?

Solution

STEP 1 **Given** 1.50 mol $C_9H_8O_4$ **Need** moles of C atoms

STEP 2 **Plan** Moles of $C_9H_8O_4$ Subscript moles of C atoms

STEP 3 **Equalities/Conversion Factors**

$$1 \text{ mol } C_9H_8O_4 = 9 \text{ mol C atoms}$$

$$\frac{9 \text{ mol C}}{1 \text{ mol } C_9H_8O_4} \quad \text{and} \quad \frac{1 \text{ mol } C_9H_8O_4}{9 \text{ mol C}}$$

STEP 4 **Set Up Problem**

$$1.50 \text{ mol } C_9H_8O_4 \times \frac{9 \text{ mol C atoms}}{1 \text{ mol } C_9H_8O_4} = 13.5 \text{ mol C atoms}$$

Study Check

How many moles of aspirin, $C_9H_8O_4$, contain 0.480 mol O atoms?

Sample Problem 6.6 **Particles of an Element in a Compound**

How many Na^+ ions are in 3.00 mol Na_2O?

Solution

STEP 1 **Given** 3.00 mol Na_2O **Need** number of Na^+ ions

STEP 2 **Plan** Use the subscript for Na in the formula to convert from moles of Na_2O to moles of Na^+ and use Avogadro's number to calculate the number of Na^+ ions.

Moles of Na_2O Subscript moles of Na^+ Avogadro's number Na^+ ions

STEP 3 **Equalities/Conversion Factors** In Na_2O, an ionic compound, there are 2 mol Na^+ ions in 1 mol Na_2O.

$$1 \text{ mol } Na_2O = 2 \text{ mol } Na^+$$

$$\frac{2 \text{ mol } Na^+ \text{ ions}}{1 \text{ mol } Na_2O} \quad \text{and} \quad \frac{1 \text{ mol } Na_2O}{2 \text{ mol } Na^+}$$

$$1 \text{ mol } Na^+ = 6.022 \times 10^{23} \text{ Na}^+ \text{ ions}$$

$$\frac{6.022 \times 10^{23} \text{ Na}^+ \text{ ions}}{1 \text{ mol } Na^+} \quad \text{and} \quad \frac{1 \text{ mol } Na^+}{6.022 \times 10^{23} \text{ Na}^+ \text{ ions}}$$

STEP 4 **Set Up Problem**

$$3.00 \text{ mol } Na_2O \times \frac{2 \text{ mol } Na^+}{1 \text{ mol } Na_2O} \times \frac{6.022 \times 10^{23} \text{ Na}^+ \text{ ions}}{1 \text{ mol } Na^+ \text{ ion}} = 3.61 \times 10^{24} \text{ Na}^+ \text{ ions}$$

Study Check

How many SO_4^{2-} ions are in 2.50 mol $Fe_2(SO_4)_3$, a compound used in water and sewage treatment?

Questions and Problems   **The Mole**

6.9 What is the mole?

6.10 What is Avogadro's number?

6.11 What is the difference in saying "1 mol of chlorine atoms" and "1 mol of chlorine molecules"?

6.12 What is the difference in saying "1 mol of nitrogen atoms" and "1 mol of nitrogen molecules"?

6.13 Calculate each of the following:
  a. number of C atoms in 0.500 mol C
  b. number of SO_2 molecules in 1.28 mol SO_2
  c. moles of Fe in 5.22×10^{22} atoms Fe
  d. moles of C_2H_5OH in 8.50×10^{24} molecules C_2H_5OH

6.14 Calculate each of the following:
  a. number of Li atoms in 4.5 mol Li
  b. number of CO_2 molecules in 0.0180 mol CO_2
  c. moles of Cu in 7.8×10^{21} atoms Cu
  d. moles of C_2H_6 in 3.754×10^{23} molecules C_2H_6

6.15 Calculate each of the following quantities in 2.00 mol H_3PO_4:
  a. moles of H   **b.** moles of O
  c. atoms of P   **d.** atoms of O

6.16 Calculate each of the following quantities in 0.185 mol $(C_3H_5)_2O$:
  a. moles of C   **b.** moles of O
  c. atoms of H   **d.** atoms of C

6.3 Molar Mass

A single atom or molecule is much too small to weigh, even on the most sensitive balance. In fact, it takes a huge number of atoms or molecules to make enough of a substance for you to see. An amount of water that contains Avogadro's number of water molecules is only a few sips. However, in the laboratory, we can use a balance to weigh out Avogadro's number of particles or 1 mol of substance.

For any element, the quantity called **molar mass** is the quantity in grams that equals the atomic mass of that element. We are counting out 6.022×10^{23} atoms of an element when we weigh out the number of grams equal to its molar mass. For example, if we need 1 mol of carbon (C) atoms, we would first find the atomic mass of 12.01 for carbon on the periodic table. Then to obtain 1 mol of carbon atoms, we would weigh out 12.01 g carbon. Thus, we can use the periodic table to determine the molar mass of any element because molar mass is numerically equal to the atomic mass in grams. One mol of sulfur has a molar mass of 32.07 grams and 1 mol of silver atoms has a molar mass of 107.9 grams.

6.022×10^{23} atoms C

1 mol C atoms

12.01 g C atoms

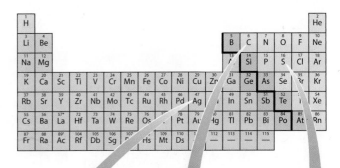

1 mol of silver atoms has a mass of 107.9 g

1 mol of carbon atoms has a mass of 12.01 g

1 mol of sulfur atoms has a mass of 32.07 g

Molar Mass of a Compound

To determine the molar mass of a compound, multiply the molar mass of each element by its subscript in the formula and add the results. For example, the molar mass of sulfur trioxide, SO_3, is obtained by adding the molar masses of 1 mol sulfur and 3 mol oxygen.

<table>
<tr><td>

Guide to Calculating Molar Mass

STEP 1
Obtain the molar mass of each element.

STEP 2
Multiply each molar mass by the number of moles (subscript) in the formula.

STEP 3
Calculate the molar mass by adding the masses of the elements.

</td></tr>
</table>

STEP 1 Using the periodic table, obtain the molar masses of sulfur and oxygen.

$$\frac{32.07 \text{ g S}}{1 \text{ mol S}} \qquad \frac{16.00 \text{ g O}}{1 \text{ mol O}}$$

STEP 2 Obtain the mass of each element in the formula by multiplying each molar mass by its number of moles (subscript) in the formula.

Grams from 1 mol S

$$1 \text{ mol S} \times \frac{32.07 \text{ g S}}{1 \text{ mol S}} = 32.07 \text{ g S}$$

Grams from 3 mol O

$$3 \text{ mol O} \times \frac{16.00 \text{ g O}}{1 \text{ mol O}} = 48.00 \text{ g O}$$

STEP 3 Obtain the molar mass of SO_3 by adding the masses of 1 S and 3 O.

Molar mass of $SO_3 = 80.07$ g

Figure 6.2 shows some 1-mol quantities of substances. Table 6.2 lists the molar mass for several 1-mol samples.

1-Mol Quantities

| S | Fe | NaCl | $K_2Cr_2O_7$ | $C_{12}H_{22}O_{11}$ |

Figure 6.2 1-mol samples: sulfur, S (32.07 g); iron, Fe (55.85 g); salt, NaCl (58.44 g); potassium dichromate, $K_2Cr_2O_7$ (294.20 g); and sugar, sucrose, $C_{12}H_{22}O_{11}$ (342.30 g).

Q How is the molar mass for $K_2Cr_2O_7$ obtained?

Table 6.2   The Molar Mass and Number of Particles in 1-Mol Quantities

Substance	Molar Mass	Number of Particles in 1 Mol
1 mol carbon (C)	12.01 g	6.022×10^{23} C atoms
1 mol sodium (Na)	22.99 g	6.022×10^{23} Na atoms
1 mol iron (Fe)	55.85 g	6.022×10^{23} Fe atoms
1 mol NaF (helps prevent cavities)	41.99 g	6.022×10^{23} NaF formula units
1 mol $C_6H_{12}O_6$ (glucose)	180.16 g	6.022×10^{23} glucose molecules
1 mol $C_8H_{10}N_4O_2$ (caffeine)	194.12 g	6.022×10^{23} caffeine molecules

Sample Problem 6.7   **Calculating the Molar Mass of a Compound**

Find the molar mass of Li_2CO_3 used to produce red color in fireworks.

Solution

STEP 1   Using the periodic table, obtain the molar masses of lithium, carbon, and oxygen.

$$\frac{6.941 \text{ g Li}}{1 \text{ mol Li}} \qquad \frac{12.01 \text{ g C}}{1 \text{ mol C}} \qquad \frac{16.00 \text{ g O}}{1 \text{ mol O}}$$

STEP 2   Obtain the mass of each element in the formula by multiplying each molar mass by its number of moles (subscript) in the formula.

Grams from 2 mol Li

$$2 \text{ mol Li} \times \frac{6.941 \text{ g Li}}{1 \text{ mol Li}} = 13.88 \text{ g Li}$$

Grams from 1 mol C

$$1 \text{ mol C} \times \frac{12.01 \text{ g C}}{1 \text{ mol C}} = 12.01 \text{ g C}$$

Grams from 3 mol O

$$3 \text{ mol O} \times \frac{16.00 \text{ g O}}{1 \text{ mol O}} = 48.00 \text{ g O}$$

STEP 3   Obtain the molar mass of Li_2CO_3 by adding the masses of 2 Li, 1 C, and 3 O.

$$\text{Molar mass of } Li_2CO_3 = 73.89 \text{ g}$$

Study Check

Calculate the molar mass of salicylic acid, $C_7H_6O_3$.

Questions and Problems   **Molar Mass**

6.17 Calculate the molar mass for each of the following:
  a. NaCl (table salt)
  b. Fe_2O_3 (rust)
  c. $C_{20}H_{23}N$ (antidepressant)
  d. $Al_2(SO_4)_3$ (antiperspirant)
  e. $KC_4H_5O_6$ (cream of tartar)
  f. $C_{16}H_{19}N_3O_5S$ (amoxicillin, an antibiotic)
6.18 Calculate the molar mass for each of the following:
  a. $FeSO_4$ (iron supplement)
  b. Al_2O_3 (absorbent and abrasive)
  c. $C_7H_5NO_3S$ (saccharin)
  d. C_3H_8O (rubbing alcohol)

  e. $(NH_4)_2CO_3$ (baking powder)
  f. $Zn(C_2H_3O_2)_2$ (zinc dietary supplement)
6.19 Calculate the molar mass of each of the following:
  a. Cl_2 **b.** $C_3H_6O_3$
  c. $Mg_3(PO_4)_2$ **d.** AlF_3
  e. $C_2H_4Cl_2$ **f.** SnF_2
6.20 Calculate the molar mass of each of the following:
  a. O_2 **b.** KH_2PO_4
  c. $Fe(ClO_4)_3$ **d.** $C_4H_8O_4$
  e. $Ga_2(CO_3)_3$ **f.** $KBrO_4$

Learning **Goal**

Given the number of moles of a substance, calculate the mass in grams; given the mass, calculate the number of moles.

WEB TUTORIAL
Stoichiometry

6.4 Calculations Using Molar Mass

The molar mass of an element or a compound is one of the most useful conversion factors in chemistry. Molar mass is used to change from moles of a substance to grams, or from grams to moles. To do these calculations, we use the molar mass as a conversion factor. For example, 1 mol magnesium has a mass of 24.31 g. To express molar mass as an equality, we can write

$$1 \text{ mol Mg} = 24.31 \text{ g Mg}$$

From this equality, two conversion factors can be written.

$$\frac{24.31 \text{ g Mg}}{1 \text{ mol Mg}} \quad \text{and} \quad \frac{1 \text{ mol Mg}}{24.31 \text{ g Mg}}$$

Conversion factors are written for compounds in the same way. For example, the molar mass of the compound H_2O is 18.02 g.

$$1 \text{ mol } H_2O = 18.02 \text{ g } H_2O$$

The conversion factors from the molar mass of H_2O are written as

$$\frac{18.02 \text{ g } H_2O}{1 \text{ mol } H_2O} \quad \text{and} \quad \frac{1 \text{ mol } H_2O}{18.02 \text{ g } H_2O}$$

We can now change from moles to grams, or grams to moles, using the conversion factors derived from the molar mass. (Remember, you must determine the molar mass of the substance first.)

Sample Problem 6.8   **Converting Moles of an Element to Grams**

Silver metal is used in the manufacture of tableware, mirrors, jewelry, and dental alloys. If the design for a piece of jewelry requires 0.750 mol silver, how many grams of silver are needed?

Solution

STEP 1   **Plan**   0.750 mol Ag   Molar mass factor   grams of Ag

Guide to Converting Moles to Grams

STEP 1
Write a plan that converts moles to grams.

STEP 2
Write the conversion factors for molar mass.

STEP 3
Set up the problem to convert moles to grams.

STEP 2 **Equalities/Conversion Factors**

$$1 \text{ mol Ag} = 107.9 \text{ g Ag}$$

$$\frac{107.9 \text{ g Ag}}{1 \text{ mol Ag}} \quad \text{and} \quad \frac{1 \text{ mol Ag}}{107.9 \text{ g Ag}}$$

STEP 3 **Set Up Problem**   Calculate the grams of silver using the molar mass factor that cancels "mol Ag."

$$0.750 \text{ mol Ag} \times \frac{107.9 \text{ g Ag}}{1 \text{ mol Ag}} = 80.9 \text{ g Ag}$$

Study Check

Calculate the number of grams of gold (Au) present in 0.124 mol gold.

Sample Problem 6.9   **Converting Mass of a Compound to Moles**

A box of salt contains 737 g NaCl. How many moles of NaCl are present in the box?

Solution

STEP 1   **Plan**   737 g NaCl   Molar mass factor    moles of NaCl

STEP 2   **Equalities/Conversion Factors**   The molar mass of NaCl is the sum of the masses of 1 mol Na^+ and 1 mol Cl^-:

$$(1 \times 22.99 \text{ g/mol}) + (1 \times 35.45 \text{ g/mol}) = 58.44 \text{ g/mol}$$

$$1 \text{ mol NaCl} = 58.44 \text{ g}$$

$$\frac{58.44 \text{ g NaCl}}{1 \text{ mol NaCl}} \quad \text{and} \quad \frac{1 \text{ mol NaCl}}{58.44 \text{ g NaCl}}$$

STEP 3   **Set Up Problem**   We calculate the moles of NaCl using the molar mass factor that cancels "g NaCl."

$$737 \text{ g NaCl} \times \frac{1 \text{ mol NaCl}}{58.44 \text{ g NaCl}} = 12.6 \text{ mol NaCl}$$

Study Check

One gel cap of an antacid contains 311 mg $CaCO_3$ and 232 mg $MgCO_3$. In a recommended dosage of two gel caps, how many moles each of $CaCO_3$ and $MgCO_3$ are present?

We can now combine the calculations from the previous problems to convert the mass in grams of a compound to the number of molecules. In these calculations, such as shown in Sample Problem 6.10, we see the central role of moles in the conversion of mass to particles.

<table>
<tr><td>

Guide to Converting Grams to Particles

STEP 1
Write a plan that converts grams to particles.

STEP 2
Write the conversion factors for molar mass and Avogadro's number.

STEP 3
Set up the problem to convert grams to moles to particles.

</td></tr>
</table>

Sample Problem 6.10   **Converting Grams to Particles**

A 10.00-lb bag of table sugar contains 4536 g sucrose, $C_{12}H_{22}O_{11}$. How many molecules of sugar are present?

Solution

STEP 1   **Plan**   We need to convert the grams of sugar to moles of sugar using the molar mass of $C_{12}H_{22}O_{11}$. Then we use Avogadro's number to calculate the number of molecules of sugar.

4536 g sugar   Molar mass factor   moles   Avogadro's number   molecules

STEP 2   **Equalities/Conversion Factors**   The molar mass of $C_{12}H_{22}O_{11}$ is the sum of the masses of 12 mol C + 22 mol H + 11 mol O:

$$(12 \times 12.01 \text{ g/mol}) + (22 \times 1.008 \text{ g/mol}) + (11 \times 16.00 \text{ g/mol}) = 342.30 \text{ g/mol}$$

$$1 \text{ mol } C_{12}H_{22}O_{11} = 342.30 \text{ g } C_{12}H_{22}O_{11}$$

$$\frac{342.30 \text{ g } C_{12}H_{22}O_{11}}{1 \text{ mol } C_{12}H_{22}O_{11}} \quad \text{and} \quad \frac{1 \text{ mol } C_{12}H_{22}O_{11}}{342.30 \text{ g } C_{12}H_{22}O_{11}}$$

$$1 \text{ mol } C_{12}H_{22}O_{11} = 6.022 \times 10^{23} \text{ molecules } C_{12}H_{22}O_{11}$$

$$\frac{6.022 \times 10^{23} \text{ molecules}}{1 \text{ mol } C_{12}H_{22}O_{11}} \quad \text{and} \quad \frac{1 \text{ mol } C_{12}H_{22}O_{11}}{6.022 \times 10^{23} \text{ molecules}}$$

STEP 3   **Set Up Problem**   Using the molar mass factor, the grams of sugar are converted to moles of sugar, which are then converted to molecules using Avogadro's number.

$$4536 \text{ g } C_{12}H_{22}O_{22} \times \frac{1 \text{ mol } C_{12}H_{22}O_{11}}{342.30 \text{ g } C_{12}H_{22}O_{11}} \times \frac{6.022 \times 10^{23} \text{ molecules}}{1 \text{ mol } C_{12}H_{22}O_{11}} = 7.980 \times 10^{24} \text{ molecules}$$

Study Check

How many moles of nitrogen are in 2.50 g caffeine $C_8H_{10}N_4O_2$?

We summarize the calculations in Figure 6.3, which gives the connections between the moles of a compound, its mass in grams, and number of molecules (or formula units if ionic), and the moles and atoms of each element in that compound.

Figure 6.3 The moles of a compound are related to its mass in grams by molar mass, to the number of molecules (or formula units) by Avogadro's number, and to the moles of each element by the subscripts in the formula.

Q What steps are needed to calculate the number of H atoms in 5.00 g of CH₄?

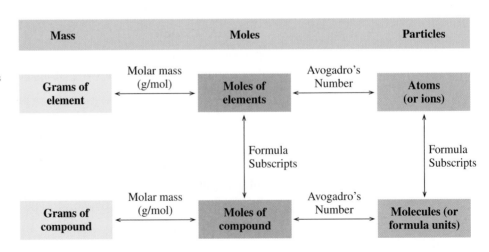

Questions and Problems   Calculations Using Molar Mass

6.21 Calculate the mass in grams for each of the following:
  a. 1.50 mol Na   **b.** 2.80 mol Ca
  c. 0.125 mol CO_2 **d.** 0.0485 mol Na_2CO_3
  e. 7.14×10^2 mol PCl_3

6.22 Calculate the mass in grams for each of the following:
  a. 5.12 mol Al   **b.** 0.75 mol Cu
  c. 3.52 mol $MgBr_2$ **d.** 0.145 mol C_2H_6O
  e. 2.08 mol $(NH_4)_2SO_4$

6.23 Calculate the mass in grams in 0.150 mol of each of the following:
  a. Ne   **b.** I_2 **c.** Na_2O
  d. $Ca(NO_3)_2$ **e.** C_6H_{14}

6.24 Calculate the mass in grams in 2.28 mol of each of the following:
  a. N_2 **b.** SO_3 **c.** $C_3H_6O_3$
  d. $Mg(HCO_3)_2$ **e.** SF_6

6.25 Calculate the number of moles in each of the following:
  a. 82.0 g Ag   **b.** 0.288 g C
  c. 15.0 g ammonia, NH_3 **d.** 7.25 g propane, C_3H_8
  e. 245 g Fe_2O_3

6.26 Calculate the number of moles in each of the following:
  a. 85.2 g Ni   **b.** 144 g K
  c. 6.4 g H_2O **d.** 308 g $BaSO_4$
  e. 252.8 g fructose, $C_6H_{12}O_6$

6.27 Calculate the number of moles in 25.0 g of each of the following:
  a. He   **b.** O_2 **c.** $Al(OH)_3$
  d. Ga_2S_3 **e.** C_4H_{10}, butane

6.28 Calculate the number of moles in 4.00 g of each of the following:
  a. He   **b.** SnO_2 **c.** $Cr(OH)_3$
  d. Ca_3N_2 **e.** $C_6H_8O_6$, vitamin C

6.29 Calculate the number of atoms of C in each of the following:
  a. 25.0 g C   **b.** 0.688 mol CO_2
  c. 275 g C_3H_8 **d.** 1.84 mol C_2H_6O
  e. 7.5×10^{24} molecules CH_4

6.30 Calculate the number of atoms of N in each of the following:
  a. 0.755 mol N_2 **b.** 0.82 g $NaNO_3$
  c. 40.0 g N_2O **d.** 6.24×10^{-3} mol NH_3
  e. 1.4×10^{22} molecules N_2O_4

6.31 Propane gas, C_3H_8, is used as a fuel for many barbecues.
  a. How many grams of the compound are in 1.50 mol propane?
  b. How many moles of the compound are in 34.0 g propane?
  c. How many grams of carbon are in 34.0 g propane?
  d. How many atoms of hydrogen are in 0.254 g propane?

6.32 Allyl sulfide $(C_3H_5)_2S$ is the substance that gives garlic its characteristic odor.
  a. How many moles of sulfur are in 23.2 g $(C_3H_5)_2S$?
  b. How many hydrogen atoms are in 0.75 mol $(C_3H_5)_2S$?
  c. How many grams of carbon are in 4.20×10^{23} molecules $(C_3H_5)_2S$?
  d. How many carbon atoms are in 15.0 g $(C_3H_5)_2S$?

6.5 Percent Composition and Empirical Formulas

We have seen that the atoms of the elements present in a compound are combined in a definite proportion. Now we can also say that the molar mass of any compound contains a definite proportion by mass of its elements. Using molar mass, we can determine the **percent composition** of a compound, which is the mass percent (mass %) of each element present.

$$\text{Mass \% of each element in a compound} = \frac{\text{mass of each element}}{\text{molar mass of compound}} \times 100\%$$

For example, let us calculate the percent composition of dinitrogen oxide, "laughing gas," N_2O.

STEP 1 **Determine the total mass of each element in the molar mass.**

$$2 \text{ mol N} \times \frac{14.01 \text{ g N}}{1 \text{ mol N}} = 28.02 \text{ g N}$$

$$1 \text{ mol O} \times \frac{16.00 \text{ g O}}{1 \text{ mol O}} = 16.00 \text{ g O}$$

$$\overline{1 \text{ mol } N_2O \qquad\qquad = 44.02 \text{ g } N_2O \text{ (molar mass)}}$$

Guide to Calculating Percent Composition

STEP 1
Determine the total mass of each element in the molar mass of a formula.

STEP 2
Divide the total mass of each element by the molar mass and multiply by 100.

STEP 2   **Calculate percent by multiplying the mass ratio by 100%.**

$$\%N = \frac{28.02 \text{ g N}}{44.02 \text{ g N}_2\text{O}} \times 100 = 63.65\%$$

$$\%O = \frac{16.00 \text{ g O}}{44.02 \text{ g N}_2\text{O}} \times 100 = 36.35\%$$

Check that the total percent is equal or very close to 100%.

$$63.65\% \text{ N} + 36.35\% \text{ O} = 100.00\%$$

Sample Problem 6.11   **Calculating Percent Composition**

The odor of pears is due to propyl acetate, $C_5H_{10}O_2$. What is its percent composition by mass?

Solution

STEP 1   **Determine the total mass of each element in the molar mass.**

$$5 \text{ mol C} \times \frac{12.01 \text{ g C}}{1 \text{ mol C}} = 60.05 \text{ g C}$$

$$10 \text{ mol H} \times \frac{1.008 \text{ g H}}{1 \text{ mol H}} = 10.08 \text{ g H}$$

$$2 \text{ mol O} \times \frac{16.00 \text{ g O}}{1 \text{ mol O}} = 32.00 \text{ g O}$$

1 mol of $C_5H_{10}O_2$ = 102.13 g (molar mass)

STEP 2   **Calculate percent by multiplying the mass ratio by 100%.**

$$\%C = \frac{60.05 \text{ g C}}{102.13 \text{ g C}_5\text{H}_{10}\text{O}_2} \times 100 = 58.80\%$$

$$\%H = \frac{10.08 \text{ g H}}{102.13 \text{ g C}_5\text{H}_{10}\text{O}_2} \times 100 = 9.870\%$$

$$\%O = \frac{32.00 \text{ g O}}{102.13 \text{ g C}_5\text{H}_{10}\text{O}_2} \times 100 = 31.33\%$$

Check that the total percent is equal to 100.0%.

$$58.80\% \text{ C} + 9.870\% \text{ O} + 31.33\% \text{ O} = 100.00\%$$

Study Check

What is the percent composition by mass of potassium carbonate K_2CO_3?

Empirical Formulas

Up to now, the formulas you have seen have been **molecular formulas,** which are the actual or true formulas of compounds. If we write a formula that represents the lowest whole-number ratio of the atoms in a compound, it is called the simplest or **empirical formula.** For example, the compound

CHEM NOTE

CHEMISTRY OF FERTILIZERS

Every year in the spring, homeowners and farmers add fertilizers to the soil to produce greener lawns and larger crops. Plants require several nutrients, but the major ones are nitrogen, phosphorus, and potassium. Nitrogen promotes green growth, phosphorus promotes strong root development for strong plants and abundant flowers, and potassium helps plants defend against diseases and weather extremes. The numbers on a package of fertilizer give the percentages each of N, P, and K by mass. For example, the numbers 30-3-4 indicate 30% nitrogen, 3% phosphorus, and 4% potassium.

The major nutrient, nitrogen, is present in huge quantities as N_2 in the atmosphere, but plants cannot utilize nitrogen in this

form. Bacteria in the soil convert atmospheric N_2 to usable forms by nitrogen fixation. To provide additional nitrogen to plants, several types of nitrogen-containing chemicals, including ammonia, nitrates, and ammonium compounds, are added to soil. The nitrates are absorbed directly, but ammonia and ammonium salts are first converted to nitrates by the soil bacteria.

The percent nitrogen in a fertilizer depends on the nitrogen compound. The percent nitrogen by mass in each type is calculated using percent composition.

Nitrogen (N) 30%
Phosphorus (P) 3%
Potassium (K) 4%

Type of Fertilizer	Percent Nitrogen by Mass	
NH_3	$\dfrac{14.01 \text{ g N}}{17.03 \text{ g NH}_3}$	$\times\ 100\% = 82.27\%$
NH_4NO_3	$\dfrac{28.02 \text{ g N}}{80.05 \text{ g NH}_4NO_3}$	$\times\ 100\% = 35.00\%$
$(NH_4)_2SO_4$	$\dfrac{28.02 \text{ g N}}{132.15 \text{ g (NH}_4)_2SO_4}$	$\times\ 100\% = 21.20\%$
$(NH_4)_2HPO_4$	$\dfrac{28.02 \text{ g N}}{132.06 \text{ g (NH}_4)_2HPO_4}$	$\times\ 100\% = 21.22\%$

The choice of a fertilizer depends on its use and convenience. A fertilizer can be prepared as crystals or a powder, in a liquid solution, or as a gas such as ammonia. The ammonia and ammonium fertilizers are water soluble and quick-acting. Other forms may be made to slow release by enclosing water-soluble ammonium salts in a thin plastic coating. The most commonly used fertilizer is NH_4NO_3 because it is easy to apply and has a high percent of N by mass.

benzene, with molecular formula C_6H_6, has the empirical formula CH. Some molecular formulas and their empirical formulas are shown in Table 6.3.

Table 6.3   Examples of Molecular and Empirical Formulas

Name	Molecular (actual)	Empirical (simplest formula)
acetylene	C_2H_2	CH
benzene	C_6H_6	CH
ammonia	NH_3	NH_3
hydrazine	N_2H_4	NH_2
ribose	$C_5H_{10}O_5$	CH_2O
glucose	$C_6H_{12}O_6$	CH_2O

The empirical formula of a compound is determined by converting the number of grams of each element to moles and finding the lowest whole-number ratio to use as subscripts, as shown in Sample Problem 6.12.

Guide to Calculating
Empirical Formula

STEP 1
Calculate the moles of
each element.

STEP 2
Divide by the smallest
number of moles.

STEP 3
Use lowest whole
number ratio of moles
as subscripts.

Sample Problem 6.12 **Calculating an Empirical Formula**

A laboratory experiment indicates that a compound contains 6.87 g iron and 13.1 g chlorine. What is the empirical formula of the compound?

Solution

STEP 1 **Calculate the number of moles of each element.**

$$6.87 \ \cancel{\text{g Fe}} \times \frac{1 \ \text{mol Fe}}{55.85 \ \cancel{\text{g Fe}}} = 0.123 \ \text{mol Fe}$$

$$13.1 \ \cancel{\text{g Cl}} \times \frac{1 \ \text{mol Cl}}{35.45 \ \cancel{\text{g Cl}}} = 0.370 \ \text{mol Cl}$$

STEP 2 **Divide each calculated number of moles by the smaller number of moles.** In this problem, the 0.123 mol Fe is the smaller amount.

$$\frac{0.123 \ \text{mol Fe}}{0.123} = 1.00 \ \text{mol Fe}$$

$$\frac{0.370 \ \text{mol Cl}}{0.123} = 3.01 \ \text{mol Cl}$$

STEP 3 **Use the lowest whole-number ratio of moles as subscripts.** The relationship of mol Fe to mol Cl is 1 to 3, which we obtain by rounding 3.01 to 3.

$$Fe_{1.00}Cl_{3.01} \longrightarrow Fe_1Cl_3, \text{ written as } FeCl_3$$

Study Check

Some yellow crystals are analyzed and found to contain 2.32 g Li and 2.68 g O. What is the empirical formula?

Often the relative amounts of the elements are given as the percent composition of a compound. Because percent means parts of one element per 100 parts of the total compound, we can express the mass of element as grams in 100 g of the compound, as shown in Sample Problem 6.13.

Sample Problem 6.13 **Calculating an Empirical Formula from Percent Composition**

Calculate the empirical formula of a compound that has 74.2% Na and 25.8% O.

Solution

STEP 1 **Calculate the number of moles of each element.** We must first change the percentages of the elements to grams by choosing a sample size of 100 g. Then the numerical values remain the same, but we have grams instead of the % signs. In exactly 100 g of this compound, there are 74.2 g Na and 25.8 g O. Then the number of moles for each element is calculated.

$$74.2 \text{ g Na} \times \frac{1 \text{ mol Na}}{22.99 \text{ g Na}} = 3.23 \text{ mol Na}$$

$$25.8 \text{ g O} \times \frac{1 \text{ mol O}}{16.00 \text{ g O}} = 1.61 \text{ mol O}$$

STEP 2 **Divide each of the calculated mole values by the smaller number of moles.**

$$\frac{3.23 \text{ mol Na}}{1.61} = 2.01 \text{ mol Na}$$

$$\frac{1.61 \text{ mol O}}{1.61} = 1.00 \text{ mol O}$$

STEP 3 **Use the lowest whole-number ratio of moles as subscripts.**

Na_2O

Study Check

Determine the empirical formula of a compound that has a percent composition of K 44.9%, S 18.4%, and O 36.7%.

Converting Decimal Numbers to Whole Numbers

Sometimes the result of dividing by the smallest number of moles gives a decimal instead of a whole number. Decimal values that are very close to whole numbers can be rounded. For example, 2.04 rounds to 2 and 6.98 rounds to 7. However, a decimal that is greater than 0.1 or less than 0.9 should not be rounded. Instead we multiply by a small integer until we obtain a whole number. Some multipliers that are typically used are listed in Table 6.4.

Let us suppose the numbers of moles we obtain give subscripts in the ratio of $C_{1.00}H_{2.33}O_{0.99}$. While 0.99 rounds to 1, we cannot round 2.33. If we multiply 2.33×2, we obtain 4.66, which is still not a whole number. If we multiply 2.33 by 3, the answer is 6.99, which rounds to 7. To complete the empirical formula, all the other subscripts must be multiplied by 3.

$$C_{(1.00 \times 3)}H_{(2.33 \times 3)}O_{(0.99 \times 3)} = C_{3.00}H_{6.99}O_{2.97} \longrightarrow C_3H_7O_3$$

Table 6.4 Some Multipliers That Convert Decimals to Whole Numbers

Decimal	Multiply by	Example		Whole Number
0.20	5	1.20×5	=	6
0.25	4	2.25×4	=	9
0.33	3	1.33×3	=	4
0.50	2	2.50×2	=	5
0.67	3	1.67×3	=	5

Sample Problem 6.14   Calculating an Empirical Formula

Ascorbic acid (vitamin C), found in citrus fruits and vegetables, contains carbon (40.9%), hydrogen (4.58%), and oxygen (54.5%). What is the empirical formula of ascorbic acid?

Solution

STEP 1 **Calculate the number of moles for each element.**   From the percent composition, we can write that 100 g ascorbic acid contains 40.9% C or 40.9 g C, 4.58% H or 4.58 g H, and 54.5% O or 54.5 g O.

$$C = 40.9\,g \qquad H = 4.58\,g \qquad O = 54.5\,g$$

Now we convert the grams of each element to moles.

$$\text{Moles of C} \qquad 40.9\,\cancel{g\,C} \;\times\; \frac{1\;mol\;C}{12.01\;\cancel{g\,C}} \;=\; 3.41\;mol\;C$$

$$\text{Moles of H} \qquad 4.58\,\cancel{g\,H} \;\times\; \frac{1\;mol\;H}{1.008\;\cancel{g\,H}} \;=\; 4.54\;mol\;H$$

$$\text{Moles of O} \qquad 54.5\,\cancel{g\,O} \;\times\; \frac{1\;mol\;O}{16.00\;\cancel{g\,O}} \;=\; 3.41\;mol\;O$$

STEP 2 **Divide by the smallest number of moles.**   For this problem, the smallest number of moles is 3.41.

$$C = \frac{3.41\;mol}{3.41\;mol} = 1.00$$

$$H = \frac{4.54\;mol}{3.41\;mol} = 1.33$$

$$O = \frac{3.41\;mol}{3.41\;mol} = 1.00$$

STEP 3 **Use the lowest whole-number ratio of moles as subscripts.** As calculated thus far, the ratio of moles gives the formula

$$C_{1.00}H_{1.33}O_{1.00}$$

However, the subscript for H is not close to a whole number, which means we cannot round it. If we multiply each of the subscripts by 3, we obtain a subscript for H that is close enough to 4 to round. Thus, we see that the empirical formula of ascorbic acid is $C_3H_4O_3$.

$$C_{(1.00\times3)}H_{(1.33\times3)}O_{(1.00\times3)} = C_{3.00}H_{3.99}O_{3.00} \longrightarrow C_3H_4O_3$$

Study Check

What is the empirical formula of glyoxylic acid, which occurs in unripe fruit, if it contains carbon 32.5%, hydrogen 2.70%, and oxygen 64.8%?

> **Questions and Problems**   **Percent Composition and Empirical Formulas**

6.33 Calculate the percent composition by mass of each of the following compounds:
  a. MgF_2, magnesium fluoride
  b. $Ca(OH)_2$, calcium hydroxide
  c. $C_4H_8O_4$, erythrose
  d. $(NH_4)_3PO_4$, ammonium phosphate
  e. $C_{17}H_{19}NO_3$, morphine

6.34 Calculate the percent composition by mass of each of the following compounds:
  a. $CaCl_2$, calcium chloride
  b. $Na_2Cr_2O_7$, sodium dichromate
  c. $C_2H_3Cl_3$, trichloroethane, a cleaning solvent
  d. $Ca_3(PO_4)_2$, calcium phosphate
  e. $C_{18}H_{36}O_2$, stearic acid

6.35 Calculate the percent by mass of N in each of the following compounds:
  a. N_2O_5, dinitrogen pentoxide
  b. NH_4NO_3, ammonium nitrate, fertilizer
  c. $C_2H_8N_2$, dimethylhydrazine, rocket fuel
  d. $C_9H_{15}N_5O$, Rogaine, stimulates hair growth
  e. $C_{14}H_{22}N_2O$, Lidocaine, local anesthetic

6.36 Calculate the percent by mass of sulfur in each of the following compounds:
  a. Na_2SO_4, sodium sulfate
  b. Al_2S_3, aluminum sulfide
  c. SO_3, sulfur trioxide
  d. C_2H_6SO, dimethylsulfoxide, topical anti-inflammatory
  e. $C_{10}H_{10}N_4O_2S$, sulfadiazine, antibacterial

6.37 Calculate the empirical formula for each of the following substances:

a. 3.57 g N and 2.04 g O
b. 7.00 g C and 1.75 g H
c. 0.175 g H, 2.44 g N, and 8.38 g O
d. 2.06 g Ca, 2.66 g Cr, and 3.28 g O

6.38 Calculate the empirical formula for each of the following substances:
  a. 2.90 g Ag and 0.430 g S
  b. 2.22 g Na and 0.774 g O
  c. 2.11 g Na, 0.0900 g H, 2.94 g S, and 5.86 g O
  d. 5.52 g K, 1.45 g P, and 3.00 g O

6.39 In an experiment, 2.51 g of sulfur combines with fluorine to give 11.44 g of a fluoride compound. What is the empirical formula of the compound?

6.40 In an experiment, 1.26 g iron combines with oxygen to give a compound that has a mass of 1.80 g. What is the empirical formula of the compound?

6.41 Calculate the empirical formula for each of the following substances:
  a. 70.9% K and 29.1% S
  b. 55.0% Ga and 45.0% F
  c. 31.0% B and 69.0% O
  d. 18.8% Li, 16.3% C, and 64.9% O
  e. 51.7% C, 6.90% H, 41.3% O

6.42 Calculate the empirical formula for each of the following substances:
  a. 55.5% Ca and 44.5% S
  b. 78.3% Ba and 21.7% F
  c. 76.0% Zn and 24.0 % P
  d. 29.1% Na, 40.6% S, and 30.3% O
  e. 19.8% C, 2.20% H, and 78.0% Cl

6.6 Molecular Formulas

Learning Goal

> Determine the molecular formula of a substance from the empirical formula and molar mass.

An empirical formula represents the lowest whole-number ratio of atoms in a compound. However, empirical formulas do not necessarily represent the actual number of atoms in a molecule. A molecular formula is related to the empirical formula by a small integer such as 1, 2, or 3.

$$\text{Molecular formula} = \text{small integer} \times \text{empirical formula}$$

For example, in Table 6.5, we see several different compounds that have the same empirical formula, CH_2O. The subscripts for the molecular formulas are related to the subscripts in the empirical formulas by small integers. The same relationship is true for the molar mass and empirical formula mass. The molar mass of each of the different compounds is related to the mass of the empirical formula (30.0 g) by the same small integer.

$$\text{Molar mass} = \text{small integer} \times \text{empirical formula mass}$$

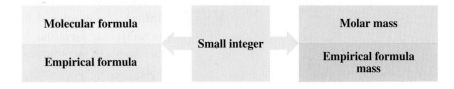

Table 6.5   Comparing the Molar Mass of Some Compounds with the Empirical Formula of CH_2O

Compound	Empirical Formula	Molecular Formula	Molar Mass (g)	Integer × Empirical Formula	Integer × Empirical Formula Mass
acetaldehyde	CH_2O	CH_2O	30.0	$(CH_2O)_1$	1(30.0)
acetic acid	CH_2O	$C_2H_4O_2$	60.0	$(CH_2O)_2$	2(30.0)
lactic acid	CH_2O	$C_3H_6O_3$	90.0	$(CH_2O)_3$	3(30.0)
erythrose	CH_2O	$C_4H_8O_4$	120.0	$(CH_2O)_4$	4(30.0)
ribose	CH_2O	$C_5H_{10}O_5$	150.0	$(CH_2O)_5$	5(30.0)

Relating Empirical and Molecular Formulas

Once we determine the empirical formula, we can calculate its mass in grams. If we are given the molar mass of the compound, we can calculate the value of the small integer.

$$\text{Small integer} = \frac{\text{molar mass of compound (g/mol)}}{\text{empirical formula mass (g/mol)}}$$

For example, when the molar mass of ribose is divided by the empirical formula mass, the integer is 5.

$$\text{Small integer} = \frac{\text{molar mass of ribose}}{\text{empirical formula mass of ribose}} = \frac{150.0 \text{ g/mol}}{30.0 \text{ g/mol}} = 5$$

Multiplying the subscripts in the empirical formula (CH_2O) by 5 gives the molecular formula of ribose, $C_5H_{10}O_5$.

$$5 \times \text{empirical formula } (CH_2O) = \text{molecular formula } (C_5H_{10}O_5)$$

Calculating a Molecular Formula

Earlier we worked out the empirical formula of ascorbic acid (vitamin C) to be $C_3H_4O_3$. If we are given a molar mass of 176.12 g for ascorbic acid, we can determine its molecular formula as follows.

Guide to Calculating a Molecular Formula from an Empirical Formula

STEP 1 Calculate the empirical formula mass.

STEP 2 Divide the molar mass by the empirical formula mass to obtain a small integer.

STEP 3 Multiply the empirical formula by an integer to obtain the molecular formula.

STEP 1 Calculate the mass of the empirical formula. The mass of the empirical formula $C_3H_4O_3$ is obtained in the same way as molar mass.

Empirical formula = 3 mol C + 4 mol H + 3 mol O
Empirical formula mass = (3 × 12.01) + (4 × 1.008) + (3 × 16.00)
= 88.06 g/mol

STEP 2 Divide the molar mass by the empirical formula mass to obtain a small integer.

$$\text{Small integer} = \frac{\text{molar mass of ascorbic acid}}{\text{empirical formula mass of } C_3H_4O_3} = \frac{176.12 \text{ g/mol}}{88.06 \text{ g/mol}} = 2$$

STEP 3 Multiply the empirical formula by the small integer to obtain the molecular formula. Multiplying all the subscripts in the empirical formula of ascorbic acid by 2 gives its molecular formula.

$$C_{(3\times2)}H_{(4\times2)}O_{(3\times2)} = C_6H_8O_6$$

Sample Problem 6.15   **Determination of a Molecular Formula**

Melamine, which is used to make plastic items such as dishes and toys, contains 28.57% carbon, 4.80% hydrogen, and 66.64% nitrogen. If the molar mass is 126.13 g, what is the molecular formula of melamine?

Solution

STEP 1   **Calculate the mass of the empirical formula.**   Write each percentage as grams in 100 grams of melamine and determine the number of moles of each.

$$C = 28.57 \text{ g} \qquad H = 4.80 \text{ g} \qquad N = 66.64 \text{ g}$$

$$\text{Moles of C} \qquad 28.57 \text{ g C} \times \frac{1 \text{ mol C}}{12.01 \text{ g C}} \quad = \quad 2.38 \text{ mol C}$$

$$\text{Moles of H} \qquad 4.80 \text{ g H} \times \frac{1 \text{ mol H}}{1.008 \text{ g H}} \quad = \quad 4.76 \text{ mol H}$$

$$\text{Moles of N} \qquad 66.64 \text{ g N} \times \frac{1 \text{ mol N}}{14.01 \text{ g N}} \quad = \quad 4.76 \text{ mol N}$$

Divide the moles of each element by the smallest number of moles, 2.38 mol, to obtain the subscripts of each element in the formula.

$$C = \frac{2.38 \text{ mol}}{2.38 \text{ mol}} = 1.00$$

$$H = \frac{4.76 \text{ mol}}{2.38 \text{ mol}} = 2.00$$

$$N = \frac{4.76 \text{ mol}}{2.38 \text{ mol}} = 2.00$$

Using these values as subscripts, $C_{1.00}H_{2.00}N_{2.00}$, we write the empirical formula of melamine as $C_1H_2N_2$ or CH_2N_2.

$$C_{1.00}H_{2.00}N_{2.00} = CH_2N_2$$

Now we calculate the molar mass of this empirical formula as follows:

Empirical formula = 1 mol C   + 2 mol H   + 2 mol N

Empirical formula mass = $(1 \times 12.01) + (2 \times 1.008) + (2 \times 14.01)$

$$= 42.05 \text{ g/mol}$$

STEP 2   **Divide the molar mass by the empirical formula mass to obtain a small integer.**

$$\text{Small integer} = \frac{\text{molar mass of melamine}}{\text{empirical formula mass of } CH_2N_2} = \frac{126.13 \text{ g/mol}}{42.05 \text{ g/mol}} = 3$$

STEP 3 **Multiply the empirical formula by the small integer to obtain the molecular formula.** Because the molar mass was three times the empirical formula mass, the subscripts in the empirical formula are multiplied by 3 to give the molecular formula.

$$C_{(1 \times 3)}H_{(2 \times 3)}N_{(2 \times 3)} = C_3H_6N_6$$

Study Check

The insecticide lindane has a percent composition of 24.78% C, 2.08% H, and 73.14% Cl. If its molar mass is about 291 g/mol, what is the molecular formula?

Questions and Problems   **Molecular Formulas**

6.43 Write the empirical formulas of each of the following substances:
   a. H_2O_2, peroxide
   b. $C_{18}H_{12}$, chrysene
   c. $C_{10}H_{16}O_2$, chrysanthemic acid in pyrethrum flowers
   d. $C_9H_{18}N_6$, altretamine
   e. $C_2H_4N_2O_2$, oxamide

6.44 Write the empirical formulas of each of the following substances:
   a. $C_6H_6O_3$, pyrogallol, a developer in photography
   b. $C_6H_{12}O_6$, galactose, a carbohydrate
   c. B_6H_{10}, hexaborane
   d. C_6Cl_6, hexachlorobenzene, a fungicide
   e. $C_{24}H_{16}O_{12}$, laccaic acid, a crimson dye

6.45 The carbohydrate fructose found in honey and fruits has an empirical formula of CH_2O. If the molar mass of fructose is 180 g, what is its molecular formula?

6.46 Caffeine has an empirical formula of $C_4H_5N_2O$. If it has a molar mass of 194.2 g, what is the molecular formula of caffeine?

6.47 Benzene and acetylene have the same empirical formula, CH. However, benzene has a molar mass of 78 g, and acetylene has a molar mass of 26 g. What are the molecular formulas of benzene and acetylene?

6.48 Glyoxyl, used in textiles; maleic acid, used to retard oxidation of fats and oils; and acontic acid, a plasticizer, all have the same empirical formula,

CHO. However, they each have a different molar mass: glyoxyl 58.0 g, maleic acid 116.1 g, and acontic acid 174.1 g. What are the molecular formulas of glyoxyl, maleic acid, and acontic acid?

6.49 Mevalonic acid is involved in the biosynthesis of cholesterol. Mevalonic acid is 48.64% C, 8.16% H, and 43.20% O. If mevalonic acid has a molar mass of 148, what is its molecular formula?

6.50 Chloral hydrate, a sedative, contains C 14.52%, H 1.83%, Cl 64.30%, and O 19.35%. If it has a molar mass of 165.42, what is the molecular formula of chloral hydrate?

6.51 Vanillic acid contains 57.14% C, 4.80% H, and 38.06% O and has a molar mass of 168.1 g. What is the molecular formula of vanillic acid?

6.52 Lactic acid is the substance that builds up in muscles during aerobic exercise. It has a molar mass of 90. g and has a composition of C 40.0%, H 6.71%, and the rest is oxygen. What is the molecular formula of lactic acid?

6.53 A sample of nicotine, a poisonous compound found in tobacco leaves, is 74.0% C, 8.7% H, and the remainder is nitrogen. If the molar mass of nicotine is 162 g, what is the molecular formula of nicotine?

6.54 Adenine is a nitrogen-containing compound found in DNA and RNA. If adenine has a composition of 44.5% C, 3.70% H, and 51.8% N and a molar mass of 135.1 g, what is the molecular formula?

Concept Map   **Chemical Quantities**

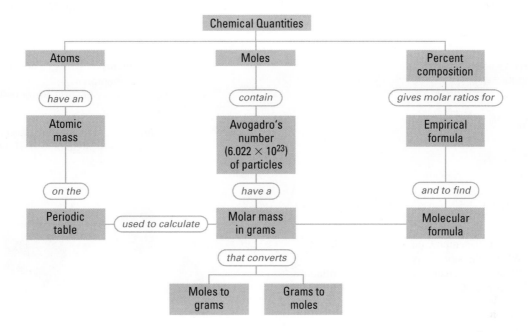

Chapter Review

6.1 Atomic Mass and Formula Mass

The formula mass of a compound is the sum of the atomic masses of the elements in the formula of the compound. The formula mass is used to convert between the number of atoms and a specified mass (amu) of an element or compound.

6.2 The Mole

One mole of an element contains 6.022×10^{23} atoms; a mole of a compound contains 6.022×10^{23} molecules or formula units.

6.3 Molar Mass

The molar mass (g/mol) of any substance is the mass in grams equal numerically to its atomic mass, or the sum of the atomic masses, which have been multiplied by their subscripts in a formula.

6.4 Calculations Using Molar Mass

The molar mass is the mass in grams of 1 mol of an element or a compound. It becomes an important conversion factor when it is used to change a quantity in grams to moles, or to change a given number of moles to grams.

6.5 Percent Composition and Empirical Formulas

The percent composition of a compound is obtained by dividing the mass in grams of each element in a compound by the molar mass of that compound. The empirical formula can be calculated by determining the mole ratio from the grams of the elements present in a sample of a compound or the percent composition.

6.6 Molecular Formulas

A molecular formula is equal to or a multiple of the empirical formula. The molar mass, which must be known, is divided by the mass of the empirical formula to obtain the multiple.

Key Terms

Avogadro's number The number of items in a mole, equal to 6.022×10^{23}.

empirical formula The simplest or smallest whole-number ratio of the atoms in a formula.

formula mass The mass in amu of a molecule or formula unit equal to the sum of the atomic masses of all the atoms in the formula.

formula unit The group of ions represented by the formula of an ionic compound.

molar mass The mass in grams of 1 mol of an element is equal numerically to its atomic mass. The molar mass of a compound is equal to the sum of the masses of the elements in the formula.

mole A group of atoms, molecules, or formula units that contains 6.022×10^{23} of these items.

molecular formula The actual formula that gives the number of atoms of each type of element in the compound.

percent composition The percent by mass of the elements in a formula.

▶Understanding the Concepts

6.55 A dandruff shampoo contains pyrithion, $C_{10}H_8N_2O_2S_2$, which acts as antibacterial and antifungal agent.

a. What is the empirical formula of pyrithion?
b. What is the molar mass of pyrithion?
c. What is the percent composition of pyrithion?
d. How many C atoms are in 25.0 g pyrithion?
e. How many moles of pyrithion contain 8.2×10^{24} nitrogen atoms?

6.56 Ibuprofen, the anti-inflammatory ingredient, has the formula $C_{13}H_{18}O_2$.

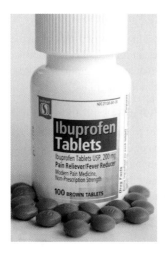

a. What is the percent by mass of oxygen in ibuprofen?
b. How many atoms of carbon are in 0.425 g ibuprofen?
c. How many grams of hydrogen are in 3.75×10^{22} molecules of ibuprofen?
d. How many atoms of hydrogen are in 0.245 g ibuprofen?

6.57 Using the following models of the molecules, determine (black = C, light blue = H, yellow = S, green = Cl)

1.

2.
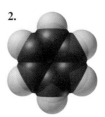

a. molecular formula   b. empirical formula
c. molar mass   d. percent composition

6.58 Using the following models of the molecules, determine (black = C, light blue = H, yellow = S, red = O)
a. molecular formula   b. empirical formula
c. molar mass   d. percent composition

1.

2.
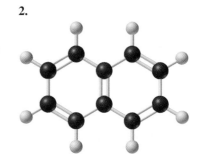

▶Additional Questions and Problems

6.59 Calculate the formula mass of each of the following:
a. $FeSO_4$, ferrous sulfate, iron supplement
b. $Ca(IO_3)_2$, calcium iodate, iodine source in table salt
c. $C_5H_8NNaO_4$, monosodium glutamate, flavor enhancer
d. $C_6H_{12}O_2$, isoamyl formate used to make artificial fruit syrups

6.60 Calculate the formula mass of each of the following:
a. $Mg(HCO_3)_2$, magnesium hydrogen carbonate

b. $Au(OH)_3$, gold(III) hydroxide, used in gold plating
c. $C_{18}H_{34}O_2$, oleic acid from olive oil
d. $C_{21}H_{26}O_5$, prednisone, anti-inflammatory

6.61 Calculate the percent composition for each of the following compounds:
a. K_2CrO_4
b. $Al(HCO_3)_3$
c. $C_6H_{12}O_6$

6.62 Determine the percent phosphorus in each of the following compounds:
   a. P_4O_{10} **b.** Mg_3P_2 **c.** $Ca_3(PO_4)_2$

6.63 During heavy exercise and workouts, lactic acid, $C_3H_6O_3$, accumulates in the muscles where it can cause pain and soreness.

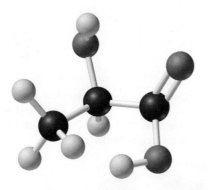

   a. What is the % oxygen in lactic acid?
   b. How many atoms of carbon are in 125 g lactic acid?
   c. How many grams of lactic acid contain 3.50 g hydrogen?
   d. What is the empirical formula of lactic acid?

6.64 Ammonium sulfate $(NH_4)_2SO_4$ is used in fertilizers.
   a. What is the % nitrogen in $(NH_4)_2SO_4$?
   b. How many hydrogen atoms are in 0.75 mol $(NH_4)_2SO_4$?
   c. How many grams of oxygen are in 4.50×10^{23} formula units $(NH_4)_2SO_4$?
   d. What mass of $(NH_4)_2SO_4$ contains 2.50 g sulfur?

6.65 Aspirin, $C_9H_8O_4$, is used to reduce inflammation and reduce fever.
   a. What is the percent composition by mass for aspirin?
   b. How many moles of aspirin contain 5.00×10^{24} carbon atoms?
   c. How many atoms of oxygen are in 7.50 g aspirin?
   d. How many molecules of aspirin contain 2.50 g hydrogen?

6.66 Rolaids is an antacid used for an upset stomach. One tablet contains 550 mg calcium carbonate, $CaCO_3$, and 110 mg magnesium hydroxide, $Mg(OH)_2$. One package contains 12 tablets.
   a. How many moles of calcium carbonate are present in one package?
   b. How many calcium ions, Ca^{2+}, would you obtain from two tablets?
   c. How many atoms of magnesium would you ingest if you take three tablets?
   d. How many hydroxide ions are in one tablet?

6.67 A mixture contains 0.250 mol Mn_2O_3 and 20.0 g MnO_2.
   a. How many atoms of oxygen are present in the mixture?
   b. How many grams of manganese are in the mixture?

6.68 A mixture contains 4.00×10^{23} molecules PCl_3 and 0.250 mol PCl_5.
   a. How many grams of Cl are present in the mixture?
   b. How many moles of P are in the mixture?

6.69 Write the empirical formula for each of the following compounds:
   a. $C_5H_5N_5$, adenine, a nitrogen compound in RNA and DNA
   b. FeC_2O_4, iron(II) oxalate, photographic developer
   c. $C_{16}H_{16}N_4$, stilbamidine, antibiotic for animals
   d. $C_6H_{14}N_2O_2$, lysine, an amino acid needed for growth

6.70 Write the empirical formula for each of the following compounds:
   a. N_2H_4, hydrazine, rocket fuel
   b. $C_{10}H_{10}O_5$, opionic acid
   c. $CrCl_3$, chromium(III) chloride, used in chrome plating
   d. $C_{16}H_{16}N_2O_2$, lysergic acid, a controlled substance from ergot

6.71 Calculate the empirical formula of each compound that contains
   a. 2.20 g S and 7.81 g F
   b. 6.35 g Ag, 0.825 g N, and 2.83 g O
   c. 89.2 g Au and 10.9 g O

6.72 Calculate the empirical formula of each compound from the percent composition.
   a. 61.0% Sn and 39.0% F
   b. 25.9% N and 74.1% O
   c. 22.1% Al, 25.4% P, and 52.5% O

6.73 Oleic acid, a component of olive oil, is 76.54% C, 12.13% H, and 11.33% O. The experimental value of the molar mass is about 282 g.
   a. What is the molecular formula of oleic acid?
   b. If oleic acid has a density of 0.895 g/mL, how many molecules of oleic acid are in 3.00 mL oleic acid?

6.74 Iron pyrite, commonly known as "fool's gold," is 46.5% Fe and 53.5% S.
   a. If a crystal of iron pyrite contains 4.58 g iron, what is the mass in grams of the crystal?
   b. If the empirical formula and the molecular formula are the same, what is the formula of the compound?
   c. How many moles of iron are in the crystal?

6.75 Succinic acid is 40.7% C, 5.12% H, and 54.2% O. If it has a molar mass of about 118 g, what are the empirical and molecular formulas?

6.76 Mercurous chlorate is 70.6% Hg, 12.5% Cl, and 16.9% O. If it has a molar mass of about 568 g, what are the empirical and molecular formulas?

6.77 A compound contains 1.65×10^{23} atoms C, 0.552 g H, and 4.39 g O. If 1 mol of the compound contains 4 mol O, what is the molecular formula and molar mass of the compound?

6.78 What is the molecular formula of a compound if 0.500 mol contains 0.500 mol Sr, 1.81×10^{24} atoms O, and 35.5 grams Cl?

Challenge Questions

6.79 A toothpaste contains 0.24% by mass sodium flouride used to prevent dental caries and 0.30% by mass triclosan, $C_{12}H_7Cl_3O_2$, a preservative and antigingivitis agent. One tube contains 119 g toothpaste.

  a. How many moles of NaF are in the tube of toothpaste?
  b. How many fluoride atoms, F^-, are in the tube of toothpaste?
  c. How many grams of sodium ion, Na^+, are in 1.50 g toothpaste?
  d. How many molecules of triclosan are in the tube of toothpaste?
  e. What is the percent composition by mass of triclosan?

6.80 Sorbic acid, an inhibitor of mold in cheese, has a percent composition of 64.27% carbon, 7.19% hydrogen, and 28.54% oxygen. If sorbic acid has a molar mass of about 112 g, what is its molecular formula?

6.81 Iron(III) chromate, a yellow powder used as a pigment in paints, contains 24.3% Fe, 33.9% Cr, and 41.8% O. If it has a molar mass of about 460 g, what are the empirical and molecular formulas?

6.82 A gold bar is 2.31 cm long, 1.48 cm wide, and 0.0758 cm thick.

  a. If gold has a density of 19.3 g/mL, what is the mass of the gold bar?
  b. How many atoms of gold are in the bar?
  c. When the same mass of gold combines with oxygen, the oxide product has a mass of 5.61 g. How many moles of oxygen atoms are combined with the gold?
  d. What is the molecular formula of the oxide product if it is the same as the empiciral formula?

Answers

Answers to Study Checks

6.1 1.0×10^5 amu

6.2 5.00×10^3 atoms Au

6.3 202.24 amu

6.4 0.432 mol H_2O

6.5 0.120 mol aspirin

6.6 4.52×10^{24} SO_4^{2-} ions

6.7 138.12 g

6.8 24.4 g Au

6.9 6.21×10^{-3} mol $CaCO_3$; 5.50×10^{-3} mol $MgCO_3$

6.10 0.0515 mol N

6.11 %K = 56.58%, %C = 8.690%, %O = 34.73%

6.12 Li_2O

6.13 K_2SO_4

6.14 $C_2H_2O_3$

6.15 $C_6H_6Cl_6$

Answers to Selected Questions and Problems

6.1
  a. 400.0 amu (4.000×10^2 amu)
  b. 7.800×10^6 amu
  c. 1.186×10^{22} amu

6.3
  a. 20 atoms Al
  b. 2.73×10^2 atoms Ag
  c. 1.7×10^3 atoms Ne

6.5
  a. 146.1 amu   **b.** 164.10 amu
  c. 46.07 amu   **d.** 180.16 amu

6.7
  a. 126.05 amu   **b.** 78.00 amu
  c. 151.92 amu   **d.** 310.1 amu

6.9 1 mol contains 6.022×10^{23} atoms, molecules of a covalent substance, or formula units of an ionic substance.

6.11 1 mol of chlorine atoms contains 6.022×10^{23} atoms chlorine and 1 mol chlorine molecules contains 6.022×10^{23} molecules chlorine, which is $2 \times 6.022 \times 10^{23}$ or 1.204×10^{24} atoms chlorine.

6.13
  a. 3.01×10^{23} C atoms
  b. 7.71×10^{23} SO_2 molecules
  c. 0.0867 mol Fe
  d. 14.1 mol C_2H_5OH

6.15
  a. 6.00 mol H   **b.** 8.00 mol O
  c. 1.20×10^{24} P atoms   **d.** 4.82×10^{24} O atoms

6.17
  a. 58.44 g   **b.** 159.7 g
  c. 277.4 g   **d.** 342.2 g
  e. 188.2 g   **f.** 365.5 g

6.19
  a. 70.90 g   **b.** 90.08 g
  c. 262.9 g   **d.** 83.98 g
  e. 98.95 g   **f.** 156.7 g

6.21
  a. 34.5 g   **b.** 112 g
  c. 5.50 g   **d.** 5.14 g
  e. 9.80×10^4 g

6.23
  a. 3.03 g   **b.** 38.1 g
  c. 9.30 g   **d.** 24.6 g
  e. 12.9 g

6.25
  a. 0.760 mol Ag   **b.** 0.0240 mol C
  c. 0.881 mol NH_3 **d.** 0.164 mol C_3H_8
  e. 1.53 mol Fe_2O_3

6.27
  a. 6.25 mol He   **b.** 0.781 mol O_2
  c. 0.320 mol $Al(OH)_3$ **d.** 0.106 mol Ga_2S_3
  e. 0.430 mol C_4H_{10}

6.29
  a. 1.25×10^{24} C atoms
  b. 4.14×10^{23} C atoms
  c. 1.13×10^{25} C atoms
  d. 2.22×10^{24} C atoms
  e. 7.5×10^{24} C atoms

6.31
  a. 66.1 g propane
  b. 0.771 mol propane
  c. 27.8 g C
  d. 2.78×10^{22} H atoms

6.33
  a. 39.01% Mg; 60.99% F
  b. 54.09% Ca; 43.18% O; 2.72% H
  c. 40.00% C; 6.71% H; 53.29% O
  d. 28.19% N; 8.12% H; 20.77% P; 42.92% O
  e. 71.56% C; 6.71% H; 4.91% N; 16.82% O

6.35
  a. 25.94% N   **b.** 35.00% N
  c. 46.62% N   **d.** 33.48% N
  e. 11.96% N

6.37
  a. N_2O **b.** CH_3
  c. HNO_3 **d.** $CaCrO_4$

6.39 SF_6

6.41
  a. K_2S **b.** GaF_3
  c. B_2O_3 **d.** Li_2CO_3
  e. $C_5H_8O_3$

6.43
  a. HO   **b.** C_3H_2
  c. C_5H_8O **d.** $C_3H_6N_2$
  e. CH_2NO

6.45 $C_6H_{12}O_6$

6.47 benzene C_6H_6; acetylene C_2H_2

6.49 $C_6H_{12}O_4$

6.51 $C_8H_8O_4$

6.53 $C_{10}H_{14}N_2$

6.55 a. C_5H_4NOS
b. 252.3 g
c. 47.60% C, 3.20% H, 11.10% N, 12.68% O, 25.42% S
d. 5.97×10^{23} C atoms
e. 6.8 mol pyrithion

6.57 1. a. S_2Cl_2 **b.** SCl
c. 135.04 g/mol **d.** 47.50% S, 52.50% Cl
2. a. C_6H_6 **b.** CH
c. 78.11 g/mol **d.** 92.25% C, 7.74% H

6.59 a. 151.92 amu **b.** 389.9 amu
c. 169.11 amu **d.** 116.16 amu

6.61 a. 40.27% K; 26.78% Cr; 32.96% O
b. 12.85% Al; 1.44% H; 17.16% C; 68.57% O
c. 40.00% C; 6.71% H; 53.29% O

6.63 a. 53.29% O **b.** 2.51×10^{24} C atoms
c. 52.1 g **d.** CH_2O

6.65 a. 60.00% C, 4.48% H, 35.52% O
b. 0.923 mol aspirin
c. 1.00×10^{23} O atoms
d. 1.87×10^{23} molecules aspirin

6.67 a. 7.29×10^{23} O atoms **b.** 40.1 g Mn

6.69 a. CHN **b.** FeC_2O_4
c. C_4H_4N **d.** C_3H_7NO

6.71 a. SF_6 **b.** $AgNO_3$
c. Au_2O_3

6.73 a. $C_{18}H_{34}O_2$
b. 5.73×10^{21} molecules oleic acid

6.75 The empirical formula is $C_2H_3O_2$; the molecular formula is $C_4H_6O_4$.

6.77 The molecular formula is $C_4H_8O_4$; the molar mass is 120.10 g/mol.

6.79 a. 0.00680 mol NaF.
b. 4.10×10^{21} F^- ions are in the tube.
c. 0.00197 g of Na^+.
d. 7.4×10^{20} triclosan molecules are in the tube.
e. 49.78% C, 2.437% H, 36.73% Cl, 11.05% O

6.81 The empirical formula is $Fe_2Cr_3O_{12}$. The molecular formula is $Fe_2Cr_3O_{12}$.

Combining Ideas
from Chapters 4–6

CI 6 The active ingredient in Tums has the percent composition Ca 40.0%, C 12.0%, and O 48.0%.

a. If the empirical and molecular formulas are the same, what is its molecular formula?
b. What is the name of the ingredient?
c. If one Tums tablet contains 500. mg of this ingredient and a person takes two tablets a day, how many calcium ions does this person obtain from the Tums tablets?

CI 7 Oxalic acid found in plants and vegetables such as rhubarb has the percent composition C 26.7%, H 2.24%, and O 71.1%. The oxalic acid present in rhubarb can be toxic when large quantities of raw or cooked leaves are ingested because oxalic acid can interfere with respiration. Oxalic acid can cause kidney or bladder stones. Rhubarb leaves contain about 0.5% oxalic acid. The lethal dose (LD$_{50}$) in rats for oxalic acid is 375 mg/kg.

a. What is the empirical formula of oxalic acid?
b. If oxalic acid has a molar mass of about 90. g/mol, what is the molecular formula?

c. Using the LD$_{50}$, how many grams of oxalic acid would be toxic for a 160-lb person?
d. How many kilograms of rhubarb leaves would the person in part c need to eat to reach the toxic level of oxalic acid?

CI 8 The compound butyric acid gives rancid butter its characteristic odor.

butryic acid

a. If black (grey) spheres are carbon atoms, light blue spheres are hydrogen atoms, and red spheres are oxygen atoms, what is the molecular formula?
b. What is the empirical formula of butyric acid?
c. What is the percent composition of butyric acid?
d. How many grams of carbon are in 0.850 g butyric acid?
e. How many grams of butyric acid contain 3.28 × 10^{23} oxygen atoms?
f. Butyric acid has a density of 0.959 g/mL at 20°C. How many moles of butyric acid are contained in 0.565 mL butyric acid?

CI 9 Tamiflu (Oseltamivir), C$_{16}$H$_{28}$N$_2$O$_4$, is an antiviral drug that is used to treat influenza. The preparation of Tamiflu begins with the extraction of shikimic acid from the seedpods of the Chinese spice star anise. However, the star anise has no antiviral activity itself.

shikimic acid

a. What is the empirical formula of Tamiflu?
b. What is the percent composition of Tamiflu?
c. What is the molecular formula of shikimic acid? (Black spheres are carbon, light blue spheres are hydrogen, and red spheres are oxygen.)
d. How many protons would be present in one molecule of shikimic acid?

CI 10 From 2.6 g of star anise, 0.13 g shikimic acid can be obtained and used to produce one capsule containing 75 mg Tamiflu. The usual adult dosage for treatment of influenza is 75 mg Tamiflu twice daily for 5 days. (See story CI 9.)

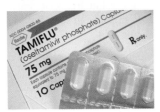

a. How many moles of shikimic acid are obtained from 1.3 g shikimic acid?
b. How many capsules containing 75 mg Tamiflu could be produced from 155 g star anise?
c. How many grams of carbon are in one dose (75 mg) of Tamiflu?
d. How many kilograms of Tamiflu would be needed to treat all the people in a city with a population of 500 000 people?
e. How many pounds of star anise would be needed to provide the Tamiflu needed in part d?

Answers to CIs

CI 7 a. CHO_2
b. $C_2H_2O_4$
c. 27 g oxalic acid
d. 5 kg of rhubarb

CI 9 a. $C_8H_{14}NO_2$
b. 61.51% C, 9.03% H, 8.97% N, 20.49% O
c. $C_7H_{10}O_5$
d. 92 protons

Chemical Reactions

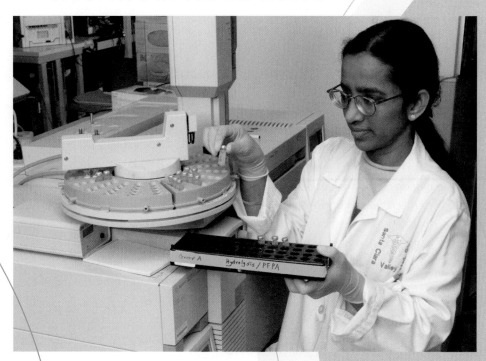

"*We use mass spectrometry to analyze and confirm the presence of drugs,*" *says Valli Vairavan, clinical lab technologist—Mass Spectrometry, Santa Clara Valley Medical Center.* "*A mass spectrometer separates and identifies compounds including drugs by mass. When we screen a urine sample, we look for metabolites, which are the products of drugs that have metabolized in the body. If the presence of one or more drugs such as heroin and cocaine is indicated, we confirm it by using mass spectrometry.*"

Drugs or their metabolites are detected in urine 24–48 hours after use. Cocaine metabolizes to benzoylecgonine and hydroxycocaine, morphine to morphine-3-glucuronide, and heroin to acetylmorphine. Amphetamines and methamphetamines are detected unchanged.

the Chemistry place

Visit **www.aw-bc.com/chemplace** for extra quizzes, interactive tutorials, career resources, PowerPoint slides for chapter review, math help, and case studies.

Chemical reactions occur everywhere. The fuel in our cars burns with oxygen to provide energy to make the cars move. When we cook our food or bleach our hair, chemical reactions take place. In our bodies chemical reactions convert food substances into molecules to build muscles and into energy to move them. In the leaves of trees and plants, carbon dioxide and water are converted into carbohydrates.

Some chemical reactions are simple, whereas others are quite complex. However, they can all be written with the chemical equations that chemists use to describe chemical reactions. In every chemical reaction, the atoms in the reacting substances, called reactants, are rearranged to give new substances called products. However, the atoms in the reactants are the same atoms as in the products, which means that matter is conserved and no matter is lost during a chemical change.

In this chapter, we will see how equations are written and how we can determine the amount of reactant or product involved in a chemical reaction. We do much the same thing at home when we use a recipe to make cookies. An auto mechanic does essentially the same thing when adjusting the fuel system of an engine to allow for the correct amounts of fuel and oxygen. In the body a certain amount of O_2 must reach the tissues for efficient metabolic reactions. When we know the chemical equation for a reaction, we can determine the amount of reactant needed or amount of product that can be produced.

7.1 Chemical Reactions

Learning **Goal**

Identify a change in a substance as a chemical or a physical change.

In a **physical change,** the appearance of a substance is altered, but not its composition. When liquid water becomes a gas, or freezes to a solid, it is still water (see Figure 7.1). If we smash a rock or tear a piece of paper, only the size of the material changes. The smaller pieces are still rock or paper because there was no change in the composition of the substances.

In a **chemical change,** the reacting substances change into new substances that have different compositions and different properties. New properties may involve a change in color, a change in temperature, or the formation of bubbles or a solid. For instance, when silver tarnishes, the bright silver metal (Ag) reacts with sulfur (S) to become the dull, black substance we call tarnish (Ag_2S) (see Figure 7.1). Table 7.1 gives some examples of some typical physical and chemical changes.

Table 7.1 Comparison of Some Chemical and Physical Changes

Chemical Changes	Physical Changes
Rusting nail	Melting ice
Bleaching a stain	Boiling water
Burning a log	Sawing a log in half
Tarnishing silver	Tearing paper
Fermenting grapes	Breaking a glass
Souring of milk	Pouring milk

Sample Problem 7.1 | **Classifying Chemical and Physical Change**

Classify each of the following changes as physical or chemical:
a. water freezing into an icicle
b. burning a match
c. breaking a chocolate bar
d. digesting a chocolate bar

Solution
a. Physical. Freezing water involves only a change from liquid water to ice. No change has occurred in the substance water.
b. Chemical. Burning a match causes the formation of new substances that have different properties.
c. Physical. Breaking a chocolate bar does not affect its composition.
d. Chemical. The digestion of the chocolate bar converts it into new substances.

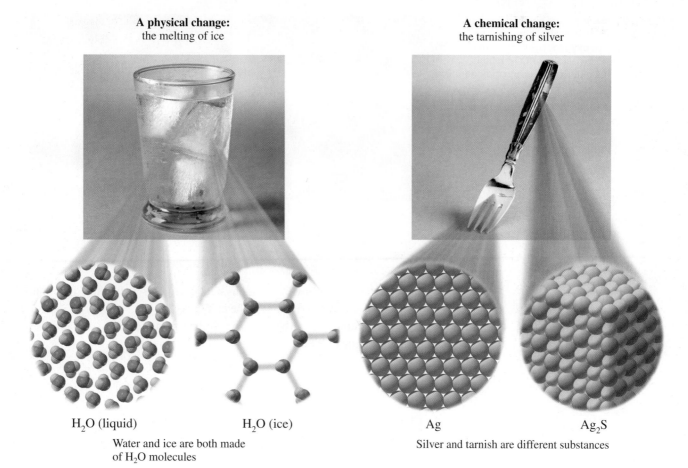

A physical change:
the melting of ice

A chemical change:
the tarnishing of silver

H_2O (liquid) H_2O (ice)

Water and ice are both made
of H_2O molecules

Ag Ag_2S

Silver and tarnish are different substances

Figure 7.1 A chemical change produces new substances; a physical change does not.

Q Why is the formation of tarnish considered a chemical change?

Study Check

Classify the following changes as physical or chemical:
a. chopping a carrot
b. developing a Polaroid picture
c. inflating a balloon

Changes During a Chemical Reaction

In a **chemical reaction,** the original substances change to new substances with different physical properties and different compositions. All of the atoms of the original substances are found in the new substances. However, some of the bonds between the atoms in the original substances have been broken and new bonds have formed between different combinations of atoms to give new substances. For example, when you light a gas burner, the molecules of methane gas (CH_4) react with oxygen (O_2) in the air to produce CO_2, H_2O, and heat. In another chemical reaction, a piece of iron (Fe) combines with oxygen (O_2) in the air to produce a new substance, rust (Fe_2O_3), which has a reddish-orange color. When an antacid tablet is placed in water, bubbles appear as sodium bicarbonate ($NaHCO_3$) and citric acid ($C_6H_8O_7$) in the tablet react to form carbon dioxide (CO_2) gas. (See Figure 7.2.) In each of these chemical reactions, new properties are visible, which are clues that tell you a chemical reaction has taken place. Table 7.2 summarizes some types of visible evidence of a chemical reaction.

Table 7.2   Types of Visible Evidence of a Chemical Reaction

1. Change in the color
2. Formation of a solid (precipitate)
3. Formation of a gas (bubbles)
4. Heat (or a flame) produced or heat absorbed

Figure 7.2 Examples of chemical reactions involve chemical change: iron (Fe) reacts with oxygen (O_2) to form rust (Fe_2O_3) and an antacid ($NaHCO_3$) tablet in water forms bubbles of carbon dioxide (CO_2).

Q What is the evidence for chemical change in these chemical reactions?

Fe Fe_2O_3 $NaHCO_3$

Sample Problem 7.2   **Evidence of a Chemical Reaction**

Identify each of the following as a physical change or a chemical reaction. If it is a chemical reaction, what is the evidence?
a. propane burning in a barbecue   **b.** chopping an onion
c. using peroxide to bleach hair

Solution

a. The production of heat during burning is evidence of a chemical reaction.
b. Chopping an onion into smaller pieces is a physical change.
c. The change in hair color is evidence of a chemical reaction.

Study Check

Striking a match is an example of a chemical reaction. What evidence would you see that indicates a chemical reaction?

Questions and Problems   **Chemical Reactions**

7.1 Classify each of the following changes as chemical or physical:
  a. grinding coffee
  b. ignition of fuel in the space shuttle
  c. drying clothes
  d. neutralizing stomach acid with an antacid tablet
7.2 Classify each of the following changes as chemical or physical:
  a. fogging the mirror during a shower
  b. tarnishing of a silver bracelet
  c. breaking a bone
  d. mending a broken bone
7.3 Identify each of the following as a physical change or a chemical reaction. If it is a chemical reaction, what is the evidence?

a. formation of snowflakes
b. an exploding dynamite stick
c. slicing a loaf of bread
d. toasting marshmallows over a campfire
7.4 Identify each of the following as a physical change or a chemical reaction. If it is a chemical reaction, what is the evidence?
  a. formation of reddish-brown smog from car exhaust
  b. tanning the skin
  c. removing tarnish from a silver bowl
  d. slicing potatoes for fries

7.2 Chemical Equations

When you build a model airplane, prepare a new recipe, or mix a medication, you follow a set of directions. These directions tell you what materials to use and the products you will obtain. In chemistry, a **chemical equation** tells us the materials we need and the products that will form in a chemical reaction.

Writing a Chemical Equation

Suppose you work in a bicycle shop, assembling wheels and bodies into bicycles. You could represent this process by a simple equation:

Equation: Wheels + Body ⟶ Bicycle

Reactants   Product

When you burn charcoal in a grill, the carbon in the charcoal combines with oxygen to form carbon dioxide. We can represent this reaction by a chemical equation that is much like the one for the bicycle:

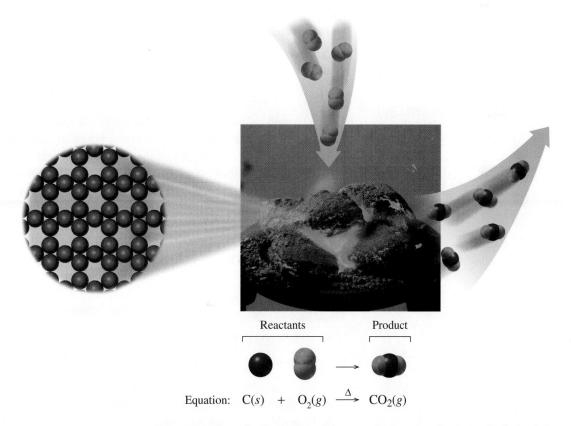

Reactants   Product

Equation: $C(s)$ + $O_2(g)$ $\xrightarrow{\Delta}$ $CO_2(g)$

In an equation, the formulas of the **reactants** are written on the left of the arrow and the formulas of the **products** on the right. When there are two or more formulas on the same side, they are separated by plus (+) signs. The delta sign (Δ) indicates that heat was used to start the reaction.

Table 7.3	Some Symbols Used in Writing Equations
Symbol	**Meaning**
+	Separates two or more formulas
⟶	Reacts to form products
Δ	The reactants are heated
(s)	Solid
(l)	Liquid
(g)	Gas
(aq)	Aqueous

Sometimes, as in the charcoal case, the formulas in an equation may include letters, in parentheses, that give the physical state of the substances as solid (*s*), liquid (*l*), or gas (*g*). If a substance is dissolved in water, it is an aqueous (*aq*) solution. Table 7.3 summarizes some of the symbols used in equations.

When a reaction takes place, the bonds between the atoms of the reactants are broken and new bonds are formed to give the products. In any chemical reaction, there must be the same number of atoms of each element present in the original substances as there are in the new substances. Atoms cannot be gained, lost, or changed into other types of atoms during a reaction. Therefore, a reaction must be written as a **balanced equation,** which shows the same number of atoms for each element on both sides of the arrow. Let's see if the preceding equation we wrote for burning carbon is balanced:

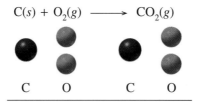

$$C(s) + O_2(g) \longrightarrow CO_2(g)$$

C   O   C   O

Reactant atoms   =   Product atoms

The answer is yes: this equation is *balanced* because there is one carbon atom and two oxygen atoms on each side of the arrow.

Now consider the reaction in which hydrogen reacts with oxygen to form water. First we write the formulas of the reactants and products:

$$H_2(g) + O_2(g) \longrightarrow H_2O(g)$$

Is the equation balanced? The answer is no; the equation is *not balanced*. There are two oxygen atoms to the left of the arrow, but only one to the right. Thus, the atoms on the left side do not match the atoms on the right side. To balance this equation, we place whole numbers called **coefficients** in front of the formulas. First we write a coefficient of 2 in front of the H_2O formula in order to have two oxygen atoms. Now the product has four hydrogen atoms, which means we must also write a coefficient of 2 in front of the formula H_2 in the reactants. Now the number of hydrogen atoms and oxygen atoms is the same in the reactants as in the products. The equation is *balanced*.

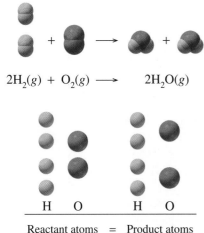

$$2H_2(g) + O_2(g) \longrightarrow 2H_2O(g)$$

H   O   H   O

Reactant atoms   =   Product atoms

Sample Problem 7.3   Chemical Equations

Hydrogen and nitrogen react to form ammonia, NH_3.

$$3H_2(g) + N_2(g) \longrightarrow 2NH_3(g)$$
Ammonia

a. What are the coefficients in the equation?

b. How many atoms of each element are in the reactants and products of the equation?

Solution

a. The coefficients are three (3) in front of H_2; one (1), which is understood, in front of N_2; and two (2) in front of NH_3.

b. In the reactants there are six hydrogen atoms and two nitrogen atoms. In the product there are also six hydrogen atoms and two nitrogen atoms.

Study Check

When ethane (C_2H_6) burns in oxygen, the products are carbon dioxide and water. The balanced equation is as follows.

$$2C_2H_6(g) + 7O_2(g) \longrightarrow 4CO_2(g) + 6H_2O(g)$$

State the total number of atoms of each element on each side of the equation.

Questions and Problems   Chemical Equations

7.5 State the number of atoms of each element on the reactant and on the product sides of the following equations:
 a. $2NO(g) + O_2(g) \longrightarrow 2NO_2(g)$
 b. $5C(s) + 2SO_2(g) \longrightarrow CS_2(g) + 4CO(g)$
 c. $2C_2H_2(g) + 5O_2(g) \longrightarrow 4CO_2(g) + 2H_2O(g)$
 d. $N_2H_4(g) + 2H_2O_2(g) \longrightarrow N_2(g) + 4H_2O(g)$

7.6 State the number of atoms of each element on the reactant and on the product sides of the following equations:
 a. $CH_4(g) + 2O_2(g) \longrightarrow CO_2(g) + 2H_2O(g)$
 b. $4P(s) + 5O_2(g) \longrightarrow P_4O_{10}(s)$
 c. $4NH_3(g) + 6NO(g) \longrightarrow 5N_2(g) + 6H_2O(g)$
 d. $6CO_2(g) + 6H_2O(l) \longrightarrow C_6H_{12}O_6(aq) + 6O_2(g)$
       Glucose

7.7 Determine whether each of the following equations is balanced or not balanced:
 a. $S(s) + O_2(g) \longrightarrow SO_3(g)$
 b. $2Al(s) + 3Cl_2(g) \longrightarrow 2AlCl_3(s)$
 c. $H_2(g) + O_2(g) \longrightarrow H_2O(g)$
 d. $C_3H_8(g) + 5O_2(g) \longrightarrow 3CO_2(g) + 4H_2O(g)$

7.8 Determine whether each of the following equations is balanced or not balanced:
 a. $PCl_3(l) + Cl_2(g) \longrightarrow PCl_5(s)$
 b. $CO(g) + 2H_2(g) \longrightarrow CH_3OH(l)$
 c. $2KClO_3(s) \longrightarrow 2KCl(s) + O_2(g)$
 d. $Mg(s) + N_2(g) \longrightarrow Mg_3N_2(s)$

7.9 All of the following are balanced equations. State the number of atoms of each element in the reactants and in the products.
 a. $2Na(s) + Cl_2(g) \longrightarrow 2NaCl(s)$
 b. $PCl_3(l) + 3H_2(g) \longrightarrow PH_3(g) + 3HCl(g)$
 c. $P_4O_{10}(s) + 6H_2O(l) \longrightarrow 4H_3PO_4(aq)$

7.10 All of the following are balanced equations. State the number of atoms of each element in the reactants and in the products.
 a. $2N_2(g) + 3O_2(g) \longrightarrow 2N_2O_3(g)$
 b. $Al_2O_3(s) + 6HCl(aq) \longrightarrow 2AlCl_3(aq) + 3H_2O(l)$
 c. $C_5H_{12}(l) + 8O_2(g) \longrightarrow 5CO_2(g) + 6H_2O(l)$

7.3 Balancing a Chemical Equation

Learning Goal

Write a balanced chemical equation from the formulas of the reactants and products for a reaction.

WEB TUTORIAL
Chemical Reactions and Equations

We have seen that a chemical equation must be balanced. In many cases, we can use a method of trial and error to balance an equation. To demonstrate the process, let us balance the reaction of the gas methane, CH_4, with oxygen to produce carbon dioxide and water. This is the principal reaction that occurs in the flame of a laboratory burner or a gas stove.

STEP 1 **Write an equation, using the correct formulas.** As a first step, we write the equation using the correct formulas for the reactants and products.

$$CH_4(g) + O_2(g) \longrightarrow CO_2(g) + H_2O(g)$$

CH_4 O_2 CO_2 H_2O

Guide to Balancing a Chemical Equation

STEP 1
Write an equation using the correct formulas of the reactants and products.

STEP 2
Count the atoms of each element in reactants and products.

STEP 3
Use coefficients to balance each element.

STEP 4
Check final equation for balance.

STEP 2 **Determine if the equation is balanced.** When we compare the atoms on the reactant side with the atoms on the product side, we see that there are more hydrogen atoms on the left side and more oxygen atoms on the right.

$$CH_4(g) + O_2(g) \longrightarrow CO_2(g) + H_2O(g)$$

1 C	1 C Balanced
4 H	2 H Not balanced
2 O	3 O Not balanced

STEP 3 **Balance the equation one element at a time.** Balance the hydrogen atoms by placing a coefficient of 2 in front of the formula for water.

$$CH_4(g) + O_2(g) \longrightarrow CO_2(g) + 2H_2O(g)$$

Then balance the oxygen atoms by placing a coefficient of 2 in front of the formula for oxygen. There are now four oxygen atoms and four hydrogen atoms in both the reactants and products.

$$CH_4(g) + 2O_2(g) \longrightarrow CO_2(g) + 2H_2O(g)$$

STEP 4 **Check to see if the equation is balanced.** Rechecking the balanced equation shows that the numbers of atoms of carbon, hydrogen, and oxygen are the same for both the reactants and the products. The equation is balanced using the lowest possible whole numbers as coefficients.

$$CH_4(g) + 2O_2(g) \longrightarrow CO_2(g) + 2H_2O(g) \quad \text{Balanced}$$

Reactants		**Products**
1 C atom	=	1 C atom
4 H atoms	=	4 H atoms
4 O atoms	=	4 O atoms

Suppose you had written the equation as follows:

$$2CH_4(g) + 4O_2(g) \longrightarrow 2CO_2(g) + 4H_2O(g) \quad \text{Incorrect}$$

Although there are equal numbers of atoms on both sides of the equation, this is not written correctly. All the coefficients must be divided by 2 to obtain the lowest possible whole numbers.

Sample Problem 7.4   Balancing Equations

Balance the following equation:

$$Na_3PO_4(aq) + MgCl_2(aq) \longrightarrow Mg_3(PO_4)_2(s) + NaCl(aq)$$

Solution

STEP 1 In the equation, the correct formulas are written.

$$Na_3PO_4(aq) + MgCl_2(aq) \longrightarrow Mg_3(PO_4)_2(s) + NaCl(aq)$$

STEP 2 When we compare the number of ions on the reactant and product sides, we find that the equation is not balanced. In this equation, it is more convenient to balance the phosphate polyatomic ion as a group instead of its individual atoms.

Reactants	Products	
$Na_3PO_4(aq) + MgCl_2(aq) \longrightarrow$	$Mg_3(PO_4)_2(s) + NaCl(aq)$	
$3\,Na^+$	$1\,Na^+$	Not balanced
$1\,PO_4^{3-}$	$2\,PO_4^{3-}$	Not balanced
$1\,Mg^{2+}$	$3\,Mg^{2+}$	Not balanced
$2\,Cl^-$	$1\,Cl^-$	Not balanced

STEP 3 We begin with the formula of $Mg_3(PO_4)_2$, which is the most complex. A 3 in front of $MgCl_2$ balances magnesium and a 2 in front of Na_3PO_4 balances the phosphate ion. Looking again at each of the ions in the reactants and products, we see that the sodium and chloride ions are not yet equal. A 6 in front of the NaCl balances the equation.

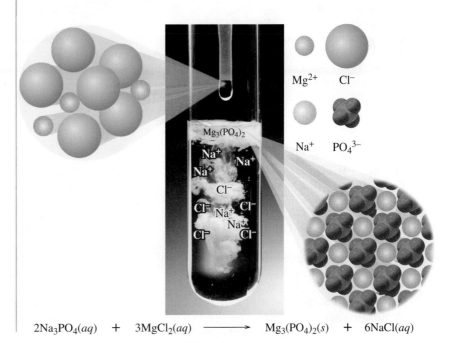

$$2Na_3PO_4(aq) + 3MgCl_2(aq) \longrightarrow Mg_3(PO_4)_2(s) + 6NaCl(aq)$$

STEP 4 A check of the atoms indicates the equation is balanced.

Reactants	Products

$$2Na_3PO_4(aq) + 3MgCl_2(aq) \longrightarrow Mg_3(PO_4)_2(s) + 6NaCl(aq)$$

Balanced

$6\,Na^+$	$=$	$6\,Na^+$
$2\,PO_4^{3-}$	$=$	$2\,PO_4^{3-}$
$3\,Mg^{2+}$	$=$	$3\,Mg^{2+}$
$6\,Cl^-$	$=$	$6\,Cl^-$

Study Check

Balance the following equation:

$$Fe(s) + O_2(g) \longrightarrow Fe_3O_4(s)$$

Questions and Problems   **Balancing a Chemical Equation**

7.11 Balance the following equations:
 a. $N_2(g) + O_2(g) \longrightarrow NO(g)$
 b. $HgO(s) \longrightarrow Hg(l) + O_2(g)$
 c. $Fe(s) + O_2(g) \longrightarrow Fe_2O_3(s)$
 d. $Na(s) + Cl_2(g) \longrightarrow NaCl(s)$
 e. $Cu_2O(s) + O_2(g) \longrightarrow CuO(s)$

7.12 Balance the following equations:
 a. $Ca(s) + Br_2(l) \longrightarrow CaBr_2(s)$
 b. $P_4(s) + O_2(g) \longrightarrow P_4O_{10}(s)$
 c. $C_4H_8(g) + O_2(g) \longrightarrow CO_2(g) + H_2O(l)$
 d. $Sb_2S_3(s) + HCl(aq) \longrightarrow SbCl_3(s) + H_2S(g)$
 e. $Fe_2O_3(s) + C(s) \longrightarrow Fe(s) + CO(g)$

7.13 Balance the following equations:
 a. $Mg(s) + AgNO_3(aq) \longrightarrow Mg(NO_3)_2(aq) + Ag(s)$
 b. $CuCO_3(s) \longrightarrow CuO(s) + CO_2(g)$
 c. $Al(s) + CuSO_4(aq) \longrightarrow Cu(s) + Al_2(SO_4)_3(aq)$
 d. $Pb(NO_3)_2(aq) + NaCl(aq) \longrightarrow$
    $PbCl_2(s) + NaNO_3(aq)$
 e. $Al(s) + HCl(aq) \longrightarrow AlCl_3(aq) + H_2(g)$

7.14 Balance the following equations:
 a. $Zn(s) + H_2SO_4(aq) \longrightarrow ZnSO_4(aq) + H_2(g)$
 b. $Al(s) + H_2SO_4(aq) \longrightarrow Al_2(SO_4)_3(aq) + H_2(g)$
 c. $K_2SO_4(aq) + BaCl_2(aq) \longrightarrow$
    $BaSO_4(s) + KCl(aq)$
 d. $CaCO_3(s) \longrightarrow CaO(s) + CO_2(g)$
 e. $Al_2(SO_4)_3(aq) + KOH(aq) \longrightarrow$
    $Al(OH)_3(s) + K_2SO_4(aq)$

7.15 Balance the following equations:
 a. $Fe_2O_3(s) + CO(g) \longrightarrow Fe(s) + CO_2(g)$
 b. $Li_3N(s) \longrightarrow Li(s) + N_2(g)$
 c. $Al(s) + HBr(aq) \longrightarrow AlBr_3(aq) + H_2(g)$
 d. $Ba(OH)_2(aq) + Na_3PO_4(aq) \longrightarrow$
    $Ba_3(PO_4)_2(s) + NaOH(aq)$
 e. $As_4S_6(s) + O_2(g) \longrightarrow As_4O_6(s) + SO_2(g)$

7.16 Balance the following equations:
 a. $K(s) + H_2O(l) \longrightarrow KOH(aq) + H_2(g)$
 b. $Cr(s) + S_8(s) \longrightarrow Cr_2S_3(s)$

 c. $BCl_3(s) + H_2O(l) \longrightarrow H_3BO_3(aq) + HCl(aq)$
 d. $Fe(OH)_3(s) + H_2SO_4(aq) \longrightarrow$
    $Fe_2(SO_4)_3(aq) + H_2O(l)$
 e. $BaCl_2(aq) + Na_3PO_4(aq) \longrightarrow$
    $Ba_3(PO_4)_2(s) + NaCl(aq)$

7.17 Write a balanced equation using the correct formulas and include conditions (*s*, *l*, *g*, or *aq*) for each of the following reactions:
 a. Lithium metal reacts with liquid water to form hydrogen gas and aqueous lithium hydroxide.
 b. Solid phosphorus reacts with chlorine gas to form solid phosphorus pentachloride.
 c. Solid iron(II) oxide reacts with carbon monoxide gas to form solid iron and carbon dioxide gas.
 d. Liquid pentene (C_5H_{10}) burns in oxygen gas to form carbon dioxide gas and water vapor.
 e. Hydrogen sulfide gas and solid iron(III) chloride react to form solid iron(III) sulfide and hydrogen chloride gas.

7.18 Write a balanced equation using the correct formulas and include conditions (*s*, *l*, *g*, or *aq*) for each of the following reactions:
 a. Solid calcium carbonate decomposes to produce solid calcium oxide and carbon dioxide gas.
 b. Nitrogen oxide gas reacts with carbon monoxide gas to produce nitrogen gas and carbon dioxide gas.
 c. Iron metal reacts with solid sulfur to produce solid iron(III) sulfide.
 d. Solid calcium reacts with nitrogen gas to produce solid calcium nitride.
 e. In the Apollo lunar module, hydrazine gas, N_2H_4, reacts with dinitrogen tetroxide gas to produce gaseous nitrogen and water vapor.

7.4 Types of Reactions

Learning Goal

Identify a reaction as a combination, decomposition, replacement, or combustion reaction.

A great number of reactions occur in nature, in biological systems, and in the laboratory. However, there are some general patterns among all reactions that help us classify reactions. Some reactions may fit into more than one reaction type.

Combination Reactions

In a **combination reaction,** two or more elements or compounds bond to form one product. For example, sulfur and oxygen combine to form the product sulfur dioxide.

Two or more reactants   combine to yield   a single product

  +

$$S(s) \; + \; O_2(g) \; \longrightarrow \; SO_2(g)$$

In Figure 7.3, the elements magnesium and oxygen combine to form a single product, magnesium oxide.

$$2Mg(s) + O_2(g) \longrightarrow 2MgO(s)$$

In other examples of combination reactions, elements or compounds combine to form a single product.

$$N_2(g) \quad + \; 3H_2(g) \longrightarrow 2NH_3(g) \qquad \text{Ammonia}$$
$$Cu(s) \quad + \; S(s) \quad \longrightarrow CuS(s)$$
$$MgO(s) \; + \; CO_2(g) \longrightarrow MgCO_3(s)$$

the **C**hemistry place

WEB TUTORIAL
Chemical Reactions and Equations

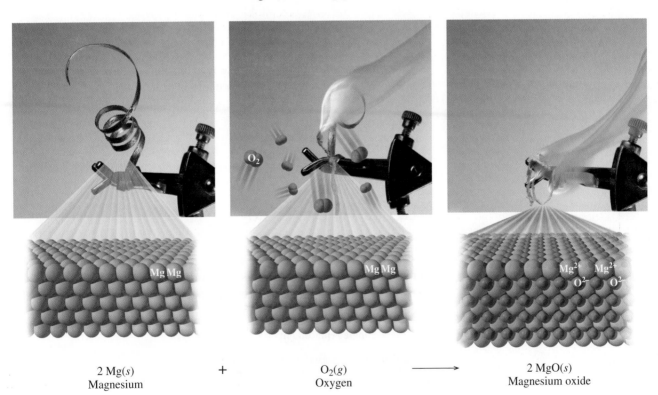

$$\begin{array}{ccccc} 2\,Mg(s) & + & O_2(g) & \longrightarrow & 2\,MgO(s) \\ \text{Magnesium} & & \text{Oxygen} & & \text{Magnesium oxide} \end{array}$$

Figure 7.3 In a combination reaction, two or more substances combine to form one substance as product.

Q What happens to the reactants in a combination reaction?

Figure 7.4 In a decomposition reaction, one reactant breaks down into two or more products.

Q How do the differences in the reactant and products classify this as a decomposition reaction?

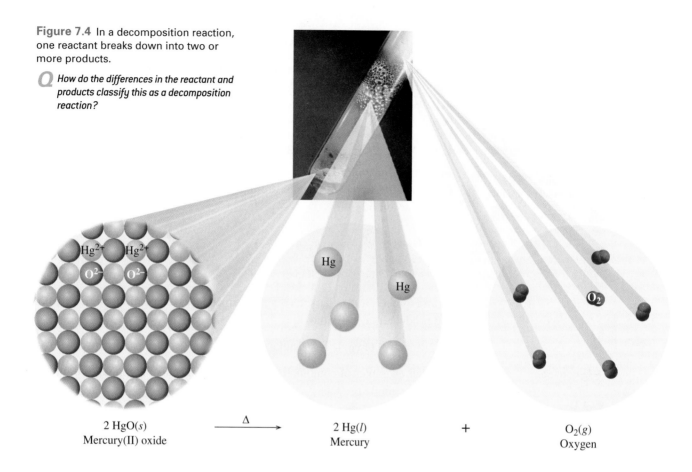

$$2\,HgO(s) \xrightarrow{\Delta} 2\,Hg(l) + O_2(g)$$

2 HgO(s)   2 Hg(l)   +   O₂(g)
Mercury(II) oxide   Mercury   Oxygen

A reactant	splits into	two or more products
A B	⟶	A + B

Decomposition Reactions

In a **decomposition reaction,** a single reactant splits into two or more products. For example, when mercury(II) oxide is heated, the products are mercury and oxygen. (See Figure 7.4.)

$$2HgO(s) \xrightarrow{\Delta} 2Hg(l) + O_2(g)$$

In another example of a decomposition reaction, calcium carbonate breaks apart into simpler compounds of calcium oxide and carbon dioxide.

$$CaCO_3(s) \xrightarrow{\Delta} CaO(s) + CO_2(g)$$

Replacement Reactions

In replacement reactions, elements in compounds are replaced by other elements. In a **single replacement reaction,** an uncombined element takes the place of an element in a compound.

Single replacement

One element   replaces   another element

A + B C ⟶ A C + B

In the single replacement reaction shown in Figure 7.5, zinc replaces hydrogen in hydrochloric acid, HCl(aq).

$$Zn(s) + 2HCl(aq) \longrightarrow ZnCl_2(aq) + H_2(g)$$

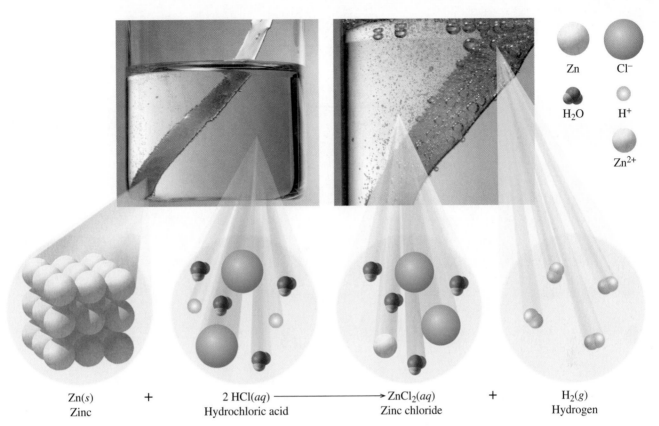

$$\underset{\text{Zinc}}{\text{Zn}(s)} \quad + \quad \underset{\text{Hydrochloric acid}}{2\,\text{HCl}(aq)} \longrightarrow \underset{\text{Zinc chloride}}{\text{ZnCl}_2(aq)} \quad + \quad \underset{\text{Hydrogen}}{\text{H}_2(g)}$$

Figure 7.5 In a single replacement reaction, an atom or ion replaces an atom or ion in a compound.

Q What changes in the formulas of the reactants identify this equation as a single replacement reaction?

In the following single replacement reaction, the halogen chlorine in Group 7A (17) replaces bromine in the compound potassium bromide.

$$\text{Cl}_2(g) + 2\text{KBr}(s) \longrightarrow 2\text{KCl}(s) + \text{Br}_2(l)$$

In a **double replacement reaction,** the positive ions in the reacting compounds switch places.

Double replacement

Two elements replace each other

A B + C D ⟶ A D + C B

For example, in the reaction shown in Figure 7.6, barium ions change places with sodium ions in the reactants to form sodium chloride and a white solid precipitate of barium sulfate.

$$\text{Na}_2\text{SO}_4(aq) + \text{BaCl}_2(aq) \longrightarrow \text{BaSO}_4(s) + 2\text{NaCl}(aq)$$

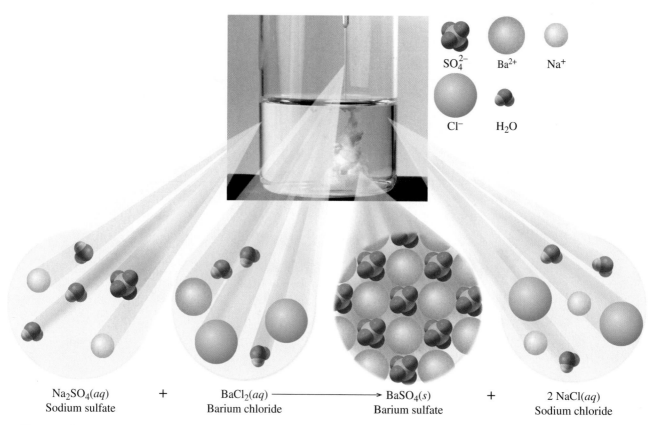

$$\text{Na}_2\text{SO}_4(aq) \quad + \quad \text{BaCl}_2(aq) \longrightarrow \text{BaSO}_4(s) \quad + \quad 2\,\text{NaCl}(aq)$$

Sodium sulfate   Barium chloride   Barium sulfate   Sodium chloride

Figure 7.6 In a double replacement reaction, the positive ions in the reactants replace each other.

Q How do the changes in the formulas of the reactants identify this equation as a double replacement reaction?

When sodium hydroxide and hydrochloric acid (HCl) react, sodium and hydrogen ions switch places, forming sodium chloride and water.

$$\text{NaOH}(aq) + \text{HCl}(aq) \longrightarrow \text{NaCl}(aq) + \text{HOH}(l)$$

Combustion Reactions

The burning of a log in a fireplace and the burning of gasoline in the engine of a car are examples of combustion reactions. In a **combustion reaction,** oxygen is required and often the reaction produces an oxide, water, and heat. For example, methane gas (CH_4) reacts with oxygen to produce carbon dioxide and water. The heat produced by this combustion reaction cooks our food and heats our homes.

$$\text{CH}_4(g) + 2\text{O}_2(g) \longrightarrow \text{CO}_2(g) + 2\text{H}_2\text{O}(g) + \text{heat}$$

Combustion reactions also occur in the cells of the body in order to metabolize food, which provides energy for the activities we want to do. We absorb oxygen (O_2) from the air to burn glucose ($C_6H_{12}O_6$) from our food, and eventually our cells produce CO_2, H_2O, and energy.

$$\text{C}_6\text{H}_{12}\text{O}_6(aq) + 6\text{O}_2(g) \longrightarrow 6\text{CO}_2(g) + 6\text{H}_2\text{O}(l) + \text{energy}$$

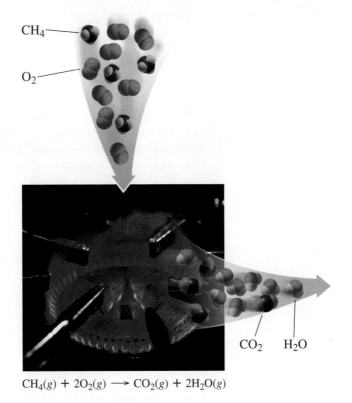

$$CH_4(g) + 2O_2(g) \longrightarrow CO_2(g) + 2H_2O(g)$$

In some combustion reactions, a single oxide product forms, which means they are also combination reactions.

$$S(s) + O_2(g) \longrightarrow SO_2(g)$$

$$2Al(s) + 3O_2(g) \longrightarrow 2Al_2O_3(s)$$

Table 7.4 summarizes the reaction types and gives examples.

Table 7.4   Summary of Reaction Types

Reaction Type	Example
Combination	
$A + B \longrightarrow AB$	$Ca(s) + Cl_2(g) \longrightarrow CaCl_2(s)$
Decomposition	
$AB \longrightarrow A + B$	$Fe_2S_3(s) \longrightarrow 2Fe(s) + 3S(s)$
Single Replacement	
$A + BC \longrightarrow AC + B$	$Cu(s) + 2AgNO_3(aq) \longrightarrow Cu(NO_3)_2(aq) + 2Ag(s)$
Double Replacement	
$AB + CD \longrightarrow AD + CB$	$BaCl_2(aq) + K_2SO_4(aq) \longrightarrow BaSO_4(s) + 2KCl(aq)$
Combustion	
$AH_4 + 2O_2 \longrightarrow AO_2 + 2H_2O + heat$	$CH_4(g) + 2O_2(g) \longrightarrow CO_2(g) + 2H_2O(g) + heat$
$B + O_2 \longrightarrow BO_2$	$S(s) + O_2(g) \longrightarrow SO_2(g)$

> **Sample Problem 7.5** ▸   **Identifying Reactions and Predicting Products**
>
> 1. Classify the following reactions as combination, decomposition, or single or double replacement:
>   a. $2Fe_2O_3(s) + 3C(s) \longrightarrow 3CO_2(g) + 4Fe(s)$
>   b. $Fe_2S_3(s) \longrightarrow 2Fe(s) + 3S(s)$
>   c. $BaCl_2(aq) + K_2SO_4(aq) \longrightarrow BaSO_4(s) + 2KCl(aq)$
> 2. Predict the product for each of the following and balance the equation:
>   a. $Al(s) + CuCl_2(aq) \longrightarrow$ (single replacement)
>   b. $K(s) + Cl_2(g) \longrightarrow$ (combination)
>
> *Solution*
>
> 1. a. In this *single replacement* reaction, a C atom replaces Fe in Fe_2O_3 to form the compound CO_2 and Fe atoms.
>   b. When one reactant breaks down into two products, the reaction is *decomposition.*
>   c. There are two reactants and two products, but the positive ions have exchanged places, which makes this a *double replacement* reaction.
> 2. a. To complete this single replacement reaction, the Cu in the $CuCl_2$ compound is replaced by Al.
>
> $$2Al(s) + 3CuCl_2(aq) \longrightarrow 2AlCl_3(aq) + 3Cu(s)$$
>
>   b. To complete this combination reaction, K and Cl are combined in a compound, which would be KCl.
>
> $$2K(s) + Cl_2(g) \longrightarrow 2KCl(s)$$
>
> *Study Check*
>
> Nitrogen gas and oxygen gas react to form nitrogen dioxide gas. Write the balanced equation and identify the reaction type.

Questions and Problems ▸   **Types of Reactions**

7.19 a. Why is the following reaction called a decomposition reaction?

$$2Al_2O_3(s) \xrightarrow{\Delta} 4Al(s) + 3O_2(g)$$

b. Why is the following reaction called a single replacement reaction?

$$Br_2(g) + BaI_2(s) \longrightarrow BaBr_2(s) + I_2(g)$$

7.20 a. Why is the following reaction called a combination reaction?

$$H_2(g) + Br_2(g) \longrightarrow 2HBr(g)$$

b. Why is the following reaction called a double replacement reaction?

$$AgNO_3(aq) + NaCl(aq) \longrightarrow AgCl(s) + NaNO_3(aq)$$

7.21 Classify each of the following reactions as a combination, decomposition, single replacement, double replacement, or combustion reaction:
a. $4Fe(s) + 3O_2(g) \longrightarrow 2Fe_2O_3(s)$
b. $Mg(s) + 2AgNO_3(aq) \longrightarrow$
$$Mg(NO_3)_2(aq) + 2Ag(s)$$
c. $CuCO_3(s) \longrightarrow CuO(s) + CO_2(g)$
d. $NaOH(aq) + HCl(aq) \longrightarrow NaCl(aq) + H_2O(l)$
e. $C_4H_8(g) + 6O_2(g) \longrightarrow 4CO_2(g) + 4H_2O(g)$
f. $ZnCO_3(s) \longrightarrow CO_2(g) + ZnO(s)$
g. $Al_2(SO_4)_3(aq) + 6KOH(aq) \longrightarrow$
$$2Al(OH)_3(s) + 3K_2SO_4(aq)$$
h. $Pb(s) + O_2(g) \longrightarrow PbO_2(s)$

7.22 Classify each of the following reactions as a combination, decomposition, single replacement, double replacement, or combustion reaction:

   a. $CuO(s) + 2HCl(aq) \longrightarrow CuCl_2(aq) + H_2O(l)$

   b. $2Al(s) + 3Br_2(g) \longrightarrow 2AlBr_3(s)$

   c. $2C_2H_2(g) + 5O_2(g) \longrightarrow 4CO_2(g) + 2H_2O(g)$

   d. $Pb(NO_3)_2(aq) + 2NaCl(aq) \longrightarrow$
$$PbCl_2(s) + 2NaNO_3(aq)$$

   e. $2Mg(s) + O_2(g) \longrightarrow 2MgO(s)$

   f. $Fe_2O_3(s) + 3C(s) \longrightarrow 2Fe(s) + 3CO(g)$

   g. $C_6H_{12}O_6(aq) \longrightarrow 2C_2H_6O(aq) + 2CO_2(g)$

   h. $BaCl_2(aq) + K_2CO_3(aq) \longrightarrow$
$$BaCO_3(s) + 2KCl(aq)$$

7.23 Try your hand at predicting the products that would result from the following types of reactions. Balance each equation you write.

   a. combination: $Mg(s) + Cl_2(g) \longrightarrow$

   b. decomposition: $HBr(g) \longrightarrow$

   c. single replacement: $Mg(s) + Zn(NO_3)_2(aq) \longrightarrow$

   d. double replacement: $K_2S(aq) + Pb(NO_3)_2(aq) \longrightarrow$

   e. combustion: $C_2H_6(g) + O_2(g) \longrightarrow$

7.24 Try your hand at predicting the products that would result from the reactions of the following. Balance each equation you write.

   a. combination: $Ca(s) + S(s) \longrightarrow$

   b. decomposition: $PbO_2(s) \longrightarrow$

   c. single replacement: $KI(s) + Cl_2(g) \longrightarrow$

   d. double replacement: $CuCl_2(aq) + Na_2S(aq) \longrightarrow$

   e. combustion: $C_2H_4(g) + O_2(g) \longrightarrow$

CHEM NOTE

SMOG AND HEALTH CONCERNS

There are two types of smog. One, photochemical smog, requires sunlight to initiate reactions that produce pollutants such as nitrogen oxides and ozone. The other type of smog, industrial or London smog, occurs in areas where coal containing sulfur is burned and the unwanted product sulfur dioxide is emitted.

Photochemical smog is most prevalent in cities where people are dependent on cars for transportation. On a typical day in Los Angeles, for example, nitrogen oxide (NO) emissions from car exhausts increase as traffic increases on the roads. The nitrogen oxide is formed when N_2 and O_2 react at high temperatures in car and truck engines.

$$N_2(g) + O_2(g) \xrightarrow{\text{Heat}} 2NO(g)$$

Then, NO reacts with oxygen in the air to produce NO_2, a reddish-brown gas that is irritating to the eyes and damaging to the respiratory tract.

$$2NO(g) + O_2(g) \longrightarrow 2NO_2(g)$$

When NO_2 is exposed to sunlight, it is converted into NO and oxygen atoms.

$$NO_2(g) \xrightarrow{\text{Sunlight}} NO(g) + \underset{\text{Oxygen atoms}}{O(g)}$$

Oxygen atoms are so reactive that they combine with oxygen molecules in the atmosphere, forming ozone.

$$O(g) + O_2(g) \longrightarrow \underset{\text{Ozone}}{O_3(g)}$$

In the upper atmosphere (the stratosphere), ozone is beneficial because it protects us from harmful ultraviolet radiation that comes from the sun. However, in the lower atmosphere, ozone irritates the eyes and respiratory tract, where it causes coughing, decreased lung function, and fatigue. It also causes deterioration of fabrics, cracks rubber, and damages trees and crops.

Industrial smog is prevalent in areas where coal with a high sulfur content is burned to produce electricity. During combustion, the sulfur is converted to sulfur dioxide:

$$S(s) + O_2(g) \longrightarrow SO_2(g)$$

The SO_2 is damaging to plants, suppresses growth, and is corrosive to metals such as steel. SO_2 is also damaging to humans and can cause lung impairment and respiratory difficulties. The SO_2 in the air reacts with more oxygen to form SO_3. SO_3 combines with water in the air to form sulfuric acid, which makes the rain acidic.

$$2SO_2(g) + O_2(g) \longrightarrow 2SO_3(g)$$
$$SO_3(g) + H_2O(l) \longrightarrow \underset{\text{Sulfuric acid}}{H_2SO_4(aq)}$$

The presence of sulfuric acid in rivers and lakes causes an increase in the acidity of the water, reducing the ability of animals and plants to survive.

Given the heat of reaction (enthalpy change), calculate the loss or gain of heat for an exothermic or endothermic reaction.

7.5 Energy in Chemical Reactions

Almost every chemical reaction involves a loss or gain of energy. To discuss energy change for a reaction, we refer to the reactants and products as the *system*. Everything else in contact with the system, such as the reaction flask and the air in the room, is called the *surroundings*.

Heat of Reaction (Enthalpy Change)

When warm food is placed in a refrigerator, heat flows out of the food and into the refrigeration unit. When a pan of water is placed on a hot burner, the heat from the burner flows into the pan and water. The **heat of reaction** or enthalpy change (symbol ΔH) is a measure of the amount of heat that flows into (absorbed) or out of (given off) a system as a reaction takes place at constant pressure. In a chemical reaction, a change of energy occurs as reactants break apart and products form. We determine a heat of reaction as the difference in the energy of the reactants and the products.

$$\Delta H = H_{\text{products}} - H_{\text{reactants}}$$

In an **endothermic reaction** (*endo* means "within"), the energy of the products is higher than the energy of the reactants. This means that an endothermic reaction requires energy. Heat from the surroundings must flow into the system to convert the reactants to products. When heat flows into the system, the ΔH value has a positive sign (+). In the equation for an endothermic reaction, the ΔH can be written as one of the reactants. For example, 570 kJ of heat are required to convert 2 mol carbon dioxide to 2 mol carbon monoxide and 1 mol oxygen.

Endothermic, Heat Flows in

$$2CO_2(g) + 570\,\text{kJ} \longrightarrow 2CO(g) + O_2(g)$$

$$2CO_2(g) \longrightarrow 2CO(g) + O_2(g) \qquad \Delta H = +570\,\text{kJ}$$

Heat Flow for Endothermic Reactions

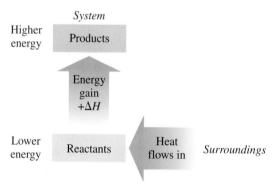

In an **exothermic reaction** (*exo* means "out"), the energy of the reactants is greater than the energy of the products. When an exothermic reaction occurs, energy is given off as products form. When heat flows out of a system to the surroundings, the ΔH value has a negative sign ($-$). In the equation for an exothermic reaction, the ΔH can be written as one of the products. For example, in the thermite reaction, the reaction of aluminum and iron(III) oxide produces a great amount of heat. The amount of heat produced during the thermite reaction is so immense that the products reach temperatures of 2500°C, forming liquids before they cool to solids. This reaction has been used to cut or weld railroad tracks.

Exothermic, Heat Given Off

$$2Al(s) + Fe_2O_3(s) \longrightarrow 2Fe(s) + Al_2O_3(s) + 850\,kJ$$

$$2Al(s) + Fe_2O_3(s) \longrightarrow 2Fe(s) + Al_2O_3(s) \qquad \Delta H = -850\,kJ$$

Heat Flow for Exothermic Reactions

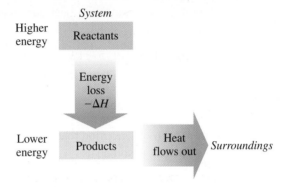

Reaction	Energy Change	Heat in the Equation	Sign of ΔH
Endothermic	Heat flows in (absorbed)	Reactant side	Positive ($+$)
Exothermic	Heat flows out (given off)	Product side	Negative ($-$)

Sample Problem 7.6   **Exothermic and Endothermic Reactions**

In the reaction of 1 mol solid carbon with oxygen gas, the energy of the carbon dioxide produced is 393 kJ lower than the energy of the reactants.
a. Is the reaction exothermic or endothermic?
b. Write the balanced equation for the reaction, including the heat of the reaction.
c. Write the value of ΔH for this reaction.

Solution

a. When the products have a lower energy than the reactants, the reaction is exothermic.
b. $C(s) + O_2(g) \longrightarrow CO_2(g) + 393\,kJ$
c. $\Delta H = -393\,kJ$

Study Check

The reaction of 1 mol hydrogen gas (H_2) with iodine gas (I_2) to form hydrogen iodide (HI) is endothermic and requires 50. kJ of heat.
a. Write a balanced equation for the reaction, including the heat of reaction.
b. Write the value of ΔH for this reaction.

Calculations of Heat in Reactions

The value of ΔH refers to the heat change for the number of moles (kJ/mol) of each substance in the balanced equation for the reaction. Consider the following decomposition reaction:

$$2H_2O(l) \longrightarrow 2H_2(g) + O_2(g) \qquad \Delta H = 572 \text{ kJ}$$
$$2H_2O(l) + 572 \text{ kJ} \longrightarrow 2H_2(g) + O_2(g)$$

For this reaction, 572 kJ are absorbed by 2 mol H_2O to produce 2 mol H_2 and 1 mol O_2. We can write heat conversion factors for each substance in this reaction.

$$\frac{572 \text{ kJ}}{2 \text{ mol } H_2O} \qquad \frac{572 \text{ kJ}}{2 \text{ mol } H_2} \qquad \frac{572 \text{ kJ}}{1 \text{ mol } O_2}$$

Moles of compound ◁ Heat of reaction, ΔH ▷ Heat (kJ)

Suppose in this reaction that 9.00 g H_2O undergoes reaction. We can calculate the heat absorbed as

$$9.00 \text{ g } H_2O \times \frac{1 \text{ mol } H_2O}{18.02 \text{ g } H_2O} \times \frac{572 \text{ kJ}}{2 \text{ mol } H_2O} = 143 \text{ kJ absorbed}$$

Guide to Calculations Using Heat of Reaction (ΔH)

STEP 1
List given and needed data for the equation.

STEP 2
Write a plan using heat of reaction and any molar mass needed.

STEP 3
Write the conversion factors including heat of reaction.

STEP 4
Set up the problem.

Sample Problem 7.7 ▷ **Calculating Heat in a Reaction**

The formation of ammonia from hydrogen and nitrogen has a $\Delta H = -92.2 \text{ kJ}$.

$$N_2(g) + 3H_2(g) \longrightarrow 2NH_3(g) \qquad \Delta H = -92.2 \text{ kJ}$$

How much heat in kilojoules is released when 50.0 g of ammonia forms?

Solution

STEP 1 **List given and needed.**

Given 50.0 g NH_3 **Need** Heat in kilojoules (kJ) to form NH_3

STEP 2 **Plan** Use conversion factors that relate the heat released to the moles of NH_3 in the balanced equation.

grams of NH_3 — Molar mass → moles of NH_3 — Heat of reaction → kilojoules

STEP 3 **Equalities/Conversion Factors**

$$1 \text{ mol NH}_3 = 17.03 \text{ g}$$

$$\frac{17.03 \text{ g}}{1 \text{ mol NH}_3} \quad \text{and} \quad \frac{1 \text{ mol NH}_3}{17.03 \text{ g NH}_3}$$

$$2 \text{ mol NH}_3 = 92.2 \text{ kJ}$$

$$\frac{92.2 \text{ kJ}}{2 \text{ mol NH}_3} \quad \text{and} \quad \frac{2 \text{ mol NH}_3}{92.2 \text{ kJ}}$$

STEP 4 **Set Up Problem**

$$50.0 \text{ g NH}_3 \times \frac{1 \text{ mol NH}_3}{17.03 \text{ g NH}_3} \times \frac{92.2 \text{ kJ}}{2 \text{ mol NH}_3} = \begin{array}{c}135 \text{ kJ} \\ \text{is released}\end{array}$$

Study Check

Mercury(II) oxide decomposes to mercury and oxygen:

$$2\text{HgO}(s) \longrightarrow 2\text{Hg}(l) + \text{O}_2(g) \qquad \Delta H = 182 \text{ kJ}$$

a. Is the reaction exothermic or endothermic?
b. How many kJ are needed to react 25.0 g mercury(II) oxide?

CHEM NOTE

HOT PACKS AND COLD PACKS

In a hospital, at a first-aid station, or at an athletic event, a *cold pack* may be used to reduce swelling from an injury, remove heat from inflammation, or decrease capillary size to lessen the effect of hemorrhaging. Inside the plastic container of a cold pack, there is a compartment containing solid ammonium nitrate (NH_4NO_3) that is separated from a compartment containing water. The pack is activated when it is hit or squeezed hard enough to break the walls between the compartments and cause the ammonium nitrate to mix with the water (shown as H_2O over the reaction arrow). In an endothermic process, each gram of NH_4NO_3 that dissolves absorbs 330 J of heat from the water. The temperature drops and the pack becomes cold and ready to use.

Endothermic Reaction in a Cold Pack

$$26 \text{ kJ} + \text{NH}_4\text{NO}_3(s) \xrightarrow{\text{H}_2\text{O}} \text{NH}_4\text{NO}_3(aq)$$

Hot packs are used to relax muscles, lessen aches and cramps, and increase circulation by expanding capillary size. Constructed in the same way as cold packs, a hot pack may contain the salt $CaCl_2$. The dissolving of the salt in water is exothermic

and releases 670 J per gram of salt. The temperature rises and the pack becomes hot and ready to use:

Exothermic Reaction in a Hot Pack

$$\text{CaCl}_2(s) \xrightarrow{\text{H}_2\text{O}} \text{CaCl}_2(aq) + 75 \text{ kJ}$$

Questions and Problems   **Energy in Chemical Reactions**

7.25 In an exothermic reaction, is the energy of the products higher or lower than the reactants?

7.26 In an endothermic reaction, is the energy of the products higher or lower than the reactants?

7.27 Classify the following as exothermic or endothermic reactions:
   a. 550 kJ is released.
   b. The energy level of the products is higher than the reactants.

   c. The metabolism of glucose in the body provides energy.

7.28 Classify the following as exothermic or endothermic reactions:
   a. The energy level of the products is lower than the reactants.
   b. In the body, the synthesis of proteins requires energy.
   c. 125 kJ is absorbed.

7.29 Classify the following as exothermic or endothermic reactions and give ΔH for each:

a. gas burning in a Bunsen burner:

$$CH_4(g) + 2O_2(g) \longrightarrow CO_2(g) + 2H_2O(g) + 890 \text{ kJ}$$

b. dehydrating limestone:

$$Ca(OH)_2(s) + 65.3 \text{ kJ} \longrightarrow CaO(s) + H_2O(l)$$

c. formation of aluminum oxide and iron from aluminum and iron(III) oxide:

$$2Al(s) + Fe_2O_3(s) \longrightarrow Al_2O_3(s) + 2Fe(s) + 850 \text{ kJ}$$

7.30 Classify the following as exothermic or endothermic reactions and give ΔH for each:

a. the combustion of propane:

$$C_3H_8(g) + 5O_2(g) \longrightarrow 3CO_2(g) + 4H_2O(g) + 2220 \text{ kJ}$$

b. the formation of "table" salt:

$$2Na(s) + Cl_2(g) \longrightarrow 2NaCl(s) + 819 \text{ kJ}$$

c. decomposition of phosphorus pentachloride:

$$PCl_5(g) + 67 \text{ kJ} \longrightarrow PCl_3(g) + Cl_2(g)$$

7.31 The equation for the formation of silicon tetrachloride from silicon and chlorine is

$$Si(s) + 2Cl_2(g) \longrightarrow SiCl_4(g) \qquad \Delta H = -657 \text{ kJ}$$

How many kilojoules are released when 125 g Cl_2 reacts with silicon?

7.32 Methanol (CH_3OH), which is used as a cooking fuel, undergoes combustion to produce carbon dioxide and water:

$$2CH_3OH(l) + 3O_2(g) \longrightarrow 2CO_2(g) + 4H_2O(l)$$
$$\Delta H = -726 \text{ kJ}$$

How many kilojoules are released when 75.0 g methanol is burned?

Concept Map   **Chemical Reactions**

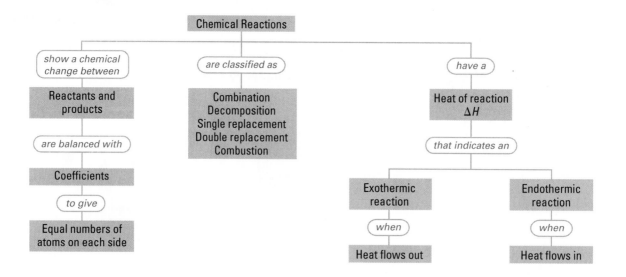

Chapter Review

7.1 Chemical Reactions

A chemical change occurs when the atoms of the initial substances rearrange to form new substances. When new substances form, a chemical reaction has taken place.

7.2 Chemical Equations

A chemical equation shows the formulas of the substances that react on the left of the reaction arrow and the formulas of the products that form on the right side of the reaction arrow.

7.3 Balancing a Chemical Equation

An equation is balanced by writing the smallest whole numbers (coefficients) in front of formulas to equalize the atoms of each element in the reactants and the products.

7.4 Types of Reactions

Many chemical reactions can be organized by reaction type: combination, decomposition, single replacement, double replacement, and/or combustion.

7.5 Energy in Chemical Reactions

In chemical reactions, the heat of reaction (ΔH) is the energy difference between the reactants and the products. In an exothermic reaction, the energy of the products is lower than that of the reactants. Heat is released and ΔH is negative. In an endothermic reaction, the energy of the products is higher than that of the reactants. Heat is absorbed and the ΔH is positive.

Key Terms

balanced equation The final form of a chemical reaction that shows the same number of atoms of each element in the reactants and products.

chemical change The formation of a new substance with a different composition and properties than the initial substance.

chemical equation A shorthand way to represent a chemical reaction using chemical formulas to indicate the reactants and products.

chemical reaction The process by which a chemical change takes place.

coefficients Whole numbers placed in front of the formulas in an equation to balance the number of atoms of each element.

combination reaction A reaction in which reactants combine to form a single product.

combustion reaction A reaction in which an element or a compound reacts with oxygen to form oxide products.

decomposition reaction A reaction in which a single reactant splits into two or more simpler substances.

double replacement reaction A reaction in which parts of two different reactants exchange places.

endothermic reaction A reaction wherein the energy of the products is greater than that of the reactants.

exothermic reaction A reaction wherein the energy of the reactants is greater than that of the products.

heat of reaction The heat (symbol ΔH) absorbed or released when a reaction takes place at constant pressure.

physical change A change in which the physical properties change but the chemical composition does not change.

products The substances formed as a result of a chemical reaction.

reactants The initial substances that undergo change in a chemical reaction.

single replacement reaction A reaction in which an element replaces a different element in a compound.

 # Understanding the Concepts

7.33 If red spheres represent oxygen atoms and blue spheres represent nitrogen atoms,
  a. write a balanced equation for the reaction.
  b. indicate the type of reaction as decomposition, combination, single replacement, or double replacement.

7.34 If purple spheres represent iodine atoms and light blue spheres represent hydrogen atoms,
  a. write a balanced equation for the reaction represented.
  b. indicate the type of reaction as decomposition, combination, single replacement, or double replacement.

Reactants   **Products**

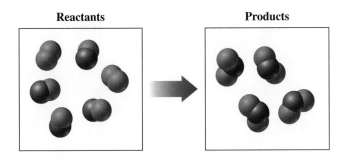

Reactants   **Products**

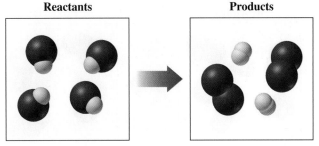

7.35 If blue spheres represent nitrogen atoms and purple spheres represent iodine atoms,
  a. write a balanced equation for the reaction represented.
  b. indicate the type of reaction as decomposition, combination, single replacement, or double replacement.

Reactants **Products**

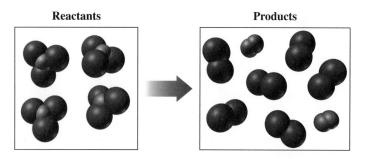

7.36 If green spheres represent chlorine atoms, yellow-green spheres represent fluorine atoms, and light blue spheres represent hydrogen atoms,
  a. write a balanced equation for the reaction represented.
  b. indicate the type of reaction as decomposition, combination, single replacement, or double replacement.

Reactants **Products**

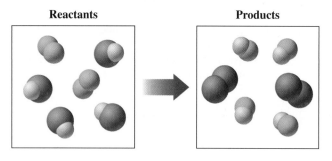

7.37 If green spheres represent chlorine atoms and red spheres represent oxygen atoms,
  a. write a balanced equation for the reaction represented.
  b. indicate the type of reaction as decomposition, combination, single replacement, or double replacement.

Reactants **Products**

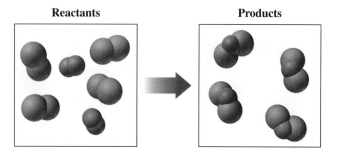

7.38 If blue spheres represent nitrogen atoms and purple spheres represent iodine atoms,
  a. write a balanced equation for the reaction represented.
  b. indicate the type of reaction as decomposition, combination, single replacement, or double replacement.

Reactants **Products**

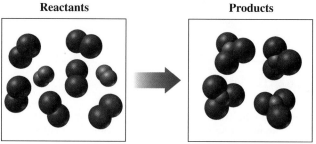

7.39 Identify some of the physical and chemical changes of a burning wax candle.

7.40 Balance each of the following by adding coefficients and identify the type of reaction for each:

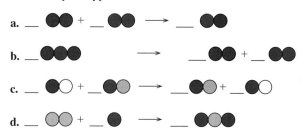

Additional Questions and Problems

7.41 Identify the type of each of the following as combination, decomposition, single replacement, double replacement, or combustion.
  a. A metal and a nonmetal element form an ionic compound.
  b. A compound containing carbon and hydrogen reacts with oxygen.
  c. Two compounds react to produce two new compounds.
  d. Heating calcium carbonate produces calcium oxide and carbon dioxide.
  e. Zinc replaces copper in $Cu(NO_3)_2$.

7.42 Identify the type of each of the following as combination, decomposition, single replacement, double replacement, or combustion.
  a. A compound breaks apart into its elements.
  b. An element replaces the ion in a compound.
  c. Copper and bromine form copper(II) bromide.
  d. Iron(II) sulfite breaks down to iron(II) oxide and sulfur dioxide.
  e. Silver ion from $AgNO_3(aq)$ forms a solid with bromide ion from $KBr(aq)$.

7.43 Balance each of the following unbalanced equations and identify the type of reaction:
  a. $NH_3(g) + HCl(g) \longrightarrow NH_4Cl(s)$
  b. $Fe_3O_4(s) + H_2(g) \longrightarrow Fe(s) + H_2O(g)$
  c. $Sb(s) + Cl_2(g) \longrightarrow SbCl_3(s)$
  d. $NI_3(s) \longrightarrow N_2(g) + I_2(g)$
  e. $KBr(aq) + Cl_2(aq) \longrightarrow KCl(aq) + Br_2(l)$
  f. $Fe(s) + H_2SO_4(aq) \longrightarrow Fe_2(SO_4)_3(aq) + H_2(g)$
  g. $Al_2(SO_4)_3(aq) + NaOH(aq) \longrightarrow$
  $Na_2SO_4(aq) + Al(OH)_3(s)$

7.44 Balance each of the following unbalanced equations and identify the type of reaction:
  a. $Li_3N(s) \longrightarrow Li(s) + N_2(g)$
  b. $Mg(s) + N_2(g) \longrightarrow Mg_3N_2(s)$
  c. $Al(s) + HCl(aq) \longrightarrow AlCl_3(aq) + H_2(g)$
  d. $Mg(s) + H_3PO_4(aq) \longrightarrow Mg_3(PO_4)_2(s) + H_2(g)$
  e. $Cr_2O_3(s) + H_2(g) \longrightarrow Cr(s) + H_2O(g)$
  f. $Al(s) + Cl_2(g) \longrightarrow AlCl_3(s)$
  g. $MgCl_2(aq) + AgNO_3(aq) \longrightarrow$
  $Mg(NO_3)_2(aq) + AgCl(s)$

7.45 Predict the products and write a balanced equation for each of the following reactions:

  a. $Zn(s) + HCl(aq) \longrightarrow$ _____ + _____
  (single replacement)
  b. $BaCO_3(s) \longrightarrow$ _____ + _____
  (decomposition)
  c. $NaOH(aq) + HCl(aq) \longrightarrow$ _____ + _____
  (double replacement)
  d. $Al(s) + F_2(g) \longrightarrow$ _____   (combination)

7.46 Predict the products and write a balanced equation for each of the following reactions.
  a. $NaCl(s) \xrightarrow{\text{Electricity}}$ _____ + _____ (decomposition)
  b. $Ca(s) + Br_2(g) \longrightarrow$ _____ (combination)
  c. $SO_2(g) + O_2(g) \longrightarrow$ _____ (combination)
  d. $HCl(aq) + NaOH(aq) \longrightarrow$ _____ + _____
  (double replacement)

7.47 Write and balance equations for each of the following reactions and identify the type of reaction:
  a. Solid potassium chlorate is heated to form solid potassium chloride and oxygen gas.
  b. Aqueous sodium chloride and aqueous silver nitrate react to form solid silver chloride and aqueous sodium nitrate.

7.48 Write and balance equations for each of the following reactions and identify the type of reaction:
  a. Sodium metal reacts with oxygen gas to form solid sodium oxide.
  b. Carbon monoxide gas and oxygen gas combine to form carbon dioxide gas.

7.49 The formation of 2 mol nitrogen oxide, NO, from $N_2(g)$ and $O_2(g)$, requires 90.2 kJ of heat.

  $$N_2(g) + O_2(g) \longrightarrow 2NO(g) \qquad \Delta H = 90.2\ kJ$$

  a. How many kJ are required to form 3.00 g NO?
  b. What is the complete equation (including heat) for the decomposition of NO?
  c. How many kJ are released when 5.00 g NO decomposes to N_2 and O_2?

7.50 The formation of rust (Fe_2O_3) from solid iron and oxygen gas releases 1.7×10^3 kJ.

  $$4Fe(s) + 3O_2(g) \longrightarrow 2Fe_2O_3(s)$$
  $$\Delta H = -1.7 \times 10^3\ kJ$$

  a. How many kJ are released when 2.00 g Fe react?
  b. How many grams of rust form when 150 kcal are released?

Challenge Questions

7.51 Write a balanced equation for each of the following reaction descriptions and identify each type of reaction:
  a. An aqueous solution of lead(II) nitrate is mixed with aqueous sodium phosphate to produce solid lead(II) phosphate and aqueous sodium nitrate.
  b. Gallium metal heated in oxygen gas forms solid gallium(III) oxide.
  c. When solid sodium nitrate is heated, solid sodium nitrite and oxygen gas are produced.

7.52 Write a balanced equation for each of the following reaction descriptions and identify each type of reaction:
  a. Solid bismuth(III) oxide and solid carbon react to form bismuth metal and carbon monoxide.
  b. Solid potassium chlorate is heated to form solid potassium chloride and oxygen gas.
  c. Butane gas (C_4H_{10}) reacts with oxygen gas to form two gaseous products: carbon dioxide and water.

7.53 In the following diagram, blue spheres are X and yellow spheres are Y.
  a. Write the formulas of the reactants and the products.
  b. Write the balanced equation for the reaction.
  c. Identify the type of reaction.

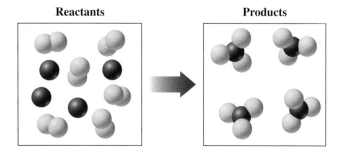

Reactants   **Products**

7.54 In the following diagram, red spheres are A, light blue spheres are B, and green spheres are C.
  a. Write the formulas of the reactants and the products.
  b. Write the balanced equation for the reaction.
  c. Identify the type of reaction.

Reactants   **Products**

7.55 When a compound containing C and H is burned in oxygen, the reaction produces 0.40 mol CO_2 and 0.60 mol H_2O.
  a. Determine the number of moles of carbon and moles of hydrogen in the compound.
  b. Determine the empirical formula of the product.
  c. If one molecule of the compound contains 6 H atoms, what is the molecular formula?
  d. Write the balanced equation for the combustion reaction.

7.56 A copper wire with a mass of 4.32 g reacts with sulfur to form 5.41 g of a copper sulfide compound.
  a. Determine the number of moles of copper and sulfur in the compound.
  b. Determine the empirical formula of the product.
  c. If the actual formula of the sulfide compound contains 1 S, what is the formula of the compound?
  d. Write the balanced equation for the combination reaction.

7.57 For the reaction: $2S(s) + 3O_2(g) \longrightarrow 2SO_3(g)$
$$\Delta H = -790 \text{ kJ}$$
  a. Is the reaction endothermic or exothermic?
  b. How many kJ are released when 1.5 mol S reacts?
  c. How many kJ are released when 125 g SO_3 is formed?
  d. What is the ΔH in kilojoules for the following reaction?
$$2SO_3(g) \longrightarrow 2S(s) + 3O_2(g)$$
  e. Is the reaction in part d endothermic or exothermic?

7.58 When peroxide (H_2O_2) is used in rocket fuels, it produces water, oxygen, and heat:
$$2H_2O_2(l) \longrightarrow 2H_2O(l) + O_2(g)$$
  a. If the reaction of 2.00 g H_2O_2 releases 5.76 kJ, what is the heat of reaction?
  b. How many kilojoules are released when 275 g peroxide react?

Answers

Answers to Study Checks

7.1 a. physical   **b.** chemical
   c. physical

7.2 The production of light and heat are evidence of a chemical reaction.

7.3 There are 4 carbon atoms, 12 hydrogen atoms, and 14 oxygen atoms on the reactant side of the equation and on the product side.

7.4 $3Fe(s) + 2O_2(g) \longrightarrow Fe_3O_4(s)$

7.5 $N_2(g) + 2O_2(g) \longrightarrow 2NO_2(g)$ combustion and combination reaction

7.6 a. $H_2(g) + I_2(g) + 50.\ kJ \longrightarrow 2\,HI(g)$
   b. $\Delta H = 50.\ kJ$

7.7 a. endothermic   **b.** $\Delta H = 10.5\ kJ$

Answers to Selected Questions and Problems

7.1 a. physical   **b.** chemical
   c. physical   **d.** chemical

7.3 a. Physical change; There is no evidence of a chemical reaction when water forms snowflakes.
   b. Light, heat, and the formation of a gas are all evidence of a chemical reaction.
   c. Physical change; There is no evidence of a change in the properties of the bread.
   d. The change in color during toasting a marshmallow is evidence of a chemical reaction.

7.5 a. Reactant side: 2 N atoms, 4 O atoms
      Product side: 2 N atoms, 4 O atoms
   b. Reactant side: 5 C atoms, 2 S atoms, 4 O atoms
      Product side: 5 C atoms, 2 S atoms, 4 O atoms
   c. Reactant side: 4 C atoms, 4 H atoms, 10 O atoms
      Product side: 4 C atoms, 4 H atoms, 10 O atoms
   d. Reactant side: 2 N atoms, 8 H atoms, 4 O atoms
      Product side: 2 N atoms, 8 H atoms, 4 O atoms

7.7 a. not balanced   **b.** balanced
   c. not balanced   **d.** balanced

7.9 a. 2 Na atoms, 2 Cl atoms
   b. 1 P atom, 3 Cl atoms, 6 H atoms
   c. 4 P atoms, 16 O atoms, 12 H atoms

7.11 a. $N_2(g) + O_2(g) \longrightarrow 2NO(g)$
   b. $2HgO(s) \longrightarrow 2Hg(l) + O_2(g)$
   c. $4Fe(s) + 3O_2(g) \longrightarrow 2Fe_2O_3(s)$
   d. $2Na(s) + Cl_2(g) \longrightarrow 2NaCl(s)$
   e. $2Cu_2O(s) + O_2(g) \longrightarrow 4CuO(s)$

7.13 a. $Mg(s) + 2AgNO_3(aq) \longrightarrow$
                       $Mg(NO_3)_2(aq) + 2Ag(s)$
   b. $CuCO_3(s) \longrightarrow CuO(s) + CO_2(g)$
   c. $2Al(s) + 3CuSO_4(aq) \longrightarrow$
                       $3Cu(s) + Al_2(SO_4)_3(aq)$
   d. $Pb(NO_3)_2(aq) + 2NaCl(aq) \longrightarrow$
                       $PbCl_2(s) + 2NaNO_3(aq)$
   e. $2Al(s) + 6HCl(aq) \longrightarrow 2AlCl_3(aq) + 3H_2(g)$

7.15 a. $Fe_2O_3(s) + 3CO(g) \longrightarrow 2Fe(s) + 3CO_2(g)$
   b. $2Li_3N(s) \longrightarrow 6Li(s) + N_2(g)$
   c. $2Al(s) + 6HBr(aq) \longrightarrow 2AlBr_3(aq) + 3H_2(g)$
   d. $3Ba(OH)_2(aq) + 2Na_3PO_4(aq) \longrightarrow$
                       $Ba_3(PO_4)_2(s) + 6NaOH(aq)$
   e. $As_4S_6(s) + 9O_2(g) \longrightarrow As_4O_6(s) + 6SO_2(g)$

7.17 a. $2Li(s) + 2H_2O(l) \longrightarrow H_2(g) + 2LiOH(aq)$
   b. $2P(s) + 5Cl_2(g) \longrightarrow 2PCl_5(s)$
   c. $FeO(s) + CO(g) \longrightarrow Fe(s) + CO_2(g)$
   d. $2C_5H_{10}(l) + 15O_2(g) \longrightarrow$
                       $10CO_2(g) + 10H_2O(g)$
   e. $3H_2S(g) + 2FeCl_3(s) \longrightarrow Fe_2S_3(s) + 6HCl(g)$

7.19 a. A single reactant splits into two simpler substances (elements).
   b. One element in the reacting compound is replaced by the other reactant.

7.21 a. combination; combustion
   b. single replacement
   c. decomposition
   d. double replacement
   e. combustion
   f. decomposition
   g. double replacement
   h. combination; combustion

7.23 a. $Mg(s) + Cl_2(g) \longrightarrow MgCl_2(s)$
   b. $2HBr(g) \longrightarrow H_2(g) + Br_2(g)$
   c. $Mg(s) + Zn(NO_3)_2(aq) \longrightarrow$
                       $Mg(NO_3)_2(aq) + Zn(s)$
   d. $K_2S(aq) + Pb(NO_3)_2(aq) \longrightarrow$
                       $PbS(g) + 2KNO_3(aq)$
   e. $2C_2H_6(g) + 7O_2(g) \longrightarrow 4CO_2(g) + 6H_2O(g)$

7.25 In exothermic reactions, the energy of the products is lower than that of the reactants.

7.27 a. exothermic   **b.** endothermic
   c. exothermic

7.29 a. exothermic, $\Delta H = -890\ kJ$
   b. endothermic, $\Delta H = 65.3\ kJ$
   c. exothermic, $\Delta H = -850\ kJ$

7.31 579 kJ

7.33 a. $2NO(g) + O_2(g) \longrightarrow 2NO_2(g)$
b. combination

7.35 a. $2NI_3(s) \longrightarrow N_2(g) + 3I_2(g)$
b. decomposition

7.37 a. $2Cl_2(g) + O_2(g) \longrightarrow 2OCl_2(g)$
b. combination

7.39 Physical changes include the following: solid wax melts to form liquid, the candle becomes shorter, the liquid wax changes to solid, the shape of the wax changes, the wick becomes shorter.

   Chemical changes include the following: heat and light are emitted, wax and the wick burn in the presence of oxygen.

7.41 a. combination   **b.** combustion
c. double replacement   **d.** decomposition
e. single replacement

7.43 a. $NH_3(g) + HCl(g) \longrightarrow NH_4Cl(s)$ combination
b. $Fe_3O_4(s) + 4H_2(g) \longrightarrow 3Fe(s) + 4H_2O(g)$
single replacement
c. $2Sb(s) + 3Cl_2(g) \longrightarrow 2SbCl_3(s)$ combination
d. $2NI_3(s) \longrightarrow N_2(g) + 3I_2(g)$ decomposition
e. $2KBr(aq) + Cl_2(aq) \longrightarrow 2KCl(aq) + Br_2(l)$
single replacement
f. $2Fe(s) + 3H_2SO_4(aq) \longrightarrow$
$Fe_2(SO_4)_3(aq) + 3H_2(g)$
single replacement
g. $Al_2(SO_4)_3(aq) + 6NaOH(aq) \longrightarrow$
$3Na_2SO_4(aq) + 2Al(OH)_3(s)$
double replacement

7.45 a. $Zn(s) + 2HCl(aq) \longrightarrow ZnCl_2(aq) + H_2(g)$
b. $BaCO_3(s) \xrightarrow{\Delta} BaO(s) + CO_2(g)$
c. $NaOH(aq) + HCl(aq) \longrightarrow NaCl(aq) + H_2O(l)$
d. $2Al(s) + 3F_2(g) \longrightarrow 2AlF_3(s)$

7.47 a. $2KClO_3(s) \longrightarrow 2KCl(s) + 3O_2(g)$ decompostion
b. $NaCl(aq) + AgNO_3(aq) \longrightarrow$
$AgCl(s) + NaNO_3(aq)$
double replacement

7.49 a. 4.51 kJ
b. $2NO(g) \longrightarrow N_2(g) + O_2(g) + 902\,kJ$
c. 7.51 kJ

7.51 a. $3Pb(NO_3)_2(aq) + 2Na_3PO_4(aq) \longrightarrow$
$Pb_3(PO_4)_2(s) + 6\,NaNO_3(aq)$
double replacement
b. $4Ga(s) + 3O_2(g) \longrightarrow 2Ga_2O_3(s)$
combination/combustion
c. $2\,NaNO_3(s) \longrightarrow 2\,NaNO_2(s) + O_2(g)$
decomposition

7.53 a. Reactants: X and Y_2; Products: XY_3
b. $2X + 3Y_2 \longrightarrow 2XY_3$
c. combination

7.55 a. 0.40 mol carbon and 1.20 mol hydrogen
b. The ratio is 0.40 mol C : 1.20 mol H, which gives the empirical formula CH_3.
c. C_2H_6
d. $2C_2H_6(g) + 7O_2(g) \longrightarrow 4CO_2(g) + 6H_2O(g)$

7.57 a. exothermic
b. 590 kJ is released.
c. 620 kJ is released.
d. $\Delta H = +790\,kJ$ is released.
e. endothermic

Chemical Quantities in Reactions

"In our food science laboratory I develop a variety of food products, from cake donuts to energy beverages," says Anne Cristofano, senior food technologist at Mattson & Company. *"When I started the donut project, I researched the ingredients, then weighed them out in the lab. I added water to make a batter and cooked the donuts in a fryer. The batter and the oil temperature make a big difference. If I don't get the right taste or texture, I adjust the ingredients, such as sugar and flour, or adjust the temperature."*

A food technologist studies the physical and chemical properties of food and develops scientific ways to process and preserve it for extended shelf life. The food products are tested for texture, color, and flavor. The results of these tests help improve the quality and safety of food.

the Chemistry place

Visit **www.aw-bc.com/chemplace** for extra quizzes, interactive tutorials, career resources, PowerPoint slides for chapter review, math help, and case studies.

When we know the chemical equation for a reaction, we can determine the amount of product that can be produced. We do much the same thing at home when we use a recipe to make cookies or add the right quantity of water to make soup. At the automotive repair shop, a mechanic adjusts the carburetor or fuel injection system of an engine to allow for the correct amounts of fuel and oxygen so the engine will run properly. In the hospital, a respiratory therapist evaluates the level of CO_2 and O_2 in the blood. A certain amount of oxygen must reach the tissues for efficient metabolic reactions. If the oxygenation of the blood is low, the therapist will oxygenate the patient and recheck the blood levels.

In this chapter, we will describe the quantity of a compound as a mole and calculate the mass related to the formula of a compound. From a balanced equation, we can determine the mass and number of moles of a reactant and calculate the amount of product. Knowing how to determine the quantitative results of a chemical reaction is important to both chemists and medical personnel such as pharmacists and respiratory therapists.

Learning Goal

Given a quantity in moles of reactant or product, use mole–mole factors from the balanced equation to calculate the moles of another substance in the reaction.

WEB TUTORIAL
Stoichiometry

8.1 Mole Relationships in Chemical Equations

In Chapter 7, we saw that equations are balanced in terms of the numbers of each type of atom in the reactants and products. However, when experiments are done in the laboratory or medications prepared in the pharmacy, samples contain so many atoms and molecules that it is impossible to count them individually. What can be measured conveniently is mass, using a balance. Because mass is related to the number of particles through the molar mass, measuring the mass is equivalent to counting the number of particles or moles.

Conservation of Mass

In any chemical reaction, the total amount of matter in the reactants is equal to the total amount of matter in the products. If all of the reactants were weighed, they would have a total mass equal to the total mass of the products. This is known as the **law of conservation of mass,** which says that there is no change in the total mass of the substances reacting in a chemical reaction. Thus, no material is lost or gained as original substances are changed to new substances.

For example, tarnish forms when silver reacts with sulfur to form silver sulfide.

$$2Ag(s) + S(s) \longrightarrow Ag_2S(s)$$

In this reaction, the number of silver atoms that react is two times the number of sulfur atoms. When 200 silver atoms react, 100 sulfur atoms are required. Normally, however, many more atoms would actually be present in this reaction. If we are dealing with molar amounts, then the coefficients in the equation can be interpreted in terms of moles. Thus, 2 mol silver react with each 1 mol sulfur. Since the molar mass of each can be determined, the quantities of silver and sulfur can also be stated in terms of mass in grams of each. Therefore, an equation for a chemical equation can be interpreted several ways, as seen in Table 8.1.

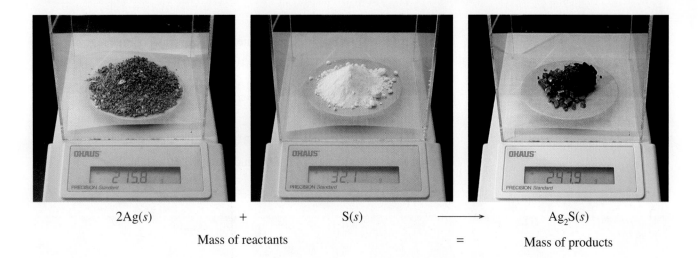

$$2\text{Ag}(s) \qquad + \qquad \text{S}(s) \qquad \longrightarrow \qquad \text{Ag}_2\text{S}(s)$$

Mass of reactants $\qquad\qquad = \qquad$ Mass of products

Table 8.1   Information Available from a Balanced Equation

	Reactants		Products
Equation	**2Ag(s)**	**+ S(s)**	$\longrightarrow$ **Ag₂S(s)**
Atoms	2 Ag atoms	+ 1 S atom	$\longrightarrow$ 1 Ag₂S formula unit
	200 Ag atoms	+ 100 S atoms	$\longrightarrow$ 100 Ag₂S formula units
Avogadro's number of atoms	$2(6.022 \times 10^{23})$ Ag atoms	$+\ 1(6.022 \times 10^{23})$ S atoms	$\longrightarrow 1(6.022 \times 10^{23})$ Ag₂S formula units
Moles	2 mol Ag	+ 1 mol S	$\longrightarrow$ 1 mol Ag₂S
Mass (g)	2(107.9 g) Ag	+ 1(32.07 g) S	$\longrightarrow$ 1(247.9 g) Ag₂S
Total mass (g)	247.9 g		$\longrightarrow$ 247.9 g

Sample Problem 8.1   **Conservation of Mass**

Calculate the total mass of reactants and products for the following equation when 1 mol CH_4 reacts:

$$CH_4(g) + 2O_2(g) \longrightarrow CO_2(g) + 2H_2O(g)$$

Solution

Interpreting the coefficients in the equation as the number of moles of each substance and multiplying by the respective molar masses gives the total mass of reactants and products.

	Reactants			Products	
Equation	$CH_4(g)$	$+ 2O_2(g)$	$\longrightarrow$	$CO_2(g)$	$+ 2H_2O(g)$
Moles	1 mol CH_4	+ 2 mol O_2	$\longrightarrow$	1 mol CO_2	+ 2 mol H_2O
Mass	16.04 g CH_4	+ 64.00 g O_2	$\longrightarrow$	44.01 g CO_2	+ 36.03 g H_2O
Total mass		80.04 g reactants	$=$	80.04 g products	

Study Check

For the following reaction, calculate the total mass of the reactants and products when 4 mol NH_3 react:

$$4NH_3(g) + 3O_2(g) \longrightarrow 2N_2(g) + 6H_2O(g)$$

Mole–Mole Factors in an Equation

When iron reacts with sulfur, the product is iron(III) sulfide.

$$2Fe(s) + 3S(s) \longrightarrow Fe_2S_3(s)$$

Iron (Fe)   +   Sulfur (S) $\longrightarrow$ Iron(III) sulfide (Fe_2S_3)
$2Fe(s)$ $3S(s)$ $Fe_2S_3(s)$

Because the equation is balanced, we know the proportions of iron and sulfur in the reaction. For this reaction, we see that 2 mol iron reacts with 3 mol sulfur to form 1 mol iron(III) sulfide. From the coefficients, we can write **mole–mole factors** between reactants and between reactants and products.

Fe and S: $\quad \dfrac{2 \text{ mol Fe}}{3 \text{ mol S}} \quad$ and $\quad \dfrac{3 \text{ mol S}}{2 \text{ mol Fe}}$

Fe and Fe_2S_3: $\quad \dfrac{2 \text{ mol Fe}}{1 \text{ mol Fe}_2S_3} \quad$ and $\quad \dfrac{1 \text{ mol Fe}_2S_3}{2 \text{ mol Fe}}$

S and Fe_2S_3: $\quad \dfrac{3 \text{ mol S}}{1 \text{ mol Fe}_2S_3} \quad$ and $\quad \dfrac{1 \text{ mol Fe}_2S_3}{3 \text{ mol S}}$

Sample Problem 8.2 ▶ **Writing Mole–Mole Factors**

Consider the following balanced equation:

$$4Na(s) + O_2(s) \longrightarrow 2Na_2O(s)$$

Write the mole–mole factors for
a. Na and O_2 **b.** Na and Na_2O

Solution

a. $\dfrac{4 \text{ mol Na}}{1 \text{ mol O}_2} \quad$ and $\quad \dfrac{1 \text{ mol O}_2}{4 \text{ mol Na}}$

b. $\dfrac{4 \text{ mol Na}}{2 \text{ mol Na}_2O} \quad$ and $\quad \dfrac{2 \text{ mol Na}_2O}{4 \text{ mol Na}}$

Study Check

Using the equation in Sample Problem 8.2, write the mole–mole factors for O_2 and Na_2O.

Using Mole–Mole Factors in Calculations

Whenever you prepare a recipe, adjust an engine for the proper mixture of fuel and air, or prepare medicines in a pharmaceutical laboratory, you need to know the proper amounts of reactants to use and how much of the product will form. Earlier, we wrote all the possible conversion factors that can be obtained from the balanced equation $2Fe(s) + 3S(s) \longrightarrow Fe_2S_3(s)$. Now we will show how mole–mole factors are used in chemical calculations.

Sample Problem 8.3   **Calculating Moles of a Reactant**

In the reaction of iron and sulfur, how many moles of sulfur are needed to react with 1.42 mol iron?

$$2Fe(s) + 3S(s) \longrightarrow Fe_2S_3(s).$$

Solution

Guide to Using Mole Factors
STEP 1 Write the given and needed moles.
STEP 2 Write a plan to convert the given to the needed moles.
STEP 3 Use coefficients to write relationships and mole–mole factors.
STEP 4 Set up problem using the mole factor that cancels given moles.

STEP 1   **Write the given and needed number of moles.**   In this problem, we need to find the number of moles of S that react with 1.42 mol Fe.

> **Given**   1.42 mol Fe   **Need**   moles of S

STEP 2   **Write the plan to convert the given to the needed.**

> moles of Fe   Mole–mole factor   moles of S

STEP 3   **Use coefficients to write relationships and mole–mole factors.** Use coefficients to write the mole–mole factors for the given and needed substances.

> 2 mol Fe = 3 mol S
>
> $$\frac{2 \text{ mol Fe}}{3 \text{ mol S}} \quad \text{and} \quad \frac{3 \text{ mol S}}{2 \text{ Mol Fe}}$$

STEP 4   **Set up problem using the mole–mole factor that cancels given moles.** Use a mole–mole factor to cancel the given moles and provide needed moles.

> $$1.42 \text{ mol Fe} \times \frac{3 \text{ mol S}}{2 \text{ mol Fe}} = 2.13 \text{ mol S}$$

The answer is given with three significant figures because the given quantity 1.42 mol Fe has 3 SFs. The values in the mole–mole factor are exact.

Study Check

Using the equation in Sample Problem 8.3, calculate the number of moles of iron needed to completely react with 2.75 mol sulfur.

Sample Problem 8.4   **Calculating Moles of a Product**

In a combustion reaction, propane (C_3H_8) reacts with oxygen. How many moles of CO_2 can be produced when 2.25 mol C_3H_8 reacts?

$$C_3H_8(g) + 5O_2(s) \longrightarrow 3CO_2(g) + 4H_2O(g)$$

Propane

Solution

STEP 1 **Write the given and needed moles.**   In this problem, we need to find the number of moles of CO_2 that can be produced when 2.25 mol C_3H_8 react.

Given   2.25 mol C_3H_8 **Need**   moles of CO_2

STEP 2 **Write the plan to convert the given to the needed.**

moles of C_3H_8 | Mole–mole factor | moles of CO_2

STEP 3 **Use coefficients to write relationships and mole–mole factors.** Use coefficients to write the mole–mole factors for the given and needed substances.

$$1 \text{ mol } C_3H_8 = 3 \text{ mol } CO_2$$

$$\frac{1 \text{ mol } C_3H_8}{3 \text{ mol } CO_2} \quad \text{and} \quad \frac{3 \text{ mol } CO_2}{1 \text{ mol } C_3H_8}$$

STEP 4 **Set up problem using the mole–mole factor that cancels given moles.**   Use a mole–mole factor to cancel the given moles and provide needed moles.

$$2.25 \text{ mol } C_3H_8 \times \frac{3 \text{ mol } CO_2}{1 \text{ mol } C_3H_8} = 6.75 \text{ mol } CO_2$$

The answer is given with three significant figures because the given quantity 2.25 mol C_3H_8 has 3 SFs. The values in the mole–mole factor are exact.

Study Check

Using the equation in Sample Problem 8.4, calculate the number of moles of oxygen that must react to produce 0.756 mol water.

Questions and Problems   **Mole Relationships in Chemical Equations**

8.1 Give an interpretation of the following equations in terms of (1) number of particles and (2) number of moles:
   a. $2SO_2(g) + O_2(g) \longrightarrow 2SO_3(g)$
   b. $4P(s) + 5O_2(g) \longrightarrow 2P_2O_5(s)$

8.2 Give an interpretation of the following equations in terms of (1) number of particles and (2) number of moles:
   a. $2Al(s) + 3Cl_2(g) \longrightarrow 2AlCl_3(s)$
   b. $4HCl(g) + O_2(g) \longrightarrow 2Cl_2(g) + 2H_2O(g)$

8.3 Calculate the total masses of the reactants and the products in each of the equations of problem 8.1.

8.4 Calculate the total masses of the reactants and the products in each of the equations of problem 8.2.

8.5 Write all of the mole–mole factors for the equations listed in problem 8.1.

8.6 Write all of the mole–mole factors for the equations listed in problem 8.2.

8.7 The reaction of hydrogen with oxygen produces water.

$$2H_2(g) + O_2(g) \longrightarrow 2H_2O(g)$$

a. How many moles of O_2 are required to react with 2.0 mol H_2?
b. If you have 5.0 mol O_2, how many moles of H_2 are needed for the reaction?
c. How many moles of H_2O form when 2.5 mol O_2 reacts?

8.8 Ammonia is produced by the reaction of hydrogen and nitrogen.

$$N_2(g) + 3H_2(g) \longrightarrow 2NH_3(g)$$
$$\text{Ammonia}$$

a. How many moles of H_2 are needed to react with 1.0 mol N_2?
b. How many moles of N_2 reacted if 0.60 mol NH_3 is produced?

c. How many moles of NH_3 are produced when 1.4 mol H_2 reacts?

8.9 Carbon disulfide and carbon monoxide are produced when carbon is heated with sulfur dioxide.

$$5C(s) + 2SO_2(g) \longrightarrow CS_2(l) + 4CO(g)$$

a. How many moles of C are needed to react with 0.500 mol SO_2?
b. How many moles of CO are produced when 1.2 mol C reacts?
c. How many moles of SO_2 are required to produce 0.50 mol CS_2?
d. How many moles of CS_2 are produced when 2.5 mol C reacts?

8.10 In the acetylene torch, acetylene gas (C_2H_2) burns in oxygen to produce carbon dioxide and water.

$$2C_2H_2(g) + 5O_2(g) \longrightarrow 4CO_2(g) + 2H_2O(g)$$

a. How many moles of O_2 are needed to react with 2.00 mol C_2H_2?
b. How many moles of CO_2 are produced when 3.5 mol C_2H_2 reacts?
c. How many moles of C_2H_2 are required to produce 0.50 mol H_2O?
d. How many moles of CO_2 are produced from 0.100 mol O_2?

8.2 Mass Calculations for Reactions

Learning Goal

Given the mass in grams of a substance in a reaction, calculate the mass in grams of another substance in the reaction.

When you perform a chemistry experiment in the laboratory, you use a laboratory balance to obtain a certain mass of reactant. From the mass in grams, you can determine the number of moles of reactant. By using mole–mole factors, you can predict the moles of product that can be produced. Then the molar mass of the product is used to convert the moles back into mass in grams.

Guide to Calculating the Masses of Reactants and Products in a Chemical Reaction

STEP 1
Use molar mass to convert grams of given to moles (if necessary).

STEP 2
Write a mole–mole factor from the coefficients in the equation.

STEP 3
Convert moles of given to moles of needed substance using mole–mole factor.

STEP 4
Convert moles of needed substance to grams using molar mass.

Sample Problem 8.5 **Mass of Products from Moles of Reactant**

In the formation of smog, nitrogen reacts with oxygen to produce nitrogen oxide. Calculate the grams of NO produced when 2.15 mol O_2 react.

$$N_2(g) + O_2(g) \longrightarrow 2NO(g)$$

Solution

STEP 1 **Given** 2.15 mol O_2 **Need** grams of NO

STEP 2 **Plan**

moles of O_2 → Mole–mole factor → moles of NO → Molar mass → grams of NO

STEP 3 **Equalities/Conversion Factors** The mole–mole factor that converts moles of O_2 to moles of NO is derived from the coefficients in the balanced equation.

$$1 \text{ mol } O_2 = 2 \text{ mol NO}$$

$$\frac{2 \text{ mol NO}}{1 \text{ mol } O_2} \text{ and } \frac{1 \text{ mol } O_2}{2 \text{ mol NO}}$$

$$1 \text{ mol NO} = 30.01 \text{ g}$$

$$\frac{30.01 \text{ g NO}}{1 \text{ mol NO}} \text{ and } \frac{1 \text{ mol NO}}{30.01 \text{ g NO}}$$

STEP 4 **Set Up Problem** First, we can change the given, 2.15 mol O_2, to moles of NO.

$$2.15 \text{ mol } O_2 \times \frac{2 \text{ mol NO}}{1 \text{ mol } O_2} = 4.30 \text{ mol NO}$$

Now, the moles of NO can be converted to grams of NO using its molar mass.

$$4.30 \text{ mol NO} \times \frac{30.01 \text{ g NO}}{1 \text{ mol NO}} = 129 \text{ g NO}$$

These two steps can also be written as a sequence of conversion factors that lead to the mass in grams of NO.

$$2.15 \text{ mol } O_2 \times \frac{2 \text{ mol NO}}{1 \text{ mol } O_2} \times \frac{30.01 \text{ g NO}}{1 \text{ mol NO}} = 129 \text{ g NO}$$

Study Check

Using the equation in Sample Problem 8.5, calculate the grams of NO that can be produced when 0.734 mol N_2 reacts.

Sample Problem 8.6   **Mass of Product from Mass of Reactant**

In a combustion reaction, acetylene (C_2H_2) burns with oxygen.

$$2C_2H_2(g) + 5O_2(g) \longrightarrow 4CO_2(g) + 2H_2O(g)$$

How many grams of carbon dioxide are produced when 54.6 g C_2H_2 is burned?

Solution

STEP 1 **Given** 54.6 g C_2H_2 **Need** grams of CO_2

STEP 2 **Plan** Once we convert grams of C_2H_2 to moles of C_2H_2 using its molar mass, we can use a mole–mole factor to find the moles of CO_2. Then the molar mass of CO_2 will give us the grams of CO_2.

grams of C_2H_2 → Molar mass → moles of C_2H_2 → Mole–mole factor → moles of CO_2 → Molar mass → grams of CO_2

STEP 3 **Equalities/Conversion Factors** We need the molar mass of C_2H_2 and CO_2. The mole–mole factor that converts moles of C_2H_2 to moles of CO_2 is derived from the coefficients in the balanced equation.

$$1 \text{ mol } C_2H_2 = 26.04 \text{ g}$$

$$\frac{26.04 \text{ g } C_2H_2}{1 \text{ mol } C_2H_2} \quad \text{and} \quad \frac{1 \text{ mol } C_2H_2}{26.04 \text{ g } C_2H_2}$$

$$2 \text{ mol } C_2H_2 = 4 \text{ mol } CO_2$$

$$\frac{4 \text{ mol } C_2H_2}{2 \text{ mol } CO_2} \quad \text{and} \quad \frac{2 \text{ mol } CO_2}{4 \text{ mol } C_2H_2}$$

$$1 \text{ mol } CO_2 = 44.01 \text{ g}$$

$$\frac{44.01 \text{ g } CO_2}{1 \text{ mol } CO_2} \quad \text{and} \quad \frac{1 \text{ mol } CO_2}{44.01 \text{ g } CO_2}$$

STEP 4 **Set Up Problem** Using our plan, we first convert grams of C_2H_2 to moles of C_2H_2.

$$54.6 \text{ g } C_2H_2 \times \frac{1 \text{ mol } C_2H_2}{26.04 \text{ g } C_2H_2} = 2.10 \text{ mol } C_2H_2$$

Then we change moles of C_2H_2 to moles of CO_2 by using the mole–mole factor.

$$2.10 \text{ mol } C_2H_2 \times \frac{4 \text{ mol } CO_2}{2 \text{ mol } C_2H_2} = 4.20 \text{ mol } CO_2$$

Finally, we can convert moles of CO_2 to grams of CO_2.

$$4.20 \text{ mol } CO_2 \times \frac{44.01 \text{ g } CO_2}{1 \text{ mol } CO_2} = 185 \text{ g } CO_2$$

The solution can be obtained using the conversion factors in sequence.

$$54.6 \text{ g } C_2H_2 \times \frac{1 \text{ mol } C_2H_2}{26.04 \text{ g } C_2H_2} \times \frac{4 \text{ mol } CO_2}{2 \text{ mol } C_2H_2} \times \frac{44.01 \text{ g } CO_2}{1 \text{ mol } CO_2} = 185 \text{ g } CO_2$$

Study Check

Using the equation in Sample Problem 8.6, calculate the grams of CO_2 that can be produced when 25.0 g O_2 reacts.

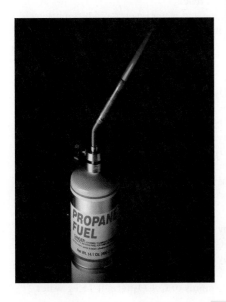

Sample Problem 8.7   **Calculating the Mass of Reactant**

Propane gas, (C_3H_8), a fuel for camp stoves and some specially equipped automobiles, reacts with oxygen to produce carbon dioxide and water. How many grams of O_2 are required to react with 22.0 g C_3H_8?

$$C_3H_8(g) + 5O_2(g) \longrightarrow 3CO_2(g) + 4H_2O(g)$$

Solution

STEP 1 **Given** 22.0 g C_3H_8 **Need** grams of O_2

STEP 2 **Plan** The 22.0 g C_3H_8 is first changed to moles of C_3H_8 using its molar mass. Then the moles of C_3H_8 are changed to moles of O_2, which can be converted to grams of O_2.

grams of C_3H_8 → Molar mass → moles of C_3H_8 → Mole–mole factor → moles of O_2 → Molar mass → grams of O_2

STEP 3 **Equalities/Conversion Factors**

$$1 \text{ mol } C_3H_8 = 44.09 \text{ g}$$

$$\frac{44.09 \text{ g } C_3H_8}{1 \text{ mol } C_3H_8} \quad \text{and} \quad \frac{1 \text{ mol } C_3H_8}{44.09 \text{ g } C_3H_8}$$

$$1 \text{ mol } C_3H_8 = 5 \text{ mol } O_2$$

$$\frac{1 \text{ mol } C_3H_8}{5 \text{ mol } O_2} \quad \text{and} \quad \frac{5 \text{ mol } O_2}{1 \text{ mol } C_3H_8}$$

$$1 \text{ mol } O_2 = 32.00 \text{ g}$$

$$\frac{32.00 \text{ g } O_2}{1 \text{ mol } O_2} \quad \text{and} \quad \frac{1 \text{ mol } O_2}{32.00 \text{ g } O_2}$$

STEP 4 **Set Up Problem**   Using a sequence of three factors converts grams of C_3H_8 to grams of O_2.

$$22.0 \text{ g } C_3H_8 \times \frac{1 \text{ mol } C_3H_8}{44.09 \text{ g } C_3H_8} \times \frac{5 \text{ mol } O_2}{1 \text{ mol } C_3H_8} \times \frac{32.00 \text{ g } O_2}{1 \text{ mol } O_2} = 79.8 \text{ g } O_2$$

Study Check

Using the equation in Sample Problem 8.7, calculate the grams of C_3H_8 that are needed to produce 15.0 g H_2O.

Questions and Problems   **Mass Calculations in Reactions**

8.11   Sodium reacts with oxygen to produce sodium oxide.

$$4Na(s) + O_2(g) \longrightarrow 2Na_2O(s)$$

   a. How many grams of Na_2O are produced when 2.50 mol Na react?
   b. If you have 18.0 g Na, how many grams of O_2 are required for reaction?
   c. How many grams of O_2 are needed in a reaction that produces 75.0 g Na_2O?

8.12   Nitrogen gas reacts with hydrogen gas to produce ammonia by the following equation:

$$N_2(g) + 3H_2(g) \longrightarrow 2NH_3(g)$$

   a. If you have 1.80 mol H_2, how many grams of NH_3 can be produced?
   b. How many grams of H_2 are needed to react with 2.80 g N_2?
   c. How many grams of NH_3 can be produced from 12.0 g H_2?

8.13   Ammonia and oxygen react to form nitrogen and water.

$$4NH_3(g) + 3O_2(g) \longrightarrow 2N_2(g) + 6H_2O(g)$$

   Ammonia

   a. How many grams of O_2 are needed to react with 8.00 mol NH_3?
   b. How many grams of N_2 can be produced when 6.50 g O_2 reacts?
   c. How many grams of water are formed from the reaction of 34.0 g NH_3?

8.14   Iron(III) oxide reacts with carbon to give iron and carbon monoxide.

$$Fe_2O_3(s) + 3C(s) \longrightarrow 2Fe(s) + 3CO(g)$$

   a. How many grams of C are required to react with 2.50 mol Fe_2O_3?
   b. How many grams of CO are produced when 36.0 g C reacts?
   c. How many grams of Fe can be produced when 6.00 g Fe_2O_3 reacts?

8.15   Nitrogen dioxide and water react to produce nitric acid, HNO_3, and nitrogen oxide.

$$3NO_2(g) + H_2O(l) \longrightarrow 2HNO_3(aq) + NO(g)$$

   a. How many grams of H_2O are required to react with 28.0 g NO_2?
   b. How many grams of NO are obtained from 15.8 g NO_2?
   c. How many grams of HNO_3 are produced from 8.25 g NO_2?

8.16   Calcium cyanamide reacts with water to form calcium carbonate and ammonia.

$$CaCN_2(s) + 3H_2O(l) \longrightarrow CaCO_3(s) + 2NH_3(g)$$

   a. How many grams of water are needed to react with 75.0 g $CaCN_2$?
   b. How many grams of NH_3 are produced from 5.24 g $CaCN_2$?
   c. How many grams of $CaCO_3$ form if 155 g water react?

8.17 When the ore lead(II) sulfide burns in oxygen, the products are lead(II) oxide and sulfur dioxide.
    a. Write the balanced equation for the reaction.
    b. How many grams of oxygen are required to react with 0.125 mol lead(II) sulfide?
    c. How many grams of sulfur dioxide can be produced when 65.0 g lead(II) sulfide react?
    d. How many grams of lead(II) sulfide are used to produce 128 g lead(II) oxide?

8.18 When the gases dihydrogen sulfide and oxygen react, they form the gases sulfur dioxide and water vapor.
    a. Write the balanced equation for the reaction.
    b. How many grams of oxygen are required to react with 2.50 g dihydrogen sulfide?
    c. How many grams of sulfur dioxide can be produced when 38.5 g oxygen react?
    d. How many grams of oxygen are required to produce 55.8 g water vapor?

8.3 Limiting Reactants

Learning **Goal**

Identify a limiting reactant when given the quantities of two or more reactants; calculate the amount of product formed from the limiting reactant.

When you make peanut butter sandwiches for lunch, you need 2 slices of bread and 1 tablespoon of peanut butter for each sandwich. As an equation, we could write:

2 slices of bread + 1 tablespoon peanut butter ⟶ 1 peanut butter sandwich

If you have 8 slices of bread and a full jar of peanut butter, you will run out of bread after you make 4 peanut butter sandwiches. You cannot make any more sandwiches once the bread is used up, even though there is a lot of peanut butter left in the jar. The number of slices of bread has limited the number of sandwiches you can make. On a different day, you might have 8 slices of bread, but only a tablespoon of peanut butter left in the peanut butter jar. You will run out of peanut butter after you make just 1 peanut butter sandwich with 6 slices of bread left over. The small amount of peanut butter available has limited the number of sandwiches you can make. This time the amount of peanut butter is limited.

8 slices bread +   make   4 peanut butter   peanut butter
1 jar peanut butter   sandwiches   left over

8 slices of bread   + 1 tablespoon   make   1 peanut butter   6 slices bread
                       of peanut butter   sandwich   left over

In a similar way, the availability of reactants in a chemical reaction can limit the amount of product that forms. In many reactions, the reactants are not combined in quantities that allow each to be used up at exactly the same time. Then one reactant is used up before the other. The reactant that is completely used up in the reaction is called the **limiting reactant.** The other reactant, called the **excess reactant,** is left over. In the last analogy, the bread was the limiting reactant and the jar of peanut butter was the excess reactant.

Bread	Peanut Butter	Sandwiches	Limiting	Excess
1 loaf (20 slices)	1 tablespoon	1	Peanut butter	Bread
4 slices	1 full jar	2	Bread	Peanut butter
8 slices	1 full jar	4	Bread	Peanut butter

Sample Problem 8.8 ▸ **Limiting Reactants**

You are going to have a dinner party. In your silverware drawer, there are 10 spoons, 8 forks, and 6 knives. If each place setting requires 1 spoon, 1 fork, and 1 knife, how many people can you serve?

Solution

You can make up place settings until you run out of one of the utensils.

$$1 \text{ spoon} + 1 \text{ fork} + 1 \text{ knife} = 1 \text{ place setting}$$

The number of place settings from each utensil is

$$10 \text{ spoons} \times \frac{1 \text{ place setting}}{1 \text{ spoon}} = 10 \text{ place settings}$$

$$8 \text{ forks} \times \frac{1 \text{ place setting}}{1 \text{ fork}} = 8 \text{ place settings}$$

$$6 \text{ knives} \times \frac{1 \text{ place setting}}{1 \text{ knife}} = 6 \text{ place settings}$$
(smallest number of settings)

The limiting utensil is 6 knives, which limits your dinner party to 6 place settings. Thus, you can have 5 guests and yourself at the dinner party.

Study Check

You have a bicycle parts store. If there are 45 wheels and 18 bicycle seats, how many bicycles can you put together?

Calculating Moles of Product from a Limiting Reactant

Our analogies of everyday limiting reactants also apply to the substances in a chemical reaction. Consider the reaction in which hydrogen and chlorine form hydrogen chloride.

$$H_2(g) + Cl_2(g) \longrightarrow 2HCl(g)$$

Suppose the reaction mixture contains 2 mol H_2 and 5 mol Cl_2. From the equation, we see that 1 mol H_2 reacts with 1 mol Cl_2 to produce 2 mol HCl. Mole–mole factors for these relationships are written as

2 mol HCl = 1 mol H_2		2 mol HCl = 1 mol Cl_2	
$\dfrac{2 \text{ mol HCl}}{1 \text{ mol } H_2}$ and $\dfrac{1 \text{ mol } H_2}{2 \text{ mol HCl}}$		$\dfrac{2 \text{ mol HCl}}{1 \text{ mol } Cl_2}$ and $\dfrac{1 \text{ mol } Cl_2}{2 \text{ mol HCl}}$	

Now we need to calculate the amount of product that is possible from each of the reactants. We are looking for the limiting reactant, which is the one that produces the smaller amount of product.

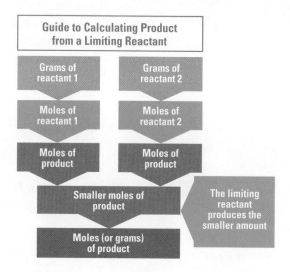

Moles of HCl from H_2

$$2 \; \cancel{\text{mol } H_2} \; \times \; \frac{2 \text{ mol HCl}}{1 \; \cancel{\text{mol } H_2}} \; = \; 4 \text{ mol HCl (smaller amount of product)}$$

Moles of HCl from Cl_2

$$5 \; \cancel{\text{mol } Cl_2} \; \times \; \frac{2 \text{ mol HCl}}{1 \; \cancel{\text{mol } Cl_2}} \; = \; 10 \text{ mol HCl (not possible)}$$

In this reaction mixture, H_2 is the limiting reactant because it reacts to produce 4 mol HCl, the smaller amount. When 2 mol H_2 are used up, the reaction stops. The excess reactant is the 3 mol Cl_2 left in the mixture that cannot react. We can show the changes in each reactant and the product as follows.

	Reactants		Product
Equation	H_2 +	Cl_2 $\longrightarrow$	2HCl
Initial moles	2 mol	5 mol	0
Moles used/formed	-2 mol	-2 mol	$+4$ mol
Moles left	0	3 mol	4 mol
Identify as	Limiting reactant	Excess reactant	Product possible

---

Sample Problem 8.9   **Moles of Product from Limiting Reactant**

Consider the reaction for the synthesis of methanol (CH_3OH).

$$CO(g) + 2H_2(g) \longrightarrow CH_3OH(g)$$

In the laboratory, 3.00 mol CO and 5.00 mol H_2 are combined. Calculate the number of moles of CH_3OH that can form and identify the limiting reactant.

Solution

STEP 1   **Given**   3.00 mol CO and 5.00 mol H_2
Need   moles of CH_3OH

STEP 2 **Plan** We can determine the limiting reactant and the excess reactant by calculating the moles of methanol that each reactant would produce if it were all used up. The actual number of moles of CH_3OH produced is from the reactant that produces the smaller number of moles.

moles of CO | Mole–mole factor | moles of CH_3OH

moles of H_2 | Mole–mole factor | moles of CH_3OH

STEP 3 **Equalities/Conversion Factors** The mole–mole factors needed are obtained from the equation.

$$1 \text{ mol CO} = 1 \text{ mol CH}_3\text{OH}$$
$$\frac{1 \text{ mol CH}_3\text{OH}}{1 \text{ mol CO}} \quad \text{and} \quad \frac{1 \text{ mol CO}}{1 \text{ mol CH}_3\text{OH}}$$

$$2 \text{ mol H}_2 = 1 \text{ mol CH}_3\text{OH}$$
$$\frac{1 \text{ mol CH}_3\text{OH}}{2 \text{ mol H}_2} \quad \text{and} \quad \frac{2 \text{ mol H}_2}{1 \text{ mol CH}_3\text{OH}}$$

STEP 4 **Set Up Problem** The moles of CH_3OH from each reactant are determined in separate calculations.

$$3.00 \text{ mol CO} \times \frac{1 \text{ mol CH}_3\text{OH}}{1 \text{ mol CO}} = 3.00 \text{ mol CH}_3\text{OH}$$

$$5.00 \text{ mol H}_2 \times \frac{1 \text{ mol CH}_3\text{OH}}{2 \text{ mol H}_2} = 2.50 \text{ mol CH}_3\text{OH} \text{ (smaller amount)}$$

The smaller amount (2.5 mol CH_3OH) is all the methanol that can be produced. Thus, H_2 is the limiting reagent and CO is in excess.

	Reactants		Product
Equation	CO +	$2H_2 \longrightarrow$	CH_3OH
Initial moles	3.0 mol	5.0 mol	0
Moles used/formed	−2.5 mol	−5.0 mol	+2.5 mol
Moles left	0.5 mol	0 mol	2.5 mol
Identify as	Excess reactant	Limiting reactant	Product possible

Study Check

Consider the reaction for the formation of copper(II) oxide from copper(I) oxide.

$$2Cu_2O(s) + O_2(g) \longrightarrow 4CuO(s)$$

Complete the following table for the quantities of reactants and products if 5.0 mol of copper(I) oxide and 2.0 mol of oxygen are allowed to react.

	Reactants		Product
Equation	$2Cu_2O$ +	$O_2 \longrightarrow$	$4CuO$
Initial moles	5.0 mol	2.0 mol	0 mol
Moles used/formed			
Moles left			
Identify as			

Calculating Mass of Product from a Limiting Reactant

The quantities of the substances in a reaction can also be given in grams. The calculations to identify the limiting reactant are the same as before, but the grams of each reactant must first be converted to moles. Once the limiting reactant is determined, the moles of possible product can be converted to grams using molar mass. This calculation is shown in Sample Problems 8.10 and 8.11.

Sample Problem 8.10 Mass of Product from a Limiting Reactant

Carbon monoxide and hydrogen gas react to form methanol, CH_3OH.

$$CO(g) + 2H_2(g) \longrightarrow CH_3OH(l)$$

If 48.0 g CO and 10.0 g H_2 react, how many grams of methanol can be produced?

Solution

STEP 1 Given 48.0 g CO and 10.0 g H_2 **Need** grams of CH_3OH

STEP 2 Plan Convert the grams of each reactant to moles and calculate the moles of CH_3OH that each reactant can produce. Then convert the number of moles of CH_3OH from the limiting reactant to grams of CH_3OH using molar mass.

48.0 g CO | Molar mass | moles CO | Mole–mole factor | moles CH_3OH (if smaller) | Molar mass | grams CH_3OH

or

10.0 g H_2 | Molar mass | moles H_2 | Mole–mole factor | moles CH_3OH (if smaller) | Molar mass | grams CH_3OH

STEP 3 Equalities and Conversion Factors

$$1 \text{ mol CO} = 28.01 \text{ g CO}$$
$$\frac{1 \text{ mol CO}}{28.01 \text{ g CO}} \text{ and } \frac{28.01 \text{ g CO}}{1 \text{ mol CO}}$$

$$1 \text{ mol CO} = 1 \text{ mol CH}_3\text{OH}$$
$$\frac{1 \text{ mol CO}}{1 \text{ mol CH}_3\text{OH}} \text{ and } \frac{1 \text{ mol CH}_3\text{OH}}{1 \text{ mol CO}}$$

$$1 \text{ mol H}_2 = 2.016 \text{ g H}_2$$
$$\frac{1 \text{ mol H}_2}{2.016 \text{ g H}_2} \text{ and } \frac{2.016 \text{ g H}_2}{1 \text{ mol H}_2}$$

$$2 \text{ mol H}_2 = 1 \text{ mol CH}_3\text{OH}$$
$$\frac{2 \text{ mol H}_2}{1 \text{ mol CH}_3\text{OH}} \text{ and } \frac{1 \text{ mol CH}_3\text{OH}}{2 \text{ mol H}_2}$$

$$1 \text{ mol CH}_3\text{OH} = 32.04 \text{ g CH}_3\text{OH}$$
$$\frac{1 \text{ mol CH}_3\text{OH}}{32.04 \text{ g CH}_3\text{OH}} \text{ and } \frac{32.04 \text{ g CH}_3\text{OH}}{1 \text{ mol CH}_3\text{OH}}$$

STEP 4 Set Up Problem The moles of CH_3OH from each reactant can now be determined in separate calculations.

Moles CH_3OH produced from CO

$$48.0 \text{ g CO} \times \frac{1 \text{ mol CO}}{28.01 \text{ g CO}} \times \frac{1 \text{ mol CH}_3\text{OH}}{1 \text{ mol CO}} = \textbf{1.17 mol CH}_3\textbf{OH}$$
(smaller number of moles)

Moles CH$_3$OH produced from H$_2$

$$10.0 \text{ g H}_2 \times \frac{1 \text{ mol H}_2}{2.016 \text{ g H}_2} \times \frac{1 \text{ mol CH}_3\text{OH}}{2 \text{ mol H}_2} = 2.48 \text{ mol CH}_3\text{OH}$$

Because CO produces the smaller number of moles of CH$_3$OH, CO is the limiting reactant. Now the grams of product CH$_3$OH from these reactants is calculated by converting the moles of CH$_3$OH obtained from CO to grams using molar mass.

$$1.71 \text{ mol CH}_3\text{OH} \times \frac{32.04 \text{ g CH}_3\text{OH}}{1 \text{ mol CH}_3\text{OH}} = 54.8 \text{ g CH}_3\text{OH}$$

Study Check

Hydrogen sulfide burns with oxygen to give sulfur dioxide and water. How many moles of sulfur dioxide are formed from the reaction of 0.250 mol H$_2$S and 0.300 mol O$_2$?

$$2H_2S(g) + 3O_2(g) \longrightarrow 2SO_2(g) + 2H_2O(g)$$

Sample Problem 8.11 **Mass of Product from a Limiting Reactant**

Ammonia and fluorine react to form dinitrogen tetrafluoride, N$_2$F$_4$, and hydrogen fluoride.

$$2NH_3(g) + 5F_2(g) \longrightarrow N_2F_4(g) + 6HF(g)$$

If 5.00 g NH$_3$ and 20.0 g F$_2$ react, how many grams of hydrogen fluoride are produced?

Solution

STEP 1 **Given** 5.00 g NH$_3$ and 20.0 g F$_2$ **Need** grams of HF

STEP 2 **Plan** We determine the limiting reactant and the excess reactant by calculating the possible number of moles of HF that each reactant can produce. We are looking for the reactant that produces the smaller number of moles of HF. In this calculation, the grams of each reactant is converted to moles and the number of moles of HF each will produce is calculated. Then the number of moles of HF from the limiting reactant, the smaller amount, is converted to the mass in grams of HF using its molar mass.

5.00 g NH$_3$ → Molar mass → moles NH$_3$ → Mole–mole factor → moles HF (if smaller) → Molar mass → grams HF

20.0 g F$_2$ → Molar mass → moles F$_2$ → Mole–mole factor → moles HF (if smaller) → Molar mass → grams HF

STEP 3 **Equalities/Conversion Factors**

$$1 \text{ mol NH}_3 = 17.03 \text{ g NH}_3$$

$$\frac{1 \text{ mol NH}_3}{17.03 \text{ g NH}_3} \quad \text{and} \quad \frac{17.03 \text{ g NH}_3}{1 \text{ mol NH}_3}$$

$$2 \text{ mol NH}_3 = 6 \text{ mol HF}$$

$$\frac{2 \text{ mol NH}_3}{6 \text{ mol HF}} \quad \text{and} \quad \frac{6 \text{ mol HF}}{2 \text{ mol NH}_3}$$

$$1 \text{ mol } F_2 = 38.00 \text{ g } F_2$$

$$\frac{1 \text{ mol } F_2}{38.00 \text{ g } F_2} \quad \text{and} \quad \frac{38.00 \text{ g } F_2}{1 \text{ mol } F_2}$$

$$6 \text{ mol HF} = 5 \text{ mol } F_2$$

$$\frac{5 \text{ mol } F_2}{6 \text{ mol HF}} \quad \text{and} \quad \frac{6 \text{ mol HF}}{5 \text{ mol } F_2}$$

$$1 \text{ mol HF} = 20.01 \text{ g HF}$$

$$\frac{1 \text{ mol HF}}{20.01 \text{ g HF}} \quad \text{and} \quad \frac{20.01 \text{ g HF}}{1 \text{ mol HF}}$$

STEP 4 **Set Up Problem** The moles of HF from each reactant are determined in separate calculations.

$$5.00 \text{ g } NH_3 \times \frac{1 \text{ mol } NH_3}{17.03 \text{ g } NH_3} \times \frac{6 \text{ mol HF}}{2 \text{ mol } NH_3} = 0.881 \text{ mol HF}$$

$$20.0 \text{ g } F_2 \times \frac{1 \text{ mol } F_2}{38.00 \text{ g } F_2} \times \frac{6 \text{ mol HF}}{5 \text{ mol } F_2} = \textbf{0.632 mol HF (smaller number of moles)}$$

The limiting reactant is F_2 because it produces the smaller number of moles of HF product. Next the quantity of the product, HF, is converted from moles to grams using its molar mass.

$$0.632 \text{ mol HF} \times \frac{20.01 \text{ g HF}}{1 \text{ mol HF}} = 12.6 \text{ g HF}$$

Study Check

When silicon dioxide (sand) and carbon are heated, the ceramic material silicon carbide, SiC, and carbon monoxide are produced. How many grams of SiC are formed from 20.0 g SiO_2 and 50.0 g C?

$$SiO_2(s) + 3C(s) \xrightarrow{\text{Heat}} SiC(s) + 2CO(g)$$

Questions and Problems **Limiting Reactants**

8.19 A taxi company has 10 taxis.
  a. On a certain day, only 8 taxi drivers show up for work. How many taxis can be used to pick up passengers?
  b. On another day, 10 taxi drivers show up for work but 3 taxis are in the repair shop. How many taxis can be driven?

8.20 A clock maker has 15 clock faces. Each clock requires 1 face and 2 hands.
  a. If the clock maker also has 42 hands, how many clocks can be produced?
  b. If the clock maker has only 8 hands, how many clocks can be produced?

8.21 Nitrogen and hydrogen react to form ammonia.

$$N_2(g) + 3H_2(g) \longrightarrow 2NH_3(g)$$

Determine the limiting reactant in each of the following mixtures of reactants:
  a. 3.0 mol N_2 and 5.0 mol H_2

  b. 8.0 mol N_2 and 4.0 mol H_2
  c. 3.0 mol N_2 and 12.0 mol H_2

8.22 Iron and oxygen react to form iron(III) oxide.

$$4Fe(s) + 3O_2(g) \longrightarrow 2Fe_2O_3(s)$$

Determine the limiting reactant in each of the following mixtures of reactants:
  a. 2.0 mol Fe and 6.0 mol O_2
  b. 5.0 mol Fe and 4.0 mol O_2
  c. 16.0 mol Fe and 20.0 mol O_2

8.23 For each of the following reactions, calculate the moles of indicated product produced when 2.00 mol of each reactant is used:
  a. $2SO_2(g) + O_2(g) \longrightarrow 2SO_3(g)$ (SO_3)
  b. $3Fe(s) + 4H_2O(l) \longrightarrow$
         $Fe_3O_4(s) + 4H_2(g)$ (Fe_3O_4)
  c. $C_7H_{16}(g) + 11O_2(g) \longrightarrow$
         $7CO_2(g) + 8H_2O(g)$ (CO_2)

8.24 For each of the following reactions, calculate the moles of indicated product produced when 3.00 mol of each reactant is used:

a. $4Li(s) + O_2(g) \longrightarrow 2Li_2O(s)$　　(Li_2O)

b. $Fe_2O_3(s) + 3H_2(g) \longrightarrow$
　　　　　　$2Fe(s) + 3H_2O(l)$　　(Fe)

c. $Al_2S_3(s) + 6H_2O(l) \longrightarrow$
　　　　　　$2Al(OH)_3(aq) + 3H_2S(g)$　　(H_2S)

8.25 For each of the following reactions, calculate the moles of indicated product produced when 20.0 g of each reactant is used:

a. $2Al(s) + 3Cl_2(g) \longrightarrow 2AlCl_3(s)$　　($AlCl_3$)

b. $4NH_3(g) + 5O_2(g) \longrightarrow$
　　　　　　$4NO(g) + 6H_2O(g)$　　(H_2O)

c. $CS_2(g) + 3O_2(g) \longrightarrow$
　　　　　　$CO_2(g) + 2SO_2(g)$　　(SO_2)

8.26 For each of the following reactions, calculate the moles of indicated product produced when 20.0 g of each reactant is used.

a. $4Al(s) + 3O_2(g) \longrightarrow 2Al_2O_3(s)$　　(Al_2O_3)

b. $3NO_2(g) + H_2O(l) \longrightarrow$
　　　　　　$2HNO_3(aq) + NO(g)$　　(HNO_3)

c. $4NH_3(g) + 5O_2(g) \longrightarrow$
　　　　　　$4NO(g) + 6H_2O(g)$　　(H_2O)

8.4 Percent Yield

Learning Goal

Given the actual quantity of product, determine the percent yield for a reaction.

Up to this point, we have done calculations as though the amount of product were the maximum quantity possible or 100%. In other words, we assumed that all of the reactants were changed completely to product. While this would be an ideal situation, it does not usually happen. As we run a reaction and transfer products from one container to another, some product is lost. There may also be side reactions that use up some of the reactants to give a different product. Thus, in a real experiment, the predicted amount of the desired product is never really obtained.

Suppose we are running a chemical reaction in the laboratory. We first measure out specific quantities of the reactants and place them in a reaction flask. Then we calculate the **theoretical yield** for the reaction, which is the amount of product we could expect if all the reactants were converted to product according to the mole ratios of the equation. The **actual yield** is the amount of product we collect when the reaction ends. Because some product is lost, the actual yield is always less than the theoretical yield. If we know the actual yield and the theoretical yield for a product, we can express the actual yield as a **percent yield.**

Guide to Calculations for Percent Yield

STEP 1
Write the given and needed quantities.

STEP 2
Write a plan to calculate the theoretical yield and the percent yield.

STEP 3
Write the molar mass for the reactant and the mole–mole factor from the balanced equation.

STEP 4
Solve for the percent yield ratio by dividing the actual yield (given) by the theoretical yield × 100.

$$\text{Percent yield (\%)} = \frac{\text{Actual yield}}{\text{Theoretical yield}} \times 100\%$$

Sample Problem 8.12 ▸ **Calculating Percent Yield**

On a spaceship, LiOH is used to absorb exhaled CO_2 from breathing air to form $LiHCO_3$.

$$LiOH(s) + CO_2(g) \longrightarrow LiHCO_3(s)$$

What is the percent yield of the reaction if 50.0 g LiOH gives 72.8 g $LiHCO_3$?

Solution

STEP 1　**Given**　50.0 g LiOH and 72.8 g $LiHCO_3$ (actually produced)
　　　　　Need　% yield $LiHCO_3$

STEP 2　**Plan**

　　　　　Calculation of theoretical yield:

50.0 g LiOH | Molar mass | → moles LiOH | Mole–mole factor | → moles LiHCO₃ | Molar mass | → grams LiHCO₃

Calculation percent yield:

$$\frac{\text{Actual yield}}{\text{Theoretical yield}} \times 100\%$$

STEP 3 **Equalities/Conversion Factors**

1 mol LiOH = 23.95 g LiOH

$$\frac{1 \text{ mol LiOH}}{23.95 \text{ g LiOH}} \quad \text{and} \quad \frac{23.95 \text{ g LiOH}}{1 \text{ mol LiOH}}$$

1 mol LiHCO₃ = 1 mol LiOH

$$\frac{1 \text{ mol LiHCO}_3}{1 \text{ mol LiOH}} \quad \text{and} \quad \frac{1 \text{ mol LiOH}}{1 \text{ mol LiHCO}_3}$$

1 mol LiHCO₃ = 67.96 g LiHCO₃

$$\frac{67.96 \text{ g LiHCO}_3}{1 \text{ mol LiHCO}_3} \quad \text{and} \quad \frac{1 \text{ mol LiHCO}_3}{67.96 \text{ g LiHCO}_3}$$

STEP 4 **Set Up Problem**

Calculation of theoretical yield:

$$50.0 \text{ g LiOH} \times \frac{1 \text{ mol LiOH}}{23.95 \text{ g LiOH}} \times \frac{1 \text{ mol LiHCO}_3}{1 \text{ mol LiOH}} \times \frac{67.96 \text{ g LiHCO}_3}{1 \text{ mol LiHCO}_3} = 142 \text{ g LiHCO}_3$$

Calculation of percent yield:

$$\frac{\text{Actual yield (given)}}{\text{Theoretical yield (calculated)}} \times 100\% = \frac{72.8 \text{ g LiHCO}_3}{142 \text{ g LiHCO}_3} \times 100\% = 51.3\%$$

A percent yield of 51.3% means that 72.8 g of the theoretical amount of 142 g LiHCO₃ was actually produced by the reaction.

Study Check

For the reaction in Sample Problem 8.12, what is the percent yield if 8.00 g CO_2 produces 10.5 g LiHCO₃?

Questions and Problems ▸ **Percent Yield**

8.27 Carbon disulfide is produced by the reaction of carbon and sulfur dioxide.

$$5C(s) + 2SO_2(g) \longrightarrow CS_2(g) + 4CO(g)$$

a. What is the percent yield for the reaction if 40.0 g carbon produces 36.0 g carbon disulfide?
b. What is the percent yield for the reaction if 32.0 g sulfur dioxide produces 12.0 g carbon disulfide?

8.28 Iron(III) oxide reacts with carbon monoxide to produce iron and carbon dioxide.

$$Fe_2O_3(s) + 3CO(g) \longrightarrow 2Fe(s) + 3CO_2(g)$$

a. What is the percent yield for the reaction if 65.0 g iron(III) oxide produces 15.0 g iron?

b. What is the percent yield for the reaction if 75.0 g carbon monoxide produces 15.0 g carbon dioxide?

8.29 Aluminum reacts with oxygen to produce aluminum oxide.

$$4Al(s) + 3O_2(g) \longrightarrow 2Al_2O_3(s)$$

The reaction of 50.0 g aluminum and sufficient oxygen has a 75.0% yield. How many grams of aluminum oxide are produced?

8.30 Propane (C_3H_8) burns in oxygen to produce carbon dioxide and water.

$$C_3H_8(g) + 5O_2(g) \longrightarrow 3CO_2(g) + 4H_2O(g)$$

Calculate the mass of CO_2 that can be produced if the reaction of 45.0 g propane and sufficient oxygen has a 60.0% yield.

8.31 When 30.0 g carbon is heated with silicon dioxide, 28.2 g carbon monoxide is produced. What is the percent yield of this reaction?

$$SiO_2(s) + 3C(s) \longrightarrow SiC(s) + 2CO(g)$$

8.32 Calcium and nitrogen react to form calcium nitride.

$$3Ca(s) + N_2(g) \longrightarrow Ca_3N_2(s)$$

If 56.6 g calcium is mixed with nitrogen gas and 32.4 g calcium nitride is produced, what is the percent yield of the reaction?

Concept Map ▶   **Chemical Quantities in Reactions**

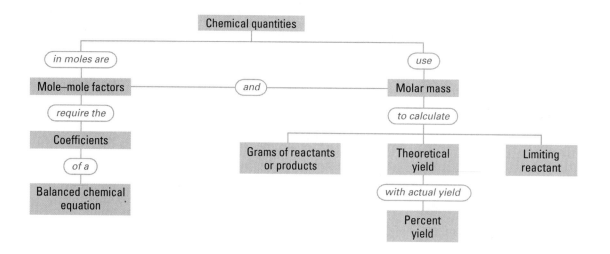

Chapter Review

8.1 Mole Relationships in Chemical Equations

In a balanced equation, the total mass of the reactants is equal to the total mass of the products. The coefficients in an equation describing the relationship between the moles of any two components are used to write mole–mole factors. When the number of moles for one substance is known, a mole-mole factor is used to find the moles of a different substance in the reaction.

8.2 Mass Calculations for Reactions

In calculations using equations, the molar masses of the substances and their mole factors are used to change the number of grams of one substance to the corresponding grams of a different substance.

8.3 Limiting Reactants

A limiting reactant is the reactant that produces the smaller amount of product while some excess reactant is left over. When the mass of two or more reactants is given, the mass of a product is calculated from the product produced by the limiting reactant.

8.4 Percent Yield

The percent yield of a reaction indicates the percent of product actually produced during a reaction. The percent yield is calculated by dividing the actual yield in grams of a product by the theoretical yield in grams.

Key Terms

actual yield The actual amount of product produced by a reaction.

excess reactant The reactant that remains when the limiting reactant is used up in a reaction.

law of conservation of mass In a chemical reaction, the total mass of the reactants is equal to the total mass of the product; matter is neither lost nor gained.

limiting reactant The reactant used up during a chemical reaction, which limits the amount of product that can form.

mole–mole factor A conversion factor that relates the number of moles of two compounds in an equation derived from their coefficients.

percent yield The ratio of the actual yield of a reaction to the theoretical yield possible for the reaction.

theoretical yield The maximum amount of product that a reaction can produce from a given amount of reactant.

Understanding the Concepts

8.33 If red spheres represent oxygen atoms and blue spheres represent nitrogen atoms,
   a. write a balanced equation for the reaction.
   b. identify the limiting reactant.

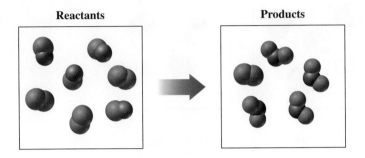

8.34 If purple spheres represent iodine atoms and light blue spheres represent hydrogen atoms,
   a. write a balanced equation for the reaction represented.
   b. identify the diagram that shows the products that result.

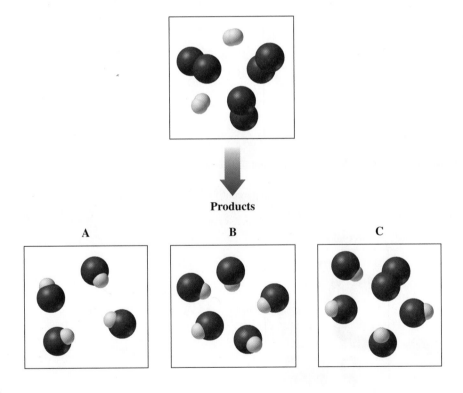

8.35 If blue spheres represent nitrogen atoms and light blue spheres represent hydrogen atoms,
    a. write a balanced equation for the reaction represented.
    b. identify the diagram that shows the products that result.

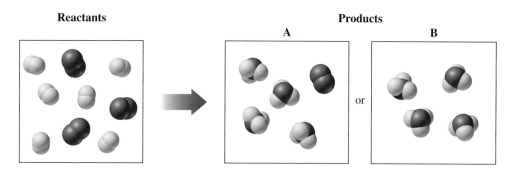

8.36 If green spheres represent chlorine atoms, yellow-green spheres represent fluorine atoms, and light blue spheres represent hydrogen atoms,
    a. write a balanced equation for the reaction represented.
    b. identify the limiting reactant.

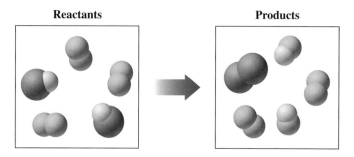

8.37 If blue spheres represent nitrogen atoms and purple spheres represent iodine atoms,
    a. write a balanced equation for the reaction represented.
    b. from the diagram of the actual products that result, calculate the percent yield for the reaction.

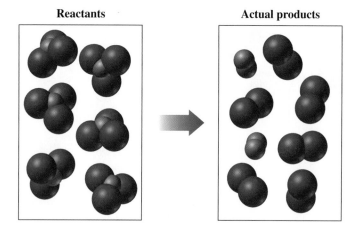

8.38 If green spheres represent chlorine atoms and red spheres represent oxygen atoms,
    a. write a balanced equation for the reaction represented.
    b. identify the limiting reactant.
    c. from the diagram of the actual products that result, calculate the percent yield for the reaction.

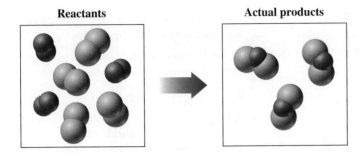

Reactants   **Actual products**

▶ Additional Questions and Problems

8.39 At a winery, glucose ($C_6H_{12}O_6$) in grapes undergoes fermentation to produce ethanol (C_2H_6O) and carbon dioxide.

$$C_6H_{12}O_6 \longrightarrow 2C_2H_6O + 2CO_2$$
Glucose   Ethanol

    a. How many moles of glucose are required to form 124 g ethanol?
    b. How many grams of ethanol would be formed from the reaction of 0.240 kg glucose?

8.40 Gasohol is a fuel containing ethanol (C_2H_6O) that burns in oxygen (O_2) to give carbon dioxide and water.
    a. State the reactants and products for this reaction in the form of a balanced equation.
    b. How many moles of O_2 are needed to completely react with 4.0 mol C_2H_6O?
    c. If a car produces 88 g CO_2, how many grams of O_2 are used up in the reaction?
    d. If you add 125 g C_2H_6O to your gas, how many grams of CO_2 and H_2O can be produced?

8.41 Balance the following equation:

$$NH_3(g) + F_2(g) \longrightarrow N_2F_4(g) + HF(g)$$

    a. How many moles of each reactant are needed to produce 4.00 mol HF?
    b. How many grams of F_2 are required to react with 1.50 mol NH_3?
    c. How many grams of N_2F_4 can be produced when 3.40 g NH_3 reacts?

8.42 Propane gas, C_3H_8, reacts with oxygen to produce water and carbon dioxide. Propane has a density of 2.02 g/L at room temperature.

$$C_3H_8(g) + 5O_2(g) \longrightarrow 3CO_2(g) + 4H_2O(l)$$
Propane

    a. How many moles of water form when 5.00 L propane gas (C_3H_8) completely reacts?

    b. How many grams of CO_2 are produced from 18.5 g oxygen gas and excess propane?
    c. How many grams of H_2O can be produced from the reaction of 8×10^{22} molecules propane gas, C_3H_8?

8.43 When a mixture of 12.8 g Na and 10.2 g Cl_2 reacts, what is the mass of NaCl that is produced?

$$2Na(s) + Cl_2(g) \longrightarrow 2NaCl(s)$$

8.44 If a mixture of 35.8 g CH_4 and 75.5 g S reacts, how many grams of H_2S are produced?

$$CH_4(g) + 4S(g) \longrightarrow CS_2(g) + 2H_2S(g)$$

8.45 Pentane gas, C_5H_{12}, reacts with oxygen to produce carbon dioxide and water.

$$C_5H_{12}(g) + 8O_2(g) \longrightarrow 5CO_2(g) + 6H_2O(g)$$

    a. How many grams of pentane must react to produce 4.0 mol water?
    b. How many grams of CO_2 are produced from 32.0 g oxygen and excess pentane?
    c. How many grams of CO_2 are formed if 44.5 g C_5H_{12} is mixed with 108 g O_2?

8.46 When nitrogen dioxide (NO_2) from car exhaust combines with water in the air it forms nitric acid (HNO_3), which causes acid rain, and nitrogen oxide.

$$3NO_2(g) + H_2O(l) \longrightarrow 2HNO_3(aq) + NO(g)$$

    a. How many molecules of NO_2 are needed to react with 0.250 mol H_2O?
    b. How many grams of HNO_3 are produced when 60.0 g NO_2 completely reacts?
    c. How many grams of HNO_3 can be produced if 225 g NO_2 is mixed with 55.2 g H_2O?

8.47 Acetylene gas C_2H_2 burns in oxygen to produce carbon dioxide and water. If 62.0 g CO_2 is produced when 22.5 g C_2H_2 reacts with sufficient oxygen, what is the percent yield for the reaction?

8.48 When 50.0 g iron(III) oxide reacts with carbon monoxide, 32.8 g iron is produced. What is the percent yield of the reaction?

$$Fe_2O_3(s) + 3CO(g) \longrightarrow 2Fe(s) + 3CO_2(g)$$

8.49 A reaction of nitrogen and sufficient hydrogen produced 30.0 g ammonia, which is a 65.0% yield for the reaction. How many grams of nitrogen reacted?

$$N_2(g) + 3H_2(g) \longrightarrow 2NH_3(g)$$

8.50 Consider the *unbalanced* equation for the decomposition of potassium chlorate.

$$KClO_3(s) \longrightarrow KCl(s) + O_2(g)$$

a. When 46.0 g $KClO_3$ is heated, 12.1 g O_2 is formed. How many grams of KCl is also formed?
b. What is the percent yield of KCl for the reaction?

8.51 Acetylene, C_2H_2, used in welders' torches, burns according to the following equation:

$$2C_2H_2(g) + 5O_2(g) \longrightarrow 4CO_2(g) + 2H_2O(g)$$

a. How many molecules of oxygen are needed to react with 22.0 g acetylene?
b. How many grams of carbon dioxide could be produced from the complete reaction of the acetylene in part a?
c. If the reaction in part a produces 64.0 g CO_2, what is the *percent yield* for the reaction?

8.52 Consider the *unbalanced* equation for the reaction of sodium and nitrogen to form sodium nitride.

$$Na(s) + N_2(g) \longrightarrow Na_3N(s)$$

a. If 80.0 g sodium is mixed with 20.0 g nitrogen gas, what mass of sodium nitride forms?
b. If the reaction in part a has a percent yield of 75.0%, how much sodium nitride is actually produced?

Challenge Questions

8.53 Carbon monoxide reacts with hydrogen to form methanol, CH_3OH.

$$CO(g) + 2H_2(g) \longrightarrow CH_3OH(l)$$

Suppose you mix 50.0 g CO and 10.0 g H_2.
a. What is the limiting reactant?
b. What is the excess reactant?
c. How many grams of methanol can be produced?
d. How many grams of the excess reactant are left over?

8.54 The gaseous hydrocarbon acetylene, C_2H_2, used in welders' torches, releases a large amount of heat when it burns according to the following equation:

$$2C_2H_2(g) + 5O_2(g) \longrightarrow 4CO_2(g) + 2H_2O(g)$$

a. How many moles of water are produced from the complete reaction of 64.0 g oxygen?
b. How many moles of oxygen are needed to react completely with 2.25×10^{24} molecules of acetylene?

c. How many grams of carbon dioxide are produced from the complete reaction of 78.0 g acetylene?
d. If the reaction in part c produces 186 g CO_2, what is the percent yield for the reaction?

8.55 Consider the following *unbalanced* equation:

$$Al(s) + O_2(g) \longrightarrow Al_2O_3(s)$$

a. Balance the equation.
b. Identify the type of reaction.
c. How many moles of oxygen react with 4.50 mol Al?
d. How many grams of aluminum oxide are produced when 50.2 g aluminum react?
e. When 0.500 mol aluminum is reacted with 8.00 g oxygen, how many grams of aluminum oxide can form?
f. If 45.0 g aluminum and 62.0 g oxygen undergo a reaction that has a 70.0% yield, what mass of aluminum oxide forms?

Answers

Answers to Study Checks

8.1 The total mass of the reactants is 164.14 g (68.14 g NH_3 + 96.00 g O_2), which equals the total mass of the products of 164.14 g (56.04 g N_2 + 108.10 g H_2O).

8.2 $\dfrac{2 \text{ mol } Na_2O}{1 \text{ mol } O_2}$ and $\dfrac{1 \text{ mol } O_2}{2 \text{ mol } Na_2O}$

8.3 1.83 mol Fe

8.4 0.945 mol O_2

8.5 44.1 g NO

8.6 27.5 g CO_2

8.7 9.18 g C_3H_8

8.8 18 bicycles

8.9

	Reactants		Product
Equation	$2Cu_2O(s)$ +	$O_2(g) \longrightarrow$	$4CuO(s)$
Initial moles	5.0 mol	2.0 mol	0 mol
Moles used/formed	−4.0 mol	−2.0 mol	8.0 mol
Moles left	1.0 mol	0	8.0 mol
Identify as	Excess reactant	Limiting reactant	Product possible

8.10 0.200 mol SO_2

8.11 13.3 g SiC

8.12 84.7%

Answers to Selected Questions and Problems

8.1 a. **(1)** Two molecules of sulfur dioxide gas react with 1 molecule oxygen gas to produce 2 molecules sulfur trioxide gas.
  b. **(2)** Two mol of sulfur dioxide gas react with 1 mol oxygen gas to produce 2 mol sulfur trioxide gas.
  c. **(1)** Four atoms of solid phosphorus react with 5 molecules of oxygen gas to produce 2 molecules of solid diphosphorus pentoxide.
  d. **(2)** Four mol solid phosphorus react with 5 mol oxygen gas to produce 2 mol solid diphosphorus pentoxide.

8.3 a. 160.14 g reactants = 160.14 g products
  b. 283.88 g reactants = 283.88 g products

8.5 a. $\dfrac{2 \text{ mol } SO_2}{1 \text{ mol } O_2}$ and $\dfrac{1 \text{ mol } O_2}{2 \text{ mol } SO_2}$

$\dfrac{2 \text{ mol } SO_2}{2 \text{ mol } SO_3}$ and $\dfrac{2 \text{ mol } SO_3}{2 \text{ mol } SO_2}$

$\dfrac{2 \text{ mol } SO_3}{1 \text{ mol } O_2}$ and $\dfrac{1 \text{ mol } O_2}{2 \text{ mol } SO_3}$

  b. $\dfrac{4 \text{ mol } P}{5 \text{ mol } O_2}$ and $\dfrac{5 \text{ mol } O_2}{4 \text{ mol } P}$

$\dfrac{4 \text{ mol } P}{2 \text{ mol } P_2O_5}$ and $\dfrac{2 \text{ mol } P_2O_5}{4 \text{ mol } P}$

$\dfrac{5 \text{ mol } O_2}{2 \text{ mol } P_2O_5}$ and $\dfrac{2 \text{ mol } P_2O_5}{5 \text{ mol } O_2}$

8.7 a. 1.0 mol O_2 **b.** 10. mol H_2
  c. 5.0 mol H_2O

8.9 a. 1.25 mol C   **b.** 0.96 mol CO
  c. 1.0 mol SO_2 **d.** 0.50 mol CS_2

8.11 a. 77.5 g Na_2O **b.** 6.26 g O_2
  c. 19.4 g O_2

8.13 a. 192 g O_2 **b.** 3.79 g N_2
  c. 54.0 g H_2O

8.15 a. 3.66 g H_2O **b.** 3.44 g NO
  c. 7.53 g HNO_3

8.17 a. $2PbS(s)$ + $3O_2(g) \longrightarrow 2PbO(s)$ + $2SO_2(g)$
  b. 6.00 g O_2
  c. 18.4 g SO_2
  d. 137 g PbS

8.19 a. 8 taxis can be used to pick up passengers
  b. 7 taxis can be driven

8.21 a. 5.0 mol H_2 **b.** 4.0 mol H_2
  c. 3.0 mol N_2

8.23 a. 2.00 mol SO_3 **b.** 0.500 mol Fe_3O_4
  c. 1.27 mol CO_2

8.25 a. 0.188 mol $AlCl_3$ **b.** 0.750 mol H_2O
  c. 0.417 mol SO_2

8.27 a. 71.0%   **b.** 63.2%

8.29 70.9 g Al_2O_3

8.31 60.5%

8.33 a. $2NO + O_2 \longrightarrow 2NO_2$
  b. NO is the limiting reactant.

8.35 a. $N_2 + 3H_2 \longrightarrow 2NH_3$ **b.** A

8.37 a. $2NI_3 \longrightarrow N_2 + 3I_2$ **b.** 67% yield

8.39 a. 1.35 mol glucose   **b.** 123 g ethanol

8.41 a. $2NH_3(g) + 5F_2(g) \longrightarrow N_2F_4(g) + 6HF(g)$
  b. 1.33 mol NH_3 and 3.33 mol F_2
  c. 143 g F_2
  d. 10.4 g N_2F_4

8.43 16.8 g NaCl

8.45 a. 48 g C_5H_{12} **b.** 27.5 g CO_2
  c. 92.8 g CO_2

8.47 81.5%

8.49 38.0 g N_2 reacted.

8.51 a. 1.27×10^{24} molecules O_2
  b. 74.4 g CO_2
  c. 86.0% yield

8.53 a. 1.79 mol CO will form 1.79 mol methanol. CO is the limiting reactant.
  b. H_2 is the excess reactant.
  c. 57.4 g methanol is formed.
  d. 2.80 g H_2 is left over.

8.55 a. $4Al(s) + 3O_2(g) \longrightarrow 2Al_2O_3(s)$
  b. This is a combination reaction.
  c. 3.38 mol oxygen.
  d. 94.9 g aluminum oxide
  e. 17.0 g aluminum oxide
  f. 59.6 g aluminum oxide

Combining Ideas from Chapters 7 and 8

CI 11 In an experiment as seen in the figure, a piece of copper is weighed and reacted with oxygen gas to produce solid copper(II) oxide.

a. If copper has a density of 8.94 g/cm^3, what is the volume (cm^3) of the copper?

b. How many copper atoms are in the sample?

c. Write the equation for the reaction.

d. How many grams of oxygen are required to completely react with the copper?

e. How many grams of copper(II) oxide will result from the reaction of 8.56 g Cu and 3.72 g oxygen?

f. How many grams of copper(II) oxide will result in part d if the percent yield for the reaction is 85%?

CI 12 Butyric acid, a compound containing carbon, hydrogen, and oxygen, contributes to the characteristic odor of parmesan cheese. When 8.81 g butyric acid undergoes combustion, the products are 17.6 g CO_2 and 7.21 g H_2O.

a. How many moles of carbon, hydrogen, and oxygen are in the sample of butyric acid?

b. What is the empirical formula of the compound?

c. If butyric acid has a molar mass of about 88, what is its molecular formula?

d. Write the balanced equation for the combustion reaction of butyric acid.

e. How many grams of oxygen are needed to completely react 1.58 g butyric acid?

CI 13 When clothes have stains, bleach is often added to the wash to react with the soil and make the stains colorless.

One brand of bleach contains 5.25% sodium hypochlorite by mass (active ingredient) with a density of 1.08 g/mL. The liquid bleach solution is prepared by bubbling chlorine gas into a solution of sodium hydroxide to produce sodium hypochlorite, sodium chloride, and water.

a. What is the formula and molar mass of sodium hypochlorite?

b. What is the percent composition of sodium hypochlorite?

c. How many hypochlorite ions are present in 1 gallon of bleach solution?

d. Write the equation for the preparation of bleach.

e. How many grams of NaOH are required to produce the mass of sodium hypochlorite for 1 gallon of bleach?

f. If 165 g Cl_2 is passed through a solution containing 275 g NaOH and 162 g sodium hypochlorite is produced, what is the percent yield for the reaction?

CI 14 Automobile exhaust is a major cause of air pollution. The pollutants formed from gasoline include nitrogen oxide, which is produced at high temperatures in the car engine from nitrogen and oxygen in the air. Once emitted into the air, nitrogen oxide reacts with oxygen to produce nitrogen dioxide, a reddish gas with a sharp, pungent odor that makes up smog. Gasoline, C_8H_{18}, has a density of 0.803 g/cm^3. Suppose a car obtains 28 mi/gal and has a total mileage for one year of 24 000 mi.

a. Write a balanced equation for the complete combustion of gasoline.

b. How many molecules of C_8H_{18} are present in 15.2 gallons of gasoline?

c. How many kilograms of carbon dioxide would this car produce in one year (assume complete combustion)?

d. Write balanced equations for the production of nitrogen oxide and nitrogen dioxide.

Answers to CIs

CI 11 a. 4.21 cm^3
  b. 3.56 $\times$ 10^{23} Cu atoms
  c. $2Cu(s) + O_2(g) \longrightarrow 2CuO(s)$
  d. 9.47 g O_2
  e. 10.7 g CuO
  f. 9.10 g

CI 13 a. NaOCl, 74.44 g/mol
  b. Na 30.88%; O 21.49%; Cl 47.62%
  c. 1.74 $\times$ 10^{24} OCl^-
  d. $2NaOH + Cl_2 \longrightarrow NaOCl + NaCl + H_2O$
  e. 231 g NaOH
  f. 93.6%

Electronic Structure and Periodic Trends

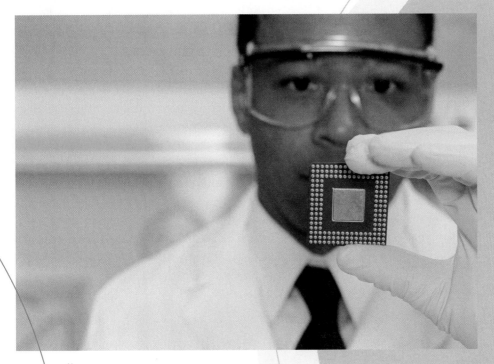

"The unique qualities of semiconducting metals make it possible for us to create sophisticated electronic circuits," says Tysen Streib, Global Product Manager, Applied Materials. *"Elements from columns 3A, 4A, and 5A of the periodic table often make good semiconductors because they readily form covalently bonded crystals. When small amounts of impurities are added, free-flowing electrons or holes can travel through the crystal with very little interference. Without these covalent bonds and loosely bound electrons, we wouldn't have any of the microchips that we use in computers, cell phones, and thousands of other devices."*

Materials scientists study the chemical properties of materials to find new uses for them in products such as cars, bridges, and clothing. They also develop materials that can be used as superconductors or in integrated-circuit chips and fuel cells. Chemistry is important in materials science because it provides information about structure and composition.

the Chemistry place

Visit **www.aw-bc.com/chemplace** for extra quizzes, interactive tutorials, career resources, PowerPoint slides for chapter review, math help, and case studies.

When sunlight passes through a prism or a drop of water, the light bends at different angles and separates into many colors. These are the colors we see in a rainbow that forms when sunlight passes through raindrops. Similar colors are also seen when certain elements are heated. You see this effect during a fireworks display. Lithium and strontium make red colors in fireworks, sodium gives yellow, magnesium is used for white, barium makes a green color, copper is blue, and mixtures of strontium and copper make purple.

After Rutherford concluded that the atom contained a nucleus and a large volume occupied only by electrons, scientists wondered how electrons move about the nucleus. Eventually, they proposed that electrons had different energies and must be arranged in specific energy levels within atoms. Further work by physicists and chemists revealed that electrons behaved both as particles and light waves. Eventually, the modern view of the atom emerged.

In this chapter, we will look at more details of atomic structure. We begin our discussion with the study of light, which is a tool scientists use to probe the behavior of electrons in atoms. Then we will look at how electrons in atoms are arranged in energy levels and how the behavior of the elements determines their position in the periodic table. Electron arrangement determines the similarity of atoms in a group and how atoms in each group of elements react to form new substances.

Learning **Goal**

Compare the wavelength of radiation with its energy.

9.1 Electromagnetic Radiation

Perhaps today you listened to a radio, cooked food with a microwave oven, or used a remote control to turn on a television. Maybe you called someone on a cellular phone. You might have seen the bright colors of a neon sign, and if it rained, you may have seen a rainbow. All of these examples of **electromagnetic radiation** involve energy that travels as waves through space.

Wavelength and Frequency

You are probably familiar with the action of waves in the ocean. If you were at a beach, you might notice that the water in each wave rises and falls as the wave comes into shore. The highest point on the wave is called a peak or a crest. On a calm day, there might be long distances between each peak. However, if there were a storm with a lot of energy, the peaks would be much closer together.

The waves of electromagnetic radiation have peaks too. The **wavelength** (symbol λ, lambda) is the distance from a peak in one wave to the peak in the next wave. (See Figure 9.1.) In some types of radiation, the peaks are spaced far apart, while in others, they are close together. When the peaks are close together, many waves pass a certain point in 1 second. When the peaks are far apart, only a few waves pass a certain point in 1 second. The **frequency** (symbol ν, nu) is the number of waves that pass a certain point in 1 second. The velocity of a wave is the distance in meters it travels in 1 second. The velocity of radiation is given as meters per second (m/s or ms^{-1}). All electromagnetic radiation travels at the speed of light (c) in a vacuum, which is

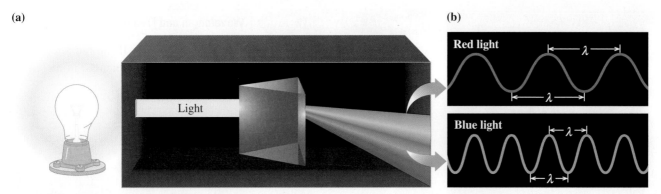

(a)   **(b)**

Figure 9.1 **(a)** Light passing through a prism is separated into a spectrum of colors we call a rainbow. **(b)** A wavelength is the distance from a point in one wave to the same point in the next. Waves with different wavelengths have different frequencies.

Q *How does the wavelength of red light compare to that of blue light?*

3.00×10^8 m/s. These characteristics of waves are related by the following equation

$$\text{Velocity}(c) = \text{Wavelength}(\lambda) \times \text{Frequency}(\nu)$$
$$c = \lambda \nu$$
$$3.00 \times 10^8 \text{ m/s} = \lambda \nu$$

When the equation is rearranged, we see that the wavelength and frequency are inversely related. Therefore, electromagnetic radiation with a long wavelength has a low frequency; radiation with a short wavelength has a high frequency.

$$\lambda = \frac{c}{\nu} = \frac{3.00 \times 10^8 \text{ m/s}}{\nu}$$

Different kinds of radiations such as radio waves, visible light, and gamma rays have different wavelengths. For example, radio waves have longer wavelengths and lower frequencies than the shorter wavelengths and higher frequencies of ultraviolet light or gamma rays.

Calculations of Frequency, Wavelength, and Velocity

The wavelengths of radiation are expressed in meters (m). Frequencies are measured in cycles per second. The unit hertz (Hz) represents one wave per second ($1/s$ or s^{-1}).

$$\text{Frequency } (\nu) = \frac{1 \text{ cycle of a wave}}{s} = 1/s \quad \text{or} \quad 1 \text{ s}^{-1} = 1 \text{ Hz}$$

You may have listened to FM radio and heard the announcer say they were broadcasting at a frequency of 91.5 megahertz. That would mean that 91.5×10^6 cycles of the FM radio waves are moving past a given point in 1 second.

$$91.5 \text{ MHz} = 91.5 \times 10^6 \text{ Hz} = 91.5 \times 10^6 \text{ s}^{-1} = 9.15 \times 10^7 \text{ s}^{-1}$$

Now we can use the equation for wavelength to determine the wavelength of the radio broadcast at 91.5 MHz.

$$\lambda \text{ (wavelength)} = \frac{3.00 \times 10^8 \text{ ms}^{-1}}{9.15 \times 10^7 \text{ s}^{-1}} = 3.28 \text{ m}$$

Sample Problem 9.1 ▸ **Wavelength and Frequency**

A student uses a microwave oven to make popcorn. If the radiation has a frequency of 2500 megahertz (MHz), what is the wavelength in meters of the microwaves?

Solution

STEP 1   **Given** $\nu = 2500\,\text{MHz}$ **Need**   m

STEP 2   **Plan**

MHz → | Metric factor | → Hz → | Frequency factor | → s^{-1} → | Wavelength equation | → m

After we convert the unit MHz to Hz and to cycles per second (s^{-1}), we can solve for wavelength using the equation for wavelength and frequency:

$$\lambda = \frac{c}{\nu} = \frac{3.00 \times 10^{8}\,\text{ms}^{-1}}{\nu}$$

STEP 3   **Equalities/Conversion Factors**

$$1\,\text{MHz} = 10^{6}\,\text{Hz}$$

$$\frac{10^{6}\,\text{Hz}}{1\,\text{MHz}} \quad\text{and}\quad \frac{1\,\text{MHz}}{10^{6}\,\text{Hz}}$$

$$1\,\text{Hz} = 1\,\text{s}^{-1}$$

$$\frac{1\,\text{Hz}}{1\,\text{s}^{-1}} \quad\text{and}\quad \frac{1\,\text{s}^{-1}}{1\,\text{Hz}}$$

STEP 4   **Set Up Problem**

$$2500\,\cancel{\text{MHz}} \times \frac{10^{6}\,\cancel{\text{Hz}}}{1\,\cancel{\text{MHz}}} \times \frac{1\,\text{s}^{-1}}{1\,\cancel{\text{Hz}}} = 2.5 \times 10^{9}\,\text{s}^{-1}$$

To solve for wavelength, substitute the calculated frequency into the equation for wavelength:

$$\lambda\,(\text{wavelength}) \quad \frac{3.00 \times 10^{8}\,\text{ms}^{\cancel{-1}}}{2.5 \times 10^{9}\,\cancel{\text{s}^{-1}}} = 1.2 \times 10^{-1}\,\text{m}$$

Study Check

If your dentist uses X rays with a wavelength of $2.2 \times 10^{-10}\,\text{m}$, what is the frequency of the X rays?

Electromagnetic Spectrum

The **electromagnetic spectrum** is an arrangement of all the various forms of electromagnetic radiation in order of decreasing wavelengths or increasing frequencies. (See Figure 9.2.) At one end are the *radio waves* with long wavelengths that are used for AM and FM radio bands, cellular phones, and TV signals. The wavelength of a typical AM radio wave can be as long as a football field. *Microwaves* have shorter wavelengths and higher frequencies than radio waves. Microwaves of about 10 mm are used in radar. *Infrared radiation* (IR) with wavelength about $10^{-5}\,\text{m}$ is responsible for the heat we feel

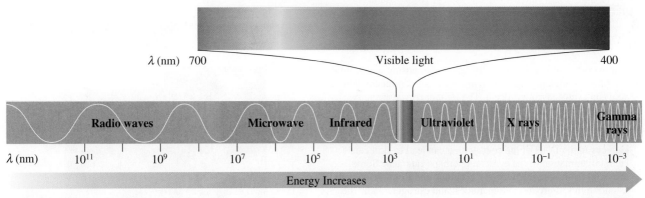

λ (nm) 700   Visible light   400

Energy Increases

Low energy   High energy

Figure 9.2 The electromagnetic spectrum contains radiation with long and short wavelengths in the visible and invisible regions.

Q How does the energy of ultraviolet light compare to that of a microwave?

from sunlight and the heat of infrared lamps used to warm food in restaurants. Certain infrared radiation is used in remote controls for TV sets and stereos.

Visible light with wavelengths from 700 to 400 nm is the only light our eyes can detect. Red light has the longest wavelength at 700 nm; orange is about 600 nm; green is about 500 nm; and violet at 400 nm has the shortest wavelength of visible light. We see objects as different colors because the objects reflect only certain wavelengths, which are absorbed by our eyes. We see white light from a light bulb because all the wavelengths of visible light are mixed together.

Ultraviolet (UV) light has shorter wavelengths and higher frequencies than violet light of the visible range. UV light in sunlight can cause sunburn, but most of the UV light from the sun is blocked by the protective ozone layer. *X rays* and *gamma rays* have shorter wavelengths than ultraviolet light, which means they have some of the highest frequencies. X rays can pass through soft substances but not metals or bone, which is why they are used to scan luggage at airports and to see images of the bones and teeth in the body. Gamma rays are produced by radioactive atoms and in nuclear processes in the sun and stars. Gamma rays are dangerous because they kill cells in the body, which is the reason gamma rays are used in the treatment of tumors and cancers.

Sample Problem 9.2   **The Electromagnetic Spectrum**

Visible light contains colors from red to violet.
a. What color has the shortest wavelength?
b. What color has the lowest frequency?

Solution

a. Violet light has the shortest wavelength.
b. Red light has the lowest frequency.

Study Check

Arrange the following in order of increasing frequencies: X rays, ultraviolet light, FM radio waves, and microwaves.

BIOLOGICAL REACTIONS TO UV LIGHT

Our everyday life depends on sunlight, but exposure to sunlight can have damaging effects on living cells, and too much exposure can even cause their death. The light energy, especially ultraviolet (UV), excites electrons and may lead to unwanted chemical reactions. The list of damaging effects of sunlight includes sunburn; wrinkling; premature aging of the skin; changes in the DNA of the cells, which can lead to skin cancers; inflammation of the eyes; and perhaps cataracts. Some drugs, like the acne medications Accutane and Retin-A, as well as antibiotics, diuretics, sulfonamides, and estrogen, make the skin extremely sensitive to light.

High-energy radiation is the most damaging biologically. Most of the radiation in this range is absorbed in the epidermis of the skin. The degree to which radiation is absorbed depends on the thickness of the epidermis, the hydration of the skin, the amount of coloring pigments and proteins of the skin, and the arrangement of the blood vessels. In light-skinned people, 85–90% of the radiation is absorbed by the epidermis, with the rest reaching the dermis layer. In dark-skinned people, 90–95% of the radiation is absorbed by the epidermis, with a smaller percentage reaching the dermis.

However, medicine does take advantage of the beneficial effect of sunlight. Phototherapy can be used to treat certain skin conditions, including psoriasis, eczema, and dermatitis. In the treatment of psoriasis, for example, oral drugs are given to make the skin more photosensitive; exposure to UV radiation

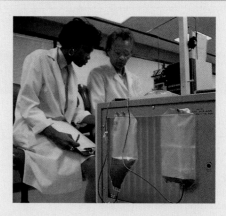

follows. Low-energy radiation is used to break down bilirubin in neonatal jaundice. Sunlight is also a factor in stimulating the immune system.

In cutaneous T-cell lymphoma, an abnormal increase in T cells causes painful ulceration of the skin. The skin is treated by photophoresis, in which the patient receives a photosensitive chemical, and then blood is removed from the body and exposed to ultraviolet light. The blood is returned to the patient, and the treated T cells stimulate the immune system to respond to the cancer cells.

Questions and Problems **Electromagnetic Radiation**

9.1 What is meant by the wavelength of UV light?

9.2 How are the wavelength and frequency of light related?

9.3 What is the difference between "white" light and blue or red light?

9.4 Why can we use X rays, but not radio waves or microwaves, to give an image of bones and teeth?

9.5 One AM radio station broadcasts news at 650 kHz and another at 980 kHz. What are the wavelengths of these AM radio waves in meters?

9.6 A wavelength of 850 nm (8.5×10^{11} m) is used for fiber-optic transmission. What is its frequency?

9.7 If orange light has a frequency of $4.8 \times 10^{14} \, \text{s}^{-1}$, what is the wavelength of orange light?

9.8 What is the frequency of green light if it has a wavelength of 5.0×10^{11} m?

9.9 Which type of electromagnetic radiation—ultraviolet light, microwaves, or X rays—has the longest wavelengths?

9.10 Of radio waves, infrared light, and UV light, which has the shortest wavelengths?

9.11 Place the following types of electromagnetic radiation in order of increasing wavelengths: the blue color in a rainbow, X rays, microwaves from an oven, infrared radiation from a heat lamp.

9.12 Place the following types of electromagnetic radiation in order of decreasing frequencies: AM music station, UV radiation from the sun, police radar.

Learning Goal

Explain how atomic spectra correlate with energy levels in atoms.

9.2 Atomic Spectra and Energy Levels

When the white light from the sun or a light bulb is passed through a prism, it produces a continuous spectrum like a rainbow. Perhaps you have seen this happen when sunlight goes through a prism or through raindrops. When atoms of elements are heated, they produce light. At night, you may have seen the yellow color of sodium lamps or the red color of neon lights.

Atomic Spectra

When the light emitted from heated elements is passed through a prism, it does not produce a continuous spectrum. Instead, an **atomic spectrum** is produced that consists of lines of different colors separated by dark areas. (See Figure 9.3.) This separation of colors indicates that only certain wavelengths of light are produced when an element is heated, which gives each element a unique atomic spectrum.

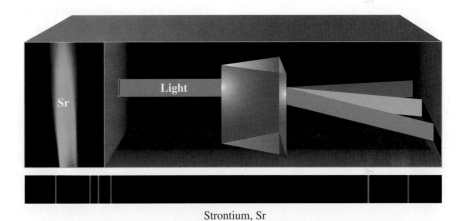

Strontium, Sr

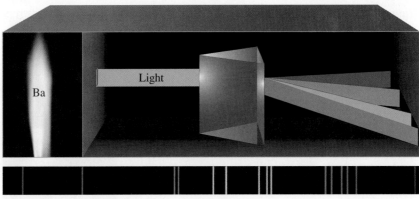

Barium, Ba

Figure 9.3 A spectrum unique to each element is produced as light emitted from the heated element passes through a prism, which separates the light into colored lines.

Q Why don't the elements form a continuous spectrum as seen with white light?

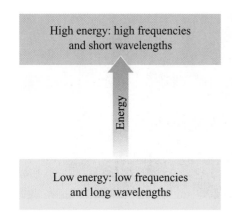

High energy: high frequencies and short wavelengths

Energy

Low energy: low frequencies and long wavelengths

Light Energy and Photons

When light is emitted, it behaves like a stream of small particles called **photons.** The energy (E) of a photon is related to the frequency (ν) of the light emitted. Using a number called *Planck's constant* (h), the energy of a photon in terms of frequency (ν) is $E = h\nu$. Thus the energy of a photon is directly proportional to its frequency. We can also express the energy of a photon as inversely proportional to its wavelength ($\nu = c/\lambda$).

$$E = \frac{hc}{\lambda}$$

In summary, high-energy radiation has high frequencies and short wavelengths, whereas low-energy radiation has low frequencies and long wavelengths.

Electron Energy Levels

With an understanding of photons, scientists could associate the lines in an atomic spectrum with changes in the energies of the electrons. In an atom, each electron has a fixed or specific energy known as its energy level. The energy of an electron is *quantized,* which means that the energy of an electron can never be between any two specific energy levels.

The energy levels are assigned values called **principal quantum numbers** (*n*), which are positive integers ($n = 1$, $n = 2\ldots$). Generally, electrons in the lower energy levels are closer to the nucleus, while electrons in the higher energy levels are farther away.

Changes in Energy Levels

By absorbing the amount of energy equal to the difference in energy levels, an electron is raised to a higher energy level. An electron loses energy when it falls to a lower energy level and emits electromagnetic radiation equal to the energy level difference. (See Figure 9.4.) If the electromagnetic radiation emitted has a wavelength in the visible range, we see a color. Ultraviolet light can be produced when electrons drop from higher energy levels to the first

Figure 9.4 Electrons can absorb a specific amount of energy to move to a higher energy level. When electrons lose energy, photons with specific energies are emitted.

Q How does the energy of a photon of green light compare to the energy of a photon of red light?

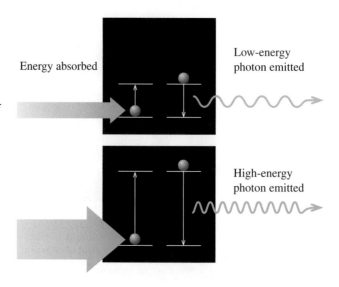

Energy absorbed

Low-energy photon emitted

High-energy photon emitted

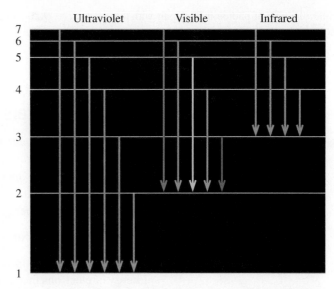

Figure 9.5 When electrons drop from a higher level to the first level, second level, and third level, photons of ultraviolet light, visible light, and infrared are emitted.

Q Why is a different photon of light emitted when an electron drops from energy level 5 to level 3 than when an electron drops from energy level 4 to 2?

($n = 1$) level. A series of colored lines of visible light can be produced when electrons fall from higher energy levels to the second ($n = 2$) level. Photons in the infrared can be produced when electrons drop to the third ($n = 3$) energy level. (See Figure 9.5.)

Sample Problem 9.3   **Electron Energy Levels**

a. How does an electron move to a higher energy level?
b. When an electron drops to a lower energy level, how is energy lost?

Solution

a. An electron moves to a higher energy level when it absorbs an amount of energy equal to the difference in levels.
b. Energy equal to the difference in levels is emitted when an electron drops to a lower level.

Study Check

Why did scientists propose that electrons occupied specific energy levels in the atom?

Questions and Problems   **Atomic Spectra and Energy Levels**

9.13 What feature of an atomic spectrum indicates that the energy emitted by heating an element is not continuous?
9.14 How can we explain the distinct lines that appear in an atomic spectrum?
9.15 Electrons can jump to higher energy levels when they (absorb/emit) a photon.

9.16 Electrons drop to lower energy levels when they (absorb/emit) a photon.
9.17 An electron drop to what energy level is likely to emit a photon in the infrared region?
9.18 An electron drop to what energy level is likely to emit a photon of red light?

9.19 Identify the type of photon in each pair with the greater energy:
   a. green light or yellow light
   b. red light or blue light

9.20 Identify the type of photon in each pair with the greater energy:
   a. orange light or violet light
   b. infrared light or ultraviolet light

Learning Goal

Describe the energy levels, sublevels, and orbitals in atoms.

9.3 Energy Levels, Sublevels, and Orbitals

As we discussed in Chapter 4, the energy levels (or *shells*) are assigned the value of a positive integer known as a *principal quantum number* (*n*). The energies of the electrons increase as the value of *n* increases. The electrons with the lowest energy are found in the first energy level (*n* = 1), while the electrons occupying higher energy levels have greater energies.

Principal Quantum Number (*n*)

$$1 < 2 < 3 < 4 < 5 < 6 < 7$$
Energy of electrons increases ⟶ .

There is a limit to the number of electrons in an energy level. Only a few electrons can occupy the lower energy levels, while more electrons can be accommodated in higher energy levels. The maximum number of electrons allowed in any energy level is calculated using the formula $2n^2$ (two times the square of the principal quantum number). In Table 9.1, we calculate the maximum number of electrons allowed in the first four energy levels.

Table 9.1   Maximum Number of Electrons Allowed in Energy Levels 1–4

Energy Level (*n*)	1	2	3	4
$2n^2$	$2(1)^2$	$2(2)^2$	$2(3)^2$	$2(4)^2$
Maximum Number of Electrons	2	8	18	32

Sublevels

Each energy level consists of one or more **sublevels** (or *subshells*), which contain electrons with identical energy. The sublevels are identified by the letters *s*, *p*, *d*, and *f*. The number of sublevels within an energy level is equal to the principal quantum number. The first energy level (*n* = 1) has only one sublevel, 1*s*. The second energy level (*n* = 2) has two sublevels, 2*s* and 2*p*. The third energy level (*n* = 3) has three sublevels, 3*s*, 3*p*, and 3*d*. The fourth energy level (*n* = 4) has four sublevels, 4*s*, 4*p*, 4*d*, and 4*f*. (See Figure 9.6.)

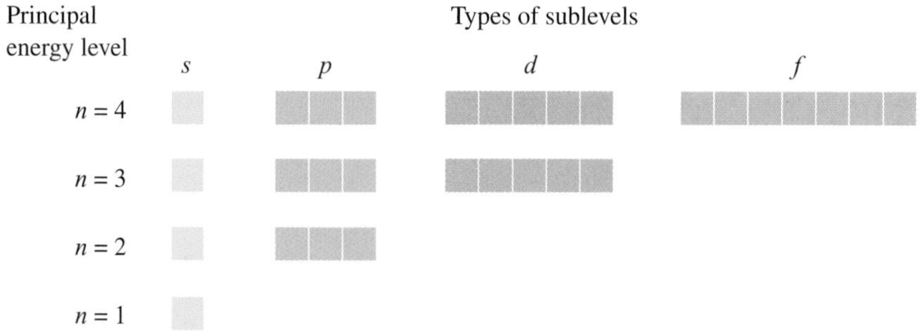

Figure 9.6 The number of sublevels in an energy level is the same as the principal quantum number *n*.

Q How many sublevels are in energy level n = 5?

Principal energy level		Types of sublevels		
	s	*p*	*d*	*f*
n = 4				
n = 3				
n = 2				
n = 1				

Within each energy level, the *s* sublevel has the lowest energy. If there are additional sublevels, the *p* sublevel has the next lowest energy, then the *d* sublevel, and finally the *f* sublevel.

Order of Increasing Energy of Sublevels Within an Energy Level

$$s < p < d < f$$

Lowest $\longrightarrow$ Highest
energy   energy

Energy levels from $n = 5$ and higher have as many sublevels as the value of *n*, but only *s, p, d,* and *f* sublevels are needed to hold the electrons of atoms of the elements known today.

Number of Electrons in Sublevels

There is a maximum number of electrons that can occupy each sublevel. An *s* sublevel holds 1 or 2 electrons. A *p* sublevel takes up to 6 electrons, a *d* sublevel can hold up to 10 electrons, and an *f* sublevel holds a maximum of 14 electrons. The total number of electrons in all the sublevels adds up to give the electrons in each energy level, as shown in Table 9.2.

Orbitals

There is no way to know the exact location of an electron in an atom. Instead, scientists describe the location of an electron in terms of probability. A region in an atom where there is the highest probability of finding an electron is called an **orbital**. Suppose you could draw an imaginary circle with a 100-m radius around your chemistry classroom. There is a high probability of finding you within that area when your chemistry class is in session. But once in a while, you may be found outside that circle because you were sick or your car did not start. The shapes of orbitals represent the three-dimensional regions in which electrons have the highest probability of being found.

Table 9.2   Electron Capacity in Sublevels for Principal Energy Levels 1–4

Principal Energy Level (Shell)	Number of Sublevels	Type of Sublevels	Maximum Number of Electrons	Maximum Total Electrons
4	4	4*f*	14	32
		4*d*	10	
		4*p*	6	
		4*s*	2	
3	3	3*d*	10	18
		3*p*	6	
		3*s*	2	
2	2	2*p*	6	8
		2*s*	2	
1	1	1*s*	2	2

Electron spinning
counterclockwise

Electron spinning
clockwise

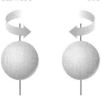

Opposite spins of
electrons in an orbital

An *s* orbital

Figure 9.7 An *s* orbital is a sphere that represents the region of highest probability of finding an *s* electron around the nucleus of an atom.

Q *Is the probability high or low of finding an s electron outside an s orbital?*

3*s*

2*s*

1*s*

Figure 9.8 All *s* orbitals have the same shape, but the volume increases when they contain electrons at higher energy levels.

Q *How would you compare the energy of the electrons in the 1s, 2s, and 3s orbitals?*

Orbital Capacity and Electron Spin

The *Pauli exclusion principle* states that an orbital can hold a maximum of 2 electrons. According to a useful model for electron behavior, an electron is seen as spinning on its axis, which generates a magnetic field. When 2 electrons are in the same orbital, they will repel each other unless their magnetic fields cancel. This happens only when the 2 electrons spin in opposite directions. We can represent the spins of the electrons in the same orbital with one arrow pointing up and the other pointing down.

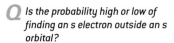

Shapes of Orbitals

Each sublevel within an energy level is composed of the same type of orbitals. There is an *s* orbital for each *s* sublevel; *p* orbitals for each *p* sublevel; *d* orbitals for each *d* sublevel; and *f* orbitals for each *f* sublevel. Each type of orbital has a unique shape. In an *s* orbital, the electrons are most likely found in a region with a spherical shape. (See Figure 9.7.) Every *s* orbital can hold 1 or 2 electrons; there is just one *s* orbital for every *s* sublevel. While the shape of every *s* orbital is spherical, there is an increase in the size of the *s* orbitals in higher energy levels. (See Figure 9.8.)

A *p* sublevel consists of three *p* orbitals, each of which has two lobes. The three *p* orbitals are arranged in three different directions (*x*, *y*, and *z* axes) around the nucleus. (See Figure 9.9.) Because each *p* orbital can hold up to 2 electrons, the three *p* orbitals can accommodate 6 electrons in a *p* sublevel. At higher energy levels, the shape of *p* orbitals is the same, but the volume increases.

A *d* sublevel consists of five *d* orbitals, which means that a *d* sublevel can hold a maximum of 10 electrons. With a total of seven *f* orbitals, an *f* sublevel can hold up to 14 electrons. The shapes of *d* orbitals and *f* orbitals are complex and we have not included them in this text.

| **Sample Problem 9.4** | **Sublevels and Orbitals** |

Indicate the type and number of orbitals in each of the following energy levels or sublevels:

a. 3*p* sublevel   **b.** *n* = 2
c. *n* = 3   **d.** 4*d* sublevel

Solution

a. The 3*p* sublevel contains three 3*p* orbitals.
b. The *n* = 2 principal energy level consists of 2*s* (one) and 2*p* (three) orbitals.

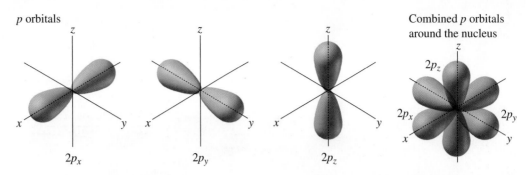

Figure 9.9 Each of the *p* orbitals has a dumbbell shape and is aligned along a different axis. Each *p* orbital holds a maximum of 2 electrons.

Q If there are three p orbitals in a p sublevel, how many total electrons are possible in a p sublevel?

  c. The $n = 3$ principal energy level consists of $3s$ (one), $3p$ (three), and $3d$ (five) orbitals.

  d. The $4d$ sublevel contains five $4d$ orbitals.

Study Check

What is similar and what is different for $1s$, $2s$, and $3s$ orbitals?

Questions and Problems   **Energy Levels, Sublevels, and Orbitals**

9.21 Describe the shape of the following orbitals:
   a. $1s$ **b.** $2p$ **c.** $5s$

9.22 Describe the shape of the following orbitals:
   a. $3p$ **b.** $6s$ **c.** $4p$

9.23 What is similar about the following?
   a. $1s$ and $2s$ orbitals
   b. $3s$ and $3p$ sublevels
   c. $3p$ and $4p$ sublevels
   d. three $3p$ orbitals

9.24 What is similar about the following?
   a. $5s$ and $6s$ orbitals
   b. $3p$ and $4p$ orbitals
   c. $3s$ and $4s$ sublevels
   d. $2s$ and $2p$ orbitals

9.25 Indicate the number of each in the following:
   a. orbitals in the $3d$ sublevel
   b. sublevels in the $n = 1$ principal energy level

   c. orbitals in the $6s$ sublevel
   d. orbitals in the $n = 3$ principal energy level

9.26 Indicate the number of each in the following:
   a. orbitals in $n = 2$ principal energy level
   b. sublevels in $n = 4$ principal energy level
   c. orbitals in the $5f$ sublevel
   d. orbitals in the $6p$ sublevel

9.27 Indicate the maximum number of electrons in the following:
   a. $2p$ orbital   **b.** $3p$ sublevel
   c. main energy level $n = 4$ **d.** $5d$ sublevel

9.28 Indicate the maximum number of electrons in the following:
   a. $3s$ sublevel   **b.** $4p$ orbital
   c. main energy level $n = 3$ **d.** $4f$ sublevel

9.4 Writing Orbital Diagrams and Electron Configurations

Learning **Goal**

> **Write the orbital diagrams and electron configurations for hydrogen to argon.**

We can now look at how electrons are arranged in the orbitals within an atom. In an **orbital diagram,** boxes (or circles) represent the orbitals. We see from the energy diagram (Figure 9.10) that the electrons in the $1s$ orbital have a lower energy level than in the $2s$ orbital. Thus, the first 2 electrons in an atom will go into the $1s$ orbital. Because the $1s$ sublevel can hold only 2 electrons,

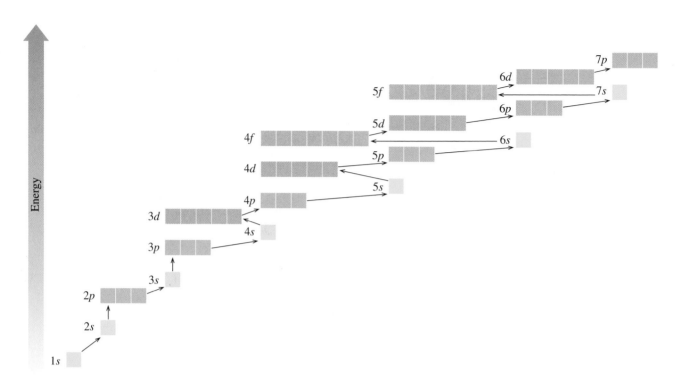

Figure 9.10 The sublevels fill in order of increasing energy beginning with 1s.

Q Why do 3d orbitals fill after the 4s orbital?

Electron arrangements in orbitals in energy levels 1 and 2 for carbon

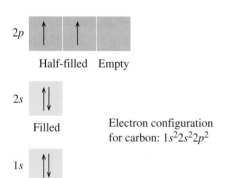

Electron configuration for carbon: $1s^2 2s^2 2p^2$

the next 2 electrons will go into the 2s orbital, which has the next lowest energy on the energy diagram.

With few exceptions (which will be noted later in this chapter), the "building" of electrons continues to the next lowest energy sublevel that is available until all the electrons are placed. For example, the atomic number of carbon is 6, which means that a carbon atom has 6 electrons. The first 2 electrons go into the 1s orbital, the next 2 electrons go into the 2s orbital. In the orbital diagram, the 2 electrons in the 1s and 2s orbitals are shown with opposite spins, the first arrow is up and the second down. The last 2 electrons in carbon begin to fill the 2p sublevel, the next lowest energy sublevel. However, there are three 2p orbitals of equal energy. Because the negatively charged electrons repel each other, they are placed in different 2p orbitals.

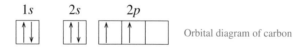

Orbital diagram of carbon

Electron Configurations

The **electron configuration** of an atom is "built up" by placing the electrons of an atom in the sublevels in order of increasing energy. Starting with the lowest energy sublevel, the electron configuration for carbon is written as

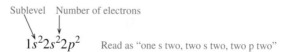

Read as "one s two, two s two, two p two"

Period 1   Hydrogen and Helium

We can begin to draw the orbital diagrams and build the electron configurations for the elements H and He in Period 1. The $1s$ orbital (which is also the $1s$ sublevel) is used first because it has the lowest energy. Hydrogen has 1 electron in the $1s$ sublevel; helium has 2. In the orbital diagram, the electrons for helium are shown with opposite spins.

Atomic Number	Element	Orbital Diagram	Electron Configuration
		$1s$	
1	H	↑	$1s^1$
2	He	↑↓	$1s^2$

Period 2   Lithium to Neon

Period 2 begins with lithium, which has 3 electrons. The first 2 electrons fill the $1s$ orbital, while the third electron goes into the $2s$ orbital, the sublevel with the next lowest energy. In beryllium, another electron is added to complete the $2s$ orbital. From boron to nitrogen, electrons add one each to different $2p$ orbitals, which gives three half-filled $2p$ orbitals. From oxygen to neon, the remaining electrons are paired up using opposite spins until the $2p$ orbitals in neon are filled. When the $2p$ orbitals are filled, the $2p$ sublevel is complete. In writing the complete electron configurations for the elements in Period 2, the electrons always begin with the $1s$ sublevel followed by the filling of the $2s$ and then the $2p$ sublevels.

Atomic Number	Element	Orbital Diagram	Electron Configuration	Abbreviated Configuration
		$1s$ $2s$		
3	Li	↑↓ ↑	$1s^2 2s^1$	[He] $2s^1$
4	Be	↑↓ ↑↓	$1s^2 2s^2$	[He] $2s^2$
		$2p$		
5	B	↑↓ ↑↓ ↑ ☐ ☐	$1s^2 2s^2 2p^1$	[He] $2s^2 2p^1$
6	C	↑↓ ↑↓ ↑ ↑ ☐	$1s^2 2s^2 2p^2$	[He] $2s^2 2p^2$
		Unpaired electrons		
7	N	↑↓ ↑↓ ↑ ↑ ↑	$1s^2 2s^2 2p^3$	[He] $2s^2 2p^3$
8	O	↑↓ ↑↓ ↑↓ ↑ ☐	$1s^2 2s^2 2p^4$	[He] $2s^2 2p^4$
9	F	↑↓ ↑↓ ↑↓ ↑↓ ↑	$1s^2 2s^2 2p^5$	[He] $2s^2 2p^5$
10	Ne	↑↓ ↑↓ ↑↓ ↑↓ ↑↓	$1s^2 2s^2 2p^6$	[He] $2s^2 2p^6$

An electron configuration can also be written in an *abbreviated configuration*. The electron configuration of the preceding noble gas is replaced by writing its symbol inside square brackets. For example, the electron configuration for lithium, $1s^2 2s^1$, can be abbreviated as $[He]2s^1$, where [He] replaces $1s^2$.

Sample Problem 9.5 **Writing Electron Configurations**

Write the orbital diagram, electron configuration, and the abbreviated electron configuration for nitrogen.

Solution

Nitrogen with atomic number 7 has 7 electrons. To write the orbital diagram, draw boxes for the 1s, 2s, and 2p orbitals.

Add 7 electrons, starting with the 1s orbital. Two electrons with opposite spins are added to the 1s and 2s orbitals. For the 2p orbitals, the 3 remaining electrons are placed in separate orbitals with parallel spins.

Orbital diagram for nitrogen (N)

The electron configuration for nitrogen shows the electrons filling the sublevels in order of increasing energy.

$1s^2 2s^2 2p^3$ Electron configuration for nitrogen (N)

The abbreviated electron configuration for nitrogen is written by substituting the symbol [He] for the $1s^2$.

$[He]\, 2s^2 2p^3$ Abbreviated electron configuration for nitrogen (N)

Study Check

Write the orbital diagram, electron configuration, and the abbreviated electron configuration for fluorine.

Period 3   Sodium to Argon

In Period 3, electrons enter the orbitals of the 3s and 3p sublevels, but not the 3d sublevel. We notice that the elements sodium to argon, which are directly below the elements lithium to neon in Period 2, have a similar pattern of filling their s and p orbitals. In sodium and magnesium, 1 and 2 electrons go into the 3s orbital. The remaining electrons for aluminum, silicon, and phosphorus go into different $3p$ orbitals so that phosphorus has a half-filled $3p$ sublevel. We can write the complete orbital diagram for phosphorus as follows:

The final electrons in sulfur, chlorine, and argon pair up with the electrons already in the $3p$ orbitals up to argon, which has a complete $3p$ sublevel. For elements in Period 3 and above, we often write the orbital diagrams only for the electrons in the highest energy levels. In Period 3, the symbol [Ne] replaces the electron configuration of neon, $1s^22s^22p^6$. The abbreviated form is convenient to use for electron configurations that contain several sublevel notations.

Atomic Number	Element	Orbital Diagram (3s and 3p orbitals only)	Electron Configuration	Abbreviated Form
11	Na		$1s^22s^22p^63s^1$	[Ne] $3s^1$
12	Mg		$1s^22s^22p^63s^2$	[Ne] $3s^2$
13	Al		$1s^22s^22p^63s^23p^1$	[Ne] $3s^23p^1$
14	Si		$1s^22s^22p^63s^23p^2$	[Ne] $3s^23p^2$
15	P		$1s^22s^22p^63s^23p^3$	[Ne] $3s^23p^3$
16	S		$1s^22s^22p^63s^23p^4$	[Ne] $3s^23p^4$
17	Cl		$1s^22s^22p^63s^23p^5$	[Ne] $3s^23p^5$
18	Ar		$1s^22s^22p^63s^23p^6$	[Ne] $3s^23p^6$

Sample Problem 9.6   **Orbital Diagrams and Electron Configurations**

For each of the following elements, write the stated type of electron notation:
a. orbital diagram for silicon
b. electron configuration for phosphorus
c. abbreviated electron configuration for chlorine

Solution

a. Silicon in Period 3 has atomic number 14, which tells us that it has 14 electrons. To write the orbital diagram, we draw boxes for the orbitals up to $3p$.

Add 14 electrons, starting with the $1s$ orbital. Show paired electrons in the same orbital with opposite spins and place the last 2 electrons in different $3p$ orbitals.

$1s$	$2s$	$2p$	$3s$	$3p$
⬆⬇	⬆⬇	⬆⬇ ⬆⬇ ⬆⬇	⬆⬇	⬆ ⬆

b. The electron configuration gives the electrons that fill the sublevel in order of increasing energy. Phosphorus is in Group 5A (15) in Period 3. In Periods 1 and 2, a total of 10 electrons fill sublevels $1s^2$, $2s^2$, and $2p^6$. In Period 3, 2 electrons go into $3s^2$. The 3 remaining electrons (total of 15) are placed in the $3p$ sublevel.

$$P \qquad 1s^2 2s^2 2p^6 3s^2 3p^3$$

c. In chlorine, the previous noble gas is neon. For the abbreviated configuration, write [Ne] for $1s^2 2s^2 2p^6$ and the electrons in the $3s$ and $3p$ sublevels.

$$[Ne]\, 3s^2 3p^5 \qquad \text{Abbreviated electron configuration for Cl}$$

Study Check

Write the complete and abbreviated electron configuration for sulfur.

Questions and Problems ▷ **Writing Orbital Diagrams and Electron Configurations**

9.29 Compare the terms *electron configuration* and *abbreviated configuration*.

9.30 Compare the terms *orbital diagram* and *electron configuration*.

9.31 Write an orbital diagram for an atom of each of the following:
- **a.** boron
- **b.** aluminum
- **c.** phosphorus
- **d.** argon

9.32 Write an orbital diagram for an atom of each of the following:
- **a.** fluorine
- **b.** sodium
- **c.** magnesium
- **d.** sulfur

9.33 Write a complete electron configuration for an atom of each of the following:
- **a.** nitrogen
- **b.** sodium
- **c.** sulfur
- **d.** boron

9.34 Write a complete electron configuration for an atom of each of the following:
- **a.** carbon
- **b.** silicon
- **c.** phosphorus
- **d.** argon

9.35 Write an abbreviated electron configuration for an atom of each of the following:
- **a.** magnesium
- **b.** sulfur
- **c.** aluminum
- **d.** nitrogen

9.36 Write an abbreviated electron configuration for an atom of each of the following:
- **a.** phosphorus
- **b.** silicon
- **c.** sodium
- **d.** oxygen

9.37 Give the symbol of the element with each of the following electron configurations:
- **a.** $1s^2 2s^1$
- **b.** $1s^2 2s^2 2p^6 3s^2 3p^4$
- **c.** $[Ne]3s^2 3p^2$
- **d.** $[He]2s^2 2p^5$

9.38 Give the symbol of the element with each of the following electron configurations:
- **a.** $1s^2 2s^2 2p^4$
- **b.** $[Ne]3s^2$
- **c.** $1s^2 2s^2 2p^6 3s^2 3p^6$
- **d.** $[Ne]3s^2 3p^1$

9.39 Give the symbol of the element that meets the following conditions:
- **a.** has 3 electrons in energy level $n = 3$
- **b.** has 2 $2p$ electrons
- **c.** completes the $3p$ sublevel
- **d.** has 2 electrons in the $2s$ sublevel

9.40 Give the symbol of the element that meets the following conditions:
- **a.** has 5 electrons in the $3p$ sublevel
- **b.** has 3 $2p$ electrons
- **c.** completes the $3s$ sublevel
- **d.** has 1 electron in the $3s$ sublevel

Write the arrangement of electrons for atoms using the sublevel blocks on the periodic table.

the
Chemistry
place

WEB TUTORIAL
Bohr's Shell Model

9.5 Electron Configurations and the Periodic Table

Up until now we have written electron configurations using the energy diagram. As configurations involve more sublevels, this becomes tedious. However, on the periodic table, the atomic numbers are in order of increasing sublevel energy. Therefore, we can "build up" atoms by reading the periodic table from left to right across each period.

Sublevel Blocks on the Periodic Table

The electron configurations of the elements are related to their position in the periodic table. Different sections or blocks within the table correspond to the *s, p, d,* and *f* sublevels. (See Figure 9.11.)

1. The *s* **block elements** include hydrogen and helium and the elements in Group 1A (1) and Group 2A (2). This means that the final 1 or 2 electrons in the elements of the *s* block are located in *s* sublevels. The period number indicates the particular *s* sublevel that is filling:1*s*, 2*s*, and so on.

2. The *p* **block elements** consists of the elements in Group 3A (13) to Group 8A (18). There are six *p* block elements in each period because each *p* sublevel with three *p* orbitals can hold up to 6 electrons. The period number indicates the particular *p* sublevel that is filling: 2*p*, 3*p*, and so on.

3. The *d* **block elements** first appear after calcium (atomic number 20) with the ten columns of elements of the transition metals. There are ten elements in the *d* block because the five *d* orbitals in each *d* sublevel can hold up to 10 electrons. The particular *d* sublevel is one less ($n - 1$) than the period number. For example, in Period 4, the first *d* block is the 3*d* sublevel. In Period 5, the second *d* block is the 4*d* sublevel.

4. The *f* **block elements** include all the elements in the two rows at the bottom of the periodic table. There are 14 elements in each *f* block because the seven *f* orbitals in an *f* sublevel can hold up to 14 electrons. Elements that have atomic numbers higher than 57 (La) have electrons in the 4*f*

Figure 9.11 Electron configuration follows the order of sublevels on the periodic table.

Q If neon is in the Group 8A, Period 2, how many electrons are in the 1s, 2s, and 2p sublevels of neon?

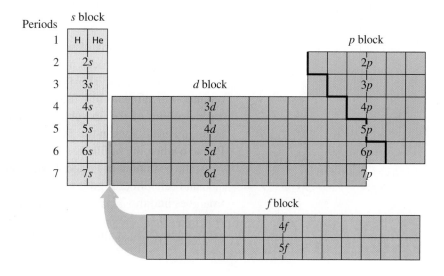

block. The particular f sublevel is two less $(n - 2)$ than the period number. For example, in Period 6, the first f block is the $4f$ sublevel. In Period 7, the second f block is the $5f$ sublevel.

Writing Electron Configurations Using Sublevel Blocks

Now we can write electron configurations using the sublevel blocks on the periodic table. As before, each configuration begins at H. But now we move across the table, writing down each block we come to until we reach the element for which we are writing an electron configuration. For example, let's write the electron configuration for chlorine (atomic number 17) from the sublevel blocks on the periodic table. For each element, we can use the following steps:

STEP 1 Locate the element on the periodic table.

STEP 2 Write the filled sublevels in order going across each period (left to right).

STEP 3 Count the number of electrons in the sublevel for the given element and complete the configuration.

Period		Sublevel Blocks Filled
1	$1s$ sublevel (H $\longrightarrow$ He)	$1s^2$
2	$2s$ sublevel (Li $\longrightarrow$ Be) then $2p$ sublevel (B $\longrightarrow$ Ne)	$2s^2 \longrightarrow 2p^6$
3	$3s$ sublevel (Na $\longrightarrow$ Mg) then $3p$ sublevel (Al $\longrightarrow$ Cl)	$3s^2 \longrightarrow 3p?$ (Cl)

Writing the sublevel blocks in order up to chlorine gives

$$1s^2 2s^2 2p^6 3s^2 3p?$$

Chlorine is the fifth element in the $3p$ block, which means that chlorine has 5 $3p$ electrons. The complete electron configuration for chlorine is written as

$$1s^2 2s^2 2p^6 3s^2 3p^5 \quad \text{Final sublevel block where Cl appears}$$

Period 4

Up until now, the filling of sublevels has progressed in order. However, if we look at the sublevel blocks in Period 4, we see that the $4s$ sublevel block fills before the $3d$ orbitals. The $4s$ orbital has a slightly lower energy than the $3d$ orbitals, which means that the $4s$ sublevel has the next lowest energy following the filling of the $3p$ sublevel at the end of Period 3. This order occurs again in Period 5 when the $5s$ orbital fills before the $4d$ orbitals, and again in Period 6 when the $6s$ fills before the $5d$.

At the beginning of Period 4, the 1 and 2 remaining electrons in potassium (19) and calcium (20) go into the $4s$ orbital. In scandium, the remaining electron goes into the $3d$ block, which continues to fill until it has 10 electrons at zinc (30). Once the $3d$ block is complete, the next 6 electrons go into the $4p$ block for elements gallium (31) to krypton (36).

Atomic Number	Element	Electron Configuration	Abbreviated Configuration
4s Block Elements			
19	K	$1s^22s^22p^63s^23p^64s^1$	$[Ar]4s^1$
20	Ca	$1s^22s^22p^63s^23p^64s^2$	$[Ar]4s^2$
3d Block Elements			
21	Sc	$1s^22s^22p^63s^23p^64s^23d^1$	$[Ar]4s^23d^1$
22	Ti	$1s^22s^22p^63s^23p^64s^23d^2$	$[Ar]4s^23d^2$
23	V	$1s^22s^22p^63s^23p^64s^23d^3$	$[Ar]4s^23d^3$
24	Cr	$1s^22s^22p^63s^23p^64s^13d^5$	$[Ar]4s^13d^5$ (exception)
25	Mn	$1s^22s^22p^63s^23p^64s^23d^5$	$[Ar]4s^23d^5$
26	Fe	$1s^22s^22p^63s^23p^64s^23d^6$	$[Ar]4s^23d^6$
27	Co	$1s^22s^22p^63s^23p^64s^23d^7$	$[Ar]4s^23d^7$
28	Ni	$1s^22s^22p^63s^23p^64s^23d^8$	$[Ar]4s^23d^8$
29	Cu	$1s^22s^22p^63s^23p^64s^13d^{10}$	$[Ar]4s^13d^{10}$(exception)
30	Zn	$1s^22s^22p^63s^23p^64s^23d^{10}$	$[Ar]4s^23d^{10}$
4p Block Elements			
31	Ga	$1s^22s^22p^63s^23p^64s^23d^{10}4p^1$	$[Ar]4s^23d^{10}4p^1$
32	Ge	$1s^22s^22p^63s^23p^64s^23d^{10}4p^2$	$[Ar]4s^23d^{10}4p^2$
33	As	$1s^22s^22p^63s^23p^64s^23d^{10}4p^3$	$[Ar]4s^23d^{10}4p^3$
34	Se	$1s^22s^22p^63s^23p^64s^23d^{10}4p^4$	$[Ar]4s^23d^{10}4p^4$
35	Br	$1s^22s^22p^63s^23p^64s^23d^{10}4p^5$	$[Ar]4s^23d^{10}4p^5$
36	Kr	$1s^22s^22p^63s^23p^64s^23d^{10}4p^6$	$[Ar]4s^23d^{10}4p^6$

Some Exceptions in Sublevel Block Order

Within the filling of the $3d$ sublevel, exceptions occur for chromium and copper. In Cr and Cu, the $3d$ sublevel is close to a half-filled or filled sublevel, which is particularly stable. Thus, chromium has only 1 electron in the $4s$ and 5 electrons in the $3d$ sublevel to give the added stability of a half-filled d sublevel.

A similar exception occurs when copper achieves a completely filled $3d$ sublevel by placing only 1 electron in the $4s$ sublevel and using 10 electrons to complete the $3d$ sublevel.

After the $4s$ and $3d$ sublevels are completed, the $4p$ sublevel fills as expected from gallium to krypton, the noble gas that completes Period 4.

**Guide to
Writing Electron Configurations
with Sublevel Blocks**

STEP 1
Locate the element on the periodic table.

STEP 2
Write the filled sublevels in order going across each period.

STEP 3
Count the number of electrons in the sublevel for the given element and complete the configuration.

Sample Problem 9.7 **Using Sublevel Blocks to Write Electron Configurations**

Use the sublevel blocks on the periodic table to write the electron configuration for

a. bromine   **b.** cesium

Solution

a. **STEP 1** Bromine is in the p block and in Period 4.

STEP 2 Beginning with $1s^2$, go across the periodic table writing each filled sublevel block as follows:

Period 1 $1s^2$
Period 2 $2s^2 \longrightarrow 2p^6$
Period 3 $3s^2 \longrightarrow 3p^6$
Period 4 $4s^2 \longrightarrow 3d^{10} \longrightarrow 4p?$

STEP 3 There are 5 electrons in the $4p$ sublevel for Br ($4p^5$), which determine the electron configuration for Br: $1s^2 2s^2 2p^6 3s^2 3p^6 4s^2 3d^{10} 4p^5$

b. **STEP 1** Cesium is in the s block and in Period 6.

STEP 2 Going across the periodic table starting with Period 1, the sublevel blocks fill as follows:

Period 1 $1s^2$
Period 2 $2s^2 \longrightarrow 2p^6$
Period 3 $3s^2 \longrightarrow 3p^6$
Period 4 $4s^2 \longrightarrow 3d^{10} \longrightarrow 4p^6$
Period 5 $5s^2 \longrightarrow 4d^{10} \longrightarrow 5p^6$
Period 6 $6s?$

STEP 3 There is 1 electron in the $6s$ sublevel for Cs ($6s^1$), which determines the electron configuration for Cs: $1s^2 2s^2 2p^6 3s^2 3p^6 4s^2 3d^{10} 4p^6 5s^2 4d^{10} 5p^6 6s^1$

Study Check

Write the electron configuration for tin.

Questions and Problems **Electron Configurations and the Periodic Table**

9.41 Write a complete electron configuration for an atom of each of the following:
a. arsenic   **b.** iron
c. palladium   **d.** iodine

9.42 Write a complete electron configuration for an atom of each of the following:
a. calcium   **b.** cobalt
c. gallium   **d.** cadmium

9.43 Write an abbreviated electron configuration for an atom of each of the following:
  a. titanium   **b.** strontium
  c. barium   **d.** lead

9.44 Write an abbreviated electron configuration for an atom of each of the following:
  a. nickel   **b.** arsenic
  c. tin   **d.** antimony

9.45 Give the symbol of the element with each of the following electron configurations:
  a. $1s^2 2s^1$ **b.** $1s^2 2s^2 2p^6 3s^2 3p^6 4s^2 3d^7$
  c. $[Ne]3s^2 3p^2$ **d.** $[Ar]4s^2 3d^{10} 4p^5$

9.46 Give the symbol of the element with each of the following electron configurations:
  a. $1s^2 2s^2 2p^4$ **b.** $[Ar]4s^2 3d^4$
  c. $1s^2 2s^2 2p^6 3s^2 3p^6$ **d.** $[Xe]6s^2 5d^{10} 4f^{14} 6p^3$

9.47 Give the symbol of the element that meets the following conditions:

  a. has 3 electrons in energy level $n = 4$
  b. has two $2p$ electrons
  c. completes the $5p$ sublevel
  d. has 2 electrons in the $4d$ sublevel

9.48 Give the symbol of the element that meets the following conditions:
  a. has 5 electrons in energy level $n = 3$
  b. has 1 electron in the $6p$ sublevel
  c. completes the $7s$ orbital
  d. has four $5p$ electrons

9.49 Give the number of electrons in the indicated orbitals for the following:
  a. $3d$ in zinc   **b.** $2p$ in sodium
  c. $4p$ in arsenic   **d.** $5s$ in rubidium

9.50 Give the number of electrons in the indicated orbitals for the following:
  a. $3d$ in manganese   **b.** $5p$ in antimony
  c. $6p$ in lead   **d.** $3s$ in magnesium

9.6 Periodic Trends of the Elements

Learning **Goal**

Use the electron configurations of elements to explain periodic trends.

The electron configurations of atoms are an important factor in the physical and chemical behavior of the elements. In this section, we will look at the *valence electrons* in atoms, the trends in the sizes of atoms, and *ionization energy*. Going across a period, there is a pattern of regular change in these properties from one group to the next. Known as *periodic properties,* each property increases or decreases across a period and then the trend is repeated in each successive period. We can use the seasonal changes in temperatures as an analogy for periodic properties. In the winter, temperatures are usually cold and become warmer in the spring. By summer, the outdoor temperatures are hot but begin to cool in the fall. By winter, we expect cold temperatures again as the pattern of decreasing and increasing temperatures repeats for another year.

Group Number and Valence Electrons

The chemical properties of representative elements are mostly due to the **valence electrons,** which are the electrons in the outermost energy levels. These valence electrons occupy the *s* and *p* sublevels with the highest quantum number *n*. The group numbers indicate the number of valence (outer) electrons for the elements in each vertical column. For example, the elements in Group 1A (1), such as lithium, sodium, and potassium, all have 1 electron in the outer energy level. Looking at the sublevel block, we can represent the valence electron in the alkali metals of Group 1A (1) as ns^1. All the elements in Group 2A (2), the alkaline earth metals, have two (2) valence electrons, ns^2. The halogens in Group 7A (17) have seven (7) valence electrons, $ns^2 np^5$.

We can see the repetition of the outermost *s* and *p* electrons for the representative elements in Periods 1 to 4 in Table 9.3. Helium is included in Group 8A (18) because it is a noble gas, but it has only 2 electrons in its complete energy level.

Table 9.3   Valence Electrons for Representative Elements in Periods 1–4

1A (1)	2A (2)	3A (13)	4A (14)	5A (15)	6A (16)	7A (17)	8A (18)
1							2
H							He
$1s^1$							$1s^2$
3	4	5	6	7	8	9	10
Li	Be	B	C	N	O	F	Ne
$2s^1$	$2s^2$	$2s^2 2p^1$	$2s^2 2p^2$	$2s^2 2p^3$	$2s^2 2p^4$	$2s^2 2p^5$	$2s^2 2p^6$
11	12	13	14	15	16	17	18
Na	Mg	Al	Si	P	S	Cl	Ar
$3s^1$	$3s^2$	$3s^2 3p^1$	$3s^2 3p^2$	$3s^2 3p^3$	$3s^2 3p^4$	$3s^2 3p^5$	$3s^2 3p^6$
19	20	31	32	33	34	35	36
K	Ca	Ga	Ge	As	Se	Br	Kr
$4s^1$	$4s^2$	$4s^2 4p^1$	$4s^2 4p^2$	$4s^2 4p^3$	$4s^2 4p^4$	$4s^2 4p^5$	$4s^2 4p^6$

Sample Problem 9.8   Using Group Numbers

Using the periodic table, write the group number and the electron configuration of the valence electrons for the following:
a. calcium   **b.** selenium   **c.** lead

Solution

a. Calcium is in Group 2A (2). It has a valence electron configuration of $4s^2$.

b. Selenium is in Group 6A (16). It has a valence electron configuration of $4s^2 4p^4$.

c. Lead is in Group 4A (14). It has a valence electron configuration of $6s^2 6p^2$. The electrons in the $5d$ and $4f$ sublevels are not valence electrons.

Study Check

What are the group numbers and valence electron arrangements for sulfur and strontium?

Atomic Size

Although there are no fixed boundaries to atoms, scientists have a good idea of the typical volume of the electron clouds in atoms. This volume can be described in terms of its **atomic radius,** which is the distance from the nucleus to the valence (outermost) electrons. For representative elements, the atomic radius increases going down each group. We would expect an increase in radius because the outermost electrons in higher energy levels are further from the nucleus. For example, in the alkali metals, Li has a valence electron of $2s^1$; Na has a valence electron of $3s^1$; K has a valence electron of $4s^1$; and Rb has a valence electron of $5s^1$. (See Figure 9.12.)

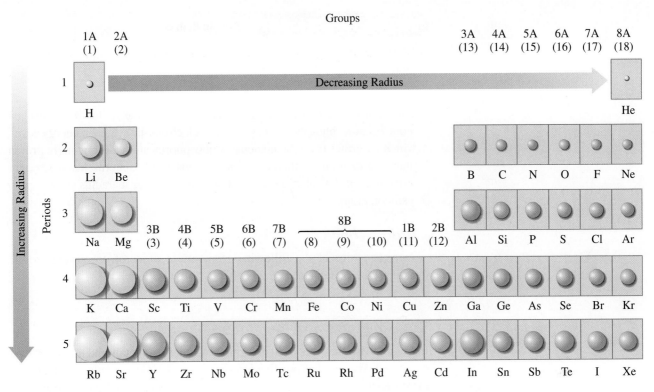

Figure 9.12 The atomic radius increases going down a group but decreases going from left to right across a period.

Q *Why does the atomic radius increase going down a group?*

The atomic radii of representative elements typically decrease from left to right on the periodic table. Going across a period, there is an increase in the number of valence electrons with a corresponding increase in protons, which causes a stronger attraction for the outermost electrons. (See Figure 9.13.) Going across a period, the valence electrons are pulled closer to the nucleus, which makes the atoms smaller.

The atomic radii of the transition elements going across a period change only slightly because electrons are adding to inner *d* sublevels. The increase in nuclear charge is balanced by an increase in the number of inner electrons. Therefore, in atoms of transition elements of the same period, the attraction for valence electrons remains about the same, as does their distance from the nucleus.

Figure 9.13 Going across Period 2, the greater nuclear charge attracts the valence electrons more strongly, pulling them closer to the nucleus and making the atoms smaller.

Q *Why does an Mg atom have a smaller atomic radius than a Na atom but a larger atomic radius than a Si atom?*

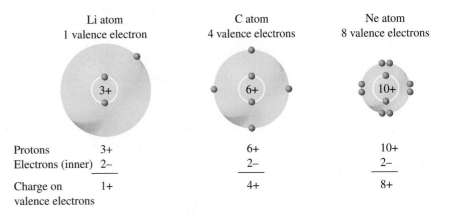

Sample Problem 9.9 — Atomic Radius

Why is the radius of a phosphorus atom larger than the radius of a nitrogen atom but smaller than the radius of a silicon atom?

Solution

The radius of a phosphorus atom is larger than the radius of a nitrogen atom because phosphorus has valence electrons in a higher energy level, which is further from the nucleus. A phosphorus atom has one more proton than a silicon atom; therefore, the nucleus in phosphorus has a stronger attraction for the valence electrons, which decreases its radius compared to a silicon atom.

Study Check

Arrange atoms of the following elements in order of decreasing atomic radius: Mg, S, and Na.

Sizes of Atoms and Their Ions

In Figure 9.14, the sizes of ions formed by atoms of metals and nonmetals are compared. We see that positive ions in Group 1A (1) are much smaller than the corresponding metal atoms. This occurs because metal atoms lose all of their valence electrons from the outermost energy level. If we look at the second and third energy levels for sodium, we see that the electron in the third energy level is lost to form the sodium ion (Na^+), which has an octet in the second energy level.

Figure 9.14 The positive ions of metals are about half the size of the corresponding metal atoms. The negative ions of nonmetals are about twice the size of the corresponding nonmetal atoms.

Q *How would you compare the size of positive ions going down a group?*

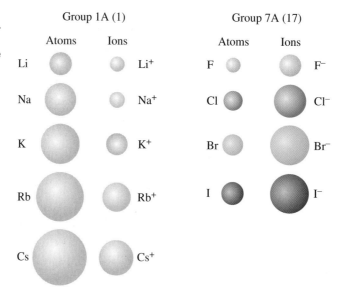

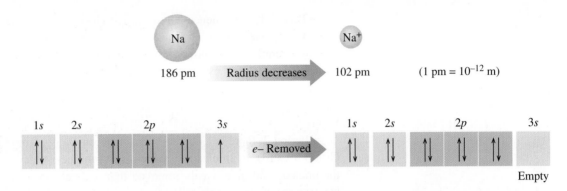

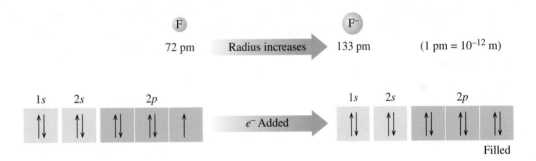

When nonmetal atoms add electrons, their size increases because of the repulsion between electrons. (See Figure 9.14.) For example, a fluoride ion is larger than a fluorine atom because a valence electron is added to the second energy level, which completes an octet.

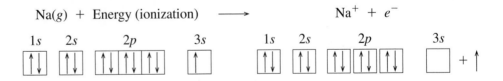

Ionization Energy

Electrons are held in atoms by their attraction to the nucleus. Therefore, energy is required to remove an electron from an atom. The **ionization energy** is the energy needed to remove the least tightly bound electron from an atom in the gaseous (*g*) state. When an electron is removed from a neutral atom, a cation with a 1+ charge is formed.

$$Na(g) \; + \; \text{Energy (ionization)} \longrightarrow \quad Na^+ \; + \; e^-$$

The ionization energy generally decreases going down a group. Less energy is needed to remove an electron as nuclear attraction for electrons farther from the nucleus decreases. (See Figure 9.15.) *Going across a period from left to right, the ionization energy generally increases.* As the attraction for outermost electrons increases across a period, more energy is required to remove an electron. In general, the ionization energy is low for the metals and high for the nonmetals. In addition to these general trends, there are specific differences in ionization energies that can be explained by the special stability of a filled or a half-filled sublevel.

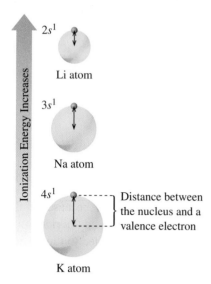

Figure 9.15 As the distance to the nucleus in Li, Na, and K atoms increases, the attraction to the nucleus decreases and less energy is required to remove the valence electron.

Q Why would Cs have a lower ionization energy than K?

In Period 1, the valence electrons are close to the nucleus and strongly held. H and He have high ionization energies because a large amount of energy is required to remove an electron. The ionization energy for He is the highest of any element because He has a full, stable $1s$ sublevel, which is disrupted by removing an electron. The high ionization energies of the noble gases indicate that their electron configurations are especially stable. (See Figure 9.16.)

In Period 2, there is a decrease in ionization energy from Group 2A (2) to Group 3A (13). A $2p$ electron, which is in a higher sublevel, is further from the nucleus and more easily removed than a $2s$ electron, and removing this electron results in a stable, filled sublevel. In Groups 4A (14) and 5A (15), the increased charge of the nucleus attracts the $2p$ electrons more strongly, which requires a greater ionization energy. Then there is a slight decrease in ionization energy for Group 6A (16) compared to Group 5A (15). For Group 6A (16), the removal of a $2p$ electron requires less energy because it results in a stable, half-filled $2p$ sublevel. Ionization energy increases for both Group 7A (17) and Group 8A (18) due to increased nuclear charge. Disrupting the stable ns^2np^6 electron configuration of the noble gases requires the highest ionization energies of all the groups.

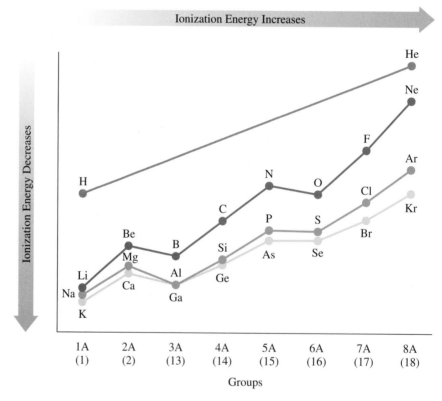

Figure 9.16 Ionization energies for the representative elements tend to decrease going down a group and increase going across a period.

Q Why is the ionization energy for Li less than for O?

Sample Problem 9.10 > **Ionization Energy**

Indicate the element in each that has the higher ionization energy and explain your choice.
a. K or Na
b. Mg or Cl
c. F or N

Solution

a. Na. In Na, an electron is removed from an orbital closer to the nucleus.
b. Cl. Nuclear charge increases the attraction for the outermost electrons for elements in the same period.
c. F. The increased nuclear charge of F requires a higher ionization energy compared to N.

Study Check

Arrange Sn, Sr, and I in order of increasing ionization energy.

Questions and Problems > **Periodic Trends of the Elements**

9.51 What does the group number of an element indicate about its electron configuration?

9.52 What is similar and different about the valence electrons of the elements in a group?

9.53 Write the group number using both A/B notation and 1–18 numbering of elements that have the following outer electron configuration:
a. $2s^2$
b. $3s^23p^3$
c. $4s^23d^5$
d. $5s^24d^{10}5p^4$

9.54 Write the group number using both A/B notation and 1–18 numbering of elements that have the following outer electron configuration:
a. $4s^24p^5$
b. $4s^1$
c. $4s^23d^8$
d. $5s^24d^{10}5p^2$

9.55 Write the valence electron configuration for each of the following (example ns^2np^4):
a. alkali metals
b. Group 4A
c. Group 13
d. Group 5A

9.56 Write the valence electron configuration for each of the following (example ns^2np^4):
a. halogens
b. Group 6A
c. Group 10
d. alkaline earth metals

9.57 Indicate the number of valence (outermost) electrons in each of the following:
a. aluminum
b. Group 5A
c. nickel
d. F, Cl, Br, and I

9.58 Indicate the number of valence (outermost) electrons in each of the following:
a. Li, Na, K, Rb, and Cs
b. zinc and cadmium
c. C, Si, Ge, Sn, and Pb
d. Group 8A

9.59 Explain the reasons that Mg and Ca are in the same group.

9.60 Explain the reasons that Cl and Br are in the same group.

9.61 Place the elements in each set in order of decreasing atomic radius.
a. Mg, Al, Si
b. Cl, Br, I
c. I, Sb, Sr
d. P, Si, Na

9.62 Place the elements in each set in order of decreasing atomic radius.
a. Cl, S, P
b. Ge, Si, C
c. Ba, Ca, Sr
d. O, S, Se

9.63 Select the larger atom in each pair.
a. Na or O
b. Na or Rb
c. Na or Mg
d. Na or Cl

9.64 Select the larger atom in each pair.
a. S or Cl
b. S or O
c. S or Se
d. S or Al

9.65 Why is a potassium ion smaller than a potassium atom?

9.66 Why is a bromide ion larger than a bromine atom?

9.67 Which is larger in each of the following?
a. Na or Na^+
b. Br or Br^-
c. S or S^{2-}

9.68 Which is smaller in each of the following?
a. I or I^-
b. Ca or Ca^{2+}
c. Rb or Rb^+

9.69 Arrange each set of elements in order of increasing ionization energy.
a. F, Cl, Br
b. Na, Cl, Al
c. Na, K, Cs
d. As, Sb, Sn

9.70 Arrange each set of elements in order of increasing ionization energy.
a. C, N, O
b. P, S, Cl
c. As, P, N
d. Al, Si, P

9.71 Select the element in each pair with the higher ionization energy.
a. Br or I
b. Mg or Al
c. S or P
d. I or Xe

9.72 Select the element in each pair with the higher ionization energy.
a. O or Ne
b. K or Br
c. Ca or Ba
d. N or O

Concept Map **Electronic Structure and Periodic Trends**

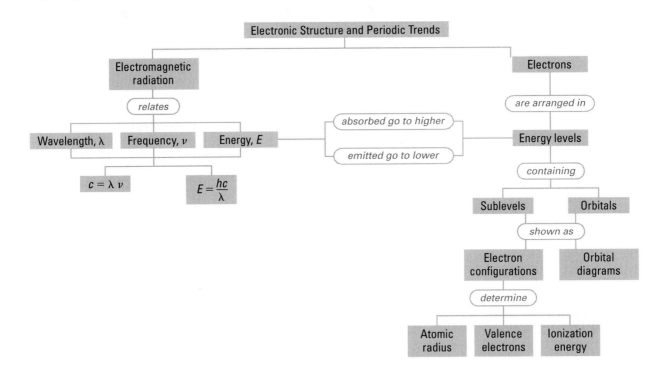

Chapter Review

9.1 Electromagnetic Radiation

Electromagnetic radiation such as radio waves and visible light is energy that travels as waves at the speed of light. Each particular type of radiation has a specific wavelength and frequency. A wavelength is the distance between a peak on one wave and a peak on the next wave. Frequency is the number of waves that pass a point in 1 second. Long-wavelength radiation has low frequencies, while short-wavelength radiation has high frequencies. Radiation with a high frequency has high energy.

9.2 Atomic Spectra and Energy Levels

The atomic spectra of elements are related to the specific energy levels occupied by electrons. When energy is absorbed, an electron moves to a higher energy level; energy is lost when the electron drops to a lower energy level and emits a photon. Each element has it own unique spectrum.

9.3 Energy Levels, Sublevels, and Orbitals

An orbital is a region around the nucleus where an electron with a specific energy is most likely to be found. Each orbital holds a maximum of 2 electrons, which must have opposite spins. In each principal energy level (n), electrons occupy orbitals within sublevels. An s sublevel contains one s orbital,

a p sublevel contains three p orbitals, a d sublevel contains five d orbitals, and an f sublevel contains seven f orbitals. Each type of orbital has a unique shape.

9.4 Writing Orbital Diagrams and Electron Configurations

Within a sublevel, electrons enter orbitals in the same energy level one at a time until all the orbitals are half-filled. Additional electrons enter with opposite spins until the orbitals in that sublevel are filled with 2 electrons each. The electrons in an atom can be written in an orbital diagram, which shows the orbitals that are occupied by paired and unpaired electrons. The electron configuration shows the number of electrons in each sublevel. An abbreviated electron configuration places the symbol of a noble gas in brackets to represent the filled sublevels.

9.5 Electron Configurations and the Periodic Table

The periodic table consists of s, p, d, and f sublevel blocks. An electron configuration can be written following the order of the sublevel blocks in the periodic table. Beginning with $1s$, an electron configuration is obtained by writing the sublevel blocks in order going across the periodic table until the element is reached.

9.6 Periodic Trends of the Elements

The properties of elements are related to the valence electrons of the atoms. With only a few minor exceptions, each group of elements has the same arrangement of valence electrons differing only in the energy level. The radius of an atom increases going down a group and decreases going across a period. The energy required to remove a valence electron is the ionization energy, which generally decreases going down a group and generally increases going across a period.

Key Terms

atomic radius The distance of the outermost electrons from the nucleus.

atomic spectrum A series of lines specific for each element produced by photons emitted by electrons dropping to lower energy levels.

_d_ block elements The block ten elements wide in Groups 3B (3) to 2B (12) in which electrons fill the five *d* orbitals in *d* sublevels.

electromagnetic radiation Forms of energy such as visible light, microwaves, radio waves, infrared, ultraviolet light, and X rays that travel as waves at the speed of light.

electromagnetic spectrum The arrangement of types of radiation from long wavelengths to short wavelengths.

electron configuration A list of the number of electrons in each sublevel within an atom, arranged by increasing energy.

_f_ block elements The block 14 elements wide in the rows at the bottom of the periodic table in which electrons fill the seven *f* orbitals in 4*f* and 5*f* sublevels.

frequency The number of times the crests of a wave pass a point in 1 second.

ionization energy The energy needed to remove the least tightly bound electron from the outermost energy level of an atom.

orbital The region around the nucleus of an atom where electrons of certain energy are most likely to be found: *s* orbitals are spherical, *p* orbitals have two lobes.

orbital diagram A diagram that shows the distribution of electrons in the orbitals of the energy levels.

_p_ block elements The elements in Groups 3A (13) to 8A (18) in which electrons fill the *p* orbitals in the *p* sublevels.

photon The smallest particle of light.

principal quantum number (*n*) The numbers ($n = 1, n = 2 \ldots$) assigned to energy levels.

_s_ block elements The elements in Groups 1A (1) and 2A (2) in which electrons fill the *s* orbitals.

sublevel A group of orbitals of equal energy within principal energy levels. The number of sublevels in each energy level is the same as the principal quantum number (*n*).

valence electrons The electrons in the outermost energy levels.

wavelength The distance between the peaks of two adjacent waves.

▶ Understanding the Concepts

Use the following diagram for Problems 9.73 and 9.74.

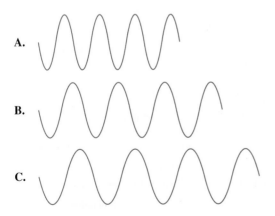

A.

B.

C.

9.73 Select diagram A, B, or C that
   a. has the longest wavelength
   b. has the shortest wavelength
   c. has the highest frequency
   d. has the lowest frequency

9.74 Select diagram A, B, or C that
   a. has the highest energy
   b. has the lowest energy
   c. would represent blue light
   d. would represent red light
   e. would represent green light

9.75 Match the following with an *s* or *p* orbital.

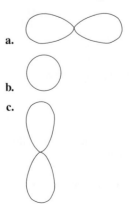

   a.

   b.

   c.

9.76 Match the following with *s* or *p* orbitals.
- **a.** two lobes
- **b.** spherical shape
- **c.** found in *n* = 2
- **d.** found in *n* = 3

9.77 Indicate if the following sections of orbital diagrams are or are not possible and explain your reason. When the section is possible, indicate the element it represents.

a.

b.

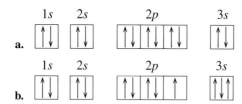

9.78 Indicate if the following sections of orbital diagrams are or are not possible and explain your reason. When the section is possible, indicate the element it represents.

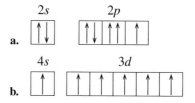

a.

b.

9.79 Match the spheres represented with atoms of Li, Na, K, and Rb.

A.   **B.**   **C.**   **D.**

9.80 Match the spheres represented with atoms of K, Ge, Ca, and Kr.

A.   **B.**   **C.**   **D.**

▶ Additional Questions and Problems

9.81 The average distance between the sun and Earth is about 1.5×10^8 km. How many minutes does it take for ultraviolet light to travel this distance?

9.82 The average distance between Mars and Earth is about 2.3×10^8 km. How many minutes does it take for red light to travel this distance?

9.83 Calculate the frequency of the electromagnetic radiation with the following wavelengths:
- **a.** yellow light at 590 nm
- **b.** ultraviolet light at 4.5×10^{-7} m
- **c.** X rays at 1.0×10^{-1} nm

9.84 Calculate the wavelength in meters of the following frequencies of electromagnetic radiation:
- **a.** heat lamp at 1×10^{13} MHz
- **b.** gamma rays from radioactive cobalt at 2.8×10^{20} s^{-1}
- **c.** an FM music station at 101.5 MHz

9.85 What is the difference between a continuous spectrum and atomic spectra?

9.86 The red light in a neon sign is given off because
- **a.** electrons are removed
- **b.** electrons jump to higher energy levels
- **c.** electrons fall to lower energy levels
- **d.** electrons are zipping around the atom at the speed of light

9.87 What is the Pauli exclusion principle?

9.88 Five unpaired electrons will enter a *d* sublevel before they start to pair up. True or false.

9.89 How are each of the following similar and how are they different?
- **a.** 2*p* and 3*p* orbitals
- **b.** 2*s* and 2*p* orbitals
- **c.** the orbitals in the 4*p* sublevel

9.90 Indicate the number of unpaired electrons in each of the following:
  a. chromium
  b. an element in Group 16
  c. an element in Group 3A
  d. F, Cl, Br, and I

9.91 Which of the following orbitals are possible in an atom: $4p$, $2d$, $3f$, and $5f$?

9.92 Which of the following orbitals are possible in an atom: $1p$, $4f$, $6s$, and $4d$?

9.93 **a.** What electron sublevel starts to fill after completion of the $3s$ sublevel?
  b. What electron sublevel starts to fill after completion of the $4p$ sublevel?
  c. What electron sublevel starts to fill after completion of the $3d$ sublevel?
  d. What electron sublevel starts to fill after completion of the $3p$ sublevel?

9.94 **a.** What electron sublevel starts to fill after completion of the $5s$ sublevel?
  b. What electron sublevel starts to fill after completion of the $4d$ sublevel?
  c. What electron sublevel starts to fill after completion of the $4f$ sublevel?
  d. What electron sublevel starts to fill after completion of the $5p$ sublevel?

9.95 **a.** How many $3d$ electrons are in Fe?
  b. How many $5p$ electrons are in Ba?
  c. How many $4d$ electrons are in I?
  d. How many $6s$ electrons are in Ba?

9.96 **a.** How many $3d$ electrons are in Zn?
  b. How many $4p$ electrons are in Br?
  c. How many $6p$ electrons are in Bi?
  d. How many $5s$ electrons are in Cd?

9.97 What do the elements Ca, Sr, and Ba have in common in terms of their electron configuration? Where are they located in the periodic chart?

9.98 What do the elements O, S, and Se have in common in terms of their electron configuration? Where are they located in the periodic chart?

9.99 Consider three elements with the following noble gas notations:

  $X = [Ar]4s^2 \qquad Y = [Ne]3s^23p^4$
  $Z = [Ar]4s^23d^{10}4p^4$

  a. Identify each element as a metal, metalloid, or nonmetal.
  b. Which element has the largest atomic radius?
  c. Which elements have similar properties?
  d. Which element has the highest ionization energy?
  e. Which element has the smallest atomic radius?

9.100 Consider three elements with the following noble gas notations:

  $X = [Ar]4s^23d^5 \qquad Y = [Ar]4s^23d^{10}4p^1$
  $Z = [Ar]4s^23d^{10}4p^6$

  a. Identify each element as a metal, metalloid, or nonmetal.
  b. Which element has the smallest atomic radius?
  c. Which elements have similar properties?
  d. Which element has the highest ionization energy?
  e. Which element has a half-filled sublevel?

9.101 Name the element that corresponds to each of the following:
  a. $1s^22s^22p^63s^23p^3$
  b. alkali metal with the smallest atomic radius
  c. $[Kr]\ 5s^24d^{10}$
  d. Group 5A element with highest ionization energy
  e. Period 3 element with largest atomic radius

9.102 Name the element that corresponds to each of the following:
  a. $1s^22s^22p^63s^23p^64s^13d^5$
  b. $[Xe]\ 6s^24f^{14}5d^{10}6p^5$
  c. halogen with the highest ionization energy
  d. Group 5A element with the smallest ionization energy
  e. Period 4 element with smallest atomic radius

9.103 An oxide ion, O^{2-}, is about twice the size of an oxygen atom. How would you explain this?

9.104 An aluminum ion, Al^{3+}, is only one-third the size of an aluminum atom. How would you explain this?

9.105 For each pair, select the smaller.
  a. Sr or Sr^{2+}
  b. Se or Se^{2-}
  c. Br^- or I^-

9.106 For each pair, select the smaller.
  a. Li^+ or Li
  b. Br or Br^-
  c. Ca^{2+} or Mg^{2+}

9.107 Why is the ionization energy of Ca higher than K but lower than Mg?

9.108 Why is the ionization energy of Cl lower than F but higher than S?

9.109 The ionization energy generally increases going from left to right across a period. Why do Group 3A elements have a lower ionization energy than Group 2A elements?

9.110 The ionization energy generally increases going from left to right across a period. Why do Group 6A elements have a lower ionization energy than Group 5A elements?

9.111 Of the elements Na, P, Cl, and F, which
  a. is a metal?
  b. has the largest atomic radius?
  c. has the highest ionization energy?
  d. loses an electron most easily?
  e. is found in Group 7A, Period 3?

9.112 Of the elements Mg, Ca, Br, Kr, which
  a. is a noble gas?
  b. has the smallest atomic radius?
  c. has the lowest ionization energy?
  d. requires the most energy to remove an electron?
  e. is found in Group 2A, Period 4?

9.113 Write the abbreviated electron configuration and group number for each of the following elements:
a. Si   **b.** Se   **c.** Mn   **d.** Sb

9.114 Write the abbreviated electron configuration and group number for each of the following elements:
a. Zn   **b.** Rh   **c.** Tc   **d.** Pb

9.115 Give the symbol of the element that has the
a. smallest atomic radius in Group 6A
b. smallest atomic radius in Period 3

c. highest ionization energy in Group 15
d. lowest ionization energy in Period 3
e. configuration $[Kr]\,5s^24d^6$

9.116 Give the symbol of the element that has the
a. largest atomic radius in Period 5
b. largest atomic radius in Group 3A
c. highest ionization energy in Group 18
d. lowest ionization energy in Period 2
e. configuration $[Ar]\,5s^24d^{10}5p^2$

Challenge Questions

9.117 Compare the speed, wavelengths, and frequencies of ultraviolet light and microwaves.

9.118 Radio waves with a frequency of $5.0 \times 10^5\,s^{-1}$ are used to communicate with satellites in space.
a. What is the wavelength in meters and nanometers of the radio waves?
b. How many minutes will it take for a message from flight control to reach a satellite that is 5.0×10^7 km from Earth?

9.119 How do scientists explain the colored lines observed in the spectra of heated atoms?

9.120 Even though H has only one electron, there are many lines in the spectrum of H. Explain.

9.121 What is meant by a principal energy level, a sublevel, and an orbital?

9.122 In some periodic charts, H is placed in Group 1A (1). In other periodic charts, H is placed in Group 7A (17). Why?

9.123 Compare O, S, and Cl in terms of atomic radius and ionization energy.

Answers

Answers to Study Checks

9.1 $1.4 \times 10^{18}\,s^{-1}$

9.2 From lowest to highest frequency: FM radio waves, microwaves, ultraviolet light, and X rays.

9.3 Because the spectra of elements consisted of only discrete, separated lines, scientists concluded that electrons occupied only certain energy levels in the atom.

9.4 The 1s, 2s, and 3s orbitals are all spherical but they increase in volume because the electron is most likely to be found further from the nucleus for higher energy levels.

9.5

1s	2s	2p	
↑↓	↑↓	↑↓ ↑ ↑	Orbital diagram for fluorine (F)

$1s^22s^22p^5$ Electron configuration for fluorine (F)

$[He]\,2s^22s^5$ Abbreviated electron configuration for fluorine (F)

9.6 $1s^22s^22p^63s^23p^4$ Electron configuration for sulfur (S)

$[Ne]3s^23p^4$ Abbreviated electron configuration for sulfur (S)

9.7 Tin has the electron configuration:

$1s^22s^22p^63s^23p^64s^23d^{10}4p^65s^24d^{10}5p^2$

9.8 Sulfur in Group 6A (16) has a $3s^2 3p^4$ valence electron configuration. Strontium in Group 2A (2) has a $5s^2$ valence electron configuration.

9.9 Going across a period, atomic radius decreases: Na is largest, then Mg, and S is the smallest.

9.10 Ionization energy increases going across a period: Sr is lowest, Sn is higher, and I is the highest of this set.

Answers to Selected Questions and Problems

9.1 The wavelength of UV light is the distance between crests of the wave.

9.3 White light has all the colors of the spectrum, including red and blue light.

9.5 650 kHz has a wavelength of 4.6×10^2 m.
980 kHz has a wavelength of 3.1×10^2 m.

9.7 Orange light has a wavelength of 6.3×10^{-7} m or 630 nm.

9.9 Microwaves have a longer wavelength than ultraviolet light or X rays.

9.11 From shortest to longest wavelength: X rays, blue light, infrared, microwaves.

9.13 Atomic spectra consist of a series of lines separated by dark sections, indicating that the energy emitted by the elements is not continuous.

9.15 absorb

9.17 A photon in the infrared region of the spectrum is emitted when an excited electron drops to the third energy level.

9.19 The photon with greater energy is
a. green light   **b.** blue light

9.21 **a.** spherical   **b.** two lobes
c. spherical

9.23 **a.** Both are spherical.
b. Both are part of the third energy level.
c. Both contain three p orbitals.
d. All have two lobes and belong in the third energy level.

9.25 **a.** There are a maximum of five orbitals in the $3d$ sublevel.
b. There is one sublevel in the $n = 1$ energy level.
c. There is one orbital in the $6s$ sublevel.
d. There are nine orbitals in the $n = 3$ energy level.

9.27 **a.** There is a maximum of 2 electrons in a $2p$ orbital.
b. There is a maximum of 6 electrons in the $3p$ sublevel.
c. There is a maximum of 32 electrons in the $n = 4$ energy level.
d. There is a maximum of 10 electrons in the $5d$ sublevel.

9.29 The electron configuration shows the number of electrons in each sublevel of an atom. The abbreviated electron configuration uses the symbol of the noble gas to show completed sublevels.

9.31 **a.**

$1s$	$2s$	$2p$		
↑↓	↑↓	↑		

b.

$3s$	$3p$		
↑↓	↑		

c.

$3s$	$3p$		
↑↓	↑	↑	↑

d.

$3s$	$3p$		
↑↓	↑↓	↑↓	↑↓

9.33 **a.** N $1s^2 2s^2 2p^3$ **b.** Na $1s^2 2s^2 2p^6 3s^1$
c. S $1s^2 2s^2 2p^6 3s^2 3p^4$ **d.** B $1s^2 2s^2 2p^1$

9.35 **a.** Mg [Ne]$3s^2$ **b.** S [Ne]$3s^2 3p^4$
c. Al [Ne]$3s^2 3p^1$ **d.** N [He]$2s^2 2p^3$

9.37 **a.** Li   **b.** S   **c.** Si   **d.** F

9.39 **a.** Al   **b.** C   **c.** Ar   **d.** Be

9.41 **a.** As $1s^2 2s^2 2p^6 3s^2 3p^6 4s^2 3d^{10} 4p^3$
b. Fe $1s^2 2s^2 2p^6 3s^2 3p^6 4s^2 3d^6$
c. Pd $1s^2 2s^2 2p^6 3s^2 3p^6 4s^2 3d^{10} 4p^6 5s^2 4d^8$
d. I $1s^2 2s^2 2p^6 3s^2 3p^6 4s^2 3d^{10} 4p^6 5s^2 4d^{10} 5p^5$

9.43 **a.** Ti [Ar] $4s^2 3d^2$ **b.** Sr [Kr] $5s^2$
c. Ba [Xe] $6s^2$ **d.** Pb [Xe] $6s^2 4f^{14} 5d^{10} 6p^2$

9.45 **a.** Li   **b.** Co   **c.** Si   **d.** Br

9.47 **a.** Ga   **b.** C   **c.** Xe   **d.** Zr

9.49 **a.** 10   **b.** 6   **c.** 3   **d.** 1

9.51 The group numbers 1A–8A or the one's digit in 1, 2, and 13–18 indicate 1–8 valence electrons. The group numbers 3–12 give the electrons in the s and d sublevel. The Group B indicates that the d sublevel is filling.

9.53 **a.** 2A (2)   **b.** 5A (15)
c. 7B (7)   **d.** 6A (16)

9.55 **a.** ns^1 **b.** $ns^2 np^2$
c. $ns^2 np^1$ **d.** $ns^2 np^3$

9.57 **a.** 3   **b.** 5   **c.** 2   **d.** 7

9.59 Mg and Ca both have 2 valence electrons, Mg $3s^2$ and Ca $4s^2$.

9.61 **a.** Mg, Al, Si   **b.** I, Br, Cl
c. Sr, Sb, I   **d.** Na, Si, P

9.63 **a.** Na   **b.** Rb   **c.** Na   **d.** Na

9.65 When potassium ion is formed, it loses the only valence electron in its outermost energy level and is smaller than a potassium atom.

9.67 **a.** Na   **b.** Br^- **c.** S^{2-}

9.69 **a.** Br, Cl, F   **b.** Na, Al, Cl
       c. Cs, K, Na   **d.** Sn, Sb, As

9.71 **a.** Br   **b.** Mg   **c.** P   **d.** Xe

9.73 **a.** C has the longest wavelength.
       b. A has the shortest wavelength.
       c. A has the highest frequency.
       d. C has the lowest frequency.

9.75 **a.** p **b.** s **c.** p

9.77 **a.** This is possible. This element is magnesium.
       b. Not possible. The $2p$ sublevel would fill before the $3s$, and only 2 electrons are allowed in an orbital.

9.79 Li is D, Na is A, K is C, and Rb is B.

9.81 8.3 min

9.83 **a.** $5.1 \times 10^{14} \, s^{-1}$ **b.** $6.7 \times 10^{14} \, s^{-1}$
       c. $3.0 \times 10^{18} \, s^{-1}$

9.85 A continuous spectrum from white light contains wavelengths of all energies. Atomic spectra are line spectra in which a series of lines correspond to energy emitted when electrons drop from a higher energy level to a lower level.

9.87 The Pauli exclusion principle states that two electrons in the same orbital must have opposite spins.

9.89 **a.** A $2p$ and a $3p$ orbital have the same spatial shape with two lobes; each p orbital can hold up to 2 electrons with opposite spins. However, the $3p$ orbital is larger because the $3p$ electron has a higher energy level and is most likely to be found further from the nucleus.
       b. A $2s$ orbital and a $2p$ orbital are found in the same energy level, $n = 2$, and each can hold up to 2 electrons with opposite spins. However, the shapes of a $2s$ orbital and a $2p$ orbital are different.
       c. The orbitals in the $4p$ sublevel all have the same energy level and shape. However, there are three $4p$ orbitals directed along the $x, y,$ and z axes around the nucleus.

9.91 A $4p$ orbital is possible because $n = 4$ has four sublevels, including a p sublevel. A $2d$ orbital is not possible because $n = 2$ has only s and p sublevels. There are no $3f$ orbitals because only $s, p,$ and d sublevels are allowed for $n = 3$. A $5f$ sublevel is possible in $n = 5$ because five sublevels are allowed.

9.93 **a.** $3p$ **b.** $5s$ **c.** $4p$ **d.** $4s$

9.95 **a.** 6   **b.** 6   **c.** 10   **d.** 2

9.97 Ca, Sr, and Ba all have 2 valence electrons, ns^2, which places them in Group 2A (2).

9.99 **a.** X is a metal; Y and Z are nonmetals.
       b. X has the largest atomic radius.
       c. Y and Z have 6 valence electrons and are in Group 6A (16).
       d. Y has the highest ionization energy.
       e. Y has the smallest atomic radius.

9.101 **a.** phosphorus   **b.** lithium (H is a nonmetal)
        c. cadmium   **d.** nitrogen
        e. sodium

9.103 When 2 electrons are added to the outermost electron level in an oxygen atom, there is an increase in electron repulsion that pushes the electrons apart and increases the size of the oxide ion compared to the oxygen atom.

9.105 **a.** Sr^{2+} **b.** Se   **c.** Br^-

9.107 Calcium has a greater number of protons than K. The least tightly bound electron in Ca is further from the nucleus than in Mg and needs less energy to remove.

9.109 In Group 3A (13), ns^2np^1, the p electron is further from the nucleus and easier to remove.

9.111 **a.** Na   **b.** Na   **c.** F   **d.** Na   **e.** Cl

9.113 **a.** [Ne] $3s^23p^2$; Group 4A (14)
        b. [Ar] $4s^23d^{10}4p^4$; Group 6A (16)
        c. [Ar] $4s^23d^5$ Group 7B (7)
        d. [Kr] $5s^24d^{10}5p^3$; Group 5A (15)

9.115 **a.** O   **b.** Ar   **c.** N   **d.** Na   **e.** Ru

9.117 Ultraviolet light and microwaves both travel at the same speed: $3.0 \times 10^8 \, ms^{-1}$. The wavelength of ultraviolet light is shorter than the wavelength of microwaves, and the frequency of ultraviolet light is higher than the frequency of microwaves.

9.119 The series of lines separated by dark sections in spectra indicate that the energy emitted by the elements is not continuous and that electrons are moving between discrete energy levels.

9.121 The principal energy level contains all the electrons with similar energy. A sublevel contains electrons with the same energy, while an orbital is the region around the nucleus where electrons of a certain energy are most likely to be found.

9.123 S has a larger atomic radius than O; S is also larger than Cl. O has a higher ionization energy than S, and Cl also has a higher ionization energy than S.

Molecular Structure: Solids and Liquids

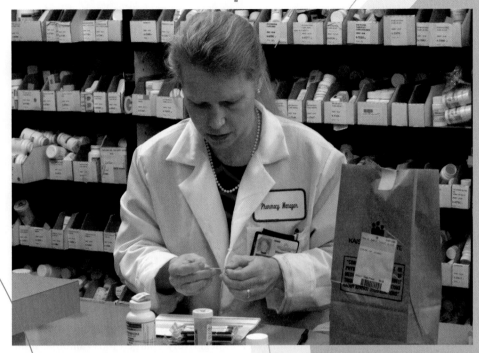

"The pharmacy is one of the many factors in the final integration of chemistry and medicine in patient care," says Dorothea Lorimer, pharmacist, Kaiser Hospital. *"If someone is allergic to a medication, I have to find out if a new medication has similar structural features. For instance, some people are allergic to sulfur. If there is sulfur in the new medication, there is a chance it will cause a reaction."*

A prescription indicates a specific amount of a medication. At the pharmacy, the chemical name, formula, and quantity in milligrams or micrograms are checked. Then the prescribed number of capsules is prepared and placed in a container. If it is a liquid medication, a specific volume is measured and poured into a bottle for liquid prescriptions.

the **Chemistry** place

Visit **www.aw-bc.com/chemplace** for extra quizzes, interactive tutorials, career resources, PowerPoint slides for chapter review, math help, and case studies.

With the information in Chapter 9 about electron configurations, we can now look more closely at how the valence electrons, those electrons in the outermost s and p sublevels, influence the reactivity of atoms. A bond is ionic or covalent depending on whether valence electrons are lost and gained or shared by the reacting atoms. However, in both types of bond, each reacting atom typically acquires the s^2p^6 electron configuration of the valence electrons of the nearest noble gas.

For covalent compounds, an electron-dot formula is used to diagram the valence electrons that are shared and not shared. The presence of single and multiple bonds can be determined as well as the possibility of resonance structures. From the electron-dot formula, the three-dimensional shapes and polarities of molecules can be determined. Then we can look at the kinds of attractive forces between ions and molecules that influence the physical properties of substances such as melting and boiling point. Finally, we will look at the physical states of solids, liquids, and gases and how changes in state take place.

10.1 Electron Configuration of Ionic Compounds

Learning **Goal**

Use the electron configurations to show the formation of ions.

In ionic compounds, metals, which have low ionization energies, tend to give up electrons, and nonmetallic elements, with high ionization energies, tend to accept electrons. This loss and gain of electrons results in the formation of positively charged ions (cations) and negatively charged ions (anions). Ionic bonds form as a result of the attraction between positively charged ions and negatively charged ions.

Electron Configurations of Ions

Now using the electron configurations, we can review how atoms of the representative elements lose or gain electrons to form ions with the electron configurations of noble gases. For example, the removal of 2 valence electrons from magnesium gives the magnesium ion the electron configuration of neon. The addition of 1 electron to fluorine gives a fluoride ion the electron configuration of neon.

Atoms			Ions		

$$
\begin{array}{ccccc}
\textbf{Mg} & + & \textbf{2F} & \longrightarrow & \text{Mg}^{2+} + \ \ 2\text{F}^- \\
\text{2A (2)} & & \text{7A (17)} & & \text{MgF}_2
\end{array}
$$

$$1s^22s^22p^63s^2 + 1s^22s^2\boldsymbol{2p^5} \longrightarrow 1s^22s^2\boldsymbol{2p^6} + 1s^22s^2\boldsymbol{2p^6}$$
$$1s^22s^2\boldsymbol{2p^5} \qquad\qquad 1s^22s^2\boldsymbol{2p^6}$$
$$1e^- \qquad\qquad \text{Ne} \qquad \text{Ne}$$

Formation of Ions by Transition Metals

We have seen that the electron configurations of the transition metals usually have 2 s electrons and from 1 to 10 d electrons. Transition metals form positively charged ions by losing the s electrons from their valence shells. Thus, many of the transition metals in Period 4 form stable ions with a 2^+ charge.

Other stable ions can form when transition metals also lose 1 or 2 d electrons. Iron forms two cations: Fe^{2+}, in which 2 $4s$ electrons are removed, and Fe^{3+}, which forms when a d electron is also lost in addition to the 2 $4s$ electrons.

Atoms   **Ions**

$$Fe \quad + \quad O \quad \longrightarrow \quad Fe^{2+} \quad + \quad O^{2-}$$
$$2e^{-} \qquad\qquad FeO$$

$$1s^2 2s^2 2p^6 3s^2 3p^6 4s^2 3d^6 + 1s^2 2s^2 2p^4 \longrightarrow 1s^2 2s^2 2p^6 3s^2 3p^6 3d^6 + 1s^2 2s^2 2p^6$$

Atoms   **Ions**

$$Fe \quad + \quad 3F \quad \longrightarrow \quad Fe^{3+} \quad + \quad 3F^{-}$$
$$1e^{-} \qquad\qquad FeF_3$$

$$1s^2 2s^2 2p^6 3s^2 3p^6 4s^2 3d^6 + 1s^2 2s^2 2p^5 \longrightarrow 1s^2 2s^2 2p^6 3s^2 3p^6 3d^5 + 1s^2 2s^2 2p^6$$
$$1e^{-}$$
$$+ 1s^2 2s^2 2p^5 \longrightarrow \qquad\qquad\qquad + 1s^2 2s^2 2p^6$$
$$1e^{-}$$
$$+ 1s^2 2s^2 2p^5 \longrightarrow \qquad\qquad\qquad + 1s^2 2s^2 2p^6$$

In Fe^{2+}, there are 6 $4d$ electrons, whereas in Fe^{3+}, there are 5 $4d$ electrons. The loss of a d electron gives Fe^{3+} a half-filled d sublevel, which is stable. As seen in these examples, cations formed by transition metals have electrons in their d sublevels. Thus, they do not acquire the electron configuration of a noble gas. Two cations with 1^+ charges are formed from copper and silver atoms. Because they use an s electron to complete their d sublevels, they lose only 1 s electron.

Sample Problem 10.1   **Formation of Ions**

Identify each of the following electron configurations of a cation with a 2^+ charge:

a. $1s^2 2s^2 2p^6 3s^2 3p^6$
b. $1s^2 2s^2 2p^6 3s^2 3p^6 3d^6$
c. $1s^2 2s^2 2p^6 3s^2 3p^6 3d^{10}$

Solution
a. Ca^{2+} **b.** Fe^{2+} **c.** Zn^{2+}

Study Check

What are the electron configurations for N^{3-} and P^{3-}?

Electron Configurations of Ionic Compounds

10.1 Show the changes in electron configurations when electrons are lost or gained by atoms to form each of the following ions:
  a. Ca^{2+} **b.** S^{2-}
  c. Ni^{2+} **d.** Ti^{2+}

10.2 Show the changes in electron configurations when electrons are lost or gained by atoms to form each of the following ions:
  a. Mn^{2+} **b.** Al^{3+}
  c. Co^{3+} **d.** O^{2-}

10.3 For each compound, write the electron configuration of the cation and indicate if it has the electron configuration of a noble gas.
  a. $ScCl_3$ **b.** BaO
  c. KCl **d.** VO

10.4 For each compound, write the electron configuration of the cation and indicate if it has the electron configuration of a noble gas.
  a. MnO_2 **b.** AlF_3
  c. CoF_3 **d.** $SrCl_2$

Learning Goal

Write the electron-dot formulas for covalent compounds or ions with multiple bonds and show resonance structures.

10.2 Electron-Dot Formulas

In covalent compounds or polyatomic ions, atoms of two or more nonmetals, which have similar and fairly high ionization energies, share valence electrons. With our understanding of electron configuration, we will describe the bonding between two nonmetal atoms as the sharing of s and p valence electrons to achieve the stability of their nearest noble gases. We will use electron-dot symbols for atoms we wrote in Chapter 4 in which dots representing the valence electrons are placed around the symbol of an element.

Drawing Electron-Dot Formulas

The electron-dot formula for a molecule shows the sequence of atoms, the bonding pairs shared between atoms, and the unshared electrons in the valence shells. We will write an electron-dot formula for PCl_3. Table 10.1 relates the group numbers of some nonmetals to the number of covalent bonds the elements typically form; for example, phosphorus forms three bonds and chlorine forms one bond.

Table 10.1 Typical Bonding Patterns of Some Nonmetals in Covalent Compounds

1A (1)	3A (13)	4A (14)	5A (15)	6A (16)	7A (17)
[a]H					
1 bond					
	[a]B	C	N	O	F
	3 bonds	4 bonds	3 bonds	2 bonds	1 bond
		Si	P	S	Cl
		4 bonds	3 bonds	2 bonds	1 bond
					Br
					1 bond
					I
					1 bond

[a]H and B do not form 8-electron octets. H atoms share one electron pair; B atoms share three electron pairs for a set of 6 electrons.

| Sample Problem 10.2 | **Writing Electron-Dot Formulas with Lone Pairs** |

Write the electron-dot formulas for phosphorus trichloride, PCl_3.

Solution

Guide to Writing Electron-Dot Formulas

STEP 1
Determine the arrangement of atoms.

STEP 2
Determine the total number of valence electrons.

STEP 3
Attach each bonded atom to the central atom with a pair of electrons.

STEP 4
Place the remaining electrons as lone pairs to complete octets (two for H, six for B).

STEP 5
If octets are not complete, form a multiple bond. Convert a lone pair to a bonding pair with the central atom.

STEP 1 **Determine the arrangement of atoms.** In PCl_3, the central atom is P.

Cl P Cl
Cl

STEP 2 **Determine the total number of valence electrons.** We can use the group numbers to determine the valence electrons for each of the atoms in the molecule.

Element	Group	Atoms	Valence Electrons	=	Total
P	5A (15)	1 P	$\times 5\,e^-$	=	$5\,e^-$
Cl	7A (17)	3 Cl	$\times 7\,e^-$	=	$21\,e^-$
		Total valence electrons for PCl_3		=	$26\,e^-$

STEP 3 **Attach the central atom to each bonded atom by a pair of electrons.**

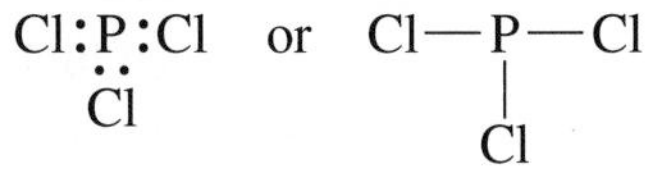

Cl:P:Cl or Cl—P—Cl
Cl |
              Cl

STEP 4 **Arrange the remaining electrons as lone pairs to complete octets.** A total of 6 electrons ($3 \times 2\,e^-$) are needed to bond the central P atom to three Cl atoms. There are 20 valence electrons left.

$$26 \text{ valence } e^- - 6 \text{ bonding } e^- = 20\,e^- \text{ remaining}$$

The remaining electrons are placed as lone pairs around the outer Cl atoms first, which uses 18 more electrons.

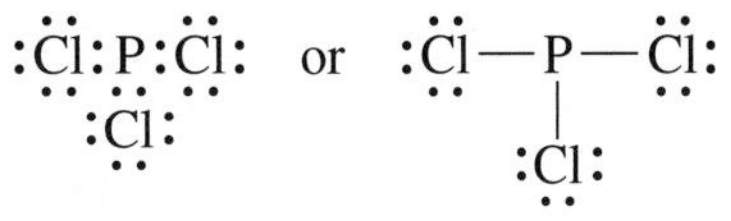

:Cl:P:Cl: or :Cl—P—Cl:
  :Cl: :Cl:

The remaining 2 electrons are used to complete the octet for the P atom. Octets are written for all the atoms using 26 valence electrons.

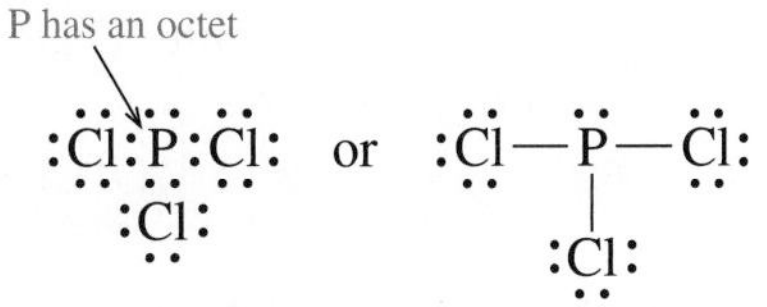

P has an octet

:Cl:P:Cl: or :Cl—P—Cl:
  :Cl: :Cl:

Study Check

Write the electron-dot formula for Cl_2O.

While the octet rule is useful, there are exceptions. We have already seen that a hydrogen (H_2) molecule requires just 2 electrons or a single bond to achieve the stability of the nearest noble gas, helium. In BF_3, the nonmetal B can share only 3 valence electrons to give a total of 6 valence electrons or three bonds. The nonmetals typically form octets. However, atoms such as P, S, Cl, Br, and I can share more of their valence electrons and expand the octets to stable valence shells of 10, 12, or even 14 electrons. In PCl_3, the P atom has an octet, but in PCl_5 the P atom has 10 valence electrons or five bonds. In H_2S, the S atom has an octet, but in SF_6, there are 12 valence electrons or six bonds to sulfur.

Sample Problem 10.3   **Writing Electron-Dot Formulas for Ions**

Write the electron dot formula for the chlorite ion, ClO_2^-.

Solution

STEP 1   **Determine the arrangement of atoms.**   In ClO_2^-, the central atom is Cl. The atoms and electrons of a polyatomic ion are placed in brackets with the charge written outside.

$$\left[\text{O} \quad \text{Cl} \quad \text{O} \right]^-$$

STEP 2   **Determine the total number of valence electrons.**   We can use the group numbers to determine the valence electrons for each of the atoms in the molecule. Because the ion has a negative charge, 1 more electron is added to the valence electrons.

Element	Group	Atoms	Valence Electrons	=	Total
O	6A (16)	2 O	$\times 6\,e^-$	=	$12\,e^-$
Cl	7A (17)	1 Cl	$\times 7\,e^-$	=	$7\,e^-$
Ionic charge (negative) add			$1\,e^-$	=	$1\,e^-$
	Total valence electrons for ClO_2^-			=	$20\,e^-$

STEP 3   **Attach the central atom to each bonded atom by a pair of electrons.**

$$\text{O:Cl:O} \quad \text{or} \quad \text{O—Cl—O}$$

STEP 4   **Arrange the remaining electrons as lone pairs to complete octets.**   A total of 4 electrons ($2 \times 2\,e^-$) are used to bond the O atoms to the Cl atom. The number of valence electrons remaining is 20 valence e^- − 4 bonding e^- = 16 e^- remaining.

The remaining electrons are placed as lone pairs around the outer O atoms first.

$$\left[:\ddot{\text{O}}:\text{Cl}:\ddot{\text{O}}: \right]^- \quad \text{or} \quad \left[:\ddot{\text{O}}\text{—Cl—}\ddot{\text{O}}: \right]^-$$

A total of 4 electrons are left, which complete the octet for the Cl atom.

$$\left[:\ddot{\text{O}}:\ddot{\text{Cl}}:\ddot{\text{O}}: \right]^- \quad \text{or} \quad \left[:\ddot{\text{O}}\text{—}\ddot{\text{Cl}}\text{—}\ddot{\text{O}}: \right]^-$$

The electron dot formula for a polyatomic ion is typically enclosed in square brackets with the charge written as a superscript in the upper right-hand corner.

Study Check

Write the electron-dot formula for NH_2^-.

Multiple Covalent Bonds and Resonance

the Chemistry place

WEB TUTORIAL

Bonds and Bond Polarities

Up to now, we have looked at covalent bonding in molecules having only single bonds. In many covalent compounds, atoms share two or three pairs of electrons to complete their octets. A **double bond** is the sharing of two pairs of electrons, and in a **triple bond,** three pairs of electrons are shared. Atoms of carbon, oxygen, nitrogen, and sulfur are most likely to form multiple bonds. Atoms of hydrogen and the halogens do not form double or triple bonds. Double and triple bonds are formed when single covalent bonds fail to complete the octets of all the atoms in the molecule. In N_2, an octet is achieved when each nitrogen atom shares 3 electrons; three covalent bonds, a triple bond, will form.

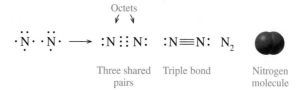

Sample Problem 10.4 Writing Electron-Dot Formulas with Double Bonds

Write the electron-dot formula for carbon dioxide, CO_2.

Solution

STEP 1 **Determine the arrangement of atoms.** In CO_2, the central atom is C.

O C O

STEP 2 **Determine the total number of valence electrons.**

Element	Group	Atoms	Valence Electrons	=	Total
O	6A (16)	2 O	$\times$ 6 e^-	=	12 e^-
C	4A (14)	1 C	$\times$ 4 e^-	=	4 e^-
	Total valence electrons for CO_2			=	16 e^-

STEP 3 **Attach the central atom to each bonded atom by a pair of electrons.**

O:C:O: or O—C—O

STEP 4 **Arrange remaining electrons as lone pairs to complete octets.** Four (4) electrons ($2 \times 2\,e^-$) were used to bond the central C atom to the O atoms. The remaining 12 electrons ($16 - 4$) are placed as six lone pairs to complete the octets for the O atoms.

$$:\overset{..}{\underset{..}{O}}:C:\overset{..}{\underset{..}{O}}: \quad \text{or} \quad :\overset{..}{\underset{..}{O}}\!-\!Cl\!-\!\overset{..}{\underset{..}{O}}:$$

STEP 5 **If octets are not complete, form one or more multiple bonds.** Although all the valence electrons are used, the central C atom does not have an octet. To provide an octet for C, use a lone pair from each O atom as a second bonding pair between C and each O.

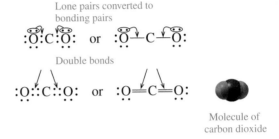

Lone pairs converted to bonding pairs

Double bonds

$$:\overset{..}{O}::C::\overset{..}{O}: \quad \text{or} \quad :\overset{..}{O}\!\!=\!\!C\!\!=\!\!\overset{..}{O}:$$

Molecule of carbon dioxide

Study Check

Write the electron-dot formula for HCN (atoms arranged as H   C   N).

Resonance Structures

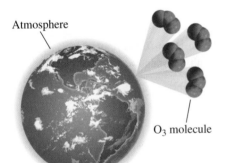

Atmosphere

O_3 molecule

When a molecule or polyatomic ion contains multiple bonds, it is often possible to write more than one electron-dot formula for the same arrangement of atoms. Suppose we want to write the electron-dot formula for ozone, O_3, a component in the stratosphere that protects us from the ultraviolet rays of the sun. Although all of the valence electrons (18) are used, one of the oxygen atoms does not have an octet. One lone pair must be moved to form a double bond. But which one should be used? One possibility is to form a double bond on the left and the other possibility is to form a double bond on the right. Both electron-dot formulas give complete octets to all the O atoms. However, experiments show that the actual bond length is equivalent to a molecule with a "one-and-a-half" bond between the central O atom and each outside O atom. In this *hybrid,* the electrons are shown spread equally over all the O atoms. When two or more electron-dot formulas can be written, they are called **resonance structures,** which are shown with a double-headed arrow. Although we will write resonance structures of some molecules and ions, the true structure is really a mix or average of the possible structures.

$$:\overset{..}{\underset{..}{O}}\!-\!\overset{..}{\underset{..}{O}}\!-\!\overset{..}{\underset{..}{O}}:$$

or

$$:\overset{..}{\underset{..}{O}}\!-\!\overset{..}{\underset{..}{O}}\!-\!\overset{..}{\underset{..}{O}}: \qquad :\overset{..}{\underset{..}{O}}\!-\!\overset{..}{\underset{..}{O}}\!-\!\overset{..}{\underset{..}{O}}:$$

$$:\overset{..}{O}\!\!=\!\!\overset{..}{\underset{..}{O}}\!-\!\overset{..}{\underset{..}{O}}: \longleftrightarrow :\overset{..}{\underset{..}{O}}\!-\!\overset{..}{\underset{..}{O}}\!\!=\!\!\overset{..}{O}: \quad \text{or} \quad :\overset{..}{\underset{..}{O}}\!-\!\overset{..}{\underset{..}{O}}\!-\!\overset{..}{\underset{..}{O}}:$$

Resonance structures   Hybrid

Sample Problem 10.5   **Writing Resonance Structures**

Write two resonance structures for sulfur dioxide, SO_2.

Solution

STEP 1   **Determine the arrangement of atoms.**   In SO_2, the S atom is the central atom.

O   S   O

STEP 2   **Determine the total number of valence electrons.**

Element	Group	Atoms	Valence Electrons	=	Total
S	6A (16)	1 S	$\times 6\,e^-$	=	$6\,e^-$
O	6A (16)	2 O	$\times 6\,e^-$	=	$12\,e^-$
		Total valence electrons for SO_2		=	$18\,e^-$

STEP 3   **Attach the central atom to each bonded atom by a pair of electrons.**   We will use a single line to represent a pair of bonding electrons.

O—S—O

STEP 4   **Arrange the remaining electrons to complete octets.**   After 4 electrons are used to write single bonds between the S atom and the O atoms, the remaining electrons are placed as lone pairs around the O atoms and the S atom.

:Ö—S̈—Ö:

STEP 5   **If octets are not complete, form a multiple bond.**   The octet for S is completed using one lone pair from an O atom as a second bonding pair and forming a double bond. Because the lone pair can come from either O atom, two resonance structures are possible.

:Ö—S̈=O: ⟷ :O=S̈—Ö:

Study Check

Write three resonance structures for SO_3.

Table 10.2 summarizes this method of writing electron-dot formulas for several molecules and ions.

Questions and Problems   **Electron-Dot Formulas**

10.5 Determine the total number of valence electrons in each of the following:
    a. H_2S **b.** I_2
    c. CCl_4 **d.** OH^-

10.6 Determine the total number of valence electrons in each of the following:
    a. SBr_2 **b.** NBr_3
    c. CH_3OH **d.** NH_4^+

10.7 Write the electron-dot formula for each of the following molecules or ions:
    a. HF   **b.** SF_2 **c.** NBr_3 **d.** BH_4^-
                                        H
    e. CH_3OH (methyl alcohol) H C O H
                                 H
                         H H
    f. N_2H_4 (hydrazine) H N N H

10.8 Write the electron-dot formula for each of the following molecules or ions:
  a. H_2O
  b. CCl_4
  c. H_3O^+
  d. SiF_4
  e. CF_2Cl_2
  f. C_2H_6 (ethane)
$$H \quad H$$
$$H\ C\ C\ H$$
$$H \quad H$$

10.9 When is it necessary to write a multiple bond in an electron-dot formula?

10.10 If the available valence electrons for a molecule or polyatomic ions do not complete all of the octets in an electron-dot formula, what should you do?

10.11 What is resonance?

10.12 When does a covalent compound have resonance?

10.13 Write the electron-dot formulas for each of the following molecules or ions:
  a. CO (carbon monoxide)
  b. H_2CCH_2 (ethylene)
  c. H_2CO (C is the central atom)

10.14 Write the electron-dot formulas for each of the following molecules or ions:
  a. HCCH (acetylene)
  b. CS_2 (C is the central atom)
  c. NO^+

10.15 Write resonance structures for each of the following molecules or ions.
  a. $ClNO_2$ (N is the central atom)
  b. OCN^-

10.16 Write resonance structures for each of the following molecules or ions.
  a. HCO_2^- (C is the central atom)
  b. N_2O (N N O)

Table 10.2   Using Valence Electrons to Write Electron-Dot Formulas

Molecule or Polyatomic Ion	Valence Electrons	Bonds to Attach Atoms	Electrons Left	Completed Octets (or H:)	Electron Check
Cl_2	$2(7) = 14$	Cl—Cl	$14 - 2 = 12$	$:\ddot{Cl}—\ddot{Cl}:$	14
HCl	$1 + 7 = 8$	H—Cl	$8 - 2 = 6$	$H—\ddot{Cl}:$	8
H_2O	$2(1) + 6 = 8$	H—O—H	$8 - 4 = 4$	$H—\ddot{O}—H$	8
PCl_3	$5 + 3(7) = 26$	Cl—P—Cl with Cl above	$26 - 6 = 20$	$:\ddot{Cl}—P—\ddot{Cl}:$ with $:\ddot{Cl}:$ above	26
ClO_3^-	$7 + 3(6) + 1 = 26$	$[O—Cl—O$ with O above$]^-$	$26 - 6 = 20$	$[:\ddot{O}—\ddot{Cl}—\ddot{O}:$ with $:\ddot{O}:$ above$]^-$	26
NO_2^-	$5 + 2(6) + 1 = 18$	$[O—N—O]^-$	$18 - 4 - 14$	$[:\ddot{O}—\ddot{N}=\ddot{O}:]^-$ $\updownarrow$ $[:\ddot{O}=\ddot{N}—\ddot{O}:]^-$	18 18

10.3 Shapes of Molecules and Ions (VSEPR Theory)

Now that we have counted valence electrons and written electron-dot formulas, we can look at the three-dimensional shapes of molecules. The shape is important in our understanding of how molecules interact with enzymes or certain antibiotics or produce our sense of taste and smell.

To predict the three-dimensional shape of a molecule, we use the **valence shell electron-pair repulsion (VSEPR) theory.** In the VSEPR theory, the shape of a molecule depends on minimizing repulsions, which means the electron groups around a central atom are arranged as far apart as possible. Once the electron-dot formulas are written, the specific shape of a molecule can be determined.

Two Electron Groups

In $BeCl_2$, there are two chlorine atoms bonded to a central beryllium atom. Because Be has a strong attraction for its valence electrons, it forms a covalent rather than ionic compound. With only two electron groups around the central atom, the electron-dot formula of $BeCl_2$ is an exception to the octet rule. The best arrangement of two electron groups for minimal repulsion is to place them on opposite sides of the Be atom. This gives a **linear** shape and a bond angle of 180° to the $BeCl_2$ molecule.

$$:\ddot{C}l - Be - \ddot{C}l:$$

Another example of a linear molecule is CO_2. In predicting shape, a *double* or *triple* bond is treated as one electron group. Thus, a multiple bond is counted the same as a single electron group in determining electron repulsion. In CO_2, the two double bonds, which are counted as two electron groups around C, are arranged on opposite sides of the C atom. With two double bonds, the shape of the CO_2 molecule is *linear* with a bond angle of 180°.

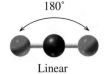

Linear

$$:\ddot{O} = C = \ddot{O}:$$

Three Electron Groups

In BF_3, the central atom B is attached to fluorine atoms by only three groups (another exception to the octet rule). The arrangement of three electron groups as far apart as possible is called **trigonal planar** and has bond angles of 120°. In BF_3, each electron group is bonded to an atom, which gives BF_3 a trigonal planar structure. Thus, the BF_3 molecule is flat with all the atoms in the same plane and 120° bond angles.

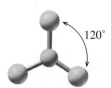

Trigonal planar

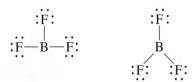

Electron-dot formula   Electron arrangement

In the electron-dot formula for SO_2, there is a single bond, a double bond, and a lone pair of electrons surrounding the S atom. Thus there are three electron groups around the S atom, which assume a trigonal planar arrangement for minimal repulsion. But only the bonded atoms attached to the central atom determine the shape and structure of a molecule. Therefore, with two O atoms bonded to the central S atom, the structure of the SO_2 molecule is a **bent** shape. In the SO_2 molecule, the bond angle is slightly affected by the lone pair but is close to 120°.

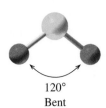

120°
Bent

Electron-dot formula   Electron arrangement

Four Electron Groups

Up to now, the shapes of molecules have been in two dimensions. However, when there are four electron groups around a central atom, the minimum repulsion is obtained by placing four electron groups at the corners of a three-dimensional tetrahedron. A regular tetrahedron consists of four sides that are equilateral triangles.

In CH_4, all four electron pairs around the central C atom are bonded to hydrogen atoms. From the electron-dot formula, CH_4 appears planar with 90° bond angles, but this is not the largest angle possible. The best arrangement for minimal repulsion is **tetrahedral,** with the electron groups going toward the corners of a tetrahedron, which creates bond angles of 109.5°. Because CH_4 has four H atoms attached to the central C, the CH_4 molecule has a tetrahedral shape with bond angles of 109.5°.

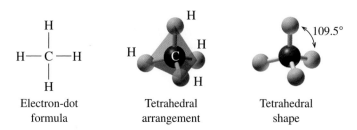

Electron-dot   Tetrahedral   Tetrahedral
formula   arrangement   shape

Now we will look at other molecules with four electron groups but only two or three attached atoms. For example, ammonia, NH_3, has four electron groups around the central nitrogen atom, which means the electron groups have a tetrahedral arrangement. The three electron groups attached to H atoms and the one lone pair occupy the corners of a tetrahedron. Because one corner has no bonded atom, the NH_3 molecule is said to have a **trigonal pyramidal** shape, which is determined by the attached H atoms. In the NH_3 molecule, the bond angles are somewhat affected by the lone pairs but are about 109.5°.

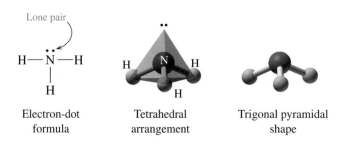

Electron-dot   Tetrahedral   Trigonal pyramidal
formula   arrangement   shape

In water, H_2O, the four electron groups have a tetrahedral arrangement around the central O atom. Two electron groups attached to H atoms and two lone pairs occupy the corners of the tetrahedron. With only two bonded atoms, the H_2O molecule has a bent shape. In the H_2O molecule, the bond angle is somewhat affected by the lone pairs but is close to 109.5°. Table 10.3 gives the molecular shapes for molecules with two, three, and four electron groups.

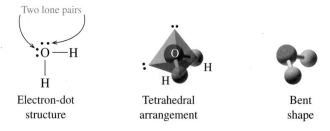

Electron-dot   Tetrahedral   Bent
structure   arrangement   shape

Table 10.3   Molecular Shapes for a Central Atom with Two, Three, and Four Electron Groups

Electron Groups	Bonded Atoms	Lone Pairs	Bond Angle	Molecular Shape	Example	
2	2	0	180°	linear	$BeCl_2$	
3	3	0	120°	trigonal planar	BF_3	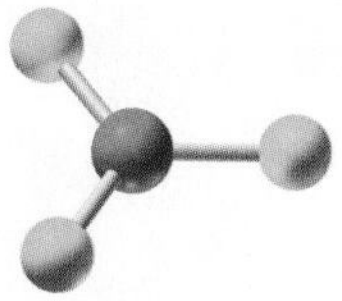
3	2	1	~120°	bent	SO_2	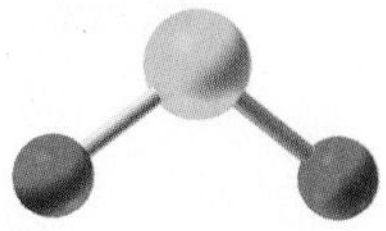
4	4	0	109.5°	tetrahedral	CH_4	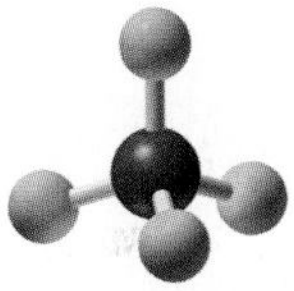
4	3	1	~109.5°	trigonal pyramidal	NH_3	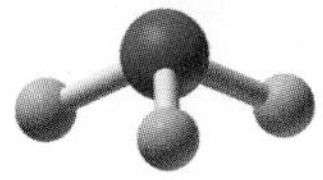
4	2	2	~109.5°	bent	H_2O	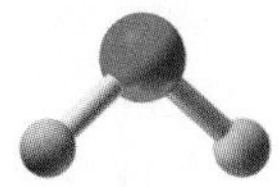

Sample Problem 10.6   Predicting Shapes

Use VSEPR theory to predict the shape of the following molecules or ions:
a. PH_3 **b.** H_2Se **c.** NO_3^-

Solution

a. PH_3

Guide to Predicting Molecular Shape (VSEPR Theory)
STEP 1 Write the electron-dot formula for the molecule.
STEP 2 Arrange the electron groups around the central atom to minimize repulsion.
STEP 3 Use the atoms bonded to the central atom to determine the molecular shape.

STEP 1 **Write the electron-dot formula.** In the electron-dot formula for PH_3, there are four electron groups.

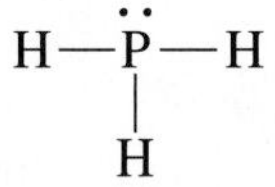

$$H-\overset{\cdot\cdot}{\underset{\underset{H}{|}}{P}}-H$$

STEP 2 **Arrange the electron groups around the central atom to minimize repulsion.** The four electron groups have a tetrahedral arrangement.

STEP 3 **Use the atoms bonded to the central atom to determine the molecular shape.** Three bonded atoms and one lone pair give PH_3 a trigonal pyramidal shape.

b. H_2Se

STEP 1 **Write the electron-dot formula.** In the electron-dot formula for H_2Se, there are four electron groups.

$$:\overset{..}{Se}-H$$
$$|$$
$$H$$

STEP 2 **Arrange the electron groups around the central atom to minimize repulsion.** The four electron groups around Se would have a tetrahedral arrangement.

STEP 3 **Use the atoms bonded to the central atom to determine the molecular shape.** Two bonded atoms and two lone pairs give H_2Se a bent shape.

c. NO_3^-

STEP 1 **Write the electron-dot formula.** For the ion NO_3^-, three electron groups surround the N atom in each resonance structure.

$$\left[:\overset{..}{\underset{..}{O}}-N=\overset{..}{O}:\right]^- \longleftrightarrow \left[:\overset{..}{\underset{..}{O}}-N-\overset{..}{\underset{..}{O}}:\right]^- \longleftrightarrow \left[:\overset{..}{O}=N-\overset{..}{\underset{..}{O}}:\right]^-$$
$$:\overset{..}{\underset{..}{O}}: \qquad\qquad :\overset{..}{\underset{..}{O}}: \qquad\qquad :\overset{..}{\underset{..}{O}}:$$

STEP 2 **Arrange the electron groups around the central atom to minimize repulsion.** A double bond and two single bonds give NO_3^- three electron groups, which would have a trigonal planar arrangement.

STEP 3 **Use the atoms bonded to the central atom to determine the molecular shape.** Three bonded atoms and no lone pairs give NO_3^- a trigonal planar shape.

Study Check

Predict the shape of ClO_2^-.

Questions and Problems   **Shapes of Molecules and Ions (VSEPR Theory)**

10.17 What is the shape of a molecule with each of the following:
   a. Two bonded atoms and no lone pairs
   b. Three bonded atoms and one lone pair

10.18 What is the shape of a molecule with each of the following:
   a. Four bonded atoms
   b. Two bonded atoms and two lone pairs

10.19 In the molecule PCl_3, the four electron groups around the phosphorus atom are arranged in a tetrahedral geometry. However, the shape of the molecule is called trigonal pyramidal. Why does the shape of the molecule have a different name from the name of the electron group geometry?

10.20 In the molecule H_2S, the four electron groups around the sulfur atom are arranged in a tetrahedral geometry. However, the shape of the molecule is called bent. Why does the shape of the molecule have a different name from the name of the electron group geometry?

10.21 Compare the electron-dot formulas of BF_3 and NF_3. Why do these molecules have different shapes?

10.22 Compare the electron-dot formulas of CH_4 and H_2O. Why do these molecules have the same angles but different names for their shapes?

10.23 Use the VSEPR theory to predict the shape of each molecule:
   a. GaH_3 **b.** OF_2 **c.** HCN
   d. CCl_4 **e.** SeO_2

10.24 Use the VSEPR theory to predict the shape of each molecule:
  a. CF_4
  b. NCl_3
  c. SCl_2
  d. CS_2
  e. $BFCl_2$

10.25 Write the electron-dot formula and predict the shape for the polyatomic ions:
  a. CO_3^{2-}
  b. SO_4^{2-}
  c. BH_4^{-}
  d. NO_2^{+}

10.26 Write the electron-dot formula and predict the shape for the polyatomic ions:
  a. NO_2^{-}
  b. PO_4^{3-}
  c. ClO_4^{-}
  d. SF_3^{+}

Learning **Goal**

> **Use electronegativity to determine the polarity of a bond or a molecule.**

the **Chemistry** place

WEB TUTORIAL
Electronegativity

10.4 Electronegativity and Polarity

The ability of an atom to attract bonding electrons to itself is called its **electronegativity.** The electronegativity values assigned to the representative elements are shown in Figure 10.1. Nonmetals have high electronegativity values compared to metals because nonmetals have a greater attraction for electrons than metals. The nonmetals with the highest electronegativity values are fluorine (4.0) at the top of Group 7A (17) and oxygen (3.5) at the top of Group 6A (16). The metals cesium and francium at the bottom of Group 1A (1) have the lowest electronegativity value (0.7). Smaller atoms tend to have higher electronegativity values because the valence electrons they share are closer to their nuclei. Thus, the values of electronegativity generally increase going from left to right across each period of the periodic table and increase going up within each group. The values of electronegativity for the transition metals are also low, but we will not include them in our discussion. Note that there are no electronegativity values for the noble gases because they do not typically form bonds.

Figure 10.1 The electronegativity of representative elements indicates the ability of atoms to attract shared electrons. Electronegativity values increase across a period and going up a group.

 What element on the periodic chart has the strongest attraction for shared electrons?

Electronegativity increases →

H
2.1

Electronegativity decreases ↓

1 Group 1A	2 Group 2A		13 Group 3A	14 Group 4A	15 Group 5A	16 Group 6A	17 Group 7A	18 Group 8A
Li 1.0	Be 1.5		B 2.0	C 2.5	N 3.0	O 3.5	F 4.0	
Na 0.9	Mg 1.2		Al 1.5	Si 1.8	P 2.1	S 2.5	Cl 3.0	
K 0.8	Ca 1.0		Ga 1.6	Ge 1.8	As 2.0	Se 2.4	Br 2.8	
Rb 0.8	Sr 1.0		In 1.7	Sn 1.8	Sb 1.9	Te 2.1	I 2.5	
Cs 0.7	Ba 0.9		Tl 1.8	Pb 1.9	Bi 1.9	Po 2.0	At 2.1	

Figure 10.2 In the nonpolar covalent bond of H_2, electrons are shared equally. In the polar covalent bond of HCl, electrons are shared unequally.

Q *H_2 has a nonpolar covalent bond, but HCl has a polar covalent bond. Explain.*

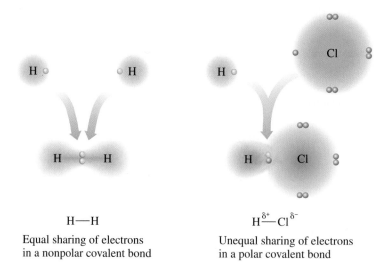

H—H

Equal sharing of electrons
in a nonpolar covalent bond

H^{δ^+}—Cl^{δ^-}

Unequal sharing of electrons
in a polar covalent bond

Polarity of Bonds

Earlier we discussed bonding as either *ionic,* in which electrons are transferred, or *covalent,* in which electrons are equally shared. The difference in the electronegativity values of two atoms gives an indication of the type of bond that forms. In H—H, the electronegativity difference is zero $(2.1 - 2.1 = 0)$, which means the bonding electrons are shared equally between the two hydrogen atoms. A bond between atoms with identical or very similar electronegativities is a **nonpolar covalent bond.** However, most covalent bonds are between different atoms with different electronegativity values. For example, in H—Cl, there is an electronegativity difference of $3.0 - 2.1 = 0.9$. (See Figure 10.2.) When the electrons are shared unequally in a covalent bond, it is called a **polar covalent bond.**

In a polar covalent bond, the shared electrons are attracted to the more electronegative atom, which makes it partially negative. At the other end of the polar bond, the atom with the lower electronegativity becomes partially positive. Because a polar covalent bond has a separation of positive and negative charges, or two poles, it is called a **dipole.** The positive and negative ends of a polar covalent bond are indicated by the lowercase Greek letter delta with a positive or negative sign, δ^+ and δ^-. An arrow pointing from the positive charge to the negative charge (⟵⟶) may be used to indicate the dipole.

Examples of Dipoles in Polar Covalent Bonds

$$\overset{\delta^+}{C}—\overset{\delta^-}{O} \qquad \overset{\delta^+}{N}—\overset{\delta^-}{O} \qquad \overset{\delta^+}{Cl}—\overset{\delta^-}{F}$$
$$\longmapsto \qquad\quad \longmapsto \qquad\quad \longmapsto$$

As the electronegativity difference increases, the shared electrons are attracted more strongly to the more electronegative atom. The **polarity,** which depends on the separation of charges, also increases. Eventually, the difference in electronegativity is great enough that the electrons are

Table 10.4   Predicting Bond Type from Electronegativity Differences

Molecule		Type of Electron Sharing	Electronegativity Difference[a]	Bond Type
H_2	H—H	Shared equally	$2.1 - 2.1 = 0$	Nonpolar covalent
Cl_2	Cl—Cl	Shared equally	$3.0 - 3.0 = 0$	Nonpolar covalent
HBr	$\overset{\delta^+}{H}—\overset{\delta^-}{Br}$	Shared unequally	$2.8 - 2.1 = 0.7$	Polar covalent
HCl	$\overset{\delta^+}{H}—\overset{\delta^-}{Cl}$	Shared unequally	$3.0 - 2.1 = 0.9$	Polar covalent
NaCl	Na^+Cl^-	Electron transfer	$3.0 - 0.9 = 2.1$	Ionic
MgO	$Mg^{2+}O^{2-}$	Electron transfer	$3.5 - 1.2 = 2.3$	Ionic

[a]Values are taken from Figure 10.1.

transferred from one atom to another, which results in an ionic bond. For example, the electronegativity difference for the ionic compound NaCl is $3.0 - 0.9 = 2.1$. Thus for large differences in electronegativity, we would predict an ionic bond. (See Table 10.4.)

The variations in bonding are continuous; there is no definite point at which one type of bond stops and the next starts. However, for purposes of discussion, we can use some general ranges for predicting the type of bond between atoms. (See Table 10.5.) When electronegativity differences are from 0.0 to 0.4, the electrons are shared about equally in a nonpolar covalent bond. For example, H—H ($2.1 - 2.1 = 0$) and C—H ($2.5 - 2.1 = 0.4$) are classified as nonpolar covalent bonds. As electronegativity differences increase, there is also an increase in the polarity of the covalent bond. For example, an O—H bond with a difference of 1.4 is much more polar than an O—F bond with a difference of only 0.5. Differences in electronegativity of 1.8 or greater generally indicate a bond that is mostly ionic. For example, we would classify K—Cl with an electronegativity difference of $3.0 - 0.8 = 2.2$ as an ionic bond resulting from a transfer of electrons.

Table 10.5   Electronegativity Difference and Types of Bonds

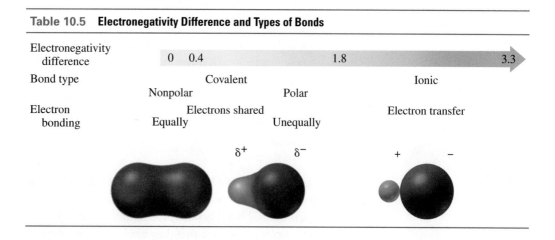

Sample Problem 10.7 ▶ **Bond Polarity**

Use electronegativity values to classify each bond as nonpolar covalent, polar covalent, or ionic.

N—N   O—H   Cl—As   O—K

Solution

For each bond, we obtain the electronegativity values and calculate the difference.

Bond	Electronegativity Difference	Type of Bond
N—N	$3.0 - 3.0 = 0.0$	nonpolar covalent
O—H	$3.5 - 2.1 = 1.4$	polar covalent
Cl—As	$3.0 - 2.0 = 1.0$	polar covalent
O—K	$3.5 - 0.8 = 2.7$	ionic

Study Check

For each of the following pairs, identify the more polar bond:
a. Si—S or Si—N
b. O—P or N—P

Polarity of Molecules

The shape of a molecule and the polarity of its bonds determine the polarity of the molecule. It is the polarity of a molecule that affects its chemical reactivity and physical properties, including solubility and melting and boiling points.

Nonpolar Molecules Diatomic molecules such as H_2 or Cl_2 are nonpolar because they contain one nonpolar covalent bond.

Nonpolar Molecules

H—H   Cl—Cl

Nonpolar covalent bonds

Molecules with two or more polar bonds can also be nonpolar if the polar bonds have a symmetrical arrangement in the molecule. If the polar bonds or dipoles cancel each other, it is a nonpolar molecule. For example, CO_2 contains two polar covalent bonds. Because the CO_2 molecule is linear, the dipoles oppose each other and cancel out. Thus, the CO_2 molecule is nonpolar.

O=C=O

⟵ + +⟶
Dipoles cancel

Another molecule that is nonpolar is CCl_4, which has four chlorine atoms at the corners of a tetrahedron surrounding the central, carbon atom. Although each C—Cl bond is polar, the arrangement of the four dipoles in the tetrahedral molecule is such that they cancel out and the CCl_4 molecule is nonpolar.

The four C—Cl dipoles cancel out

Polar Molecules In a polar molecule, one end of the molecule is more negative than the other end. Polarity in a molecule occurs when the dipoles from the polar bonds do not cancel each other. Polarity depends on the type of atoms and the shape of the electron groups around the central atom. For example, HCl is a polar molecule because it has one polar covalent bond.

Polar molecule

H—Cl

⟶

Dipole does not cancel

In molecules with more atoms, the shape of the molecule determines whether the dipoles cancel or not. For example, H_2O is a polar molecule because its bent shape does not allow the H—O dipoles to cancel.

More negative end of molecule

More positive end of molecule

H_2O is a polar molecule because dipoles do not cancel

The NH_3 molecule, which has a trigonal pyramidal shape, is a polar molecule because the N—H dipoles do not cancel.

More negative end of molecule

More positive end of molecule

NH_3 is a polar molecule because dipoles do not cancel

Sample Problem 10.8   **Polarity of Molecules**

Indicate if each of the following molecules is polar or nonpolar:
a. BF_3 **b.** CH_3F

Solution

a. The molecule BF_3 consists of a central atom B with three polar B—F bonds and no lone pairs. Because it has a trigonal planar shape, the B—F dipoles would cancel out. Thus, BF_3 is a nonpolar molecule.

b. The molecule CH_3F consists of a central atom C with three nonpolar C—H bonds and one polar C—F bond. Although it has a tetrahedral shape, the C—F dipole does not cancel. Thus, CH_3F is a polar molecule.

Study Check

Is the PCl_3 molecule polar or nonpolar?

Questions and Problems >   **Electronegativity and Polarity**

10.27 Describe the trend in electronegativity going across a period.

10.28 Describe the trend in electronegativity going down a group.

10.29 Approximately what electronegativity difference would you expect for a nonpolar covalent bond?

10.30 Approximately what electronegativity difference would you expect for a polar covalent bond?

10.31 Using the periodic table only, arrange the atoms in each of the following sets in order of increasing electronegativity:
  a. Li, Na, K
  b. Na, P, Cl
  c. Ca, Br, O

10.32 Using the periodic table only, arrange the atoms in each of the following sets in order of increasing electronegativity:
  a. F, Cl, Br
  b. B, N, O
  c. Mg, S, F

10.33 For each of the following bonds, indicate the positive end with δ^+ and the negative end with δ^-. Write an arrow to show the dipole for each.
  a. N—F   **b.** Si—P
  c. C—O   **d.** P—Br
  e. B—Cl

10.34 For each of the following bonds, indicate the positive end with δ^+ and the negative end with δ^-. Write an arrow to show the dipole for each.
  a. Si—Br   **b.** Se—F
  c. Br—F   **d.** N—H
  e. N—P

10.35 Predict whether each of the following bonds is ionic, polar covalent, or nonpolar covalent:
  a. Si—Br   **b.** Li—F
  c. Br—F   **d.** Br—Br
  e. N—P   **f.** C—P

10.36 Predict whether each of the following bonds is ionic, polar covalent, or nonpolar covalent:
  a. Si—O   **b.** K—Cl
  c. S—F   **d.** P—Br
  e. Li—O   **f.** N—P

10.37 Why is F_2 a nonpolar molecule, but HF is a polar molecule?

10.38 Why is CBr_4 a nonpolar molecule, but NBr_3 is a polar molecule?

10.39 Identify the following molecules as polar or nonpolar:
  a. CS_2 **b.** NF_3
  c. Br_4 **d.** SO_3

10.40 Identify the following molecules as polar or nonpolar:
  a. H_2S **b.** PBr_3
  c. $SiCl_4$ **d.** SO_2

10.5 Attractive Forces in Compounds

Learning **Goal**

Describe the attractive forces between ions, polar molecules, and nonpolar molecules.

the **Chemistry** place

WEB TUTORIAL

Intermolecular Forces

In gases, the interactions between particles are minimal, which allows gas molecules to move far apart from each other. In solids and liquids, there are sufficient interactions between the particles to hold them close together, although some solids have low melting points while others have very high melting points. Such differences in properties are explained by looking at the various kinds of attractive forces between particles.

In ionic compounds, the positive and negative ions are held together by ionic bonds. These are strong attractive forces that require large amounts of energy to pull the ions apart and melt the ionic solid. As a result, ionic solids have high melting points. For example, the ionic solid NaCl melts at 801°C. In compounds held together by covalent bonds, there are other types of attractive forces that exist in the solid. These attractive forces, which are weaker than ionic bonds, include dipole–dipole attractions, hydrogen bonding, and dispersion forces.

Dipole–Dipole and Hydrogen Bonds

For polar molecules, there are attractive forces, called **dipole–dipole attractions,** that occur between the positive end of one molecule and the negative end of another. For example, in a sample of HCl, the positive hydrogen end of one dipole attracts the negative chlorine atom in another molecule.

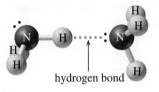

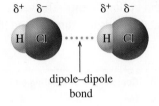

dipole–dipole
bond

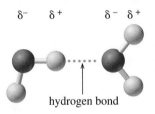

hydrogen bond

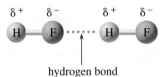

hydrogen bond

In certain polar molecules, strong dipoles occur when a hydrogen atom is attached to an atom of fluorine, oxygen, or nitrogen, all of which have high electronegativity values. In a special type of dipole–dipole attraction called a **hydrogen bond,** a strong attractive force occurs between the partially positive hydrogen atom and a lone pair of electrons on another N, O, or F atom. Hydrogen bonds are a major factor in the formation and structure of biological molecules such as proteins and DNA.

Hydrogen bonds are the strongest type of attractive forces between polar molecules. When we compare compounds with similar masses in Table 10.6, we see that polar compounds have higher melting and boiling points than nonpolar compounds. The polar O—H bond in ethanol provides hydrogen bonding between the ethanol molecules. As a result, ethanol requires more energy and higher temperatures to form a liquid and a gas than propane. In formic acid, two oxygen atoms provide more polarity to the molecule and more opportunities for hydrogen bonding. Thus, formic acid has the highest melting point and boiling point of the three compounds.

Table 10.6   Comparison of Boiling Points of Polar and Nonpolar Covalent Molecules

Compound	Formula	Molar Mass	Type of Compound	Attractive Force	Boiling Point (°C)
Propane	$CH_3—CH_2—CH_3$	44	Nonpolar	Dispersion	−42
Methyl chloride	$CH_3—Cl$	50	Polar	Dipole–dipole	24
Ethanol	$CH_3—CH_2—O—H$	46	Polar	Hydrogen bond	79
Formic acid	$H—\overset{\overset{O}{\|\|}}{C}—OH$	46	Polar	Hydrogen bonds	101

Table 10.7	Melting Points of Selected Substances

Substance	Melting Point(°C)
Ionic Bonds	
MgF_2	1248
KF	860
NaCl	801
Hydrogen bonds	
H_2O	0
NH_3	−78
HF	−83
Dipole–dipole interactions	
HI	−51
HBr	−89
HCl	−115
Dispersion forces	
I_2	114
Br_2	−7
Cl_2	−101
F_2	−220
H_2	−259
C_5H_{12}	−130
CH_4	−182

A comparison of the melting points and types of attractive forces of some other substances is shown in Table 10.7.

Dispersion Forces

Nonpolar compounds do form solids, but at low temperatures. Very weak attractions called **dispersion forces** occur when temporary dipoles form within nonpolar molecules. Usually, the electrons in a nonpolar molecule are distributed symmetrically. However, at any moment, the motion of the electrons may cause more electrons to be present at one end of the molecule, which forms a temporary dipole. Although dispersion forces are very weak, they make it possible for nonpolar molecules to form liquids and solids. As the size of the nonpolar compound increases, there are more electrons that produce stronger temporary dipoles. Larger nonpolar molecules have higher melting and boiling points. The various types of attractions between particles in solids and liquids are summarized in Table 10.8.

Table 10.8   Comparison of Bonding and Attractive Forces

Type of Force	Particle Arrangement	Energy (kJ/mol)	Example
Between atoms or ions			
Ionic bond		500–5000	$Na^+ \cdots Cl^-$
Covalent bond (X = nonmetal)		100–1000	$Cl-Cl$
Between molecules			
Hydrogen bond (X = F, O, or N)		10–40	$H-F \cdots H-F$
Dipole–dipole (X and Y = different nonmetals)		5–20	$Br-Cl \cdots Br-Cl$
Dispersion (Temporary shift of electrons in nonpolar bonds)		1–10	$F-F \cdots F-F$

Sample Problem 10.9   Attractive Forces Between Particles

Indicate the major type of molecular interaction
a. dipole–dipole   **b.** hydrogen bonding
c. dispersion forces

expected of each of the following:
1. $H-F$ **2.** $F-F$ **3.** PCl_3

Solution

1. **(b)** H—F is a polar molecule that interacts with other H—F molecules by hydrogen bonding.
2. **(c)** Because F—F is nonpolar, only dispersion forces provide attractive forces.
3. **(a)** The polarity of the PCl_3 molecules provides dipole–dipole interactions.

Study Check

Why is the boiling point of H_2S lower than that of H_2O?

Questions and Problems   **Attractive Forces in Compounds**

10.41 Identify the major type of interactive force in each of the following substances:
  a. BrF   **b.** KCl
  c. CCl_4 **d.** HF
  e. Cl_2

10.42 Identify the major type of interactive force in each of the following substances:
  a. HCl   **b.** MgF_2
  c. PBr_3 **d.** Br_2
  e. NH_3

10.43 Identify the strongest attractive forces between each of the following:
  a. CH_3OH **b.** Cl_2
  c. HCl   **d.** CCl_4
  e. CH_3CH_3

10.44 Identify the strongest attractive forces between each of the following:
  a. O_2 **b.** HF
  c. CH_3Cl **d.** H_2O
  e. NH_3

10.6 Matter and Changes of State

Learning **Goal**

Identify the states of matter as solid, liquid, or gas and calculate the energy involved in changes of state.

We can now use the information on the types of attractions between particles to discuss the states of matter: typically, *gases, liquids,* and *solids.* Solids change to liquids and liquids change to gases when sufficient energy is added to overcome the attractive forces between the particles. In a **solid,** very strong attractive forces hold the particles close together. They are arranged in such a rigid pattern that they can only vibrate slowly in their fixed positions. This gives a solid a definite shape and volume. For many solids, this rigid structure produces a crystal, as seen in amethyst.

In a **liquid,** the particles have enough energy to move freely in random directions. They are still close to each other with strong attractions to maintain a definite volume, but there is no rigid structure. Thus, when oil, water, or vinegar is poured from one container to another, the liquid maintains its own volume but takes the shape of the new container (see Figure 10.3).

The air you breathe is made of gases, mostly nitrogen and oxygen. In a **gas,** the molecules move at high speeds, creating great distances between molecules. This behavior allows gases to fill their container. Gases have no definite shape or volume of their own; they take the shape and volume of their container. Table 10.9 compares some of the properties of the three states of matter.

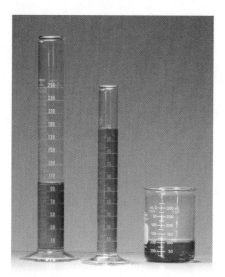

Figure 10.3 A liquid with a volume of 100 mL takes the shape of its container.

Q Why does a liquid have a definite volume, but not a definite shape?

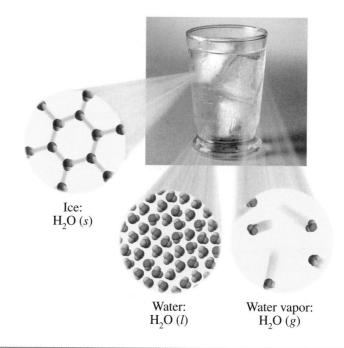

Ice:
H_2O (s)

Water:
H_2O (l)

Water vapor:
H_2O (g)

Table 10.9   **Some Properties of Solids, Liquids, and Gases**

Property	Solid	Liquid	Gas
Shape	Has a definite shape	Takes the shape of the container	Takes the shape of its container
Volume	Has a definite volume	Has a definite volume	Fills the volume of the container
Arrangement of particles	Fixed, very close	Random, close	Random, far apart
Interaction between particles	Very strong	Strong	Essentially none
Movement of particles	Very slow	Moderate	Very fast
Examples	Ice, salt, iron	Water, oil, vinegar	Water vapor, helium, air

Changes of State

Matter undergoes a **change of state** when it is converted from one state to another. When an ice cube melts in a drink, a change of state has occurred.

When heat is added to a solid, the particles in the rigid structure move faster. At a temperature called the **melting point (mp),** the particles in the solid gain sufficient energy to overcome the energy of the attractive forces that hold them together. The particles in the solid separate and move about in random patterns. The substance is **melting,** changing from a solid to a liquid.

If the temperature is lowered, the reverse process takes place. Kinetic energy is lost, the particles slow down, and attractive forces pull the particles closer together. The substance is **freezing.** A liquid changes to a solid at the **freezing point (fp),** which is the same temperature as the melting point. Every substance has its own freezing (melting) point: water freezes at 0°C; ice melts at 0°C; gold freezes (melts) at 1064°C; nitrogen freezes (melts) at −210°C.

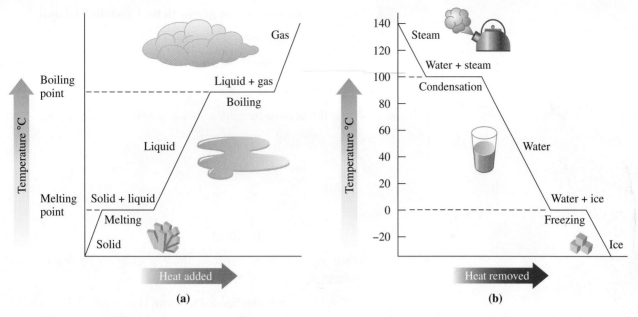

(a)   **(b)**

Figure 10.7 (a) A heating curve diagrams the temperature increases and changes in state as heat is added. **(b)** A cooling curve for water.

Q *What does the plateau at 100°C represent on the cooling curve for water?*

Steps on a Cooling Curve When heat is removed from a substance such as water vapor, the gas cools and condenses. A **cooling curve** is a diagram of this cooling process. (See Figure 10.7b.) On the cooling curve, the temperature is plotted on the vertical axis and the removal of heat on the horizontal axis. As a gas is cooled, heat is lost and the temperature drops. At the condensation point (same as the boiling point), the gas begins to condense, forming liquid. This process is indicated by a flat line (plateau) on the cooling curve at the condensation point.

After all of the gas has changed into liquid, the particles within the liquid cool, as indicated by the downward-sloping line that shows the temperature decrease. At the freezing point, the particles in the liquid slow so much that a solid begins to form. A second flat line at the freezing point indicates the change of state from liquid to solid (freezing). Once all of the substance is frozen, more heat can be lost, which lowers the solid's temperature below its freezing point.

Combining Energy Calculations

Up to now, we have calculated one step in a heating or cooling curve. However, many problems require a combination of steps that include a temperature change as well as a change of state. The heat is calculated for each step separately and then added together to find the total energy, as seen in Sample Problem 10.13.

Sample Problem 10.13 ⟩ **Combining Heat Calculations**

Calculate the total heat in joules needed to convert 15.0 g liquid water at 25.0°C to steam at 100°C.

Solution

STEP 1 **List grams of substance and change of state.**

Given 15.0 g water at 25°C

Need Heat (J) needed to warm water and change to steam.

STEP 2   **Write the plan to convert grams to heat and desired unit.**

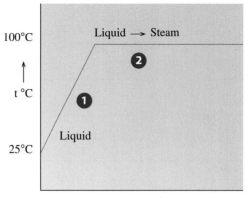

When several changes occur, draw a diagram of heating and changes of state.

Total heat = joules needed to warm H_2O from 25°C to 100°C
+ joules to change liquid to steam at 100°C

STEP 3   **Write the heat conversion factors needed.**

Equalities/Conversion Factors

$$SH_{H_2O} = 4.184 \frac{J}{g \, °C}$$

$$\frac{4.184 \, J}{g \, °C} \quad \text{and} \quad \frac{g \, °C}{4.184 \, J}$$

$$1 \, g \, H_2O \, (l \rightarrow g) = 2260 \, J$$

$$\frac{2260 \, J}{1 \, g \, H_2O} \quad \text{and} \quad \frac{1 \, g \, H_2O}{2260 \, J}$$

STEP 4   **Set up problem with factors.**

Problem Setup:

$$\Delta T = 100°C - 25°C = 75°C$$

Heat needed to warm H_2O (25°C) to H_2O (100°C):

$$15.0 \, g \times 75°C \times \frac{4.184 \, J}{g \, °C} = 4700 \, J$$

Heat needed to change H_2O (liquid) to H_2O (gas) at 100°C:

$$15.0 \, g \times \frac{2260 \, J}{1 \, g} = 33\,900 \, J$$

Calculate the total heat:

Heating water	4 700 J
Changing liquid to steam (100°C)	33 900 J
Total heat needed	38 600 J

Study Check

How many kilojoules (kJ) are released when 75.0 g steam at 100°C condenses, cools to 0°C, and freezes? (*Hint:* The solution will require three energy calculations.)

Questions and Problems   **Matter and Changes of State**

10.45 Indicate whether each of the following describes a gas, a liquid, or a solid:
  a. This substance has no definite volume or shape.
  b. The particles in this substance do not interact strongly with each other.
  c. The particles of this substance are held in a definite structure.

10.46 Indicate whether each of the following describes a gas, a liquid, or a solid:
  a. The substance has a definite volume but takes the shape of the container.
  b. The particles of this substance are very far apart.
  c. This substance occupies the entire volume of the container.

10.47 Calculate the heat needed at 0°C in each of the following and indicate whether heat was absorbed or released:
  a. joules to melt 65.0 g ice
  b. joules to melt 17.0 g ice
  c. kilojoules to freeze 225 g water
  d. kilojoules to freeze 50.0 g water

10.48 Calculate the heat needed at 0°C in each of the following and indicate whether heat was absorbed or released:
  a. joules to freeze 35.2 g water
  b. joules to freeze 275 g water
  c. kilojoules to melt 145 g ice
  d. kilojoules to melt 5.00 kg ice

10.49 For each of the following problems, calculate the heat change at 100°C and indicate whether heat was absorbed or released:
  a. joules to vaporize 10.0 g water
  b. kilojoules to vaporize 50.0 g water
  c. joules to condense 8.00 kg steam
  d. kilojoules to condense 175 g steam

10.50 For each of the following problems, calculate the heat change at 100°C and indicate whether heat was absorbed or released:

  a. joules to condense 10.0 g steam
  b. kilojoules to condense 76.0 g steam
  c. joules to vaporize 44.0 g water
  d. kilojoules to vaporize 5.0 kg water

10.51 Using the values for the heat of fusion, specific heat of water, and/or heat of vaporization, calculate the amount of heat energy in each of the following:
  a. joules needed to warm 20.0 g water at 15°C to 72°C
  b. joules needed to melt 50.0 g ice at 0°C and to warm the liquid to 65.0°C
  c. kilojoules released when 15.0 g steam condenses at 100°C and the liquid cools to 0°C
  d. kilojoules need to melt 24.0 g ice at 0°C, warm the liquid to 100°C, and change it to steam at 100°C

10.52 Using the values for the heat of fusion, specific heat of water, and/or heat of vaporization, calculate the amount of heat energy in each of the following:
  a. joules to condense 125 g steam at 100°C and to cool the liquid to 15.0°C
  b. joules needed to melt a 525-g ice sculpture at 0°C and to warm the liquid to 15.0°C
  c. kilojoules released when 85.0 g steam condenses at 100°C, and to cool the liquid and freeze it at 0°C
  d. joules to warm 55.0 mL water (density = 1.00g/mL) from 10.0°C to 100°C and vaporize it at 100°C

10.53 An ice bag containing 275 g ice at 0°C was used to treat sore muscles. When the bag was removed, the ice had melted and the liquid water had a temperature of 24.0°C. How many kilojoules of heat were absorbed?

10.54 A 115-g sample of steam at 100°C is emitted from a volcano. It condenses, cools, and falls as snow at 0°C. How many kilojoules of heat were released?

Concept Map   **Molecular Structure: Solids and Liquids**

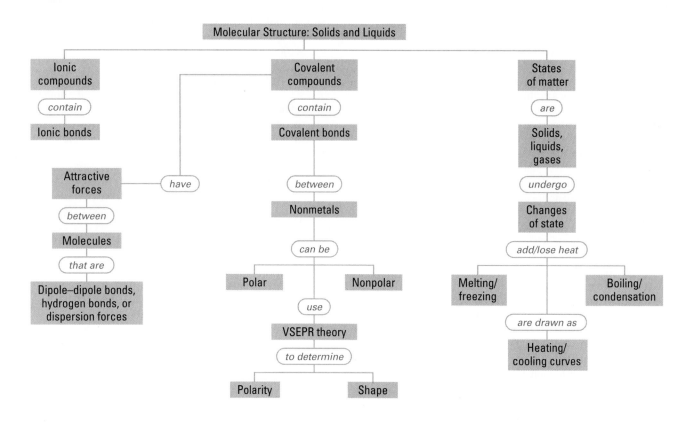

Chapter Review

10.1 Electron Configurations of Ionic Compounds

In ionic compounds, metals of representative elements lose electrons and nonmetals gain electrons to give a $s^2 p^6$ configuration for the valence electrons. The transition metals first lose their s valence electrons to form $2+$ ions, but can also lose 1 or more d electrons to form ions with charges of $3+$ and higher. Thus, the cations of transition metals do not typically acquire the electron configuration of a noble gas.

10.2 Electron-Dot Formulas

In a covalent bond, atoms of nonmetals share valence electrons. In most covalent compounds, atoms achieve a noble gas configuration. The number of valence electrons are added up for all the atoms in the molecule or ion. Any negative charge is added to the total valence electrons, while any positive charge is subtracted. In the electron-dot formulas, a bonding pair is placed between the central atom and attached atoms. The remaining valence electrons are used as lone pairs to complete the octets of the surrounding atoms and then the central atom.

When octets are not completed, lone pairs are converted to bonding pairs forming double or triple bonds. Resonance

structures are possible when two or more electron-dot formulas can be drawn for a molecule or ion with a multiple bond.

10.3 Shapes of Molecules and Ions (VSEPR Theory)

The shape of a molecule is determined from the electron-dot formula and the number of bonded atoms and lone pairs. The electron arrangement of two electron groups around a central atom is linear; three electron groups are trigonal planar; and four are tetrahedral. When all the electron groups are bonded to atoms, the shape has the same name as the electron arrangement. A central atom with two bonded atoms and one or two lone pairs has a bent shape. A central atom with three bonded atoms and one lone pair has a trigonal pyramidal shape.

10.4 Electronegativity and Polarity

Electronegativity is the ability of an atom to attract the electrons it shares with another atom. In general, the electronegativities of metals are low, while nonmetals have high electronegativities. In a nonpolar covalent bond, atoms share electrons equally. In a polar covalent bond, the electrons are unequally shared because they are attracted to the more electronegative atom. The atom in a polar bond with the lower electronegativity is partially positive (δ^+) and the atom with the higher electronegativity is partially negative (δ^-). Atoms that form ionic bonds have large differences in electronegativity values.

Nonpolar molecules contain nonpolar covalent bonds or have an arrangement of bonded atoms that causes the dipoles to cancel out. In polar molecules, the dipoles do not cancel because there are nonidentical bonded atoms or lone pairs on the central atom.

10.5 Attractive Forces in Compounds

In ionic solids, oppositely charged ions are held in a rigid structure by ionic bonds. Attractive forces called dipole–dipole attractions and hydrogen bonds hold the solid and liquid states of polar covalent compounds together. Nonpolar compounds form solids and liquids by temporary dipoles called dispersion forces.

10.6 Matter and Changes of State

The three states of matter are solid, liquid, and gas. A substance undergoes a physical change when its shape, size, or state changes, but its identity does not change. Melting occurs when the particles in a solid absorb enough energy to break apart and form a liquid. The amount of energy required to convert exactly 1 g solid to liquid is called the heat of fusion. For water, 334 J are needed to melt 1 g ice or must be removed to freeze 1 g water. Sublimation is a process whereby a solid changes directly to a gas.

Evaporation occurs when particles in a liquid state absorb enough energy to break apart and form gaseous particles. Boiling is the vaporization of liquid at its boiling point. The heat of vaporization is the amount of heat needed to convert exactly 1 g liquid to vapor. For water, 2260 J are needed to vaporize 1 g water or must be removed to condense 1 g steam.

A heating or cooling curve illustrates the changes in temperature and state as heat is added to or removed from a substance. Plateaus on the graph indicate changes of state. The total heat absorbed or removed from a substance undergoing temperature changes and changes of state is the sum of energy calculations for change(s) of state and change(s) in temperature.

Key Terms

bent The shape of a molecule with two bonded atoms and one lone pair or two lone pairs.

boiling The formation of bubbles of gas throughout a liquid.

boiling point (bp) The temperature at which a substance exists as a liquid and gas; liquid changes to gas (boils), and gas changes to liquid (condenses).

change of state The transformation of one state of matter to another, for example, solid to liquid, liquid to solid, liquid to gas.

condensation The change of state of a gas to a liquid.

cooling curve A diagram that illustrates temperature changes and changes of states for a substance as heat is removed.

deposition The change of a gas directly to a solid; the reverse of sublimation.

dipole The separation of positive and negative charge in a polar bond indicated by an arrow that is drawn from the more positive atom to the more negative atom.

dipole–dipole attractions Attractive forces between oppositely charged ends of polar molecules.

dispersion forces Weak dipole bonding that results from a momentary polarization of nonpolar molecules in a substance.

double bond A sharing of two pairs of electrons by two atoms.

electronegativity The relative ability of an element to attract electrons in a bond.

evaporation The formation of a gas (vapor) by the escape of high-energy molecules from the surface of a liquid.

freezing A change of state from liquid to solid.

freezing point (fp) The temperature at which the solid and liquid forms of a substance are in equilibrium; a liquid changes to a solid (freezes), a solid changes to a liquid (melts).

gas A state of matter characterized by no definite shape or volume. Particles in a gas move rapidly.

heat of fusion The energy required to melt exactly 1 g of a substance. For water, 334 J are needed to melt 1 g ice; 334 J are released when 1 g water freezes.

heat of vaporization The energy required to vaporize 1 g of a substance. For water, 2260 J are needed to vaporize exactly 1 g liquid; 1 g steam gives off 2260 J when it condenses.

heating curve A diagram that shows the temperature changes and changes of state of a substance as it is heated.

hydrogen bond The attraction between a partially positive H and a strongly electronegative atom of F, O, or N.

linear The shape of a molecule that has two bonded atoms and no lone pair.

liquid A state of matter that takes the shape of its container but has a definite volume.

melting The conversion of a solid to a liquid.

melting point (mp) The temperature at which a solid becomes a liquid (melts). It is the same temperature as the freezing point.

nonpolar covalent bond A covalent bond in which the electrons are shared equally.

polar covalent bond A covalent bond in which the electrons are shared unequally.

polarity A measure of the unequal sharing of electrons, indicated by the difference in electronegativity values.

resonance structures Two or more electron-dot formulas that can be written for a molecule or ion by placing a multiple bond between different atoms.

solid A state of matter that has its own shape and volume.

sublimation The change of state in which a solid is transformed directly to a gas without forming a liquid.

tetrahedral The shape of a molecule with four bonded atoms.

trigonal planar The shape of a molecule with three bonded atoms and no lone pair.

trigonal pyramidal The shape of a molecule that has three bonded atoms and one lone pair.

triple bond A sharing of three pairs of electrons by two atoms.

valence shell electron-pair repulsion (VSEPR) theory A theory that predicts the shape of a molecule by moving the electron groups on a central atom as far apart as possible to minimize the repulsion of the negative regions.

Understanding the Concepts

10.55 Identify the major attractive force present in each of the following:
   a. Br_2 **b.** HF   **c.** HBr   **d.** Kr

10.56 Identify the major attractive force present in each of the following:
   a. Cl_2
   b. CH_3—OH
   c. CH_3—CH_3
   d. CH_3—CH_2—Cl

10.57 Why would BCl_3 be a nonpolar molecule when PCl_3 is a polar molecule?

10.58 Why would CO_2 be a nonpolar molecule when SO_2 is a polar molecule?

10.59 Use your knowledge of changes of states to explain the following.
   a. How does perspiration during heavy exercise cool the body?
   b. Why do towels dry more quickly on a hot summer day than on a cold winter day?
   c. Why do wet clothes stay wet in a plastic bag?

10.60 Use your knowledge of changes of states to explain the following.

   a. Why is a spray that evaporates quickly, such as ethyl chloride, used to numb a sports injury during a game?
   b. Why does water in a wide, flat, shallow dish evaporate more quickly than the same amount of water in a tall, narrow vase?
   c. Why does a sandwich on a plate dry out faster than a sandwich in plastic wrap?

10.61 Draw a heating curve for a sample of ice that is heated from $-20°C$ to $130°C$. Indicate the segment of the graph that corresponds to each of the following:
   a. solid   **b.** melting point
   c. liquid   **d.** boiling point
   e. gas

10.62 Draw a cooling curve for a sample of steam that cools from $110°C$ to $-10°C$. Indicate the segment of the graph that corresponds to each of the following:
   a. solid   **b.** melting point
   c. liquid   **d.** boiling point
   e. gas

10.63 The following is a heating curve for chloroform, a solvent for fats, oils, and waxes.

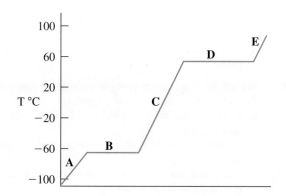

a. What is the melting point of chloroform?
b. What is the boiling point of chloroform?
c. On the heating curve, identify the segments A, B, C, D, and E as solid, liquid, gas, melting, or boiling.

d. At the following temperatures, is chloroform a solid, liquid, or gas: $-80°C$; $-40°C$; $25°C$; $80°C$?

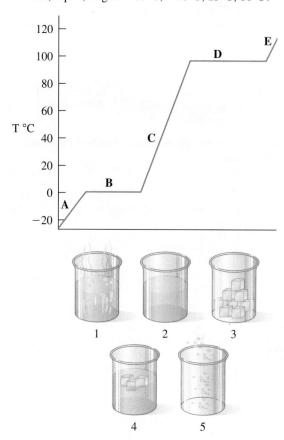

Additional Questions and Problems

10.65 Determine the total number of valence electrons in each of the following:
a. CS_2 **b.** CH_3CHO
c. PH_4^+ **d.** BCl_3
e. SO_3^{2-}

10.66 Determine the total number of valence electrons in each of the following:
a. $COCl_2$ **b.** N_2O
c. ClO_2^- **d.** $SeCl_2$
e. PBr_3

10.67 Write the electron-dot formula for each of the following:
a. BF_4^- **b.** Cl_2O
c. H_2NOH (N is the central atom)   **d.** NO_2^+
e. H_2CCCl_2

10.68 Write the electron-dot formula for each of the following:
a. H_3COCH_3 (the atoms are in the order C O C)
b. CS_2 (the atoms are in the order S C S)
c. ClO_2^-
d. SF_3^+
e. H_2CCHCN (the atoms are in the order C C C)

10.69 Write resonance structures for each of the following:
a. N_2O (the atoms are bonded in the order N N O)
b. NO_2^+ **c.** CNS^-

10.70 Write resonance structures for each of the following:
a. NO_3^- **b.** CO_3^{2-} **c.** OCN^-

10.71 Write the electron-dot formula and determine the shape for each of the following:
a. NF_3 **b.** $SiBr_4$
c. $BeCl_2$ **d.** SO_2

10.72 Write the electron-dot formula and determine the shape for each of the following:
a. NH_4^+ **b.** O_2^{2-}
c. $COCl_2$ (C is the central atom)   **d.** BCl_3

10.73 Use the periodic table to arrange the following atoms in order of increasing electronegativity:
a. I , F, Cl   **b.** Li, K, S, Cl
c. Mg, Sr, Ba, Be

10.74 Use the periodic table to arrange the following atoms in order of increasing electronegativity:
a. Cl, Br, Se   **b.** Na, Cs, O, S
c. O, F, B, Li

10.75 Select the more polar bond in each of the following pairs:
- **a.** C—N or C—O
- **b.** N—F or N—Br
- **c.** Br—Cl or S—Cl
- **d.** Br—Cl or Br—I
- **e.** N—S or N—O

10.76 Select the more polar bond in each of the following pairs:
- **a.** C—C or C—O
- **b.** P—Cl or P—Br
- **c.** Si—S or Si—Cl
- **d.** F—Cl or F—Br
- **e.** P—O or P—S

10.77 Show the dipole arrow for each of the following bonds:
- **a.** Si—Cl
- **b.** C—N
- **c.** F—Cl
- **d.** C—F
- **e.** N—O

10.78 Show the dipole arrow for each of the following bonds:
- **a.** C—O
- **b.** N—F
- **c.** O—Cl
- **d.** S—Cl
- **e.** P—F

10.79 Classify each of the following bonds as nonpolar covalent, polar covalent, or ionic:
- **a.** Si—Cl
- **b.** C—C
- **c.** Na—Cl
- **d.** C—H
- **e.** F—F

10.80 Classify each of the following bonds as nonpolar covalent, polar covalent, or ionic:
- **a.** C—N
- **b.** Cl—Cl
- **c.** K—Br
- **d.** H—H
- **e.** N—F

10.81 Use the electron-dot formula to determine the shape for each of the following molecules and ions:
- **a.** BrO_2^-
- **b.** H_2O
- **c.** CO_3^{2-}
- **d.** CF_4
- **e.** CS_2
- **f.** PO_3^{3-}
- **g.** NH_4^+

10.82 Use the electron-dot formula to determine the shape for each of the following molecules and ions:
- **a.** PH_3
- **b.** NO_3^-
- **c.** HCN
- **d.** ClO_2^-
- **e.** SO_3^{2-}
- **f.** SF_3^+
- **g.** ClO_4^-

10.83 Classify the following molecules as polar or nonpolar:
- **a.** HBr
- **b.** SiO_2
- **c.** NCl_3
- **d.** CH_3Cl
- **e.** NF_3
- **f.** H_2O

10.84 Classify the following molecules as polar or nonpolar:
- **a.** GeH_4
- **b.** Br_2
- **c.** CF_3Cl
- **d.** PCl_3
- **e.** BF_3
- **f.** SCl_2

10.85 Predict the shape and polarity of each of the following molecules. Assume all bonds are polar.
- **a.** a central atom with three identical bonded atoms and no lone pair
- **b.** a central atom with two bonded atoms and one lone pair
- **c.** a central atom with two identical atoms and no lone pairs

10.86 Predict the shape and polarity of each of the following molecules. Assume all bonds are polar.
- **a.** a central atom with four identical bonded atoms and no lone pairs
- **b.** a central atom with three identical bonded atoms and one lone pair
- **c.** a central atom with four bonded atoms that are not identical and no lone pair

10.87 Indicate the major type of attractive force—(1) ionic, (2) dipole–dipole, (3) hydrogen bond, (4) dispersion forces—that occurs between particles of the following substances:
- **a.** NH_3
- **b.** HF
- **c.** CH_4
- **d.** $CHCl_3$
- **e.** H_2O
- **f.** LiCl

10.88 Describe the type of compound that could have each of the following types of attractive forces:
- **a.** dipole–dipole
- **b.** hydrogen bonds
- **c.** dispersion forces

10.89 When it rains or snows, the air temperature seems warmer. Explain.

10.90 **a.** Water is sprayed on the ground of an orchard when temperatures are near freezing to keep the fruit from freezing. Explain.
- **b.** How many kilojoules of energy are released if 5.0 kg water at 15°C is sprayed on the ground and cools and freezes at 0°C?

10.91 An ice cube tray holds 325 g water. If the water initially has a temperature of 25°C, how many kilojoules of heat must be removed to cool and freeze the water at 0°C?

10.92 An ice cube at 0°C with a mass of 115 g is added to H_2O in a beaker that has a temperature of 64.0°C. If the final temperature of the mixture is 24.0°C, what was the initial mass of the warm water?

Challenge Questions

10.93 Complete the electron-dot formula for each of the following:

a. H—N—H with H above N and O above C, C—H
(structure: H—N—C—H with H on N, O double/single on C)

b. Cl—C—C—N with H above and H below the first C

c. H—N—N—H

d. Cl—C—O—C—H with O above first C and H above and below second C

10.94 Identify the errors in each of the following electron-dot formulas and draw the correct formula:

a. :C̈l=O=C̈l:

b. H—C—H with :Ö: above C

c. H—N̈=Ö—H with H below N

10.95 Predict the shape of each of the following molecules or ions:

a. NH_2Cl **b.** PH_4^+
c. SCN^- **d.** SO_3

10.96 Classify each of the following molecules as polar or nonpolar:

a. BF_3 **b.** N_2
c. CS_2 **d.** NH_2Cl

10.97 A 3.0-kg block of lead is taken from a furnace at 300.°C and placed on a large block of ice at 0°C. The specific heat of lead is 0.13 J/g°C. If all the heat given up by the lead is used to melt ice, how much ice is melted when the temperature of the lead drops to 0°C?

10.98 The melting point of benzene is 5.5°C and its boiling point is 80.1°C. Sketch a heating curve for benzene from 0°C to 100°C.
a. What is the state of benzene at 15°C?
b. What happens on the curve at 5.5°C?
c. What is the state of benzene at 63°C?
d. What is the state of benzene at 98°C?
e. At what temperature will both liquid and gas be present?

10.99 A 45.0-g piece of ice at 0.0°C is added to a sample of water at 8.0°C All of the ice melts and the temperature of the water decreases to 0.0°C. How many grams of water were in the sample?

10.100 Identify the most important type of attractions between molecules for each of the following and identify the compounds with the highest and the lowest boiling point:
a. propane, C_3H_8 **b.** methanol, CH_3OH
c. bromine, Br_2 **d.** HBr
e. iodine bromide, IBr

Answers

Answers to Study Checks

10.1 $N^{3-}:1s^22s^22p^6$; $P^{3-}:1s^22s^22p^63s^23p^6$

10.2 :C̈l:Ö:C̈l: or   :C̈l—Ö—C̈l:

10.3 $\left[H:\ddot{N}:H \right]^-$ or $\left[H—\ddot{N}—H \right]^-$

10.4 H:C:::N: or   H—C≡N: In HCN, there is a triple bond between C and N atoms.

10.5 :Ö—S̈=Ö: ⟷ :Ö—S—Ö: ⟷ :Ö=S—Ö:
with :Ö: below each S

10.6 ClO_2^- is bent.

10.7 a. Si—N
b. O—P

10.8 The PCl_3 molecule is trigonal pyramidal, which makes it a polar molecule.

10.9 H_2O forms hydrogen bonds.

10.10 41 800 J removed

10.11 Gas changes to liquid.

10.12 56.5 kJ released

10.13 226 kJ released

Answers to Selected Questions and Problems

10.1 a. Ca: $1s^22s^22p^63s^23p^64s^2 \longrightarrow Ca^{2+}: 1s^22s^22p^63s^23p^6$
b. S: $1s^22s^22p^63s^23p^4 \longrightarrow S^{2-}: 1s^22s^22p^63s^23p^6$
c. Ni: $1s^22s^22p^63s^23p^64s^23d^8 \longrightarrow Ni^{2+}: 1s^22s^22p^63s^23p^63d^8$
d. Ti: $1s^22s^22p^63s^23p^64s^23d^2 \longrightarrow Ti^{2+}: 1s^22s^22p^63s^23p^63d^2$

10.3 a. $Sc^{3+}: 1s^22s^22p^63s^23p^6$; Ar
b. $Ba^{2+}: 1s^22s^22p^63s^23p^64s^23d^{10}4p^65s^24d^{10}5p^6$; Xe
c. $K^+: 1s^22s^22p^63s^23p^6$; Ar
d. $V^{2+}: 1s^22s^22p^63s^23p^63d^3$; none

10.5 a. 8 valence electrons
b. 14 valence electrons
c. 32 valence electrons
d. 8 valence electrons

10.7 **a.** HF (8 e^-) H:$\ddot{\text{F}}$: or H—$\ddot{\text{F}}$:

b. SF_2 (20 e^-) :$\ddot{\text{F}}$:$\ddot{\text{S}}$:$\ddot{\text{F}}$: or :$\ddot{\text{F}}$—$\ddot{\text{S}}$—$\ddot{\text{F}}$:

c. NBr_3 (26 e^-) :$\ddot{\text{Br}}$:$\ddot{\text{N}}$:$\ddot{\text{Br}}$: or :$\ddot{\text{Br}}$—N—$\ddot{\text{Br}}$:
with :$\ddot{\text{Br}}$: above

d. BH_4^- (8 e^-) $\left[\begin{array}{c}\text{H}\\\text{H:}\ddot{\text{B}}\text{:H}\\\text{H}\end{array}\right]^-$ or $\left[\begin{array}{c}\text{H}\\\text{H—B—H}\\\text{H}\end{array}\right]^-$

e. CH_3OH (14 e^-) H:$\ddot{\text{C}}$:$\ddot{\text{O}}$:H or H—C—$\ddot{\text{O}}$—H (with H above and below C)

f. N_2H_4 (14 e^-) H:$\ddot{\text{N}}$:$\ddot{\text{N}}$:H or H—N—N—H (with H H above and below)

10.9 When using all the valence electrons does not give complete octets, it is necessary to write multiple bonds.

10.11 Resonance occurs when we can write two or more electron-dot formulas for the same molecule or ion.

10.13 **a.** CO (10 e^-) :C:::O: or :C≡O:

b. H_2CCH_2 (12 e^-) H:$\ddot{\text{C}}$::$\ddot{\text{C}}$:H or
H—C=C—H (with H H above and below)

c. H_2CO (12 e^-) H:$\ddot{\text{C}}$:H or H—C—H with :O: above (double bond)

10.15

a. $ClNO_2$:$\ddot{\text{Cl}}$—N—$\ddot{\text{O}}$: ⟷ :$\ddot{\text{Cl}}$—N=$\ddot{\text{O}}$: (with :O: above)

b. OCN^- $\left[:\ddot{\text{O}}=\text{C}=\ddot{\text{N}}:\right]^-$ ⟷ $\left[:\ddot{\text{O}}—\text{C}≡\text{N}:\right]^-$

10.17 **a.** linear
b. trigonal pyramidal

10.19 The four electron pairs in PCl_3 have a tetrahedral arrangement, but three bonded atoms and one lone pair around a central atom give a trigonal pyramidal shape.

10.21 In BF_3, the central atom B has three bonded atoms and no lone pairs, which gives BF_3 a trigonal planar shape. In NF_3, the central atom N has three bonded atoms and one lone pair, which gives NF_3 a trigonal pyramidal shape.

10.23 **a.** trigonal planar **b.** bent (109°)
c. linear **d.** tetrahedral
e. bent (120°)

10.25 **a.** CO_3^{2-} (24 e^-) $\left[:\ddot{\text{O}}—\text{C}=\text{O}:\right]^{2-}$ (with :O: above) trigonal planar

b. SO_4^{2-} (32 e^-) $\left[:\ddot{\text{O}}—\text{S}—\ddot{\text{O}}:\right]^{2-}$ (with :O: above and :O: below) tetrahedral

c. BH_4^- (8 e^-) $\left[\begin{array}{c}\text{H}\\\text{H—B—H}\\\text{H}\end{array}\right]^-$ tetrahedral

d. NO_2^+ (16 e^-) $\left[:\ddot{\text{O}}=\text{N}=\ddot{\text{O}}:\right]^+$ linear

10.27 The electronegativity increases going across a period.

10.29 A nonpolar covalent bond would have an electronegativity difference of 0.0 to 0.4.

10.31 **a.** K, Na, Li **b.** Na, P, Cl
c. Ca, Br, O

10.33 **a.** $\overset{\delta^+\ \ \delta^-}{\text{N—F}}$ ⟶ **b.** $\overset{\delta^+\ \ \delta^-}{\text{Si—P}}$ ⟶

c. $\overset{\delta^+\ \ \delta^-}{\text{C—O}}$ ⟶ **d.** $\overset{\delta^+\ \ \delta^-}{\text{P—Br}}$ ⟶

e. $\overset{\delta^+\ \ \delta^-}{\text{B—Cl}}$ ⟶

10.35 **a.** polar covalent **b.** ionic
c. polar covalent **d.** nonpolar covalent
e. polar covalent **f.** nonpolar covalent

10.37 Electrons are shared equally between two identical atoms and unequally between nonidentical atoms.

10.39 **a.** nonpolar **b.** polar
c. nonpolar **d.** nonpolar

10.41 **a.** dipole–dipole **b.** ionic
c. dispersion **d.** hydrogen bond
e. dispersion

10.43 **a.** hydrogen bonding
b. dispersion forces
c. dipole–dipole attraction
d. dispersion forces
e. dispersion forces

10.45 **a.** gas **b.** gas **c.** solid

10.47 **a.** 21 700 J; absorbed **b.** 5680 J; absorbed
c. 75.2 kJ; released **d.** 16.7 kJ; released

10.49 a. 22 600 J; heat is absorbed
b. 113 kJ; heat is absorbed
c. 1.81×10^7 J; heat is released
d. 396 kJ; heat is released

10.51 a. 4800 J   b. 30 300 J
c. 40.2 kJ   d. 72.3 kJ

10.53 119.5 kJ

10.55 a. dispersion forces
b. hydrogen bonds
c. dipole–dipole interactions
d. dispersion forces

10.57 BCl_3 is trigonal planar; all dipoles cancel and BCl_3 is a nonpolar molecule. PCl_3 is trigonal pyramidal; the dipoles do not cancel and PCl_3 is a polar molecule.

10.59 a. The heat from the skin is used to evaporate the water (perspiration). Therefore the skin is cooled.
b. On a hot day, there are more molecules with sufficient energy to become water vapor.
c. In a closed bag, some molecules evaporate but they cannot escape and will condense back to liquid; the clothes will not dry.

10.61

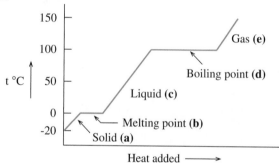

10.63 a. −60°C   b. 60°C
c. A represents the solid state. B represents the change from solid to liquid or melting of the substance. C represents the liquid state as temperature increases. D represents the change from liquid to gas or boiling of the liquid. E represents the gas state.
d. At −80°C, solid; at −40°C, liquid; at 25°C, liquid; 80°C, gas

10.65 a. 4 + 2(6) = 16 valence electrons
b. 2(4) + 4(1) + 6 = 18 valence electrons
c. 5 + 4(1) − 1 = 8 valence electrons
d. 3 + 3(7) = 24 valence electrons
e. 6 + 3(6) + 2 = 26 valence electrons

10.67 a. BF_4^- (32 e^-)

$$\left[:\ddot{F}:\!\ddot{B}:\!\ddot{F}:\right]^- \quad or \quad \left[:\ddot{F}-\overset{:\ddot{F}:}{\underset{:\ddot{F}:}{B}}-\ddot{F}:\right]^-$$

b. Cl_2O (20 e^-) $:\ddot{C}l:\ddot{O}:\ddot{C}l:$ or $:\ddot{C}l-\ddot{O}-\ddot{C}l:$

c. H_2NOH (14 e^-) $H:\ddot{N}:\ddot{O}:H$ or $H-\overset{H}{\underset{:\ddot{N}:}{N}}-\ddot{O}-H$

d. NO_2^+ (16 e^-) $\left[:\ddot{O}::N::\ddot{O}:\right]^+$ or $\left[:\ddot{O}=N=\ddot{O}:\right]^+$

e. H_2CCCl_2 (24 e^-) $H:\ddot{C}::\ddot{C}:\ddot{C}l:$ or

$$H-\overset{H}{\underset{}{C}}=\overset{:\ddot{C}l:}{\underset{}{C}}-\ddot{C}l:$$

10.69 a. 16 valence electrons $:\ddot{N}::N::\ddot{O}:$ or

$:\ddot{N}=N=\ddot{O}: \longleftrightarrow :N\equiv N-\ddot{O}:$

b. 16 valence electrons $\left[:\ddot{O}::N::\ddot{O}:\right]^+$ or

$\left[:\ddot{O}=N=\ddot{O}:\right]^+ \longleftrightarrow \left[:O\equiv N-\ddot{O}:\right]^+$

c. 16 valence electrons $\left[:C:::N:\ddot{S}:\right]^-$ or

$\left[:C\equiv N-\ddot{S}:\right]^- \longleftrightarrow \left[:\ddot{C}=N=\ddot{S}:\right]^-$

10.71 a. NF_3 $:\ddot{F}-\overset{}{\underset{:\ddot{F}:}{N}}-\ddot{F}:$ trigonal pyramidal

b. $SiBr_4$ $:\ddot{B}r-\overset{:\ddot{B}r:}{\underset{:\ddot{B}r:}{Si}}-\ddot{B}r:$ tetrahedral

c. $BeCl_2$ $:\ddot{C}l-Be-\ddot{C}l:$ linear

d. SO_2 $\left[:\ddot{O}=\ddot{S}-\ddot{O}:\right] \longleftrightarrow$

$\left[:\ddot{O}-\ddot{S}=\ddot{O}:\right]$ bent (120°)

10.73 a. I, Cl, F
b. K, Li, S, Cl
c. Ba, Sr, Mg, Be

10.75 a. C—O   b. N—F
c. S—Cl   d. Br—I
e. N—O

10.77 a. Si—Cl   b. N—C
c. F—Cl   d. C—F
e. N—O

10.79 **a.** polar covalent **b.** nonpolar covalent
c. ionic **d.** nonpolar covalent
e. nonpolar covalent

10.81 **a.** bent (109.5°) **b.** bent (109.5°)
c. trigonal planar **d.** tetrahedral
e. linear **f.** trigonal pyramidal
g. tetrahedral

10.83 **a.** polar **b.** nonpolar
c. polar **d.** polar
e. polar **f.** polar

10.85 **a.** trigonal planar, nonpolar
b. bent, polar
c. linear, nonpolar

10.87 **a.** 3 **b.** 3 **c.** 4 **d.** 2 **e.** 3 **f.** 1

10.89 When water vapor condenses or liquid water freezes, heat is released, which warms the air.

10.91 34 kJ + 109 kJ = 143 kJ removed

10.93 **a.**

$$H-\overset{\displaystyle H}{\underset{}{N}}-\overset{\displaystyle :O:}{\underset{}{C}}-H$$

b.

$$:\overset{..}{\underset{..}{Cl}}-\overset{\displaystyle H}{\underset{\displaystyle H}{C}}-C\equiv N:$$

c. $H-\overset{..}{N}=\overset{..}{N}-H$

d.

$$:\overset{..}{\underset{..}{Cl}}-\overset{\displaystyle :O:}{\underset{}{C}}-\overset{..}{\underset{..}{O}}-\overset{\displaystyle H}{\underset{\displaystyle H}{C}}-H$$

10.95 **a.** trigonal pyramidal
b. tetrahedral
c. linear
d. trigonal planar

10.97 350 g of ice will be melted.

10.99 450 g

Combining Ideas from Chapters 9 and 10

CI 15 The following reaction occurs between a metal and a nonmetal.

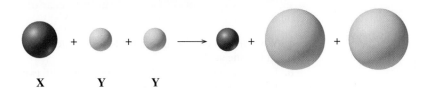

X   Y   Y

   a. Which spheres represent a metal? A nonmetal?
   b. Which reactant has the higher electronegativity?
   c. What are the ionic charges of X and Y in the product?
   d. If these elements are both in Period 3,
      1. write the electron configuration of the atoms.
      2. write the electron configuration of their ions.
      3. give the names of the noble gases with the same electron configuration as the ions.
      4. write the formula and name of the product.
   e. If these elements are both in Period 4,
      1. write the electron configuration of the atoms.
      2. write the electron configuration of their ions.
      3. write the formula and name of the product

CI 16 Using the electronegativity values in Figure 10.1, determine the most polar bond in each of the following. Show the dipoles for the most polar bond by assigning δ^+ and δ^- to the atoms in each.
   a. C—H or C—Cl or C—F
   b. N—S or N—F or N—H
   c. O—Cl or I—F or O—F
   d. Al—P or Al—S or Al—Cl
   e. C—C or C—S or C—Si

CI 17 Draw the electron-dot formula, state the number of lone pairs around the central atom, and predict the shape for each of the following:
   a. NH_2^-
   b. PCl_4^+
   c. HOBr
   d. BF_4^-

CI 18 Ethanol is obtained from renewable crops such as corn, which use the sun as their source of energy. In the United States, automobiles can now use an ethanol fuel, which is gasoline that contains 10% ethanol (known as E10) and 90% unleaded gasoline.

Ethanol, $CH_3—CH_2—OH$, has a melting point of $-115°C$, a boiling point of $78°C$, a heat of fusion of 4.60 kJ/mol, a heat of vaporization of 38.6 kJ/mol, and a specific heat of 2.46 J/g°C.
   a. Draw a heating curve for ethanol from $-150°C$ to $100°C$.

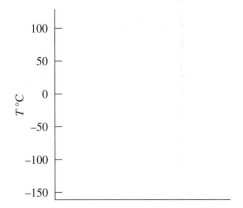

   b. When 20.0 g ethanol at $-62°C$ is heated and vaporized at $78°C$, how much energy (kJ) is required?
   c. Write the balanced chemical equation for the combustion reaction of ethanol if the products are carbon dioxide and water vapor.
   d. If a 15-gal gas tank contains 10% ethanol and 90% gasoline by volume, how many liters of ethanol are in the gas tank?

e. If the density of ethanol is 0.796 g/mL, how many kilograms of CO_2 are produced by the complete combustion of the ethanol in a full 15-gal gas tank?

CI 19 Chloral hydrate, a sedative and hypnotic, was the first drug used to treat insomnia. Chloral hydrate has a melting point of 57°C. At its boiling point of 98°C, it breaks down to chloral and water.

$$
\begin{array}{ccc}
& \text{Cl} & \text{O—H} \\
& | & | \\
\text{Cl—} & \text{C—} & \text{C—O—H} \\
& | & | \\
& \text{Cl} & \text{H}
\end{array}
\longrightarrow
\begin{array}{ccc}
& \text{Cl} & \text{O} \\
& | & \| \\
\text{Cl—} & \text{C—} & \text{C—H} + H_2O \\
& | & \\
& \text{Cl} &
\end{array}
$$

Chloral hydrate   Chloral

a. Write the electron-dot formulas for chloral hydrate and chloral.
b. What are the empirical formulas of chloral hydrate and chloral?
c. What is the % by mass chlorine in chloral hydrate?

CI 20 Ethylene glycol, $C_2H_6O_2$, used as a coolant and antifreeze, has a density of 1.11 g/mL. As a sweet-tasting syrup, it can be appealing to pets and small children, but it is toxic with an LD_{50} of 4700 mg/kg. Its accidental ingestion can cause difficulty with breathing and kidney damage. In the body, ethylene glycol is converted to another toxic substance, oxalic acid, $H_2C_2O_4$.

a. What are the empirical formulas of ethylene glycol and oxalic acid?
b. If ethylene glycol has a C—C single bond with two H atoms attached to each C atom, what is its electron-dot formula?
c. If oxalic acid has a C—C single bond, but no C—H bonds, what is its electron-dot formula?
d. Write the chemical equation for the reaction of ethylene glycol and oxygen (O_2) to give oxalic acid and water.
e. How many milliliters of ethylene glycol could be toxic for a 11-lb cat?

Answers to CIs

CI 15 a. X is a metal. Y is a nonmetal.
  b. Y has the higher electronegativity.
  c. X^{2+}, Y^-
  d. 1. $X = 1s^2 2s^2 2p^6 3s^2$
      $Y = 1s^2 2s^2 2p^6 3s^2 3p^5$
    2. $X^{2+} = 1s^2 2s^2 2p^6$
      $Y^- = 1s^2 2s^2 2p^6 3s^2 3p^6$
    3. X^{2+} has the same electron configuration as Ne
      Y^- has the same electron configuration as Ar
    4. $MgCl_2$, magnesium chloride
  e. 1. $X = 1s^2 2s^2 2p^6 3s^2 3p^6 4s^2$
      $Y = 1s^2 2s^2 2p^6 3s^2 3p^6 4s^2 3d^{10} 4p^5$
    2. $X^{2+} = 1s^2 2s^2 2p^6 3s^2 3p^6$
      $Y^- = 1s^2 2s^2 2p^6 3s^2 3p^6 4s^2 3d^{10} 4p^6$
    3. $CaBr_2$, calcium bromide

CI 17 a. $\left[\text{H—}\overset{..}{\text{N}}\text{—H} \right]^-$ With 2 bonding atoms and 2 lone pairs, the shape is bent.

  b.
$$
\left[
\begin{array}{c}
:\overset{..}{\underset{..}{\text{Cl}}}: \\
| \\
:\overset{..}{\underset{..}{\text{Cl}}}\text{—P—}\overset{..}{\underset{..}{\text{Cl}}}: \\
| \\
:\overset{..}{\underset{..}{\text{Cl}}}:
\end{array}
\right]^+
$$
  With 4 bonding atoms and no lone pairs, the shape is tetrahedral.

c. $\text{H—}\overset{..}{\underset{..}{\text{O}}}\text{—}\overset{..}{\underset{..}{\text{Br}}}:$ With 2 bonding atoms and 2 lone pairs, the shape is bent.

d.
$$
\left[
\begin{array}{c}
:\overset{..}{\underset{..}{\text{F}}}: \\
| \\
:\overset{..}{\underset{..}{\text{F}}}\text{—B—}\overset{..}{\underset{..}{\text{F}}}: \\
| \\
:\overset{..}{\underset{..}{\text{F}}}:
\end{array}
\right]^-
$$
  With 4 bonding atoms and no lone pairs, the shape is tetrahedral.

CI 19 a.
$$
\begin{array}{ccc}
:\overset{..}{\underset{..}{\text{Cl}}}: & :\overset{..}{\underset{..}{\text{O}}}\text{—H} \\
| & | \\
:\overset{..}{\underset{..}{\text{Cl}}}\text{—C—C—}\overset{..}{\underset{..}{\text{O}}}\text{—H} \\
| & | \\
:\overset{..}{\underset{..}{\text{Cl}}}: & \text{H}
\end{array}
$$

$$
\begin{array}{ccc}
:\overset{..}{\underset{..}{\text{Cl}}}: & :\overset{..}{\text{O}}: \\
| & \| \\
:\overset{..}{\underset{..}{\text{Cl}}}\text{—C—C—H} \\
| & \\
:\overset{..}{\underset{..}{\text{Cl}}}: &
\end{array}
$$

  b. chloral hydrate: $C_2H_3O_2Cl_3$
    chloral: C_2HOCl_3
  c. 64.30% %Cl (by mass)

Gases

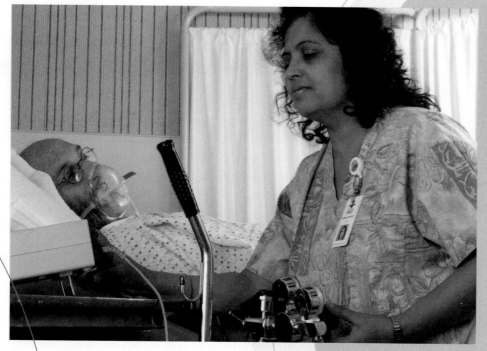

"*When oxygen levels in the blood are low, the cells in the body don't get enough oxygen,*" *says Sunanda Tripathi, registered nurse, Santa Clara Valley Medical Center.* "*We use a nasal cannula to give supplemental oxygen to a patient. At a flow rate of 2 liters per minute, a patient breathes in a gaseous mixture that is about 28% oxygen compared to 21% in ambient air.*"

When a patient has a breathing disorder, the flow and volume of oxygen into and out of the lungs is measured. A ventilator may be used if a patient has difficulty breathing. When pressure is increased, the lungs expand. When the pressure of the incoming gas is reduced, the lung volume contracts to expel carbon dioxide. These relationships—known as gas laws—are an important part of ventilation and breathing.

Visit **www.aw-bc.com/chemplace** for extra quizzes, interactive tutorials, career resources, PowerPoint slides for chapter review, math help, and case studies.

We all live at the bottom of a sea of gases called the atmosphere. The most important of these gases is oxygen, which constitutes about 21% of the atmosphere. Without oxygen, life on this planet would be impossible: oxygen is vital to all life processes of plants and animals. Ozone (O_3), formed in the upper atmosphere by the interaction of oxygen with ultraviolet light, absorbs some of the harmful radiation before it can strike Earth's surface. The other gases in the atmosphere include nitrogen (78% of the atmosphere), argon, carbon dioxide (CO_2), and water vapor. Carbon dioxide gas, a product of combustion and metabolism, is used by plants in photosynthesis, which produces the oxygen that is essential for humans and animals.

The atmosphere has become a dumping ground for other gases, such as methane, chlorofluorohydrocarbons (CFCs), sulfur dioxide, and nitrogen oxides. The chemical reactions of these gases with sunlight and oxygen in the air are contributing to air pollution, ozone depletion, global warming, and acid rain. Such chemical changes can seriously affect our health and the way all of us live. A knowledge of gases and some of the laws that govern gas behavior can help us understand the nature of matter and allow us to make decisions concerning important environmental and health issues.

Learning **Goal**

Describe the kinetic theory of gases and the properties of gases.

WEB TUTORIAL
Properties of Gases

11.1 Properties of Gases

The behavior of gases is quite different from that of liquids and solids. Gas particles are far apart, whereas particles of both liquids and solids are held close together. A gas has no definite shape or volume and will completely fill any container. Because there are great distances between its particles, a gas is less dense than a solid or liquid and can be compressed. A model for the behavior of a gas, called the **kinetic molecular theory of gases,** helps us understand gas behavior.

Kinetic Molecular Theory of Gases

1. **A gas consists of small particles (atoms or molecules) that move randomly with rapid velocities.** Gas molecules moving in all directions at high speeds cause a gas to fill the entire volume of a container.
2. **The attractive forces between the particles of a gas are usually very small.** Gas particles move far apart and fill a container of any size and shape.
3. **The actual volume occupied by gas molecules is extremely small compared to the volume that the gas occupies.** The volume of the gas is considered equal to the volume of the container. Most of the volume of a gas is empty space, which allows gases to be easily compressed.
4. **The average kinetic energy of gas molecules is proportional to the Kelvin temperature.** Gas particles move faster as the temperature increases. At higher temperatures, gas particles hit the walls of the container with more force, which produces higher pressures.
5. **Gas particles are in constant motion, moving rapidly in straight paths.** When gas particles collide, they rebound and travel in new directions. When they collide with the walls of the container, they exert gas pressure. An increase in the number or force of collisions against the walls of the container causes an increase in the pressure of the gas.

The kinetic theory helps explain some of the characteristics of gases. For example, we can quickly smell perfume from a bottle that is opened on the other side of a room, because its particles move rapidly in all directions. They move faster at higher temperatures and more slowly at lower temperatures. Sometimes tires and gas-filled containers explode when temperatures are too high. From the kinetic theory, we know that gas particles move faster when heated, hit the walls of a container with more force, and cause a buildup of pressure inside a container.

When we talk about a gas, we describe it in terms of four properties: pressure, volume, temperature, and the amount of gas.

Pressure (*P*)

Gas particles are extremely small. However, when billions and billions of gas particles hit against the walls of a container, they exert a force known as the pressure of the gas. The more molecules that strike the wall, the greater the pressure. (See Figure 11.1.) If we heat the container to make the molecules move faster, they smash into the walls of the container more often and with increased force, which increases the pressure. The gas molecules of oxygen and nitrogen in the air around us are exerting pressure on us all the time. We call the pressure exerted by the air **atmospheric pressure.** (See Figure 11.2.)

Figure 11.1 Gas particles move in straight lines within a container. The gas particles exert pressure when they collide with the walls of the container.

Q Why does heating the container increase the pressure of the gas within it?

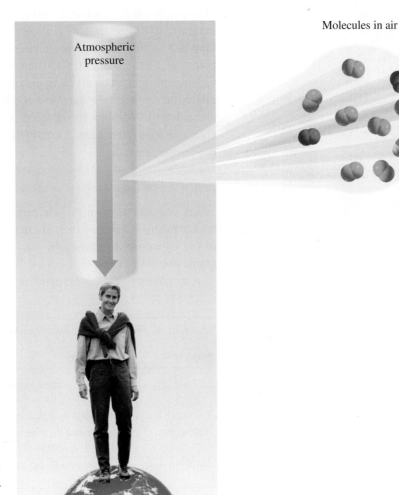

Molecules in air

Atmospheric pressure

Figure 11.2 A column of air extending from the upper atmosphere to the surface of Earth produces a pressure on each of us of about 1 atmosphere. While there is a lot of pressure on the body, it is balanced by the pressure inside the body.

Q Why is there less pressure at higher altitudes?

As you go to higher altitudes the atmospheric pressure is less because there are fewer molecules of oxygen and nitrogen in the air. The most common units used for gas measurement are the atmosphere (atm) and millimeters of mercury (mm Hg). On the TV weather report, you may hear or see the atmospheric pressure given in inches of mercury, or kilopascals in countries other than the United States. In a chemistry lab the unit torr may be used.

Volume (V)

The volume of gas equals the size of the container in which the gas is placed. When you inflate a tire or a basketball, you are adding more gas particles, which increases the number of particles hitting the walls of the tire or basketball, and its volume increases. Sometimes, on a cool morning, a tire looks flat. The volume of the tire has decreased because a lower temperature decreases the speed of the molecules, which reduces the force of their impacts on the walls of the tire. The most common units for volume measurement are liters (L) and milliliters (mL).

Temperature (T)

The temperature of a gas is related to the kinetic energy of its particles. For example, if we have a gas at 200 K in a rigid container and heat it to a temperature of 400 K, the gas particles will have twice the kinetic energy that they did at 200 K. That also means that the gas at 400 K exerts twice the pressure of the gas at 200 K. Although you measure gas temperature using a Celsius thermometer, all comparisons of gas behavior and all calculations related to temperature must use the Kelvin temperature. No one has quite created the conditions for absolute zero (K), but we predict that the particles will have zero kinetic energy, and the gas will exert zero pressure at this temperature.

Amount of Gas (n)

When you add air to a bicycle tire, you are increasing the amount of gas, which results in a higher pressure in the tire. Usually, we measure the amount of gas by its mass (grams). In gas law calculations, we need to change the grams of gas to moles.

A summary of the four properties of a gas are given in Table 11.1.

Table 11.1 Properties That Describe a Gas

Property	Description	Unit(s) of Measurement
Pressure (P)	The force exerted by gas against the walls of the container	atmosphere (atm); mm Hg; torr; pascal
Volume (V)	The space occupied by the gas	liter (L); milliliter (mL)
Temperature (T)	Determines the kinetic energy and rate of motion of the gas particles	Celsius (°C); Kelvin (K) *required in calculations*
Amount (n)	The quantity of gas present in a container	grams (g); moles (n) *required in calculations*

CHEM NOTE

MEASURING BLOOD PRESSURE

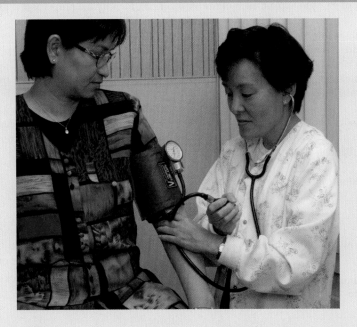

The measurement of your blood pressure is one of the important measurements a doctor or nurse makes during a physical examination. Acting like a pump, the heart contracts to create the pressure that pushes blood through the circulatory system. During contraction, the blood pressure is called systolic and is at its highest. When the heart muscles relax, the blood pressure is called diastolic and falls. Normal range for systolic pressure is 100–120 mm Hg, and for diastolic pressure, 60–80 mm Hg, usually expressed as a ratio such as 100/80. These values are somewhat higher in older people. When blood pressures are elevated, such as 140/90, there is a greater risk of stroke, heart attack, or kidney damage. Low blood pressure prevents the brain from receiving adequate oxygen, causing dizziness and fainting.

The blood pressures are measured by a sphygmomanometer, an instrument consisting of a stethoscope and an inflatable cuff connected to a tube of mercury called a manometer. After the cuff is wrapped around the upper arm, it is pumped up with air until it cuts off the flow of blood through the arm. With the stethoscope over the artery, the air is slowly released from the cuff. When the pressure equals the systolic pressure, blood starts to flow again, and the noise it makes is heard through the stethoscope. As air continues to be released, the cuff deflates until no sound in the artery is heard. That second pressure reading is noted as the diastolic pressure, the pressure when the heart is not contracting.

Sample Problem 11.1 ▶ Properties of Gases

Identify the property of a gas that is described by each of the following:
a. increases the kinetic energy of gas particles
b. the force of the gas particles hitting the walls of the container
c. the space that is occupied by a gas

Solution

a. temperature   **b.** pressure   **c.** volume

Study Check

When helium is added to a balloon, the number of grams of gas increases. What property of a gas is described?

11.1 Use the kinetic theory of gases to explain each of the following:
a. Gases move faster at higher temperatures.
b. Gases can be compressed much more easily than liquids or solids.
c. Gases have low densities.

11.2 Use the kinetic molecular theory of gases to explain each of the following:
a. A container of nonstick cooking spray explodes when thrown into a fire.
b. The air in a hot-air balloon is heated to make the balloon rise.
c. You can smell an odor from far away.

11.3 Identify the property of a gas that is measured in each of the following:
a. 350 K
b. space occupied by a gas
c. 2.00 g O_2
d. force of gas particles striking the walls of the container

11.4 Identify the property of a gas that is measured in each of the following measurements:
a. determines the kinetic energy of the gas particles
b. 1.0 atm
c. 10.0 L
d. 0.50 mol He

11.2 Gas Pressure

Describe the units of measurement used for pressure and change from one unit to another.

When water boils in a pan covered with a lid, the collisions of the molecules of steam (gas) lift up the lid. The gas molecules exert **pressure,** which is defined as a force acting on a certain area.

$$\text{Pressure } (P) = \frac{\text{force}}{\text{area}}$$

The air that covers the surface of Earth, the atmosphere, contains vast numbers of gas particles. Because the air particles have mass, they are pulled toward Earth by gravity, where they exert an *atmospheric pressure.*

The atmospheric pressure can be measured using a barometer, as shown in Figure 11.3. At a pressure of exactly 1 atmosphere (atm), the mercury column would be exactly 760 mm high. We say that the atmospheric pressure is 760 mm Hg (millimeters of mercury), which is an **atmosphere (atm)**. One atmosphere of pressure may also be expressed as 760 **torr,** a pressure unit named to honor Evangelista Torricelli, the inventor of the barometer. Because they are equal, units of torr and mm Hg are used interchangeably.

1 atm = 760 mm Hg = 760 torr
1 mm Hg = 1 torr

In SI units, pressure is measured in pascals (Pa); 1 atm is equal to 101 325 Pa. Because a pascal is a very small unit, pressures are usually reported in kilopascals.

1 atm = 1.01325×10^5 Pa = 101.325 kPa

The American equivalent of 1 atm is 14.7 pounds per square inch (lb/in.^{2}). When you use a pressure gauge to check the air pressure in the tires of a car, it may read 30–35 psi. Table 11.2 summarizes the various units used in the measurement of pressure.

If you have a barometer in your home, it probably gives pressure in inches of mercury. One atmosphere is equal to the pressure of a column of mercury that is 29.9 inches high. Atmospheric pressure changes with variations in weather and altitude. On a hot, sunny day, the air is more dense. The mercury column rises, which indicates a higher atmospheric pressure. On a rainy day, the atmosphere exerts less pressure, which causes the mercury column to fall. In the weather

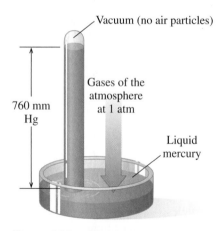

Figure 11.3 A barometer: the pressure exerted by the gases in the atmosphere is equal to the downward pressure of a mercury column in a closed glass tube. The height of the mercury column measured in mm Hg is called atmospheric pressure.

Vacuum (no air particles)
Gases of the atmosphere at 1 atm
760 mm Hg
Liquid mercury

Q *Why does the height of the mercury column change from day to day?*

Table 11.2 Units for Measuring Pressure

Unit	Abbreviation	Unit Equivalent to 1 atm
Atmosphere	atm	1 atm (exact)
Millimeters of Hg	mm Hg	760 mm Hg
Torr	torr	760 torr
Inches of Hg	in. Hg	29.9 in. Hg
Pounds per square inch	lb/in.^{2} (psi)	14.7 lb/in.^{2}
Pascal	Pa	101 325 Pa
Kilopascal	kPa	101 325 kPa

report, this type of weather is called a low-pressure system. Atmospheric pressure is greatest at sea level. Above sea level, the density of the gases in the air decreases, which causes lower atmospheric pressures; the atmospheric pressure is greater in Death Valley because it is below sea level. (See Table 11.3).

Table 11.3 Altitude and Atmospheric Pressure

Location	Altitude (km)	Atmospheric Pressure (mm Hg)
Sea level	0	760
Los Angeles	0.09	752
Las Vegas	0.70	700
Denver	1.60	630
Mount Whitney	4.00	467
Mount Everest	9.30	270

CASE STUDY

Scuba Diving and Blood Gases

Divers must be concerned about increasing pressures on their ears and lungs when they dive below the surface of the ocean. Because water is denser than air, the pressure on a diver increases rapidly as the diver descends. At a depth of 33 ft below the surface of the ocean, an additional 1 atmosphere pressure is exerted by the water on a diver, for a total of 2 atm. At 100 ft, there is a total pressure of 4 atm on a diver. The air tanks a diver carries continuously adjust the pressure of the breathing mixture to match the increase in pressure.

Sample Problem 11.2   Units of Pressure

A sample of neon gas has a pressure of 0.50 atm. Give the pressure of the neon in

a. millimeters of Hg   **b.** inches of Hg

Solution

a. The equality 1 atm = 760 mm Hg can be written as conversion factors:

$$\frac{760 \text{ mm Hg}}{1 \text{ atm}} \quad \text{or} \quad \frac{1 \text{ atm}}{760 \text{ mm Hg}}$$

CHEM NOTE

HYPERBARIC CHAMBERS

A burn patient may undergo treatment for burns and infections in a hyperbaric chamber, a device in which pressures can be obtained that are two to three times greater than atmospheric pressure. A greater oxygen pressure increases the level of dissolved oxygen in the blood and tissues, where it fights bacterial infections. High levels of oxygen are toxic to many strains of bacteria. The hyperbaric chamber may also be used during surgery, to help counteract carbon monoxide (CO) poisoning, and to treat some cancers.

The blood is normally capable of dissolving up to 95% of the oxygen. Thus, if the pressure of the oxygen is 2280 mm Hg (3 atm), 95% of that or 2170 mm Hg of oxygen can dissolve in the blood, where it saturates the tissues. In the case of carbon monoxide poisoning, this oxygen can replace the carbon monoxide that has attached to the hemoglobin.

A patient undergoing treatment in a hyperbaric chamber must also undergo decompression (reduction of pressure) at a rate that slowly reduces the concentration of dissolved oxygen in the blood. If decompression is too rapid, the oxygen dissolved in the blood may form gas bubbles in the circulatory system.

If divers do not decompress slowly, they suffer a similar condition called the bends. While below the surface of the ocean, divers breathe air at higher pressures. At such high pressures, nitrogen gas will dissolve in their blood. If they ascend to

the surface too quickly, the dissolved nitrogen forms bubbles in the blood that can produce life-threatening blood clots. The gas bubbles can also appear in the joints and tissues of the body and be quite painful. A diver suffering from the bends is placed immediately in a decompression chamber where pressure is first increased and then slowly decreased. The dissolved nitrogen can then diffuse through the lungs until atmospheric pressure is reached.

Using the appropriate conversion factor, the problem is set up as

$$0.50 \; \cancel{atm} \times \frac{760 \text{ mm Hg}}{1 \; \cancel{atm}} = 380 \text{ mm Hg}$$

b. One atm is equal to 29.9 in. Hg. Using this equality as a conversion factor in the problem setup, we obtain

$$0.50 \; \cancel{atm} \times \frac{29.9 \text{ in. Hg}}{1 \; \cancel{atm}} = 15 \text{ in. Hg}$$

Study Check

What is the pressure in atmospheres for a gas that has a pressure of 655 torr?

Questions and Problems   Gas Pressure

11.5 What units are used to measure the pressure of a gas?

11.6 Which of the following statement(s) describes the pressure of a gas?
   a. the force of the gas particles on the walls of the container
   b. the number of gas particles in a container
   c. the volume of the container
   d. 3.00 atm
   e. 750 torr

11.7 An oxygen tank contains oxygen (O_2) at a pressure of 2.00 atm. What is the pressure in the tank in terms of the following units?
   a. torr   **b.** lb/in.^{2} **c.** mm Hg   **d.** kPa

11.8 On a climb up Mt. Whitney, the atmospheric pressure is 467 mm Hg. What is the pressure in terms of the following units?
   a. atm   **b.** torr   **c.** in. Hg   **d.** Pa

11.3 Pressure and Volume (Boyle's Law)

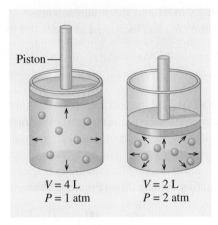

$V = 4$ L
$P = 1$ atm

$V = 2$ L
$P = 2$ atm

Figure 11.4 Boyle's law: As volume decreases, gas molecules become more crowded, which causes the pressure to increase. Pressure and volume are inversely related.

Q *If the volume of a gas increases, what will happen to its pressure?*

Imagine that you can see air particles hitting the walls inside a bicycle tire pump. What happens to the pressure inside the pump as we push down on the handle? As the volume decreases, there is a decrease in the surface area of the container. The pressure within the container increases because there are more collisions per unit area.

When a change in one property (in this case, volume) causes a change in another property (in this case, pressure), those properties are related. If the changes occur in opposite directions, such as a decrease in volume causing an increase in pressure, the properties have an **inverse relationship.** In equation form this is shown as: $P \propto \frac{1}{V}$, where $\propto$ means "proportional to." The relationship between the pressure and volume of a gas is known as **Boyle's law.** The law states that the volume (V) of a sample of gas changes inversely with the pressure (P) of the gas as long as there has been no change in the temperature (T) or amount of gas (n), as illustrated in Figure 11.4.

If we change the volume or pressure of a gas sample without any change occurring in the temperature or in the amount of the gas, the new pressure and volume will give the same product as the initial pressure and volume. Because the PV product has the same value under both conditions, we can set the initial and final PV values equal to each other.

Boyle's Law

$$P_1V_1 = P_2V_2 \qquad \text{No change in number of moles and temperature}$$

Sample Problem 11.3   Calculating Pressure When Volume Changes

A sample of hydrogen gas (H_2) has a volume of 5.0 L and a pressure of 1.0 atm. What is the new pressure if the volume is decreased to 2.0 L at constant temperature?

Solution

STEP 1   **Organize the data in a table.**   In this problem, we want to know the final pressure (P_2) for the change in volume. In calculations with gas laws, it is helpful to organize the data in a table. Because the volume decreases, we can predict that the pressure will increase.

Conditions 1	Conditions 2	Know	Predict
$V_1 = 5.0$ L	$V_2 = 2.0$ L	V decreases	
$P_1 = 1.0$ atm	$P_2 = ?$		P increases

STEP 2   **Rearrange the gas law for the unknown.**   For a PV relationship, we use Boyle's law and solve for P_2 by dividing both sides by V_2.

$$P_1V_1 = P_2V_2$$

$$\frac{P_1V_1}{V_2} = \frac{P_2 \cancel{V_2}}{\cancel{V_2}}$$

$$P_2 = \frac{P_1V_1}{V_2}$$

STEP 3 **Substitute values into the gas law to solve for the unknown.** From the table, we see that the volume has decreased. Because pressure and volume are inversely related, the pressure must increase.

$$P_2 = \frac{1.0 \text{ atm} \times 5.0 \, \cancel{L}}{2.0 \, \cancel{L}} = 2.5 \text{ atm}$$

Volume factor increases pressure

Note that the units of volume cancel and the final pressure is in atmospheres. The final pressure (P_2) has increased as predicted in STEP 1.

Study Check

A sample of helium gas has a volume of 312 mL at 648 torr. If the volume expands to 825 mL at constant temperature, what is the new pressure in torr?

Sample Problem 11.4 ▸ Calculating Volume When Pressure Changes

The gauge on a 12-L tank of compressed oxygen reads 3800 mm Hg. How many liters would this same gas occupy at a pressure of 0.75 atm at constant temperature?

Solution

STEP 1 **Organize the data in a table.**   First, we need to match the units for the initial and final pressure.

$$0.75 \, \cancel{\text{atm}} \times \frac{760 \text{ mm Hg}}{1 \, \cancel{\text{atm}}} = 570 \text{ mm Hg}$$

The pressures could also be changed to atm.

$$3800 \, \cancel{\text{mm Hg}} \times \frac{1 \text{ atm}}{760 \, \cancel{\text{mm Hg}}} = 5.0 \text{ atm}$$

Because the pressure decreases, we can predict that the volume should increase. Placing our information in a table gives the following:

Conditions 1	Conditions 2	Know	Predict
P_1 = 3800 mm Hg (5.0 atm)	P_2 = 570 mm Hg (0.75 atm)	P decreases	
V_1 = 12 L	V_2 = ?		V increases

STEP 2 **Rearrange the gas law for the unknown.**   Using Boyle's law, we solve for V_2. According to Boyle's law, a decrease in the pressure will cause an increase in the volume.

$$V_2 = \frac{V_1 P_1}{P_2}$$

CHEM NOTE

PRESSURE–VOLUME RELATIONSHIP IN BREATHING

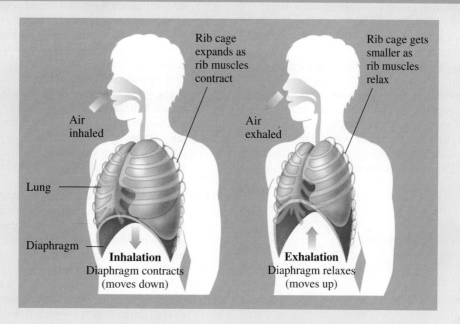

Rib cage expands as rib muscles contract

Rib cage gets smaller as rib muscles relax

Air inhaled

Air exhaled

Lung

Diaphragm

Inhalation
Diaphragm contracts (moves down)

Exhalation
Diaphragm relaxes (moves up)

The importance of Boyle's law becomes more apparent when you consider the mechanics of breathing. Our lungs are elastic, balloon-like structures contained within an airtight chamber called the thoracic cavity. The diaphragm, a muscle, forms the flexible floor of the cavity.

Inspiration

The process of taking a breath of air begins when the diaphragm flattens, and the rib cage expands, causing an increase in the volume of the thoracic cavity. The elasticity of the lungs allows them to expand when the thoracic cavity expands. According to Boyle's law, the pressure inside the lungs will decrease when their volume increases. This causes the pressure inside the lungs to fall below the pressure of the atmosphere. This difference in pressures produces a *pressure gradient* between the lungs and the atmosphere.

In a pressure gradient, molecules flow from an area of greater pressure to an area of lower pressure. Thus, we inhale as air flows into the lungs (*inspiration*), until the pressure within the lungs becomes equal to the pressure of the atmosphere.

Expiration

Expiration, or the exhalation phase of breathing, occurs when the diaphragm relaxes and moves back up into the thoracic cavity to its resting position. This reduces the volume of the thoracic cavity, which squeezes the lungs and decreases their volume. Now the pressure in the lungs is greater than the pressure of the atmosphere, so air flows out of the lungs. Thus, breathing is a process in which pressure gradients are continuously created between the lungs and the environment as a result of the changes in the volume and pressure.

STEP 3   **Substitute values into the gas law to solve for the unknown.**

$$V_2 = 12\,L \times \frac{3800\ \cancel{mm\,Hg}}{570\ \cancel{mm\,Hg}} = 80.\,L$$

$$V_2 = 12\,L \times \frac{5.0\ \cancel{atm}}{0.75\ \cancel{atm}} = 80.\,L$$

Pressure factor increases volume

The final pressure (P_2) has increased as predicted in STEP 1.

Study Check

A sample of methane gas (CH_4) has a volume of 125 mL at 0.600 atm pressure and 25°C. How many milliliters will it occupy at a pressure of 1.50 atm and 25°C?

Questions and Problems   **Pressure and Volume (Boyle's Law)**

11.9 Why do scuba divers need to exhale air when they ascend to the surface of the water?

11.10 Why does a sealed bag of chips expand when you take it to a higher altitude?

11.11 What happens to the volume of your lungs during expiration?

11.12 How do respirators (or CPR) help a person obtain oxygen?

11.13 The air in a cylinder with a piston has a volume of 220 mL and a pressure of 650 mm Hg.
a. If a change results in a higher pressure inside the cylinder, does cylinder A or B represent the final volume? Explain your choice.

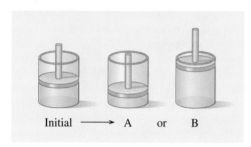

Initial ⟶ A or B

b. If the pressure inside the cylinder increases to 1.2 atm, what is the final volume of the cylinder? Complete the following table:

Property	Conditions 1	Conditions 2	Know	Predict
Pressure (P)				
Volume (V)				

11.14 A balloon is filled with helium gas. When the following changes are made at constant temperature, which of these diagrams (A, B, or C) shows the new volume of the balloon?

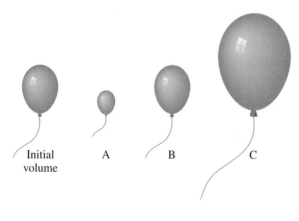

Initial volume   A   B   C

a. The balloon floats to a higher altitude where the outside pressure is lower.
b. The balloon is taken inside the house, but the atmospheric pressure remains the same.
c. The balloon is put in a hyperbaric chamber in which the pressure is increased.

11.15 A gas with a volume of 4.0 L is contained in a closed vessel. Indicate the changes in its pressure when the volume undergoes the following changes at constant temperature.
a. The volume is compressed to 2 L.
b. The volume is allowed to expand to 12 L.
c. The volume is compressed to 0.40 L.

11.16 A gas at a pressure of 2.0 atm is contained in a closed vessel. Indicate the changes in its volume when the pressure undergoes the following changes at constant temperature.
a. The pressure increases to 6.0 atm.
b. The pressure drops to 1.0 atm.
c. The pressure drops to 0.40 atm.

11.17 A 10.0-L balloon contains He gas at a pressure of 655 mm Hg. What is the new pressure of the He gas at each of the following volumes if there is no change in temperature?
a. 20.0 L   b. 2.50 L
c. 1500 mL   d. 120 mL

11.18 The air in a 5.00-L tank has a pressure of 1.20 atm. What is the new pressure of the air when the air is placed in tanks that have the following volumes, if there is no change in temperature?
a. 1.00 L   b. 2500 mL
c. 750 mL   d. 8.0 L

11.19 A sample of nitrogen (N_2) gas has a volume of 4.5 L at a pressure of 760 mm Hg. What is the new pressure if the gas sample is compressed to a volume of 2.0 L if there is no change in temperature?

11.20 A tank of oxygen holds 20.0 L of oxygen (O_2) at a pressure of 15.0 atm. When the gas is released, it provides 300.0 L oxygen. What is the pressure of this same gas at a volume of 300.0 L and constant temperature?

11.21 A sample of nitrogen (N_2) has a volume of 50.0 L at a pressure of 760. mm Hg. What is the volume of the gas at each of the following pressures if there is no change in temperature?
a. 1500 mm Hg   b. 2.0 atm
c. 0.500 atm   d. 850 torr

11.22 A sample of methane (CH_4) has a volume of 25 mL at a pressure of 0.80 atm. What is the volume of the gas at each of the following pressures if there is no change in temperature?
a. 0.40 atm   b. 2.00 atm
c. 2500 mm Hg   d. 80.0 torr

11.4 Temperature and Volume (Charles' Law)

Learning **Goal**

Use the temperature–volume relationship (Charles' law) to determine the new temperature or volume of a certain amount of gas at a constant pressure.

WEB TUTORIAL
Properties of Gases

When preparing a hot-air balloon for a flight, the air in the balloon is heated with a propane heater. As the air warms, its volume increases. The resulting decrease in density allows the balloon to rise.

To study the effect of changing temperature on the volume of a gas, we must not change the pressure or the amount of the gas. Suppose we increase the Kelvin temperature of a gas sample. The kinetic theory shows that the activity (kinetic energy) of the gas will also increase. To keep pressure constant, the volume of the container must increase. (See Figure 11.5.) By contrast, if the temperature of the gas is lowered, the volume of the container must be reduced to maintain the same pressure.

Suppose that you are going to take a ride in a hot-air balloon. The captain turns on a propane burner to heat the air inside the balloon. As the temperature rises, the air particles move faster and spread out, causing the volume of the balloon to increase. Eventually, the air in the balloon becomes less dense than the air outside, and the balloon and its passengers rise. In fact, it was in 1787 that Jacques Charles, a balloonist as well as a physicist, proposed that the volume of a gas is related to the temperature. This became **Charles' law,** which states that the volume (V) of a gas is directly related to the temperature (K) when there is no change in the pressure (P) or amount (n) of gas, $V \propto T$. A **direct relationship** is one in which the related properties increase or decrease together. For two conditions, initial and final, we can write Charles' law as follows.

Charles' Law

$$\frac{V_1}{T_1} = \frac{V_2}{T_2} \qquad \text{No change in number of moles and pressure}$$

Remember that all temperatures used in gas law calculations must be converted to their corresponding Kelvin (K) temperatures.

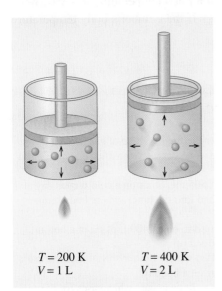

$T = 200\ \text{K}$ $T = 400\ \text{K}$
$V = 1\ \text{L}$ $V = 2\ \text{L}$

Figure 11.5 Charles' law: The Kelvin temperature of a gas is directly related to the volume of the gas when there is no change in the pressure. When the temperature increases, making the molecules move faster, the volume must increase to maintain constant pressure.

Q *If the temperature of a gas decreases at a constant pressure, how will the volume change?*

Sample Problem 11.5 ▶ **Calculating Volume When Temperature Changes**

A sample of neon gas has a volume of 5.40 L and a temperature of 15°C. Find the new volume of the gas after the temperature has been increased to 42°C at constant pressure.

Solution

STEP 1 **Organize the data in a table.** When the temperatures are given in degrees Celsius, they must be changed to Kelvins.

$$T_1 = 15°C + 273 = 288.\ \text{K}$$
$$T_2 = 42°C + 273 = 315.\ \text{K}$$

Conditions 1	Conditions 2	Know	Predict
$T_1 = 288\ \text{K}$	$T_2 = 315\ \text{K}$	T increases	
$V_1 = 5.40\ \text{L}$	$V_2 = ?$		V increases

STEP 2   **Rearrange the gas law for the unknown.**   In this problem, we want to know the final volume (V_2) when the temperature increases. Using Charles' law, we solve for V_2 by multiplying both sides by T_2.

$$\frac{V_1}{T_1} = \frac{V_2}{T_2}$$

$$\frac{V_1}{T_1} \times T_2 = \frac{V_2}{T_2} \times T_2$$

$$V_2 = \frac{V_1 T_2}{T_1}$$

STEP 3   **Substitute values into the gas law to solve for the unknown.** From the table, we see that the temperature has increased. Because temperature is directly related to volume, the volume must increase. When we substitute in the values, we see that the ratio of the temperatures (temperature factor) is greater than 1, which increases the volume, as predicted.

$$V_2 = 5.40\,\text{L} \times \frac{315\,K}{288\,K} = 5.91\,\text{L}$$

Temperature factor
increases volume

Study Check

A mountain climber inhales 486 mL of air at a temperature of $-8°C$. What volume will the air occupy in the lungs if the climber's body temperature is 37°C?

Questions and Problems   **Temperature and Volume (Charles' Law)**

11.23 Select the diagram that shows the new volume of a balloon when the following changes are made at constant pressure:

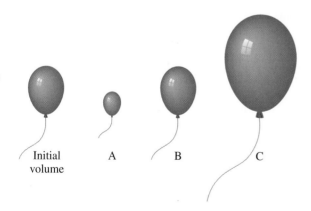

Initial volume   A   B   C

  a. The temperature is changed from 100 K to 300 K.
  b. The balloon is placed in a freezer.
  c. The balloon is first warmed, and then returned to its starting temperature.

11.24 Indicate whether the final volume of gas in each of the following is the same, larger, or smaller than the initial volume:
  a. A volume of 505 mL of air on a cold winter day at 5°C is breathed into the lungs, where body temperature is 37°C.
  b. The heater used to heat 1400 L of air in a hot-air balloon is turned off.
  c. A balloon filled with helium at the amusement park is left in a car on a hot day.

11.25 What change in volume occurs when gases for hot-air balloons are heated prior to their ascent?

11.26 On a cold, wintry morning, the tires on a car appear flat. How has their air volume changed overnight?

11.27 A balloon contains 2500 mL of helium gas at 75°C. What is the new volume of the gas when the temperature changes to the following, if n and P are not changed?
  a. 55°C   **b.** 680. K   **c.** $-25°C$ **d.** 240. K

11.28 A gas has a volume of 4.00 L at 0°C. What final temperature in degrees Celsius is needed to cause the volume of the gas to change to the following, if n and P are not changed?
  a. 10.0 L   **b.** 1200 mL   **c.** 2.50 L   **d.** 50.0 mL

11.5 Temperature and Pressure (Gay-Lussac's Law)

If we could watch the molecules of a gas as the temperature rises, we would notice that they move faster and hit the sides of the container more often and with greater force. If we keep the volume of the container the same, we would observe an increase in the pressure. A temperature–pressure relationship, also known as **Gay-Lussac's law,** states that the pressure of a gas is directly related to its Kelvin temperature. This means that an increase in temperature increases the pressure of a gas, and a decrease in temperature decreases the pressure of the gas, provided the volume and number of moles of the gas remain the same $P \propto T$. (See Figure 11.6.) The ratio of pressure (P) to temperature (T) is the same under all conditions as long as volume (V) and amount of gas (n) do not change.

Gay-Lussac's Law

$$\frac{P_1}{T_1} = \frac{P_2}{T_2}$$ No change in number of moles and volume

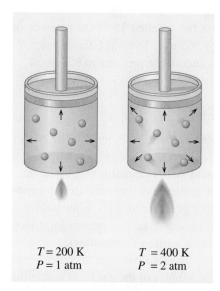

$T = 200 \text{ K}$ $T = 400 \text{ K}$
$P = 1 \text{ atm}$ $P = 2 \text{ atm}$

Figure 11.6 Gay-Lussac's law: the pressure of a gas is directly related to the temperature of the gas. When the Kelvin temperature of a gas is doubled, the pressure is doubled at constant volume.

Q *How does a decrease in the temperature of a gas affect its pressure at constant volume?*

Sample Problem 11.6 > **Calculating Pressure When Temperature Changes**

Aerosol containers can be dangerous if they are heated because they can explode. Suppose a container of hair spray with a pressure of 4.0 atm at a room temperature of 25°C is thrown into a fire. If the temperature of the gas inside the aerosol can reaches 402°C, what will be its pressure? The aerosol container may explode if the pressure inside exceeds 8.0 atm. Would you expect it to explode?

Solution

STEP 1 **Organize the data in a table.** We must first change the temperatures to kelvins.

$$T_1 = 25°C + 273 = 298 \text{ K}$$
$$T_2 = 402°C + 273 = 675 \text{ K}$$

Conditions 1	Conditions 2	Know	Predict
$P_1 = 4.0 \text{ atm}$	$P_2 = ?$		P increases
$T_1 = 298 \text{ K}$	$T_2 = 675 \text{ K}$	T increases	

STEP 2 **Rearrange the gas law for the unknown.** Using Gay-Lussac's law, we can solve for P_2.

$$\frac{P_1}{T_1} = \frac{P_2}{T_2}$$

$$\frac{P_1}{T_1} \times T_2 = \frac{P_2}{\cancel{T_2}} \times \cancel{T_2}$$

$$P_2 = \frac{P_1 T_2}{T_1}$$

STEP 3 **Substitute values into the gas law to solve for the unknown.** From the table, we see that the temperature has increased.

Because pressure and temperature are directly related, the pressure must increase. When we substitute in the values, we see the ratio of the temperatures (temperature factor) is greater than 1, which increases pressure.

$$P_2 = 4.0 \text{ atm} \times \frac{675 \text{ } K}{298 \text{ } K} = 9.1 \text{ atm}$$

Temperature factor increases volume

Because the calculated pressure exceeds 8.0 atm, we expect the can to explode.

Study Check

In a storage area where the temperature has reached 55°C, the pressure of oxygen gas in a 15.0-L steel cylinder is 965 torr. To what Celsius temperature would the gas have to be cooled to reduce the pressure to 850. torr?

Vapor Pressure and Boiling Point

In Chapter 10, we learned that liquid molecules with sufficient kinetic energy can break away from the surface of the liquid as they become gas particles or vapor. In an open container, all the liquid will eventually evaporate. In a closed container, the vapor accumulates and creates pressure called **vapor pressure.** Each liquid exerts its own vapor pressure at a given temperature. As temperature increases, more vapor forms, and vapor pressure increases. Table 11.4 lists the vapor pressure of water at various temperatures.

A liquid reaches its boiling point when its vapor pressure becomes equal to the external pressure. As boiling occurs, bubbles of the gas form within the liquid and quickly rise to the surface. For example, at an atmospheric pressure of 760 mm Hg, water will boil at 100°C, the temperature at which its vapor pressure reaches 760 mm Hg.

Table 11.4 Vapor Pressure of Water

Temperature (C°)	Vapor Pressure (mm Hg)
0	5
10	9
20	18
30	32
37	47[a]
40	55
50	93
60	149
70	234
80	355
90	528
100	760

[a] At body temperature.

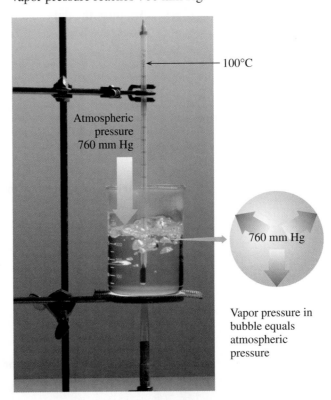

100°C

Atmospheric pressure 760 mm Hg

760 mm Hg

Vapor pressure in bubble equals atmospheric pressure

Table 11.5	Pressure and the Boiling Point of Water
Pressure (mm Hg)	Boiling Point (°C)
270	70
467	87
630	93
752	99
760	100
800	100.4
1075	110
1520 (2 atm)	120
2026	130
7600 (10 atm)	180

At higher altitudes, atmospheric pressures are lower and the boiling point of water is lower than 100°C. Earlier, we saw that the typical atmospheric pressure in Denver is 630 mm Hg. This means that water in Denver needs a vapor pressure of 630 mm Hg to boil. Because water has a vapor pressure of 630 mm Hg at 95°C, water boils at 95°C in Denver.

People who live at high altitudes often use pressure cookers to obtain higher temperatures when preparing food. When the external pressure is greater than 1 atm, a temperature higher than 100°C is needed to boil water. Laboratories and hospitals use devices called autoclaves to sterilize laboratory and surgical equipment. An autoclave, like a pressure cooker, is a closed container that increases the total pressure above the liquid so it will boil at higher temperatures. Table 11.5 shows how the boiling point of water increases as pressure increases.

Questions and Problems

Temperature and Pressure (Gay-Lussac's Law)

11.29 Why do aerosol cans explode if heated?

11.30 How can the tires on a car look flat on a cold, winter morning, but have a blowout when the car is driven on hot pavement in the desert?

11.31 Solve for the new pressure when each of the following temperature changes occurs, with n and V constant:
   a. A gas sample has a pressure of 1200 torr at 155°C. What is the final pressure of the gas after the temperature has dropped to 0°C?
   b. An aerosol can has a pressure of 1.40 atm at 12°C. What is the final pressure in the aerosol can if it is used in a room where the temperature is 35°C?

11.32 Solve for the new temperature in degrees Celsius when pressure is changed.
   a. A 10.0-L container of helium gas has a pressure of 250 torr at 0°C. To what Celsius temperature does the sample need to be heated to obtain a pressure of 1500 torr?
   b. A 500.0-mL sample of air at 40.°C and 740 mm Hg is cooled to give a pressure of 680 mm Hg.

11.33 Match the terms *vapor pressure, atmospheric pressure,* and *boiling point* to the following descriptions:
   a. the temperature at which bubbles of vapor appear within the liquid

   b. the pressure exerted by a gas above the surface of its liquid
   c. the pressure exerted on Earth by the particles in the air
   d. the temperature at which the vapor pressure of a liquid becomes equal to the external pressure

11.34 In which pair(s) would boiling occur?

Atmospheric Pressure	Vapor Pressure
a. 760 mm Hg	700 mm Hg
b. 480 torr	480 mm Hg
c. 1.2 atm	912 mm Hg
d. 1020 mm Hg	760 mm Hg
e. 740 torr	1.0 atm

11.35 Give an explanation for the following observations:
   a. Water boils at 87°C on the top of Mt. Whitney.
   b. Food cooks more quickly in a pressure cooker than in an open pan.

11.36 Give an explanation for the following observations:
   a. Boiling water at sea level is hotter than boiling water in the mountains.
   b. Water used to sterilize surgical equipment is heated to 120°C at 2.0 atm in an autoclave.

11.6 The Combined Gas Law

All pressure–volume–temperature relationships for gases that we have studied may be combined into a single relationship called the **combined gas law.** This expression is useful for studying the effect of changes in two of these variables on the third as long as the amount of gas (number of moles) remains constant.

Combined Gas Law

$$\frac{P_1 V_1}{T_1} = \frac{P_2 V_2}{T_2}$$ No change in moles of gas

Table 11.6 Summary of Gas Laws

Combined Gas Law	Properties Held Constant		Relationship	
$\dfrac{P_1}{T_1} = \dfrac{P_2 V_2}{T_2}$	T, n		$P_1 V_1 = P_2 V_2$	Boyle's law
$\dfrac{P_1 V_1}{T_1} = \dfrac{P_2 V_2}{T_2}$	P, n		$\dfrac{V_1}{T_1} = \dfrac{V_2}{T_2}$	Charles' law
$\dfrac{P_1 V_1}{T_1} = \dfrac{P_2 V_2}{T_2}$	V, n		$\dfrac{P_1}{T_1} = \dfrac{P_2}{T_2}$	Gay-Lussac's law

By remembering the combined gas law, we can derive any of the gas laws by omitting those properties that do not change. Table 11.6 summarizes the pressure–volume–temperature relationships of gases.

Sample Problem 11.7 **Using the Combined Gas Law**

A 25.0-mL bubble is released from a diver's air tank at a pressure of 4.00 atm and a temperature of 11°C. What is the volume (mL) of the bubble when it reaches the ocean surface, where the pressure is 1.00 atm and the temperature is 18°C?

Solution

STEP 1 **Organize the data in a table.** We must first change the temperature to kelvins.

$$T_1 = 11°C + 273 = 284 \text{ K}$$
$$T_2 = 18°C + 273 = 291 \text{ K}$$

Conditions 1	Conditions 2	Know	Predict
$P_1 = 4.00$ atm	$P_2 = 1.00$ atm	P decreases	
$V_1 = 25.0$ mL	$V_2 = ?$		V increases
$T_1 = 284$ K	$T_2 = 291$ K	T increases	

STEP 2 **Rearrange the gas law for the unknown.** Because the pressure and temperature are both changing, we must use the combined gas law to solve for V_2.

$$\frac{P_1 V_1}{T_1} = \frac{P_2\ V_2}{T_2}$$

$$\frac{P_1 V_1}{T_1} \times \frac{T_2}{P_2} = \frac{P_2\ V_2 \times T_2}{T_2 \times P_2}$$

$$V_2 = \frac{P_1 V_1 T_2}{T_1 P_2}$$

STEP 3 **Substitute the values into the gas law to solve for the unknown.**
From the data table, we determine that the pressure decrease and
the temperature increase will both increase the volume.

$$V_2 = 25.0 \text{ mL} \times \frac{4.00 \text{ atm}}{1.00 \text{ atm}} \times \frac{291 \text{ K}}{284 \text{ K}} = 102 \text{ mL}$$

Pressure
factor
increases
volume

Temperature
factor
increases
volume

Study Check

A weather balloon is filled with 15.0 L of helium at a temperature of 25°C
and a pressure of 685 mm Hg. What is the pressure (mm Hg) of the helium
in the balloon in the upper atmosphere when the temperature is −35°C and
the volume becomes 34.0 L?

Questions and Problems ▶ **The Combined Gas Law**

11.37 Write the expression for the combined gas law. What
gas laws are combined to make the combined gas law?

11.38 Rearrange the variables in the combined gas law to give
an expression for the following:
a. V_2 **b.** P_2

11.39 A sample of helium gas has a volume of 6.50 L at a pressure of 845 mm Hg and a temperature of 25°C. What is
the pressure of the gas in atm when the volume and temperature of the gas sample are changed to the following?
a. 1850 mL and 325 K **b.** 2.25 L and 12°C
c. 12.8 L and 47°C

11.40 A sample of argon gas has a volume of 735 mL at a pressure of 1.20 atm and a temperature of 112°C. What is the

volume of the gas in milliliters when the pressure and temperature of the gas sample are changed to the following?
a. 658 mm Hg and 281 K **b.** 0.55 atm and 75°C
c. 15.4 atm and −15°C

11.41 A 100.0-mL bubble of hot gases at 225°C and 1.80 atm
escapes from an active volcano. What is the new volume of the bubble outside the volcano where the temperature is −25°C and the pressure is 0.80 atm?

11.42 A scuba diver 40 ft below the ocean surface inhales
50.0 mL of compressed air in a scuba tank at a pressure
of 3.00 atm and a temperature of 8°C. What is the pressure of air in the lungs if the gas expands to 150.0 mL
at a body temperature of 37°C?

11.7 Volume and Moles (Avogadro's Law)

Learning **Goal**

Describe the relationship between
the amount of a gas and its
volume and use this relationship
in calculations.

In our study of the gas laws, we have looked at changes in properties for a
specified amount (n) of gas. Now we will consider how the properties of a gas
change when there is a change in number of moles or grams. For example,
when you blow up a balloon, its volume increases because you add more air
molecules. If a basketball gets a hole in it, and some of the air leaks out, its
volume decreases. In 1811, Amedeo Avogadro stated that the volume of a gas
is directly related to the number of moles of a gas when temperature and pressure are not changed. We refer to this statement as **Avogadro's law.** If the
number of moles of a gas are doubled, then the volume will double as long as
we do not change the pressure or the temperature. (See Figure 11.7.) For two
conditions, we can write

Avogadro's Law

$$\frac{V_1}{n_1} = \frac{V_2}{n_2}$$ No change in pressure or temperature.

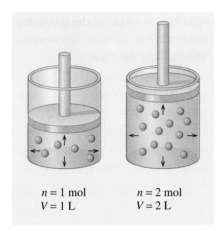

Figure 11.7 **Avogadro's law:** The volume of a gas is directly related to the number of moles of the gas. If the number of moles is doubled, the volume must double at constant temperature and pressure.

Q *If a balloon has a leak, what happens to its volume?*

Sample Problem 11.8   **Calculating Volume for a Change in Moles**

A balloon with a volume of 220 mL is filled with 2.0 mol helium. To what volume (mL) will the balloon expand if 3.0 mol helium are added, to give a total of 5.0 mol helium (the pressure and temperature do not change)?

Solution

STEP 1 **Organize the data in a table.**   A data table for our given information can be set up as follows:

Conditions 1	Conditions 2	Know	Predict
$V_1 = 220\ \text{mL}$	$V_2 = ?$		V increases
$n_1 = 2.0\ \text{mol}$	$n_2 = 5.0\ \text{mol}$	n increases	

STEP 2 **Rearrange the gas law for the unknown.**   Using Avogadro's law, we can solve for V_2.

$$\frac{V_1}{n_1} = \frac{V_2}{n_2}$$

$$n_2 \times \frac{V_1}{n_1} = \frac{V_2}{n_2} \times n_2$$

$$V_2 = \frac{V_1 n_2}{n_1}$$

STEP 3 **Substitute the values into the gas law to solve for the unknown.**   From the table, we see that the number of moles has increased. Because the number of moles and volume are directly related, the volume must increase at constant pressure and temperature. When we substitute in the values, we see the ratio of the moles (mole factor) is greater than 1, which increases volume.

$$V_2 = 220\ \text{mL} \times \frac{5.0\ \cancel{\text{mol}}}{2.0\ \cancel{\text{mol}}} = 550\ \text{mL}$$

New volume   Initial volume   Mole factor that increases volume

Study Check

At a certain temperature and pressure, 8.00 g oxygen has a volume of 5.00 L. What is the volume (L) after 4.00 g oxygen is added to the oxygen in the balloon?

STP and Molar Volume

Using Avogadro's law, we can say that any two gases will have equal volumes if they contain the same number of moles of gas at the same temperature and pressure. To help us make comparisons between different gases, *standard conditions* called standard temperature (273 K) and pressure (1 atm) (abbreviated **STP**) were selected:

STP Conditions

Standard temperature = 0°C (273 K)

Standard pressure = 1 atm (760 mm Hg)

Figure 11.8 Avogadro's law indicates that 1 mole of any gas at STP has a molar volume of 22.4 L.

Q *What volume of gas is occupied by 16.0 g methane gas, CH₄, at STP?*

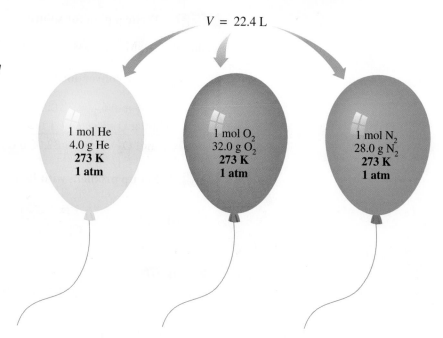

$V = 22.4$ L

1 mol He
4.0 g He
273 K
1 atm

1 mol O₂
32.0 g O₂
273 K
1 atm

1 mol N₂
28.0 g N₂
273 K
1 atm

At STP, 1 mole of any gas occupies a volume of 22.4 L. This value is known as the *molar volume* of a gas. Suppose we have three containers, one filled with 1 mol oxygen gas (O_2), another filled with 1 mol nitrogen gas (N_2), and one filled with 1 mol helium gas (He). When the gases are at STP conditions (1 atm and 273 K), each has a volume of 22.4 L. Thus, the volume of 1 mole of any gas at STP is 22.4 L, the **molar volume** of a gas. (See Figure 11.8.)

Molar Volume

The volume of 1 mol of gas at STP = 22.4 L

As long as a gas is at STP conditions, its molar volume can be used as a conversion factor to convert between the number of moles of gas and its volume.

Moles of gas	**Molar volume 22.4 L/mol**	**Volume (L) of gas**

Molar Volume Conversion Factors

$$\frac{1 \text{ mol gas (STP)}}{22.4 \text{ L}} \quad \text{and} \quad \frac{22.4 \text{ L}}{1 \text{ mol gas (STP)}}$$

Guide to Using Molar Volume

STEP 1
Identify given and needed.

STEP 2
Write a plan.

STEP 3
Write conversion factors.

STEP 4
Set up problem with factors to cancel units.

Sample Problem 11.9 **Using Molar Volume to Find Volume at STP**

What is the volume in liters of 64.0 g O_2 gas at STP?

Solution

Once we convert the mass of O_2 to moles O_2, the molar volume of a gas at STP can be used to calculate the volume (L) of O_2.

STEP 1 **Given** 64.0 g $O_2(g)$ at STP **Need** Volume in liters (L)

STEP 2   **Write a plan for solution.**

grams of O_2 Molar mass   moles of O_2 Molar volume   = liters of O_2

STEP 3   **Write conversion factors.**

$$1 \text{ mol } O_2 = 32.00 \text{ g}$$

$$\frac{32.00 \text{ g } O_2}{1 \text{ mol } O_2} \quad \text{and} \quad \frac{1 \text{ mol } O_2}{32.00 \text{ g } O_2}$$

$$1 \text{ mol } O_2 \text{ (STP)} = 22.4 \text{ L}$$

$$\frac{22.4 \text{ L } O_2}{1 \text{ mol } O_2} \quad \text{and} \quad \frac{1 \text{ mol } O_2}{22.4 \text{ L } O_2}$$

STEP 4   **Set up problem with factors to cancel units.**

$$64.0 \text{ g } O_2 \times \frac{1 \text{ mol } O_2}{32.00 \text{ g } O_2} \times \frac{22.4 \text{ L } O_2}{1 \text{ mol } O_2} = 44.8 \text{ L } O_2 \text{ (STP)}$$

Study Check

What is the volume (L) of 5.10 g of He at STP?

Density of a Gas at STP

We have seen that at the same temperature and pressure, 1 mole of a gas occupies the same volume. Thus, the density ($D = g/L$) of any gas depends on its molar mass. For example, at STP, oxygen, O_2, has a density of 1.43 g/L. The carbon dioxide, CO_2, that we exhale has a density of 1.96 g/L and settles to the ground because the density of CO_2 is greater than the density of air. On the other hand, a balloon filled with helium rises in air because helium has a density of 0.179 g/L, which makes it less dense than air. For any gas at STP, we can calculate density (g/L) using the molar mass and the molar volume, as shown in the next sample problem.

Sample Problem 11.10   **Density of a Gas at STP**

What is the density of nitrogen gas (N_2) at STP?

Solution

The moles of gas provide the grams for the density expression and molar volume will provide the volume of the gas.

STEP 1   **Given** $N_2 (g)$ **Need**   Density (g/L) of N_2 at STP

STEP 2   **Write a plan.**   At STP, the density (g/L) of any gas can be calculated by dividing its molar mass by the molar volume.

$$\text{Density} = \frac{\text{Molar mass}}{\text{Molar volume}} = \frac{\text{g/mol}}{\text{L/mol}} = \frac{\text{g}}{\text{L}}$$

STEP 3   **Write conversion factors.**

$$1 \text{ mol N}_2 = 28.02 \text{ g N}_2$$

$$\frac{28.02 \text{ g N}_2}{1 \text{ mol N}_2} \quad \text{and} \quad \frac{1 \text{ mol N}_2}{28.02 \text{ g N}_2}$$

$$1 \text{ mol N}_2 \text{ (STP)} = 22.4 \text{ L}$$

$$\frac{22.4 \text{ L N}_2}{1 \text{ mol N}_2} \quad \text{and} \quad \frac{1 \text{ mol N}_2}{22.4 \text{ L N}_2}$$

STEP 4   **Set up the problem with factors to cancel units.**

$$\text{Density (g/L) of N}_2 = \frac{\text{Mass}}{\text{Volume}} \quad \frac{\dfrac{28.02 \text{ g N}_2}{1 \text{ mol N}_2}}{\dfrac{22.4 \text{ L N}_2}{1 \text{ mol N}_2}} = 1.25 \text{ g/L}$$

Study Check

What is the density of hydrogen gas (H_2) at STP?

Questions and Problems   **Volume and Moles (Avogadro's Law)**

11.43 What happens to the volume of a bicycle tire or a basketball when you use a pump to add air?

11.44 Sometimes when you blow up a balloon and release it, it flies around the room. What is happening to the air that was in the balloon and its volume?

11.45 A sample containing 1.50 mol neon gas has a volume of 8.00 L. What is the new volume of the gas in liters when the following changes occur in the quantity of the gas at constant pressure and temperature?
   a. A leak allows one-half of the neon atoms to escape.
   b. A sample of 25.0 g neon is added to the neon gas already in the container.
   c. A sample of 3.50 mol Ne is added to the neon gas already in the container.

11.46 A sample containing 4.80 g O_2 gas has a volume of 15.0 L. Pressure and temperature remain constant.
   a. What is the new volume if 0.50 mol O_2 gas is added?
   b. Oxygen is released until the volume is 10.0 L. How many mol O_2 is removed?
   c. What is the volume after 4.00 g He is added to the O_2 gas already in the container?

11.47 Use the molar volume of a gas to solve the following at STP:
   a. the number of moles of O_2 in 44.8 L O_2 gas
   b. the number of moles of CO_2 in 4.00 L CO_2 gas
   c. the volume (L) of 6.40 g O_2
   d. the volume (mL) occupied by 50.0 g neon

11.48 Use molar volume to solve the following problems at STP:
   a. the volume (L) occupied by 2.5 mol N_2
   b. the volume (mL) occupied by 0.420 mol He
   c. the number of grams of neon contained in 11.2 L Ne gas
   d. the number of moles of H_2 in 1600 mL H_2 gas

11.49 Calculate the densities of each of the following gases in g/L at STP:
   a. F_2 **b.** CH_4
   c. Ne   **d.** SO_2

11.50 Calculate the densities of each of the following gases in g/L at STP:
   a. C_3H_8 (propane)   **b.** NH_3
   c. Cl_2 **d.** Ar

Learning **Goal**

Use the ideal gas law to solve for *P, V, T,* or *n* of a gas when given three of the four values in the ideal gas equation. Calculate density, molar mass, or volume of a gas in a chemical reaction.

the
Chemistry
place

WEB TUTORIAL

The Ideal Gas Law

11.8 The Ideal Gas Law

The four properties used in the measurement of a gas—pressure (*P*), volume (*V*), temperature (*T*), and amount of a gas (*n*)—can be combined to give a single expression called the **ideal gas law,** which is written as follows:

Ideal Gas Law

$$PV = nRT$$

Rearranging the ideal gas law shows that the four gas properties equal a constant, *R*.

$$\frac{PV}{nT} = R$$

To calculate the value of *R*, we substitute the STP conditions for molar volume into the expression: 1 mole of any gas occupies 22.4 L at STP (273 K and 1 atm).

$$R = \frac{(1.00 \text{ atm})(22.4 \text{ L})}{(1.00 \text{ mol})(273 \text{ K})} = \frac{0.0821 \text{ L} \cdot \text{atm}}{\text{mol} \cdot \text{K}}$$

The value for *R*, the **universal gas constant,** is 0.0821 L·atm per mol·K. If we use 760 mm Hg for the pressure, we obtain another useful value for *R* of 62.4 L·mm Hg per mol·K.

$$R = \frac{(760 \text{ mm Hg})(22.4 \text{ L})}{(1.00 \text{ mol})(273 \text{ K})} = \frac{62.4 \text{ L} \cdot \text{mm Hg}}{\text{mol} \cdot \text{K}}$$

The ideal gas law is a useful expression when you are given the values for any three of the four properties of a gas.

In working problems using the ideal gas law, the units of each variable must match the units in the *R* you select.

Property	Unit
Pressure (*P*)	atm or mm Hg
Volume (*V*)	L
Amount (*n*)	moles
Temperature (*T*)	K

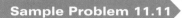

Sample Problem 11.11 **Using the Ideal Gas Law**

Dinitrogen oxide, N_2O, which is used in dentistry, is an anesthetic also called "laughing gas." What is the pressure in atmospheres of 0.350 mol N_2O at 22°C in a 5.00-L container?

Solution

STEP 1 **Organize the data, including *R*, in a table.** When three of the four quantities (*P, V, n,* and *T*) are known, we use the ideal gas law to solve for the unknown quantity. The units must match the units of the gas constant *R*. The temperature is converted from Celsius to Kelvin.

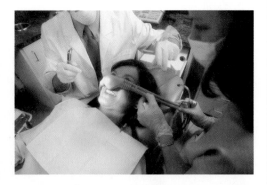

Guide to Using the Ideal Gas

STEP 1
Organize data given for the gas.

STEP 2
Solve the ideal gas law for the unknown.

STEP 3
Substitute gas data and calculate unknown value.

$P = ?$

$V = 5.00 \text{ L}$

$n = 0.350 \text{ mol}$

$R = 0.0821 \dfrac{\text{L} \cdot \text{atm}}{\text{mol} \cdot \text{K}}$

$T = 22°\text{C} + 273 = 295 \text{ K}$

STEP 2 **Rearrange the ideal gas law to solve for the unknown.** By dividing both sides of the ideal gas law by V, we solve for pressure, P:

$$P\ V = nRT \qquad \text{Ideal gas law}$$

$$P\ \frac{\cancel{V}}{\cancel{V}} = \frac{nRT}{V}$$

$$P = \frac{nRT}{V}$$

STEP 3 Substitute values from the table to calculate the unknown.

$$P = \frac{0.350 \ \cancel{\text{mol}} \times 0.0821 \ \dfrac{\text{L} \cdot \text{atm}}{\cancel{\text{mol}} \cdot K} \times 295 \ K}{5.00 \ \cancel{\text{L}}} = 1.70 \text{ atm}$$

Study Check

Chlorine gas, Cl_2, is used to purify the water in swimming pools. How many moles of chlorine gas are in a 7.00-L tank if the gas has a pressure of 865 mm Hg and a temperature of 24°C?

Sample Problem 11.12 **Calculating Mass Using the Ideal Gas Law**

Butane, C_4H_{10}, is used as a fuel for barbecues and as an aerosol propellant. If you have 108 mL butane at 715 mm Hg and 25°C, what is the mass (g) of the butane?

Solution

STEP 1 **Organize the data, including R, in a table.** When three of the four quantities (P, V, n, and T) are known, we use the ideal gas law to solve for the unknown quantity. Because the pressure is given in mm Hg, we will use the R in mm Hg. The volume given in milliliters (mL) is converted to a volume in liters (L). The temperature is converted from Celsius to Kelvin.

Initial Values	Adjusted for Units in Gas Constant R
$P = 715 \text{ mm Hg}$	715 mm Hg
$V = 108 \text{ mL}$	$108 \ \cancel{\text{mL}} \times \dfrac{1 \text{ L}}{1000 \ \cancel{\text{mL}}} = 0.108 \text{ L}$
$n = ? \text{ mol of } C_4H_{10}$	$? \text{ mol of } C_4H_{10}$
$R = \dfrac{62.4 \text{ L} \cdot \text{mm Hg}}{\text{mol} \cdot \text{K}}$	$\dfrac{62.4 \text{ L} \cdot \text{mm Hg}}{\text{mol} \cdot \text{K}}$
$T = 25°\text{C}$	$25°\text{C} + 273 = 298 \text{ K}$

STEP 2 **Rearrange the ideal gas law to solve for the unknown.** By dividing both sides of the ideal gas law by RT, we solve for moles, n:

$$PV = \boxed{n} \; RT \qquad \text{Ideal gas equation}$$

$$\frac{PV}{RT} = n \; \frac{\cancel{RT}}{\cancel{RT}}$$

$$\boxed{n} = \frac{PV}{RT}$$

STEP 3 **Substitute values from the table to calculate the unknown.**

$$n = \frac{715 \; \cancel{\text{mm Hg}} \times 0.108 \; \cancel{L}}{\dfrac{62.4 \; \cancel{L} \cdot \cancel{\text{mm Hg}}}{\text{mol} \cdot \cancel{K}} \times 298 \; \cancel{K}} = 0.00415 \; \text{mol} \; (4.15 \times 10^{-3} \, \text{mol})$$

Now we convert the moles of butane to grams using its molar mass (58.12 g/mol).

$$0.00415 \; \cancel{\text{mol C}_4\text{H}_{10}} \times \frac{58.12 \; \text{g C}_4\text{H}_{10}}{1 \; \cancel{\text{mol C}_4\text{H}_{10}}} = 0.241 \; \text{g C}_4\text{H}_{10}$$

Study Check

What is the volume of 1.20 g carbon monoxide at 8°C if it has a pressure of 724 mm Hg?

Molar Mass of a Gas

Another use of the ideal gas law is to determine the molar mass of a gas or a liquid that easily forms a gas. If the mass of the gas is known, the number of moles is calculated using the molar volume at STP or the ideal gas law. Then molar mass (g/mol) can be determined. Suppose that 0.357 g of a gas occupies a volume of 0.500 L at STP. Under STP conditions, we find the moles of the gas using the molar volume.

$$n = 0.500 \; \cancel{L} \times \frac{1 \; \text{mol}}{22.4 \; \cancel{L}} = \boxed{0.0223 \; \text{mol}}$$

$$\text{Molar mass} = \frac{\text{Mass}}{\text{Moles}} = \frac{0.357 \; \text{g}}{\boxed{0.0223 \; \text{mol}}} = 16.0 \; \text{g/mol}$$

Sample Problem 11.13 ▷ **Molar Mass of a Gas at STP**

What is the molar mass of a gas if a 2.21-g sample occupies a volume of 0.425 L at STP?

Solution

STEP 1 **Given** 2.21 g gas, V = 0.425 L **Need** Molar mass (g/mol)

STEP 2   **Write a plan.**   The molar mass is the grams of the gas divided by the moles. We can determine the moles of a gas at STP using molar volume.

liters of gas   Molar volume   moles of gas ⟶ Molar mass

STEP 3   **Write conversion factors.**

$$1 \text{ mol gas (STP)} = 22.4 \text{ L}$$

$$\frac{22.4 \text{ L gas}}{1 \text{ mol gas}} \quad \text{and} \quad \frac{1 \text{ mol gas}}{22.4 \text{ L gas}}$$

The moles of the gas are calculated as

$$n = 0.425 \text{ L gas} \times \frac{1 \text{ mol gas}}{22.4 \text{ L gas}} = 0.0190 \text{ mol}$$

STEP 4   **Set up calculation for molar mass (g/mol).**   The molar mass of the gas is obtained by dividing the mass by the moles of gas.

$$\text{Molar mass} = \frac{\text{Mass}}{\text{Moles}} = \frac{2.21 \text{ g}}{0.0190 \text{ mol}} = 116 \text{ g/mol}$$

Study Check

What is the molar mass if 4.88 g of an unknown gas has a volume of 1.50 L at STP?

Sample Problem 11.14   **Molar Mass of a Gas Using the Ideal Gas Law**

What is the molar mass of a gas if a 3.16-g sample of gas at 0.750 atm at 45°C occupies a volume of 2.05 L?

Solution

STEP 1   **Given**   3.16 g gas, $P = 0.750$ atm, $V = 2.05$ L, $T = 45°C + 273 = 318$ K   **Need**   Molar mass (g/mol)

STEP 2   **Write a plan.**   The molar mass is the grams of the gas divided by the moles. In this problem, we need to use the ideal gas law to determine the moles of the gas.

P, V, T Ideal Gas Law   moles (n) gas

$$\frac{3.16 \text{ g sample}}{\text{moles } (n) \text{ gas}} = \text{molar mass (g/mol)}$$

STEP 3   **Write conversion factors.**

$$PV = n\,RT \qquad \text{Ideal gas law}$$

$$\frac{PV}{RT} = n \frac{RT}{RT}$$

$$n = \frac{PV}{RT}$$

The moles of the gas are calculated as

$$n = \frac{0.750 \text{ atm} (2.05 \text{ L})}{0.0821 \frac{\text{L} \cdot \text{atm}}{\text{mol} \cdot \text{K}} (318 \text{ K})} = 0.0589 \text{ mol}$$

STEP 4 **Set up calculation for molar mass (g/mol).** The molar mass of the gas is obtained by dividing the mass by the moles of gas.

$$\text{Molar mass} = \frac{\text{Mass}}{\text{Moles}} = \frac{3.16 \text{ g}}{0.0589 \text{ mol}} = 53.7 \text{ g/mol}$$

Study Check

What is the molar mass of an unknown gas in a 1.50-L container if 0.488 g has a pressure of 0.0750 atm at 19.0°C?

Questions and Problems **The Ideal Gas Law**

11.51 Calculate the pressure, in atmospheres, of 2.00 mol helium gas in a 10.0-L container at 27°C.

11.52 What is the volume in liters of 4.0 mol methane gas, CH_4, at 18°C and 1.40 atm?

11.53 An oxygen gas container has a volume of 20.0 L. How many grams of oxygen are in the container if the gas has a pressure of 845 mm Hg at 22°C?

11.54 A 10.0-g sample of krypton has a temperature of 25°C at 575 mm Hg. What is the volume, in milliliters, of the krypton gas?

11.55 A 25.0-g sample of nitrogen, N_2, has a volume of 50.0 L and a pressure of 630. mm Hg. What is the temperature of the gas?

11.56 A 0.226-g sample of carbon dioxide, CO_2, has a volume of 525 mL and a pressure of 455 mm Hg. What is the temperature of the gas?

11.57 Determine the molar mass of each of the following gases:
 a. 0.84 g of a gas that occupies 450 mL at STP
 b. a gas with a density of 1.28 g/L at STP
 c. 1.48 g of a gas that occupies 1.00 L at 685 mm Hg and 22°C
 d. 2.96 g of a gas that occupies 2.30 L at 0.95 atm and 24°C

11.58 Determine the molar mass of each of the following gases:
 a. 11.6 g of a gas that occupies 2.00 L at STP
 b. a gas with a density of 0.715 g/L at STP
 c. 0.726 g of a gas that occupies 855 mL at 1.20 atm and 18°C
 d. 2.32 g of a gas that occupies 1.23 L at 685 mm Hg and 25°C

11.9 Gas Laws and Chemical Reactions

Learning **Goal**

Determine the mass or volume of a gas that reacts or forms in a chemical reaction.

Gases are involved as reactants and products in many chemical reactions. For example, we have seen that the combustion of carbon-based fuels and oxygen gas gives carbon dioxide gas and water vapor. In combination reactions, we have seen that hydrogen gas and nitrogen gas react to form ammonia gas, and hydrogen gas and oxygen gas produce water vapor. Typically, the information given for a gas in a reaction is its pressure (P), volume (V), and temperature (T). We need to use the ideal gas law or molar volume at STP to determine the moles of a gas in a reaction. Once we know the moles of one of the gases in a reaction, we can use a mole factor to determine the moles of any other substance, as we have done before.

> **Sample Problem 11.15** ▷ **Gases in Chemical Reactions at STP**

When potassium metal reacts with chlorine gas, the product is solid potassium chloride.

$$2K(s) + Cl_2(g) \longrightarrow 2KCl(s)$$

How many grams of potassium chloride are produced when 7.25 L chlorine gas at STP reacts with potassium?

Solution

Guide to Problem Solving Reactions Involving Gases

STEP 1
Find moles of gas A using molar volume or ideal gas law.

STEP 2
Determine moles of substance B using-mole factor.

STEP 3
Convert moles of substance B to grams or volume.

STEP 1 **Find moles of gas A using molar volume or ideal gas law.** At STP, we can use molar volume (22.4 L/mol) to determine moles of Cl_2 gas.

$$7.25 \text{ L } Cl_2 \times \frac{1 \text{ mol } Cl_2}{22.4 \text{ L } Cl_2} = 0.324 \text{ mol } Cl_2$$

STEP 2 **Determine moles of substance B using the mole–mole factor from the balanced equation.**

$$1 \text{ mol } Cl_2 = 2 \text{ mol } KCl$$

$$\frac{2 \text{ mol } KCl}{1 \text{ mol } Cl_2} \quad \text{and} \quad \frac{1 \text{ mol } Cl_2}{2 \text{ mol } KCl}$$

$$0.324 \text{ mol } Cl_2 \times \frac{2 \text{ mol } KCl}{1 \text{ mol } Cl_2} = 0.648 \text{ mol } KCl$$

STEP 3 **Convert moles of substance B to grams or volume.** Using the molar mass of KCl, we can determine the grams of KCl.

$$1 \text{ mol } KCl = 74.55 \text{ g } KCl$$

$$\frac{1 \text{ mol } KCl}{74.55 \text{ g } KCl} \quad \text{and} \quad \frac{74.55 \text{ g } KCl}{1 \text{ mol } KCl}$$

$$0.648 \text{ mol } KCl \times \frac{74.55 \text{ g } KCl}{1 \text{ mol } KCl} = 48.3 \text{ g } KCl$$

These steps can also be set up as a continuous solution.

$$7.25 \text{ L } Cl_2 \times \frac{1 \text{ mol } Cl_2}{22.4 \text{ L } Cl_2} \times \frac{2 \text{ mol } KCl}{1 \text{ mol } Cl_2} \times \frac{74.55 \text{ g } KCl}{1 \text{ mol } KCl} = 48.3 \text{ g } KCl$$

Study Check

Hydrogen gas forms when zinc metal reacts with aqueous HCl.

$$Zn(s) + 2HCl(aq) \longrightarrow ZnCl_2(aq) + H_2(g)$$

How many liters of hydrogen gas at STP are produced when 15.8 g zinc reacts?

Sample Problem 11.16   **Ideal Gas Law and Chemical Equations**

Limestone ($CaCO_3$) reacts with HCl to produce aqueous calcium chloride and carbon dioxide gas.

$$CaCO_3(s) + 2HCl(aq) \longrightarrow CaCl_2(aq) + CO_2(g) + H_2O(l)$$

How many liters of CO_2 are produced at 752 mm Hg and 24°C from a 25.0-g sample of limestone?

Solution

STEP 1   **Find the moles of substance A using molar mass.**   The moles of limestone require the molar mass of limestone.

$$1 \text{ mol } CaCO_3 = 100.09 \text{ g } CaCO_3$$

$$\frac{100.09 \text{ g } CaCO_3}{1 \text{ mol } CaCO_3} \quad \text{and} \quad \frac{1 \text{ mol } CaCO_3}{100.09 \text{ g } CaCO_3}$$

$$25.0 \text{ g } CaCO_3 \times \frac{1 \text{ mol } CaCO_3}{100.09 \text{ g } CaCO_3} = 0.250 \text{ mol } CaCO_3$$

STEP 2   **Determine moles of substance B using the mole–mole factor from the balanced equation.**

$$1 \text{ mol } CaCO_3 = 1 \text{ mol } CO_2$$

$$\frac{1 \text{ mol } CaCO_3}{1 \text{ mol } CO_2} \quad \text{and} \quad \frac{1 \text{ mol } CO_2}{1 \text{ mol } CaCO_3}$$

$$0.250 \text{ mol } CaCO_3 \times \frac{1 \text{ mol } CO_2}{1 \text{ mol } CaCO_3} = 0.250 \text{ mol } CO_2$$

Steps 1 and 2 can be combined as

$$25.0 \text{ g } CaCO_3 \times \frac{1 \text{ mol } CaCO_3}{100.09 \text{ g } CaCO_3} \times \frac{1 \text{ mol } CO_2}{1 \text{ mol } CaCO_3} = 0.250 \text{ mol } CO_2$$

STEP 3   **Convert moles of gas B to volume.**   Now the moles of CO_2 can be placed in the ideal gas law to solve for the volume (L) of gas. The ideal gas law solved for volume is

$$V = \frac{nRT}{P}$$

$$V = \frac{(0.250 \text{ mol } CO_2)\left(\dfrac{62.4 \text{ L} \cdot \text{mm Hg}}{K \cdot \text{mol}}\right)(297 \text{ K})}{752 \text{ mm Hg}} = 6.16 \text{ L}$$

Study Check

If 12.8 g aluminum reacts with HCl, how many liters of H_2 would be formed at 715 mm Hg and 19°C?

$$2Al(s) + 6HCl(aq) \longrightarrow 2AlCl_3(aq) + 3H_2(g)$$

11.59 Mg metal reacts with HCl to produce hydrogen gas:

$$Mg(s) + 2HCl(aq) \longrightarrow MgCl_2(aq) + H_2(g)$$

a. What volume of hydrogen at STP is released when 8.25 g Mg reacts?
b. How many grams of magnesium are needed to prepare 5.00 L H_2 at 735 mm Hg and 18°C?

11.60 When heated to 350°C at 0.950 atm, ammonium nitrate decomposes to produce nitrogen, water, and oxygen gases:

$$2NH_4NO_3(s) \longrightarrow 2N_2(g) + 4H_2O(g) + O_2(g)$$

a. How many liters of water vapor are produced when 25.8 g NH_4NO_3 decomposes?
b. How many grams of NH_4NO_3 are needed to produce 10.0 L oxygen?

11.61 Butane is used to fill gas tanks for heating. The following equation describes its combustion:

$$2C_4H_{10}(g) + 13O_2(g) \longrightarrow 8CO_2(g) + 10H_2O(g)$$

If a tank contains 55.2 g butane, what volume in liters of oxygen is needed to burn all the butane at 0.850 atm and 25°C?

11.62 What volume, in liters, of O_2 at 35°C and 1.19 atm can be produced from the decomposition of 50.0 g KNO_3?

$$2KNO_3(s) \longrightarrow 2KNO_2(s) + O_2(g)$$

11.63 Aluminum oxide is formed from its elements.

$$4Al(s) + 3O_2(g) \longrightarrow 2Al_2O_3(s)$$

How many liters of oxygen at STP are needed to react 5.4 g aluminum?

11.64 Nitrogen dioxide reacts with water to produce oxygen and ammonia.

$$4NO_2(g) + 6H_2O(g) \longrightarrow 7O_2(g) + 4NH_3(g)$$

At a temperature of 415°C and a pressure of 725 mm Hg, how many grams of NH_3 can be produced when 4.00 L NO_2 react?

11.10 Partial Pressures (Dalton's Law)

Learning Goal

Use partial pressures to calculate the total pressure of a mixture of gases.

Many gas samples are a mixture of gases. For example, the air you breathe is a mixture of mostly oxygen and nitrogen gases. For gas mixtures, we use the gas laws we have studied because particles of all gases behave in the same way. Therefore, the total pressure of the gases in a mixture is a result of the collisions of the gas particles regardless of what type of gas they are.

In a gas mixture, each gas exerts its **partial pressure,** which is the pressure it would exert if it were the only gas in the container. **Dalton's law** states that the total pressure of a gas mixture is the sum of the partial pressures of the gases in the mixture.

$$P_{total} = P_1 + P_2 + P_3 + \cdots$$

Total pressure of a gas mixture = Sum of the partial pressures of the gases in the mixture

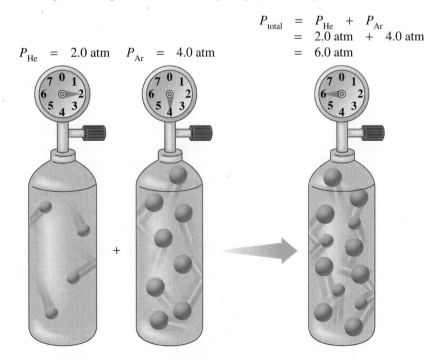

$$P_{He} = 2.0 \text{ atm} \qquad P_{Ar} = 4.0 \text{ atm}$$

$$
\begin{aligned}
P_{total} &= P_{He} + P_{Ar} \\
&= 2.0 \text{ atm} + 4.0 \text{ atm} \\
&= 6.0 \text{ atm}
\end{aligned}
$$

Suppose we have two separate tanks, one filled with helium at 2.0 atm and the other filled with argon at 4.0 atm. When the gases are combined in a single tank with the same volume and temperature, the number of gas molecules, not the type of gas, determines the pressure in a container. There the pressure of the gas mixture would be 6.0 atm, which is the sum of their individual or partial pressures.

Sample Problem 11.17   **Calculating the Total Pressure of a Gas Mixture**

A 10-L gas tank contains propane (C_3H_8) gas at a pressure of 300. torr. Another 10-L gas tank contains methane (CH_4) gas at a pressure of 500. torr. In preparing a gas fuel mixture, the gases from both tanks are combined in a 10-L container at the same temperature. What is the pressure of the gas mixture?

Solution

Using Dalton's law of partial pressure, we find that the total pressure of the gas mixture is the sum of the partial pressures of the gases in the mixture.

$$P_{total} = P_{propane} + P_{methane}$$
$$= 300. \text{ torr} + 500. \text{ torr}$$
$$= 800. \text{ torr}$$

Therefore, when both propane and methane are placed in the same container, the total pressure of the mixture is 800. torr.

Study Check

A gas mixture consists of helium with a partial pressure of 315 mm Hg, nitrogen with a partial pressure of 204 mm Hg, and argon with a partial pressure of 422 mm Hg. What is the total pressure in atmospheres?

Air Is a Gas Mixture

The air you breathe is a mixture of gases. What we call the atmospheric pressure is actually the sum of the partial pressures of the gases in the air. Table 11.7 lists partial pressures for the gases in air on a typical day.

Table 11.7   Typical Composition of Air

Gas	Partial Pressure (mm Hg)	Percentage (%)
Nitrogen, N_2	594.0	78
Oxygen, O_2	160.0	21
Carbon dioxide, CO_2	0.3 ⎫	
Water vapor, H_2O	5.7 ⎭	1
Total air	760.0	100

Sample Problem 11.18   **Partial Pressure of a Gas in a Mixture**

A mixture of oxygen and helium is prepared for a scuba diver who is going to descend 200 ft below the ocean surface. At that depth, the diver breathes a gas mixture that has a total pressure of 7.0 atm. If the partial pressure of the oxygen in the tank at that depth is 1140 mm Hg, what is the partial pressure (atm) of the helium?

Solution

STEP 1   **Write the equation for the sum of the partial pressure.**   From Dalton's law of partial pressures, we know that the total pressure is equal to the sum of the partial pressures.

$$P_{total} = P_{O_2} + P_{He}$$

Guide to Solving for Partial Pressure

STEP 1
Write the equation for sum of partial pressures.

STEP 2
Solve for the unknown pressure.

STEP 3
Substitute known pressures and calculate unknown.

STEP 2   **Solve for the unknown pressure.**

$$P_{total} = P_{O_2} + P_{He}$$
$$P_{He} = P_{total} - P_{O_2}$$

Convert units to match.

$$P_{O_2} = 1140 \ \text{mm Hg} \times \frac{1 \ \text{atm}}{760 \ \text{mm Hg}} = 1.50 \ \text{atm}$$

STEP 3   **Substitute known pressures and calculate the unknown.**

$$P_{He} = P_{total} + P_{O_2}$$
$$P_{He} = 7.0 \ \text{atm} - 1.50 \ \text{atm} = 5.5 \ \text{atm}$$

Study Check

An anesthetic consists of a mixture of cyclopropane gas, C_3H_6, and oxygen gas, O_2. If the mixture has a total pressure of 825 torr, and the partial pressure of the cyclopropane is 73 torr, what is the partial pressure of the oxygen in the anesthetic?

Gases Collected over Water

In the laboratory, gases are often collected by bubbling them through water into a container. (See Figure 11.9.) Suppose we allow magnesium (Mg) to react with HCl to form $MgCl_2$ and H_2 gas.

$$Mg(s) + 2HCl(aq) \longrightarrow MgCl_2(aq) + H_2(g)$$

The H_2 displaces water in the container. Because some water evaporates at the same temperature, there is also water vapor in the container. When we determine the pressure of the collected gas, it is actually equal to the sum of the partial pressure of the hydrogen gas and water vapor. Because we need to determine the pressure of the dry gas, we must find the vapor pressure of water (Table 11.4) at the experimental temperature and subtract it from the total gas pressure. Once we know the pressure of the dry H_2 gas, we can use the V of the container and T to determine the moles or grams of the hydrogen gas we collected.

Sample Problem 11.19   **Moles of Gas Collected over Water**

When magnesium reacts with HCl, a volume of 355 mL hydrogen gas is collected over water at 26°C.

$$Mg(s) + 2HCl(aq) \longrightarrow MgCl_2(aq) + H_2(g)$$

If the pressure of the gases is 752 mm Hg, how many moles of $H_2(g)$ were collected? The vapor pressure of water at 26°C is 25 mm Hg.

Solution

STEP 1   **Given** $P_{total} = 752 \ \text{mm Hg}, V = 355 \ \text{mL} \ (0.355 \ \text{L})$,
$T = 26°C + 273 = 299 \ \text{K}$ **Need**   Moles (n) of H_2

Figure 11.9 A gas from a reaction is collected by bubbling through water. Due to evaporation of water, the total pressure is equal to the partial pressure of the gas and the vapor pressure of water.

Q How is the pressure of the dry gas determined?

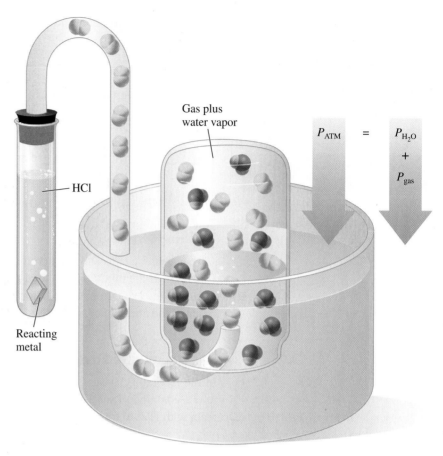

Gas plus water vapor

HCl

Reacting metal

P_{ATM} = P_{H_2O} + P_{gas}

Guide to Gases Collected over Water

STEP 1
Obtain vapor pressure of water.

STEP 2
Subtract vapor pressure from total *P* of gas mixture to give partial pressure of gas.

STEP 3
Use ideal gas law to convert P_{gas} to moles or grams of gas collected.

STEP 2 **Rearrange the ideal gas law to solve for the unknown.** By dividing both sides of the ideal gas law by *RT*, we solve for moles, *n*:

$$PV = n\ RT \qquad \text{Ideal gas law}$$

$$\frac{PV}{RT} = n\ \frac{\cancel{RT}}{\cancel{RT}}$$

$$n = \frac{PV}{RT}$$

STEP 3 **Substitute values from the table to calculate the unknown.** First, we determine the partial pressure of H_2 using Dalton's law of partial pressures.

$$P_{total} = 752\ \text{mm Hg} = P_{H_2} + P_{H_2O} = P_{H_2} + 25\ \text{mm Hg}$$

Solving for the partial pressure of H_2 gives

$$P_{H_2} = 752\ \text{mm Hg} - 25\ \text{mm Hg} = 727\ \text{mm Hg}$$

Substituting the P_{H_2} into the ideal gas law, we calculate moles of H_2 collected.

$$n_{H_2} = \frac{P_{H_2}V}{RT}$$

$$n_{H_2} = \frac{727\ \cancel{\text{mm Hg}} \times 0.355\ \cancel{L}}{62.4\ \cancel{L} \cdot \dfrac{\cancel{\text{mm Hg}}}{\text{mol} \cdot \cancel{K}} \times 299\ \cancel{K}} = 0.0138\ \text{mol}\ (1.38 \times 10^{-2}\ \text{mol})$$

Study Check

A 456-mL sample of oxygen gas (O_2) was collected over water at a pressure of 744 mm Hg and a temperature of 15°C. How many grams of O_2 were collected? At 15°C, the vapor pressure of water is 13 mm Hg.

CHEM NOTE

BLOOD GASES

Our cells continuously use oxygen and produce carbon dioxide. Both gases move in and out of the lungs through the membranes of the alveoli, the tiny air sacs at the ends of the airways in the lungs. An exchange of gases occurs in which oxygen from the air diffuses into the lungs and into the blood, while carbon dioxide produced in the cells is carried to the lungs to be exhaled. In Table 11.8, partial pressures are given for the gases in air that we inhale (inspired air), air in the alveoli, and the air that we exhale (expired air). The partial pressure of water vapor increases within the lungs because the vapor pressure of water is 47 mm Hg at body temperature.

At sea level, oxygen normally has a partial pressure of 100 mm Hg in the alveoli of the lungs. Because the partial pressure of oxygen in venous blood is 40 mm Hg, oxygen diffuses from the alveoli into the bloodstream. The oxygen combines with hemoglobin, which carries it to the tissues of the body where the partial pressure of oxygen can be very low, less than 30 mm Hg. Oxygen diffuses from the blood, where the partial pressure of O_2 is high, into the tissues, where O_2 pressure is low.

As oxygen is used in the cells of the body during metabolic processes, carbon dioxide is produced, so the partial pressure of CO_2 may be as high as 50 mm Hg or more. Carbon dioxide diffuses from the tissues into the bloodstream and is carried to the lungs. There it diffuses out of the blood, where CO_2 has a partial pressure of 46 mm Hg, into the alveoli, where the CO_2 is at 40 mm Hg and is exhaled. Table 11.9 gives the partial pressures of blood gases in the tissues and in oxygenated and deoxygenated blood.

Table 11.8 Partial Pressures of Gases During Breathing

Gas	Partial Pressure (mm Hg)		
	Inspired Air	Alveolar Air	Expired Air
Nitrogen, N_2	594.0	573	569
Oxygen, O_2	160.0	100	116
Carbon dioxide, CO_2	0.3	40	28
Water vapor, H_2O	5.7	47	47
Total	760.0	760	760

Table 11.9 Partial Pressures of Oxygen and Carbon Dioxide in Blood and Tissues

Gas	Partial Pressure (mm Hg)		
	Oxygenated Blood	Deoxygenated Blood	Tissues
O_2	100	40	30 or less
CO_2	40	46	50 or greater

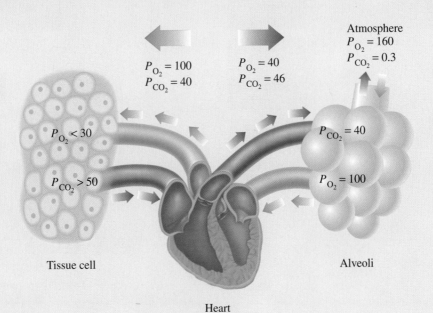

Questions and Problems   **Partial Pressures (Dalton's Law)**

11.65 When solid $KClO_3$ is heated, it decomposes to give solid KCl and O_2 gas. A volume of 256 mL of gas is collected over water at a total pressure of 765 mm Hg and 24°C. The vapor pressure of water at 24°C is 22 mm Hg.

$$2KClO_3(s) \longrightarrow 2KCl(s) + 3O_2(g)$$

a. What was the partial pressure of the O_2 gas?
b. How many moles of O_2 gas were in the gas sample?

11.66 When solid $CaCO_3$ is heated, it decomposes to give solid CaO and O_2 gas. A volume of 425 mL of gas is collected over water at a total pressure of 758 mm Hg and 16°C. The vapor pressure of water at 16°C is 14 mm Hg.

$$CaCO_3(s) \longrightarrow CaO(s) + O_2(g)$$

a. What was the partial pressure of the O_2 gas?
b. How many moles of O_2 gas were in the gas sample?

11.67 A typical air sample in the lungs contains oxygen at 100 mm Hg, nitrogen at 573 mm Hg, carbon dioxide at 40 mm Hg, and water vapor at 47 mm Hg. Why are these pressures called partial pressures?

11.68 Suppose a mixture contains helium and oxygen gases. If the partial pressure of helium is the same as the partial pressure of oxygen, what do you know about the number of helium atoms compared to the number of oxygen molecules? Explain.

11.69 In a gas mixture, the partial pressures are: nitrogen 425 torr, oxygen 115 torr, and helium 225 torr. What is the total pressure (torr) exerted by the gas mixture?

11.70 In a gas mixture, the partial pressures are: argon 415 mm Hg, neon 75 mm Hg, and nitrogen 125 mm Hg. What is the total pressure (atm) exerted by the gas mixture?

11.71 A gas mixture containing oxygen, nitrogen, and helium exerts a total pressure of 925 torr. If the partial pressures are oxygen 425 torr and helium 75 torr, what is the partial pressure (torr) of the nitrogen in the mixture?

11.72 A gas mixture containing oxygen, nitrogen, and neon exerts a total pressure of 1.20 atm. If helium added to the mixture increases the pressure to 1.50 atm, what is the partial pressure (atm) of the helium?

Concept Map   **Gases**

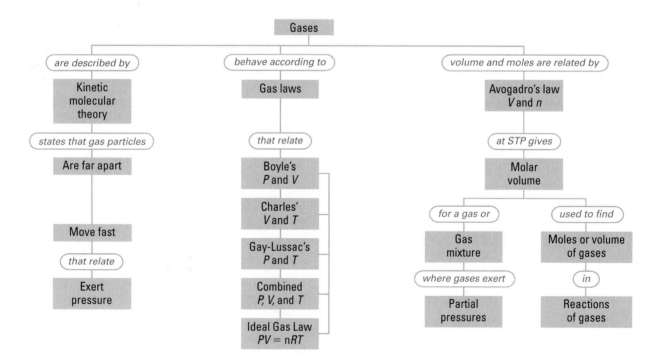

Chapter Review

11.1 Properties of Gases

In a gas, particles are far apart and moving very fast. A gas is described by the physical properties of pressure (P), volume (V), temperature (T), and amount in moles (n).

11.2 Gas Pressure

A gas exerts pressure, the force of the gas particles striking the surface of a container. Gas pressure is measured in units of torr, mm Hg, atm, and pascal.

11.3 Pressure and Volume (Boyle's Law)

The volume (V) of a gas changes inversely with the pressure (P) of the gas if there is no change in the amount and temperature: $P_1V_1 = P_2V_2$. This means that the pressure increases if volume decreases; pressure decreases if volume increases.

11.4 Temperature and Volume (Charles' Law)

The volume (V) of a gas is directly related to its Kelvin temperature (T) when there is no change in the amount and pressure of the gas:

$$\frac{V_1}{T_1} = \frac{V_2}{T_2}$$

Therefore, if temperature increases, the volume of the gas increases; if temperature decreases, volume decreases.

11.5 Temperature and Pressure (Gay-Lussac's Law)

The pressure (P) of a gas is directly related to its Kelvin temperature (T).

$$\frac{P_1}{T_1} = \frac{P_2}{T_2}$$

This means that an increase in temperature (T) increases the pressure of a gas, or a decrease in temperature decreases the pressure, as long as the amount and volume stay constant. Vapor pressure is the pressure of the gas that forms when a liquid evaporates. At the boiling point of a liquid, the vapor pressure equals the external pressure.

11.6 The Combined Gas Law

Gas laws combine into a relationship of pressure (P), volume (V), and temperature (T).

$$\frac{P_1V_1}{T_1} = \frac{P_2V_2}{T_2}$$

This expression is used to determine the effect of changes in two of the variables on the third.

11.7 Volume and Moles (Avogadro's Law)

The volume (V) of a gas is directly related to the number of moles (n) of the gas when the pressure and temperature of the gas do not change.

$$\frac{V_1}{n_1} = \frac{V_2}{n_2}$$

If the moles of gas are increased, the volume must increase, or if the moles of gas are decreased, the volume decreases. At standard temperature (273 K) and pressure (1 atm) abbreviated STP, 1mol of any gas has a volume of 22.4 L. The density of a gas at STP is the ratio of the molar mass to the molar volume.

11.8 The Ideal Gas Law

The ideal gas law gives the relationship of all the quantities P, V, n, and T that describe and measure a gas. $PV = nRT$. Any of the four variables can be calculated if the other three are known. The molar mass of a gas can be calculated using molar volume at STP or the ideal gas law.

11.9 Gas Laws and Chemical Reactions

The ideal gas law is used to convert the quantities (P, V, and T) of gases to moles in a chemical reaction. The moles of gases can be used to determine the number of moles of other substances in the reaction. The pressure and volume of other gases in the reaction can be calculated using the ideal gas law.

11.10 Partial Pressures (Dalton's Law)

In a mixture of two or more gases, the total pressure is the sum of the partial pressures of the individual gases.

$$P_{total} = P_1 + P_2 + P_3 + \cdots$$

The partial pressure of a gas in a mixture is the pressure it would exert if it were the only gas in the container. For gases collected over water, the vapor pressure of water is subtracted from the total pressure of the gas mixture to obtain the partial pressure of the dry gas.

Key Terms

atmosphere (atm) A unit equal to the pressure exerted by a column of mercury 760 mm high.

atmospheric pressure The pressure exerted by the atmosphere.

Avogadro's law A gas law that states that the volume of a gas is directly related to the number of moles of gas in the sample when pressure and temperature do not change.

Boyle's law A gas law stating that the pressure of a gas is inversely related to the volume when temperature and moles of the gas do not change; that is, if volume decreases, pressure increases.

Charles' law A gas law stating that the volume of a gas changes directly with a change in Kelvin temperature when pressure and moles of the gas do not change.

combined gas law A relationship that combines several gas laws relating pressure, volume, and temperature.

$$\frac{P_1 V_1}{T_1} = \frac{P_2 V_2}{T_2}$$

Dalton's law A gas law stating that the total pressure exerted by a mixture of gases in a container is the sum of the partial pressures that each gas would exert alone.

direct relationship A relationship in which two properties increase or decrease together.

Gay-Lussac's law A gas law stating that the pressure of a gas changes directly with a change in temperature when the number of moles of a gas and its volume are held constant.

ideal gas law A law that combines the four measured properties of a gas in the equation $PV = nRT$.

inverse relationship A relationship in which two properties change in opposite directions.

kinetic molecular theory of gases A model used to explain the behavior of gases.

molar volume A volume of 22.4 L occupied by 1 mol of a gas at STP conditions of 0°C (273 K) and 1 atm.

partial pressure The pressure exerted by a single gas in a gas mixture.

pressure The force exerted by gas particles that hit the walls of a container.

STP Standard conditions of 0°C (273 K) temperature and 1 atm pressure used for the comparison of gases.

torr A unit of pressure equal to 1 mm Hg; 760 torr = 1 atm.

universal gas constant R A numerical value that relates the quantities P, V, n, and T in the ideal gas law, $PV = nRT$.

vapor pressure The pressure exerted by the particles of vapor above a liquid.

Understanding the Concepts

11.73 At 100°C, which of the following gases exerts
  a. the lowest pressure?
  b. the highest pressure?

1.   2.   3.

11.74 Indicate which diagram represents the volume of the gas sample in a flexible container when each of the following changes takes place:

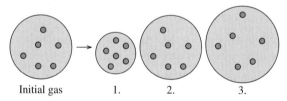

Initial gas   1.   2.   3.

  a. Temperature increases at constant pressure.
  b. Temperature decreases at constant pressure.
  c. Pressure decreases at constant temperature.
  d. Doubling the pressure and doubling the Kelvin temperature.

11.75 A balloon is filled with helium gas with a pressure of 1.00 atm and neon gas with a pressure of 0.50 atm. For each of the following changes of the initial balloon, select the diagram (A, B or C) that shows the final (new) volume of the balloon:

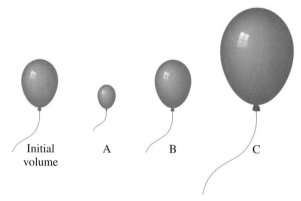

Initial volume   A   B   C

  a. The balloon is put in a cold storage unit (P and n constant).
  b. The balloon floats to a higher altitude where the pressure is less (n, T constant).
  c. All of the helium gas is removed (T and P constant).
  d. The Kelvin temperature doubles and 1/2 of the gas atoms leak out (P constant).
  e. 2.0 mol O_2 gas is added at constant T and P.

11.76 Indicate if pressure increases, decreases, or stays the same in each of the following:

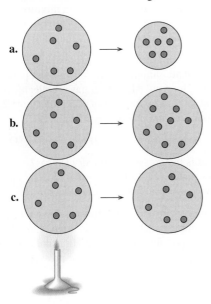

a.

b.

c.

11.77 Your space ship has docked at a space station above Mars. The temperature inside the space station is a carefully controlled 24°C at a pressure of 745 mm Hg. A balloon with a volume of 425 mL drifts into the airlock where the temperature is −95°C and the pressure is 0.115 atm. What is the new volume of the balloon? Assume that the balloon is very elastic.

11.78 At a restaurant, a customer chokes on a piece of food. You put your arms around the person's waist and use your fists to push up on the person's abdomen, an action called the Heimlich maneuver.
   a. How would this action change the volume of the chest and lungs?
   b. Why does it cause the person to expel the food item from the airway?

Additional Questions and Problems

11.79 A gas sample has a volume of 4250 mL at 15°C and 745 mm Hg. What is the new temperature (°C) after the sample is transferred to a new container with a volume of 2.50 L and a pressure of 1.20 atm?

11.80 A weather balloon has a volume of 750 L when filled with helium at 8°C at a pressure of 380 torr. What is the new volume of the balloon, where the pressure is 0.20 atm and the temperature is −45°C?

11.81 How many molecules of CO_2 are in 35.0 L $CO_2(g)$ at 1.2 atm and 5°C?

11.82 A steel cylinder with a volume of 15.0 L is filled with 50.0 g nitrogen gas at 25°C. What is the pressure of the N_2 gas in the cylinder?

11.83 A 2.00-L container is filled with methane gas, CH_4, at a pressure of 2500 mm Hg and a temperature of 18°C. How many grams of methane are in the container?

11.84 A container is filled with 4.0×10^{22} O_2 molecules at 5°C and 845 mm Hg. What is the volume in mL of the container?

11.85 A 1.00-g sample of dry ice (CO_2) is placed in a container that has a volume of 4.60 L and a temperature of 24.0°C. Calculate the pressure in mm Hg inside the container after all the dry ice changes to a gas.

$$CO_2(s) \longrightarrow CO_2(g)$$

11.86 Nitrogen (N_2) is prepared and a sample of 250 mL is collected over water at 30°C and a total pressure of 745 mm Hg.

   a. Using the vapor pressure of water in Table 11.4, what is the partial pressure of the nitrogen?
   b. What is the mass of nitrogen collected?

11.87 What is the density (g/L) of oxygen gas at STP?

11.88 What is the molar mass of a gas if 1.15 g of the gas has a volume of 225 mL at STP?

11.89 A sample of gas with a mass of 1.62 g occupies a volume of 941 mL at a pressure of 748 torr and a temperature of 20.0°C. What is the molar mass of the gas?

11.90 At STP, 762 mL of a gas has a mass of 1.02 g. If the gas has an empirical formula of CH_3, what is the molecular formula of the compound?

11.91 How many liters of H_2 gas can be produced at STP from 25.0 g Zn?

$$Zn(s) + 2HCl(aq) \longrightarrow ZnCl_2(aq) + H_2(g)$$

11.92 Hydrogen gas can be produced in the laboratory through the reaction of magnesium metal with hydrochloric acid.

$$Mg(s) + 2HCl(aq) \longrightarrow MgCl_2(aq) + H_2(g)$$

What is the volume, in liters, of H_2 gas produced at 24°C and 835 mm Hg, from the reaction of 12.0 g Mg?

11.93 Nitrogen dioxide reacts with water to produce oxygen and ammonia.

$$4NO_2(g) + 6H_2O(g) \longrightarrow 7O_2(g) + 4NH_3(g)$$

   a. How many liters of O_2 at STP are produced when 2.5×10^{23} molecules NO_2 react?

b. A 5.00-L sample of $H_2O(g)$ reacts at a temperature of 375°C and a pressure of 725 mm Hg. How many grams of NH_3 can be produced?

11.94 A sample of an unknown gas with a mass of 3.24 g occupies a volume of 1.88 L at a pressure of 748 mm Hg and a temperature of 20.0°C.
a. What is the molar mass of the gas?
b. If the unknown gas is composed of 2.78 g carbon and the rest is hydrogen, what is its molecular formula?

11.95 A gaseous compound has an empirical formula CH_2. When the gas is at 23°C and 752 mm Hg, a volume of 782 mL of the gas has a mass of 2.23 g. What is the molar mass and the molecular formula of the gas?

11.96 In the formation of smog, nitrogen and oxygen gas react to form nitrogen dioxide. How many grams of NO_2 will be produced when 2.0 L nitrogen at 840 mm Hg and 24°C are completely reacted?

$$N_2(g) + 2O_2(g) \longrightarrow 2NO_2(g)$$

11.97 A gas mixture with a total pressure of 2400 torr is used by a scuba diver. If the mixture contains 2.0 mol helium and 6.0 mol oxygen, what is the partial pressure of each gas in the sample?

11.98 What is the total pressure in mm Hg of a gas mixture containing argon gas at 0.25 atm, helium gas at 350 mm Hg, and nitrogen gas at 360 torr?

11.99 A gas mixture contains oxygen and argon at partial pressures of 0.60 atm and 425 mm Hg. If nitrogen gas added to the sample increases the total pressure to 1250 torr, what is the partial pressure in torr of the nitrogen added?

11.100 A gas mixture contains helium and oxygen at partial pressures of 255 torr and 0.450 atm. What is the total pressure in mm Hg of the mixture after it is placed in a container one-half the volume of the original container?

11.101 Solid aluminum reacts with aqueous H_2SO_4 to form H_2 gas and aluminum sulfate. When a sample of Al is allowed to react, 415 mL of gas is collected over water at 23°C, at a pressure of 755 mm Hg. At 23°C, the vapor pressure of water is 21 mm Hg.

$$2Al(s) + 3H_2SO_4(aq) \longrightarrow 3H_2(g) + Al_2(SO_4)_3(aq)$$

a. What is the pressure of the dry H_2 gas?
b. How many moles of H_2 were produced?
c. How many grams of Al were reacted?

11.102 When heated, solid $KClO_3$ forms solid KCl and O_2 gas. A sample of $KClO_3$ is heated and 226 mL of gas with a pressure of 744 mm Hg is collected over water, at 26°C. At 26°C, the vapor pressure of water is 25 mm Hg.

$$2KClO_3(s) \longrightarrow 2KCl(s) + 3O_2(g)$$

a. What is the pressure of the dry O_2 gas?
b. How many moles of O_2 were produced?
c. How many grams of $KClO_3$ were reacted?

Challenge Questions

11.103 Two flasks of equal volume and at the same temperature contain different gases. One flask contains 10.0 g Ne and the other flask contains 10.0 g He. Which of the following statements are correct? If not, explain.
a. Both flasks contain the same number of atoms.
b. The pressures in the flasks are the same.
c. The flask that contains helium has a higher pressure than the flask that contains neon.
d. The densities of the gases are the same.

11.104 A 92.0-g sample of a liquid is placed in a 25.0-L flask. At 140°C, the liquid evaporates completely to give a pressure of 0.900 atm.
a. What is the molar mass of the gas?
b. If the flask can withstand pressures up to 1.30 atm, calculate the maximum temperature that the gas can be heated to without breaking.

11.105 A sample of a carbon–hydrogen compound is found to contain 9.60 g carbon and 2.42 g hydrogen. At STP, 762 mL of the gas has a mass of 1.02 g. What is the molecular formula for the compound?

11.106 When sensors in a car detect a collision, they cause the reaction of sodium azide, NaN_3, which generates nitrogen gas to fill the air bags within 0.03 second.

$$2NaN_3(s) \longrightarrow 2Na(s) + 3N_2(g)$$

How many liters of N_2 are produced at STP if the air bag contains 132 g NaN_3?

11.107 Glucose, $C_6H_{12}O_6$, is metabolized in living systems according to the reaction

$$C_6H_{12}O_6(s) + 6O_2(g) \longrightarrow 6CO_2(g) + 6H_2O(l)$$

How many grams of water can be produced from the reaction of 18.0 g glucose and 7.50 L O_2 at 1.00 atm and 37°C?

11.108 2.00L N_2, at 25°C and 1.08 atm, is mixed with 4.00 L O_2, at 25°C and 0.118 atm, and the mixture allowed to react. How much NO, in grams, is produced?

$$N_2(g) + O_2(g) \longrightarrow 2NO(g)$$

Answers

Answers to Study Checks

11.1 The mass in grams gives the amount of gas.

11.2 0.862 atm

11.3 245 torr

11.4 50.0 mL

11.5 569 mL

11.6 16°C

11.7 241 mm Hg

11.8 7.50 L

11.9 28.5 L He

11.10 0.0900 g/L

11.11 0.327 mol Cl_2

11.12 1.04 L CO

11.13 72.9 g/mol

11.14 104 g/mol

11.15 5.41 L H_2

11.16 18.1 L H_2

11.17 1.24 atm

11.18 752 torr

11.19 0.594 g O_2

Answers to Selected Questions and Problems

11.1 **a.** At a higher temperature, gas particles have greater kinetic energy, which makes them move faster.
  b. Because there are great distances between the particles of a gas, they can be pushed closer together and still remain a gas.
  c. Gas particles are very far apart, which means that the mass of a gas in a certain volume is very small, resulting in a low density.

11.3 **a.** temperature   **b.** volume
  c. amount   **d.** pressure

11.5 atmospheres (atm), mm Hg, torr, lb/in.^{2}, pascals, kilopascals, in. Hg

11.7 **a.** 1520 torr   **b.** 29.4 lb/in.^{2}
  c. 1520 mm Hg   **d.** 203 KPa

11.9 As a diver ascends to the surface, external pressure decreases. If the air in the lungs were not exhaled, its volume would expand and severely damage the lungs. The pressure in the lungs must adjust to changes in the external pressure.

11.11 In order to exhale (expiration), the pressure in the lungs must increase, which is done when the volume of the lungs is decreased as the diaphragm relaxes.

11.13 **a.** The pressure is greater in cylinder A. According to Boyle's law, a decrease in volume pushes the gas particles closer together, which will cause an increase in the pressure.
  b.

Property	Conditions 1	Conditions 2	Know	Predict
Pressure (*P*)	650 mm Hg	1.2 atm	*P* increases	
Volume (*V*)	220 mL	160 mL		*V* decreases

11.15 **a.** The pressure doubles.
  b. The pressure falls to one-third the initial pressure.
  c. The pressure increases to ten times the original pressure.

11.17 **a.** 328 mm Hg   **b.** 2620 mm Hg
  c. 4370 mm Hg   **d.** 54 600 mm Hg

11.19 1700 mm Hg

11.21 **a.** 25 L   **b.** 25 L
  c. 100. L   **d.** 45 L

11.23 **a.** C   **b.** A
  c. B

11.25 When a gas is heated at constant pressure, the volume of the gas increases to fill the hot-air balloon.

11.27 **a.** 2400 mL   **b.** 4900 mL
  c. 1800 mL   **d.** 1700 mL

11.29 An increase in temperature increases the pressure inside the can. When the pressure exceeds the pressure limit of the can, it explodes.

11.31 **a.** 770 torr   **b.** 1.51 atm

11.33 **a.** boiling point   **b.** vapor pressure
  c. atmospheric pressure   **d.** boiling point

11.35 **a.** On top of a mountain, water boils below 100°C because the atmospheric (external) pressure is less than 1 atm.
  b. Because the pressure inside a pressure cooker is greater than 1 atm, water boils above 100°C. At a higher temperature, food cooks faster.

11.37 $\dfrac{P_1 V_1}{T_1} = \dfrac{P_2 V_2}{T_2}$

Boyle's, Charles', and Gay-Lussac's laws are combined to make this law.

11.39 **a.** 4.26 atm   **b.** 3.07 atm
  c. 0.606 atm

11.41 110 mL

11.43 The volume increases because the amount of gas particles is increased.

11.45 **a.** 4.00 L   **b.** 14.6 L
  c. 26.7 L

11.47 **a.** 2.00 mol O_2 **b.** 0.179 mol CO_2
c. 4.48 L   **d.** 55 500 mL

11.49 **a.** 1.70 g/L   **b.** 0.716 g/L
c. 0.901 g/L   **d.** 2.86 g/L

11.51 4.93 atm

11.53 29.4 g O_2

11.55 566 K (293°C)

11.57 **a.** 42 g/mol   **b.** 28.7 g/mol
c. 39.8 g/mol   **d.** 33 g/mol

11.59 **a.** 7.60 L H_2 **b.** 4.92 g Mg

11.61 178 L O_2

11.63 3.4 L O_2

11.65 **a.** 743 mm Hg   **b.** 0.0103 mol O_2

11.67 In a gas mixture, the pressure that each gas exerts as part of the total pressure is called the partial pressure of that gas. Because the air sample is a mixture of gases, the total pressure is the sum of the partial pressures of each gas in the sample.

11.69 765 torr

11.71 425 torr

11.73 **a.** 2   Fewest number of gas particles exerts the lowest pressure.
b. 1   Greatest number of gas particles exerts the highest pressure.

11.75 **a.** A: Volume decreases when temperature decreases.
b. C: Volume increases when pressure decreases.
c. A: Volume decreases when the moles of gas decrease.
d. B: Doubling temperature doubles the volume, but losing half the gas particles decreases the volume by half. The two effects cancel and no change in volume occurs.
e. C: Increasing the moles increases the volume to keep T and P constant.

11.77 2170 mL

11.79 207 K (-66°C)

11.81 1.1×10^{24} molecules CO_2

11.83 4.4 g

11.85 91.5 mm Hg

11.87 1.43 g/L

11.89 42.1 g/mol

11.91 8.57 L H_2

11.93 **a.** 16 L O_2 **b.** 1.02 g NH_3

11.95 70.1 g/mol; C_5H_{10}

11.97 He 600 torr, O_2 1800 torr

11.99 370 torr

11.101 **a.** 734 mm Hg
b. 0.0165 mol H_2
c. 0.297 g Al

11.103 **a.** False. The flask containing helium has more moles of helium and thus more helium atoms.
b. False. There are different numbers of moles in the flasks, which means the pressures are different.
c. True. There are more moles of helium, which makes the pressure of helium greater than that of neon.
d. True. The mass and volume of each are the same, which means the mass/volume ratio or density is the same in both flasks.

11.105 $0.762 \, \text{L} \times \dfrac{1 \, \text{mol}}{22.4 \, \text{L}} = 0.0340 \, \text{mol}$

$\dfrac{1.02 \, \text{g}}{0.0340 \, \text{mol}} = 30.0 \, \text{g/mol}$

$9.60 \, \text{g C} \times \dfrac{1 \, \text{mol C}}{12.01 \, \text{g C}} = 0.799 \, \text{mol C}/0.799$
$\qquad\qquad\qquad\qquad\qquad = 1.00 \, \text{mol C}$

$2.42 \, \text{g H} \times \dfrac{1 \, \text{mol H}}{1.008 \, \text{g H}} = 2.40 \, \text{mol H}/0.799$
$\qquad\qquad\qquad\qquad\qquad = 3.00 \, \text{mol H}$

Empirical formula $= CH_3$
Empirical mass $= 12.01 + 3(1.008) = 15.03 \, \text{g/EF}$

$\dfrac{30.0 \, \text{g/mol}}{15.03 \, \text{g/EF}} = 2 \qquad (CH_3)_2 = C_2H_6$

11.107 $n = \dfrac{PV}{RT} = \dfrac{(1.00 \, \text{atm})(7.50 \, \text{L})}{\left(\dfrac{0.0821 \, \text{L} \cdot \text{atm}}{\text{mol} \cdot \text{K}}\right)(310 \, \text{K})}$

$= 0.295 \, \text{mol } O_2 \times \dfrac{6 \, \text{mol } H_2O}{6 \, \text{mol } O_2} = 0.295 \, \text{mol } H_2O$

$18.0 \, \text{g } C_6H_{12}O_6 \times \dfrac{1 \, \text{mol } C_6H_{12}O_6}{180.2 \, \text{g } C_6H_{12}O_6} =$
$\qquad = 0.100 \, \text{mol } C_6H_{12}O_6 \times \dfrac{6 \, \text{mol } H_2O}{1 \, \text{mol } C_6H_{12}O_6}$
$\qquad\qquad\qquad\qquad = 0.600 \, \text{mol } H_2O$

0.295 mol H_2O is the smaller number of moles of product. Thus, O_2 is the limiting reactant.

$0.2954 \, \text{mol } H_2O \times \dfrac{18.02 \, \text{g } H_2O}{1 \, \text{mol } H_2O} = 5.31 \, \text{g } H_2O$

Solutions

"Chemistry is very important when taking care of patients in the hospital," says Dr. Denise Gee, physician, Boston Medical Center. "Blood tests can tell us the amount of various cations and anions in the body. This includes sodium, potassium, chloride, and bicarbonate, among others. An abnormal value can sometimes help diagnose disease or may signal that a patient is getting sicker. In the healthcare setting, chemistry is essential in monitoring overall patient health."

Doctors who are internists, family physicians, or pediatricians are directly involved in caring for people. Research doctors develop new therapies for cancer, genetic disorders, and infectious diseases. Other doctors teach medical students or work for pharmaceutical or health insurance companies.

LOOKING AHEAD

the Chemistry place

Visit **www.aw-bc.com/chemplace** for extra quizzes, interactive tutorials, career resources, PowerPoint slides for chapter review, math help, and case studies.

Solutions are everywhere around us. Most of the gases, liquids, and solids we see are mixtures of at least one substance dissolved in another. The air we breathe is a solution of oxygen and nitrogen gases. Carbon dioxide gas dissolved in water makes our carbonated drinks. When we make solutions of coffee or tea, we use hot water to dissolve substances from coffee beans or tea leaves.

Because the individual components in any mixture are not bonded to each other, the composition of those components can vary. Also, some of the physical properties of the individual components are still noticeable. For example, in ocean water, we detect the dissolved sodium chloride by the salty taste. The flavor we associate with coffee is due to the dissolved components. There are different types of solution. In a homogeneous solution, the components cannot be distinguished one from the other. Syrup is a homogeneous solution of sugar and water: the sugar cannot be distinguished from the water. However, in an aquarium, a heterogeneous mixture, all the components are observable including the sand on the bottom, the fish, the plants, and the water.

12.1 Solutions

A **solution** is a mixture in which one substance called the **solute** is uniformly dispersed in another substance called the **solvent.** Because the solute and the solvent do not react with each other, they can be mixed in varying proportions. A little salt dissolved in water tastes slightly salty. When more salt is added, the water tastes very salty. The solute (in this case, salt) is the substance present in the smaller amount, whereas the solvent (in this case, water) is present in the larger amount. In a solution, the particles of the solute are evenly dispersed among the molecules of the solvent. (See Figure 12.1.)

Learning Goal

Define solute and solvent; describe the formation of a solution.

Figure 12.1 A solution of copper(II) sulfate ($CuSO_4$) forms as particles of solute dissolve, move away from the crystals, and become evenly dispersed among the solvent (water) molecules.

Q *What does the uniform blue color indicate about the $CuSO_4$ solution?*

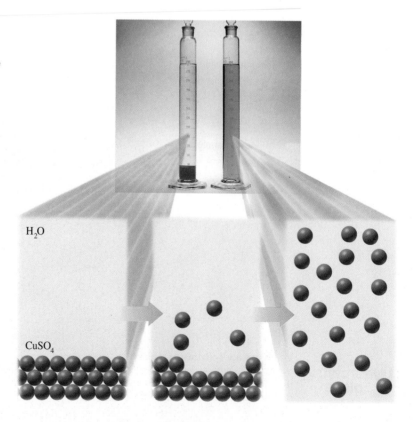

H_2O

$CuSO_4$

Solute: The substance present in lesser amount

Salt

Water

Solvent: The substance present in greater amount

WEB TUTORIAL

Hydrogen Bonding

Types of Solutes and Solvents

Solutes and solvents may be solids, liquids, or gases. The solution that forms has the same physical state as the solvent. When sugar is dissolved in a glass of water, a liquid sugar solution forms. Sugar is the solute, and water is the solvent. Soda water and soft drinks are prepared by dissolving CO_2 gas in water. The CO_2 gas is the solute, and water is the solvent. Table 12.1 lists some solutes and solvents and their solutions.

Water as a Solvent

Water is one of the most common substances in nature. In the H_2O molecule, an oxygen atom shares electrons with two hydrogen atoms. Because oxygen is much more electronegative than hydrogen, the O—H bonds are polar. In each polar bond, the oxygen atom has a partial negative (δ^-) charge, and the hydrogen atom has a partial positive δ^+ charge. Because the water molecule has a bent shape, water is a *polar substance.*

Hydrogen bonds occur between molecules where a partially positive hydrogen is attached to the strongly electronegative atoms O, N, or F. In water, hydrogen bonds are formed by the attraction between the oxygen atom of one water molecule and a hydrogen atom in another water molecule. In the diagram, hydrogen bonds are shown as dots between the water molecules. Although hydrogen bonds are much weaker than covalent or ionic bonds, there are many of them linking water molecules together. As a result, hydrogen bonding plays an important role in the properties of water and biological compounds such as proteins and DNA.

Table 12.1   Some Examples of Solutions

Type	Example	Solute	Solvent
Gas Solutions			
Gas in a gas	Air	Oxygen (gas)	Nitrogen (gas)
Liquid Solutions			
Gas in a liquid	Soda water	Carbon dioxide (gas)	Water (liquid)
	Household ammonia	Ammonia (gas)	Water (liquid)
Liquid in a liquid	Vinegar	Acetic acid (liquid)	Water (liquid)
Solid in a liquid	Seawater	Sodium chloride (solid)	Water (liquid)
	Tincture of iodine	Iodine (solid)	Alcohol (liquid)
Solid Solutions			
Liquid in a solid	Dental amalgam	Mercury (liquid)	Silver (solid)
Solid in a solid	Brass	Zinc (solid)	Copper (solid)
	Steel	Carbon (solid)	Iron (solid)

Partial negative charge

Partial positive charge

Hydrogen bonds

Formation of Solutions

An ionic compound such as sodium chloride, NaCl, is held together by ionic bonds between positive Na^+ ions and negative Cl^- ions. It dissolves in water because water is a polar solvent. When NaCl crystals are placed in water, the process of dissolution begins as the ions on the surface of the crystal come in contact with water molecules. (See Figure 12.2.) The negatively charged oxygen atom at one end of a water molecule attracts the positive Na^+ ions. The positively charged hydrogen atoms at the other end of a water molecule attract the negative Cl^- ions. The attractive forces of many water molecules provide the energy to break the ionic bonds between the Na^+ and Cl^- ions in the NaCl crystal. As the water molecules pull the ions into solution, a new surface of the NaCl crystal is exposed to the solvent. During a process called **hydration,** the dissolved Na^+ and Cl^- ions are surrounded by water molecules, which diminishes their attraction to other ions and helps keep them in solution. Later in this chapter (Section 12.3), we will look at ionic compounds that have such strong attractions between ions that they do not dissolve in water.

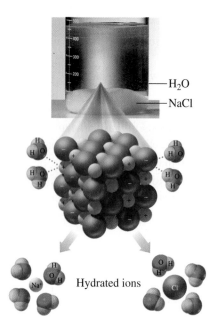

—H_2O

—NaCl

Hydrated ions

Figure 12.2 Ions on the surface of a crystal of NaCl dissolve in water as they are attracted to the polar water molecules that pull the ions into solution and surround them.

Q What helps keep the Na^+ and Cl^- ions in solution?

Like Dissolves Like

Gases form solutions easily because their particles are moving so rapidly that they are far apart and attractions to other gas particles are not important. When solids or liquids form solutions, there must be an attraction between the solute particles and the solvent particles. Then the particles of the solute and solvent will mix together. If there is no attraction between a solute and a solvent, their particles do not mix and no solution forms.

A salt such as NaCl will form a solution with water because the Na^+ and Cl^- ions in the salt are attracted to the positive and negative parts of the individual water molecules. You can dissolve methanol, CH_3OH, in water because the molecule has a polar —OH group that attracts water molecules. (See Figure 12.3.)

However, nonpolar molecules such as iodine (I_2), oil, or grease do not dissolve well in water because water is polar. Nonpolar solutes require nonpolar solvents for a solution to form. The expression "like dissolves like" is a way of saying that the polarities of a solute and a solvent must be similar in order to form a solution. Figure 12.4 illustrates the formation of some polar and nonpolar solutions.

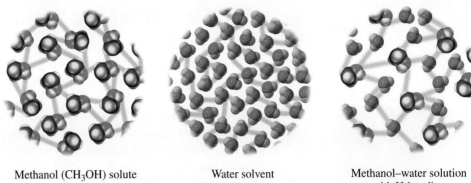

Methanol (CH₃OH) solute   Water solvent   Methanol–water solution
                                                    with H-bonding

Figure 12.3 The polar methanol, CH₃OH, forms hydrogen bonds with polar H₂O to form a methanol–water solution.

Q Why is the methanol–water solution an example of "like dissolves like"?

Figure 12.4 Like dissolves like. **(a)** The test tubes contain an upper layer of water (polar) and a lower layer of CH₂Cl₂ (nonpolar). **(b)** The nonpolar solute I₂ dissolves in the nonpolar layer. **(c)** The ionic solute Ni(NO₃)₂ dissolves in the water.

(a)   (b)   (c)

Q Which layer would dissolve polar molecules of sugar?

Sample Problem 12.1   **Polar and Nonpolar Solutes**

Indicate whether each of the following substances will dissolve in water. Explain.
a. KCl
b. octane, C₈H₁₈, a compound in gasoline
c. ethanol, C₂H₅OH, a substance in mouthwash

Solution
a. Yes. KCl is an ionic compound.
b. No. C₈H₁₈ is a nonpolar substance.
c. Yes. C₂H₅OH is a polar substance.

Study Check
Will Br₂, a nonpolar substance, dissolve in hexane, a nonpolar solvent?

Questions and Problems   **Solutions**

12.1 Identify the solute and the solvent in each solution composed of the following:
  a. 10.0 g NaCl and 100.0 g H₂O
  b. 50.0 mL ethanol, C₂H₅OH(*l*), and 10.0 mL H₂O
  c. 0.20 L O₂ and 0.80 L N₂
12.2 Identify the solute and the solvent in each solution composed of the following:
  a. 50.0 g silver and 4.0 g mercury
  b. 100.0 mL water and 5.0 g sugar
  c. 1.0 g I₂ and 50.0 mL alcohol
12.3 Water is a polar solvent; CCl₄ is a nonpolar solvent. In which solvent are each of the following more likely to be soluble?
  a. NaNO₃, ionic   **b.** I₂, nonpolar
  c. sugar, polar   **d.** gasoline, nonpolar

12.4 Water is a polar solvent; hexane is a nonpolar solvent. In which solvent are each of the following more likely to be soluble?
  a. vegetable oil, nonpolar
  b. benzene, nonpolar
  c. LiNO₃, ionic
  d. Na₂SO₄, ionic
12.5 Describe the formation of an aqueous KI solution.
12.6 Describe the formation of an aqueous LiBr solution.

Identify solutes as electrolytes or nonelectrolytes.

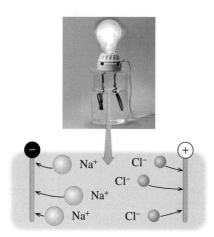

Strong electrolyte

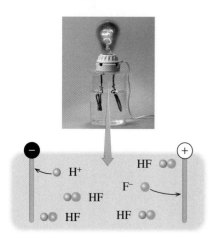

Weak electrolyte

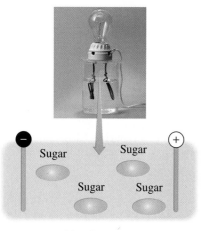

Nonelectrolyte

12.2 Electrolytes and Nonelectrolytes

Solutes can be classified by their ability to conduct an electrical current. When solutes called **electrolytes** dissolve in water, they separate into ions, which are able to conduct electricity. When solutes called **nonelectrolytes** dissolve in water, they do not separate into ions and their solutions do not conduct electricity.

To test solutions for ions, we can use an apparatus that consists of a battery and a pair of electrodes connected by wires to a light bulb. The light bulb glows when electricity can flow, which only happens when electrolytes provide ions to complete the circuit.

Strong Electrolytes

A **strong electrolyte** is a solute that dissociates completely into ions when it dissolves in water. For example, when sodium chloride (NaCl) dissolves in water, the sodium and chloride ions are attracted to water molecules. In a process called **dissociation,** the ions separate from the solid. As ions separate from the solid, they are hydrated by surrounding water molecules. In the equation for the dissociation of NaCl in water, the H_2O over the arrow indicates that water is needed for the dissociation process, but is not a reactant.

$$NaCl(s) \xrightarrow{H_2O} Na^+(aq) + Cl^-(aq)$$

Other soluble salts dissolve in a similar way. When we write the equation for the dissociation, the electrical charges must balance. For example, magnesium nitrate dissolves in water to give one magnesium ion for every two nitrate ions. However, only the ionic bonds between Mg^{2+} and NO_3^- are broken, not the covalent bonds within the polyatomic ion. The dissociation for $Mg(NO_3)_2$ is written as follows:

$$Mg(NO_3)_2(s) \xrightarrow{H_2O} Mg^{2+}(aq) + 2NO_3^-(aq)$$

Weak Electrolytes

A **weak electrolyte** is a solute that dissolves in water mostly as whole molecules. Only a few of the dissolved molecules separate, which produces a small number of ions in solution. Thus solutions of weak electrolytes do not conduct electrical current as well as solutions of strong electrolytes. For example, an aqueous solution of HF, a weak electrolyte, consists of mostly HF molecules and a few H^+ and F^- ions. First we show the formation of an aqueous HF solution.

$$HF(g) \xrightarrow{H_2O} HF(aq)$$

Within the solution, a few HF molecules dissociate into ions. As more H^+ and F^- ions form, some recombine to give HF molecules, which is indicated by a backwards arrow. Eventually, the rate of formation of ions is equal to the rate at which they recombine. The use of two arrows indicates that the forward and reverse reactions are taking place at the same time.

$$HF(aq) \underset{\text{Recombination}}{\overset{\text{Dissociation}}{\rightleftharpoons}} H^+(aq) + F^-(aq)$$

Nonelectrolytes

Solutes that are nonelectrolytes dissolve in water as molecules and do not separate into ions; thus, solutions of nonelectrolytes do not conduct electricity. For example, sucrose (sugar) is a nonelectrolyte that dissolves in water as whole molecules only.

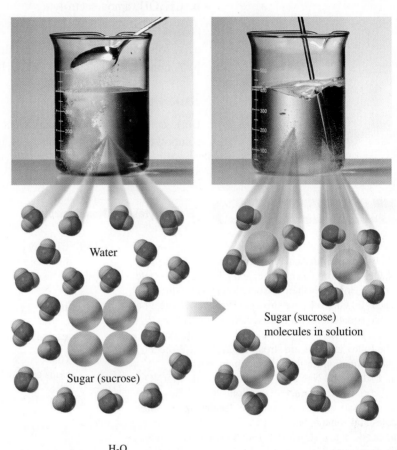

Water

Sugar (sucrose) molecules in solution

Sugar (sucrose)

$$C_{12}H_{22}O_{11}(s) \xrightarrow{H_2O} C_{12}H_{22}O_{11}(aq)$$

Table 12.2 summarizes the classification of solutes in aqueous solutions.

Table 12.2   Classification of Solutes in Aqueous Solutions

Types of Solute	Dissociation	Contained in Solution	Conducts Electricity	Examples
Strong electrolyte	Completely	Ions only	Yes	Ionic compounds such as NaCl, KBr, MgCl$_2$, NaNO$_3$; bases such as NaOH, KOH; acids such as HCl, HBr, HNO$_3$, HClO$_4$
Weak electrolyte	Partially	Mostly molecules and a few ions	Yes, but poorly	HF, H$_2$O, NH$_3$, CH$_3$COOH (acetic acid)
Nonelectrolyte	None	Molecules only	No	Carbon compounds such as CH$_3$OH, C$_2$H$_5$OH, C$_{12}$H$_{22}$O$_{11}$

Sample Problem 12.2 — **Solutions of Electrolytes and Nonelectrolytes**

Indicate whether aqueous solutions of each of the following contain ions, molecules, or both:

a. Na_2SO_4, a strong electrolyte

b. CH_3OH, a nonelectrolyte

Solution

a. A solution of Na_2SO_4 contains the ions of the salt, Na^+ and SO_4^{2-}.

b. A nonelectrolyte such as CH_3OH dissolves in water as molecules.

Study Check

Boric acid, H_3BO_3, is a weak electrolyte. Would you expect a boric acid solution to contain ions, molecules, or both?

Questions and Problems — **Electrolytes and Nonelectrolytes**

12.7 KF is a strong electrolyte, and HF is a weak electrolyte. How does their dissociation in water differ?

12.8 NaOH is a strong electrolyte, and CH_3OH is a nonelectrolyte. How does their dissociation in water differ?

12.9 The following salts are strong electrolytes. Write a balanced equation for their dissociation in water.

a. KCl   **b.** $CaCl_2$ **c.** K_3PO_4 **d.** $Fe(NO_3)_3$

12.10 The following salts are strong electrolytes. Write a balanced equation for their dissociation in water.

a. LiBr   **b.** $NaNO_3$ **c.** $FeCl_3$ **d.** $Mg(NO_3)_2$

12.11 Indicate whether aqueous solutions of the following solutes will contain ions only, molecules only, or molecules and some ions:

a. acetic acid ($HC_2H_3O_2$), found in vinegar, a weak electrolyte

b. NaBr, a salt

c. fructose ($C_6H_{12}O_6$), a nonelectrolyte

12.12 Indicate whether aqueous solutions of the following solutes will contain ions only, molecules only, or molecules and some ions:

a. Na_2SO_4, a salt

b. ethanol, C_2H_5OH, a nonelectrolyte

c. HCN, hydrocyanic acid, a weak electrolyte

12.13 Classify each solute represented in the following equations as a strong, weak, or nonelectrolyte.

a. $K_2SO_4(s) \xrightarrow{H_2O} 2K^+(aq) + SO_4^{2-}(aq)$

b. $NH_3(g) + H_2O(l) \rightleftharpoons NH_4^+(aq) + OH^-(aq)$

c. $C_6H_{12}O_6(s) \xrightarrow{H_2O} C_6H_{12}O_6(aq)$

12.14 Classify each solute represented in the following equations as a strong, weak, or nonelectrolyte:

a. $CH_3OH(l) \xrightarrow{H_2O} CH_3OH(aq)$

b. $MgCl_2(s) \xrightarrow{H_2O} Mg^{2+}(aq) + 2Cl^-(aq)$

c. $HClO(aq) \rightleftharpoons H^+(aq) + ClO^-(aq)$

12.3 Solubility

Learning Goal

Define solubility; distinguish between an unsaturated and a saturated solution. Identify an insoluble salt.

The term **solubility** is used to describe the amount of a solute that can dissolve in a given amount of solvent. Many factors, such as the type of solute, the type of solvent, and temperature, affect a solute's solubility. Solubility, usually expressed in grams of solute in 100 grams of solvent, is the maximum amount of solute that can be dissolved at a certain temperature. If a solute readily dissolves when added to the solvent, the solution does not contain the maximum amount of solute. We call the solution an **unsaturated solution.** When a solution contains all the solute that can dissolve, it is a **saturated**

solution. If we try to add more solute, undissolved solute will remain on the bottom of the container. For example, 36 g NaCl can dissolve in 100 g water at 20°C. If we add 15 g NaCl to 100 g water at 20°C, it all dissolves; the solution is unsaturated. The NaCl solution becomes saturated when we dissolve a total of 36 g NaCl.

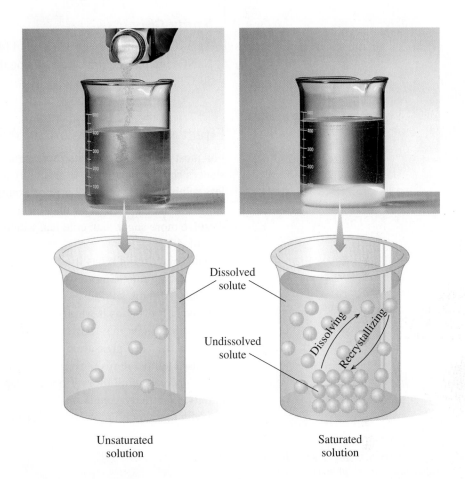

Dissolved solute

Undissolved solute

Unsaturated solution

Saturated solution

A solution becomes saturated when the rate of dissolving the solute becomes equal to the rate of recrystallizing the solute. Then there is no further change in the amounts of dissolved and solid solute. The arrows indicate that the rate at which solute dissolves is equal to the rate at which solute in solution crystallizes out as a solid.

$$\text{Solid solute} \underset{\text{Crystallizes}}{\overset{\text{Dissolves}}{\rightleftarrows}} \text{saturated solution}$$

Sample Problem 12.3 **Saturated Solutions**

At 20°C, the solubility of KCl is 34 g/100 g water. In the laboratory, a student mixes 45 g KCl with 100 g water at a temperature of 20°C.

a. How much of the KCl will dissolve?

b. Is the solution saturated?

c. What is the mass in grams of any solid KCl on the bottom of the container?

Solution

a. A total of 34 g KCl will dissolve because that is its solubility and, therefore, the maximum amount of KCl in solution at 20°C.

b. Yes, the solution is saturated.

c. The mass of the solid KCl is 11 g.

Study Check

At 50°C, the solubility of $NaNO_3$ is 114 g/100 g water. How many grams of $NaNO_3$ are needed to make a saturated $NaNO_3$ solution with 50 g water at 50°C?

Effect of Temperature on Solubility

For most solids, solubilities increase as temperature increases. A few substances show little change in solubility at higher temperatures, and a few are less soluble. (See Figure 12.5.) For example, when you add sugar to iced tea, a layer of undissolved sugar may form on the bottom of the glass. Hot tea can dissolve more sugar than cold tea, which is why sugar is added before the tea is cooled.

The solubility of a gas in water decreases as the temperature increases. At higher temperatures, more gas molecules have the energy to escape from the solution. Perhaps you have observed the bubbles escaping from a cold carbonated soft drink as it warms. At high temperatures, bottles containing carbonated solutions may burst as more gas molecules leave the solution and increase the gas pressure inside the bottle. Biologists have found that increased temperatures in rivers and lakes cause the amount of dissolved oxygen to decrease until the warm water can no longer support a biological community. Electricity-generating plants are required to have their own ponds to use with their cooling towers to lessen the threat of thermal pollution.

Figure 12.5 In water, most common solids are more soluble as the temperature increases.

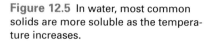

Compare the solubility of $NaNO_3$ at 20°C and 60°C.

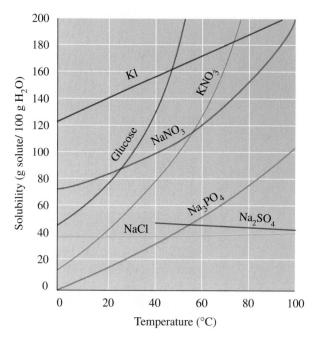

Henry's Law

Henry's law states that the solubility of gas in a liquid is directly related to the pressure of that gas above the liquid. At higher pressures, there are more gas molecules available to enter and dissolve in the liquid. A can of soda is carbonated by using CO_2 gas at high pressure to increase the solubility of the CO_2 in the beverage. When you open the can at atmospheric pressure, the pressure of the CO_2 drops, which decreases the solubility of CO_2. As a result, bubbles of CO_2 rapidly escape from the solution. The burst of bubbles is even more noticeable when you open a warm can of soda.

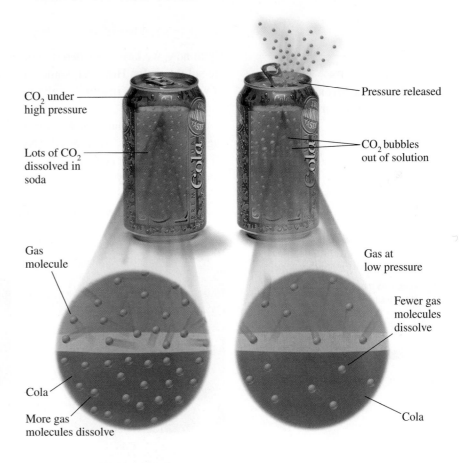

CO_2 under high pressure

Lots of CO_2 dissolved in soda

Gas molecule

Cola

More gas molecules dissolve

Pressure released

CO_2 bubbles out of solution

Gas at low pressure

Fewer gas molecules dissolve

Cola

Sample Problem 12.4 ▸ Factors Affecting Solubility

Indicate whether the solubility of the solute will increase or decrease in each of the following situations:

a. dissolving sugar using 80°C water instead of 25°C water
b. effect on dissolved O_2 in a lake as it warms

Solution

a. An increase in the temperature increases the solubility of the sugar.
b. An increase in the temperature decreases the solubility of O_2 gas.

Study Check

At 10°C, the solubility of KNO_3 is 20 g/100 g H_2O. Would the value of 5 g/100 g H_2O or 80. g/100 g H_2O be the more likely solubility at 40°C? Explain.

Table 12.3 Solubility Rules for Ionic Solids in Water

Soluble if Salt Contains		Insoluble if Salt Contains
NH_4^+, Li^+, Na^+, K^+ NO_3^-, acetate $C_2H_3O_2^-$	← but are soluble with	CO_3^{2-}, S^{2-} PO_4^{3-}, OH^-
Cl^-, Br^-, I^-	but are not soluble with →	$Ag^+, Pb^{2+},$ or Hg_2^{2+}
SO_4^{2-}	but are not soluble with →	$Ba^{2+}, Pb^{2+}, Ca^{2+}, Sr^{2+}$

the Chemistry place

WEB TUTORIAL
Solubility

Soluble and Insoluble Salts

Up to now, we have considered ionic compounds that dissolve in water; they are **soluble salts.** However, some ionic compounds do not separate into ions in water. They are **insoluble salts** that remain as solids even in contact with water.

Salts that are soluble in water typically contain at least one of the following ions: Li^+, Na^+, K^+, NH_4^+, NO_3^-, or $C_2H_3O_2^-$. Most salts containing Cl^- are soluble, but $AgCl$, $PbCl_2$, or Hg_2Cl_2 are not; they are insoluble chloride salts. Similarly, most salts containing SO_4^{2-} are soluble, but a few are insoluble, as shown in Table 12.3. Most other salts are insoluble and do not dissolve in water. (See Figure 12.6.) In an insoluble salt, attractions between

Figure 12.6 Mixing certain aqueous solutions produces insoluble salts.

Q What ions make each of these salts insoluble in water?

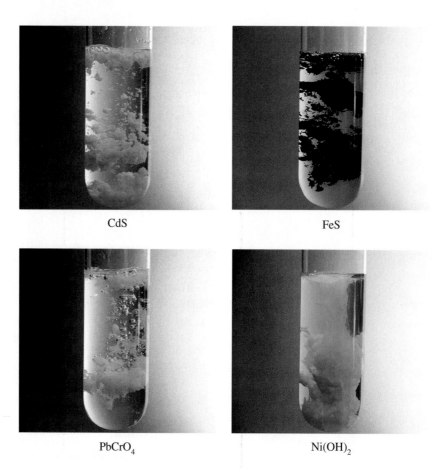

CdS

FeS

PbCrO$_4$

Ni(OH)$_2$

Table 12.4 Using Solubility Rules

Ionic Compound	Solubility in Water	Reasoning
K_2S	Soluble	Contains K^+
$Ca(NO_3)_2$	Soluble	Contains NO_3^-
$PbCl_2$	Insoluble	Is an insoluble chloride
NaOH	Soluble	Contains Na^+
$AlPO_4$	Insoluble	Contains no soluble ions

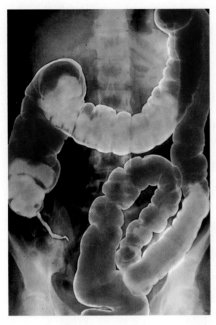

Figure 12.7 A barium sulfate enhanced X ray of the abdomen shows the large intestine.

Q Is $BaSO_4$ a soluble or an insoluble substance?

its positive and negative ions are too strong for the polar water molecules to break. We can use the solubility rules to predict whether a salt (a solid ionic compound) would be expected to dissolve in water. Table 12.4 illustrates the use of these rules.

In medicine, the insoluble salt $BaSO_4$ is used as an opaque substance to enhance X rays of the gastrointestinal tract. $BaSO_4$ is so insoluble that it does not dissolve in gastric fluids. (See Figure 12.7.) Other barium salts cannot be used because they would dissolve in water, releasing Ba^{2+}, which is poisonous.

Sample Problem 12.5 Soluble and Insoluble Salts

Predict whether each of the following salts is soluble in water:
a. Na_3PO_4 **b.** $CaCO_3$ **c.** K_2SO_4

Solution
a. Soluble. Salts containing Na^+ are soluble.
b. The salt $CaCO_3$ is not soluble. Most salts containing CO_3^{2-} are not soluble.
c. Soluble. Salts containing K^+ are soluble.

Study Check

Would you expect the following salts to be soluble in water? Why?
a. $PbCl_2$ **b.** K_3PO_4 **c.** $FeCO_3$

Formation of a Solid

We can use solubility rules to predict whether a solid called a *precipitate* forms when two solutions of ionic compounds are mixed. A solid forms when two ions of an insoluble salt come in contact with one another. For example, when a solution of $AgNO_3$ (Ag^+ and NO_3^-) is mixed with a solution of NaCl (Na^+ and Cl^-), the white insoluble salt AgCl is produced. We can write the reaction as a double replacement equation. However, the molecular equation does not show the individual ions to help us decide which, if any, insoluble salt would form. To help us determine any insoluble salt, we can first write the reactants to show all the ions present when the two solutions are mixed.

$$Ag^+(aq) + NO_3^-(aq) + Na^+(aq) + Cl^-(aq) \longrightarrow$$

Guide to Writing Net Ionic Equations for Formation of an Insoluble Salt
STEP 1   Write the ions of the reactants.
STEP 2   Write the new combinations of ions and determine if any are insoluble.
STEP 3   Write the ionic equation including any solid.
STEP 4   Write the net ionic equation by removing spectator ions.

Then we look at possible new combinations of cations and anions to see if any would form an insoluble salt. The new combination of AgCl would form an insoluble salt.

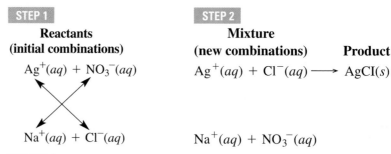

STEP 1

Reactants (initial combinations)

$$Ag^+(aq) + NO_3^-(aq)$$

$$Na^+(aq) + Cl^-(aq)$$

STEP 2

Mixture (new combinations)   **Product**

$$Ag^+(aq) + Cl^-(aq) \longrightarrow AgCl(s)$$

$$Na^+(aq) + NO_3^-(aq)$$

STEP 3   Now we can write an **ionic equation** to show that a precipitate of AgCl forms while the ions Na^+ and NO_3^- remain in solution.

$$Ag^+(aq) + NO_3^-(aq) + Na^+(aq) + Cl^-(aq) \longrightarrow$$
$$AgCl(s) + Na^+(aq) + NO_3^-(aq)$$

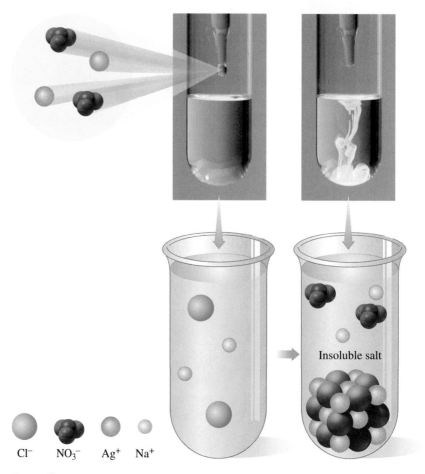

Cl^- NO_3^- Ag^+ Na^+

Insoluble salt

Type of Equation

Molecular	$AgNO_3(aq)$	$+ NaCl(aq) \longrightarrow$	$AgCl(s) + NaNO_3(aq)$
Ionic	$Ag^+(aq) + NO_3^-(aq)$	$+ Na^+(aq) + Cl^-(aq) \longrightarrow$	$AgCl(s) + Na^+(aq) + NO_3^-(aq)$
Net ionic	$Ag^+(aq)$	$+ Cl^-(aq) \longrightarrow$	$AgCl(s)$

STEP 4 Now we can remove the Na^+ and NO_3^- ions, known as *spectator ions* because they are unchanged during the reaction.

$$Ag^+(aq) + \cancel{NO_3^-(aq)} + \cancel{Na^+(aq)} + Cl^-(aq) \longrightarrow$$
$$AgCl(s) + \cancel{Na^+(aq)} + \cancel{NO_3^-(aq)}$$

Finally, a **net ionic equation** can be written that gives the chemical reaction that occurred.

The Na^+ and NO_3^- ions, the spectator ions, are removed from the ionic equation we wrote above.

$$Ag^+(aq) + Cl^-(aq) \longrightarrow AgCl(s)$$

Sample Problem 12.6   **Formation of an Insoluble Salt**

Solutions of $BaCl_2$ and K_2SO_4 are mixed and a white solid forms.
a. Write the net ionic equation.
b. What is the white solid that forms?

Solution

a. STEP 1 $Ba^{2+}(aq) + Cl^-(aq) + K^+(aq) + SO_4^{2-}(aq)$

STEP 2 $BaSO_4(s)$ is insoluble.

STEP 3 $Ba^{2+}(aq) + 2Cl^-(aq) + 2K^+(aq) + SO_4^{2-}(aq) \longrightarrow$
$BaSO_4(s) + 2Cl^-(aq) + 2K^+(aq)$

STEP 4 $Ba^{2+}(aq) + SO_4^{2-}(aq) \longrightarrow BaSO_4(s)$

b. $BaSO_4$ is the white solid.

Study Check

Predict whether a solid might form in each of the following mixtures of solutions. If so, write the net ionic equation for the reaction.
a. $NH_4Cl(aq) + Ca(NO_3)_2(aq)$
b. $Pb(NO_3)_2(aq) + KCl(aq)$

the
Chemistry
place

CASE STUDY
Kidney Stones and Saturated Solutions

Questions and Problems   **Solubility**

12.15 State whether each of the following refers to a saturated or unsaturated solution:
a. A crystal added to a solution does not change in size.
b. A sugar cube completely dissolves when added to a cup of coffee.
12.16 State whether each of the following refers to a saturated or unsaturated solution:
a. A spoonful of salt added to boiling water dissolves.
b. A layer of sugar forms on the bottom of a glass of tea as ice is added.

Use this table for problems 12.17–12.20.

Substance	Solubility (g/100 g H_2O)	
	20°C	50°C
KCl	34.0	42.6
$NaNO_3$	88.0	114.0
$C_{12}H_{22}O_{11}$ (sugar)	203.9	260.4

GOUT AND KIDNEY STONES: A PROBLEM OF SATURATION IN BODY FLUIDS

The conditions of gout and kidney stones involve compounds in the body that exceed their solubility levels and form solid products. Gout affects adults, primarily men, over the age of 40. Attacks of gout may occur when the concentration of uric acid in blood plasma exceeds its solubility, which is 7 mg/100 mL of plasma at 37°C. Insoluble deposits of needle-like crystals of uric acid can form in the cartilage, tendons, and soft tissues, where they cause painful gout attacks. They may also form in the tissues of the kidneys, where they can cause renal damage. High levels of uric acid in the body can be caused by an increase in uric acid production, failure of the kidneys to remove uric acid, or by a diet with an overabundance of foods containing purines, which are metabolized to uric acid in the body. Foods in the diet that contribute to high levels of uric acid include certain meats, sardines, mushrooms, asparagus, and beans. Drinking alcoholic beverages may also significantly increase uric acid levels and bring about gout attacks.

Treatment for gout involves diet changes and drugs. Depending on the levels of uric acid, a medication, such as probenecid, can be used to help the kidneys eliminate uric acid, or allopurinol, which blocks the production of uric acid by the body.

Kidney stones are solid materials that form in the urinary tract. Most kidney stones are composed of calcium phosphate and calcium oxalate, although they can be solid uric acid. The excessive ingestion of minerals and insufficient water intake can cause the concentration of mineral salts to exceed the solubility of the mineral salts and lead to the formation of kidney stones. When a kidney stone passes through the urinary tract, it causes considerable pain and discomfort, necessitating the use of painkillers and surgery. Sometimes ultrasound is used to break up kidney stones. Persons prone to kidney stones are advised to drink six to eight glasses of water every day to prevent saturation levels of minerals in the urine.

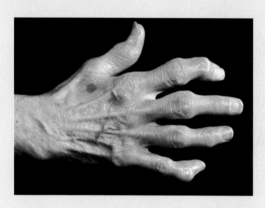

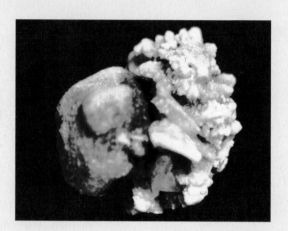

12.17 Using the table, determine whether each of the following solutions will be saturated or unsaturated at 20°C:
  a. adding 25.0 g KCl to 100 g H_2O
  b. adding 12.0 g $NaNO_3$ to 25 g H_2O
  c. adding 400.0 g sugar to 125 g H_2O

12.18 Using the table, determine whether each of the following solutions will be saturated or unsaturated at 50°C:
  a. adding 25.0 g KCl to 50 g H_2O
  b. adding 150.0 g $NaNO_3$ to 75 g H_2O
  c. adding 80.0 g sugar to 25 g H_2O

12.19 A solution containing 80.0 g KCl in 200 g H_2O at 50°C is cooled to 20°C.
  a. How many grams of KCl remain in solution at 20°C?
  b. How many grams of solid KCl came out of solution after cooling?

12.20 A solution containing 80.0 g $NaNO_3$ in 75.0 g H_2O at 50°C is cooled to 20°C.
  a. How many grams of $NaNO_3$ remain in solution at 20°C?
  b. How many grams of solid $NaNO_3$ came out of solution after cooling?

12.21 Explain the following observations:
  a. More sugar dissolves in hot tea than in iced tea.
  b. Champagne in a warm room goes flat.
  c. A warm can of soda has more spray when opened than a cold one.

12.22 Explain the following observations:
  a. An open can of soda loses its "fizz" quicker at room temperature than in the refrigerator.
  b. Chlorine gas in tap water escapes as the sample warms to room temperature.
  c. Less sugar dissolves in iced coffee than in hot coffee.

12.23 Predict whether each of the following ionic compounds is soluble in water:
  a. LiCl   **b.** AgCl
  c. $BaCO_3$ **d.** K_2O
  e. $Fe(NO_3)_3$

12.24 Predict whether each of the following ionic compounds is soluble in water:
  a. PbS   **b.** NaI
  c. Na_2S **d.** Ag_2O
  e. $CaSO_4$

12.25 Determine whether a solid forms when solutions containing the following salts are mixed. If so, write the molecular equation (double replacement) and the net ionic equation for the reaction.
  a. KCl and Na_2S
  b. $AgNO_3$ and K_2S
  c. $CaCl_2$ and Na_2SO_4

12.26 Determine whether a solid forms when solutions containing the following salts are mixed. If so, write the molecular equation (double replacement) and the net ionic equation for the reaction.
  a. Na_3PO_4 and $AgNO_3$
  b. K_2SO_4 and Na_2CO_3
  c. $Pb(NO_3)_2$ and Na_2CO_3

12.4 Percent Concentration

Learning Goal

Calculate the percent concentration of a solute in a solution; use percent concentration to calculate the amount of solute or solution.

The amount of solute dissolved in a certain amount of solution is called the **concentration** of the solution. Although there are many ways to express a concentration, they all specify a certain amount of solute in a given amount of solution.

$$\text{Concentration of a solution} = \frac{\text{Amount of solute}}{\text{Amount of solution}}$$

Mass Percent

The **mass percent** (% m/m) concentration of a solution describes the mass of the solute in every 100 grams of solution. The mass in grams of the solution is the sum of the mass of the solute and the mass of the solvent.

$$\text{Mass percent (\% m/m)} = \frac{\text{Mass of solute (g)}}{\text{Mass of solute (g)} + \text{Mass of solvent (g)}} \times 100$$

$$= \frac{\text{Mass of solute (g)}}{\text{Mass of solution (g)}} \times 100$$

Suppose we prepared a solution by mixing 8.00 g KCl (solute) with 42.00 g water (solvent). Together, the mass of the solute and mass of the solvent give the mass of the solution (8.00 g + 42.00 g = 50.00 g). The mass % is calculated by substituting in the values into the mass percent expression.

$$\frac{8.00 \text{ g KCl}}{50.00 \text{ g solution}} \times 100\% = 16.0\% \text{ (m/m)}$$

$$\underbrace{8.00 \text{ g KCl} + 42.00 \text{ g H}_2\text{O}}$$
(Solute + Solvent)

Add 8.00 g of KCl

Add water until the solution weighs 50.00 g

Sample Problem 12.7 ▶ **Mass Percent Concentration**

What is the mass percent of a solution prepared by dissolving 30.0 g NaOH in 120.0 g H_2O?

Solution

STEP 1 **Determine quantities of solute and solution.**

Given 30.0 g NaOH and 120.0 g H_2O = 150.0 g solution
Need mass percent (% m/m) of NaOH

STEP 2 **Write the % concentration expression.**

$$\text{Mass percent (\% m/m)} = \frac{\text{Grams of solute}}{\text{Grams of solution}} \times 100$$

**Guide to Calculating
Solution Concentration**

STEP 1
Determine quantities of
solute and solution.

STEP 2
Write the % concentration expression.

STEP 3
Substitute solute and solution
quantities in expression.

STEP 3 **Substitute the mass of the solute and the solution quantities into the expression.**

$$\text{Mass percent (\% m/m)} = \frac{30.0 \text{ g NaOH}}{150.0 \text{ g solution}} \times 100$$

$$= 20.0\% \text{ (m/m) NaOH}$$

Study Check

What is the mass percent of NaCl in a solution made by dissolving 2.0 g NaCl in 56.0 g H_2O?

Volume Percent

Because the volumes of liquids or gases are easily measured, the concentrations of their solutions are often expressed as **volume percent** (% v/v). The units of volume used in the ratio must be the same, for example, both in milliliters or both in liters.

$$\text{Volume percent (\% v/v)} = \frac{\text{Volume of solute}}{\text{Volume of solution}} \times 100\%$$

We interpret a volume/volume percent as the volume of solute in 100 mL of solution. In the wine industry, a label that reads 12% (v/v) means 12 mL of alcohol in 100 mL of wine.

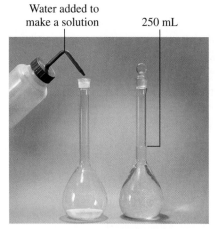

Water added to
make a solution   250 mL

5.0 mL   2.0% (v/v)
C_2H_5OH C_2H_5OH
              solution

Sample Problem 12.8   **Calculating Volume Percent Concentration**

A student prepared a solution by adding water to 5.0 mL ethanol (C_2H_5OH) to give a final volume of 250.0 mL. What is the volume percent (% v/v) of the ethanol solution?

Solution

STEP 1 **Determine quantities of solute and solution.**

Given   5.0 mL C_2H_5OH in 250.0 mL solution
Need   volume percent (% v/v) of C_2H_5OH

STEP 2 **Write the % concentration expression.**

$$\text{Volume percent (\% v/v)} = \frac{\text{Volume solute}}{\text{Volume solution}} \times 100$$

STEP 3 **Substitute solute and solution quantities into the expression.**

$$\text{Volume percent (\% v/v)} = \frac{5.0 \text{ mL } C_2H_5OH}{250.0 \text{ mL solution}} \times 100$$

$$\text{Volume percent (\% v/v)} = 2.0\% \text{ (v/v) } C_2H_5OH$$

Study Check

What is the volume percent, % v/v, of Br_2 in a solution prepared by dissolving 12 mL bromine (Br_2) in enough carbon tetrachloride to make 250 mL solution?

Table 12.5 Conversion Factors from Percent Concentrations

Percent Concentration	Meaning	Conversion Factors		
10% (m/m) KCl	There are 10 g KCl in 100 g solution.	$\dfrac{10 \text{ g KCl}}{100 \text{ g solution}}$	and	$\dfrac{100 \text{ g solution}}{10 \text{ g KCl}}$
12% (v/v) ethanol	There are 12 mL ethanol in 100 mL solution.	$\dfrac{12 \text{ mL ethanol}}{100 \text{ mL solution}}$	and	$\dfrac{100 \text{ mL solution}}{12 \text{ mL ethanol}}$

Percent Concentrations as Conversion Factors

In the preparation of solutions, we often need to calculate the amount of solute or solution. Then the percent concentration is useful as a conversion factor. Some examples of percent concentrations, their meanings, and possible conversion factors are given in Table 12.5.

Sample Problem 12.9 **Using Mass Percent to Find Mass of Solute**

An antibiotic ointment is 3.5% (m/m) neomycin. How many grams of neomycin are in a tube containing 64 grams of ointment?

Solution

Guide to Using Concentration to Calculate Mass or Volume

STEP 1 State the given and needed quantities.

STEP 2 Write a plan to calculate mass or volume.

STEP 3 Write equalities and conversion factors including concentration.

STEP 4 Set up problem to calculate mass or volume.

STEP 1 **Given** 3.5% (m/m) neomycin **Need** grams of neomycin

STEP 2 **Plan** grams of ointment → Mass % factor → grams of neomycin

STEP 3 **Equalities/Conversion Factors** The mass percent (% m/m) indicates the grams of a solute in every 100 grams of a solution.

$$\text{Mass percent (\% m/m)} = \frac{\text{g solute}}{100 \text{ g solution}}$$

The mass percent (3.5% m/m) of neomycin can be written as two conversion factors.

$$100 \text{ g ointment} = 3.5 \text{ g neomycin}$$

$$\frac{3.5 \text{ g neomycin}}{100 \text{ g ointment}} \quad \text{and} \quad \frac{100 \text{ g ointment}}{3.5 \text{ g neomycin}}$$

STEP 4 **Set Up Problem** Now we use mass percent to convert the grams of solution to grams of solute. The mass of the solution is converted to mass of solute using the conversion factor.

$$64 \text{ g ointment} \times \frac{3.5 \text{ g neomycin}}{100 \text{ g ointment}} = 2.2 \text{ g neomycin}$$

Study Check

Calculate the grams of KCl and grams of water in 225 g of an 8.00% (m/m) KCl solution.

Questions and Problems   **Percent Concentration**

12.27 How would you prepare 250. g of a 5.00% (m/m) glucose solution?

12.28 What is the difference between a 10% (v/v) methyl alcohol (CH_3OH) solution and a 10% (m/m) methyl alcohol solution?

12.29 Calculate the mass percent, % m/m, for the solute in each of the following solutions:
  a. 25 g KCl and 125 g H_2O
  b. 8.0 g $CaCl_2$ in 80.0 g solution
  c. 12 g sugar in 225 g tea (solution)

12.30 Calculate the mass percent, % m/m, for the solute in each of the following solutions:
  a. 75 g NaOH in 325 g solution
  b. 2.0 g KOH in 20.0 g H_2O
  c. 48.5 g Na_2CO_3 in 250.0 g solution

12.31 Calculate the amount of solute (g or mL) needed to prepare the following solutions:
  a. 50.0 g of a 5.0% (m/m) KCl solution
  b. 1250 g of a 4.0% (m/m) NH_4Cl solution
  c. 250 mL of a 10.0% (v/v) acetic acid solution

12.32 Calculate the amount of solute (g or mL) needed to prepare the following solutions:
  a. 150. g of a 40.0% (m/m) $LiNO_3$ solution
  b. 450 g of a 2.0% (m/m) KCl solution
  c. 225 mL of a 15% (v/v) isopropyl alcohol solution

12.33 A mouthwash contains 22.5% alcohol by volume. If the bottle of mouthwash contains 355 mL, what is the volume in milliliters of the alcohol?

12.34 A bottle of champagne is 11% alcohol by volume. If there are 750 mL of champagne in the bottle, how many milliliters of alcohol are present?

12.35 Calculate the amount of solution (g or mL) that contains each of the following amounts of solute:
  a. 5.0 g $LiNO_3$ from a 25% (m/m) $LiNO_3$ solution
  b. 40.0 g KOH from a 10.0% (m/m) KOH solution
  c. 2.0 mL acetic acid from a 10.0% (v/v) acetic acid solution

12.36 Calculate the amount of solution (g or mL) that contains each of the following amounts of solute:
  a. 7.50 g NaCl from a 2.0% (m/m) NaCl solution
  b. 4.0 g NaOH from a 25% (m/m) NaOH solution
  c. 20.0 g KBr from an 8.0% (m/m) KBr solution

12.5 Molarity and Dilutions

Learning **Goal**

> **Calculate the molarity of a solution; use molarity as a conversion factor to calculate the moles of solute or the volume needed to prepare a solution.**

When the solutes of solutions take part in reactions, chemists are interested in the number of reacting particles. For this purpose, chemists use **molarity (M),** a concentration that states the number of moles of solute in exactly 1 liter of solution. The molarity of a solution can be calculated knowing the moles of solute and the volume of solution.

$$\text{Molarity (M)} = \frac{\text{Moles of solute}}{\text{Liters of solution}} = \frac{\text{mol solute}}{\text{L soln}}$$

For example, if 1.0 mol NaCl were dissolved in enough water to prepare 1.0 L of solution, the resulting NaCl solution has a molarity of 1.0 M. The abbreviation M indicates the units of moles per liter (mol/L).

$$M = \frac{\text{Moles of solute}}{\text{Liters of solution}} = \frac{1.0 \text{ mol NaCl}}{1.0 \text{ L solution}} = 1.0 \text{ M NaCl}$$

Guide to Calculating Molarity

> **STEP 1**
> State the given quantities.

> **STEP 2**
> Write a plan to calculate molarity.

> **STEP 3**
> Write equalities and conversion factors needed.

> **STEP 4**
> Set up problem to calculate molarity.

Sample Problem 12.10   **Calculating Molarity**

What is the molarity (M) of 60.0 g NaOH in 0.250 L solution?

Solution

STEP 1   **Given**   60.0 g NaOH in 0.250 L solution
          Need   molarity (mol/L)

STEP 2   **Plan**   The calculation of molarity requires the moles of NaOH and the volume of the solution in liters.

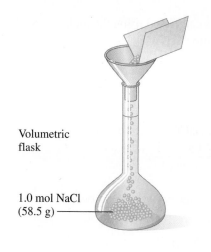

Volumetric
flask

1.0 mol NaCl
(58.5 g)

Add water until
1.0-L mark
is reached.

Mix

A 1.0 molar (M) NaCl solution

$$\text{Molarity (M)} = \frac{\text{Moles of solute}}{\text{Liters of solution}}$$

$$\text{g NaOH} \quad \boxed{\text{Molar mass}} \quad \frac{\text{mol NaOH}}{\text{Volume (L)}} = \text{M NaOH solution}$$

STEP 3 **Equalities/Conversion Factors**

$$1 \text{ mol NaOH} = 40.01 \text{ g NaOH}$$

$$\frac{1 \text{ mol NaOH}}{40.01 \text{ g NaOH}} \quad \text{and} \quad \frac{40.01 \text{ g NaOH}}{1 \text{ mol NaOH}}$$

STEP 4 **Set Up Problem**

$$\text{mol NaOH} = 60.0 \text{ g NaOH} \times \frac{1 \text{ mol NaOH}}{40.01 \text{ g NaOH}} = 1.50 \text{ mol NaOH}$$

The molarity is calculated by dividing the moles of NaOH by the volume in liters.

$$\frac{1.50 \text{ mol NaOH}}{0.250 \text{ L}} = \frac{6.00 \text{ mol NaOH}}{1 \text{ L}} = 6.00 \text{ M NaOH}$$

Study Check

What is the molarity of a solution that contains 75.0 g KNO_3 dissolved in 0.350 L solution?

Molarity as a Conversion Factor

When we need to calculate the moles of solute or the volume of solution, the molarity is used as a conversion factor. Examples of conversion factors from molarity are given in Table 12.6.

Table 12.6 **Some Examples of Molar Solutions**

Molarity	Meaning	Conversion Factors		
6.0 M HCl	6.0 mol HCl in 1 L solution	$\dfrac{6.0 \text{ mol HCl}}{1 \text{ L}}$	and	$\dfrac{1 \text{ L}}{6.0 \text{ mol HCl}}$
0.20 M NaOH	0.20 mol NaOH in 1 L solution	$\dfrac{0.20 \text{ mol NaOH}}{1 \text{ L}}$	and	$\dfrac{1 \text{ L}}{0.20 \text{ mol NaOH}}$

Using the molarity of the solution with the molar mass of the solute, we can calculate the volume of solution needed, as illustrated in Sample Problem 12.11.

Sample Problem 12.11 **Using Molarity to Find Volume**

How many liters of a 2.00 M NaCl solution are needed to provide 67.3 g NaCl?

Solution

STEP 1 **Given**   67.3 NaCl from a 2.00 M NaCl solution
Need   liters NaCl

STEP 2 **Plan** The volume of NaCl is calculated using the moles of NaCl and molarity of the NaCl solution.

g NaCl | Molar mass | mol NaCl | Molarity | L NaCl

STEP 3 **Equalities/Conversion Factors**

$$1 \text{ mol NaCl} = 58.44 \text{ g NaCl}$$

$$\frac{1 \text{ mol NaCl}}{58.44 \text{ g NaCl}} \quad \text{and} \quad \frac{58.44 \text{ g NaCl}}{1 \text{ mol NaCl}}$$

The molarity of any solution can be written as two conversion factors.

$$1 \text{ L NaCl} = 2.00 \text{ mol NaCl}$$

$$\frac{1 \text{ L NaCl}}{2.00 \text{ mol NaCl}} \quad \text{and} \quad \frac{2.00 \text{ mol NaCl}}{1 \text{ L NaCl}}$$

STEP 4 **Set Up Problem**

$$\text{L of NaCl} = 67.3 \text{ g NaCl} \times \frac{1 \text{ mol NaCl}}{58.44 \text{ g NaCl}} \times \frac{1 \text{ L NaCl}}{2.00 \text{ mol NaCl}}$$

$$= 0.576 \text{ L NaCl}$$

Study Check

How many milliliters of 2.25 M HCl solution will provide 4.12 g HCl?

To prepare a solution, we must convert the number of moles of solute needed into grams. Using the volume and the molarity of the solution with the molar mass of the solute, we can calculate the number of grams of solute necessary. This type of calculation is illustrated in Sample Problem 12.12.

Sample Problem 12.12 **Using Molarity**

How many grams of KCl would you need to weigh out to prepare 0.250 L of a 2.00 M KCl solution?

Solution

STEP 1 **Given** 0.250 L of 2.00 M KCl solution **Need** g KCl

STEP 2 **Plan** The grams of KCl are calculated by finding mol KCl using the volume and molarity of the KCl solution.

L KCl | Molarity | mol KCl | Molar mass | g KCl

STEP 3 **Equalities/Conversion Factors**

$$1 \text{ L KCl} = 2.00 \text{ mol KCl}$$

$$\frac{1 \text{ L KCl}}{2.00 \text{ mol KCl}} \quad \text{and} \quad \frac{2.00 \text{ mol KCl}}{1 \text{ L KCl}}$$

$$1 \text{ mol KCl} = 74.55 \text{ g KCl}$$

$$\frac{1 \text{ mol KCl}}{74.55 \text{ g KCl}} \quad \text{and} \quad \frac{74.55 \text{ g KCl}}{1 \text{ mol KCl}}$$

STEP 4 **Set Up Problem**

$$\text{mol KCl} = 0.250 \text{ L-solution} \times \frac{2.00 \text{ mol KCl}}{1 \text{ L-KCl}} = 0.500 \text{ mol KCl}$$

The grams of KCl is calculated by multiplying the moles of KCl by the molar mass.

$$\text{g KCl} = 0.500 \text{ mol KCl} \times \frac{74.55 \text{ g KCl}}{1 \text{ mol KCl}} = 37.3 \text{ g KCl}$$

Combining the steps, we can write the problem setup as follows:

$$0.250 \text{ L-solution} \times \frac{2.00 \text{ mol KCl}}{1 \text{ L-KCl}} \times \frac{74.55 \text{ g KCl}}{1 \text{ mol KCl}} = 37.3 \text{ g KCl}$$

Study Check

How many grams of $NaHCO_3$ are in 325 mL of a 4.50 M $NaHCO_3$ solution?

Dilutions

In chemistry, we often need to prepare a dilute solution from a more concentrated solution. In a process called **dilution,** we add water to a solution to make a larger volume. For example, you might prepare some orange juice by adding three cans of water to the original can of concentrated orange juice.

mix

1 can orange   +   3 cans water   =   4 cans of orange juice
juice concentrate

Adding water to a solution increases the volume and decreases the concentration of solute. The solution has been diluted. However, the number of moles of solute has not changed, only the volume of the solution.

Moles of solute   =   Moles of solute
(initial solution)   (diluted solution)

We know from the discussion of molarity that the moles of solute are obtained from the volume and molarity.

Moles of solute   =   Molarity × Volume
        Moles   =   M × V

Figure 12.8 When water is added to a concentrated solution, there is no change in the number of particles, but the solute particles can spread out as the volume of the diluted solution increases.

Q *What is the concentration of the diluted solution after an equal volume of water is added to a sample of 6 M HCl?*

Therefore, we can express the number of moles for the initial solutions as M_1V_1 and the number of moles in the diluted solution as M_2V_2:

Moles of solute	=	Moles of solute
(initial solution)		(diluted solution)
M_1V_1	=	M_2V_2

If we are given any 3 of the 4 variables, we can rearrange the expression to solve for the unknown quantity as seen in Sample Problem 12.13.

Sample Problem 12.13 **Molarity of a Diluted Solution**

What is the molarity of a solution prepared when 75.0 mL of a 4.00 M KCl solution is diluted to a volume of 0.500 L?

Solution

STEP 1 **Given Data in a Table** We make a table of the molar concentrations and volumes of the initial and diluted solutions.

Initial: $M_1 = 4.00 \, \text{M KCL}$ $V_1 = 75.0 \, \text{mL} = 0.0750 \, \text{L}$

Diluted: $M_2 = ? \, \text{M KCL}$ $V_2 = 0.500 \, \text{L}$

STEP 2 **Plan** The unknown molarity can be calculated by solving the dilution expression for M_2.

$$M_1V_1 = M_2 \, V_2$$

$$\frac{M_1V_1}{V_2} = M_2 \, \frac{\cancel{V_2}}{\cancel{V_2}}$$

$$M_2 = M_1 \times \frac{V_1}{V_2}$$

STEP 3 **Set Up Problem** The diluted concentration is calculated by placing the values from the table into the dilution expression.

$$= 4.00 \, \text{M} \times \frac{0.0750 \, \cancel{L}}{0.500 \, \cancel{L}} = 0.600 \, \text{M KCl} \qquad \text{(diluted solution)}$$

Guide to Calculating Dilution Quantities

STEP 1
Prepare a table of the initial and diluted volumes and concentrations.

STEP 2
Write a plan that solves the dilution expression for the unknown quantity.

STEP 3
Set up problem by placing known quantities in dilution expression.

Study Check

You need to prepare 600. mL of 2.00 M NaOH solution from a 10.0 M NaOH solution. What volume of the 10.0 M NaOH solution do you use?

Sample Problem 12.14 **Volume of a Diluted Solution**

What volume (mL) of a 0.20 M $SrCl_2$ solution can be prepared by diluting 50.0 mL of a 1.0 M $SrCl_2$ solution? How many mL of water must be added?

Solution

STEP 1 **Give Data in a Table** We make a table of the molar concentrations and volumes of the initial and diluted solutions.

Initial: $M_1 = 1.0 \text{ M } SrCl_2$ $V_1 = 50.0 \text{ mL}$

Diluted: $M_2 = 0.20 \text{ M}$ $V_2 = ? \text{ mL}$

STEP 2 **Plan** The volume of the dilute solution can be calculated by solving the dilution expression for V_2.

$$M_1 V_1 = M_2\ V_2$$

$$\frac{M_1 V_1}{M_2} = \frac{\cancel{M_2}}{\cancel{M_2}}\ V_2$$

$$V_2 = \frac{M_1 V_1}{M_2}$$

STEP 3 **Set Up Problem** Place the values from the table into the dilution expression solved for V_2.

$$V_2 = \frac{1.0\ \cancel{\text{M } SrCl_2}}{0.20\ \cancel{\text{M } SrCl_2}} \times 50.0 \text{ mL} = 250 \text{ mL} \qquad \text{(diluted } SrCl_2 \text{ solution)}$$

A volume of 200 mL of water must be added to the initial 50.0 mL of solution.

Study Check

What volume (mL) of an 8.00 M HCl solution is needed to prepare 1.00 L of a 0.500 M HCl solution?

Questions and Problems **Molarity and Dilutions**

12.37 Calculate the molarity of each of the following solutions:
 a. 2.00 mol glucose in 4.00 L solution
 b. 5.85 g NaCl in 40.0 mL solution
 c. 4.00 g KOH in 2.00 L solution

12.38 Calculate the molarity of each of the following solutions:
 a. 0.500 mol sucrose in 0.200 L solution
 b. 30.4 g LiBr in 350. mL solution
 c. 73.0 g HCl in 2.00 L solution

12.39 Calculate the grams of solute needed to prepare each of the following solutions:
 a. 2.00 L of a 1.50 M NaOH solution
 b. 125 mL of a 0.200 M KCl solution
 c. 25.0 mL of a 3.50 M HCl solution

12.40 Calculate the grams of solute needed to prepare each of the following solutions:
 a. 2.00 L of a 5.00 M NaOH solution
 b. 325 mL of a 0.100 M $CaCl_2$ solution
 c. 15.0 mL of a 0.500 M $LiNO_3$ solution

12.41 Calculate the volume in milliliters of each of the following solutions that provides the given amount of solute:
   a. 12.5 g Na_2CO_3 from a 0.120 M solution
   b. 0.850 mol $NaNO_3$ from a 0.500 M solution
   c. 30.0 g LiOH from a 2.70 M solution

12.42 Calculate the volume in liters of each of the following solutions that provides the given amount of solute:
   a. 5.00 mol NaOH from a 12.0 M solution
   b. 15.0 g Na_2SO_4 from a 4.00 M solution
   c. 28.0 g $NaHCO_3$ from a 1.50 M solution

12.43 Calculate final concentration of the solution in each of the following dilutions:
   a. Water is added to 0.150 L of a 6.00 M HCl solution to give a volume of 0.500 L.
   b. A 10.0-mL sample of 2.50 M KCl solution is diluted with water to 0.250 L.
   c. Water is added to 0.250 L of a 12.0 M KBr solution to give a volume of 1.00 L.

12.44 Calculate final concentration of the solution in each of the following dilutions:
   a. Water added to 10.0 mL of a 3.50 M KNO_3 solution gives a volume of 0.250 L.
   b. A 5.00-mL sample of 18.0 M sucrose solution is diluted with water to 100. mL.
   c. Water is added to 0.250 L of a 12.0 M KBr solution to give a volume of 1.00 L.

12.45 Determine the final volume (mL) for each of the following dilutions:
   a. diluting 50.0 mL of a 12.0 M NH_4Cl solution to give a 2.00 M NH_4Cl solution
   b. diluting 18.0 mL of a 15.0 M $NaNO_3$ solution to give a 1.50 M $NaNO_3$ solution
   c. diluting 4.50 mL of an 18.0 M H_2SO_4 solution to give a 2.50 M H_2SO_4 solution

12.46 Determine the final volume (mL) for each of the following dilutions:
   a. diluting 2.50 mL of an 8.00 M KOH solution to give a 2.00 M KOH solution
   b. diluting 50.0 mL of a 12.0 M NH_4Cl solution to give a 2.00 M NH_4Cl solution
   c. diluting 75.0 mL of a 6.00 M HCl solution to give a 0.200 M HCl solution

12.47 Determine the volume (mL) required to prepare each of the following dilutions:
   a. 255 mL of a 0.200 M HNO_3 solution from a 4.00 M HNO_3 solution
   b. 715 mL of a 0.100 M $MgCl_2$ solution using a 6.00 M $MgCl_2$ solution
   c. 0.100 L of a 0.150 M KCl solution using an 8.00 M KCl solution

12.48 Determine the volume (mL) required to prepare each of the following dilutions:
   a. 20.0 mL of a 0.250 M KNO_3 solution from a 6.00 M KNO_3 solution
   b. 25.0 mL of 2.50 M H_2SO_4 solution using a 12.0 M H_2SO_4 solution
   c. 0.500 L of a 1.50 M NH_4Cl solution using a 10.0 M NH_4Cl solution

12.49 You need to dilute 25.0 mL of a 3.00 M HCl solution to make a 0.150 M HCl solution. What is the volume of diluted solution after you add water?

12.50 You need to dilute 30.0 mL of a 2.50 NaCl solution to make a 0.500 M NaCl solution. What is the volume of diluted solution after you add water?

12.6 Solutions in Chemical Reactions

Calculations are also done for the substances in chemical reactions that take place in aqueous solutions. For a balanced equation, we use the molarity and volume of a solution to determine the moles of a substance required or produced in a chemical reaction. We can also use molarity and the number of moles of a solute to determine the volume of a solution.

Sample Problem 12.15   **Volume of a Solution in a Reaction**

Zinc reacts with HCl to produce $ZnCl_2$ and hydrogen gas, H_2.

$$Zn(s) + 2HCl(aq) \longrightarrow ZnCl_2(aq) + H_2(g)$$

How many liters of a 1.50 M HCl solution completely react with 5.32 g of zinc?

Solution

STEP 1   **Given**   5.32 g Zn and a 1.50 M HCl solution
              Need   liters of HCl solution

STEP 2 **Plan** We can use the molar mass of Zn to find the moles of Zn and the mole–mole factor from the balanced equation to convert moles of Zn to moles of HCl. Since the concentration of the HCl solution is given, the molarity can be used to convert moles to volume in liters.

g Zn | Molar mass | mol Zn | Mole–mole factor | mol HCl | Molarity | L HCl

STEP 3 **Equalities/Conversion Factors**

Molar mass of Zn

$$1 \text{ mol Zn} = 65.41 \text{ g Zn}$$

$$\frac{1 \text{ mol Zn}}{65.41 \text{ g Zn}} \quad \text{and} \quad \frac{65.41 \text{ g Zn}}{1 \text{ mol Zn}}$$

Mole-mole factor

$$1 \text{ mol Zn} = 2 \text{ mol HCl}$$

$$\frac{1 \text{ mol Zn}}{2 \text{ mol HCl}} \quad \text{and} \quad \frac{2 \text{ mol HCl}}{1 \text{ mol Zn}}$$

Molarity of HCl solution

$$1 \text{ L HCl} = 1.50 \text{ mol HCl}$$

$$\frac{1 \text{ L HCl}}{1.50 \text{ mol HCl}} \quad \text{and} \quad \frac{1.50 \text{ mol HCl}}{1 \text{ L HCl}}$$

STEP 4 **Set Up Problem** The problem is set up as seen in our plan.

$$5.32 \text{ g Zn} \times \frac{1 \text{ mol Zn}}{65.41 \text{ g Zn}} \times \frac{2 \text{ mol HCl}}{1 \text{ mol Zn}} \times \frac{1 \text{ L HCl}}{1.50 \text{ mol HCl}} = 0.108 \text{ L HCl}$$

Study Check

Using the reaction in Sample Problem 12.15, how many grams of zinc can react with 225 mL of 0.200 M HCl?

Guide to Calculations Involving Solutions in Chemical Reactions

STEP 1
State the given and needed quantities.

STEP 2
Write a plan to calculate needed quantity or concentration.

STEP 3
Write equalities and conversion factors including mole–mole and concentration factors.

STEP 4
Set up problem to calculate needed quantity or concentration.

Sample Problem 12.16 **Volume of a Reactant**

How many mL of 0.250 M $BaCl_2$ is needed to react with 32.5 mL of a 0.160 M Na_2SO_4 solution?

$$Na_2SO_4(aq) + BaCl_2(aq) \longrightarrow BaSO_4(s) + 2NaCl(aq)$$

Solution

STEP 1 **Given** 32.5 mL (0.0325 L) of 0.160 M Na_2SO_4 and 0.250 M $BaCl_2$ **Need** mL $BaCl_2$

STEP 2 **Plan** We use the volume and molarity of the Na_2SO_4 solution to determine the moles of Na_2SO_4 and then moles of $BaCl_2$ using the mole–mole factor from the equation. Use the molarity of $BaCl_2$ to calculate the volume of $BaCl_2$ in liters and milliliters.

L Na_2SO_4 | Molarity | mol Na_2SO_4 | Mole–mole factor | mol $BaCl_2$ | Molarity | L $BaCl_2$ | Metric factor | mL $BaCl_2$

STEP 3 **Equalities/Conversion Factors**

Molarity of Na₂SO₄

$$1 \text{ L Na}_2\text{SO}_4 = 0.160 \text{ mol Na}_2\text{SO}_4$$

$$\frac{1 \text{ L Na}_2\text{SO}_4}{0.160 \text{ mol Na}_2\text{SO}_4} \quad \text{and} \quad \frac{0.160 \text{ mol Na}_2\text{SO}_4}{1 \text{ L Na}_2\text{SO}_4}$$

Mole–mole factor

$$1 \text{ mol Na}_2\text{SO}_4 = 1 \text{ mol BaCl}_2$$

$$\frac{1 \text{ mol Na}_2\text{SO}_4}{1 \text{ mol BaCl}_2} \quad \text{and} \quad \frac{1 \text{ mol BaCl}_2}{1 \text{ mol Na}_2\text{SO}_4}$$

Molarity of BaCl₂

$$1 \text{ L BaCl}_2 = 0.250 \text{ mol BaCl}_2$$

$$\frac{1 \text{ L BaCl}_2}{0.250 \text{ mol BaCl}_2} \quad \text{and} \quad \frac{0.250 \text{ mol BaCl}_2}{1 \text{ L BaCl}_2}$$

STEP 4 **Set Up Problem** The problem is set up using the plan and appropriate conversion factors.

$$0.0325 \text{ L Na}_2\text{SO}_4 \times \frac{0.160 \text{ mol Na}_2\text{SO}_4}{1 \text{ L Na}_2\text{SO}_4} \times \frac{1 \text{ mol BaCl}_2}{1 \text{ mol Na}_2\text{SO}_4} \times \frac{1 \text{ L BaCl}_2}{0.250 \text{ mol BaCl}_2} \times \frac{1000 \text{ mL BaCl}_2}{1 \text{ L BaCl}_2}$$

$$= 20.8 \text{ mL BaCl}_2$$

Study Check

For the reaction in Sample Problem 12.16, how many milliliters of 0.330 M Na₂SO₄ is needed to react with 26.8 mL of a 0.216 M BaCl₂ solution?

Sample Problem 12.17 ▶ **Volume of a Gas from a Solution**

Acid rain results from the reaction of nitrogen dioxide with water in the air.

$$3NO_2(g) + H_2O(l) \longrightarrow 2HNO_3(aq) + NO(g)$$

At STP, how many liters of NO₂ gas are required to produce 0.275 L of 0.400 M HNO₃?

Solution

STEP 1 **Given** 0.275 L of 0.400 M HNO₃
Need Liters of NO₂ gas at STP

STEP 2 **Plan** We will use the volume and molarity of the HNO₃ solution to determine the moles of HNO₃ and convert to moles of NO₂ using the balanced equation. Then we can calculate the volume of NO₂ using the molar volume (22.4 L/mol) at STP.

L HNO₃ │ Molarity │ moles HNO₃ │ Mole–mole factor │ moles NO₂ │ Molar volume │ L NO₂

STEP 3 **Equalities/Conversion Factors**

Molarity of HNO₃

$$1 \text{ L HNO}_3 = 0.400 \text{ mol HNO}_3$$

$$\frac{1 \text{ L HNO}_3}{0.400 \text{ mol HNO}_3} \quad \text{and} \quad \frac{0.400 \text{ mol HNO}_3}{1 \text{ L HNO}_3}$$

Mole–mole factor

$$3 \text{ mol NO}_2 = 2 \text{ mol HNO}_3$$

$$\frac{2 \text{ mol HNO}_3}{3 \text{ mol NO}_2} \quad \text{and} \quad \frac{3 \text{ mol NO}_2}{2 \text{ mol HNO}_3}$$

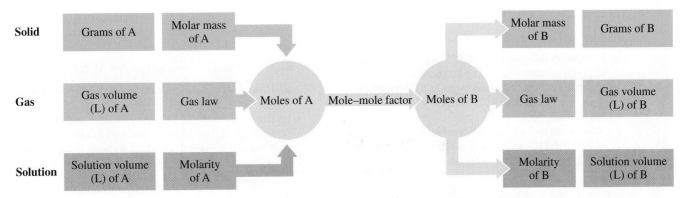

Figure 12.9 In calculations involving chemical reactions, substance A is converted to moles of A using molar mass (if solid), gas laws (if gas), or molarity (if solution). Then moles of A are converted to moles of substance B, which are converted to grams of solid, liters of gas, or liters of solution, as needed.

Q What sequence of conversion factors would you use to calculate the number of grams of $CaCO_3$ needed to react with 1.50 L 2.00 M HCl in the reaction:
$$2HCl(aq) + CaCO_3(s) \longrightarrow CaCl_2(aq) + CO_2(g) + H_2O(l)?$$

Molar volume NO_2

$$1 \text{ mol } NO_2 = 22.4 \text{ L } NO_2$$

$$\frac{22.4 \text{ L } NO_2}{1 \text{ mol } NO_2} \quad \text{and} \quad \frac{1 \text{ mol } NO_2}{22.4 \text{ L } NO_2}$$

STEP 4   **Set Up Problem**   We use the molarity of HNO_3 to find moles of HNO_3, then convert to moles of NO_2, and finally to liters of NO_2 gas at STP.

$$0.275 \text{ L } HNO_3 \times \frac{0.400 \text{ mol } HNO_3}{1 \text{ L } HNO_3} \times \frac{3 \text{ mol } NO_2}{2 \text{ mol } HNO_3} \times \frac{22.4 \text{ L } NO_2}{1 \text{ mol } NO_2} = 3.70 \text{ L } NO_2$$

Study Check

Using the equation in Sample Problem 12.17, determine the volume of NO produced at STP when 2.20 L of 1.50 HNO_3 is produced.

Figure 12.9 gives a summary of the pathways and conversion factors needed for substances including solutions involved in chemical reactions.

Questions and Problems   **Solutions in Chemical Reactions**

12.51 Given the reaction

$$Pb(NO_3)_2(aq) + 2KCl(aq) \longrightarrow PbCl_2(s) + 2KNO_3(aq)$$

   a. How many grams of $PbCl_2$ will be formed from 50.0 mL of 1.50 M KCl?
   b. How many milliliters of 2.00 M $Pb(NO_3)_2$ will react with 50.0 mL of 1.50 M KCl?
   c. What is the molarity of 20.0 mL of KCl solution that reacts with 30.0 mL of 0.400 M $Pb(NO_3)_2$?

12.52 In the reaction

$$NiCl_2(aq) + 2NaOH(aq) \longrightarrow Ni(OH)_2(s) + 2NaCl(aq)$$

   a. How many milliliters of 0.200 M NaOH are needed to react with 18.0 mL of 0.500 M $NiCl_2$?
   b. How many grams of $Ni(OH)_2$ are produced from the reaction of 35.0 mL of 1.75M NaOH and excess $NiCl_2$?
   c. What is the molarity of 30.0 mL of $NiCl_2$ solution if this volume of solution reacts completely with 10.0 mL of 0.250 M NaOH solution?

12.53 In the reaction

$$Mg(s) + 2HCl(aq) \longrightarrow MgCl_2(aq) + H_2(g)$$

a. How many milliliters of a 6.00 M HCl solution are required to react with 15.0 g magnesium?

b. How many liters of hydrogen gas at STP can form when 0.500 L of 2.00 M HCl reacts with excess magnesium?

c. What is the molarity of a HCl solution if the reaction of 45.2 mL HCl with excess magnesium produces 5.20 L H_2 gas at 735 mm Hg and 25°C?

12.54 The calcium carbonate in limestone reacts with HCl to produce a calcium chloride solution, carbon dioxide, and water.

$$CaCO_3(s) + 2HCl(aq) \longrightarrow CaCl_2(aq) + H_2O(l) + CO_2(g)$$

a. How many milliliters of 0.200 M HCl can react with 8.25 g $CaCO_3$?

b. How many liters of CO_2 gas can form at STP when 15.5 mL of 3.00 M HCl react with excess $CaCO_3$?

c. What is the molarity of a HCl solution if the reaction of 200. mL HCl with excess $CaCO_3$ produces 12.0 L CO_2 gas at 725 mm Hg and 18°C?

Concept Map   **Solutions**

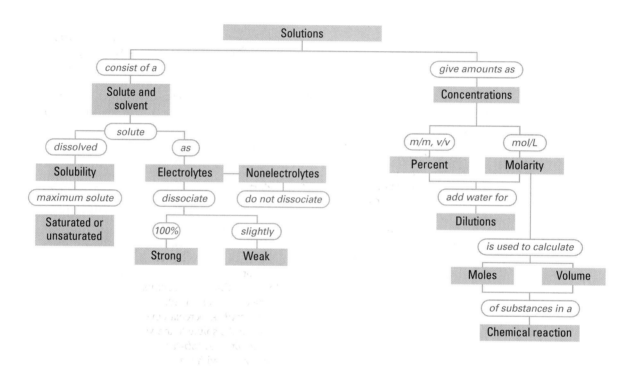

Chapter Review

12.1 Solutions

A solution forms when a solute dissolves in a solvent. The expression "like dissolves like" means that a polar or ionic solute dissolves in a polar solvent and a nonpolar solute requires a nonpolar solvent. The polar O—H bond leads to hydrogen bonding between water molecules.

12.2 Electrolytes and Nonelectrolytes

An ionic solute dissolves in water, a polar solvent, because the polar water molecules attract and pull the ions into solution, where they become hydrated. Substances that release ions are called electrolytes because the solution will conduct an electrical current. Strong electrolytes are completely ionized, whereas weak electrolytes are only partially ionized. Nonelectrolytes are substances that dissolve in water to produce molecules and solutions that cannot conduct electrical currents.

12.3 Solubility

A solution that contains the maximum amount of dissolved solute is a saturated solution. The solubility of a solute is the maximum amount of a solute that can dissolve in 100 g of solvent. A solution containing less than the maximum amount of dissolved solute is unsaturated. An increase in temperature increases the solubility of most solids in water but decreases the solubility of gases in water. Salts that are soluble in water usually contain Li^+, Na^+, K^+, NH_4^+, NO_3^-, or acetate, $C_2H_3O_2^-$. An ionic equation consists of writing all the dissolved substances in an equation for the formation of an insoluble salt as individual ions. A net ionic equation is written by removing all the ions not involved in the chemical change (spectator ions) from the ionic equation.

12.4 Percent Concentration

The concentration of a solution is the amount of solute dissolved in a certain amount of solution. Mass percent expresses the ratio of the mass of solute to the mass of solution multiplied by 100. Percent concentration can also be expressed as a volume/volume ratio. In calculations of grams or milliliters of solute or solution, the percent concentration is used as a conversion factor.

12.5 Molarity and Dilutions

Molarity is the moles of solute per liter of solution. Units of molarity, moles/liter, are used in conversion factors to solve for moles of solute or volume of solution. When water is added to a solution, the volume increases. The solute is now distributed throughout a larger volume, which dilutes the solution and decreases the concentration.

12.6 Solutions in Chemical Reactions

When solutions are involved in chemical reactions, the moles of a substance in solution can be determined from the volume and molarity of the solution. If the mass, gas volume, or solution volume and molarity of substances in a reaction are given, the balanced equation can be used to determine the quantities or concentrations of any of the other substances in the reaction.

Key Terms

concentration A measure of the amount of solute that is dissolved in a specified amount of solution.

dilution A process by which water (solvent) is added to a solution to increase the volume and decrease (dilute) the solute concentration.

dissociation The separation of a solute into ions when the solute is dissolved in water.

electrolyte A substance that produces ions when dissolved in water; its solution conducts electricity.

Henry's law The solubility of a gas in a liquid is directly related to the pressure of that gas above the liquid.

hydration The process of surrounding dissolved ions by water molecules.

insoluble salt An ionic compound that does not dissolve in water.

ionic equation An equation for a reaction in solution that gives all the individual ions, both reacting ions and spectator ions.

mass percent The grams of solute in exactly 100 g of solution.

molarity (M) The number of moles of solute in exactly 1 L of solution.

net ionic equation An equation for a reaction that gives only the reactants undergoing chemical change and leaves out spectator ions.

nonelectrolyte A substance that dissolves in water as molecules; its solution will not conduct an electrical current.

saturated solution A solution containing the maximum amount of solute that can dissolve at a given temperature. Any additional solute will remain undissolved in the container.

solubility The maximum amount of solute that can dissolve in exactly 100 g of solvent, usually water, at a given temperature.

soluble salt An ionic compound that dissolves in water.

solute A substance that is the smaller amount uniformly dispersed in another substance called the solvent.

solution A homogeneous mixture in which the solute is made up of small particles (ions or molecules).

solvent The substance in which the solute dissolves; usually the component present in greatest amount.

strong electrolyte A compound that ionizes completely when it dissolves in water. Its solution is a good conductor of electricity.

unsaturated solution A solution that contains less solute than can be dissolved.

volume percent A percent concentration that relates the volume of the solute to the volume of the solution.

weak electrolyte A substance that produces only a few ions along with many molecules when it dissolves in water. Its solution is a weak conductor of electricity.

▶ Understanding the Concepts

12.55 Select the diagram that represents the solution formed by a solute 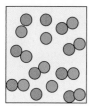 that is a
  a. nonelectrolyte   **b.** weak electrolyte
  c. strong electrolyte

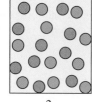
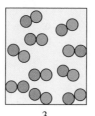

1.   2.   3.

12.56 Match the diagrams with
  a. a polar solute and a polar solvent
  b. a nonpolar solute and a polar solvent
  c. a nonpolar solute and a nonpolar solvent

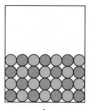
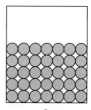

1.   2.

12.57 Select the container that represents the dilution of a 4% (m/m) KCl solution to each of the following:
a. 2% (m/m) KCl
b. 1% (m/m) KCl

12.58 Do you think solution (1) has undergone heating or cooling to give the solid shown in (2) and (3)?

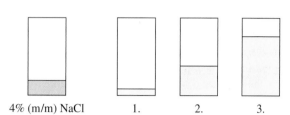

4% (m/m) NaCl   1.   2.   3.

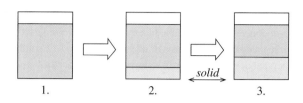

1.   2.   3.   *solid*

Use the following beakers and solutions for questions 12.59 and 12.60.

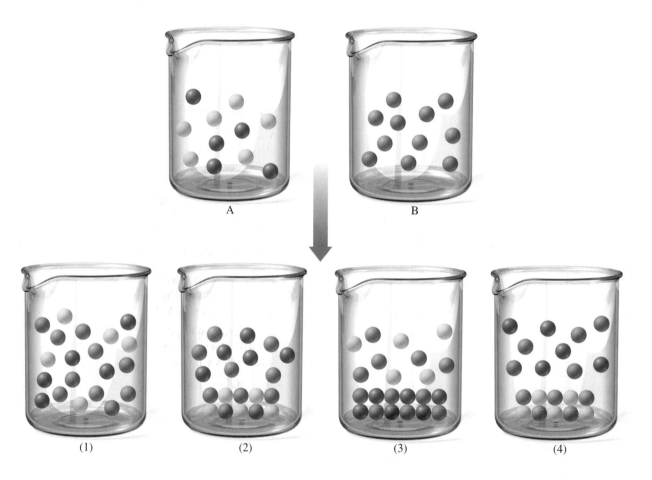

A   B

(1)   (2)   (3)   (4)

12.59 Use the following:

Na⁺   Cl⁻   Ag⁺   NO₃⁻

a. Select the beaker (1, 2, 3, or 4) that contains the products after the solutions in beakers A and B are mixed.
b. If an insoluble salt forms, write the ionic equation.
c. If a reaction occurs, write the net ionic equation.

12.60 Use the following:

K⁺   NO₃⁻   NH₄⁺   Br⁻

a. Select the beaker (1, 2, 3, or 4) that contains the products after the solutions in beakers A and B are mixed.
b. If an insoluble salt forms, write the ionic equation.
c. If a reaction occurs, write the net ionic equation.

Additional Questions and Problems

12.61 Why does iodine dissolve in hexane, but not in water?

12.62 How do temperature and pressure affect the solubility of solids and gases in water?

12.63 If NaCl has a solubility of 36.0 g at 20°C, how many grams of water are needed to prepare a saturated solution containing 80.0 g NaCl?

12.64 If the solid NaCl in a saturated solution of NaCl continues to dissolve, why is there no change in the concentration of the NaCl solution?

12.65 Potassium nitrate has a solubility of 34 g KNO_3 in 100 g H_2O at 20°C. State if each of the following forms an unsaturated or saturated solution at 20°C:
   a. 34 g KNO_3 and 200 g H_2O
   b. 17 g KNO_3 and 50 g H_2O
   c. 68 g KNO_3 and 150 g H_2O

12.66 Potassium fluoride has a solubility of 92 g KF in 100 g H_2O at 18°C. State if each of the following forms an unsaturated or saturated solution at 18°C:
   a. 46 g KF and 100 g H_2O
   b. 46 KF and 50 g H_2O
   c. 184 KF and 150 g H_2O

12.67 Why would a solution made by mixing solutions of $NaNO_3$ and KCl be clear, while a combination of KCl and $Pb(NO_3)_2$ solution produces a solid?

12.68 Indicate whether each of the following is soluble in water:
   a. KCl   **b.** $MgSO_4$
   c. PbS   **d.** $AgNO_3$
   e. $Ca(OH)_2$

12.69 Write the net ionic equation to show the formation of a precipitate (insoluble salt) when the following solutions are mixed. Write *none* if there is no precipitate.
   a. $AgNO_3(aq)$ and $NaCl(aq)$
   b. $NaCl(aq)$ and $KNO_3(aq)$
   c. $Na_2SO_4(aq)$ and $BaCl_2(aq)$

12.70 Write the net ionic equation to show the formation of a precipitate (insoluble salt) when the following solutions are mixed. Write *none* if there is no precipitate.
   a. $Ca(NO_3)_2(aq)$ and $Na_2S(aq)$
   b. $Na_3PO_4(aq)$ and $Pb(NO_3)_2(aq)$
   c. $FeCl_3(aq)$ and $NH_4NO_3(aq)$

12.71 How many milliliters of a 12% (v/v) propyl alcohol solution would you take to obtain 4.5 mL of propyl alcohol?

12.72 An 80-proof brandy is 40.0% (v/v) ethyl alcohol. The "proof" is twice the percent concentration of alcohol in the beverage. How many milliliters of alcohol are present in 750. mL of brandy?

12.73 A solution is prepared with 70.0 g HNO_3 and 130.0 g H_2O. It has a density of 1.21 g/mL.
   a. What is the mass percent of the HNO_3 solution?
   b. What is the total volume of the solution?
   c. What is its molarity (M)?

12.74 A solution is prepared by dissolving 22.0 g NaOH in 118.0 g water. The NaOH solution has a density of 1.15 g/mL.
   a. What is the % (m/m) concentration of the NaOH solution?
   b. What is the total volume (mL) of the solution?
   c. What is the molarity (M) of the solution?

12.75 How many liters of a 2.50 M KNO_3 solution can be prepared from 60.0 g KNO_3?

12.76 How many liters of 4.00 M NaCl solution will provide 25.0 g NaCl?

12.77 If you were in the laboratory, how would you prepare 250 mL of a 2.00 M KCl solution?

12.78 What is the molarity of a solution containing 15.6 g of KCl in 274 mL solution?

12.79 How many grams of solute are in each of the following solutions?
   a. 2.52 L of a 3.00 M KNO_3 solution
   b. 75.0 mL of a 0.506 M Na_2SO_4 solution
   c. 45.2 mL of a 1.80 M HCl solution

12.80 How many grams of solute are in each of the following solutions?
   a. 428 mL of a 0.450 M Na_2SO_4 solution
   b. 10.5 mL of a 2.50 M $AgNO_3$ solution
   c. 28.4 mL of a 6.00 M H_3PO_4 solution

12.81 The antacid Amphogel contains aluminum hydroxide $Al(OH)_3$. How many milliliters of 6.00 M HCl are required to react with 60.0 mL of 2.00 M $Al(OH)_3$?

$$Al(OH)_3(s) + 3HCl(aq) \longrightarrow AlCl_3(aq) + 3H_2O(aq)$$

12.82 A 255-mL sample of HCl solution reacts with excess Mg to produce 14.0 L H_2 gas at STP. What is the molarity of the HCl solution?

$$Mg(s) + 2HCl(aq) \longrightarrow MgCl_2(aq) + H_2(g)$$

12.83 A 355-mL sample of the HCl solution reacts with excess Mg to produce 4.20 L H_2 gas measured at 745 mm Hg and 35°C. What is the molarity of the HCl solution?

$$Mg(s) + 2HCl(aq) \longrightarrow MgCl_2(aq) + H_2(g)$$

12.84 Calcium carbonate, $CaCO_3$, reacts with stomach acid (HCl, hydrochloric acid) according to the following equation:

$$CaCO_3(s) + 2HCl(aq) \longrightarrow$$
$$CaCl_2(aq) + H_2O(l) + CO_2(g)$$

One tablet of Tums, an antacid, contains 500.0 mg $CaCO_3$. If one tablet of Tums is added to 20.0 mL of 0.100 M HCl, how many liters of CO_2 gas are produced at STP?

12.85 Calculate the molarity of the solution when water is added to prepare each of the following solutions:
a. 25.0 mL of 0.200 M NaBr diluted to 50.0 mL
b. 15.0 mL of 1.20 M K_2SO_4 diluted to 40.0 mL
c. 75.0 mL of 6.00 M NaOH diluted to 255 mL

12.86 Calculate the molarity of the solution when water is added to prepare each of the following solutions:
a. 25.0 mL of 18.0 M HCl diluted to 500. mL
b. 50.0 mL of 1.50 M NaCl diluted to 125 mL
c. 4.50 mL of 8.50 M KOH diluted to 75.0 mL

12.87 What is the final volume in mL when 25.0 mL of 5.00 M HCl is diluted to each of the following concentrations?
a. 2.50 M HCl
b. 1.00 M HCl
c. 0.500 M HCl

12.88 What is the final volume in mL when 5.00 mL of 12.0 M NaOH is diluted to each of the following concentrations?
a. 0.600 M
b. 1.00 M
c. 2.50 M

Challenge Questions

12.89 Indicate whether each of the following ionic compounds is soluble (S) or insoluble (I) in water:
a. Na_3PO_4 b. $PbBr_2$
c. KCl   d. $(NH_4)_2S$
e. $MgCO_3$ f. $FePO_4$

12.90 Write the net ionic equation to show the formation of a precipitate (insoluble salt) when the following solutions are mixed. Write *none* if no insoluble salt forms.
a. $AgNO_3$ + Na_2SO_4
b. KCl + $Pb(NO_3)_2$
c. $CaCl_2$ + $Mg_3(PO_4)_2$
d. Na_2SO_4 + $BaCl_2$

12.91 In a laboratory experiment, a 10.0-mL sample of NaCl solution is poured into an evaporating dish with a mass of 24.10 g. The combined mass of the evaporating dish and NaCl is 36.15 g. After heating, the evaporating dish and dry NaCl have a combined mass of 25.50 g.
a. What is the % (m/m) of the NaCl solution?
b. What is the molarity (M) of the NaCl solution?
c. If water is added to 10.0 mL of the initial NaCl solution to give a final volume of 60.0 mL, what is the molarity of the dilute NaCl solution?

12.92 A solution contains 4.56 g KCl in 175 mL of solution. If the density of the KCl solution is 1.12 g/mL, what are the % (m/m) and molarity, M, for the potassium chloride solution?

12.93 How many milliliters of a 1.75 M LiCl solution contain 15.2 g of LiCl?

12.94 How many grams of NaBr are contained in 75.0 mL of a 1.50 M NaBr solution?

12.95 Magnesium reacts with HCl to produce magnesium chloride and hydrogen gas.

$$Mg(s) + 2HCl(aq) \longrightarrow MgCl_2(aq) + H_2(g)$$

What is the molarity of the HCl solution if 250. mL of the HCl solution reacts with magnesium to produce 4.20 L of H_2 gas measured at STP?

12.96 How many liters of NO gas can be produced at STP from 80.0 mL of 4.00 M HNO_3 and 10.0 g of Cu?

$$3Cu(s) + 8HNO_3(aq) \longrightarrow$$
$$3Cu(NO_3)_2(aq) + 4H_2O(l) + 2NO(g)$$

Answers

Answers to Study Checks

12.1 Yes. Both the solute and solvent are nonpolar substances; "like dissolves like."

12.2 A solution of a weak electrolyte will contain mostly molecules and a few ions.

12.3 57 g $NaNO_3$

12.4 The value of 80. g/100 g H_2O is more likely because the solubility of most solids increases when the temperature increases.

12.5 a. No; $PbCl_2$ is an insoluble chloride.
b. Yes; a salt containing K^+ ion is soluble.
c. No; $FeCO_3$ is insoluble.

12.6 a. No solid forms.
b. $Pb^{2+}(aq) + 2Cl^-(aq) \longrightarrow PbCl_2(s)$

12.7 3.4% (m/m) NaCl solution

12.8 4.8% (v/v) Br_2 in CCl_4

12.9 18 g KCl and 207 g H_2O

12.10 2.12 M KNO_3

12.11 50.2 mL

12.12 123 g $NaHCO_3$

12.13 120. mL of 10.0 M NaOH solution

12.14 62.5 mL of 8.00 M HCl

12.15 1.47 g Zn

12.16 17.5 mL Na_2SO_4 solution

12.17 37.0 L NO

Answers to Selected Questions and Problems

12.1 **a.** NaCl, solute; water, solvent
b. water, solute; ethanol, solvent
c. oxygen, solute; nitrogen, solvent

12.3 **a.** water   **b.** CCl_4 **c.** water   **d.** CCl_4

12.5 The polar water molecules pull the K^+ and I^- ions away from the solid and into solution, where they are hydrated.

12.7 In a solution of KF, only the ions of K^+ and F^- are present in the solvent. In an HF solution, there are a few ions of H^+ and F^- present but mostly dissolved HF molecules.

12.9 **a.** $KCl \xrightarrow{H_2O} K^+ + Cl^-$

b. $CaCl_2 \xrightarrow{H_2O} Ca^{2+} + 2Cl^-$

c. $K_3PO_4 \xrightarrow{H_2O} 3K^+ + PO_4^{3-}$

d. $Fe(NO_3)_3 \xrightarrow{H_2O} Fe^{3+} + 3NO_3^-$

12.11 **a.** mostly molecules and a few ions
b. ions only
c. molecules only

12.13 **a.** strong electrolyte   **b.** weak electrolyte
c. nonelectrolyte

12.15 **a.** saturated   **b.** unsaturated

12.17 **a.** unsaturated   **b.** unsaturated
c. saturated

12.19 **a.** 68.0 g KCl   **b.** 12.0 g KCl

12.21 **a.** The solubility of solid solutes typically increases as temperature increases.
b. The solubility of a gas is less at a higher temperature.
c. Gas solubility is less at a higher temperature and the CO_2 pressure in the can is increased.

12.23 **a.** soluble   **b.** insoluble
c. insoluble   **d.** soluble
e. soluble

12.25 **a.** no solid forms
b. $2Ag^+(aq) + S^{2-}(aq) \longrightarrow Ag_2S(s)$
c. $Ca^{2+}(aq) + SO_4^{2-}(aq) \longrightarrow CaSO_4(s)$

12.27 12.5 g glucose is added to 237.5 g water to make 250. g of 5.0% (m/m) glucose solution.

12.29 **a.** 17%   **b.** 10.%
c. 5.3%

12.31 **a.** 2.5 g KCl   **b.** 50. g NH_4Cl
c. 25 mL acetic acid

12.33 79.9 mL alcohol

12.35 **a.** 20. g   **b.** 400. g
c. 20. mL

12.37 **a.** 0.500 M glucose   **b.** 2.50 M NaCl
c. 0.0356 M KOH

12.39 **a.** 120. g NaOH   **b.** 1.86 g KCl
c. 3.19 g HCl

12.41 **a.** 983 mL   **b.** 1700 mL (1.70×10^3)
c. 464 mL

12.43 **a.** 1.80 M HCl   **b.** 0.100 M KCl
c. 3.00 M KBr

12.45 **a.** 300. mL NH_4Cl solution
b. 180. mL $NaNO_3$ solution
c. 32.4 mL H_2SO_4 solution

12.47 **a.** 12.8 mL HNO_3 solution
b. 11.9 mL $MgCl_2$ solution
c. 0.00188 mL KCl solution

12.49 500. mL HCl solution

12.51 **a.** 10.4 g $PbCl_2$
b. 18.8 mL $Pb(NO_3)_2$ solution
c. 1.20 M KCl

12.53 **a.** 206 mL HCl solution   **b.** 11.2 L H_2 gas
c. 9.09 M HCl

12.55 **a.** 3 (no dissociation)
b. 1 (some dissociation, a few ions)
c. 2 (all ionized)

12.57 **a.** 2; to halve the % concentration, the volume would double.
b. 3; to go to one-fourth the % concentration, the volume would be four times the initial volume.

12.59 **a.** beaker 3
b. $Na^+ + Cl^- + Ag^+ + NO_3^- \longrightarrow$
$AgCl + Na^+ + NO_3^-$
c. $Ag^+ + Cl^- \longrightarrow AgCl$

12.61 Because iodine is a nonpolar molecule, it will dissolve in hexane, a nonpolar solvent. Iodine does not dissolve in water because water is a polar solvent.

12.63 222 g water

12.65 **a.** unsaturated solution   **b.** saturated solution
c. saturated solution

12.67 When solutions of $NaNO_3$ and KCl are mixed, no insoluble products are formed. All the combinations of salts are soluble. When KCl and $Pb(NO_3)_2$ solutions are mixed, the insoluble salt $PbCl_2$ forms.

12.69 **a.** $Ag^+(aq) + Cl^-(aq) \longrightarrow AgCl(s)$
b. none
c. $Ba^{2+}(aq) + SO_4^{2-}(aq) \longrightarrow BaSO_4(s)$

12.71 38 mL solution

12.73 **a.** 35.0% HNO_3 **b.** 165 mL
c. 6.73 M

12.75 **a.** 0.237 L KNO_3 solution

12.77 To make a 2.00 M KCl solution, weigh out 37.3 g KCl (0.500 mol) and place in a volumetric flask. Add water to dissolve the KCl and give a final volume of 0.250 L.

12.79 **a.** 764 g KNO_3 **b.** 5.39 g Na_2SO_4
c. 2.97 g HCl

12.81 60.0 mL HCl

12.83 0.917 M HCl solution

12.85 **a.** 0.100 M NaBr   **b.** 0.450 M K_2SO_4
c. 1.76 M NaOH

12.87 **a.** 50.0 mL HCl   **b.** 125 mL HCl
c. 250. mL HCl

12.89 **a.** Na^+ salts are soluble.
b. The halide salt containing Pb^{2+} is insoluble.
c. K^+ salts are soluble.
d. Salts containing NH_4^+ ions are soluble.
e. Salts containing CO_3^{2-} are usually insoluble.
f. Salts containing PO_4^{3-} and Fe^{3+} are insoluble.

12.91 **a.** 11.6% NaCl (m/m)   **b.** 2.40 M
c. 0.400 M

12.93 205 mL

12.95 1.50 M HCl

Chemical Equilibrium

"*I use radioactive isotopes to understand the cycling of elements like carbon and phosphorus in the ocean,*" *explains Claudia Benitez-Nelson, a chemical oceanographer and assistant professor of geological sciences at the University of South Carolina.* "*For example, I use thorium-234 to trace how and when particles are formed and transported to the bottom of the ocean. I also examine the biological consumption of the nutrient phosphorus by measuring fluctuations over time in the levels of the naturally occurring radioactive isotopes of phosphorus. My knowledge of chemistry is essential for understanding nutrient biogeochemistry and carbon sequestration in the oceans.*"

Oceanographers study the oceans and the plants and animals that live in the ocean. They study marine life, the chemical compounds in the ocean, the shape and composition of the ocean floor, and the effects of waves and tides.

the
Chemistry
place

Visit **www.aw-bc.com/chemplace** for extra quizzes, interactive tutorials, career resources, PowerPoint slides for chapter review, math help, and case studies.

Earlier we looked at chemical reactions and determined the amounts of substances that react and the products that form. Now we are interested in how fast a reaction goes. If we know how fast a medication acts on the body, we can adjust the time over which the medication is taken. In construction, substances are added to cement to make it dry faster so work can continue. Some reactions such as explosions or the formation of precipitates in a solution are very fast. We know that when we roast a turkey or bake a cake that the reaction is slower. Some reactions such as the tarnishing of silver and the aging of the body are much slower. (See Figure 13.1.) We will see that some reactions need energy to keep running while other reactions produce energy. We burn gasoline in our automobile engines to produce energy to make our cars move. We will also look at the effect of changing the concentrations of reactants or products on the rate of reaction.

Up to now, we have considered a reaction as proceeding in a forward direction from reactants to products. However, in many reactions a reverse reaction also takes place as products collide to re-form reactants. When the forward and reverse reactions take place at the same rate, the amounts of reactants and products stay the same. When this balance in forward and reverse rate is reached, we say that the reaction has reached *equilibrium*. At equilibrium both reactants and products are present, though some reaction mixtures contain mostly reactants and form only a few products, while others contain mostly products and few reactants.

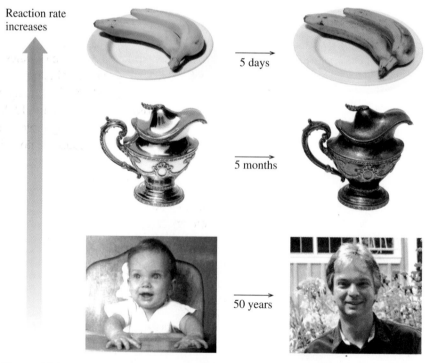

Figure 13.1 Reaction rates vary greatly for everyday processes. A banana ripens in a few days, silver tarnishes in a few months, while the aging process of humans takes many years.

Q How would you compare the rates of the reaction that forms sugars in plants by photosynthesis with the reactions that digest sugars in the body?

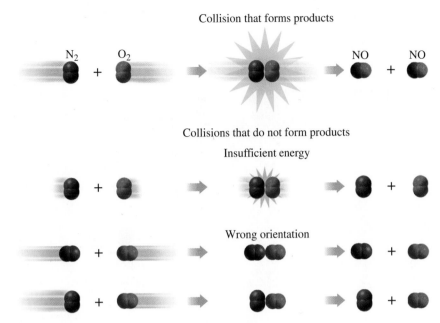

Figure 13.2 Reacting molecules must collide, have a minimum amount of energy, and have the proper orientation to form products.

Q *What happens when reacting molecules collide with the minimum energy but don't have the proper orientation?*

13.1 Rates of Reactions

Learning Goal

Describe how temperature, concentration, and catalysts affect the rate of a reaction.

For a chemical reaction to take place, the molecules of the reactants must come in contact with each other. The **collision theory** indicates that a reaction takes place only when molecules collide with the proper orientation, and sufficient energy. Many collisions can occur, but only a few actually lead to the formation of product. For example, consider the reaction of nitrogen and oxygen molecules. (See Figure 13.2.) To form NO product, the collisions between the N_2 and the O_2 molecules must place the atoms in the proper alignment. If the molecules are not aligned properly, no reaction takes place.

Activation Energy

Even when a collision has the proper orientation, there still must be sufficient energy to break the bonds between the atoms of the reactants. The amount of energy required to break the bonds between atoms of reactants is the **activation energy**. In Figure 13.3, this appears as an energy hill. The concept of activation energy is analogous to climbing a hill. To reach a destination on the other side, we must have the energy needed to climb to the top of the hill. Once we are at the top, we can run down the other side. The energy needed to get us from our starting point to the top of the hill would be the activation energy.

In the same way, a collision must provide enough energy to push the reactants to the top of the energy hill. Then the reactants may be converted to products. If the energy provided by the collision is less than the activation energy, the molecules simply bounce apart and no reaction occurs. The features that lead to a successful reaction are summarized next.

Figure 13.3 The activation energy is the energy needed to convert the colliding molecules into product.

Q *What happens in a collision of reacting molecules that have the proper orientation, but not the energy of activation?*

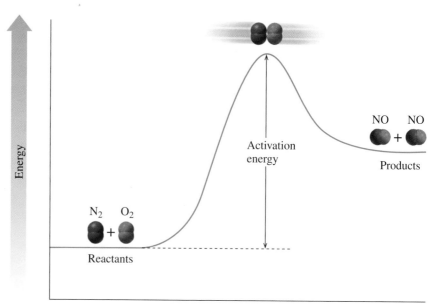

Three Conditions Required for a Reaction to Occur

1. **Collision**   The reactants must collide.
2. **Orientation**   The reactants must align properly to break and form bonds.
3. **Energy**   The collision must provide the energy of activation.

Reaction Rates

The **rate** (or speed) **of reaction** is determined by measuring the amount of a reactant used up, or the amount of a product formed, in a certain period of time.

$$\text{Rate of reaction} = \frac{\text{Change in concentration}}{\text{Change in time}}$$

Perhaps we can describe the rate of reaction by the analogy of eating a pizza. When we start to eat, we have a whole pizza. As time goes by, there are fewer slices of pizza left. If we know how long it took to eat the pizza, we could determine the rate at which the pizza was consumed. Let's assume 4 slices are eaten every 8 minutes. That gives a rate of $\frac{1}{2}$ slice per minute. After 16 minutes, all 8 slices are gone.

Rate at Which Pizza Slices Are Eaten

Slices eaten	0	4 slices	6 slices	8 slices
Time (min)	0	8 min	12 min	16 min

$$\text{Rate} = \frac{4 \text{ slices}}{8 \text{ min}} = \frac{1 \text{ slice}}{2 \text{ min}} = \frac{\frac{1}{2} \text{ slice}}{1 \text{ min}}$$

Factors That Affect the Rate of a Reaction

Some reactions go very fast, while others are very slow. For any reaction, the rate is affected by changes in temperature, changes in the concentration of the reactants, and the addition of catalysts.

Temperature At higher temperatures, the increase in kinetic energy makes the reacting molecules move faster. As a result, more collisions occur and more colliding molecules have sufficient energy to react and form products. If we want food to cook faster, we use more heat to raise the temperature. When body temperature rises, there is an increase in the pulse rate, rate of breathing, and metabolic rate. On the other hand, we slow down reactions by lowering the temperature. We refrigerate perishable foods to retard spoilage and make them last longer. For some injuries, we apply ice to lessen the bruising process.

Concentrations of Reactants The rate of a reaction increases when the concentrations of the reactants increases. When there are more reacting molecules, more collisions can occur, and the reaction goes faster. (See Figure 13.4.) For example, a person having difficulty breathing may be given oxygen. The increase in the number of oxygen molecules in the lungs increases the rate at which oxygen combines with hemoglobin and helps the person breathe more easily.

Catalysts Another way to speed up a reaction is to lower the energy of activation. We saw that the energy of activation is the energy needed to break apart the bonds of the reacting molecules. If a collision provides less than the activation energy, the bonds do not break and the molecules bounce apart. A **catalyst** speeds up a reaction by providing an alternative pathway that has a

Reactants in   Possible
reaction flask   collisions

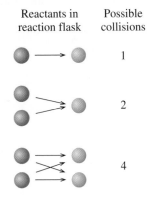

Figure 13.4 Increasing the concentration of a reactant increases the number of collisions that are possible.

Q How many collisions are possible if one more red reactant is added?

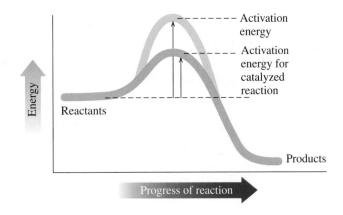

Table 13.1 **Factors That Increase Reaction Rate**

Factor	Reason
More reactants	More collisions
Higher temperature	More collisions, more collisions with energy of activation
Adding a catalyst	Lowers energy of activation

lower energy of activation. When activation energy is lowered, more collisions provide sufficient energy for reactants to form product. During a reaction, a catalyst is not changed or used up.

Catalysts have many uses in industry. In the manufacturing of margarine, hydrogen (H_2) is added to vegetable oils. Normally, the reaction is very slow because it has a high activation energy. However, when platinum (Pt) is used as a catalyst, the reaction occurs rapidly. In the body, biocatalysts called enzymes make most metabolic reactions proceed at rates necessary for proper cellular activity. A summary of the factors affecting reaction rates is given in Table 13.1.

Sample Problem 13.1 **Factors That Affect the Rate of Reaction**

Indicate whether the following changes will increase, decrease, or have no effect upon the rate of reaction:
a. increase in temperature
b. decrease in the number of reactants
c. adding a catalyst

Solution
a. increase
b. decrease
c. increase

Study Check

How does the lowering of temperature affect the rate of reaction?

Questions and Problems **Rates of Reactions**

13.1 **a.** What is meant by the rate of a reaction?
b. Why does bread grow mold more quickly at room temperature than in the refrigerator?

13.2 **a.** How does a catalyst affect the activation energy?
b. Why is pure oxygen used in respiratory distress?

13.3 In the following reaction, what happens to the number of collisions when more Br_2 molecules are added?

$$H_2(g) + Br_2(g) \longrightarrow 2HBr(g)$$

13.4 In the following reaction, what happens to the number of collisions when the temperature of the reaction is decreased?

$$H_2(g) + Br_2(g) \longrightarrow 2HBr(g)$$

13.5 How would each of the following change the rate of the reaction shown here?

$$2SO_2(g) + O_2(g) \longrightarrow 2SO_3(g)$$

a. adding SO_2
b. raising the temperature
c. adding a catalyst
d. removing some O_2

13.6 How would each of the following change the rate of the reaction shown here?

$$2NO(g) + 2H_2(g) \longrightarrow N_2(g) + 2H_2O(g)$$

a. adding more NO
b. lowering the temperature
c. removing some H_2
d. adding a catalyst

CHEM NOTE

CATALYTIC CONVERTERS

For many years, manufacturers have been required to include catalytic converters on automobile engines. When gasoline burns, the products found in the exhaust of a car contain high levels of pollutants. These include carbon monoxide (CO) from incomplete combustion, hydrocarbons such as C_7H_{16} from unburned fuel, and nitrogen oxide (NO) from the reaction of N_2 and O_2 at the high temperatures reached within the engine. Carbon monoxide is toxic, and nitrogen oxide is involved in the formation of smog and acid rain.

The purpose of a catalytic converter is to lower the activation energy for reactions that convert each of these pollutants into substances such as CO_2, N_2, O_2, and H_2O, which are already present in the atmosphere.

$$2CO(g) + O_2(g) \longrightarrow 2CO_2(g)$$
$$C_7H_{16}(g) + 11O_2(g) \longrightarrow 7CO_2(g) + 8H_2O(g)$$
$$2NO(g) \longrightarrow N_2(g) + O_2(g)$$

A catalytic converter consists of solid-particle catalysts, such as platinum (Pt) and palladium (Pd), on a ceramic honeycomb that provides a large surface area and facilitates contact with pollutants. As the pollutants pass through the converter, they react with the catalysts. Today, we all use unleaded gasoline because

lead interferes with the ability of the Pt and Pd catalysts in the converter to react with the pollutants.

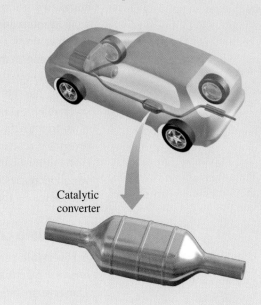

Catalytic converter

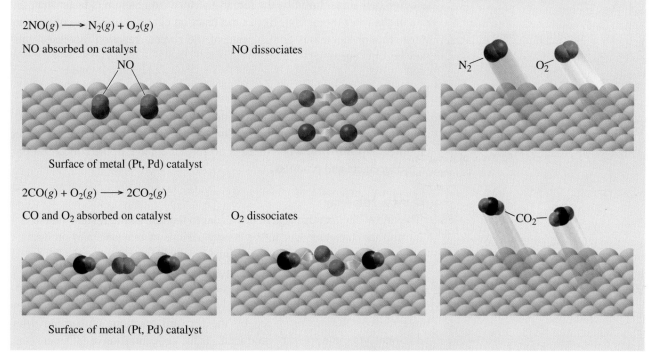

$2NO(g) \longrightarrow N_2(g) + O_2(g)$

NO absorbed on catalyst

NO

Surface of metal (Pt, Pd) catalyst

NO dissociates

N_2 O_2

$2CO(g) + O_2(g) \longrightarrow 2CO_2(g)$

CO and O_2 absorbed on catalyst

O_2 dissociates

CO_2

Surface of metal (Pt, Pd) catalyst

13.2 Chemical Equilibrium

Learning Goal

Use the concept of reversible reactions to explain chemical equilibrium.

In earlier chapters, we considered the *forward reaction* in an equation and assumed that all of the reactants were converted to products. However, most of the time reactants are not completely converted to products because a *reverse reaction* takes place in which products come together and form the reactants. When a reaction consists of both a forward and reverse direction, it is said to be reversible. We have looked at other reversible processes. For example, the melting of solids to form liquids and the freezing of liquids to solids is a reversible physical change. Even in our daily life we have reversible events. We go from home to school and we return from school to home. We go up an escalator and come back down. We put money in our bank account and take money out.

An analogy for a forward and reverse reaction can be found in the phrase "I am going to the grocery store." Although we mention our trip in one direction, we know that we will also return home from the grocery store. Because our trip has both a forward and reverse direction, we can say the trip is reversible. It is not very likely that we would stay at the grocery store forever.

A trip to the grocery store can be used to illustrate another aspect of reversible reactions. Perhaps the grocery store is close and we usually walk. However, we can change our rate. Suppose one day that we drive to the store, which increases our rate and gets us to the store faster. Correspondingly, a car also increases the rate at which we return home.

Reversible Chemical Reactions

A **reversible reaction** consists of both a forward and a reverse reaction. The *forward reaction* begins as collisions occur between the reactant molecules. The *reverse reaction* begins once there are sufficient product molecules to undergo collisions. Initially, the rate of the forward reaction is faster than the rate of the reverse reaction. But as the reaction continues, the rate of the forward reaction decreases and the rate of the reverse reaction increases until they become equal.

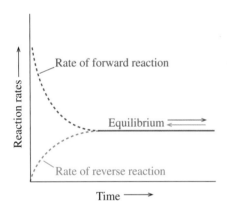

Equilibrium

Eventually, the rates of the forward and reverse reactions are equal; the reactants form products as often as products form reactants. A reaction reaches **chemical equilibrium** when there is no further change in the concentrations of the reactants and products.

> At equilibrium:
> The rate of the forward reaction is equal to the rate of the reverse reaction. No further changes occur in the concentrations of reactants and products, even though the two reactions continue at equal but opposite rates.

Let us look at the process as the reaction of N_2 and O_2 proceeds to equilibrium. (See Figure 13.5.) Initially, only N_2 and O_2 are present. Soon, a few molecules of NO are produced by the forward reaction. With more time, additional NO molecules are produced. As the concentration of NO increases, more NO molecules collide and react in the reverse direction.

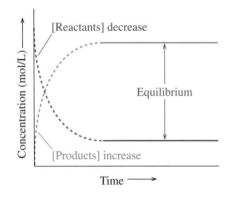

Forward reaction: $N_2(g) + O_2(g) \longrightarrow 2NO(g)$

Reverse reaction: $2NO(g) \longrightarrow N_2(g) + O_2(g)$

As NO product builds up, the rate of the reverse reaction increases while the rate of the forward reaction decreases. Eventually, the rates become equal, which means the reaction has reached equilibrium. Even though the concentrations remain constant at equilibrium, the forward and reverse reactions

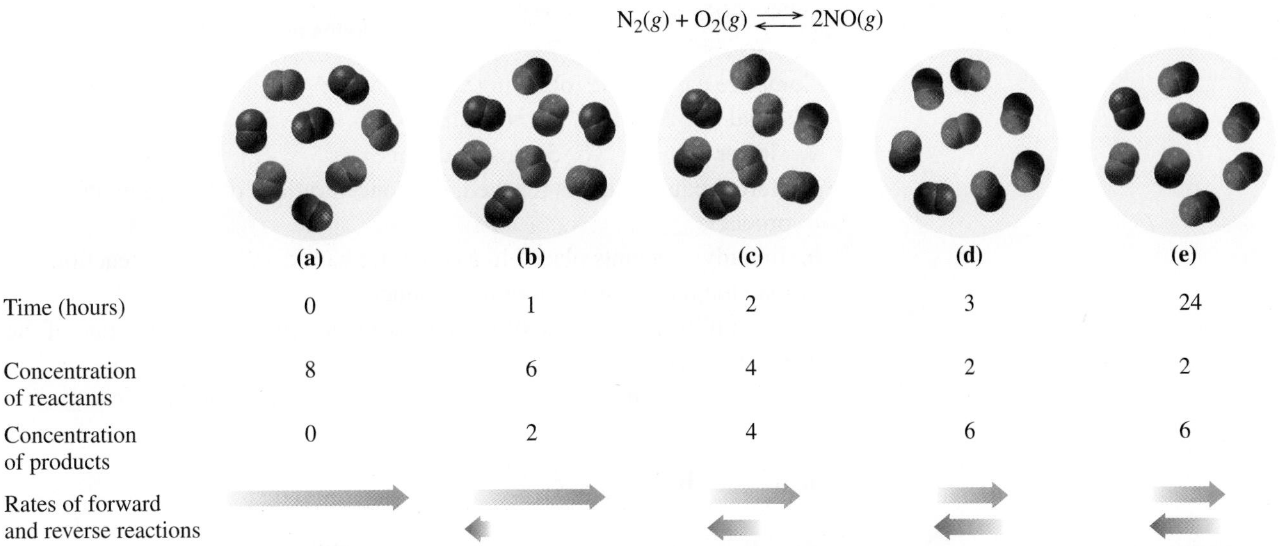

$$N_2(g) + O_2(g) \rightleftharpoons 2NO(g)$$

	(a)	(b)	(c)	(d)	(e)
Time (hours)	0	1	2	3	24
Concentration of reactants	8	6	4	2	2
Concentration of products	0	2	4	6	6
Rates of forward and reverse reactions					

Figure 13.5 (a) Initially, the reaction flask contains only the reactants N_2 and O_2. **(b)** The forward reaction between O_2 and N_2 begins to produce 2NO. **(c)** As the reaction proceeds, there are fewer molecules of O_2 and N_2 and more molecules of NO, which increases the rate of the reverse reaction. **(d)** At equilibrium, the concentrations of reactants N_2 and O_2 and product NO are constant. **(e)** The reaction continues with the rate of the forward reaction equal to the rate of the reverse reactions.

Q How do the rates of the forward and reverse reactions compare once a chemical reaction reaches equilibrium?

continue to occur. The forward and reverse reactions are usually shown together in a single equation by using a double arrow. A reversible reaction is two opposing reactions that occur at the same time.

$$N_2(g) + O_2(g) \quad \underset{\text{Reverse reaction}}{\overset{\text{Forward reaction}}{\rightleftharpoons}} \quad 2NO(g)$$

Sample Problem 13.2 Reversible Reactions

Write the forward and reverse reactions for each of the following:
a. $N_2(g) + 3H_2(g) \rightleftharpoons 2NH_3(g)$ **b.** $2CO(g) + O_2 \rightleftharpoons 2CO_2(g)$

Solution
The equations are separated into forward and reverse reactions.
a. Forward reaction: $N_2(g) + 3H_2(g) \longrightarrow 2NH_3(g)$
  Reverse reaction: $2NH_3(g) \longrightarrow N_2(g) + 3H_2(g)$
b. Forward reaction: $2CO(g) + O_2(g) \longrightarrow 2CO_2(g)$
  Reverse reaction: $2CO_2(g) \longrightarrow O_2(g) + 2CO(g)$

Study Check

Write the equation for the reaction that contains the following reverse reaction:

$$2HBr(g) \longrightarrow H_2(g) + Br_2(g)$$

Sample Problem 13.3 ▸ Reaction Rates and Equilibrium

Complete each of the following with

1. equal
2. not equal
3. faster
4. slower
5. do not change
6. change

a. Before equilibrium is reached, the concentrations of the reactants and products ____.

b. Initially, reactants placed in a container have a ____ rate of reaction than the rate of reaction of the products.

c. At equilibrium, the rate of the forward reaction is ____ to the rate of the reverse reaction.

d. At equilibrium, the concentrations of the reactants and products ____.

Solution

a. 6   **b.** 3   **c.** 1   **d.** 5

Study Check

Using the choice of answers in Sample Problem 13.3, complete the following:

As reactants are used up and products accumulate, the rate of the forward reaction becomes ____, while the rate of the reverse reaction becomes ____.

Questions and Problems ▸ Chemical Equilibrium

13.7 What is meant by the term *reversible reaction*?

13.8 When does a reversible reaction reach equilibrium?

13.9 Which of the following processes are reversible?
   a. breaking a glass   **b.** melting ice
   c. heating a pan

13.10 Which of the following processes are at equilibrium?
   a. Opposing rates of reaction are equal.
   b. The rate of the forward reaction is faster than the rate of the reverse reaction.
   c. The concentrations of reactants and products do not change.

13.3 Equilibrium Constants

Learning Goal

Calculate the equilibrium constant for a reversible reaction given the concentrations of reactants and products at equilibrium.

At equilibrium, reactions occur in opposite directions at the same rate, which means the concentrations of the reactants and products remain constant. We can use a ski lift as an analogy. Early in the morning, skiers at the bottom of the mountain begin to ride the ski lift up to the slopes. As skiers reach the top of the mountain, they ski down. Eventually, the number of people riding up the ski lift becomes equal to the number of people skiing down the mountain. There is no further change in the number of skiers on the slopes; the system is at equilibrium.

Equilibrium Constant Expression

Because the concentrations in a reaction at equilibrium no longer change, they can be used to set up a relationship between the products and the reactants. Suppose we write a general equation for reactants A and B that form products C and D. The small italic letters are the coefficients in the balanced equation.

$$a\text{A} + b\text{B} \rightleftharpoons c\text{C} + d\text{D}$$

An **equilibrium constant expression** for the reaction multiplies the concentrations of the products together and divides by the concentrations of the reactants. Each concentration is raised to a power that is its coefficient in the balanced chemical reaction. The square bracket around each substance indicates the concentration is expressed in moles per liter (M). The **equilibrium constant**, K_c, is the numerical value obtained by substituting molar concentrations at equilibrium into the expression. For our general reaction, the equilibrium constant expression is

Equilibrium   Equilibrium constant
constant   expression

$$K_c = \frac{\text{Products}}{\text{Reactants}} \quad \frac{[C]^c\,[D]^d}{[A]^a\,[B]^b} \quad \text{Coefficients}$$

We can write the equilibrium constant expression for the reaction of H_2 and I_2 using the balanced equation

$$H_2(g) + I_2(g) \rightleftharpoons 2HI(g)$$

Guide to Writing the K_c Expression

STEP 1
Write the balanced equilibrium equation.

STEP 2
Write the products in brackets as the numerator and reactants in brackets as the denominator. Do not include pure solids or liquids.

STEP 3
Write the coefficient of each substance in the equation as an exponent.

STEP 1 **Write the balanced equilibrium equation.**

$$H_2(g) + I_2(g) \rightleftharpoons 2HI(g)$$

STEP 2 **Write in brackets the products as the numerator and reactants as the denominator.**

$$\frac{\text{Products}}{\text{Reactants}} \quad \frac{[HI]}{[H_2][I_2]}$$

STEP 3 **Write the coefficient of each substance as an exponent.**

$$K_c = \frac{[HI]^2}{[H_2][I_2]}$$

Sample Problem 13.4   **Writing Equilibrium Constant Expressions**

Write the equilibrium constant expression for the following:

$$2SO_2(g) + O_2(g) \rightleftharpoons 2SO_3(g)$$

Solution

STEP 1 **Write the balanced equilibrium equation.**

$$2SO_2(g) + O_2(g) \rightleftharpoons 2SO_3(g)$$

STEP 2 **Write in brackets the products as the numerator and reactants as the denominator.**

$$\frac{\text{Products}}{\text{Reactants}} \quad \frac{[SO_3]}{[SO_2][O_2]}$$

STEP 3 **Write the coefficient of each substance as an exponent.**

$$K_c = \frac{[SO_3]^2}{[SO_2]^2[O_2]}$$

$$CaCO_3(s) \rightleftharpoons CaO(s) + CO_2(g)$$

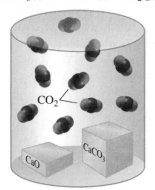

$T = 800°C$

$T = 800°C$

Figure 13.6 At equilibrium at constant temperature, the concentration of CO_2 is the same regardless of the amounts of $CaCO_3(s)$ and $CaO(s)$ in the container.

Q *Why are the concentrations of CaO(s) and CaCO₃(s) not included in K꜀ for the decomposition of CaCO₃?*

Study Check

Write the balanced chemical equation that would give the following equilibrium constant expression:

$$K_c = \frac{[NO_2]^2}{[NO]^2[O_2]}$$

Heterogeneous Equilibrium

Up to now, our examples have been reactions that involve only gases. A reaction in which all the reactants and products are in the same state reaches **homogenous equilibrium.** When the reactants and products are in two or more states, the equilibrium is termed a **heterogeneous equilibrium.** For example, in the following reaction, the decomposition of calcium carbonate reaches heterogeneous equilibrium with calcium oxide and carbon dioxide. (See Figure 13.6.)

$$CaCO_3(s) \rightleftharpoons CaO(s) + CO_2(g)$$

In contrast to gases, the concentrations of pure solids and pure liquids in a heterogeneous equilibrium are constant; they do not change. Therefore, pure solids and liquids are not included in the equilibrium constant expression. For this heterogeneous equilibrium, the K_c expression does not include the concentration of $CaCO_3(s)$ or $CaO(s)$. It is written as

$$K_c = [CO_2]$$

Sample Problem 13.5 ▶ **Heterogeneous Equilibrium Constant Expression**

Write the equilibrium constant expression for the following heterogeneous equilibria:
a. $Si(s) + 2Cl_2(g) \rightleftharpoons SiCl_4(g)$
b. $2Mg(s) + O_2(g) \rightleftharpoons 2MgO(s)$

Solution

In the equilibrium constant expressions for heterogeneous reactions, the concentrations of the pure solids are not included.

a. $K_c = \dfrac{[SiCl_4]}{[Cl_2]^2}$ **b.** $K_c = \dfrac{1}{[O_2]}$

Study Check

Solid iron(II) oxide and carbon monoxide gas are in equilibrium with solid iron and carbon dioxide gas. Write the equation and the equilibrium constant expression for the reaction.

Calculating Equilibrium Constants

The numerical value of the equilibrium constant is calculated from the equilibrium constant expression by substituting experimentally measured concentrations of the reactants and products at equilibrium into the equilibrium constant expression. For example, the equilibrium constant expression for the reaction of H_2 and I_2 is written

$$H_2(g) + I_2(g) \rightleftharpoons 2HI(g) \qquad K_c = \frac{[HI]^2}{[H_2][I_2]}$$

Suppose in an experiment we measured the molar concentrations for the reactants and products at equilibrium as $[H_2] = 0.10$ M, $[I_2] = 0.20$ M, and $[HI] = 1.04$ M. When we substitute these values into the equilibrium constant expression, we obtain the numerical value of the equilibrium constant.

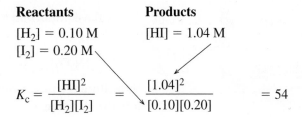

Reactants
$[H_2] = 0.10$ M
$[I_2] = 0.20$ M

Products
$[HI] = 1.04$ M

$$K_c = \frac{[HI]^2}{[H_2][I_2]} = \frac{[1.04]^2}{[0.10][0.20]} = 54$$

Suppose we look at different equilibrium concentrations of reactants and products for the H_2, I_2, and HI system at equilibrium at 700. K (427°C). When the concentrations of reactants and products are measured in each equilibrium sample and used to calculate the K_c for the reaction, the same value of K_c is obtained for each. (See Table 13.2.) Thus, a reaction at a specific temperature can have only one value for the equilibrium constant.

Table 13.2 Equilibrium Constant for $H_2(g) + I_2(g) \rightleftharpoons 2HI(g)$ at 427°C

Experiment	$[H_2]$	$[I_2]$	$[HI]$	$K_c = \dfrac{[HI]^2}{[H_2][I_2]}$
1	0.10 M	0.20 M	1.04 M	54
2	0.20 M	0.20 M	1.47	54
3	0.30 M	0.17 M	1.66	54

The units of K_c depend on the specific equation. In the example of $[H_2]$, $[I_2]$, and $[HI]$, K_c has the units of $[M]^2/[M]^2$, which results in a value with no units. However, in the following example, the $[M]$ units would not cancel because $[M]^2/[M]^4 = 1/[M]^2 = [M]^{-2}$. However, we usually do not attach any units to the value of K_c. At 500. K, the value of K_c for the following reaction is 1.7×10^2.

$$N_2(g) + 3H_2(g) \rightleftharpoons 2NH_3(g)$$

$$K_c = \frac{[NH_3]^2}{[N_2][H_2]^3} = 1.7 \times 10^2\,[M]^{-2} \quad \text{or usually written as } 1.7 \times 10^2$$

Sample Problem 13.6   **Calculating an Equilibrium Constant**

The decomposition of dinitrogen tetroxide forms nitrogen dioxide.

$$N_2O_4(g) \rightleftharpoons 2NO_2(g)$$

What is the value of K_c at 100.°C if a reaction mixture at equilibrium contains $[N_2O_4] = 0.45$ M and $[NO_2] = 0.31$ M?

Solution

Given   reactant: $[N_2O_4] = 0.45$ M   product: $[NO_2] = 0.31$ M
Need K_c value

Guide to Calculating the K_c Value
STEP 1 Write the K_c expression for the equilibrium.
STEP 2 Substitute equilibrium (molar) concentrations and calculate K_c.

STEP 1   **Write the equilibrium expression.**

$$K_c = \frac{[NO_2]^2}{[N_2O_4]}$$

STEP 2   **Substitute the molar concentrations at equilibrium and calculate the K_c value.**

$$K_c = \frac{[0.31]^2}{[0.45]} = \frac{0.31 \text{ M} \times 0.31 \text{ M}}{0.45 \text{ M}} = 0.21 \text{ M, or } 0.21$$

Study Check

Ammonia decomposes when heated to give nitrogen and hydrogen.

$$2NH_3(g) \rightleftharpoons 3H_2(g) + N_2(g)$$

Calculate the equilibrium constant if an equilibrium mixture contains $[NH_3] = 0.040$ M, $[N_2] = 0.20$ M, and $[H_2] = 0.60$ M.

Questions and Problems   **Equilibrium Constants**

13.11 Write the equilibrium constant expression, K_c, for each of the following reactions:
   a. $CH_4(g) + 2H_2S(g) \rightleftharpoons CS_2(g) + 4H_2(g)$
   b. $2NO(g) \rightleftharpoons N_2(g) + O_2(g)$
   c. $2SO_3(g) + CO_2(g) \rightleftharpoons CS_2(g) + 4O_2(g)$

13.12 Write the equilibrium constant expression K_c for each of the following reactions:
   a. $2HBr(g) \rightleftharpoons H_2(g) + Br_2(g)$
   b. $CO(g) + 2H_2(g) \rightleftharpoons CH_3OH(g)$
   c. $CH_4(g) + H_2O(g) \rightleftharpoons CO(g) + 3H_2(g)$

13.13 Identify each of the following as a homogeneous or heterogeneous equilibrium:
   a. $2O_3(g) \rightleftharpoons 3O_2(g)$
   b. $2NaHCO_3(s) \rightleftharpoons Na_2CO_3(s) + CO_2(g) + H_2O(g)$
   c. $CH_4(g) + H_2O(g) \rightleftharpoons 3H_2(g) + CO(g)$
   d. $4HCl(g) + O_2(g) \rightleftharpoons 2H_2O(l) + 2Cl_2(g)$

13.14 Identify each of the following as a homogeneous or heterogeneous equilibrium:
   a. $CO(g) + H_2(g) \rightleftharpoons C(s) + H_2O(g)$
   b. $CO(g) + 2H_2(g) \rightleftharpoons CH_3OH(l)$
   c. $CS_2(g) + 4H_2(g) \rightleftharpoons CH_4(g) + 2H_2S(g)$
   d. $Br_2(g) + Cl_2(g) \rightleftharpoons 2BrCl(g)$

13.15 Write the equilibrium constant expression for each of the reactions in problem 13.13.

13.16 Write the equilibrium constant expression for each of the reactions in problem 13.14.

13.17 What is the value of K_c for the following equilibrium

$$N_2O_4(g) \rightleftharpoons 2NO_2(g)$$

if $[NO_2] = 0.21$ M and $[N_2O_4] = 0.030$ M?

13.18 What is the value of K_c for the following equilibrium

$$CO_2(g) + H_2(g) \rightleftharpoons CO(g) + H_2O(g)$$

if $[CO] = 0.20$ M, $[H_2O] = 0.30$ M, $[CO_2] = 0.30$ M, and $[H_2] = 0.033$ M?

13.19 What is the value of K_c for the following equilibrium at 1000°C

$$CO(g) + 3H_2(g) \rightleftharpoons CH_4(g) + H_2O(g)$$

if $[H_2] = 0.30$ M, $[CO] = 0.51$ M, $[CH_4] = 1.8$ M, and $[H_2O] = 2.0$ M?

13.20 What is the value of K_c for the following equilibrium at 500°C

$$N_2(g) + 3H_2(g) \rightleftharpoons 2NH_3(g)$$

if $[H_2] = 0.40$ M, $[N_2] = 0.44$ M, $[NH_3] = 2.2$ M?

13.4 Using Equilibrium Constants

We have seen that the values of K_c can be large or small. We can now look at the K_c values to predict how far the reaction proceeds to products at equilibrium. When a K_c is large, the numerator (products) is greater than the concentrations of the reactants in the denominator.

$$\frac{[\text{Products}]}{[\text{Reactants}]} = \text{Large } K_c$$

When a K_c is small, the numerator is smaller than the denominator, which means that the reaction favors the reactants.

$$\frac{[\text{Products}]}{[\text{Reactants}]} = \text{Small } K_c$$

Using a general reaction and its equilibrium constant expression, we can look at the relative concentrations of reactant A and product B.

$$A(g) \rightleftharpoons B(g) \qquad K_c = \frac{[B]}{[A]}$$

For a large K_c, [B] is greater than [A]. For example, if the K_c is 1×10^3, or 1000, [B] would be 1000 times greater than [A] at equilibrium.

$$K_c = \frac{[B]}{[A]} = 1000 \quad \text{or rearranged} \quad [B] = 1000\,[A]$$

For a small K_c, [A] is greater than [B]. For example, if the K_c is 1×10^{-2}, [A] is 100 times greater than [B] at equilibrium.

$$K_c = \frac{[B]}{[A]} = \frac{1}{100} \quad \text{or rearranged} \quad [A] = 100\,[B]$$

Equilibrium mixtures with different values of K_c are illustrated in Figure 13.7.

$$A \rightleftharpoons B \qquad K_c = \frac{B}{A}$$

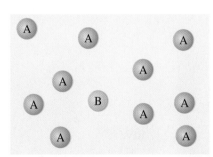

Small $K_c = 0.1$

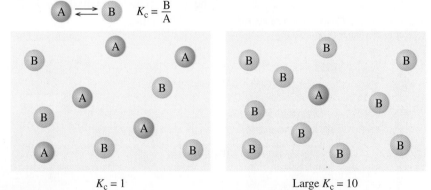

$K_c = 1$ Large $K_c = 10$

Figure 13.7 A reaction with $K_c < 1$ contains a higher concentration of reactants [A] than products [B]. A reaction with K_c of about 1.0 has about the same concentrations of products [B] as reactants [A]. A reaction with $K_c > 1$ has a higher concentration of products [B] than reactants [A].

Q *Does a reaction in which [A] = 100[B] at equilibrium have a K_c greater than, about equal to, or less than 1?*

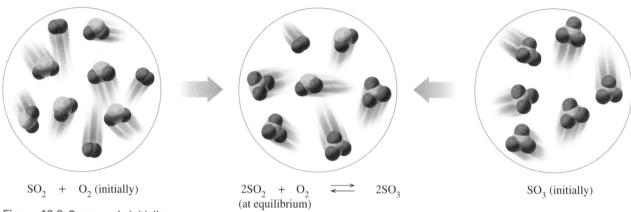

SO$_2$ + O$_2$ (initially)

2SO$_2$ + O$_2$ $\rightleftharpoons$ 2SO$_3$
(at equilibrium)

SO$_3$ (initially)

Figure 13.8 One sample initially contains SO$_2$ and O$_2$, while another sample contains only SO$_3$. At equilibrium, mostly SO$_3$ and only small amounts of SO$_2$ and O$_2$ are present in both equilibrium mixtures.

Q *Why is the same equilibrium mixture obtained from reactants as from products?*

Equilibrium with a Large K_c

A reaction with a large K_c forms a substantial amount of product by the time equilibrium is established. The greater the value of K_c, the more the equilibrium favors the products. A reaction with a very large K_c essentially goes to completion to give mostly products. Consider the reaction of SO$_2$ and O$_2$, which has a large K_c. At equilibrium, the reaction mixture contains mostly product and very little reactant.

$$2SO_2(g) + O_2(g) \rightleftharpoons 2SO_3(g)$$

$$K_c = \frac{[SO_3]^2}{[SO_2]^2[O_2]} \quad \frac{\text{Mostly product}}{\text{Little reactant}} = 3.4 \times 10^2 \quad \text{\small Reaction favors products}$$

We can start the reaction with only the reactants SO$_2$ and O$_2$ or we can start the reaction with just the products SO$_3$. (See Figure 13.8.) In one reaction, SO$_2$ and O$_2$ form SO$_3$ and in the other, SO$_3$ reacts to form SO$_2$ and O$_2$. However, in both equilibrium mixtures, the concentration of SO$_3$ is much higher than the concentrations of SO$_2$ and O$_2$. (See Figure 13.9.) Because there is more product than reactant at equilibrium, the energy of activation for the forward reaction must be lower than the energy of activation for the reverse reaction.

Figure 13.9 In the reaction of SO$_2$ and O$_2$, the equilibrium favors the formation of product SO$_3$, which results in a large K_c.

Q *Why is an equilibrium mixture obtained after starting with pure SO$_3$?*

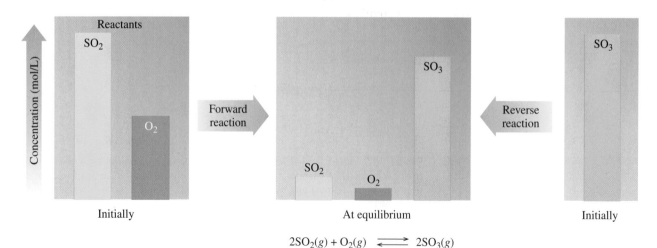

2SO$_2$(g) + O$_2$(g) $\rightleftharpoons$ 2SO$_3$(g)

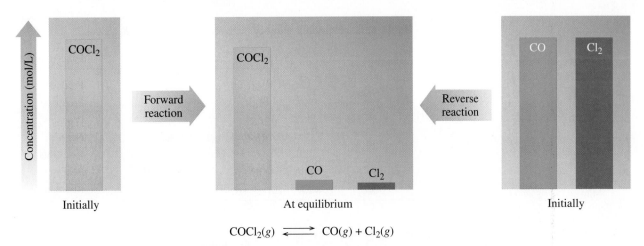

$$COCl_2(g) \rightleftharpoons CO(g) + Cl_2(g)$$

Figure 13.10 At equilibrium, the reaction $COCl_2 \rightleftharpoons CO + Cl_2$ favors the reactant and the reaction mixture at equilibrium contains mostly $COCl_2$, which results in a small K_c.

Q *Starting with only $COCl_2$ in a closed container, how do the forward and reverse reactions change as equilibrium is reached?*

Equilibrium with a Small K_c

For a reaction with a small K_c, the equilibrium mixture contains very small concentrations of products. Consider the reaction for the decomposition of $COCl_2$, which has a small K_c. (See Figure 13.10.)

$$COCl_2(g) \rightleftharpoons CO(g) + Cl_2(g)$$

$$K_c = \frac{[CO][Cl_2]}{[COCl_2]} \quad \frac{\text{Little product}}{\text{Mostly reactant}} = 2.2 \times 10^{-10} \quad \text{Reaction favors reactants}$$

Whether the reaction begins with only the reactant, $COCl_2$, or with the products, CO and Cl_2, the equilibrium mixture contains mostly reactant and very little product. The energy of activation for the forward reaction would be much greater than the energy of activation for the reverse reaction. Reactions with very small K_c produce essentially no product.

Reactions with equilibrium constants close to 1 have about the same concentrations of reactants and products. (See Figure 13.11.) Table 13.3 lists some equilibrium constants and the extent of their reaction.

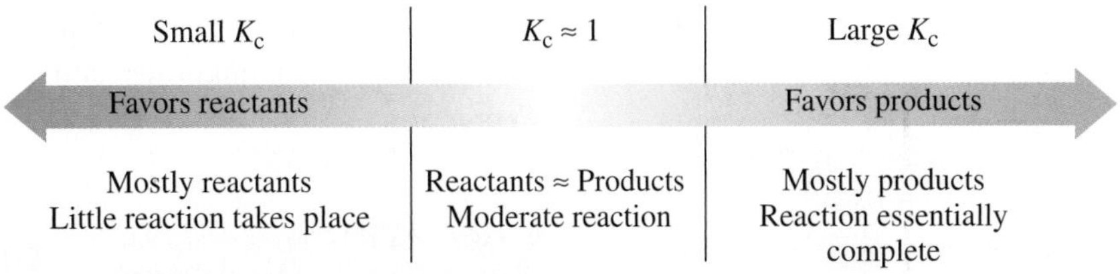

Small K_c	$K_c \approx 1$	Large K_c
Favors reactants		Favors products
Mostly reactants Little reaction takes place	Reactants $\approx$ Products Moderate reaction	Mostly products Reaction essentially complete

Figure 13.11 The equilibrium constant K_c indicates how far a reaction goes to products. A reaction with a large K_c contains mostly products; a reaction with a small K_c contains mostly reactants.

Q *Does a reaction with a $K_c = 1.2 \times 10^{15}$ contain mostly reactants or products at equilibrium?*

Table 13.3   Examples of Reactions with Large and Small K_c Values

Reactants		Products	K_c	Equilibrium Favors
$2CO(g) + O_2(g)$	$\rightleftharpoons$	$2CO_2(g)$	2×10^{11}	Products
$2H_2(g) + S_2(g)$	$\rightleftharpoons$	$2H_2S(g)$	1.1×10^7	Products
$N_2(g) + 3H_2(g)$	$\rightleftharpoons$	$2NH_3(g)$	1.6×10^2	Products
$H_2O(g) + CH_4(g)$	$\rightleftharpoons$	$CO(g) + 3H_2(g)$	4.7	Neither
$PCl_5(g)$	$\rightleftharpoons$	$PCl_3(g) + Cl_2(g)$	1.2×10^{-2}	Reactants
$N_2(g) + O_2(g)$	$\rightleftharpoons$	$2NO(g)$	2×10^{-9}	Reactants
$COCl_2(g)$	$\rightleftharpoons$	$CO(g) + Cl_2(g)$	2.2×10^{-10}	Reactants

---

Sample Problem 13.7 ▶ Extent of Reaction

Predict whether the equilibrium favors the reactants or products for each of the following reactions:

a. $2H_2(g) + O_2(g) \rightleftharpoons 2H_2O(g)$ $K_c = 2.9 \times 10^{82}$ at 25°C
b. $PCl_5(g) \rightleftharpoons PCl_3(g) + Cl_2(g)$ $K_c = 1.2 \times 10^{-2}$ at 225°C

Solution

a. A large K_c indicates that the equilibrium favors the products.
b. A small K_c indicates that the equilibrium favors the reactants.

Study Check

The equilibrium constant for the reaction

$$2CO_2(g) \rightleftharpoons 2CO(g) + O_2(g)$$

is 6.4×10^{-7}. In the reaction mixture, is the concentration of the reactants much greater, much smaller, or about the same as the products?

Calculating Concentrations at Equilibrium

When a reaction goes essentially all to products, we can use the mole factors we studied earlier to calculate the quantity of a product. However, many reactions reach equilibrium without using up all the reactants. If this is the case, then we need to use the equilibrium constant to calculate the amount of product that can be formed in the reaction. For example, if we know the equilibrium constant for a reaction and all the concentrations except one, we can calculate the unknown concentration using the equilibrium constant expression.

Sample Problem 13.8 ▶ Calculating Concentration Using an Equilibrium Constant

For the reaction of carbon dioxide and hydrogen, the equilibrium concentrations are $[CO_2] = 0.25$ M, $[H_2] = 0.80$ M, and $[H_2O] = 0.50$ M. What is the equilibrium concentration of $CO(g)$?

$$CO_2(g) + H_2(g) \rightleftharpoons CO(g) + H_2O(g)   K_c = 0.11$$

Guide to Using K_c Value

STEP 1
Write the K_c expression for the equilibrium equation.

STEP 2
Solve the K_c expression for the unknown concentration.

STEP 3
Substitute the known values into the rearranged K_c expression.

STEP 4
Check answer by using calculated concentrations in the K_c expression.

Solution

STEP 1 **Write the K_c expression for the equilibrium.** From the equation, the equilibrium constant expression is written as

$$K_c = \frac{[CO][H_2O]}{[CO_2][H_2]}$$

STEP 2 **Solve the K_c expression for the unknown concentration.** To rearrange the K_c for [CO], multiply both sides by $[CO_2][H_2]$ and cancel.

$$K_c \times [CO_2][H_2] = \frac{[CO][H_2O]}{[\cancel{CO_2}][\cancel{H_2}]} \times [\cancel{CO_2}][\cancel{H_2}]$$

$$K_c[CO_2][H_2] = [CO][H_2O]$$

Dividing both sides by $[H_2O]$ gives [CO].

$$\frac{K_c[CO_2][H_2]}{[H_2O]} = \frac{[CO][\cancel{H_2O}]}{[\cancel{H_2O}]}$$

$$[CO] = \frac{K_c[CO_2][H_2]}{[H_2O]}$$

STEP 3 **Substitute the known values into the rearranged K_c expression.** Substitute the K_c value and the concentrations given at equilibrium: $[CO_2] = 0.25$ M, $[H_2] = 0.80$ M, and $[H_2O] = 0.50$ M to solve for [CO].

$$[CO] = \frac{K_c[CO_2][H_2]}{[H_2O]} = \frac{0.11[0.25][0.80]}{[0.50]}$$

$$[CO] = 0.044 \text{ M}$$

STEP 4 **Check answer by using the calculated concentrations in the K_c.**

$$K_c = \frac{[CO][H_2O]}{[CO_2][H_2]} = \frac{[0.044][0.50]}{[0.25][0.80]}$$

$$K_c = 0.11$$

Study Check

Ethanol can be produced by reacting ethylene (C_2H_4) with water vapor. At 327°C, the K_c is 9×10^3.

$$C_2H_4(g) + H_2O(g) \rightleftharpoons C_2H_5OH(g)$$

If an equilibrium mixture has concentrations of $[C_2H_4] = 0.020$ M and $[H_2O] = 0.015$ M, what is the equilibrium concentration of C_2H_5OH?

Using Equilibrium Constants

13.21 Indicate whether each of the following equilibrium mixtures contain mostly products or mostly reactants:
a. $Cl_2(g) + NO(g) \rightleftharpoons 2NOCl(g)$ $K_c = 3.7 \times 10^8$
b. $2H_2(g) + S_2(g) \rightleftharpoons 2H_2S(g)$ $K_c = 1.1 \times 10^7$
c. $3O_2(g) \rightleftharpoons 2O_3$ $K_c = 1.7 \times 10^{-56}$

13.22 Indicate whether each of the following equilibrium mixtures contain mostly products or mostly reactants:
a. $CO(g) + Cl_2(g) \rightleftharpoons COCl_2(g)$ $K_c = 5.0 \times 10^{-9}$
b. $2HF(g) \rightleftharpoons H_2(g) + F_2(g)$ $K_c = 1.0 \times 10^{-95}$
c. $2NO(g) + O_2 \rightleftharpoons 2NO_2(g)$ $K_c = 6.0 \times 10^{13}$

13.23 The equilibrium constant, K_c, for the equilibrium

$$H_2(g) + I_2(g) \rightleftharpoons 2HI(g)$$

is 54 at 425°C. If the equilibrium mixture contains 0.030 M HI and 0.015 M I_2, what is the equilibrium concentration of H_2?

13.24 The equilibrium constant, K_c, for the equilibrium

$$N_2O_4(g) \rightleftharpoons 2NO_2(g)$$

is 4.6×10^{-3}. If the equilibrium mixture contains 0.050 M NO_2, what is the concentration of N_2O_4?

13.25 The K_c at 100°C is 2.0 for the reaction

$$2NOBr(g) \rightleftharpoons 2NO(g) + Br_2(g)$$

If the system at equilibrium contains [NO] = 2.0 M and $[Br_2]$ = 1.0 M, what is the [NOBr]?

13.26 An equilibrium mixture at 225°C contains 0.14 M NH_3 and 0.18 M H_2 for the reaction

$$3H_2(g) + N_2(g) \rightleftharpoons 2NH_3(g)$$

If the K_c at this temperature is 1.7×10^2, what is the equilibrium concentration of N_2?

13.5 Changing Equilibrium Conditions: Le Châtelier's Principle

Learning Goal

Use Le Châtelier's principle to describe the changes made in equilibrium concentrations when reaction conditions change.

We have seen that when a reaction reaches equilibrium, the rates of the forward and reverse reactions are equal and the concentrations remain constant. Now we will look at what happens to a system at equilibrium when changes occur in reaction conditions, such as changes in temperature, concentration, and pressure.

Le Châtelier's Principle

In the previous section, we saw that a system at equilibrium consists of forward and reverse reactions occurring at equal rates. Thus, at equilibrium, the concentrations of the substances do not change. However, any changes that occur in the reaction conditions will disturb the equilibrium. Concentrations can be changed by adding or removing one of the substances, the volume (pressure) can be changed, or the temperature can be changed. When we alter any of the conditions of a system at equilibrium, the rates of the forward and reverse reaction will no longer be equal. We say that a *stress* is placed on the equilibrium. We use Le Châtelier's principle to determine the direction that the equilibrium must shift to relieve that stress and reestablish equilibrium.

Le Châtelier's Principle

When a stress (change in conditions) is placed on a reaction at equilibrium, the equilibrium shifts in the direction that relieves the stress.

Effect of Concentration Changes

We will use the equilibrium for the reaction of PCl_5 to illustrate the stress caused by a change in concentration and how the system reacts to the stress. Consider the following reaction, which has a K_c of 0.042 at 250°C:

$$PCl_5(g) \rightleftharpoons PCl_3(g) + Cl_2(g)$$

For a reaction at a given temperature, there is only one equilibrium constant. Even if there are changes in the concentrations of the components, the K_c value does not change. What will change are the concentrations of the other components in the reaction in order to relieve the stress. For example, we can see that an equilibrium mixture that contains 1.20 M PCl_5, 0.20 M PCl_3, and 0.25 M Cl_2 has a K_c of 0.042.

$$K_c = \frac{[PCl_3][Cl_2]}{[PCl_5]} = \frac{[0.20][0.25]}{[1.20]} = 0.042$$

Suppose that now we add PCl_5 to the equilibrium mixture to increase $[PCl_5]$ to 2.00 M. If we substitute the concentrations into the equilibrium expression at this point, the ratio of products to reactants is 0.025, which is smaller than the K_c of 0.042.

$$\frac{\text{Products}}{\text{Reactants (added)}} \qquad \frac{[PCl_3][Cl_2]}{[PCl_5]} = \frac{[0.20][0.25]}{[2.00]} = 0.025 < K_c$$

Because a K_c cannot change for a reaction at a given temperature, adding more PCl_5 places a stress on the system. (See Figure 13.12.) Forming more of the products can relieve this stress. According to Le Châtelier's principle, adding reactants causes the equilibrium to *shift* toward the products.

Add PCl_5

$$PCl_5(g) \rightleftharpoons PCl_3(g) + Cl_2(g)$$

In our experiment, equilibrium is reestablished with new concentrations of $[PCl_5] = 1.94$ M, $[PCl_3] = 0.26$, and $[Cl_2] = 0.31$ M. The resulting equilibrium mixture now contains more reactants and products, but their new concentrations in the equilibrium expression are once again equal to the K_c.

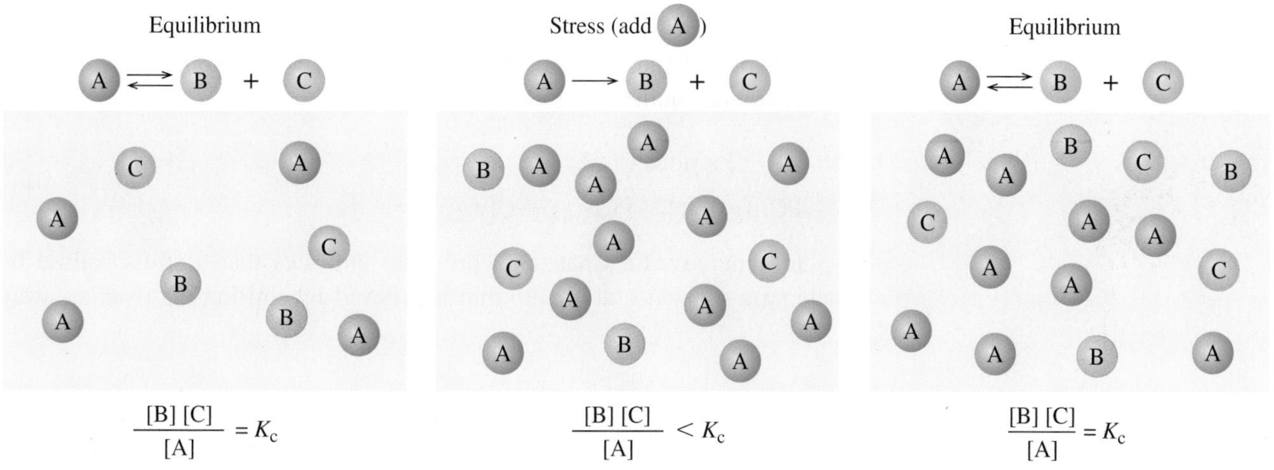

Figure 13.12 The addition of A places stress on the equilibrium of A $\rightleftharpoons$ B + C. To relieve the stress, the forward reaction converts some A to B + C and the equilibrium is reestablished.

Q *When C is added, does the equilibrium shift to products or reactants? Why?*

Equilibrium	Stress (add A)	Equilibrium
A $\rightleftharpoons$ B + C	A $\longrightarrow$ B + C	A $\rightleftharpoons$ B + C
$\dfrac{[B][C]}{[A]} = K_c$	$\dfrac{[B][C]}{[A]} < K_c$	$\dfrac{[B][C]}{[A]} = K_c$

$$K_c = \frac{[PCl_3][Cl_2]}{[PCl_5]} = \frac{[0.26][0.31]}{[1.94]} = 0.042 = K_c \qquad \text{New higher concentrations}$$

Suppose that in another experiment some PCl_5 is removed from the original equilibrium mixture, which lowers the $[PCl_5]$ to 0.76 M. Now the ratio of the products to the reactants is greater than the K_c value of 0.042. The removal of some of the reactant has placed a stress on the equilibrium.

$$\frac{\text{Product}}{\text{Reactant (removed)}} \qquad \frac{[PCl_3][Cl_2]}{[PCl_5]} = \frac{[0.20][0.25]}{[0.76]} = 0.066 > K_c$$

In this case, the stress is relieved as the reverse reaction converts some products to reactants. Using Le Châtelier's principle, we see that removing some reactant *shifts* the equilibrium toward the reactants.

> Remove PCl_5

$$PCl_5(g) \rightleftharpoons PCl_3(g) + Cl_2(g)$$

In this experiment, equilibrium is reestablished with new concentrations of $[PCl_5] = 0.80$ M, $[PCl_3] = 0.16$ M, and $[Cl_2] = 0.21$ M. The resulting equilibrium mixture now contains lower concentrations of reactants and products, but their new concentrations in the equilibrium expression once again are equal to the K_c.

$$K_c = \frac{[PCl_3][Cl_2]}{[PCl_5]} = \frac{[0.16][0.21]}{[0.80]} = 0.042 = K_c \qquad \text{New lower concentrations}$$

There can also be changes in the concentrations of other components in this reaction. We could add or remove one of the products in this reaction. Suppose the $[Cl_2]$ is doubled, which makes the product/reactant ratio greater than K_c.

$$\frac{[PCl_3][Cl_2]}{[PCl_5]} = \frac{[0.20][0.50]}{[1.2]} = 0.083 > K_c$$

With an increase in the concentration of Cl_2, the rate of the reverse reaction increases and converts some of the products to reactants. Using Le Châtelier's principle, we see that the addition of a product causes a *shift* toward the reactants.

> Add Cl_2

$$PCl_5(g) \rightleftharpoons PCl_3(g) + Cl_2(g)$$

On the other hand, we could remove some Cl_2, which would decrease $[Cl_2]$ and *shift* the equilibrium toward the products.

> Remove Cl_2

$$PCl_5(g) \rightleftharpoons PCl_3(g) + Cl_2(g)$$

In summary, Le Châtelier's principle indicates that a stress caused by adding a substance at equilibrium is relieved by shifting the reaction away

Table 13.4   Effect of Concentration Changes on Equilibrium
$PCl_5(g) \rightleftharpoons PCl_3(g) + Cl_2(g)$

Stress	Shift	Equilibrium Changes		
		$PCl_5(g)$	$PCl_3(g)$	$Cl_2(g)$
Increase PCl_5	Toward products	added	more	more
Decrease PCl_5	Toward reactants	removed	less	less
Increase PCl_3	Toward reactants	more	added	less
Decrease PCl_3	Toward products	less	removed	more
Increase Cl_2	Toward reactants	more	less	added
Decrease Cl_2	Toward products	less	more	removed

from that substance. When a substance is removed, the equilibrium shifts toward that substance. These features of Le Châtelier's principle are summarized in Table 13.4.

Catalysts

Sometimes a catalyst is added to a reaction. Earlier we showed that a catalyst speeds up a reaction by lowering the activation energy. As a result, the rates of the forward and reverse reactions both increase. The time required to reach equilibrium is shorter, but the same ratios of products and reactants are attained. Therefore, a catalyst speeds up the forward and reverse reactions, but it has no effect on the equilibrium constant.

Sample Problem 13.9   Effect of Changes in Concentrations

What is the effect of each of the following on the reaction at equilibrium?
a. increasing [CO]
b. increasing $[H_2]$
c. decreasing $[H_2O]$
d. decreasing $[CO_2]$
e. adding a catalyst

$$CO(g) + H_2O(g) \rightleftharpoons CO_2(g) + H_2(g)$$

Solution
According to Le Châtelier's principle, equilibrium shifts to relieve the stress.
a. Increasing [CO] shifts the equilibrium to products.
b. Increasing $[H_2]$ shifts the equilibrium to reactants.
c. Decreasing $[H_2O]$ shifts the equilibrium to reactants.
d. Decreasing $[CO_2]$ shifts the equilibrium to products.
e. Adding a catalyst does not cause a shift in the equilibrium.

Study Check

What is the effect of increasing $[CO_2]$ on the equilibrium for the reaction in Sample Problem 13.9?

OXYGEN–HEMOGLOBIN EQUILIBRIUM AND HYPOXIA

The transport of oxygen involves an equilibrium between hemoglobin (Hb), oxygen, and oxyhemoglobin.

$$Hb + O_2 \rightleftharpoons HbO_2$$

When the O_2 level is high in the alveoli of the lung, the reaction favors the product HbO_2. In the tissues where O_2 concentration is low, the reverse reaction releases the oxygen from the hemoglobin. The equilibrium expression is written

$$K_c = \frac{[HbO_2]}{[Hb][O_2]}$$

At normal atmospheric pressure, oxygen diffuses into the blood because the partial pressure of oxygen in the alveoli is higher than that in the blood. At altitudes above 8000 feet, the decrease in the amount of oxygen in the air results in a significant reduction of oxygen to the blood and body tissues. At an altitude of 18 000 feet, a person will obtain 29% less oxygen. When oxygen levels are lowered, a person may experience hypoxia, which has symptoms that include increased respiratory rate, headache, decreased mental acuteness, fatigue, decreased physical coordination, nausea, vomiting, and cyanosis. A similar problem occurs in persons with a history of lung disease that impairs gas diffusion in the alveoli or in persons that have a reduced number of red blood cells, which occurs in smokers.

From the equilibrium expression, we see that a decrease in oxygen will shift the equilibrium to the reactants. Such a shift depletes the concentration of HbO_2 and causes the hypoxia condition.

$$Hb + O_2 \longleftarrow HbO_2$$

Immediate treatment of altitude sickness includes hydration, rest, and if necessary, descending to a lower altitude. The adaptation to lowered oxygen levels requires about ten days. During this time the bone marrow increases red blood cell production, providing more red blood cells and hemoglobin. A person living at a high altitude can have 50% more red blood cells than someone at sea level. This increase in hemoglobin causes a shift in the equilibrium back to HbO_2 product. Eventually, the higher concentration of HbO_2 will provide more oxygen to the tissues and the symptoms of hypoxia will lessen.

$$Hb + O_2 \longrightarrow HbO_2$$

For some who climb high mountains, it is important to stop and acclimatize for several days at increasing altitudes. At very high altitudes, it may be necessary to use an oxygen tank.

Effect of Volume (Pressure) Changes on Equilibrium

The gases involved in a reaction exert pressure. Although the volume and therefore pressure can change, the value of the equilibrium constant does not change at a given temperature. Using the gas laws, we know that increasing the pressure of the container decreases the volume, while decreasing the pressure increases the volume.

According to Le Châtelier's principle, decreasing the number of moles of gas relieves the stress of increased pressure. This means that the reaction shifts toward the fewer number of moles of gas. Let's look at the effect of decreasing the volume of the equilibrium mixture that originally contained 1.20 M PCl_5, 0.20 M PCl_3, and 0.25 M Cl_2 with a K_c of 0.042.

$$PCl_5(g) \rightleftharpoons PCl_3(g) + Cl_2(g)$$

$$K_c = \frac{[PCl_3][Cl_2]}{[PCl_5]} = \frac{[0.20][0.25]}{[1.20]} = 0.042$$

If we decrease the volume by half, all the molar concentrations are doubled. In the equation there are more moles of products than reactants, so there is an increase in the product/reactant ratio.

$$\frac{[PCl_3][Cl_2]}{[PCl_5]} = \frac{[0.40][0.50]}{[2.40]} = 0.083 > K_c$$

To relieve the stress, the equilibrium shifts toward the reactants, which will reduce the mol/L of the products and increase the mol/L of the reactant. (See Figure 13.13.)

Decrease V

$$PCl_5(g) \rightleftharpoons PCl_3(g) + Cl_2(g)$$
1 mol   2 mol

When equilibrium is reestablished, the new concentrations are $[PCl_5] =$ 2.52 M, $[PCl_3] = 0.28$ M, and $[Cl_2] = 0.38$ M. The resulting equilibrium mixture contains new concentrations of reactants and products that are now equal to the K_c value.

$$K_c = \frac{[PCl_3][Cl_2]}{[PCl_5]} = \frac{[0.28][0.38]}{[2.52]} = 0.042$$

On the other hand, when volume increases and pressure decreases, the reaction shifts toward the greater number of moles of gas. Suppose that the volume is doubled. Then the molar concentrations of all the gases decrease by half. Because there are more moles of products than reactants, there is a decrease in the product/reactant ratio.

$$\frac{[PCl_3][Cl_2]}{[PCl_5]} = \frac{[0.100][0.125]}{[0.600]} = 0.021 < K_c$$

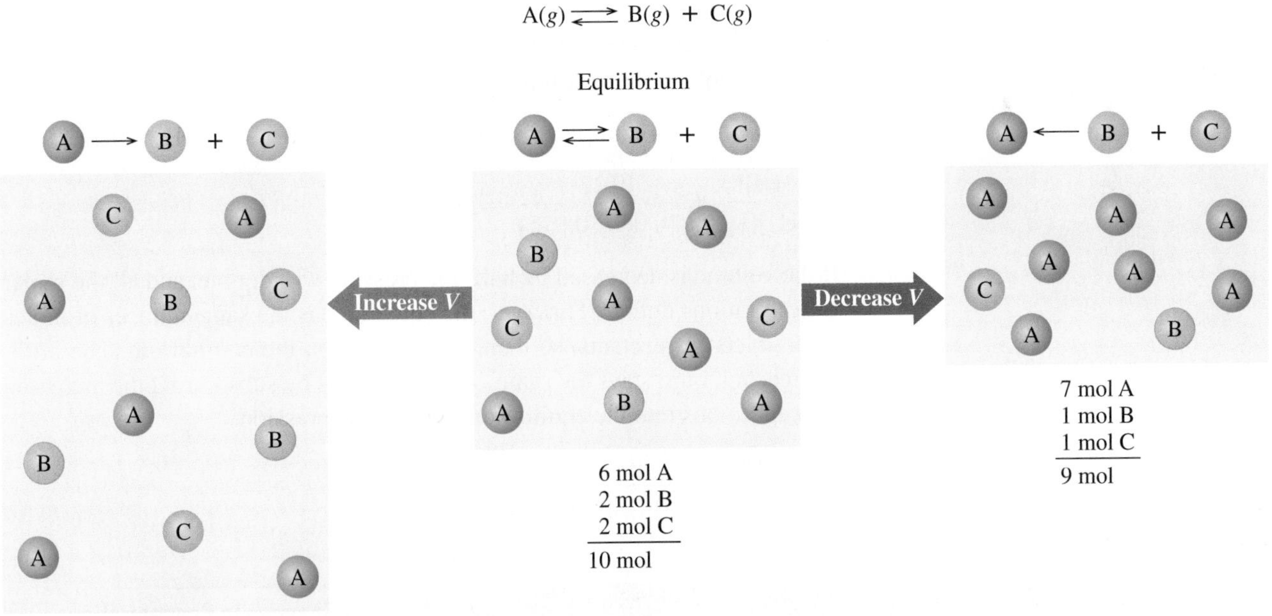

$$A(g) \rightleftharpoons B(g) + C(g)$$

Equilibrium

Figure 13.13 The decrease in the volume of the container places stress on the equilibrium: $A(g) \rightleftharpoons B(g) + C(g)$. To relieve the stress, the reverse reaction converts some products to reactants, which gives a smaller number of moles of gas, reduces pressure, and reestablishes the equilibrium. When the volume increases, the forward reaction converts reactants to product to increase the moles of gas and relieve the stress.

Q *If you want to increase the products, would you increase or decrease the volume of the reaction container?*

Now the equilibrium has to shift toward the products to relieve the stress; this will increase the concentrations (mol/L) of the products.

> Increase V

$$PCl_5(g) \; \rightleftharpoons \; PCl_3(g) + Cl_2(g)$$
$$\text{1 mol} \qquad\qquad \text{2 mol}$$

When equilibrium is reestablished, the new concentrations are $[PCl_5] = 0.56$ M, $[PCl_3] = 0.14$ M, and $[Cl_2] = 0.17$ M. The resulting equilibrium mixture contains new concentrations of reactants and products that are now equal to the K_c value.

At Volume (1)	At Volume (2)
$[PCl_3] = 0.20$ M	0.14 M
$[Cl_2] \;\; = 0.25$ M	0.17 M
$[PCl_5] = 1.20$ M	0.56 M

$$K_c = \frac{[PCl_3][Cl_2]}{[PCl_5]} = \frac{[0.14][0.17]}{[0.56]} = 0.042$$

When a reaction has the same number of moles of gases in the reactants as products, a volume change does not affect the equilibrium. There is no effect on equilibrium because the molar concentrations of the reactants and products change in the same way. Consider the reaction of H_2 and I_2 to form HI, which has a K_c of 54.

$$H_2(g) + I_2(g) \; \rightleftharpoons \; 2HI(g)$$
$$\text{2 mol} \qquad\qquad \text{2 mol}$$

Suppose we start with $[H_2] = 0.060$ M, $[I_2] = 0.015$ M, $[HI] = 0.22$ M.

$$\frac{[HI]^2}{[H_2][I_2]} = \frac{[0.22]^2}{[0.060][0.015]} = 54$$

If the volume is decreased by half, the pressure will double, and all the molar concentrations double. However, the equation has the same number of moles of products as reactants, so there is no effect on the equilibrium. The product/reactant ratio stays the same. We can see this by substituting the increased concentrations into the equilibrium constant expression.

$$\frac{[HI]^2}{[H_2][I_2]} = \frac{[0.44]^2}{[0.12][0.030]} = 54$$

Sample Problem 13.10   **Effect of Changes in Concentrations**

Indicate whether the effect of decreasing the volume for each of the following equilibria causes the number of moles of product to increase, decrease, or not change:

a. $C_2H_2(g) + 2H_2(g) \rightleftharpoons C_2H_6(g)$

b. $2NO_2(g) \rightleftharpoons 2NO(g) + O_2(g)$

c. $CO(g) + H_2O(g) \rightleftharpoons CO_2(g) + H_2(g)$

Solution

To relieve the stress of decreasing the volume, the equilibrium shifts toward the side with the fewer moles of gaseous components.

a. The equilibrium shifts to C_2H_6 product to reduce the number of moles of gas. The number of moles of product is increased.

$$C_2H_2(g) + 2H_2(g) \longrightarrow C_2H_6(g)$$
$$\text{3 mol gas} \qquad\qquad \text{1 mol gas}$$

b. The equilibrium shifts to NO_2 reactant to reduce the number of moles of gas. The number of moles of product is decreased.

$$2NO_2(g) \longleftarrow 2NO(g) + O_2(g)$$
$$\text{2 mol gas} \qquad\quad \text{3 mol gas}$$

c. There is no shift in equilibrium because there is no change in the number of moles; the moles of reactant are equal to the moles of product. The number of moles of product does not change.

$$CO(g) + H_2O(g) \rightleftarrows CO_2(g) + H_2(g)$$
$$\text{2 mol gas} \qquad\qquad \text{2 mol gas}$$

Study Check

Suppose you want to increase the yield of product in the following reaction. Would you increase or decrease the volume of the reaction container?

$$CO(g) + 2H_2(g) \rightleftarrows CH_3OH(g)$$

Effect of a Change in Temperature on Equilibrium

As we have seen, the effects of changes on equilibrium are to shift equilibrium to reestablish the same value of the equilibrium constant. However, if we change the temperature of a system at equilibrium, we change the value of K_c. When the temperature of an equilibrium system increases, the reaction that is favored is the one that removes heat. When heat is added to an endothermic reaction, the equilibrium shifts to the products to use up heat. The value of K_c increases because the shift increases the product concentration and decreases the reactant concentration.

Increase T; increase K_c

$$N_2(g) + O_2(g) + \text{heat} \rightleftarrows 2NO(g)$$

If the temperature is lowered, the equilibrium shifts to increase the concentrations of the reactants, and the value of K_c decreases. (See Table 13.5.)

Decrease T; decrease K_c

$$N_2(g) + O_2(g) + \text{heat} \rightleftarrows 2NO(g)$$

Table 13.5   Equilibrium Shifts for Temperature Changes in an Endothermic Reaction

K_c	Temperature Change	Equilibrium Shift	Change in K_c Value
$\dfrac{[NO]^2}{[N_2][O_2]}$	Increase	More product $\dfrac{[NO]^2}{[N_2][O_2]}$ Less reactant	⬆ Increases
$\dfrac{[NO]^2}{[N_2][O_2]}$	Decrease	Less product $\dfrac{[NO]^2}{[N_2][O_2]}$ More reactant	⬇ Decreases

For an exothermic reaction, the addition of heat favors the reverse reaction, which uses up heat. The value of K_c for an exothermic reaction decreases when the temperature increases.

Increase T; decrease K_c

$$2SO_2(g) + O_2(g) \rightleftharpoons 2SO_3(g) + \text{heat}$$

If heat is removed, the equilibrium of an exothermic reaction favors the products, which provides heat. (See Table 13.6.)

Decrease T; increase K_c

$$2SO_2(g) + O_2(g) \rightleftharpoons 2SO_3(g) + \text{heat}$$

Table 13.6   Equilibrium Shifts for Temperature Changes in an Exothermic Reaction

K_c	Temperature Change	Equilibrium Shift	Change in K_c Value
$\dfrac{[SO_3]^2}{[SO_2]^2[O_2]}$	Increase	Less product $\dfrac{[SO_3]^2}{[SO_2]^2[O_2]}$ More reactant	⬇ Decreases
$\dfrac{[SO_3]^2}{[SO_2]^2[O_2]}$	Decrease	More product $\dfrac{[SO_3]^2}{[SO_2]^2[O_2]}$ Less reactant	⬆ Increases

| **Sample Problem 13.11** | **Effect of Temperature Change on Equilibrium** |

Indicate the change in the concentration of products and the K_c when the temperature of each of the following reactions at equilibrium is increased:

a. $N_2(g) + 3H_2(g) \rightleftharpoons 2NH_3(g) + 92\text{ kJ}$ $(\Delta H = -92\text{ kJ})$

b. $N_2(g) + O_2(g) + 180\text{ kJ} \rightleftharpoons 2NO(g)$ $(\Delta H = +180\text{ kJ})$

Solution

a. The addition of heat shifts an exothermic reaction to reactants, which decreases the concentration of the products. The K_c will decrease.

b. The addition of heat shifts an endothermic reaction to products, which increases the concentration of the products. The K_c will increase.

Study Check

Indicate the change in the concentration of reactants and the K_c when there is a decrease in the temperature of each of the reactions at equilibrium in Sample Problem 13.11.

Table 13.7 summarizes the ways we can use Le Châtelier's principle to determine the shift in equilibrium that relieves a stress caused by change in a condition.

Table 13.7   **Effects of Condition Changes on Equilibrium**

Condition	Change (Stress)	Reaction to Remove Stress
Concentration	Add reactant	Forward
	Remove reactant	Reverse
	Add product	Reverse
	Remove product	Forward
Volume (container)	Decrease	Toward fewer moles in the gas phase
	Increase	Toward more moles in the gas phase
Temperature	**Endothermic reaction**	
	Raise T	Forward, larger value for K_c
	Lower T	Reverse, smaller value for K_c
	Exothermic reaction	
	Raise T	Reverse, smaller value for K_c
	Lower T	Forward, larger value for K_c
Catalyst	Increases rates equally	No effect

CHEM NOTE

HOMEOSTASIS: REGULATION OF BODY TEMPERATURE

In a physiological system of equilibrium called homeostasis, changes in our environment are balanced by changes in our bodies. It is crucial to our survival that we balance heat gain with heat loss. If we do not lose enough heat, our body temperature rises. At high temperatures, the body can no longer regulate our metabolic reactions. If we lose too much heat, body temperature drops. At low temperatures, essential functions proceed too slowly.

The skin plays an important role in the maintenance of body temperature. When the outside temperature rises, receptors in the skin send signals to the brain. The temperature-regulating part of the brain stimulates the sweat glands to produce perspiration. As perspiration evaporates from the skin, heat is removed and the body temperature is lowered.

In cold temperatures, epinephrine is released, causing an increase in metabolic rate, which increases the production of heat. Receptors on the skin signal the brain to contract the blood vessels. Less blood flows through the skin, and heat is conserved. The production of perspiration stops to lessen the heat lost by evaporation.

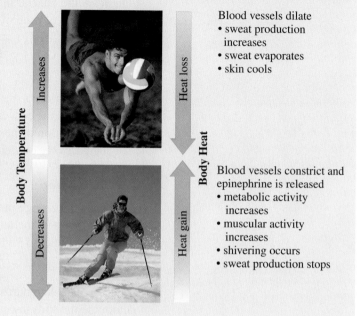

Blood vessels dilate
• sweat production increases
• sweat evaporates
• skin cools

Blood vessels constrict and epinephrine is released
• metabolic activity increases
• muscular activity increases
• shivering occurs
• sweat production stops

Questions and Problems   **Changing Equilibrium Conditions: Le Châtelier's Principle**

13.27 a. Does the addition of reactant to an equilibrium mixture cause the product/reactant ratio to be higher or lower than the K_c?
   b. According to Le Châtelier's principle, how is equilibrium in part a established?

13.28 a. What is the effect on the K_c when the temperature of an exothermic reaction is lowered?
   b. According to Le Châtelier's principle, how is equilibrium in part a established?

13.29 In the lower atmosphere, oxygen is converted to ozone (O_3) by the energy provided from lightning.

$$3O_2(g) + \text{heat} \rightleftharpoons 2O_3(g)$$

For each of the following changes at equilibrium, indicate whether the equilibrium shifts to products or reactants or does not shift:
   a. adding $O_2(g)$
   b. adding $O_3(g)$
   c. raising the temperature
   d. decreasing the volume of the container
   e. adding a catalyst

13.30 Ammonia is produced by reacting nitrogen gas and hydrogen gas.

$$N_2(g) + 3H_2(g) \rightleftharpoons 2NH_3(g) + 92 \text{ kJ}$$

For each of the following changes at equilibrium, indicate whether the equilibrium shifts to products or reactions or does not shift:
   a. removing $N_2(g)$
   b. lowering the temperature

   c. adding $NH_3(g)$
   d. adding $H_2(g)$
   e. increasing the volume of the container

13.31 Hydrogen chloride can be made by reacting hydrogen gas and chlorine gas.

$$H_2(g) + Cl_2(g) + \text{heat} \rightleftharpoons 2HCl(g)$$

For each of the following changes at equilibrium, indicate whether the equilibrium shifts to products or reactants, or does not shift:
   a. adding $H_2(g)$
   b. increasing the temperature
   c. removing HCl (g)
   d. adding a catalyst
   e. removing $Cl_2(g)$

13.32 When heated, carbon reacts with water to produce carbon monoxide and hydrogen.

$$C(s) + H_2O(g) + \text{heat} \rightleftharpoons CO(g) + H_2(g)$$

For each of the following changes at equilibrium, indicate whether the equilibrium shifts to products or reactants, or does not shift:
   a. increasing the temperature
   b. adding $C(s)$
   c. removing $CO(g)$ as its forms
   d. adding $H_2O(g)$
   e. decreasing the volume of the container

13.6 Equilibrium in Saturated Solutions

Learning **Goal**

Calculate the solubility product for a saturated solution; use the solubility product to calculate molar ion concentrations.

Up until now, we have primarily looked at the equilibrium that is reached with reactants and products that are gases. However, there are also equilibrium systems that involve aqueous solutions, some of which are saturated solutions or contain insoluble salts. Everyday examples of solubility equilibrium in solution are found in tooth decay and kidney stones. When bacteria in the mouth react with sugars in food, acids are produced that dissolve the enamel of a tooth, which is made of a mineral called hydroxyapatite, $Ca_5(PO_4)_3OH$. Kidney stones are composed of calcium salts such as calcium oxalate, CaC_2O_4, and calcium phosphate, $Ca_3(PO_4)_2$, which are rather insoluble. When Ca^{2+} ions and oxalate $C_2O_4^{2-}$ ions exceed their solubility in the kidneys, they will precipitate, forming solid CaC_2O_4.

$$Ca^{2+}(aq) + C_2O_4^{2-}(aq) \rightleftharpoons CaC_2O_4(s)$$
$$3Ca^{2+}(aq) + 2PO_4^{2-}(aq) \rightleftharpoons Ca_3(PO_4)_2(s)$$

To understand the role of solubility in biology and the environment, we can look at the equilibrium that occurs in saturated solutions.

Solubility Product Constant

In Chapter 12, we learned that a saturated solution contains some undissolved solute in contact with the maximum amount of dissolved solute. A saturated solution is a dynamic system in which the rate of dissolution for a solute has become equal to the rate of solute recrystallization out of solution. As long as the temperature remains constant, the concentration of the ions in the saturated solution is constant. Let us look at the equilibrium equation for CaC_2O_4, which is written with the solid solute on the left and the ions in solution on the right.

$$CaC_2O_4(s) \rightleftharpoons Ca^{2+}(aq) + C_2O_4^{2-}(aq)$$

At equilibrium, the concentrations of the Ca^{2+} and $C_2O_4^{2-}$ are constant. We can represent the solubility of CaC_2O_4 by an equilibrium expression called the **solubility product constant** or **solubility product, (K_{sp})**.

$$K_{sp} = [Ca^{2+}][C_2O_4^{2-}]$$

You may notice that the K_{sp} is similar to other equilibrium constants we have written. As in heterogeneous equilibrium, the concentration of the solid is constant and not included in the K_{sp} expression. In another example, we consider the equilibrium of solid calcium fluoride and its ions Ca^{2+} and F^-.

$$CaF_2(s) \rightleftharpoons Ca^{2+}(aq) + 2F^-(aq)$$

At equilibrium, the rate of dissolving for CaF_2 is equal to the rate of its recrystallization, which means the concentrations of the ions remain constant. The solubility product for this solubility is written as the product of the ion concentrations. As with other equilibrium expressions, the $[F^-]$ is raised to the power of 2 because there is a coefficient of 2 in the equilibrium equation.

$$K_{sp} = [Ca^{2+}][F^-]^2$$

Sample Problem 13.12   Writing Solubility Product Constants

For each of the following slightly soluble salts, write the equilibrium equation and the solubility product expression:

a. AgBr   **b.** $PbCl_2$ **c.** Li_2CO_3

Solution

The equilibrium equation gives the solid salt on the left and the ions in solution on the right. The K_{sp} gives the molar concentrations of the ions raised to a power equal to the coefficients in the balanced equation.

a. $AgBr(s) \rightleftharpoons Ag^+(aq) + Br^-(aq)$ $K_{sp} = [Ag^+][Br^-]$

b. $PbCl_2(s) \rightleftharpoons Pb^{2+}(aq) + 2Cl^-(aq)$ $K_{sp} = [Pb^{2+}][Cl^-]^2$

c. $Li_2CO_3(s) \rightleftharpoons 2Li^+(aq) + CO_3^{2-}(aq)$ $K_{sp} = [Li^+]^2[CO_3^{2-}]$

Study Check

If a salt has the solubility product expression $K_{sp} = [Fe^{3+}][OH^-]^3$, what is the equilibrium equation for the solubility?

Calculating Solubility Product Constant

Experiments in the lab can measure the concentrations of ions in a saturated solution. For example, we can make a saturated solution of $CaCO_3$ by adding solid $CaCO_3$ to water and stirring until equilibrium is reached. Then we would measure the concentrations of Ca^{2+} and CO_3^{2-} in solution. Suppose that a saturated solution of $CaCO_3$ has $[Ca^{2+}] = 7.1 \times 10^{-5}$ M and $[CO_3^{2-}] = 7.1 \times 10^{-5}$ M.

STEP 1 Write the equilibrium equation for solubility.

$$CaCO_3(s) \rightleftharpoons Ca^{2+}(aq) + CO_3^{2-}(aq)$$

STEP 2 Write the solubility product constant (K_{sp}).

$$K_{sp} = [Ca^{2+}][CO_3^{2-}]$$

STEP 3 Substitute the molar concentrations of the ions into the K_{sp} expression.

$$K_{sp} = [7.1 \times 10^{-5} \text{ M}][7.1 \times 10^{-5} \text{ M}] = 5.0 \times 10^{-9}$$

Table 13.8 gives values of K_{sp} for a selected group of ionic compounds at 25°C.

Guide to Calculating K_{sp}

STEP 1
Write the equilibrium equation for the dissociation of the ionic compound.

STEP 2
Write the K_{sp} expression with the molarity of each ion raised to a power equal to its coefficient.

STEP 3
Substitute the molarity of each ion into the K_{sp} and calculate.

Table 13.8   Solubility Product Constants (K_{sp}) for Selected Ionic Compounds (25°C)

Formula	K_{sp}
AgCl	1.8×10^{-10}
Ag_2SO_4	1.2×10^{-5}
$BaCO_3$	2.0×10^{-9}
$BaSO_4$	1.1×10^{-10}
CaF_2	3.2×10^{-11}
$Ca(OH)_2$	6.5×10^{-6}
$CaSO_4$	2.4×10^{-5}
$PbCl_2$	1.5×10^{-6}
$PbCO_3$	7.4×10^{-14}

Sample Problem 13.13   Calculating the Solubility Product Constant

A saturated solution of strontium fluoride, SrF_2, has $[Sr^{2+}] = 8.7 \times 10^{-4}$ M and $[F^-] = 1.7 \times 10^{-3}$ M. What is the value of K_{sp} for SrF_2?

Solution

STEP 1 Write the equilibrium equation for solubility.

$$SrF_2(s) \rightleftharpoons Sr^{2+}(aq) + 2F^-(aq)$$

STEP 2 **Write the solubility product constant (K_{sp}).**

$$K_{sp} = [Sr^{2+}][F^-]^2$$

STEP 3 **Substitute the molar concentrations of the ions into the K_{sp} expression.**

$$K_{sp} = [8.7 \times 10^{-4}\,M][1.7 \times 10^{-3}\,M]^2 = 2.5 \times 10^{-9}$$

Study Check

What is the K_{sp} of silver bromide, AgBr, if a saturated solution has $[Ag^+] = 7.1 \times 10^{-7}$ and $[Br^-] = 7.1 \times 10^{-7}$?

Calculating Molar Solubility (S)

If we know the K_{sp} of an ionic compound, we can use it to calculate the molar solubility of the compound. The molar solubility (S) is the number of moles of solute dissolved in 1 liter of solution. Using the equilibrium equation and the K_{sp}, we can calculate the molarity of each ion and determine the molar solubility. For example, the molar solubility of CdS is 1×10^{-12} mol per liter. This means that 1×10^{-12} mol CdS dissociates into Cd^{2+} and S^{2-} ions to give $[Cd^{2+}] = 1 \times 10^{-12}$ M and $[S^{2-}] = 1 \times 10^{-12}$ M.

$$CdS(s) \rightleftharpoons Cd^{2+}(aq) + S^{2-}(aq)$$

$S = 1 \times 10^{-12}$ mol/L 1×10^{-12} M and 1×10^{-12} M

We can use the equilibrium equation and the K_{sp} of a salt to calculate the molarity of each ion and determine the molar solubility.

Guide to Calculating Molar Solubility from K_{sp}

STEP 1
Write the equilibrium equation for the dissociation of the ionic compound.

STEP 2
Write the K_{sp} expression.

STEP 3
Substitute S for the molarity of each ion into the K_{sp}.

STEP 4
Calculate the molar solubility (S).

Sample Problem 13.14   **Calculating the Molar Solubility**

Calculate the molar solubility(S) of PbSO$_4$ if the K_{sp} is 1.6×10^{-8}.

Solution

STEP 1 **Write the equilibrium equation for solubility.**

$$PbSO_4(s) \rightleftharpoons Pb^{2+}(aq) + SO_4^{2-}(aq)$$

STEP 2 **Write the solubility product constant (K_{sp}).**

$$K_{sp} = [Pb^{2+}][SO_4^{2-}]$$

STEP 3 **Show the molarity of the ions as S in the equation with the known value of K_{sp}.**

$$K_{sp} = [S][S] = 1.6 \times 10^{-8}$$

STEP 4 **Solve for the solubility S.**

$$S \times S = S^2 = 1.6 \times 10^{-8}$$
$$S = \sqrt{1.6 \times 10^{-8}} = 1.3 \times 10^{-4}\,\text{mol/L}$$

The molar solubility of PbSO$_4$ is 1.3×10^{-4} mol/L. This means that when 1.3×10^{-4} mol PbSO$_4$ dissolves in 1 liter of solution, $[Pb^{2+}] = 1.3 \times 10^{-4}$ M and $[SO_4^{2-}] = 1.3 \times 10^{-4}$ M.

$$PbSO_4(s) \quad \rightleftharpoons \quad Pb^{2+}(aq) \quad + \quad SO_4^{2-}(aq)$$

$$1.3 \times 10^{-4} \, M \qquad 1.3 \times 10^{-4} \, M$$

Study Check

Calculate the molar solubility(S) of CdS if CdS has a K_{sp} of 4×10^{-30}.

Effect of Adding One of the Ions

We have seen when a slightly soluble salt such as $MgCO_3$ dissolves in water that small amounts of Mg^{2+} and CO_3^{2-} ions are produced in equal quantities.

$$MgCO_3(s) \rightleftharpoons Mg^{2+}(aq) + CO_3^{2-}(aq) \quad K_{sp} = 3.5 \times 10^{-8}$$

Then the solubility of $MgCO_3$ is $1.9 \times 10^{-4} \, M$ and the concentrations of both Mg^{2+} and CO_3^{2-} are the same ($1.9 \times 10^{-4} \, M$).

However, we can change the concentration of Mg^{2+} or CO_3^{2-} by adding a soluble salt containing one of those ions. Suppose $MgCl_2$, a soluble salt, is added to the above solution. The soluble salt provides Mg^{2+} and 2 Cl^-, which increases the concentration of Mg^{2+}. Because the product of the ions is now greater than the K_{sp}, some Mg^{2+} combines with CO_3^{2-} to from solid $MgCO_3$. Some CO_3^{2-} is removed from solution, lowering the concentration of CO_3^{2-}. However, the K_{sp} does not change. The product of the increased concentration of Mg^{2+} and the decreased concentration of CO_3^{2-} is still equal to the K_{sp}. Thus, the solubility of a slightly soluble salt decreases when one of its ions is present.

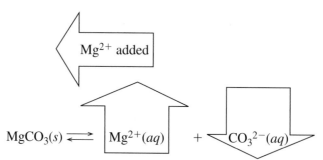

The same thing will happen if Na_2CO_3 is added to the solution. Solid $MgCO_3$ would form, the concentration of Mg^{2+} would be lower, and the concentration of CO_3^{2-} would be higher.

Earlier we described kidney stones as crystals composed of calcium salts such as calcium oxalate CaC_2O_4. In the body, the molar concentrations of Ca^{2+} and $C_2O_4^{2-}$ are not equal. But when the product of their concentrations exceeds the K_{sp}, solid CaC_2O_4 will form kidney stones. If we know the $C_2O_4^{2-}$ concentration in body fluid, we can calculate the maximum concentration of Ca^{2+} for a saturated CaC_2O_4 solution as seen in Sample Problem 13.15.

Sample Problem 13.15   **Calculating the Concentration of an Ion**

CaC_2O_4 has a K_{sp} of 2.7×10^{-9}. What is the maximum concentration of Ca^{2+} if $[C_2O_4^{2-}]$ is 3.5×10^{-5} M?

Solution

STEP 1   **Write the equilibrium equation for solubility.**

$$CaC_2O_4(s) \rightleftharpoons Ca^{2+}(aq) + C_2O_4^{2-}(aq)$$

STEP 2   **Write the solubility product constant (K_{sp}).**

$$K_{sp} = [Ca^{2+}][C_2O_4^{2-}]$$

STEP 3   **Substitute the known concentration into the equation with the known value of K_{sp}.**

$$K_{sp} = [Ca^{2+}][3.5 \times 10^{-5}] = 2.7 \times 10^{-9}$$

STEP 4   **Solve for the unknown molar concentration.**

$$[Ca^{2+}] = \frac{2.7 \times 10^{-9}}{3.5 \times 10^{-5}} = 7.7 \times 10^{-5}$$

The maximum concentration of Ca^{2+} is 7.7×10^{-5} mol/L.

Study Check

Nickel(II) carbonate $NiCO_3$ has a K_{sp} of 1.3×10^{-7}. What is the maximum concentration of Ni^{2+} if $[CO_3^{2-}]$ is 4.2×10^{-5} M?

Kidney stones form when $[Ca^{2+}][C_2O_4^{2-}]$ is equal to or greater than the solubility product. Normally, the urine contains substances such as magnesium and citrate that prevent the formation of kidney stones. Some of the factors that contribute to the formation of kidney stones are drinking too little water, limited activity, consuming foods with high levels of oxalate, and some metabolic diseases. Prevention includes drinking large quantities of water and decreasing consumption of foods high in oxalate such as spinach, rhubarb, and soybean products.

Questions and Problems   **Equilibrium in Saturated Solutions**

13.33 For each of the following salts, write the equilibrium equation and the solubility product expression:
   a. $MgCO_3$ **b.** CaF_2
   c. Ag_3PO_4

13.34 For each of the following salts, write the equilibrium equation and the solubility product expression:
   a. $PbSO_4$ **b.** $Al(OH)_3$
   c. BaF_2

13.35 A saturated solution of barium sulfate, $BaSO_4$, has $[Ba^{2+}] = 1 \times 10^{-5}$ M and $[SO_4^{2-}] = 1 \times 10^{-5}$ M. What is the value of K_{sp} for $BaSO_4$?

13.36 A saturated solution of silver bromide, $AgBr$, has $[Ag^+] = 7.1 \times 10^{-7}$ M and $[Br^-] = 7.1 \times 10^{-7}$ M. What is the value of K_{sp} for $AgBr$?

13.37 A saturated solution of silver carbonate, Ag_2CO_3, has $[Ag^+] = 2.6 \times 10^{-4}$ M and $[CO_3^{2-}] = 1.3 \times 10^{-4}$ M. What is the value of K_{sp} for Ag_2CO_3?

13.38 A saturated solution of barium fluoride, BaF_2, has $[Ba^{2+}] = 3.6 \times 10^{-3}$ M and $[F^-] = 7.2 \times 10^{-3}$ M. What is the value of K_{sp} for BaF_2?

13.39 What are $[Cu^+]$ and $[I^-]$ in a saturated CuI solution if the K_{sp} of CuI is 1×10^{-12}?

13.40 What are $[Sn^{2+}]$ and $[S^{2-}]$ in a saturated SnS solution if the K_{sp} of SnS is 1×10^{-26}?

13.41 If a saturated solution of AgCl has $[Ag^+] = 2.0 \times 10^{-7}$ M, what is $[Cl^-]$? (See Table 13.8 for the K_{sp}.)

13.42 If a saturated solution of $PbCO_3$ has $[CO_3^{2-}] = 3.0 \times 10^{-8}$ M, what is $[Pb^{2+}]$? (See Table 13.8 for the K_{sp}.)

Concept Map ▸ **Chemical Equilibrium**

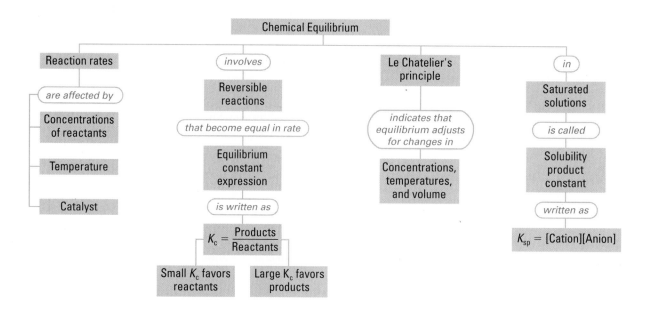

Chapter Review

13.1 Rates of Reactions

The rate of a reaction is the speed at which the reactants are converted to products. Increasing the concentrations of reactants, raising the temperature, or adding a catalyst can increase the rate of a reaction.

13.2 Chemical Equilibrium

Chemical equilibrium occurs in a reversible reaction when the rate of the forward reaction becomes equal to the rate of the reverse reaction. At equilibrium, no further change occurs in the concentrations of the reactants and products as the forward and reverse reactions continue.

13.3 Equilibrium Constants

An equilibrium constant, K_c, is the ratio of the concentrations of the products to the concentrations of the reactants with each concentration raised to a power equal to its coefficient in the chemical equation. For heterogeneous solid–gas reactions, only gases are placed in the equilibrium expression.

13.4 Using Equilibrium Constants

A large value of K_c indicates the equilibrium favors the products and could go nearly to completion, whereas a small value

of K_c shows that the equilibrium favors the reactants. Equilibrium constants can be used to calculate the concentration of a component in the equilibrium mixture.

13.5 Changing Equilibrium Conditions: Le Châtelier's Principle

The addition of reactants or removal of products favors the forward reaction. Removal of reactants or addition of products favors the reverse reaction. Changing the volume of a reaction container changes the pressure of gases at equilibrium, causing a shift to the side with the fewer number of moles. Raising or lowering the temperature for exothermic and endothermic reactions changes the value of K_c and shifts the equilibrium for a reaction.

13.6 Equilibrium in Saturated Solutions

In a saturated solution of a slightly soluble salt, the rate of dissolving the solute is equal to the rate of recrystallization. In a saturated solution the concentrations of the ions from the solute are constant and can be used to calculate the solubility product constant (K_{sp}) for the salt. If the K_{sp} is known, the solubility of the salt can be calculated.

Key Terms

activation energy The energy that must be provided by a collision to break apart the bonds of the reacting molecules.

catalyst A substance that increases the rate of reaction by lowering the activation energy.

chemical equilibrium The point at which the forward and reverse reactions take place at the same rate so that there is no further change in concentrations of reactants and products.

collision theory A model for a chemical reaction that states that molecules must collide with sufficient energy in order to form products.

equilibrium constant, K_c The numerical value obtained by substituting the equilibrium concentrations of the components into the equilibrium constant expression.

equilibrium constant expression The ratio of the concentrations of products to the concentrations of reactants with each component raised to an exponent equal to the coefficient of that compound in the chemical equation.

heterogeneous equilibrium An equilibrium system in which the components are in different states.

homogenous equilibrium An equilibrium system in which all components are in the same state.

Le Châtelier's principle When a stress is placed on a system at equilibrium, the equilibrium shifts to relieve that stress.

rate of reaction The speed at which reactants are used to form product(s).

reversible reaction A reaction in which a forward reaction occurs from reactants to products, and a reverse reaction occurs from products back to reactants.

solubility product, K_{sp} The product of the concentrations of the ions in a saturated solution of a slightly soluble salt with each concentration raised to a power equal to its coefficient in the equilibrium equation.

▶ Understanding the Concepts

13.43 Write the equilibrium constant expression for each of the following reactions:
   a. $CH_4(g) + 2O_2(g) \rightleftharpoons CO_2(g) + 2H_2O(g)$
   b. $4NH_3(g) + 3O_2(g) \rightleftharpoons 2N_2(g) + 6H_2O(g)$
   c. $C(s) + 2H_2(g) \rightleftharpoons CH_4(g)$

13.44 Write the equilibrium constant expression for each of the following reactions:
   a. $2C_2H_6(g) + 7O_2(g) \rightleftharpoons 4CO_2(g) + 6H_2O(g)$
   b. $2NaHCO_3(s) \rightleftharpoons$
   $$Na_2CO_3(s) + CO_2(g) + H_2O(g)$$
   c. $4NH_3(g) + 5O_2(g) \rightleftharpoons 4NO(g) + 6H_2O(g)$

13.45 Use the initial and equilibrium diagrams to determine if the reaction has a large or small equilibrium constant.

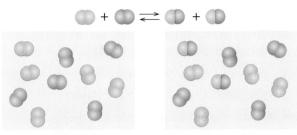

Initial   Equilibrium

13.46 Use the initial and equilibrium diagrams to determine if the reaction has a large or small equilibrium constant.

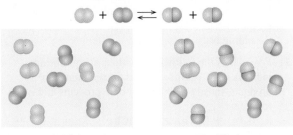

Initial   Equilibrium

13.47 Use the reaction and equilibrium diagrams below for the following:
   a. Would T_2 be higher or lower than T_1?
   b. Would K_c for T_2 be larger or smaller than the K_c for T_1?

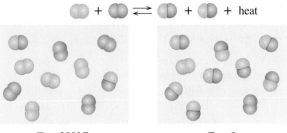

$T_1 = 300°C$ $T_2 = ?$

13.48 Use the reaction and equilibrium diagrams opposite for the following:

a. Would the reaction be exothermic or endothermic?

b. To increase K_c for this reaction, would you raise or lower the temperature?

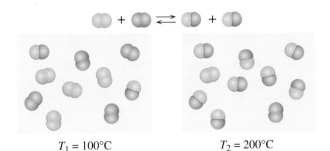

$T_1 = 100°C$ $T_2 = 200°C$

▶ Additional Questions and Problems

13.49 Consider the following reaction at equilibrium:

$$C_2H_4(g) + Cl_2(g) \rightleftharpoons C_2H_4Cl_2(g) + heat$$

Indicate how each of the following will shift the equilibrium:

a. raising the temperature of the reaction

b. decreasing the volume of the reaction container

c. adding a catalyst

d. adding Cl_2

13.50 Consider the following reaction at equilibrium:

$$N_2(g) + O_2(g) + heat \rightleftharpoons 2NO(g)$$

Indicate how each of the following will shift the equilibrium:

a. raising the temperature of the reaction

b. decreasing the volume of the reaction container

c. adding a catalyst

d. adding N_2

13.51 For each of the following reactions at equilibrium, indicate if the equilibrium mixture contains mostly products, mostly reactants, or both products and reactants:

a. $H_2(g) + Cl_2(g) \rightleftharpoons 2HCl(g)$ $K_c = 1.3 \times 10^{34}$

b. $2NOBr(g) \rightleftharpoons 2NO(g) + Br_2(g)$ $K_c = 2.0$

c. $2NOCl(g) \rightleftharpoons Cl_2(g) + 2NO(g)$

                                 $K_c = 2.7 \times 10^{-9}$

13.52 For each of the following reactions at equilibrium, indicate if the equilibrium mixture contains mostly products, mostly reactants, or both products and reactants:

a. $2H_2O(g) \rightleftharpoons 2H_2(g) + O_2(g)$ $K_c = 4 \times 10^{-48}$

b. $N_2(g) + 3H_2(g) \rightleftharpoons 2NH_3(g)$ $K_c = 0.30$

c. $2SO_2(g) + O_2(g) \rightleftharpoons 2SO_3(g)$ $K_c = 1.2 \times 10^9$

13.53 Write the equation for each of the following equilibrium constant expressions:

a. $K_c = \dfrac{[SO_2][Cl_2]}{[SO_2Cl_2]}$ b. $K_c = \dfrac{[BrCl]^2}{[Br_2][Cl_2]}$

c. $K_c = \dfrac{[CH_4][H_2O]}{[CO][H_2]^3}$ d. $K_c = \dfrac{[N_2O][H_2O]^3}{[O_2]^2[NH_3]^2}$

13.54 Write the equation for each of the following equilibrium constant expressions:

a. $K_c = \dfrac{[CO_2][H_2]}{[CO][H_2O]}$ b. $K_c = \dfrac{[H_2][F_2]}{[HF]^2}$

c. $K_c = \dfrac{[O_2][HCl]^4}{[Cl_2]^2[H_2O]^2}$ d. $K_c = \dfrac{[CS_2][H_2]^4}{[CH_4][H_2S]^2}$

13.55 Consider the reaction

$$2NH_3(g) \rightleftharpoons N_2(g) + 3H_2(g)$$

a. Write the equilibrium constant expression for K_c.

b. What is the K_c for the reaction if at equilibrium the concentrations are $[NH_3] = 0.20$ M, $[H_2] = 0.50$ M, and $[N_2] = 3.0$ M?

13.56 Consider the reaction

$$2SO_2(g) + O_2(g) \rightleftharpoons 2SO_3(g)$$

a. Write the equilibrium constant expression for K_c.

b. What is the K_c for the reaction if at equilibrium the concentrations are $[SO_2] = 0.10$ M, $[O_2] = 0.12$ M, $[SO_3] = 0.60$ M.

13.57 The equilibrium constant for the following reaction is 5.0 at 100°C. If an equilibrium mixture contains $[NO_2] = 0.50$ M, what is the $[N_2O_4]$?

$$2NO_2(g) \rightleftharpoons N_2O_4(g)$$

13.58 The equilibrium constant for the following reaction is 0.20 at 1000°C. If an equilibrium mixture contains solid carbon, $[H_2O] = 0.40$ M, and $[CO] = 0.40$ M, what is the $[H_2]$?

$$C(s) + H_2O(g) \rightleftharpoons CO(g) + H_2(g)$$

13.59 According to Le Châtelier's principle, does the equilibrium shift to products or reactants when O_2 is added to the equilibrium mixture of each of the following reactions?

a. $3O_2(g) \rightleftharpoons 2O_3(g)$

b. $2CO_2(g) \rightleftharpoons 2CO(g) + O_2(g)$

c. $P_4(g) + 5O_2(g) \rightleftharpoons P_4O_{10}(s)$

d. $2NO_2(g) \rightleftharpoons N_2(g) + 2O_2(g)$

13.60 According to Le Châtelier's principle, what is the effect on the products when N_2 is added to the equilibrium mixture of each of the following reactions:
  a. $2NH_3(g) \rightleftharpoons 3H_2(g) + N_2(g)$
  b. $N_2(g) + O_2(g) \rightleftharpoons 2NO(g)$
  c. $2NO_2(g) \rightleftharpoons N_2(g) + 2O_2(g)$
  d. $4NH_3(g) + 3O_2(g) \rightleftharpoons 2N_2(g) + 6H_2O(g)$

13.61 Would decreasing the volume of the equilibrium mixture of each of the following reactions cause the equilibrium to shift toward products, toward reactants, or no shift?
  a. $3O_2(g) \rightleftharpoons 2O_3(g)$
  b. $2CO_2(g) \rightleftharpoons 2CO(g) + O_2(g)$
  c. $P_4(g) + 5O_2(g) \rightleftharpoons P_4O_{10}(s)$
  d. $2NO_2(g) \rightleftharpoons N_2(g) + 2O_2(g)$

13.62 Would increasing the volume of the equilibrium mixture of each of the following reactions cause the equilibrium to shift toward products, toward reactants or no shift?
  a. $2NH_3(g) \rightleftharpoons 3H_2(g) + N_2(g)$
  b. $N_2(g) + O_2(g) \rightleftharpoons 2NO(g)$
  c. $2NO_2(g) \rightleftharpoons N_2(g) + 2O_2(g)$
  d. $4NH_3(g) + 3O_2(g) \rightleftharpoons 2N_2(g) + 6H_2O(g)$

13.63 For each of the following salts, write the equilibrium equation and the solubility product expression:
  a. $CuCO_3$ **b.** PbF_2 **c.** $Fe(OH)_3$

13.64 For each of the following salts, write the equilibrium equation and the solubility product expression:
  a. CuS **b.** Ag_2SO_4 **c.** $Zn(OH)_2$

13.65 A saturated solution of iron(II) sulfide, FeS, has $[Fe^{2+}] = 7.7 \times 10^{-10}$ M and $[S^{2-}] = 7.7 \times 10^{-10}$ M. What is the value of K_{sp} for FeS?

13.66 A saturated solution of copper(I) chloride, CuCl, has $[Cu^+] = 1.1 \times 10^{-3}$ M and $[Cl^-] = 1.1 \times 10^{-3}$ M. What is the value of K_{sp} for CuCl?

13.67 A saturated solution of $Mn(OH)_2$ has $[Mn^{2+}] = 3.7 \times 10^{-5}$ M and $[OH^-] = 7.4 \times 10^{-5}$ M. What is the value of K_{sp} for $Mn(OH)_2$?

13.68 A saturated solution of silver chromate, Ag_2CrO_4, has $[Ag^+] = 1.3 \times 10^{-4}$ M and $[CrO_4^{2-}] = 6.5 \times 10^{-5}$ M. What is the value of K_{sp} for Ag_2CrO_4?

13.69 What are $[Cd^{2+}]$ and $[S^{2-}]$ in a saturated CdS solution if the K_{sp} of CdS is 1.0×10^{-24}?

13.70 What are $[Cu^{2+}]$ and $[CO_3^{2-}]$ in a saturated $CuCO_3$ solution if the K_{sp} of $CuCO_3$ is 1×10^{-26}?

13.71 If a saturated solution of $BaSO_4$ has $[Ba^{2+}] = 1.0 \times 10^{-3}$ M, what is the $[SO_4^{2-}]$? (See Table 13.8 for the K_{sp}.)

13.72 If a saturated solution of AgCl has $[Ag^+] = 2.0 \times 10^{-2}$ M, what is the $[Cl^-]$? (See Table 13.8 for the K_{sp}.)

Challenge Questions

13.73 For each of the following K_c values, indicate whether the equilibrium mixture contains mostly reactants, mostly products, or similar amounts of reactants and products:
  a. $N_2(g) + O_2(g) \rightleftharpoons 2NO(g)$ $K_c = 1 \times 10^{-30}$
  b. $H_2(g) + Br(g) \rightleftharpoons 2HBr(g)$ $K_c = 2.0 \times 10^{19}$

13.74 The K_c at 250°C is 4.2×10^{-2} for the reaction

$$PCl_5(g) \rightleftharpoons PCl_3(g) + Cl_2(g)$$

  a. Write the equilibrium constant expression.
  b. Initially, 0.60 mol PCl_5 are placed in a 1.00-L flask. At equilibrium, there is 0.16 mol PCl_3 in the flask. What are the equilibrium concentrations of the PCl_5 and Cl_2?
  c. What is the equilibrium constant for the reaction?
  d. If 0.20 mol Cl_2 is added to the equilibrium mixture, will $[PCl_5]$ increase or decrease?

13.75 The K_c at 100°C is 2.0 for the reaction

$$2NOBr(g) \rightleftharpoons 2NO(g) + Br_2(g)$$

In an experiment, 1.0 mol of each substance is placed in a 1.0-L container.
  a. What is the equilibrium constant expression for the reaction?
  b. Is the system at equilibrium?

  c. If not, will the rate of the forward or reverse reaction initially speed up?
  d. Which concentrations will increase and which will decrease when the system has come to equilibrium?

13.76 For the reaction

$$C(s) + CO_2(g) \rightleftharpoons 2CO(g)$$

the equilibrium mixture contains solid carbon, $[CO] = 0.030$ M, and $[CO_2] = 0.060$ M.
  a. What is the value of K_c for the reaction at this temperature?
  b. What is the effect of adding more CO_2 to the equilibrium mixture?
  c. What is the effect of decreasing the volume of the container?

13.77 The antacid milk of magnesia, which contains $Mg(OH)_2$, is used to neutralize excess stomach acid. If the solubility of $Mg(OH)_2$ in water is 9.7×10^{-3} g/L, what is the K_{sp}?

13.78 In a saturated solution of CaF_2, the $[F^-] = 2.2 \times 10^{-3}$ M. What is $[Ca^{2+}]$? (See Table 13.8 for the K_{sp}.)

Answers

Answers to Study Checks

13.1 Lowering the temperature will decrease the rate of reaction.

13.2 $H_2(g) + Br_2(g) \rightleftharpoons 2HBr(g)$

13.3 (4) slower; (3) faster

13.4 $2NO(g) + O_2(g) \rightleftharpoons 2NO_2(g)$

13.5 $FeO(s) + CO(g) \rightleftharpoons Fe(s) + CO_2(g)$

$$K_c = \frac{[CO_2]}{[CO]}$$

13.6 $K_c = 27$

13.7 The concentration of the reactants would be much greater than the concentration of the products.

13.8 $[C_2H_5OH] = 2.7 \text{ M}$

13.9 Increasing $[CO_2]$ will shift the equilibrium toward the reactants.

13.10 Decreasing the volume of the reaction container will shift the equilibrium to the product side, which has fewer moles of gas.

13.11 **a.** A decrease in temperature will decrease the concentration of reactants and increase the K_c value.
b. A decrease in temperature will increase the concentration of reactants and decrease the K_c value.

13.12 $Fe(OH)_3(s) \rightleftharpoons Fe^{3+}(aq) + 3OH^-(aq)$

13.13 $K_{sp} = 5.0 \times 10^{-13}$

13.14 Molar solubility $(S) = 2 \times 10^{-15} \text{ M}$

13.15 The maximum Ni^{2+} concentration is $3.1 \times 10^{-3} \text{ M}$.

Answers to Selected Questions and Problems

13.1 **a.** The rate of the reaction indicates how fast the products form.
b. At room temperature, the reactions involved in the growth of bread mold will proceed at a faster rate than in the lower temperature of the refrigerator.

13.3 The number of collisions will increase when the number of Br_2 molecules is increased.

13.5 **a.** increase **b.** increase
c. increase **d.** decrease

13.7 A reversible reaction is one in which a forward reaction converts reactants to products, whereas a reverse reaction converts products to reactants.

13.9 **a.** not reversible **b.** reversible
c. reversible

13.11 **a.** $K_c = \dfrac{[CS_2][H_2]^4}{[CH_4][H_2S]^2}$ **b.** $K_c = \dfrac{[N_2][O_2]}{[NO]^2}$

c. $K_c = \dfrac{[CS_2][O_2]^4}{[SO_3]^2[CO_2]}$

13.13 **a.** homogeneous equilibrium
b. heterogeneous equilibrium
c. homogeneous equilibrium
d. heterogeneous equilibrium

13.15 **a.** $K_c = \dfrac{[O_2]^3}{[O_3]^2}$ **b.** $K_c = [CO_2][H_2O]$

c. $K_c = \dfrac{[H_2]^3[CO]}{[CH_4][H_2O]}$ **d.** $K_c = \dfrac{[Cl_2]^2}{[HCl]^4[O_2]}$

13.17 $K_c = 1.5$

13.19 $K_c = 260$

13.21 **a.** mostly products **b.** mostly products
c. mostly reactants

13.23 $[H_2] = 1.1 \times 10^{-3} \text{ M}$

13.25 $[NOBr] = 1.4 \text{ M}$

13.27 **a.** When more reactant is added to an equilibrium mixture, the product/reactant ratio is initially less than K_c.
b. According to Le Châtelier's principle, equilibrium is reestablished when the forward reaction forms more products to make the product/reactant ratio equal the K_c again.

13.29 **a.** Equilibrium shifts to products.
b. Equilibrium shifts to reactants.
c. Equilibrium shifts to products.
d. Equilibrium shifts to products.
e. No shift in equilibrium occurs.

13.31 **a.** Equilibrium shifts to products.
b. Equilibrium shifts to products.
c. Equilibrium shifts to products.
d. No shift in equilibrium occurs.
e. Equilibrium shifts to reactants.

13.33 **a.** $MgCO_3(s) \rightleftharpoons Mg^{2+}(aq) + CO_3^{2-}(aq)$;
$K_{sp} = [Mg^{2+}][CO_3^{2-}]$
b. $CaF_2(s) \rightleftharpoons Ca^{2+}(aq) + 2F^-(aq)$;
$K_{sp} = [Ca^{2+}][F^-]^2$
c. $Ag_3PO_4(s) \rightleftharpoons 3Ag^+(aq) + PO_4^{3-}(aq)$;
$K_{sp} = [Ag^+]^3[PO_4^{3-}]$

13.35 $K_{sp} = 1 \times 10^{-10}$

13.37 $K_{sp} = 8.8 \times 10^{-12}$

13.39 $[Cu^+] = 1 \times 10^{-6}$; $[I^-] = 1 \times 10^{-6}$

13.41 $[Cl^-] = 9.0 \times 10^{-4}$

13.43 **a.** $K_c = \dfrac{[CO_2][H_2O]^2}{[CH_4][O_2]^2}$ **b.** $K_c = \dfrac{[N_2]^2[H_2O]^6}{[NH_3]^4[O_2]^3}$

c. $K_c = \dfrac{[CH_4]}{[H_2]^2}$

13.45 The reaction would have a small value of the equilibrium constant.

13.47 **a.** T_2 is lower than T_1.
b. K_c for T_2 is larger than K_c for T_1.

13.49 **a.** shift toward reactants
b. shift toward products
c. no change
d. shift toward products

13.51 **a.** mostly products
b. both products and reactants
c. mostly reactants

13.53 **a.** $SO_2Cl_2(g) \rightleftarrows SO_2(g) + Cl_2(g)$
b. $Br_2(g) + Cl_2(g) \rightleftarrows 2BrCl(g)$
c. $CO(g) + 3H_2(g) \rightleftarrows CH_4(g) + H_2O(g)$
d. $2O_2(g) + 2NH_3(g) \rightleftarrows N_2O(g) + 3H_2O(g)$

13.55 **a.** $K_c = \dfrac{[N_2][H_2]^3}{[NH_3]^2}$ **b.** $K_c = 9.4$

13.57 $[N_2O_4] = 1.3M$

13.59 **a.** Equilibrium shifts to products.
b. Equilibrium shifts to reactants.
c. Equilibrium shifts to products.
d. Equilibrium shifts to reactants.

13.61 **a.** Equilibrium shifts to products.
b. Equilibrium shifts to reactants.
c. Equilibrium shifts to products.
d. Equilibrium shifts to reactants.

13.63 **a.** $CuCO_3(s) \rightleftarrows Cu^{2+}(aq) + CO_3^{2-}(aq)$;
$K_{sp} = [Cu^{2+}][CO_3^{2-}]$
b. $PbF_2(s) \rightleftarrows Pb^{2+}(aq) + 2F^-(aq)$;
$K_{sp} = [Pb^{2+}][F^-]^2$
c. $Fe(OH)_3(s) \rightleftarrows Fe^{3+}(aq) + 3OH^-(aq)$;
$K_{sp} = [Fe^{3+}][OH^-]^3$

13.65 $K_{sp} = 5.9 \times 10^{-19}$

13.67 $K_{sp} = 2.0 \times 10^{-13}$

13.69 $[Cd^{2+}] = 1.0 \times 10^{-12}$; $[S^{2-}] = 1.0 \times 10^{-12}$

13.71 $[SO_4^{2-}] = 1.1 \times 10^{-7}$

13.73 **a.** A small K_c indicates that the equilibrium mixture contains mostly reactants.
b. A large K_c indicates that the equilibrium mixture contains mostly products.

13.75 **a.** $K_c = \dfrac{[Br_2][NO]^2}{[NOBr]^2} = 2$
b. When the concentrations are placed in the expression, the result is 1, which is not equal to K_c. The system is not at equilibrium.
c. Since the value in part b is less than K_c, the forward reaction will speed up.
d. The $[Br_2]$ and $[NO]$ will increase and the $[NOBr]$ will decrease.

13.77 The solubility is
$$\frac{9.7 \times 10^{-3}\, \cancel{g}}{L} \times \frac{1\ mol}{58.33\, \cancel{g}} = \frac{1.7 \times 10^{-4}\ mol}{L}.$$
The $[Mg^{2+}] = 1.7 \times 10^{-4}$ mol/L and the
$[OH^-] = 2 \times 1.7 \times 10^{-4}$ mol/L $= 3.4 \times 10^{-4}$ mol/L.
$$K_{sp} = [Mg^{2+}] \times [OH^-]^2 = [1.7 \times 10^{-4}][3.4 \times 10^{-4}]^2$$
$$= 2.0 \times 10^{-11}$$

Acids and Bases

"*In our toxicology lab, we measure the drugs in samples of urine or blood,*" *says Penny Peng, assistant supervisor of chemistry, Toxicology Lab, Santa Clara Valley Medical Center.* "*But first we extract the drugs from the fluid and concentrate them so they can be detected in the machine we use. We extract the drugs by using different organic solvents such as methanol, ethyl acetate, or methylene chloride, and by changing the pH. We evaporate most of the organic solvent to concentrate any drugs it may contain. A small sample of the concentrate is placed into a machine called a gas chromatograph. As the gas moves over a column, the drugs in it are separated. From the results, we can identify as many as 10 to 15 different drugs from one urine sample.*"

the Chemistry place

Visit **www.aw-bc.com/chemplace** for extra quizzes, interactive tutorials, career resources, PowerPoint slides for chapter review, math help, and case studies.

$\overset{\mathsf{A}}{}$cids and bases are important substances in health, industry, and the environment. One of the most common characteristics of acids is their sour taste. Lemons and grapefruits are sour because they contain citric and ascorbic acid (vitamin C). Vinegar tastes sour because it contains acetic acid. When we exercise, lactic acid forms in our muscles, causing fatigue and soreness. Acid from bacteria turns milk sour in the production of yogurt or cottage cheese. We have hydrochloric acid in our stomach that helps us digest food. Sometimes we take antacids, which are bases such as sodium bicarbonate or milk of magnesia, to neutralize the effects of too much stomach acid.

Acids and bases have many uses in the chemical industry. Sulfuric acid, H_2SO_4, is used to produce fertilizers and plastics, to manufacture detergents, and to conduct electricity in lead-acid storage batteries for automobiles. Sulfuric acid is the world's most widely produced chemical. Sodium hydroxide is used in the production of pulp and paper, the manufacture of soaps, in the textile industries, in oven and drain cleaners, and in the manufacture of glass.

In the environment, the acidity, or pH, of rain, water, and soil can have significant effects. When rain becomes too acidic, it can dissolve marble statues and accelerate the corrosion of metals. In lakes and ponds, the acidity can affect the ability of fish to survive. The acidity of soil around plants affects their growth. If the soil pH is too acidic or too basic, the roots of the plant cannot take up some nutrients. Most plants thrive in soil with a nearly neutral pH, although certain plants such as orchids, camellias, and blueberries require a more acidic soil.

14.1 Acids and Bases

Learning **Goal**

Describe and name acids and bases using the Arrhenius and the Brønsted–Lowry concepts.

The term *acid* comes from the Latin word *acidus,* which means "sour." We are familiar with the sour tastes of vinegar and lemons and other common acids in foods.

In the nineteenth century, Arrhenius was the first to describe **acids** as substances that produce hydrogen ions (H^+) when they dissolve in water. For example, hydrogen chloride ionizes in water to give hydrogen ions, H^+, and chloride ions, Cl^-. The hydrogen ions, H^+, give acids a sour taste, change blue litmus to red, and corrode some metals.

$$HCl(g) \xrightarrow{\;H_2O\;} H^+(aq) + Cl^-(aq)$$

Polar covalent   Ionization
compound   in water

Naming Acids

When an acid dissolves in water to produce a hydrogen ion and an anion, the prefix *hydro* is used before the name of the nonmetal and its *ide* ending is changed to *ic acid.* For example, hydrogen chloride (HCl) dissolves in water to form HCl(*aq*), which is named hydrochloric acid.

Most acids are *oxo acids,* which means they also contain oxygen. When an oxo acid dissolves in water, it produces H^+ and a polyatomic ion. The name of an oxo acid comes from the name of its polyatomic ion. When a polyatomic ion ends in *ate,* the ending is replaced by *ic acid,* which is the ending for the most common form of oxo acids. When the name of the polyatomic ion ends in *ite,* the acid name ends in *ous acid.* Thus, HNO_3, which produces

Table 14.1   **Naming Common Acids**

Acid	Name of Acid	Anion	Name of Anion
HCl	**hydro**chloric **acid**	Cl^-	chlor**ide**
HBr	**hydro**bromic **acid**	Br^-	brom**ide**
HNO_3	nitr**ic acid**	NO_3^-	nitr**ate**
HNO_2	nitr**ous acid**	NO_2^-	nitr**ite**
H_2SO_4	sulfur**ic acid**	SO_4^{2-}	sulf**ate**
H_2SO_3	sulfur**ous acid**	SO_3^{2-}	sulf**ite**
H_2CO_3	carbon**ic acid**	CO_3^{2-}	carbon**ate**
H_3PO_4	phosphor**ic acid**	PO_4^{3-}	phosph**ate**
$HClO_3$	chlor**ic acid**	ClO_3^-	chlor**ate**
$HClO_2$	chlor**ous acid**	ClO_2^-	chlor**ite**
CH_3COOH	acet**ic acid**	CH_3COO^-	acet**ate**

the nitrate ion (NO_3^-), is named nitric acid; HNO_2, which produces the nitrite ion (NO_2^-), is named nitrous acid.

$$HNO_3(l) \xrightarrow{\text{H}_2\text{O}} H^+(aq) + NO_3^-(aq)$$
Nitric acid   Nitrate ion

$$HNO_2(l) \xrightarrow{\text{H}_2\text{O}} H^+(aq) + NO_2^-(aq)$$
Nitrous acid   Nitrite ion

As we can see in the formulas, the *ous acid* form has one oxygen less than the common *ic acid* form. The names of some common acids and their anions, including polyatomic ions, are listed in Table 14.1.

In Group 7A (17), additional oxo acids are possible. The names of the oxo acids of chlorine, for example, are shown in Table 14.2. When the acid formula contains one more oxygen than the common *ic* form of the acid, the prefix *per* is used; $HClO_4$ is named *per*chlor*ic acid*. When the acid formula has one oxygen less than the common *ic* form of the acid, the suffix *ous* is used, as in chlor*ous acid*. The prefix *hypo* is used for the acid that has two oxygen atoms fewer than the common *ic* form of the acid $HClO_3$. Thus $HClO$ is named *hypo*chlor*ous acid*.

Bases

You may be familiar with some bases such as antacids, drain openers, and oven cleaners. According to the Arrhenius theory, **bases** are ionic compounds that dissociate into a metal ion and hydroxide ions (OH^-) when they

Table 14.2   **Names of Oxo Acids and Anions of Chlorine**

O in Acid	Acid	Name	Anion	Name
One O more	$HClO_4$	**per**chlor**ic acid**	ClO_4^-	**per**chlor**ate**
Common form	$HClO_3$	chlor**ic acid**	ClO_3^-	chlor**ate**
One O less	$HClO_2$	chlor**ous acid**	ClO_2^-	chlor**ite**
Two O less	$HClO$	**hypo**chlor**ous acid**	ClO^-	**hypo**chlor**ite**

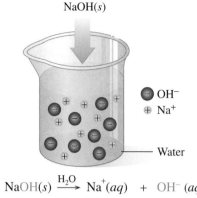

NaOH(s)

● OH⁻
⊕ Na⁺

Water

$$NaOH(s) \xrightarrow{H_2O} Na^+(aq) + OH^-(aq)$$

Ionic   Dissociation   Hydroxide
compound   ion

dissolve in water. For example, sodium hydroxide is an Arrhenius base that dissociates in water to give sodium ions, Na^+, and hydroxide ions, OH^-.

Most Arrhenius bases are from Groups 1A (1) and 2A (2), such as NaOH, KOH, LiOH, and $Ca(OH)_2$. There are other bases such as $Al(OH)_3$ and $Fe(OH)_3$, but they are fairly insoluble. The hydroxide ions (OH^-) produced by Arrhenius bases give these bases common characteristics such as a bitter taste and soapy feel.

Naming Bases

Typical Arrhenius bases are named as hydroxides.

Bases	Name
NaOH	sodium **hydroxide**
KOH	potassium **hydroxide**
$Ca(OH)_2$	calcium **hydroxide**
$Al(OH)_3$	aluminum **hydroxide**

Sample Problem 14.1   **Names of Acids and Bases**

Name the following as acids or bases:
   a. H_3PO_4 **b.** NaOH   **c.** HNO_2 **d.** $HBrO_2$

Solution
   a. phosphoric acid   **b.** sodium hydroxide
   c. nitrous acid   **d.** bromous acid

Study Check

Give the names for H_2SO_4 and KOH.

Questions and Problems   **Acids and Bases**

14.1 Indicate whether each of the following statements indicates an acid or a base:
   a. has a sour taste
   b. neutralizes bases
   c. produces H_3O^+ ions in water
   d. is named potassium hydroxide

14.2 Indicate whether each of the following statements indicates an acid or a base:
   a. neutralizes acids
   b. produces OH^- in water
   c. has a soapy feel
   d. turns litmus red

14.3 Name each of the following as an acid or base:
   a. HCl   **b.** $Ca(OH)_2$
   c. H_2CO_3 **d.** HNO_3
   e. H_2SO_3 **f.** $HBrO_3$

14.4 Name each of the following as an acid or base:
   a. $Al(OH)_3$ **b.** HBr
   c. H_2SO_4 **d.** KOH
   e. HNO_2 **f.** $HClO_2$

14.5 Write formulas for the following acids and bases:
   a. magnesium hydroxide
   b. hydrofluoric acid
   c. phosphoric acid
   d. lithium hydroxide
   e. ammonium hydroxide
   f. periodic acid

14.6 Write formulas for the following acids and bases:
   a. barium hydroxide
   b. hydriodic acid
   c. nitric acid
   d. potassium hydroxide
   e. sodium hydroxide
   f. hypochlorous acid

14.2 Brønsted–Lowry Acids and Bases

Early in the twentieth century, Brønsted and Lowry expanded the definition of acids and bases. A **Brønsted–Lowry acid** donates a proton (hydrogen ion, H⁺) to another substance, and a **Brønsted–Lowry base** accepts a proton.

A Brønsted–Lowry acid is a proton (H⁺) donor.

A Brønsted–Lowry base is a proton (H⁺) acceptor.

A free, dissociated proton (H⁺) does not actually exist in water. Its attraction to polar water molecules is so strong that the proton bonds to a water molecule and forms a **hydronium ion, H_3O^+**.

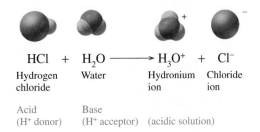

Water   Proton   Hydronium ion

We write the formation of a hydrochloric acid solution as a transfer of a proton from hydrogen chloride to water. By accepting a proton in the reaction, water is acting as a base according to the Brønsted–Lowry concept.

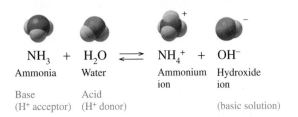

HCl + H_2O ⟶ H_3O^+ + Cl⁻
Hydrogen   Water   Hydronium   Chloride
chloride   ion   ion

Acid   Base
(H⁺ donor)   (H⁺ acceptor)   (acidic solution)

In another reaction, ammonia (NH_3) reacts with water. Because the nitrogen of NH_3 has a stronger attraction for a proton than the oxygen of water, water acts as an acid and donates a proton.

NH_3 + H_2O ⇌ NH_4^+ + OH⁻
Ammonia   Water   Ammonium   Hydroxide
                        ion   ion

Base   Acid
(H⁺ acceptor) (H⁺ donor)   (basic solution)

Table 14.3 compares some characteristics of acids and bases.

Table 14.3 Some Characteristics of Acids and Bases

Characteristic	Acids	Bases
Reaction, Arrhenius	Produce H^+	Produce OH^-
Reaction, Brønsted–Lowry	Donate H^+	Accept H^+
Electrolytes	Yes	Yes
Taste	Sour	Bitter, chalky
Feel	May sting	Soapy, slippery
Litmus	Red	Blue
Phenolphthalein	Colorless	Red
Neutralization	Neutralize bases	Neutralize acids

Sample Problem 14.2 ▶ **Acids and Bases**

In each of the following equations, identify the reactant that is an acid (H^+ donor) and the reactant that is a base (H^+ acceptor).

a. $HBr(aq) + H_2O(l) \longrightarrow H_3O^+(aq) + Br^-(aq)$
b. $H_2O(l) + CN^-(aq) \rightleftharpoons HCN(aq) + OH^-(aq)$

Solution
a. HBr, acid; H_2O, base
b. H_2O, acid; CN^-, base

Study Check

When HNO_3 reacts with water, water acts as a base (H^+ acceptor). Write the equation for the reaction.

Conjugate Acid–Base Pair

Donates H^+

HF → F^-

Conjugate Acid–Base Pair

Accepts H^+

H_2O → H_3O^+

According to the Brønsted–Lowry theory, the reaction between an acid and base involves proton transfer. When an acid (HA) donates H^+ to a base (B), the products are A^- and BH^+. Because the products are also an acid and a base, a reverse reaction can occur in which the acid BH^+ donates H^+ to the base (A^-). When a pair of molecules or ions are related by the loss or gain of one H^+, they are called a **conjugate acid–base pair.** Because protons are transferred in both a forward and reverse reaction, each acid–base reaction contains two conjugate acid–base pairs. In this general reaction, the acid HA has a conjugate base, A^-, and the base B has a conjugate acid, BH^+.

Conjugate acid–base pair

HA	+	**B**	⇌	**A$^-$**	+	**BH$^+$**
Acid 1		**Base 2**		**Base 1**		**Acid 2**
H^+ donor		H^+ acceptor		H^+ acceptor		H^+ donor

Conjugate acid–base pair

Now we can identify the conjugate acid–base pairs in a reaction such as hydrofluoric acid and water. Because the reaction is reversible, the conjugate acid H_3O^+ can transfer a proton to the conjugate base F^- and re-form the acid HF. Using the relationship of loss and gain of one H^+, we identify the conjugate acid–base pairs as HF and F^- along with H_3O^+ and H_2O.

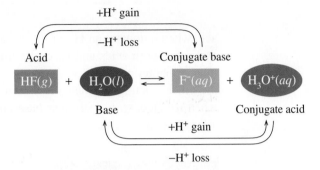

In another proton-transfer reaction, ammonia, NH_3, accepts H^+ from H_2O to form the conjugate acid NH_4^+ and conjugate base OH^-. Each of these conjugate acid–base pairs, NH_3/NH_4^+ and H_2O/OH^-, are related by the loss and gain of one H^+. Table 14.4 gives some more examples of conjugate acid–base pairs. In these two examples, we see that water can act as an

Table 14.4   Some Conjugate Acid–Base Pairs

Acid		Conjugate Base	
Strong Acids			
Perchloric acid	$HClO_4$	Perchlorate ion	ClO_4^-
Sulfuric acid	H_2SO_4	Hydrogen sulfate ion	HSO_4^-
Hydroiodic acid	HI	Iodide ion	I^-
Hydrobromic acid	HBr	Bromide ion	Br^-
Hydrochloric acid	HCl	Chloride ion	Cl^-
Nitric acid	HNO_3	Nitrate ion	NO_3^-
Weak Acids			
Hydronium ion	H_3O^+	Water	H_2O
Hydrogen sulfate ion	HSO_4^-	Sulfate ion	SO_4^{2-}
Phosphoric acid	H_3PO_4	Dihydrogen phosphate ion	$H_2PO_4^-$
Nitrous acid	HNO_2	Nitrite ion	NO_2^-
Hydrofluoric acid	HF	Fluoride ion	F^-
Acetic acid	CH_3COOH	Acetate ion	CH_3COO^-
Extremely Weak Acids			
Carbonic acid	H_2CO_3	Bicarbonate ion	HCO_3^-
Hydrosulfuric acid	H_2S	Hydrogen sulfide ion	HS^-
Ammonium ion	NH_4^+	Ammonia	NH_3
Hydrocyanic acid	HCN	Cyanide ion	CN^-
Bicarbonate ion	HCO_3^-	Carbonate ion	CO_3^{2-}
Hydrogen sulfide ion	HS^-	Sulfide ion	S^{2-}
Water	H_2O	Hydroxide ion	OH^-

Increasing acid strength

Increasing base strength

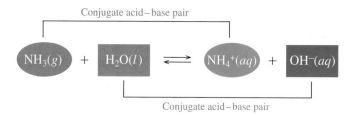

Conjugate acid–base pair

$$NH_3(g) + H_2O(l) \rightleftharpoons NH_4^+(aq) + OH^-(aq)$$

Conjugate acid–base pair

acid when it donates an H^+ or as a base when it accepts H^+. Substances that can act as both acids and bases are **amphoteric.** For water, the most common amphoteric substance, the acidic or basic behavior depends on whether the other reacting substance is a stronger acid or base. Water donates H^+ when it reacts with a stronger base and accepts H^+ when it reacts with a stronger acid.

Sample Problem 14.3   **Conjugate Acid–Base Pairs**

Write the formula of the conjugate base of each of the following Brønsted–Lowry acids:
a. $HClO_3$ **b.** H_2CO_3 **c.** $H_2PO_4^-$

Solution

The conjugate base forms when the acid donates a proton.
a. ClO_3^- is the conjugate base of $HClO_3$.
b. HCO_3^- is the conjugate base of H_2CO_3.
c. HPO_4^{2-} is the conjugate base of $H_2PO_4^-$.

Study Check

Write the formula of the conjugate acid of each of the following Brønsted–Lowry bases:
a. HS^- **b.** Cl^- **c.** NO_2^-

Sample Problem 14.4   **Identifying Conjugate Acid–Base Pairs**

Write the Brønsted–Lowry equation for the reaction of the acid HBr and the base NH_3. Identify the conjugate acid–base pairs.

Solution

In the reaction, HBr donates H^+ to NH_3. The resulting Br^- is the conjugate base and NH_4^+ is the conjugate acid.

$$HBr + NH_3 \rightleftharpoons Br^- + NH_4^+$$

The conjugate acid–base pairs are HBr/Br^- along with NH_3/NH_4^+.

Study Check

In the following reaction, identify the conjugate acid–base pairs:

$$HCN + SO_4^{2-} \rightleftharpoons CN^- + HSO_4^-$$

14.7 In each of the following equations, identify the acid (proton donor) and base (proton acceptor) for the reactants:
   a. $HI(aq) + H_2O(l) \longrightarrow H_3O^+(aq) + I^-(aq)$
   b. $F^-(aq) + H_2O(l) \rightleftharpoons HF(aq) + OH^-(aq)$

14.8 In each of the following equations, identify the acid (proton donor) and base (proton acceptor) for the reactants:
   a. $CO_3^{2-}(aq) + H_2O(l) \rightleftharpoons HCO_3^-(aq) + OH^-(aq)$
   b. $H_2SO_4(aq) + H_2O(l) \longrightarrow H_3O^+(aq) + HSO_4^-(aq)$

14.9 Write the formula and name of the conjugate base for each of the following acids:
   a. HF   **b.** H_2O
   c. H_2CO_3 **d.** HSO_4^-
   e. $HClO_2$

14.10 Write the formula and name of the conjugate base for each of the following acids:
   a. HCO_3^- **b.** H_3O^+
   c. HPO_4^{2-} **d.** HNO_2
   e. HBrO

14.11 Write the formula and name of the conjugate acid for each of the following bases:
   a. CO_3^{2-} **b.** H_2O
   c. $H_2PO_4^-$ **d.** Br^-
   e. ClO_4^-

14.12 Write the formula and name of the conjugate acid for each of the following bases:
   a. SO_4^{2-} **b.** CN^-
   c. OH^- **d.** ClO_2^-
   e. HS^-

14.13 Identify the acid and base on the left side of the following equations and identify their conjugate species on the right side:
   a. $H_2CO_3(aq) + H_2O(l) \rightleftharpoons$
                          $H_3O^+(aq) + HCO_3^-(aq)$
   b. $NH_4^+(aq) + H_2O(l) \rightleftharpoons H_3O^+(aq) + NH_3(aq)$
   c. $HCN(aq) + NO_2^-(aq) \rightleftharpoons$
                          $CN^-(aq) + HNO_2(aq)$

14.14 Identify the acid and base on the left side of the following equations and identify their conjugate species on the right side:
   a. $H_3PO_4(aq) + H_2O(l) \rightleftharpoons$
                          $H_3O^+(aq) + H_2PO_4^-(aq)$
   b. $CO_3^{2-}(aq) + H_2O(l) \rightleftharpoons$
                          $OH^-(aq) + HCO_3^-(aq)$
   c. $H_3PO_4(aq) + NH_3(aq) \rightleftharpoons$
                          $NH_4^+(aq) + H_2PO_4^-(aq)$

14.15 When ammonium chloride dissolves in water, the ammonium ion NH_4^+ donates a proton to water. Write a balanced equation for the reaction of the ammonium ion with water.

14.16 When sodium carbonate dissolves in water, the carbonate ion CO_3^{2-} acts as a base. Write a balanced equation for the reaction of the carbonate ion with water.

14.3 Strengths of Acids and Bases

Write equations for the dissociation of strong and weak acids; identify the direction of reaction.

The *strength* of acids is determined by the moles of H_3O^+ that are produced for each mole of acid that dissolves. The *strength* of bases is determined by the moles of OH^- that are produced for each mole of base that dissolves. In the process called **dissociation,** an acid or base separates into or produces ions in water. Acids and bases vary greatly in their ability to produce H_3O^+ or OH^-. Strong acids and strong bases dissociate completely. In water, weak acids and weak bases dissociate only slightly, leaving most of the initial acid or base undissociated.

Strong and Weak Acids

Strong acids give up protons so easily that their dissociation in water is virtually complete. For example, when HCl, one of the strong acids, dissociates in water by transferring a proton to H_2O, the resulting HCl solution contains only the dissolved ions H_3O^+ and Cl^-. We consider the reaction of HCl in water as going 100% to products. Therefore, the equation for a strong acid such as HCl is written with a single arrow to the products.

$$HCl(g) + H_2O(l) \longrightarrow H_3O^+(aq) + Cl^-(aq)$$

Figure 14.1 A strong acid such as HCl is completely dissociated ($\approx$100%), whereas a weak acid such as CH$_3$COOH contains mostly molecules and a few ions.

Q What is the difference between a strong acid and a weak acid?

Most acids are weak acids. **Weak acids** dissociate slightly in water, which means that only a small percentage of the weak acid transfers a proton to water to produce a small amount of H_3O^+ ions and anions. (See Figure 14.1.) Many of the products you drink or use at home contain weak acids. In carbonated soft drinks, CO_2 dissolves in water to form carbonic acid, H_2CO_3, which remains mostly undissociated in solution. In a weak acid such as H_2CO_3, there is equilibrium between the undissociated H_2CO_3 molecules and the dissociation products, H_3O^+ and HCO_3^-. Therefore, the reaction for a weak acid in water is written with a double arrow. Sometimes a longer reverse arrow is used to indicate that the equilibrium of a weak acid favors the formation of the undissociated reactants.

$$H_2CO_3(aq) + H_2O(l) \underset{\longleftarrow}{\overset{\rightarrow}{}} H_3O^+(aq) + HCO_3^-(aq)$$

Citric acid is a weak acid found in fruits and fruit juices such as lemons, oranges, and grapefruit. Vinegar contains another weak acid known as acetic acid, CH_3COOH. In the vinegar used on salads, acetic acid is present as a 5% acetic acid solution.

$$CH_3COOH(aq) + H_2O(l) \underset{\longleftarrow}{\overset{\rightarrow}{}} H_3O^+(aq) + CH_3COO^-(aq)$$

Strong and Weak Bases

Strong bases are dissociated virtually completely in water. For example, when KOH, a strong base, dissociates in water, the solution consists only of the ions K^+ and OH^-. Essentially no undissociated KOH remains. Therefore, the equation for a strong base such as KOH is written with a single arrow to products.

$$KOH(s) \xrightarrow{H_2O} K^+(aq) + OH^-(aq)$$

The Arrhenius bases of Groups 1A (1) and 2A (2) such as LiOH, KOH, NaOH, and $Ca(OH)_2$ are strong bases. Sodium hydroxide, NaOH (also known as lye), is used in household products to remove grease in ovens and to clean drains.

Weak bases are poor acceptors of protons and remain mostly undissociated in water. Baking soda ($NaHCO_3$) contains bicarbonate ion, HCO_3^-, which acts as a weak base in water. Detergents, which contain anions of weak acids, act as bases in water. A typical weak base, ammonia, NH_3, is found in window cleaners. In water, only a few ammonia molecules accept protons to form NH_4^+ and OH^-.

$$NH_3(g) + H_2O(l) \rightleftharpoons NH_4^+(aq) + OH^-(aq)$$

Direction of Reaction

There is a relationship between the components in each conjugate acid–base pair. The strong acids that donate protons easily have weak conjugate bases that do not readily accept protons. As the strength of the acid decreases, the strength of its conjugate base increases. Weak acids have strong conjugate bases.

In any acid–base reaction, there are two acids and two bases. However, one acid is stronger than the other acid and one base is stronger than the other base. By comparing their relative strengths, we can determine the direction of the reaction. For example, the strong acid H_2SO_4 gives up protons to water. The hydronium ion, H_3O^+, produced is a weaker acid than H_2SO_4, and the conjugate base, HSO_4^-, is a weaker base than water.

$$H_2SO_4(aq) + H_2O(l) \longrightarrow H_3O^+(aq) + HSO_4^-(aq) \quad \text{Strongly favors products}$$

Stronger acid   Stronger base   Weaker acid   Weaker base

Let's look at another reaction in which water donates a proton to carbonate, CO_3^{2-}, to form HCO_3^- and OH^-. From Table 14.4, page 461, we see that HCO_3^- is a stronger acid than H_2O. We also see that OH^- is a stronger base than CO_3^{2-}. The equilibrium favors the weaker acid and base reactants, as shown by the long arrow for the reverse reaction.

$$CO_3^{2-}(aq) + H_2O(l) \rightleftharpoons HCO_3^-(aq) + OH^-(aq) \quad \text{Strongly favors reactants}$$

Weaker base   Weaker acid   Stronger acid   Stronger base

Sample Problem 14.5   **Strengths of Acids and Bases**

For the following questions select from HCO_3^-, HSO_4^-, or HNO_2.
a. Which is the strongest acid?
b. Which acid has the strongest conjugate base?

Solution

As derived from the information in Table 14.4,
a. The strongest acid in this group is HSO_4^-.
b. The weakest acid in this group, HCO_3^-, has the strongest conjugate base CO_3^{2-}.

Study Check

Which is the stronger base: F^- or NH_3?

Does equilibrium favor the reactants or products in the following reaction?

$$HF(aq) + H_2O(l) \rightleftharpoons H_3O^+(aq) + F^-(aq)$$

Solution

From Table 14.4, we see that HF is a weaker acid than H_3O^+ and H_2O is a weaker base than F^-. Equilibrium favors the reverse direction and therefore the reactants.

$$HF(aq) + H_2O(l) \mathrel{\substack{\longrightarrow \\ \longleftarrow}} H_3O^+(aq) + F^-(aq)$$
Weaker acid   Weaker base   Stronger acid   Stronger base

Study Check

Does the reaction of nitric acid and water favor the reactants or the products?

Questions and Problems **Strengths of Acids and Bases**

14.17 What is meant by the statement "A strong acid has a weak conjugate base"?

14.18 What is meant by the statement "A weak acid has a strong conjugate base"?

14.19 Identify the stronger acid in each pair.
 a. HBr or HNO_2
 b. H_3PO_4 or HSO_4^-
 c. HCN or H_2CO_3

14.20 Identify the stronger acid in each pair.
 a. NH_4^+ or H_3O^+
 b. H_2SO_4 or HCN
 c. H_2O or H_2CO_3

14.21 Identify the weaker acid in each pair.
 a. HCl or HSO_4^-
 b. HNO_2 or HF
 c. HCO_3^- or NH_4^+

14.22 Identify the weaker acid in each pair.
 a. HNO_3 or HCO_3^-
 b. HSO_4^- or H_2O
 c. H_2SO_4 or H_2CO_3

14.23 Predict whether the equilibrium for each of the following reactions favors the reactants or the products:
 a. $H_2CO_3(aq) + H_2O(l) \rightleftharpoons$
 $\qquad\qquad\qquad\qquad H_3O^+(aq) + HCO_3^-(aq)$
 b. $NH_4^+(aq) + H_2O(l) \rightleftharpoons H_3O^+(aq) + NH_3(aq)$
 c. $HCl(aq) + NH_3(aq) \rightleftharpoons Cl^-(aq) + NH_4^+(aq)$

14.24 Predict whether the equilibrium for each of the following reactions favors the reactants or the products:
 a. $H_3PO_4(aq) + H_2O(l) \rightleftharpoons$
 $\qquad\qquad\qquad\qquad H_3O^+(aq) + H_2PO_4^-(aq)$
 b. $CO_3^{2-}(aq) + H_2O(l) \rightleftharpoons$
 $\qquad\qquad\qquad\qquad OH^-(aq) + HCO_3^-(aq)$
 c. $HS^-(aq) + H_2O(l) \rightleftharpoons H_3O^+(aq) + S^{2-}(aq)$

14.25 Write an equation for the acid–base reaction between ammonium ion and sulfate ion. Why does the equilibrium favor the reactants?

14.26 Write an equation for the acid–base reaction between nitrous acid and sulfate ion. Why does the equilibrium favor the reactants?

14.4 Dissociation Constants

Learning Goal

Write the equilibrium expression for a weak acid or weak base.

We have seen that reactions of weak acids in water reach equilibrium. If HA is a weak acid, the concentration of H_3O^+ and A^- will be small, which means that the equilibrium will favor the reactants. (See Figure 14.2.)

$$HA(aq) + H_2O(l) \rightleftharpoons H_3O^+(aq) + A^-(aq)$$

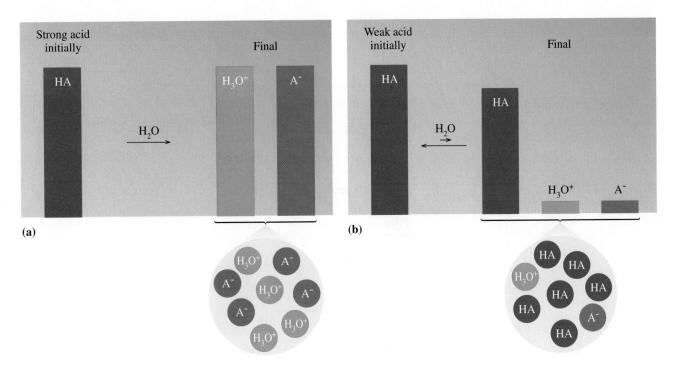

Figure 14.2 After dissociation in water, **(a)** a strong acid (HA) has a high concentration of H_3O^+ and A^-, and **(b)** a weak acid (HA) has a high concentration of HA and low concentrations of H_3O^+ and A^-.

Q How does the height of H_3O^+ and A^- in the bar diagram change for a weak acid?

Dissociation Constants for Weak Acids and Weak Bases

As we have seen, acids and bases have different strengths depending on how much they dissociate in water. Because the dissociation of strong acids in water is essentially complete, the reaction is not considered to be an equilibrium situation. However, because weak acids in water dissociate only slightly, the ion products reach an equilibrium with the undissociated weak acid molecules. Thus, an equilibrium expression can be written for weak acids that gives the ratio of the concentrations of products to the weak-acid reactants. As with other equilibrium constants, the molar concentration of the products is divided by the molar concentration of the reactants.

$$\frac{[H_3O^+][A^-]}{[HA][H_2O]}$$

Because water is a pure liquid, its concentration, which is constant, is omitted from the equilibrium constant, called the **acid dissociation constant, K_a** (or acid ionization constant). Thus for a weak acid HA, the K_a is written

$$K_a = \frac{[H_3O^+][A^-]}{[HA]} \quad \text{Acid dissociation constant}$$

Let's consider the equilibrium of carbonic acid, which dissociates in water to form bicarbonate ion and hydronium ion.

$$H_2CO_3(aq) + H_2O(l) \rightleftharpoons HCO_3^-(aq) + H_3O^+(aq)$$

The K_a expression for carbonic acid is

$$K_a = \frac{[H_3O^+][HCO_3^-]}{[H_2CO_3]} = 4.3 \times 10^{-7}$$

Table 14.5 K_a and K_b Values for Selected Weak Acids and Bases

Acids

Phosphoric acid	H_3PO_4	7.5×10^{-3}
Hydrofluoric acid	HF	7.2×10^{-4}
Nitrous acid	HNO_2	4.5×10^{-4}
Formic acid	HCOOH	1.8×10^{-4}
Acetic acid	CH_3COOH	1.8×10^{-5}
Carbonic acid	H_2CO_3	4.3×10^{-7}
Dihydrogen phosphate	$H_2PO_4^-$	6.2×10^{-8}
Hydrocyanic acid	HCN	4.9×10^{-10}
Hydrogen phosphate	HPO_4^{2-}	2.2×10^{-13}

Bases

Carbonate	CO_3^{2-}	2.2×10^{-4}
Ammonia	NH_3	1.8×10^{-5}

The K_a measured for carbonic acid at 25°C is quite small, which confirms that the equilibrium of carbonic acid in water favors the reactants. (Recall that usually the concentration units are omitted in the values given for equilibrium constants.)

We can also consider a weak base such as ammonia in water.

$$NH_3(aq) + H_2O(l) \rightleftharpoons NH_4^+(aq) + OH^-(aq)$$

We can also write a **base dissociation constant, K_b.**

$$K_b = \frac{[NH_4^+][OH^-]}{[NH_3]} = 1.8 \times 10^{-5}$$

The K_b for ammonia is small because the equilibrium favors the reactants. The smaller the K_a or K_b value, the weaker the acid or base. On the other hand, strong acids and bases, which are essentially 100% dissociated, would have large K_a and K_b values, although these values are not usually measured. Table 14.5 gives some K_a and K_b values for selected weak acids and bases.

We have described strong and weak acids in several ways. Table 14.6 summarizes the characteristics of acids in terms of strengths and equilibrium position.

Table 14.6 Characteristics of Acids

Characteristic	Strong Acids	Weak Acids
Equilibrium position	Toward ionized products	Toward reactants
K_a	Large	Small
$[H_3O^+]$ and $[A^-]$	≈100% of [HA]	Small percent of [HA]
Conjugate bases	Weak	Strong

Sample Problem 14.7 Acid Dissociation Constants

Write the expression for the acid dissociation constant for nitrous acid.

Solution

The equation for the dissociation of nitrous acid is written

$$HNO_2(aq) + H_2O(l) \rightleftharpoons H_3O^+(aq) + NO_2^-(aq)$$

The acid dissociation constant is written as the concentrations of the products divided by the concentration of the undissociated weak acid.

$$K_a = \frac{[H_3O^+][NO_2^-]}{[HNO_2]} = 4.5 \times 10^{-4}$$

Study Check

Which is the stronger acid, nitrous acid or carbonic acid? Why?

Questions and Problems Dissociation Constants

14.27 Consider the following acids and their dissociation constants:

$$H_2SO_3(aq) + H_2O(l) \rightleftharpoons H_3O^+(aq) + HSO_3^-(aq)$$
$$K_a = 1.2 \times 10^{-2}$$
$$HS^-(aq) + H_2O(l) \rightleftharpoons H_3O^+(aq) + S^{2-}(aq)$$
$$K_a = 1.3 \times 10^{-19}$$

   a. Which is the stronger acid, H_2SO_3 or HS^-?
   b. What is the conjugate base of H_2SO_3?
   c. Which acid has the weaker conjugate base?
   d. Which acid has the stronger conjugate base?
   e. Which acid produces more ions?

14.28 Consider the following acids and their dissociation constants:

$$HPO_4^{2-}(aq) + H_2O(l) \rightleftharpoons H_3O^+(aq) + PO_4^{3-}(aq)$$
$$K_a = 2.2 \times 10^{-13}$$
$$HCOOH(aq) + H_2O(l) \rightleftharpoons H_3O^+(aq) + HCOO^-(aq)$$
$$K_a = 1.8 \times 10^{-4}$$

   a. Which is the weaker acid, HPO_4^{2-} or $HCOOH$?
   b. What is the conjugate base of HPO_4^{2-}?
   c. Which acid has the weaker conjugate base?
   d. Which acid has the stronger conjugate base?
   e. Which acid produces more ions?

14.29 Phosphoric acid dissociates to form dihydrogen phosphate and hydronium ion. Phosphoric acid has a K_a of 7.5×10^{-3}. Write the equation and equilibrium constant for the dissociation of the acid.

14.30 Aniline, $C_6H_5NH_2$, a weak base with a K_b of 4.0×10^{-10}, has a conjugate acid, $C_6H_5NH_3^+$. Write the equation and equilibrium constant for the reaction of the base with water.

14.5 Ionization of Water

We have seen that in some acid–base reactions water is amphoteric and acts either as an acid or as a base. Does that mean water can be both an acid and a base? Yes, this is exactly what happens with water molecules in pure water. Let's see how this happens. In water, one water molecule donates a proton to another to produce H_3O^+ and OH^-, which means that water can behave as both an acid and a base. Let's take a look at the conjugate acid–base pairs of water.

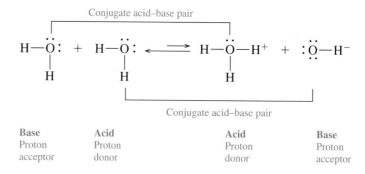

In the ionization of water, there is both a forward and reverse reaction.

$$H_2O(l) + H_2O(l) \rightleftharpoons H_3O^+(aq) + OH^-(aq)$$

In pure water, the transfer of a proton between two water molecules produces equal numbers of H_3O^+ and OH^-. Experiments have determined that, in pure water, the concentrations of H_3O^+ and OH^- at 25°C are each 1.0×10^{-7} M. Square brackets around the symbols indicate their concentrations in moles per liter (M).

Pure water $[H_3O^+] = [OH^-] = 1.0 \times 10^{-7}$ M

When we multiply these concentrations, it forms the **ion-product constant of water**, K_w, which is 1.0×10^{-14}. The concentration units are omitted in the K_w value.

$$K_w = [H_3O^+] \times [OH]$$
$$= (1.0 \times 10^{-7} \text{ M})(1.0 \times 10^{-7} \text{ M}) = 1.0 \times 10^{-14}$$

The K_w value of 1.0×10^{-14} is important because it applies to any aqueous solution: all aqueous solutions have H_3O^+ and OH^-.

When the $[H_3O^+]$ and $[OH^-]$ in a solution are equal, the solution is **neutral.** However, most solutions are not neutral and have different concentrations of $[H_3O^+]$ and $[OH^-]$. If acid is added to water, there is an increase in $[H_3O^+]$ and a decrease in $[OH^-]$, which makes an acidic solution. If base is added, $[OH^-]$ increases and $[H_3O^+]$ decreases, which makes a basic solution. (See Figure 14.3.) However, for any aqueous solution, whether it is neutral, acidic, or basic, the product $[H_3O^+] \times [OH^-]$ is equal to K_w (1.0×10^{-14}).

Figure 14.3 In a neutral solution, $[H_3O^+]$ and $[OH^-]$ are equal. In acidic solutions, the $[H_3O^+]$ is greater than the $[OH^-]$. In basic solutions, the $[OH^-]$ is greater than the $[H_3O^+]$.

Q *Is a solution that has a 1.0×10^{-3} M $[H_3O^+]$ acidic, basic, or neutral?*

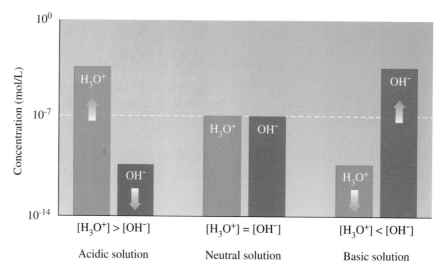

Table 14.7   **Examples of $[H_3O^+]$ and $[OH^-]$ in Neutral, Acidic, and Basic Solutions**

Type of Solution	$[H_3O^+]$	$[OH^-]$	K_w
Neutral	1.0×10^{-7} M	1.0×10^{-7} M	1.0×10^{-14}
Acidic	1.0×10^{-2} M	1.0×10^{-12} M	1.0×10^{-14}
Acidic	2.5×10^{-5} M	4.0×10^{-10} M	1.0×10^{-14}
Basic	1.0×10^{-8} M	1.0×10^{-6} M	1.0×10^{-14}
Basic	5.0×10^{-11} M	2.0×10^{-4} M	1.0×10^{-14}

Therefore, if the $[H_3O^+]$ is given, K_w can be used to calculate the $[OH^-]$. Or if the $[OH^-]$ is given, K_w can be used to calculate the $[H_3O^+]$. (See Table 14.7.)

$$K_w = [H_3O^+] \times [OH^-]$$

$$[OH^-] = \frac{K_w}{[H_3O^+]} \qquad [H_3O^+] = \frac{K_w}{[OH^-]}$$

To illustrate these calculations, let us determine the $[H_3O^+]$ for a solution that has an $[OH^-] = 1.0 \times 10^{-6}$ M.

Guide to Calculating $[H_3O^+]$ and $[OH^-]$ in Aqueous Solutions

STEP 1
Write the K_w for water.

STEP 2
Solve the K_w for the unknown $[H_3O^+]$ or $[OH^-]$.

STEP 3
Substitute the known $[H_3O^+]$ or $[OH^-]$ and calculate.

STEP 1   **Write the K_w for water.**

$$K_w = [H_3O^+] [OH^-] = 1.0 \times 10^{-14}$$

STEP 2   **Arrange the K_w to solve for the unknown.**   Dividing through by the $[OH^-]$ gives

$$\frac{K_w}{[OH^-]} = \frac{[H_3O^+] \times [OH^-]}{[OH^-]} = \frac{1.0 \times 10^{-14}}{[OH^-]}$$

$$[H_3O^+] = \frac{1.0 \times 10^{-14}}{[OH^-]}$$

STEP 3   **Substitute the $[OH^-]$ and calculate the $[H_3O^+]$.**

$$[H_3O^+] = \frac{1.0 \times 10^{-14}}{[1.0 \times 10^{-6}]} = 1.0 \times 10^{-8} \text{ M}$$

Because the $[OH^-]$ of 1.0×10^{-6} M is larger than the $[H_3O^+]$ of 1.0×10^{-8} M, the solution is basic.

Sample Problem 14.8   **Calculating $[H_3O^+]$ and $[OH^-]$ in Solution**

A vinegar solution has a $[H_3O^+] = 2.0 \times 10^{-3}$ M at 25°C. What is the $[OH^-]$ of the vinegar solution? Is the solution acidic, basic, or neutral?

Solution

STEP 1   **Write the K_w for water.**

$$K_w = [H_3O^+] \times [OH^-] = 1.0 \times 10^{-14}$$

STEP 2 **Arrange the K_w to solve for the unknown.** Rearranging the K_w for OH^- gives

$$\frac{K_w}{[H_3O^+]} = \frac{[\cancel{H_3O^+}] \times [OH^-]}{[\cancel{H_3O^+}]} = \frac{1.0 \times 10^{-14}}{[H_3O^+]}$$

$$[OH^-] = \frac{1.0 \times 10^{-14}}{[H_3O^+]}$$

STEP 3 **Substitute the known $[H_3O^+]$ and calculate.**

$$[OH^-] = \frac{1.0 \times 10^{-14}}{[2.0 \times 10^{-3}]} = 5.0 \times 10^{-12} \text{ M}$$

Because the $[H_3O^+]$ of 2.0×10^{-3} M is much larger than the $[OH^-]$ of 5.0×10^{-12} M, the solution is acidic.

Study Check

What is the $[H_3O^+]$ of an ammonia cleaning solution with an $[OH^-]$ = 4.0×10^{-4} M? Is the solution acidic, basic, or neutral?

Questions and Problems   Ionization of Water

14.31 Why are the concentrations of H_3O^+ and OH^- equal in pure water?

14.32 What is the meaning and value of K_w at 25°C?

14.33 In an acidic solution, how does the concentration of H_3O^+ compare to the concentration of OH^-?

14.34 If a base is added to pure water, why does the $[H_3O^+]$ decrease?

14.35 Indicate whether the following are acidic, basic, or neutral solutions:
a. $[H_3O^+] = 2.0 \times 10^{-5}$ M
b. $[H_3O^+] = 1.4 \times 10^{-9}$ M
c. $[OH^-] = 8.0 \times 10^{-3}$ M
d. $[OH^-] = 3.5 \times 10^{-10}$ M

14.36 Indicate whether the following are acidic, basic, or neutral solutions:
a. $[H_3O^+] = 6.0 \times 10^{-12}$ M
b. $[H_3O^+] = 1.4 \times 10^{-4}$ M
c. $[OH^-] = 5.0 \times 10^{-12}$ M
d. $[OH^-] = 4.5 \times 10^{-2}$ M

14.37 Calculate the $[H_3O^+]$ of each aqueous solution with the following $[OH^-]$:
a. coffee, 1.0×10^{-9} M
b. soap, 1.0×10^{-6} M

c. cleanser, 2.0×10^{-5} M
d. lemon juice, 4.0×10^{-13} M

14.38 Calculate the $[H_3O^+]$ of each aqueous solution with the following $[OH^-]$:
a. NaOH, 1.0×10^{-2} M
b. aspirin, 1.8×10^{-11} M
c. milk of magnesia, 1.0×10^{-5} M
d. seawater, 2.5×10^{-6} M

14.39 Calculate the $[OH^-]$ of each aqueous solution with the following $[H_3O^+]$:
a. vinegar, 1.0×10^{-3} M
b. urine, 5.0×10^{-6} M
c. ammonia, 1.8×10^{-12} M
d. NaOH, 4.0×10^{-13} M

14.40 Calculate the $[OH^-]$ of each aqueous solution with the following $[H_3O^+]$:
a. baking soda, 1.0×10^{-8} M
b. orange juice, 2.0×10^{-4} M
c. milk, 5.0×10^{-7} M
d. bleach, 4.8×10^{-12} M

14.6 The pH Scale

Learning Goal

Calculate pH from $[H_3O^+]$; given the pH, calculate $[H_3O^+]$ and $[OH^-]$ of a solution.

Many kinds of careers, such as respiratory therapy, food processing, medicine, agriculture, spa cleaning, and soap manufacturing, require personnel to measure the $[H_3O^+]$ and $[OH^-]$ of solutions. The proper levels of acidity are necessary for soil to support plant growth and prevent algae in swimming pool water. Measuring the acidity levels of blood and urine checks the function of the kidneys.

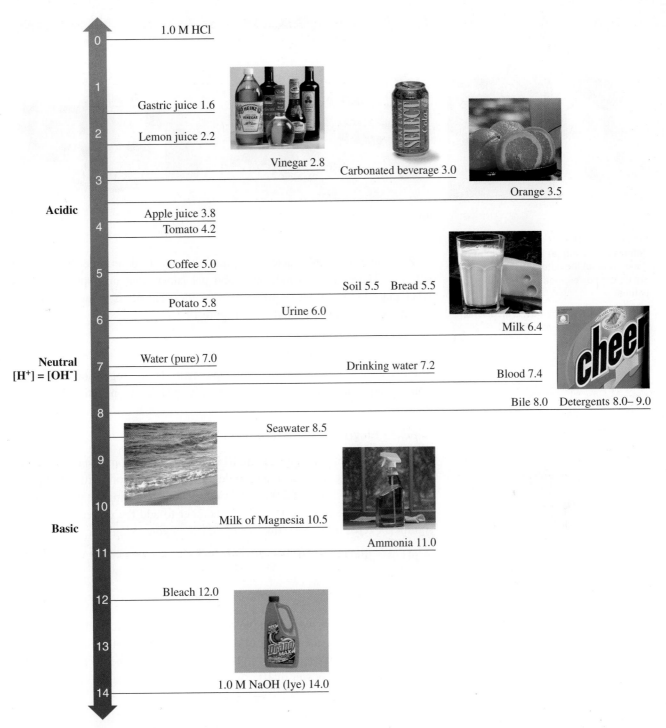

Figure 14.4 On the pH scale, values below 7 are acidic, a value of 7 is neutral, and values above 7 are basic.

Q *Is apple juice an acidic, basic, or a neutral solution?*

On the pH scale, a number between 0 and 14 represents the H_3O^+ concentration. A pH value less than 7 corresponds to an acidic solution; a pH value greater than 7 indicates a basic solution. (See Figure 14.4.)

Acidic solution	pH < 7	$[H_3O^+] > 1.0 \times 10^{-7}\,M$
Neutral solution	pH = 7	$[H_3O^+] = 1.0 \times 10^{-7}\,M$
Basic solution	pH > 7	$[H_3O^+] < 1.0 \times 10^{-7}\,M$

(a)

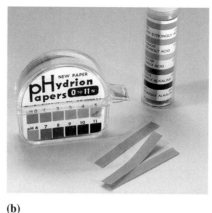

(b)

(c)

Figure 14.5 The pH of a solution can be determined using **(a)** a pH meter, **(b)** pH paper, and **(c)** indicators that turn different colors corresponding to different pH values.

Q *If a pH meter reads 4.00, is the solution acidic, basic, or neutral?*

CASE STUDY
Hyperventilation and Blood pH

WEB TUTORIAL
The pH Scale

In the laboratory, a pH meter is commonly used to determine the pH of a solution. There are also indicators and pH papers that turn specific colors when placed in solutions of different pH values. The pH is found by comparing the colors to a color chart. (See Figure 14.5.)

Calculating the pH of Solutions

The pH scale is a log scale that corresponds to the hydrogen-ion concentrations of aqueous solutions. Mathematically, **pH** is the negative logarithm (base 10) of the H_3O^+ concentration.

$$pH = -\log[H_3O^+]$$

Essentially, the negative powers of 10 in the concentrations are converted to positive numbers. For example, a lemon juice solution with $[H_3O^+] = 1.0 \times 10^{-2}$ M has a pH of 2.00. This can be calculated using the pH equation. For whole numbers in the $[H_3O^+]$, remember to add the correct number of significant zeros to the resulting pH obtained on your calculator.

$$pH = -\log[1.0 \times 10^{-2}]$$
$$pH = -(-2.00)$$
$$= 2.00$$

Let's look at how we determine the number of significant figures in the pH. For a logarithm, the number of decimal places in the pH value is equal to the number of significant figures in the $[H_3O^+]$. The number to the left of the decimal point is the power of 10.

$$[H_3O^+] = 1.0 \times 10^{-2} \qquad pH = 2.00$$

2 significant figures   2 decimal places

Steps for a pH Calculation

The pH of a solution is determined using the *log* key and *changing sign*. For example, to calculate the pH of a vinegar solution with $[H_3O^+] = 2.4 \times 10^{-3}$ M you can use the following steps:

Guide to Calculating pH of an Aqueous Solution

STEP 1
Enter the [H$_3$O$^+$] value.

STEP 2
Press the *log* key and *change sign*.

STEP 3
Adjust significant figures to the *right* of the decimal point to equal SFs in the coefficient.

STEP 1 **Enter the [H$_3$O$^+$] value.** Enter 2.4 and press EE or EXP .

Display Shows

2.4^{03} or   *2.4 03*

Enter 3 and press +/− to change the power to −3. (For calculators without a change sign key, consult the instructions for the calculator.)

2.4$^{−03}$ or   *2.4 −03*

STEP 2 **Press the** log **key.**

−2.619789

Change the sign by pressing the *change sign* key.

2.619789

The steps can be combined to give the calculator sequence as follows:

$$pH = -\log[2.4 \times 10^{-3}] = 2.4 \;\text{EE or EXP}\; 3 \;[+/-]\; [\log] \;[+/-]$$
$$= 2.619789$$

Be sure to check the instructions for your calculator. On some calculators, the log key is used first, followed by the concentration.

STEP 3 **Adjust significant figures.** In a pH value, the number to the *left* of the decimal point is an *exact* number derived from the power of 10. The number of digits to the *right* of the decimal point is equal to the number of significant figures in the coefficient.

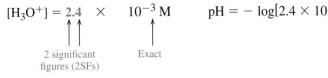

Coefficient   Power of 10

$[H_3O^+] = \mathbf{2.4} \times 10^{-3}$ M $pH = -\log[2.4 \times 10^{-3}] = \mathbf{2.62}$

2 significant figures (2SFs)   Exact   Exact   2 decimal places

Because pH is a log scale, a change of one pH unit corresponds to a tenfold change in [H$_3$O$^+$]. It is important to note that the pH decreases as the [H$_3$O$^+$] increases. For example, a solution with a pH of 2.00 has a [H$_3$O$^+$] 10 times higher than a solution with a pH of 3.00 and 100 times higher than a solution with a pH of 4.00.

Sample Problem 14.9 **Calculating pH**

Determine the pH for the following solutions:
a. [H$_3$O$^+$] = 1.0 × 10^{-5} M   **b.** [H$_3$O$^+$] = 5 × 10^{-8} M

Solution

a. **STEP 1** **Enter the [H$_3$O$^+$] using the *change sign* key.**

Display

1.0 EE or EXP 5 +/−   *1.0*$^{−05}$ or   *1.0 −05*

STEP 2 **Press the *log* key and *change sign* key.**

log +/−   *5*

STEP 3 **Adjust significant figures to the *right* of the decimal point to equal the number of significant figures in the coefficient.**

1.0 × 10^{-5} M   pH = **5.00**

2 SFs   2 SFs to the *right* of the decimal point

b. STEP 1 **Enter the concentration using the *change sign* key.**

5 $\boxed{\text{EE or EXP}}$ $\boxed{+/-}$ 5^{-08} or $5-08$

STEP 2 **Press the *log* key and then the *change sign* key.**

$\boxed{\text{log}}$ $\boxed{+/-}$ 7.301029

STEP 3 **Adjust significant figures to the *right* of the decimal point to equal the significant figures in the coefficient.**

$5 \times 10^{-8}\,M$ pH = 7.3

1 SF   1 SF to the *right* of the decimal point

Study Check

What is the pH of bleach with $[H_3O^+] = 4.2 \times 10^{-12}\,M$?

CASE STUDY
Hyperventilation and Blood pH

Sample Problem 14.10   **Calculating pH from [OH⁻]**

What is the pH of an ammonia solution with $[OH^-] = 3.7 \times 10^{-3}\,M$?

Solution

STEP 1 **Enter the $[H_3O^+]$ and press the *change sign* key.** Because $[OH^-]$ is given for the ammonia solution, we have to calculate $[H_3O^+]$ using the ion product of water, K_w. Dividing through by $[OH^-]$ gives $[H_3O^+]$.

$$\frac{K_w}{[OH^-]} = \frac{[H_3O^+]\,[\cancel{OH^-}]}{[\cancel{OH^-}]} = \frac{1.0 \times 10^{-14}}{[OH^-]}$$

$$[H_3O^+] = \frac{1.0 \times 10^{-14}}{[3.7 \times 10^{-3}\,M]} = 2.7 \times 10^{-12}\,M$$

$$pH = -\log[2.7 \times 10^{-12}] = 2.7 \boxed{\text{EE or EXP}} 12 \boxed{+/-} = \quad 2.7^{-12} \text{ OR } 2.7-12$$

STEP 2 **Press the *log* key, and then the *change sign* key.**

$\boxed{\text{log}}$ $\boxed{+/-}$ 11.56863

STEP 3 **Adjust significant figures to the *right* of the decimal point to equal the SFs in the coefficient.**

$2.7 \times 10^{-12}\,M$ pH = 11.57

2 SFs   2 SFs *after* the decimal point

Study Check

Calculate the pH of a sample of acid rain that has $[OH^-] = 2 \times 10^{-10}\,M$.

pOH

The **pOH** scale is similar to the pH scale except that pOH is associated with the $[OH^-]$ of an aqueous solution.

$$pOH = -\log[OH^-]$$

Solutions with high $[OH^-]$ have low pOH value; solutions with low $[OH^-]$ have high pOH values. In any aqueous solution, the sum of the pH and pOH is equal to 14.00, which is the negative logarithm of the K_w.

$$pH + pOH = 14.00$$

For example, if the pH of a solution is 3.50, the pOH can be calculated as follows:

$$pH + pOH = 14.00$$
$$pOH = 14.00 - pH = 14.00 - 3.50 = 10.50$$

Calculating $[H_3O^+]$ from pH

In another type of calculation, we may be given the pH of the solution and asked to determine the $[H_3O^+]$. This is a reverse of the pH calculation and may require the use of the 10^x key, which is usually a 2nd function key. On some calculators, this operation is done using the inverse key and the log key.

$$[H_3O^+] = 10^{-pH}$$

Sample Problem 14.11 | **Calculating $[H_3O^+]$ from pH**

Calculate $[H_3O^+]$ for each of the following solutions:
a. coffee, pH of 5.0 **b.** baking soda, pH of 8.25

Solution
a. coffee, pH of 5.0

> STEP 1 **Enter the pH value and press the *change sign* key.**
>
> **Display**
>
> 5.0 (+/-)   −5.0

> STEP 2 **Convert −pH to concentration.** Press the 2nd *function* key and then the *10^x* key.
>
> (2nd) (10ˣ)   0.00001 OR 1.⁻⁰⁵ OR 1−05
>
> Or press the *inverse* key and then the *log* key.
>
> (inv) (log)   0.00001 OR 1.⁻⁰⁵ OR 1−05
>
> Write the display in scientific notation with units of concentration.
>
> $1 \times 10^{-5}\,M$

> STEP 3 **Adjust the significant figures in the coefficient.** The pH value of 5.0 has only one digit to the *right* of the decimal point, which means the $[H_3O^+]$ is written with only one significant figure.
>
> $[H_3O^+] = 1 \times 10^{-5}\,M$

b. baking soda, pH of 8.25

> **STEP 1** **Enter the pH value and press the *change sign* key.**
>
> **Display**
>
> 8.25 $\boxed{+/-}$ -8.25

> **STEP 2** **Convert $-$pH to concentration.** Press the *2nd function* key and then the *10^x* key.
>
> $\boxed{2^{nd}}$ $\boxed{10^x}$ 5.62341^{-09} OR $5.62341-09$
>
> Or press the *inverse* key and then the *log* key.
>
> $\boxed{inv}$ $\boxed{log}$ 5.62341^{-09} OR $5.62341-09$
>
> Write the display in scientific notation with units of concentration.
>
> $$5.62341 \times 10^{-9}\,M$$

> **STEP 3** **Adjust the significant figures in the coefficient.** Because the pH value of 8.25 has two digits to the *right* of the decimal point, the $[H_3O^+]$ is written with two significant figures.
>
> $$[H_3O^+] = 5.6 \times 10^{-9}\,M$$

Study Check

What is the $[H_3O^+]$ and $[OH^-]$ of a beer that has a pH of 4.50?

A comparison of $[H_3O^+]$, $[OH^-]$, and their corresponding pH and pOH values is given in Table 14.8.

Table 14.8 **A Comparison of $[H_3O^+]$, $[OH^-]$, and Corresponding pH Values at 25°C**

$[H_3O^+]$	pH	$[OH^-]$	pOH	
10^0	0	10^{-14}	14	
10^{-1}	1	10^{-13}	13	
10^{-2}	2	10^{-12}	12	
10^{-3}	3	10^{-11}	11	Acidic
10^{-4}	4	10^{-10}	10	
10^{-5}	5	10^{-9}	9	
10^{-6}	6	10^{-8}	8	
10^{-7}	7	10^{-7}	7	Neutral
10^{-8}	8	10^{-6}	6	
10^{-9}	9	10^{-5}	5	
10^{-10}	10	10^{-4}	4	
10^{-11}	11	10^{-3}	3	Basic
10^{-12}	12	10^{-2}	2	
10^{-13}	13	10^{-1}	1	
10^{-14}	14	10^0	0	

> **Sample Problem 14.12** ▶ **Calculating pOH and pH**
>
> **a.** What is the pOH and pH of seawater if the [OH⁻] is 1.0×10^{-6} M?
> **b.** What is the pOH and pH of a sample of wine if [H₃O⁺] is 1.5×10^{-3} M?
>
> *Solution*
> **a.** $pOH = -\log[OH^-] = -\log[1.0 \times 10^{-6}] = 6.00$
> From the pK_w, we know that pH + pOH = 14.00. Solving for pH gives pH = 14.00 − pOH = 14.00 − 6.00 = 8.00.
> **b.** $pH = -\log[H_3O^+] = -\log[1.5 \times 10^{-3}] = 2.82$
> From the pK_w, we know that pH + pOH = 14.00. Solving for pOH gives pOH = 14.00 − pH = 14.00 − 2.82 = 11.18.
>
> *Study Check*
>
> What is the pOH and pH of a solution of milk of magnesia with [OH⁻] of 5.0×10^{-4} M?

CHEM NOTE
ACID RAIN

Rain typically has a pH of 6.2. It is slightly acidic because carbon dioxide in the air combines with water to form carbonic acid. However, in many parts of the world, rain has become considerably more acidic, with pH values as low as 3 being reported. One cause of acid rain is the sulfur dioxide (SO_2) gas produced when coal that contains sulfur is burned.

In the air, the SO_2 gas reacts with oxygen to produce SO_3, which then combines with water to form sulfuric acid, H_2SO_4, a strong acid.

$$S + O_2 \longrightarrow SO_2$$
$$2SO_2 + O_2 \longrightarrow 2SO_3$$
$$SO_3 + H_2O \longrightarrow H_2SO_4$$

In parts of the United States, acid rain has made lakes so acidic they are no longer able to support fish and plant life. Limestone ($CaCO_3$) is sometimes added to these lakes to neutralize the acid. In Eastern Europe, acid rain has brought about an environmental disaster. Nearly 40% of the forests in Poland have been severely damaged, and some parts of the land are so acidic that crops will not grow. Throughout Europe and the United States, monuments made of marble (a form of $CaCO_3$) are deteriorating as acid rain dissolves the marble.

$$2H^+ + CaCO_3 \longrightarrow Ca^{2+} + H_2O + CO_2$$

1935   1994
Marble statue in Washington Square Park

Efforts to slow or stop the damaging effects of acid rain include the reduction of sulfur emissions. This will require installation of expensive equipment in coal-burning plants to absorb more of the SO_2 gases before they are emitted. In some outdated plants, this may be impossible, and they will need to be closed. It is a difficult problem for engineers and scientists, but one that must be solved.

Questions and Problems ▸ The pH Scale

14.41 Why does a neutral solution have a pH of 7.00?

14.42 If you know the $[OH^-]$, how can you determine the pH of a solution?

14.43 State whether each of the following solutions is acidic, basic, or neutral:
 a. blood, pH 7.38
 b. vinegar, pH 2.8
 c. drain cleaner, pOH 2.8
 d. coffee, pH 5.52

14.44 State whether each of the following solutions is acidic, basic, or neutral:
 a. soda, pH 3.22
 b. shampoo, pOH 8.3
 c. laundry detergent, pOH 4.56
 d. rain, pH 5.8

14.45 A solution with a pH of 3 is 10 times more acidic than a solution with pH 4. Explain.

14.46 A solution with a pH of 10 is 100 times more basic than a solution with pH 8. Explain.

14.47 Calculate the pH of each solution given the following $[H_3O^+]$ or $[OH^-]$ values.
 a. $[H_3O^+] = 1.0 \times 10^{-4}$ M
 b. $[H_3O^+] = 3.0 \times 10^{-9}$ M
 c. $[OH^-] = 1.0 \times 10^{-5}$ M
 d. $[OH^-] = 2.5 \times 10^{-11}$ M

14.48 Calculate the pH of each solution given the following $[H_3O^+]$ or $[OH^-]$ values.
 a. $[H_3O^+] = 1.0 \times 10^{-8}$ M
 b. $[H_3O^+] = 5.0 \times 10^{-6}$ M
 c. $[OH^-] = 4.0 \times 10^{-2}$ M
 d. $[OH^-] = 8.0 \times 10^{-3}$ M

14.49 Complete the following table:

$[H_3O^+]$	$[OH^-]$	pH	pOH	Acidic, Basic, or Neutral?
	1.0×10^{-6} M			
		3.00		
2.8×10^{-5} M				
			2.00	

14.50 Complete the following table:

$[H_3O^+]$	$[OH^-]$	pH	pOH	Acidic, Basic, or Neutral?
		10.00		
				Neutral
			5.00	
6.4×10^{-12} M				

14.7 Reactions of Acids and Bases

Learning Goal

Write balanced equations for reactions of acids and bases.

Typical reactions of acids and bases include the reactions of acids with metals, bases, and carbonate or bicarbonate ions. For example, when you drop an antacid tablet in water, the bicarbonate ion and citric acid in the tablet react to produce carbon dioxide bubbles, a salt, and water. A **salt** is an ionic compound that does not have H^+ as the cation or OH^- as the anion. $NaCl$, CaF_2, and NH_4NO_3 are examples of some salts.

Acids and Metals

Acids react with certain metals known as *active metals* to produce hydrogen gas (H_2) and a salt of that metal. Active metals include potassium, sodium, calcium, magnesium, aluminum, zinc, iron, and tin. In these single replacement reactions, the metal loses electrons and the metal ion replaces the hydrogen in the acid.

$$Mg(s) + 2HCl(aq) \longrightarrow MgCl_2(aq) + H_2(g)$$
Metal   Acid   Salt   Hydrogen

$$Zn(s) + 2HCl(aq) \longrightarrow ZnCl_2(aq) + H_2(g)$$
Metal   Acid   Salt   Hydrogen

Acids and Carbonates and Bicarbonates

When strong acids are added to a carbonate or bicarbonate, the reaction produces bubbles of carbon dioxide gas, a salt, and water. In the reaction, H^+ is transferred to the carbonate to give carbonic acid, H_2CO_3, which breaks down rapidly to CO_2 and H_2O. The net ionic equation is written by omitting the metal ions and chloride ions that are not reacting.

$$HCl(aq) + NaHCO_3(aq) \longrightarrow CO_2(g) + H_2O(l) + NaCl(aq)$$

$$2HCl(aq) + Na_2CO_3(aq) \longrightarrow CO_2(g) + H_2O(l) + 2NaCl(aq)$$

Acids and Hydroxides: Neutralization

Neutralization is a reaction between an acid and a base to produce a salt and water. The cation in the salt comes from the base and the anion comes from the acid. In the neutralization reactions of strong acids and strong bases, water is always one of the products. We can write the following equation for the neutralization of HCl and NaOH:

$$HCl(aq) + NaOH(aq) \longrightarrow NaCl(aq) + H_2O(l)$$
Acid   Base   Salt   Water

To see the actual reaction, we write the strong acid and strong base as individual ions.

$$\mathbf{H^+}(aq) + Cl^-(aq) + Na^+(aq) + \mathbf{OH^-}(aq) \longrightarrow$$
$$Na^+(aq) + Cl^-(aq) + \mathbf{H_2O}(l) \quad \text{Ionic equation}$$

In the neutralization reaction, H^+ from the acid reacts with OH^- from the base to form water, leaving the spectator ions from the salt (Na^+ and Cl^-) in solution. When we omit the spectator ions from the ionic equation, we obtain the *net ionic* equation, which allows us to see that the net reaction for neutralization is the reaction of H^+ and OH^-.

$$\mathbf{H^+}(aq) + \cancel{Cl^-(aq)} + \cancel{Na^+(aq)} + \mathbf{OH^-}(aq) \longrightarrow$$
$$\cancel{Na^+(aq)} + \cancel{Cl^-(aq)} + \mathbf{H_2O}(l) \quad \text{Ionic equation}$$
$$H^+(aq) + OH^-(aq) \longrightarrow H_2O(l) \quad \text{Net ionic equation}$$

Balancing Neutralization Equations

In a neutralization reaction, one H^+ always reacts with one OH^-. Therefore, the coefficients in the neutralization equation must be chosen so that the H^+ from the acid is balanced by the OH^- provided by the base. We balance the neutralization of HCl and $Ba(OH)_2$ as follows:

Guide to Balancing an Equation for Neutralization
STEP 1 Write the reactants and products.
STEP 2 Balance the H^+ in the acid with the OH^- in the base.
STEP 3 Balance the H_2O with the H^+ and the OH^-.
STEP 4 Write the salt from the remaining ions.

STEP 1 **Write the reactants and products.**

$$HCl(aq) + Ba(OH)_2(s) \longrightarrow H_2O(l) + \text{salt}$$

STEP 2 **Balance the H^+ in the acid with the OH^- in the base.** Placing a 2 in front of the HCl provides two H^+ for the two OH^- in $Ba(OH)_2$.

$$\mathbf{2}HCl(aq) + Ba(OH)_2(s) \longrightarrow H_2O(l) + \text{salt}$$

STEP 3 **Balance the H_2O with the H^+ and OH^-.**

$$2HCl(aq) + Ba(OH)_2(s) \longrightarrow 2H_2O(l) + \text{Salt}$$

STEP 4 **Write the salt from the remaining ions in the acid and base.**

$$\mathbf{2HCl}(aq) + \mathbf{Ba(OH)_2}(s) \longrightarrow 2H_2O(l) + \mathbf{BaCl_2}(aq)$$

Sample Problem 14.13   **Reactions of Acids**

Write a balanced equation for the reaction of $HCl(aq)$ with each of the following:

a. $Al(s)$ **b.** $K_2CO_3(ag)$ **c.** $Mg(OH)_2(s)$

Solution

a. Al

STEP 1 **Write the reactants and products.** When a metal reacts with an acid, the products are H_2 gas and a salt.

$$Al(s) + HCl(aq) \longrightarrow H_2(g) + \text{salt}$$

STEP 2 **Determine the formula of the salt.** When $Al(s)$ dissolves, it forms Al^{3+}, which is balanced by three Cl^- from HCl.

$$Al(s) + HCl(aq) \longrightarrow H_2(g) + AlCl_3(aq)$$

STEP 3 **Balance the equation.**

$$2Al(s) + 6HCl(aq) \longrightarrow 3H_2(g) + 2AlCl_3(aq)$$

b. K_2CO_3

STEP 1 **Write the reactants and products.** When a carbonate reacts with an acid, the products are $CO_2(g)$, $H_2O(l)$, and a salt.

$$K_2CO_3(s) + HCl(aq) \longrightarrow CO_2(g) + H_2O(l) + \text{salt}$$

STEP 2 **Determine the formula of the salt.** When $K_2CO_3(s)$ dissolves, it forms K^+, which is balanced by one Cl^- from HCl.

$$\mathbf{K_2CO_3}(s) + HCl(aq) \longrightarrow$$
$$CO_2(g) + H_2O(l) + \mathbf{KCl}(aq)$$

STEP 3 **Balance the equation.**

$$K_2CO_3(aq) + \mathbf{2}HCl(aq) \longrightarrow$$
$$CO_2(g) + H_2O(l) + \mathbf{2}KCl(aq)$$

c. $Mg(OH)_2$

STEP 1 **Write the reactants and products.** When a base reacts with an acid, the products are $H_2O(l)$ and a salt.

$$Mg(OH)_2(s) + HCl(aq) \longrightarrow H_2O(l) + \text{salt}$$

STEP 2 **Balance the H^+ in the acid with the OH^- in the base.** Placing a 2 in front of the HCl provides two H^+ for two OH^- in $Mg(OH)_2$.

$$Mg(OH)_2(s) + 2HCl(aq) \longrightarrow H_2O(l) + \text{salt}$$

ANTACIDS

Antacids are substances used to neutralize excess stomach acid (HCl). Some antacids are mixtures of aluminum hydroxide and magnesium hydroxide. These hydroxides are not very soluble in water, so the levels of available OH^- are not damaging to the intestinal tract. However, aluminum hydroxide has the side effects of producing constipation and binding phosphate in the intestinal tract, which may cause weakness and loss of appetite. Magnesium hydroxide has a laxative effect. These side effects are less likely when a combination of the antacids is used.

$$Al(OH)_3(aq) + 3HCl(aq) \longrightarrow AlCl_3(aq) + 3H_2O(l)$$
$$Mg(OH)_2(s) + 2HCl(aq) \longrightarrow MgCl_2(aq) + 2H_2O(l)$$

Some antacids use calcium carbonate to neutralize excess stomach acid. About 10% of the calcium is absorbed into the bloodstream, where it elevates the levels of serum calcium. Calcium carbonate is not recommended for patients who have peptic ulcers or a tendency to form kidney stones.

$$CaCO_3(s) + 2HCl(aq) \longrightarrow$$
$$H_2O(l) + CO_2(g) + CaCl_2(aq)$$

Still other antacids contain sodium bicarbonate. This type of antacid has a tendency to increase blood pH and elevate sodium levels in the body fluids. It also is not recommended in the treatment of peptic ulcers.

$$NaHCO_3(s) + HCl(aq) \longrightarrow$$
$$NaCl(aq) + CO_2(g) + H_2O(l)$$

The neutralizing substances in some antacid preparations are given in Table 14.9.

Table 14.9   Basic Compounds in Some Antacids

Antacid	Base(s)
Amphojel	$Al(OH)_3$
Milk of magnesia	$Mg(OH)_2$
Mylanta, Maalox, Di-Gel, Gelusil, Riopan	$Mg(OH)_2$, $Al(OH)_3$
Bisodol	$CaCO_3$, $Mg(OH)_2$
Titralac, Tums, Pepto-Bismol	$CaCO_3$
Alka-Seltzer	$NaHCO_3$, $KHCO_3$

STEP 3 **Balance the H_2O with the H^+ and OH^-.**

$$Mg(OH)_2(s) + 2HCl(aq) \longrightarrow 2H_2O(l) + salt$$

STEP 4 **Write the salt from the remaining ions in the acid and base.**

$$Mg(OH)_2(s) + 2HCl(aq) \longrightarrow 2H_2O(l) + MgCl_2(aq)$$

Study Check

Write the balanced equation for the reaction between H_2SO_4 and $NaHCO_3$.

Questions and Problems   **Reactions of Acids and Bases**

14.51 Complete and balance the equations for the following reactions:
a. $ZnCO_3(s) + HBr(aq) \longrightarrow$
b. $Zn(s) + HCl(aq) \longrightarrow$
c. $HCl(aq) + NaHCO_3(s) \longrightarrow$
d. $H_2SO_4(aq) + Mg(OH)_2(s) \longrightarrow$

14.52 Complete and balance the equations for the following reactions:
a. $KHCO_3(s) + HCl(aq) \longrightarrow$
b. $Ca(s) + H_2SO_4(aq) \longrightarrow$
c. $H_2SO_4(aq) + Al(OH)_3(s) \longrightarrow$
d. $Na_2CO_3(s) + H_2SO_4(aq) \longrightarrow$

14.53 Balance each of the following neutralization reactions:
a. $HCl(aq) + Mg(OH)_2(s) \longrightarrow$
$$MgCl_2(aq) + H_2O(l)$$
b. $H_3PO_4(aq) + LiOH(aq) \longrightarrow$
$$Li_3PO_4(aq) + H_2O(l)$$

14.54 Balance each of the following neutralization reactions:
a. $HNO_3(aq) + Ba(OH)_2(s) \longrightarrow$
$$Ba(NO_3)_2(aq) + H_2O(l)$$
b. $H_2SO_4(aq) + Al(OH)_3(s) \longrightarrow$
$$Al_2(SO_4)_3(aq) + H_2O(l)$$

14.55 Write a balanced equation for the neutralization of each of the following:
a. $H_2SO_4(aq)$ and $NaOH(aq)$
b. $HCl(aq)$ and $Fe(OH)_3(s)$
c. $H_2CO_3(aq)$ and $Mg(OH)_2(s)$

14.56 Write a balanced equation for the neutralization of each of the following:
a. $H_3PO_4(aq)$ and $NaOH(aq)$
b. $HI(aq)$ and $LiOH(aq)$
c. $HNO_3(aq)$ and $Ca(OH)_2(s)$

14.8 Acid–Base Titration

Suppose we need to find the molarity of an HCl solution of unknown concentration. We can do this by a laboratory procedure called **titration** in which we neutralize an acid sample with a known amount of base. In our titration, we first place a measured volume of the acid in a flask and add a few drops of an **indicator,** such as phenolphthalein. In an acidic solution, phenolpthalein is colorless. Then we fill a buret with a NaOH solution of known molarity and carefully add NaOH to the acid in the flask, as shown in Figure 14.6.

In the titration, we neutralize the acid by adding a volume of base that contains a matching number of moles of OH^-. We know that neutralization has taken place when the phenolphthalein in the solution changes from colorless to pink. This is called the neutralization **endpoint.** From the volume added and molarity of the NaOH, we can calculate the number of moles of NaOH and then the concentration of the acid.

**Guide to Calculations
for an Acid–Base Titration**

STEP 1
State the given and needed
quantities and concentrations.

STEP 2
Write a plan to calculate
molarity or volume.

STEP 3
State equalities and conversion
factors including concentration.

STEP 4
Set up problem to
calculate needed quantity.

Sample Problem 14.14 **Titration of an Acid**

A 25.0-mL sample of an HCl solution is placed in a flask with a few drops of phenolphthalein (indicator). If 32.6 mL of a 0.185 M NaOH is needed to reach the endpoint, what is the concentration (M) of the HCl solution?

$$NaOH(aq) + HCl(aq) \longrightarrow NaCl(aq) + H_2O(l)$$

Solution

STEP 1 **Given:**   32.6 mL of 0.185 M NaOH;
                         25.0 mL HCl = 0.0250 L HCl

Need: Molarity of HCl

STEP 2 **Plan**

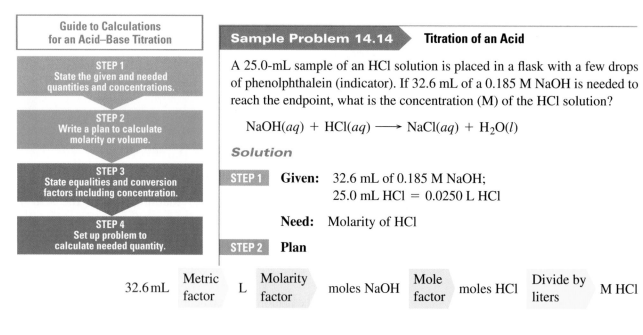

32.6 mL → Metric factor → L → Molarity factor → moles NaOH → Mole factor → moles HCl → Divide by liters → M HCl

---

Figure 14.6 The titration of an acid. A known volume of an acid is placed in a flask with an indicator and titrated with a measured volume of a base, such as NaOH, to the neutralization point.

Q What data is needed to determine the molarity of the acid in the flask?

STEP 3 **Equalities/Conversion Factors**

$$1 \text{ L NaOH} = 1000 \text{ mL NaOH}$$
$$\frac{1 \text{ L}}{1000 \text{ mL}} = \frac{1000 \text{ mL}}{1 \text{ L}}$$

$$1 \text{ L NaOH} = 0.185 \text{ mol NaOH}$$
$$\frac{1 \text{ L}}{0.185 \text{ mol NaOH}} = \frac{0.185 \text{ mol NaOH}}{1 \text{ L}}$$

$$1 \text{ mol HCl} = 1 \text{ mol NaOH}$$
$$\frac{1 \text{ mol HCl}}{1 \text{ mol NaOH}} = \frac{1 \text{ mol NaOH}}{1 \text{ mol HCl}}$$

STEP 4 **Set up problem.**

$$32.6 \text{ mL NaOH} \times \frac{1 \text{ L NaOH}}{1000 \text{ mL NaOH}} \times \frac{0.185 \text{ mol NaOH}}{1 \text{ L NaOH}} \times$$

$$\frac{1 \text{ mol HCl}}{1 \text{ mol NaOH}} = 0.00603 \text{ mol HCl}$$

$$\text{Molarity HCl} = \frac{0.00603 \text{ mol HCl}}{0.0250 \text{ L HCl}} = 0.241 \text{ M}$$

Study Check

What is the molarity of an HCl solution, if 28.6 mL of a 0.175 M NaOH solution is needed to neutralize a 25.0-mL sample of the HCl solution?

Sample Problem 14.15 **Volume of Base for Titration**

What volume in milliliters of 0.115 M NaOH would neutralize 25.0 mL of a 0.106 M H_2SO_4 solution?

$$2NaOH(aq) + H_2SO_4(aq) \longrightarrow Na_2SO_4(aq) + 2H_2O(l)$$

Solution

STEP 1 **Given:** 25.0 mL of 0.106 M H_2SO_4; 0.115 M NaOH

Need: volume (mL) of NaOH

STEP 2 **Plan**

mL → Metric factor → L → Molarity factor → Moles of H_2SO_4 → Mole-mole factor → Moles of NaOH → Molarity factor → vol (mL)

STEP 3 **Equalities/Conversion Factors**

$$1 \text{ L } H_2SO_4 = 1000 \text{ mL } H_2SO_4$$
$$\frac{1 \text{ L } H_2SO_4}{1000 \text{ mL } H_2SO_4} = \frac{1000 \text{ mL } H_2SO_4}{1 \text{ L } H_2SO_4}$$

$$1 \text{ L } H_2SO_4 = 0.106 \text{ mol } H_2SO_4$$
$$\frac{1 \text{ L } H_2SO_4}{0.106 \text{ mol } H_2SO_4} = \frac{0.106 \text{ mol } H_2SO_4}{1 \text{ L } H_2SO_4}$$

$$2 \text{ mol NaOH} = 1 \text{ mol } H_2SO_4$$
$$\frac{1 \text{ mol } H_2SO_4}{2 \text{ mol NaOH}} = \frac{2 \text{ mol NaOH}}{1 \text{ mol } H_2SO_4}$$

$$1000 \text{ mL NaOH} = 0.115 \text{ mol NaOH}$$
$$\frac{1000 \text{ mL NaOH}}{0.115 \text{ mol NaOH}} = \frac{0.115 \text{ mol NaOH}}{1000 \text{ mL NaOH}}$$

STEP 4   **Set up problem.**

$$25.0 \text{ mL } H_2SO_4 \times \frac{1 \text{ L } H_2SO_4}{1000 \text{ mL } H_2SO_4} \times \frac{0.106 \text{ mol } H_2SO_4}{1 \text{ L } H_2SO_4} \times \frac{2 \text{ mol } NaOH}{1 \text{ mol } H_2SO_4} \times \frac{1000 \text{ mL } NaOH}{0.115 \text{ mol } NaOH}$$

$$= 46.1 \text{ mL } NaOH$$

Study Check

What volume in milliliters of 0.158 M KOH is needed to neutralize 50.0 mL of a 0.212 M HCl solution?

Questions and Problems   **Acid–Base Titration**

14.57 If you need to determine the molarity of a formic acid solution, HCOOH, how would you proceed?

$$HCOOH + H_2O \xrightleftharpoons{} H_3O^+ + HCOO^-$$

14.58 If you need to determine the molarity of an acetic acid solution, CH_3COOH, how would you proceed?

$$CH_3COOH + H_2O \xrightleftharpoons{} H_3O^+ + CH_3COO^-$$

14.59 What is the molarity of an HCl solution if 5.00 mL HCl solution is titrated with 28.6 mL of 0.145 M NaOH solution?

$$HCl(aq) + NaOH(aq) \longrightarrow NaCl(aq) + H_2O(l)$$

14.60 If 29.7 mL of 0.205 M KOH is required to completely neutralize 25.0 mL of a CH_3COOH solution, what is the molarity of the acetic acid solution?

$$CH_3COOH(aq) + KOH(aq) \longrightarrow$$
$$CH_3COOK(aq) + H_2O(l)$$

14.61 If 38.2 mL of 0.163 M KOH is required to neutralize completely 25.0 mL H_2SO_4 solution, what is the molarity of the acid solution?

$$H_2SO_4(aq) + 2KOH(aq) \longrightarrow$$
$$K_2SO_4(aq) + 2H_2O(l)$$

14.62 A solution of 0.162 M NaOH is used to neutralize 25.0 mL H_2SO_4 solution. If 32.8 mL NaOH solution is required to reach the endpoint, what is the molarity of the H_2SO_4 solution?

$$H_2SO_4(aq) + 2NaOH(aq) \longrightarrow$$
$$Na_2SO_4(aq) + 2H_2O(l)$$

14.63 A solution of 0.204 M NaOH is used to neutralize 50.0 mL H_3PO_4 solution. If 16.4 mL of NaOH solution is required to reach the endpoint, what is the molarity of the H_3PO_4 solution?

$$H_3PO_4(aq) + 3NaOH(aq) \longrightarrow$$
$$Na_3PO_4(aq) + 3H_2O(l)$$

14.64 A solution of 0.312 M KOH is used to neutralize 15.0 mL H_3PO_4 solution. If 28.3 mL KOH solution is required to reach the endpoint, what is the molarity of the H_3PO_4 solution?

$$H_3PO_4(aq) + 3KOH(aq) \longrightarrow$$
$$K_3PO_4(aq) + 3H_2O(l)$$

Learning Goal

Predict whether a salt will form an acidic, basic, or neutral solution.

14.9 Acid–Base Properties of Salt Solutions

When a salt dissolves in water, it dissociates into cations and anions. Solutions of salts can be acidic, basic, or neutral. Anions and cations from strong acids and bases do not affect pH; however, anions from weak acids and cations from weak bases change the pH of an aqueous solution.

Salts That Form Neutral Solutions

A solution of a salt containing the cation from a strong base and the anion from a strong acid will be neutral. For example, a salt such as $NaNO_3$ forms a neutral solution.

$$NaNO_3(s) \xrightarrow{H_2O} Na^+(aq) + NO_3^-(aq)$$

Does not change H^+ Does not attract H^+ from water   Neutral solution (pH = 7.0)

The cation, Na^+, from a strong base does not change H^+, and the anion, NO_3^-, from a strong acid does not attract H^+ from water. Thus there is no effect on the pH of water; the solution is neutral with a pH of 7.0. Other salts such as NaCl, KCl, $NaNO_3$, KNO_3, and KBr also contain cations from strong acids and anions from strong bases and also form neutral solutions.

Some Components of Neutral Salt Solutions

Cations of strong bases: Group 1A (1): Li^+, Na^+, K^+

Group 2A (2): Ca^{2+}, Mg^{2+}, Sr^{2+}, Ba^{2+}

Anions of strong acids: Cl^-, Br^-, I^-, NO_3^-, ClO_4^-

Salts That Form Basic Solutions

A salt solution containing the cation from a strong base and the anion from a weak acid produces a basic solution. Suppose we have a solution of the salt NaF, which contains Na^+ and F^- ions.

$$NaF(s) \xrightarrow{H_2O} Na^+(aq) + F^-(aq)$$

Does not change H^+ Attracts H^+ from water

The metal ion Na^+ has no effect on the pH of the solution. However, F^- is the conjugate base of the weak acid HF. Thus, F^- will attract a proton from water and leave OH^- in solution, which makes it basic.

$$F^-(aq) + H_2O(l) \rightleftharpoons HF(aq) + OH^-(aq)$$ Basic solution (pH > 7.0)

Other salts with anions from weak acids such as NaCN, KNO_2, and Na_2SO_4 also produce basic solutions.

Some Components of Basic Salt Solutions

Cations of strong bases: Group 1A (1): Li^+, Na^+, K^+

Group 2A (2): Ca^{2+}, Mg^{2+}, Sr^{2+}, Ba^{2+}

Anions of weak acids: F^-, NO_2^-, CN^-, CO_3^{2-}, SO_4^{2-}, CH_3COO^-, S^{2-}, PO_4^{3-}

Salts That Form Acidic Solutions

A salt solution containing a cation from a weak base and an anion from a strong acid produces an acidic solution. Suppose we have a solution of the salt NH_4Cl, which contains NH_4^+ and Cl^- ions.

$$NH_4Cl(s) \xrightarrow{H_2O} NH_4^+(aq) + Cl^-(aq)$$

Donates H^+ to water   Does not attract H^+ from water

The anion Cl^- has no effect on the pH of the solution. However, as a weak acid, the cation NH_4^+ donates a proton to water, which produces H_3O^+.

$$NH_4^+(aq) + H_2O(l) \rightleftharpoons NH_3(aq) + H_3O^+(aq)$$ Acidic solution (pH > 7.0)

Table 14.10 Cations and Anions of Salts in Neutral, Basic, and Acidic Salt Solutions

Type of Solution	Cations	Anions	pH
Neutral	From strong bases Group 1A (1): Li^+, Na^+, K^+ Group 2A (2): Ca^{2+}, Mg^{2+}, Sr^{2+}, Ba^{2+} (but not Be^{2+})	From strong acids Cl^-, Br^-, I^-, NO_3^- ClO_4^-	7.0
Basic	From strong bases Group 1A (1): Li^+, Na^+, K^+ Group 2A (2): Ca^{2+}, Mg^{2+}, Sr^{2+}, Ba^{2+} (but not Be^{2+})	From weak acids F^-, NO_2^-, CN^-, CO_3^{2-}, SO_4^{2-}, CH_3COO^-, S^{2-}, PO_4^{3-}	>7.0
Acidic	From weak bases NH_4^+, Be^{2+}, Al^{3+}, Zn^{2+}, Cr^{3+}, Fe^{3+} (small, highly charged metal ions)	From strong acids Cl^-, Br^-, I^-, NO_3^-, ClO_4^-	<7.0

Some Components of Acidic Salt Solutions

Cations of weak bases: NH_4^+ and Be^{2+}, Al^{3+}, Zn^{2+}, Cr^{3+}, Fe^{3+}
(small, highly charged metal ions)

Anions of strong acids: Cl^-, Br^-, I^-, NO_3^-, ClO_4^-

Table 14.10 summarizes the cations and anions of salts that form neutral, basic, and acidic solutions. Table 14.11 summarizes the acid–base properties of some typical salts in water.

Sometimes a salt contains the cation of a weak base and the anion of a weak acid. For example, when NH_4F dissociates in water, it produces NH_4^+ and F^-. We have seen that NH_4^+ forms an acidic solution and F^- forms a basic solution. The ion that reacts to a greater extent with water determines whether the solution is acidic or basic. The salt solution will be neutral only if the ions react with water to the same extent. The determination of these reactions is complex and will not be considered in this text.

Sample Problem 14.16   **Predicting the Acid–Base Properties of Salt Solutions**

Predict whether solutions of each of the following salts would be acidic, basic, or neutral:
a. KCN
b. NH_4Br
c. $NaNO_3$

Table 14.11 Acid–Base Properties of Some Salt Solutions

Typical Salts	Types of Ions	pH	Solution
NaCl, $MgBr_2$, KNO_3	Cation from a strong base Anion from a strong acid	7.0	Neutral
NaF, $MgCO_3$, KNO_2	Cation from a strong base Anion from a weak acid	>7.0	Basic
NH_4Cl, $FeBr_3$, $Al(NO_3)_3$	Cation from a weak base Anion from a strong acid	<7.0	Acidic

Solution

a. KCN

There are only six strong acids; all other acids are weak. Bases with cations from Groups 1A (1) and 2A (2) are strong; all other bases are weak. For the salt, KCN, the cation is from a strong base (KOH), but the anion is from a weak acid (HCN).

$$KCN(s) \xrightarrow{H_2O} K^+(aq) + CN^-(aq)$$

The cation K^+ has no effect on pH. However, the anion CN^- will attract protons from water to produce a basic solution.

$$CN^-(aq) + H_2O(l) \rightleftharpoons HCN(aq) + OH^-(aq)$$

b. NH₄Br

In the salt NH_4Br, the cation is from a weak base, but the anion is from a strong acid.

$$NH_4Br(s) \xrightarrow{H_2O} NH_4^+(aq) + Br^-(aq)$$

The anion Br^- has no effect on pH because it is from HBr, a strong acid. However, the cation NH_4^+ will donate protons to water to produce an acidic solution.

$$NH_4^+(aq) + H_2O(l) \rightleftharpoons NH_3(aq) + H_3O^+(aq)$$

c. NaNO₃

The salt $NaNO_3$ contains a cation from a strong base (NaOH) and an anion from a strong acid (HNO_3). Thus, there is no change of pH; the salt solution is neutral.

$$NaNO_3(s) \xrightarrow{H_2O} Na^+(aq) + NO_3^-(aq)$$

Study Check

Would a solution of Na_3PO_4 be acidic, basic, or neutral?

Questions and Problems　　　**Acid–Base Properties of Salt Solutions**

14.65 Why does a salt containing a cation from a strong base and an anion from a weak acid form a basic solution?

14.66 Why does a salt containing a cation from a weak base and an anion from a strong acid form an acidic solution?

14.67 Predict whether each of the following salts will form an acidic, basic, or neutral solution. For acidic and basic solutions, write an equation for the reaction that takes place.
   a. $MgCl_2$ 　　　　**b.** NH_4NO_3
   c. Na_2CO_3 　　　**d.** K_2S

14.68 Predict whether each of the following salts will form an acidic, basic, or neutral solution. For acidic and basic solutions, write an equation for the reaction that takes place.
   a. Na_2SO_4 　　　**b.** KBr
   c. $BaCl_2$ 　　　　**d.** NH_4I

14.10 Buffers

When a small amount of acid or base is added to pure water, the pH changes drastically. However, if a solution is buffered, there is little change in pH. A **buffer solution** resists a change in pH when small amounts of acid or base are added. For example, blood is buffered to maintain a pH of about 7.4. If the pH of the blood goes even slightly above or below this value, changes in our uptake of oxygen and our metabolic process can be drastic enough to cause death. Even though we are constantly obtaining acids and bases from foods and biological processes, the buffers in the body so effectively absorb those compounds that blood pH remains essentially unchanged. (See Figure 14.7.)

In a buffer, an acid must be present to react with any OH^- that is added, and a base must be available to react with any added H_3O^+. However, that acid and base must not be able to neutralize each other. Therefore, a combination of an acid–base conjugate pair is used in buffers. Most buffer solutions consist of nearly equal concentrations of a weak acid and a salt containing its conjugate base. (See Figure 14.8.) Buffers may also contain a weak base and the salt of the weak base, which contains its conjugate acid.

For example, a typical buffer contains acetic acid (CH_3COOH) and a salt such as sodium acetate (CH_3COONa). As a weak acid, acetic acid dissociates slightly in water to form H_3O^+ and a very small amount of CH_3COO^-. The presence of the salt provides a much larger concentration of acetate ion (CH_3COO^-), which is necessary for its buffering capability.

$$CH_3COOH(aq) + H_2O(l) \rightleftharpoons H_3O^+(aq) + CH_3COO^-(aq)$$
Large amount   Large amount

Figure 14.7 Adding an acid or a base to water changes the pH drastically, but a buffer resists pH change when small amounts of acid or base are added.

Q Why does the pH change several pH units when acid is added to water, but not when acid is added to a buffer?

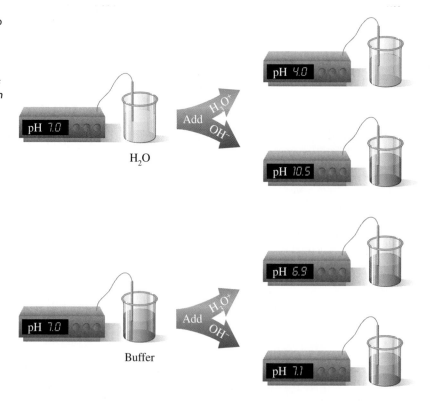

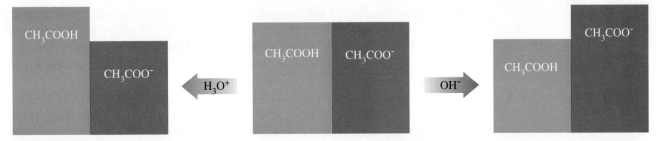

Figure 14.8 The buffer described here consists of about equal concentrations of acetic acid (CH_3COOH) and its conjugate base of acetate ion (CH_3COO^-). Adding H_3O^+ to the buffer uses up some CH_3COO^-, whereas adding OH^- neutralizes some CH_3COOH. The pH of the solution is maintained as long as the added amounts of acid or base are small compared to the concentrations of the buffer components.

❓ *How does this acetic acid/acetate ion buffer maintain pH?*

Let's see how this buffer solution maintains the H_3O^+ concentration. When a small amount of acid is added, it will combine with the acetate ion (anion) as the equilibrium shifts to the reactant acetic acid. There will be a small decrease in the $[CH_3COO^-]$ and a small increase in $[CH_3COOH]$, but the $[H_3O^+]$ will not change very much.

$$CH_3COOH(aq) + H_2O(l) \longleftarrow H_3O^+(aq) + CH_3COO^-(aq)$$

If a small amount of base is added to this buffer solution, it is neutralized by the acetic acid and water is produced. The concentration of $[CH_3COOH]$ decreases slightly and the $[CH_3COO^-]$ increases slightly, but again the $[H_3O^+]$ does not change very much.

$$CH_3COOH(aq) + OH^-(aq) \longrightarrow H_2O(aq) + CH_3COO^-(aq)$$

Sample Problem 14.17 ▷ **Identifying Buffer Solutions**

Indicate whether each of the following would make a buffer solution:
a. HCl (a strong acid) and NaCl
b. H_3PO_4 (a weak acid)
c. HF (a weak acid) and NaF

Solution
a. No. A solution of a strong acid and its salt is completely ionized.
b. No. A weak acid is not sufficient for a buffer; the salt of the weak acid is also needed.
c. Yes. This mixture contains a weak acid and its salt.

Study Check

Will a mixture of NaCl and Na_2CO_3 make a buffer solution? Explain.

Calculating the pH of a Buffer

By rearranging the K_a expression to give $[H_3O^+]$ we can obtain the ratio of the acetic acid/acetate buffer.

$$K_a = \frac{[H_3O^+][CH_3COO^-]}{[CH_3COOH]}$$

$$[H_3O^+] = K_a \times \frac{[CH_3COOH]}{[CH_3COO^-]}$$

Because K_a is a constant, the $[H_3O^+]$ is determined by the $[CH_3COOH]/[CH_3COO^-]$ ratio. As long as the addition of small amounts of either acid or base changes the ratio of $[CH_3COOH]/[CH_3COO^-]$ only slightly, the changes in $[H_3O^+]$ will be small and the pH will be maintained. It is important to note that the amount of acid or base that is added must be small compared to the supply of the buffer components CH_3COOH and CH_3COO^- from the salt CH_3COONa. If a large amount of acid or base is added, the buffering capacity of the system may be exceeded.

Buffers are also prepared from other conjugate acid–base pairs such as $H_2PO_4^-/HPO_4^{2-}$, HPO_4^{2-}/PO_4^{3-}, HCO_3^-/CO_3^{2-}, $/NH_4^+/NH_3$. The pH of the buffer solution will depend on the acid–base pair chosen.

Sample Problem 14.18   pH of a Buffer

The K_a for acetic acid, CH_3COOH, is 1.8×10^{-5}. What is the pH of a buffer prepared with 1.0 M CH_3COOH and 1.0 M CH_3COO^- (from 1.0 M CH_3COONa)?

Solution

$$CH_3COOH(aq) + H_2O(l) \rightleftharpoons H_3O^+(aq) + CH_3COO^-(aq)$$

Guide to Calculating pH of a Buffer

STEP 1
Write the K_a or K_b.

STEP 2
Rearrange the K_a for $[H_3O^+]$.

STEP 3
Substitute in the [HA] and [A⁻].

STEP 4
Use $[H_3O^+]$ to calculate pH.

STEP 1   Write the K_a expression.

$$K_a = \frac{[H_3O^+]\,[CH_3COO^-]}{[CH_3COOH]}$$

STEP 2   Rearrange K_a for $[H_3O^+]$.

$$[H_3O^+] = K_a \times \frac{[CH_3COOH]}{[CH_3COO^-]}$$

STEP 3   Substitute [HA] and [A⁻]. Substituting these values in the expression for $[H_3O^+]$ gives

$$[H_3O^+] = 1.8 \times 10^{-5} \times \frac{[1.0\,M]}{[1.0\,M]}$$

$$[H_3O^+] = 1.8 \times 10^{-5}$$

STEP 4   Use $[H_3O^+]$ to calculate pH. Using the concentration of $[H_3O^+]$ in the pH expression gives the pH of the buffer.

$$pH = -\log\,[1.8 \times 10^{-5}] = 4.74$$

Study Check

One of the conjugate acid–base pairs that buffers the blood is $H_2PO_4^-/HPO_4^{2-}$ with a K_a of 6.2×10^{-8}. What is the pH of a buffer that is 0.50 M $H_2PO_4^-$ and 0.25 M HPO_4^{2-}?

Questions and Problems   Buffers

14.69 Which of the following represent a buffer system? Explain.
  a. NaOH and NaCl
  b. H_2CO_3 and $NaHCO_3$
  c. HF and KF
  d. KCl and NaCl

14.70 Which of the following represent a buffer system? Explain.
  a. H_3PO_4
  b. $NaNO_3$
  c. CH_3COOH and CH_3COONa
  d. HCl and NaOH

14.71 Consider the buffer system of hydrofluoric acid, HF, and its salt, NaF.

$$HF(aq) + H_2O(l) \rightleftharpoons H_3O^+(aq) + F^-(aq)$$

a. What is the purpose of the buffer system?
b. Why is a salt of the acid needed?
c. How does the buffer react when some H_3O^+ is added?
d. How does the buffer react when some OH^- is added?

14.72 Consider the buffer system of nitrous acid, HNO_2, and its salt, $NaNO_2$.

$$HNO_2(aq) + H_2O(l) \rightleftharpoons H_3O^+(aq) + NO_2^-(aq)$$

a. What is the purpose of a buffer system?
b. What is the purpose of $NaNO_2$ in the buffer?
c. How does the buffer react when some H_3O^+ is added?
d. How does the buffer react when some OH^- is added?

14.73 Nitrous acid has a K_a of 4.5×10^{-4}. What is the pH of a buffer solution containing 0.10 M HNO_2 and 0.10 M NO_2^-?

14.74 Acetic acid has a K_a of 1.8×10^{-5}. What is the pH of a buffer solution containing 0.15 M CH_3COOH (acetic acid) and 0.15 M CH_3COO^-?

14.75 Compare the pH of a HF buffer that contains 0.10 M HF and 0.10 M NaF with another HF buffer that contains 0.060 M HF and 0.120 M NaF. See Table 14.5, page 468.

14.76 Compare the pH of a H_2CO_3 buffer that contains 0.10 M H_2CO_3 and 0.10 M $NaHCO_3$ with another H_2CO_3 buffer that contains 0.15 M H_2CO_3 and 0.050 M $NaHCO_3$. See Table 14.5, page 468.

CHEM NOTE

BUFFERS IN THE BLOOD

The arterial blood has a normal pH of 7.35–7.45. If changes in H_3O^+ lower the pH below 6.8 or raise it above 8.0, cells cannot function properly and death may result. In our cells, CO_2 is continually produced as an end product of cellular metabolism. Some CO_2 is carried to the lungs for elimination, and the rest dissolves in body fluids such as plasma and saliva, forming carbonic acid. As a weak acid, carbonic acid dissociates to give bicarbonate and H_3O^+. More of the anion HCO_3^- is supplied by the kidneys to give an important buffer system in the body fluid: the H_2CO_3/HCO_3^- buffer.

$$CO_2 + H_2O \rightleftharpoons H_2CO_3 \rightleftharpoons H_3O^+ + HCO_3^-$$

Excess H_3O^+ entering the body fluids reacts with the HCO_3^- and excess OH^- reacts with the carbonic acid.

$$H_2CO_3 + H_2O \longleftarrow H_3O^+ + HCO_3^-$$

Equilibrium shifts left

$$H_2CO_3 + OH^- \longrightarrow H_2O + HCO_3^-$$

Equilibrium shifts right

For the carbonic acid, we can write the equilibrium expression as

$$K_a = \frac{[H_3O^+][HCO_3^-]}{[H_2CO_3]}$$

To maintain the normal blood pH (7.35–7.45), the ratio of H_2CO_3/HCO_3^- needs to be about 1 to 10, which is obtained by typical concentrations in the blood of 0.0024 M H_2CO_3 and 0.024 M HCO_3^-.

$$[H_3O^+] = K_a \times \frac{[H_2CO_3]}{[HCO_3^-]}$$

$$= 4.3 \times 10^{-7} \times \frac{[0.0024]}{[0.024]} = 4.3 \times 10^{-7} \times 0.10 = 4.3 \times 10^{-8}$$

$$pH = -\log(4.3 \times 10^{-8}) = 7.37$$

In the body, the concentration of carbonic acid is closely associated with the partial pressure of CO_2. Table 14.12 lists the normal values for arterial blood. If the CO_2 level rises, producing more H_2CO_3, the equilibrium produces more H_3O^+, which lowers the pH. This condition is called *acidosis*. Difficulty with ventilation or gas diffusion can lead to respiratory acidosis, which can happen in emphysema or when the medulla of the brain is affected by an accident or depressive drugs.

A lowering of the CO_2 level leads to a high blood pH, a condition called *alkalosis*. Excitement, trauma, or a high temperature may cause a person to hyperventilate, which expels large amounts of CO_2. As the partial pressure of CO_2 in the blood falls below normal, the equilibrium shifts from H_2CO_3 to CO_2 and H_2O. This shift decreases the $[H_3O^+]$ and raises the pH. The kidneys also regulate H_3O^+ and HCO_3^- components, but more slowly than the adjustment made by the lungs through ventilation.

Table 14.12	Normal Values for Blood Buffer in Arterial Blood
P_{CO_2}	40 mm Hg
H_2CO_3	2.4 mmol/L of plasma
HCO_3^-	24 mmol/L of plasma
pH	7.35–7.45

Concept Map ▶ **Acids and Bases**

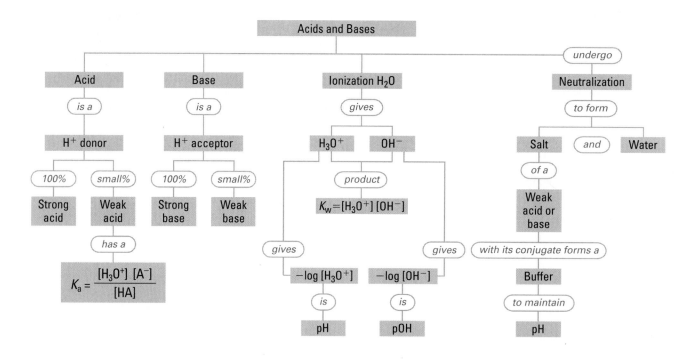

Chapter Review

14.1 Acids and Bases

According to the Arrhenius theory, an acid produces H^+ and a base produces OH^- in aqueous solutions. According to the Brønsted–Lowry theory, acids are proton (H^+) donors and bases are proton acceptors.

14.2 Brønsted–Lowry Acids and Bases

Two conjugate acid–base pairs are present in an acid–base reaction. Each acid–base pair is related by the loss or gain of one H^+. For example, when the acid HF donates a H^+, the F^- it forms is its conjugate base because F^- is capable of accepting a H^+. The other acid–base pair would be H_2O and H_3O^+.

14.3 Strengths of Acids and Bases

In strong acids, all the H^+ in the acid is donated to H_2O; in a weak acid, only a small percentage of acid molecules produce H_3O^+. Strong bases are hydroxides of Groups 1A (1) and 2A (2) that dissociate completely in water. An important weak base is ammonia, NH_3.

14.4 Dissociation Constants

In water, weak acids and weak bases produce only a few ions when equilibrium is reached. The reaction for a weak acid can be written as $HA + H_2O \rightleftharpoons H_3O^+ + A^-$. The acid dissociation constant is written as $K_a = \dfrac{[H_3O^+][A^-]}{[HA]}$. For a weak base, $B + H_2O \rightleftharpoons BH^+ + OH^-$, the base dissociation constant K_b is written as $K_b = \dfrac{[BH^+][OH^-]}{[B]}$.

14.5 Ionization of Water

In pure water, a few molecules transfer protons to other water molecules, producing small, but equal, amounts of $[H_3O^+]$ and $[OH^-]$ such that each has a concentration of 1.0×10^{-7} mol/L. The ion product, K_w, $[H_3O^+][OH^-] = 1.0 \times 10^{-14}$, applies to all aqueous solutions. In acidic solutions, the $[H_3O^+]$ is greater than the $[OH^-]$. In basic solutions, the $[OH^-]$ is greater than the $[H_3O^+]$.

14.6 The pH Scale

The pH scale is a range of numbers from 0 to 14 related to the $[H_3O^+]$ of the solution. A neutral solution has a pH of 7. In acidic solutions, the pH is below 7, and in basic solutions the pH is above 7. Mathematically, pH is the negative logarithm of the hydronium ion concentration ($-\log[H_3O^+]$). The pOH is the negative log of the hydroxide ion concentration ($pOH = -\log[OH^-]$). The sum of the pH + pOH is 14.00.

14.7 Reactions of Acids and Bases

When an acid reacts with a metal, hydrogen gas and a salt are produced. The reaction of an acid with a carbonate or bicarbonate produces carbon dioxide, a salt, and water. In neutralization, an acid reacts with a base to produce a salt and water.

14.8 Acid–Base Titration

In a laboratory procedure called titration, an acid sample is neutralized with a known amount of a base. From the volume and molarity of the base, the concentration of the acid is calculated.

14.9 Acid–Base Properties of Salt Solutions

A salt of a weak acid contains an anion that removes protons from water and makes the solution basic. A salt of a weak base contains an ion that donates a proton to water, producing an acidic solution. Salts of strong acids and bases produce neutral solutions because they contain ions that do not affect the pH.

14.10 Buffers

A buffer solution resists changes in pH when small amounts of an acid or a base are added. A buffer contains either a weak acid and its salt or a weak base and its salt. The weak acid picks up added OH^-, and the anion of the salt picks up added H^+.

Key Terms

acid A substance that dissolves in water and produces hydrogen ions (H^+), according to the Arrhenius theory. All acids are proton donors, according to the Brønsted–Lowry theory.

acid dissociation constant (K_a) The product of the ions from the dissociation of a weak acid divided by the concentration of the weak acid.

amphoteric Substances that can act as either an acid or a base in water.

base A substance that dissolves in water and produces hydroxide ions (OH^-) according to the Arrhenius theory. All bases are proton acceptors, according to the Brønsted–Lowry theory.

base dissociation constant (K_b) The product of the ions from the dissociation of a weak base divided by the concentration of the weak base.

Brønsted–Lowry acid An acid is a proton donor.

Brønsted–Lowry base A base is a proton acceptor

buffer solution A mixture of a weak acid or a weak base and its salt that resists changes in pH when small amounts of an acid or a base are added.

conjugate acid–base pair An acid and base that differ by one H^+. When an acid donates a proton, the product is its conjugate base, which is capable of accepting a proton in the reverse reaction.

dissociation The separation of an acid or a base into ions in water.

endpoint The point at which an indicator changes color. For the indicator phenolphthalein, the color change occurs when the number of moles of OH^- is equal to the number of moles of H_3O^+ in the sample.

hydronium ion, H_3O^+ The H_3O^+ ion formed by the attraction of a proton (H^+) to an H_2O molecule.

indicator A substance added to a titration sample that changes color when the acid or base is neutralized.

ion-product constant of water, K_w The product of $[H_3O^+]$ and $[OH^-]$ in solution; $K_w = [H_3O^+][OH^-]$.

neutral A solution with equal concentrations of $[H_3O^+]$ and $[OH^-]$.

neutralization A reaction between an acid and a base to form a salt and water.

pH A measure of the $[H_3O^+]$ in a solution; $pH = -\log[H_3O^+]$.

pOH A measure of the $[OH^-]$ in a solution; $pOH = -\log[OH^-]$.

salt An ionic compound that contains a metal ion or NH_4^+ and a nonmetal or polyatomic ion other than OH^-.

strong acid An acid that completely ionizes in water.

strong base A base that completely ionizes in water.

titration The addition of base to an acid sample to determine the concentration of the acid.

weak acid An acid that ionizes only slightly in solution.

weak base A base that ionizes only slightly in solution.

▶ Understanding the Concepts

14.77 In each of the following diagrams of acid solutions, determine if each diagram represents a strong acid or a weak acid. The acid has the formula HX.

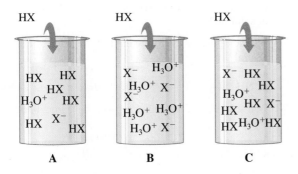

14.78 Adding a few drops of a strong acid to water will lower the pH appreciably. However, adding the same number of drops to a buffer does not appreciably alter the pH. Why?

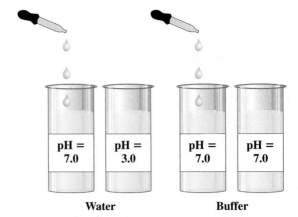

14.79 Sometimes, during stress or trauma, a person can start to hyperventilate. Then the person might breathe into a paper bag to avoid fainting.
- **a.** What changes occur in the blood pH during hyperventilation?
- **b.** How does breathing into a paper bag help return blood pH to normal?

14.80 In the blood plasma, pH is maintained by the carbonic acid–bicarbonate buffer system.
- **a.** How is pH maintained when acid is added to the buffer system?
- **b.** How is pH maintained when base is added to the buffer system?

▶ Additional Questions and Problems

14.81 Name each of the following:
- **a.** H_2SO_4
- **b.** KOH
- **c.** $Ca(OH)_2$
- **d.** HCl
- **e.** HNO_2

14.82 Are the following examples of body fluids acidic, basic, or neutral?
- **a.** saliva, pH 6.8
- **b.** urine, pH 5.9
- **c.** pancreatic juice, pH 8.0
- **d.** bile, pH 8.4
- **e.** blood, pH 7.45

14.83 What are some similarities and differences between strong and weak acids?

14.84 What are some ingredients found in antacids? What do they do?

14.85 One ingredient in some antacids is $Mg(OH)_2$.
- **a.** If the base is not very soluble in water, why is it considered a strong base?
- **b.** What is the neutralization reaction of $Mg(OH)_2$ with stomach acid, HCl?

14.86 Acetic acid, which is the acid in vinegar, is a weak acid. Why?

14.87 Using Table 14.4, page 459, determine which is the stronger acid in each of the following pairs:
- **a.** HF or HCN
- **b.** H_3O^+ or NH_4^+
- **c.** HNO_2 or CH_3COOH
- **d.** H_2O or HCO_3^-

14.88 Using Table 14.4, page 459, determine which is the stronger base in each of the following pairs:
- **a.** H_2O or Cl^-
- **b.** OH^- or NH_3
- **c.** SO_4^{2-} or NO_2^-
- **d.** CO_3^{2-} or H_2O

14.89 Determine the pH and pOH for the following solutions:
- **a.** $[H_3O^+] = 2.0 \times 10^{-8}$ M
- **b.** $[H_3O^+] = 5.0 \times 10^{-2}$ M
- **c.** $[OH^-] = 3.5 \times 10^{-4}$ M
- **d.** $[OH^-] = 0.0054$ M

14.90 Are the solutions in problem 14.89 acidic, basic, or neutral?

14.91 What are the $[H_3O^+]$ and $[OH^-]$ for a solution with the following pH values?
- **a.** 3.00
- **b.** 6.48
- **c.** 8.85
- **d.** 11.00

14.92 Solution A has a pH of 4.5 and solution B has a pH of 6.7.
- **a.** Which solution is more acidic?
- **b.** What is the $[H_3O^+]$ in each?
- **c.** What is the $[OH^-]$ in each?

14.93 What is the $[OH^-]$ in a solution that contains 0.225 g NaOH in 0.250 L of solution?

14.94 What is the $[H_3O^+]$ in a solution that contains 1.54 g HNO_3 in 0.500 L of solution?

14.95 What is the pH and pOH of a solution prepared by dissolving 2.5 g HCl in water to make 425 mL of solution?

14.96 What is the pH and pOH of a solution prepared by dissolving 1.00 g $Ca(OH)_2$ in water to make 875 mL of solution?

14.97 Will solutions of the following salts be acidic, basic, or neutral?
- **a.** KF
- **b.** NaCN
- **c.** NH_4NO_3
- **d.** NaBr

14.98 Will solutions of the following salts be acidic, basic, or neutral?
- **a.** K_2SO_4
- **b.** KNO_2
- **c.** $MgCl_2$
- **d.** NH_4Cl

14.99 A buffer is made by dissolving H_3PO_4 and NaH_2PO_4 in water.
  a. Write an equation that shows how this buffer neutralizes small amounts of acids.
  b. Write an equation that shows how this buffer neutralizes small amounts of base.
  c. Calculate the pH of this buffer if it is 0.10 M H_3PO_4 and 0.10 M $H_2PO_4^-$; the K_a for H_3PO_4 is 7.5×10^{-3}.

14.100 A buffer is made by dissolving CH_3COOH and CH_3COONa in water.
  a. Write an equation that shows how this buffer neutralizes small amounts of acid.
  b. Write an equation that shows how this buffer neutralizes small amounts of base.
  c. Calculate the pH of this buffer if it is 0.10 M CH_3COOH and 0.10 M CH_3COO^-; the K_a for CH_3COOH is 1.8×10^{-5}.

14.101 Calculate the volume (mL) of a 0.150 M NaOH solution that will completely neutralize the following:
  a. 25.0 mL of a 0.288 M HCl solution
  b. 10.0 mL of a 0.560 M H_2SO_4 solution

14.102 How many milliliters of 0.215 M NaOH solution are needed to completely neutralize 2.50 mL of 0.825 M H_2SO_4 solution?

14.103 A solution of 0.205 M NaOH is used to neutralize 20.0 mL H_2SO_4. If 45.6 mL NaOH is required to reach the endpoint, what is the molarity of the H_2SO_4 solution?

$$H_2SO_4(aq) + 2NaOH(aq) \longrightarrow Na_2SO_4(aq) + 2H_2O(l)$$

14.104 A 10.0-mL sample of vinegar, which is an aqueous solution of acetic acid, CH_3COOH, requires 16.5 mL of 0.500 M NaOH to reach the endpoint in a titration. What is the molarity of the acetic acid solution?

$$CH_3COOH(aq) + NaOH(aq) \longrightarrow CH_3COONa(aq) + H_2O(l)$$

▶ Challenge Questions

14.105 Consider the following:
  (1) H_2S (2) H_3PO_4 (3) HCO_3^-
  a. For each, write the formula of the conjugate base.
  b. For each, write the K_a expression.
  c. Write the formula of the weakest acid.
  d. Write the formula of the strongest acid.

14.106 Identify the conjugate acid–base pairs in each of the following equations and whether the equilibrium mixture contains mostly products or mostly reactants:
  a. $NH_3(aq) + HNO_3(aq) \rightleftharpoons NH_4^+(aq) + NO_3^-(aq)$
  b. $H_2O(l) + HBr(aq) \rightleftharpoons H_3O^+(aq) + Br^-(aq)$
  c. $HNO_2(aq) + HS^-(aq) \rightleftharpoons H_2S(g) + NO_2^-(aq)$
  d. $Cl^-(aq) + H_2O(l) \rightleftharpoons OH^-(aq) + HCl(aq)$

14.107 Complete and balance each of the following:
  a. $ZnCO_3(s) + H_2SO_4(aq) \longrightarrow$
  b. $Al(s) + HCl(aq) \longrightarrow$
  c. $H_3PO_4(aq) + Ca(OH)_2(s) \longrightarrow$
  d. $KHCO_3(s) + HNO_3(aq) \longrightarrow$

14.108 Predict whether a solution of each of the following salts is acidic, basic, or neutral. For salts that form acidic or basic solutions, write a balanced equation for the reaction.
  a. NH_4Br **b.** KNO_2
  c. $Mg(NO_3)_2$ **d.** BaF_2
  e. K_2S

14.109 Determine each of the following for a 0.050 M KOH solution:
  a. $[H_3O^+]$ **b.** pH   **c.** pOH
  d. products when reacted with H_3PO_4
  e. milliliters required to neutralize 40.0 mL of 0.035 M H_2SO_4

14.110 Consider the reaction of KOH and HNO_2.
  a. Write the balanced chemical equation.
  b. Calculate the milliliters of 0.122 M KOH required to neutralize 36.0 mL of 0.250 M HNO_2.
  c. Determine whether the final solution would be acidic, basic, or neutral.

Answers

Answers to Study Checks

14.1 Sulfuric acid; potassium hydroxide

14.2 $HNO_3(aq) + H_2O(l) \longrightarrow H_3O^+(aq) + NO_3^-(aq)$

14.3 A base accepts a proton to form its conjugate acid.
  a. H_2S **b.** HCl   **c.** HNO_2

14.4 The conjugate acid–base pairs are HCN/CN^- and SO_4^{2-}/HSO_4^-.

14.5 NH_3

14.6 $HNO_3 + H_2O \rightleftharpoons H_3O^+ + NO_3^-$
  The products are favored because HNO_3 is a stronger acid than H_3O^+.

14.7 Nitrous acid has a larger K_a than carbonic acid, it dissociates more in H_2O, forms more $[H_3O^+]$, and is a stronger acid.

14.8 $[H_3O^+] = 2.5 \times 10^{-11}$ M; basic

14.9 11.38

14.10 $[H_3O^+] = 5 \times 10^{-5}$ M; pH $= 4.3$

14.11 $[H_3O^+] = 3.2 \times 10^{-5}$ M; $[OH^-] = 3.1 \times 10^{-10}$ M

14.12 pOH $= 3.30$; pH $= 10.70$

14.13 $H_2SO_4(aq) + 2NaHCO_3(s) \longrightarrow$
$$Na_2SO_4(aq) + 2CO_2(g) + 2H_2O(l)$$

14.14 0.200 M HCl

14.15 67.1 mL

14.16 The anion PO_4^{3-} attracts protons from H_2O, forming the weak acid HPO_4^{2-} and OH^-, which makes the solution basic.

14.17 No. Both substances are salts; the mixture has no weak acid present.

14.18 pH $= 6.91$

Answers to Selected Questions and Problems

14.1 **a.** acid   **b.** acid   **c.** acid
  d. base

14.3 **a.** hydrochloric acid   **b.** calcium hydroxide
  c. carbonic acid   **d.** nitric acid
  e. sulfurous acid   **f.** bromic acid

14.5 **a.** $Mg(OH)_2$ **b.** HF
  c. H_3PO_4 **d.** LiOH
  e. NH_4OH **f.** HIO_4

14.7 **a.** HI is the acid (proton donor) and H_2O is the base (proton acceptor).
  b. H_2O is the acid (proton donor) and F^- is the base (proton acceptor).

14.9 **a.** F^-, fluoride ion
  b. OH^-, hydroxide ion
  c. HCO_3^-, bicarbonate ion
  d. SO_4^{2-}, sulfate ion
  e. ClO_2^-, chlorite ion

14.11 **a.** HCO_3^-, bicarbonate ion
  b. H_3O^+, hydronium ion
  c. H_3PO_4, phosphoric acid
  d. HBr, hydrobromic acid
  e. $HClO_4$, perchloric acid

14.13 **a.** acid H_2CO_3; conjugate base HCO_3^- base H_2O; conjugate acid H_3O^+
  b. acid NH_4^+; conjugate base NH_3 base H_2O; conjugate acid H_3O^+
  c. acid HCN; conjugate base CN^- base NO_2^-; conjugate acid HNO_2

14.15 $NH_4^+(aq) + H_2O(l) \rightleftarrows NH_3(aq) + H_3O^+(aq)$

14.17 A strong acid is a good proton donor, whereas its conjugate base is a poor proton acceptor.

14.19 **a.** HBr   **b.** HSO_4^- **c.** H_2CO_3

14.21 **a.** HSO_4^- **b.** HF   **c.** HCO_3^-

14.23 **a.** reactants   **b.** reactants   **c.** products

14.25 The reactants are favored because NH_4^+ is a weaker acid than HSO_4^-.
$$NH_4^+(aq) + SO_4^{2-}(aq) \rightleftarrows$$
$$NH_3(aq) + HSO_4^-(aq)$$

14.27 **a.** H_2SO_3 **b.** HSO_3^- **c.** H_2SO_3
  d. HS^- **e.** H_2SO_3

14.29 $H_3PO_4(aq) + H_2O(l) \rightleftarrows$
$$H_3O^+(aq) + H_2PO_4^-(aq) \qquad K_a = \frac{[H_3O^+][H_2PO_4^-]}{[H_3PO_4]}$$

14.31 In pure water, $[H_3O^+] = [OH^-]$ because one of each is produced every time a proton transfers from one water to another.

14.33 In an acid solution, the $[H_3O^+]$ is greater than the $[OH^-]$.

14.35 **a.** acidic   **b.** basic   **c.** basic   **d.** acidic

14.37 **a.** 1.0×10^{-5} M   **b.** 1.0×10^{-8} M
  c. 5.0×10^{-10} M   **d.** 2.5×10^{-2} M

14.39 **a.** 1.0×10^{-11} M   **b.** 2.0×10^{-9} M
  c. 5.6×10^{-3} M   **d.** 2.5×10^{-2} M

14.41 In a neutral solution, the $[H_3O^+]$ is 1.0×10^{-7} M and the pH is 7.00, which is the negative value of the power of 10.

14.43 **a.** basic   **b.** acidic   **c.** basic   **d.** acidic

14.45 An increase or decrease of 1 pH unit changes the $[H_3O^+]$ by a factor of 10. Thus a pH of 3 (10^{-3} M) is 10 times more acid than a pH of 4 (10^{-4} M).

14.47 **a.** 4.00   **b.** 8.52   **c.** 9.00   **d.** 3.40

14.49

$[H_3O^+]$	$[OH^-]$	pH	pOH	Acidic, Basic, or Neutral?
1.0×10^{-8} M	1.0×10^{-6} M	8.00	6.00	Basic
1.0×10^{-3} M	1.0×10^{-11} M	3.00	11.00	Acidic
2.8×10^{-5} M	3.6×10^{-10} M	4.55	9.45	Acidic
1.0×10^{-12} M	1.0×10^{-2} M	12.00	2.00	Basic

14.51 a. $ZnCO_3(aq) + 2HBr(aq) \longrightarrow$
$$ZnBr_2(aq) + CO_2(g) + H_2O(l)$$
b. $Zn(s) + 2HCl(aq) \longrightarrow ZnCl_2(aq) + H_2(g)$
c. $HCl(g) + NaHCO_3(s) \longrightarrow$
$$NaCl(aq) + H_2O(l) + CO_2(g)$$
d. $H_2SO_4(aq) + Mg(OH)_2(s) \longrightarrow$
$$MgSO_4(aq) + 2H_2O(l)$$

14.53 a. $2HCl(aq) + Mg(OH)_2(s) \longrightarrow$
$$MgCl_2(aq) + 2H_2O(l)$$
b. $H_3PO_4(aq) + 3LiOH(aq) \longrightarrow$
$$Li_3PO_4(aq) + 3H_2O(l)$$

14.55 a. $H_2SO_4(aq) + 2NaOH(aq) \longrightarrow$
$$Na_2SO_4(aq) + 2H_2O(l)$$
b. $3HCl(aq) + Fe(OH)_3(s) \longrightarrow$
$$FeCl_3(aq) + 3H_2O(l)$$
c. $H_2CO_3(aq) + Mg(OH)_2(s) \longrightarrow$
$$MgCO_3(s) + 2H_2O(l)$$

14.57 To a known volume of formic acid, add a few drops of indicator. Place a NaOH solution of known molarity in a buret. Add base to acid until one drop changes the color of the solution. Use the volume and molarity of NaOH and the volume of formic acid to calculate the concentration of the formic acid in the sample.

14.59 0.829 M HCl

14.61 0.125 M H_2SO_4

14.63 0.0223 M H_3PO_4

14.65 The anion from the weak acid removes a proton from H_2O to make a basic solution.

14.67 a. neutral
b. acidic $NH_4^+(aq) + H_2O(l) \rightleftarrows$
$$NH_3(aq) + H_3O^+(aq)$$
c. basic $CO_3^{2-}(aq) + H_2O(l) \rightleftarrows$
$$HCO_3^-(aq) + OH^-(aq)$$
d. basic $S^{2-}(aq) + H_2O(l) \rightleftarrows$
$$HS^-(aq) + OH^-(aq)$$

14.69 (b) and (c) are buffer systems. (b) contains the weak acid H_2CO_3 and its salt $NaHCO_3$. (c) contains HF, a weak acid and its salt KF.

14.71 a. A buffer system keeps the pH constant.
b. To neutralize any H^+ added.
c. The added H_3O^+ reacts with F^- from NaF.
d. The added OH^- is neutralized by the HF.

14.73 $[H_3O^+] = 4.5 \times 10^{-4} \times \dfrac{[0.10\ M]}{[0.10\ M]} = 4.5 \times 10^{-4}$

$pH = -\log[4.5 \times 10^{-4}] = 3.35$

14.75 The pH of the 0.10 M HF/0.10 M NaF buffer is 3.14. The pH of the 0.060 M HF/0.120 M NaF buffer is 3.44.

14.77 a. This diagram represents a weak acid; only a few HX molecules separate into H_3O^+ and X^- ions.
b. This diagram represents a strong acid; all the HX molecules separate into H_3O^+ and X^- ions.
c. This diagram represents a weak acid; only a few HX molecules separate into H_3O^+ and X^- ions.

14.79 a. During hyperventilation, a person will lose CO_2 and the blood pH will rise.
b. Breathing into a paper bag will increase the CO_2 concentration and lower the blood pH.

14.81 a. sulfuric acid **b.** potassium hydroxide
c. calcium hydroxide **d.** hydrochloric acid
e. nitrous acid

14.83 Both strong and weak acids produce H_3O^+ in water. Weak acids are only slightly ionized, whereas a strong acid exists as ions in solution.

14.85 a. The $Mg(OH)_2$ that dissolves is completely dissociated, making it a strong base.
b. $Mg(OH)_2(aq) + 2HCl(aq) \longrightarrow$
$$MgCl_2(aq) + 2H_2O(l)$$

14.87 a. HF **b.** H_3O^+ **c.** HNO_2 **d.** HCO_3^-

14.89 a. pH = 7.70; pOH 6.30
b. pH = 1.30; pOH 12.70
c. pH = 10.54; pOH 3.46
d. pH = 11.73; pOH = 2.27

14.91 a. $[H_3O^+] = 1.0 \times 10^{-3}\ M$;
$[OH^-] = 1.0 \times 10^{-11}\ M$
b. $[H_3O^+] = 3.3 \times 10^{-7}\ M$;
$[OH^-] = 3.0 \times 10^{-8}\ M$
c. $[H_3O^+] = 1.4 \times 10^{-9}\ M$;
$[OH^-] = 7.1 \times 10^{-6}\ M$
d. $[H_3O^+] = 1.0 \times 10^{-11}\ M$;
$[OH^-] = 1.0 \times 10^{-3}\ M$

14.93 $[OH^-] = 0.0225\ M$

14.95 pH = 0.80; pOH = 13.20

14.97 a. basic **b.** basic **c.** acidic **d.** neutral

14.99 a. acid: $H_2PO_4^-(aq) + H_3O^+(aq) \longrightarrow$
$$H_3PO_4(aq) + H_2O(l)$$
b. base: $H_3PO_4(aq) + OH^-(aq) \longrightarrow$
$$H_2PO_4^-(aq) + H_2O(l)$$
c. pH = 2.12

14.101 a. 48.0 mL NaOH   **b.** 74.7 mL NaOH

14.103 0.234 M H_2SO_4

14.105 a. (1) HS^- (2) $H_2PO_4^-$ (3) CO_3^{2-}

b. (1) $\dfrac{[H_3O^+][HS^-]}{[H_2S]}$ (2) $\dfrac{[H_3O^+][H_2PO_4^-]}{[H_3PO_4]}$

(3) $\dfrac{[CO_3^{2-}][H_3O^+]}{[HCO_3^-]}$

c. H_2S

d. H_3PO_4

14.107 a. $ZnCO_3(s) + H_2SO_4(aq) \longrightarrow ZnSO_4(aq) + CO_2(g) + H_2O(l)$

b. $2Al(s) + 6HCl(aq) \longrightarrow 2AlCl_3(aq) + 3H_2(g)$

c. $2H_3PO_4(aq) + 3Ca(OH)_2(s) \longrightarrow Ca_3(PO_4)_2(aq) + 6H_2O(l)$

d. $KHCO_3(s) + HNO_3(aq) \longrightarrow KNO_3(aq) + CO_2(g) + H_2O(l)$

14.109 $KOH \longrightarrow K^+ + OH^-$

$[OH^-] = 0.050 \text{ M} = 5.0 \times 10^{-2} \text{ M}$

a. $[H_3O^+] = \dfrac{1.0 \times 10^{-14} \text{ M}}{[5.0 \times 10^{-2} \text{ M}]} = 2.0 \times 10^{-13} \text{ M}$

b. $pH = -\log[2.0 \times 10^{-13}] = 12.70$

c. $pOH = -\log[5.0 \times 10^{-2}] = 1.30$ or $pOH = 14.00 - 12.70 = 1.30$

d. $3KOH(aq) + H_3PO_4(aq) \longrightarrow K_3PO_4(aq) + 3H_2O(l)$

e. $40.0 \text{ mL} \times \dfrac{1 \text{ L}}{1000 \text{ mL}} \times \dfrac{0.035 \text{ mol}}{1 \text{ L}} \times \dfrac{2 \text{ mol KOH}}{1 \text{ mol H}_2\text{SO}_4} \times \dfrac{1000 \text{ mL KOH}}{0.050 \text{ mol KOH}}$

Combining Ideas from Chapters 11–14

CI 21 Methane is a hydrocarbon containing only carbon and hydrogen that is the major component of purified natural gas used for heating and cooking. Methane gas has a density of 0.715 g/L at STP. For transport, natural gas is cooled to -163 degrees Celsius to give liquified natural gas (LNG) that has a volume that is 1/600th its volume at STP. A tank on a ship can hold 7.0 million gallons LNG, which has a density of 0.45 g/mL. The heat of combustion of methane gas is -883 kJ/mol.

a. What is the molecular formula of methane?
b. What is the mass in kilograms of LNG (assume that LNG is all methane) transported in the tank?
c. What volume of methane gas from one tank will be available to homes at STP?

d. Write the balanced combustion equation for methane that occurs in a gas burner.

e. How many kilograms of oxygen are needed to react the methane in one tank?
f. How much energy in kilojoules is provided by the tank of methane?

CI 22 A mixture of 25.0 g CS_2 gas and 30.0 g O_2 gas is placed in 10.0-L closed container and heated to 125°C. The products of complete reaction are carbon dioxide gas and sulfur dioxide gas.

a. Write a balanced equation for the reaction.
b. How many grams of CO_2 are produced?
c. What is the partial pressure of the remaining reactant?
d. What is the final pressure in the container?

CI 23 Consider the reaction at equilibrium

$$2H_2(g) + S_2(g) \rightleftharpoons 2H_2S(g) + \text{heat}$$

In a 10.0-L container, an equilibrium mixture contains 2.02 g H_2, 10.3 g S_2, and 68.2 g H_2S.

a. What is the K_c value for this equilibrium mixture?
b. If H_2 is added to the mixture, how will the equilibrium shift?

c. How will the equilibrium shift if the mixture is placed in a 5.00-L container with no change in temperature?

d. If a 5.00-L container has an equilibrium mixture of 0.30 mol H_2 and 2.5 mol H_2S, what is the equilibrium concentration of S_2 if temperature is the same?

e. Will an increase in temperature increase or decrease the K_c value?

CI 24   A bottle of vintage port wine has a volume of 750 mL and contains 18% ethanol (C_2H_6O) by volume. Ethanol has a density of 0.789 g/mL. At 20°C, port wine has a density of 0.990 g/mL. The alcohol in port wine is made when grape sugar ($C_6H_{12}O_6$) undergoes fermentation (no oxygen) to ethanol and carbon dioxide. The weight of 150 grapes is 1.5 lb and contains 26 g grape sugar (1 ton = 2 000 lb).

a. Calculate the percent concentration ethanol by mass.

b. Write the balanced equation for the fermentation reaction of grape sugar.

c. How many grams of grape sugar are required to produce one bottle of port wine?

d. How many grapes are needed to make one bottle of port wine?

e. How many bottles of port wine can be produced from 1.0 ton of grapes?

CI 25   A metal M with a mass of 0.420 g completely reacts with 34.8 mL of 0.520 M HCl to form MCl_3 and H_2 gas.

a. What volume in mL of H_2 at 720. mm Hg and 24°C is produced?

b. How many moles of metal M reacted?

c. What is the molar mass and name of metal M?

d. Write the balanced equation for the reaction.

e. What is the electron configuration of the metal M and its cation?

CI 26   A saturated solution of cobalt(II) hydroxide has a pH of 9.36.

a. Write the solubility product constant for cobalt(II) hydroxide.

b. Calculate the K_{sp} value for cobalt(II) hydroxide.

c. How many grams of cobalt(II) hydroxide will dissolve in 2.0 L water?

d. How many grams of cobalt(II) hydroxide will dissolve in 50.0 mL of 0.0100 M NaOH?

Answers to CIs

CI 21
a. CH_4
b. 1.2×10^7 kg
c. 1.7×10^{10} L
d. $CH_4(g) + 2\,O_2(g) \longrightarrow CO_2(g) + 2\,H_2O(g)$
e. 4.8×10^7 kg
f. 6.6×10^{11} kJ

CI 23
a. $K_c = 62.1$
b. If H_2 is added, the equilibrium will shift to the right.
c. If the volume decreases, the equlibrium shifts to the right.
d. $[S_2] = 1.1$ mol/L
e. An increase in temperature will decrease the value of K_c.

CI 25
a. 233 mL
b. 6.03×10^{-3} mol M
c. Molar mass 69.7; gallium
d. $2\,Ga(s) + 6\,HCl(aq) \longrightarrow 2\,GaCl_3(aq) + 3\,H_2(g)$
e. Ga: $1s^2\,2s^2\,2p^6\,3s^2\,3p^6\,4s^2\,3d^{10}\,4p^1$
   Ga^{3+}: $1s^2\,2s^2\,2p^6\,3s^2\,3p^6\,3d^{10}$

Oxidation–Reduction: Transfer of Electrons

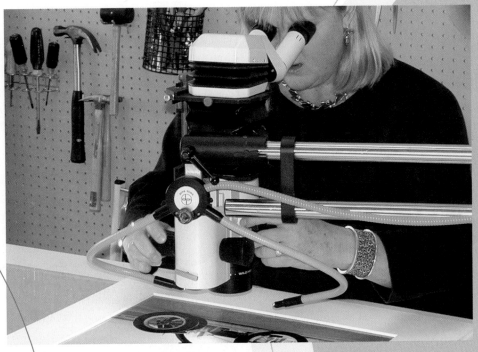

"As a conservator of photographic materials, it is essential to have an understanding of the chemical reactions of many different photographic processes," says Theresa Andrews, Conservator of Photographs at the San Francisco Museum of Modern Art. "For example, the creation of the latent image in many photographs is based upon the light sensitivity of silver halides. Photolytic silver 'prints out' when exposed to a light source such as the sun and filamentary silver 'develops out' when an exposed photographic paper is placed in a bath with reducing agents. Photolytic silver particles are much smaller than filamentary silver particles making them more vulnerable to abrasion and image loss. This knowledge is critical when making recommendations for light levels and for the protection of photographs when they are on exhibition. Conservation treatments require informed decisions based on the reactivity of the materials within the photograph and also the compatibility of materials that might be required for repair or preservation of the photograph."

the Chemistry place

Visit **www.aw-bc.com/chemplace** for extra quizzes, interactive tutorials, career resources, PowerPoint slides for chapter review, math help, and case studies.

Perhaps you have never heard of oxidation and reduction reactions. Yet this type of reaction has many important applications in our everyday lives. When you see a rusty nail, tarnish on a silver spoon, or corrosion on metal, you are observing oxidation. Historically, the term *oxidation* was used for reactions of the elements with oxygen to form oxides.

$$4Fe(s) + 3O_2(g) \longrightarrow 2Fe_2O_3(s) \qquad \text{Fe is oxidized}$$
Rust

When you turn the lights on in your car, an oxidation–reduction reaction within your car battery provides the electricity. On a cold, wintry day, you might build a wood fire. As the wood burns, oxygen combines with carbon and hydrogen to produce carbon dioxide, water, and heat. Burning wood is an oxidation–reduction reaction. When you eat foods with starches in them, you digest the starches to give glucose, which is oxidized in your cells to give you energy along with carbon dioxide and water. Every breath you take provides oxygen to carry out oxidation in your cells.

$$C_6H_{12}O_6(aq) + 6O_2(g) \longrightarrow 6CO_2(g) + 6H_2O(l) + \text{energy}$$

The term *reduction* was originally used for reactions that removed oxygen from compounds. Metal oxides in ores are reduced to obtain the pure metal. For example, iron metal is obtained by reducing the iron in iron ore with carbon.

$$2Fe_2O_3(s) + 3C(g) \longrightarrow 3CO_2(g) + 4Fe(s) \qquad \text{Fe in Fe}_2O_3 \text{ is reduced}$$

As more was learned about oxidation and reduction reactions, scientists found that they do not always involve oxygen. Today, an oxidation–reduction reaction is any reaction that involves the transfer of electrons from one substance to another.

15.1 Oxidation and Reduction

Learning Goal

Identify what is oxidized and reduced in an oxidation–reduction reaction.

In every **oxidation–reduction reaction** (abbreviated *redox*), electrons are transferred from one substance to another. If one substance loses electrons, another substance must gain electrons. **Oxidation** is defined as the *loss* of electrons; **reduction** is the *gain* of electrons. One way to remember these definitions is to use one of the following:

LEO GER
Loss of **E**lectrons is **O**xidation
Gain of **E**lectrons is **R**eduction

OIL RIG
Oxidation **I**s **L**oss of electrons
Reduction **I**s **G**ain of electrons

In general, atoms of metals lose electrons to form positive ions; whereas nonmetals gain electrons to form negative ions. In terms of oxidation and reduction, atoms of a metal are oxidized and atoms of a nonmetal are reduced. Let's look at the formation of the ionic compound CaS.

$$Ca(s) + S(s) \longrightarrow CaS(s)$$

Figure 15.1 In this single replacement reaction Zn(s) is oxidized to Zn^{2+} when it provides two electrons to reduce Cu^{2+} to Cu(s): $Zn(s) + Cu^{2+}(aq) \longrightarrow Cu(s) + Zn^{2+}(aq)$

In the oxidation, does Zn(s) lose or gain electrons?

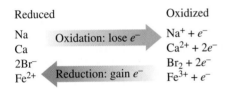

Reduced		Oxidized
Na	Oxidation: lose e^-	$Na^+ + e^-$
Ca		$Ca^{2+} + 2e^-$
$2Br^-$	Reduction: gain e^-	$Br_2 + 2e^-$
Fe^{2+}		$Fe^{3+} + e^-$

The calcium atom loses 2 electrons to form calcium ion (Ca^{2+}); calcium is oxidized.

$$Ca \longrightarrow Ca^{2+} + 2e^- \qquad \text{Oxidation: loss of electrons by Ca}$$

At the same time, the sulfur atom gains 2 electrons to form sulfide ion (S^{2-}); sulfur is reduced.

$$S + 2e^- \longrightarrow S^{2-} \qquad \text{Reduction: gain of electrons by S}$$

Therefore, the formation of CaS involves two reactions that occur simultaneously, one an oxidation and the other a reduction.

$$Ca(s) + S(s) \longrightarrow Ca^{2+} + S^{2-} = CaS(s)$$

Every time a reaction involves an oxidation and a reduction, the number of electrons lost is equal to the number of electrons gained.

Consider the reaction of zinc and copper(II) sulfate. (See Figure 15.1.)

$$Zn(s) + CuSO_4(aq) \longrightarrow ZnSO_4(aq) + Cu(s)$$

By writing the ionic equation, the atoms and ions that react can be identified.

$$\mathbf{Zn(s) + Cu^{2+}(aq) + SO_4{}^{2-}(aq) \longrightarrow Zn^{2+}(aq) + SO_4{}^{2-}(aq) + Cu(s)}$$

In this reaction, Zn atoms undergo oxidation to form Zn^{2+} ions by losing 2 electrons. At the same time, Cu^{2+} ions undergo reduction to Cu atoms by gaining 2 electrons. The $SO_4{}^{2-}$ ions are spectator ions and do not change.

$$Zn(s) \longrightarrow Zn^{2+}(aq) + 2e^- \qquad \text{Oxidation of Zn}$$
$$Cu^{2+}(aq) + 2e^- \longrightarrow Cu(s) \qquad \text{Reduction of } Cu^{2+}$$

Sample Problem 15.1 **Identification of Oxidation–Reduction Reactions**

Determine if each of the following is an oxidation–reduction reaction:
a. $Ca(s) + Cl_2(g) \longrightarrow CaCl_2(aq)$
b. $2Mg(s) + O_2(g) \longrightarrow 2MgO(s)$
c. $BaCl_2(aq) + Na_2CO_3(aq) \longrightarrow BaCO_3(s) + 2NaCl(aq)$

Solution

a. Yes. A Ca atom loses electrons to form a Ca^{2+} ion; Ca is oxidized.

$$Ca \longrightarrow Ca^{2+} + \mathbf{2e^-}$$

Each Cl atom in Cl_2 gains 1 electron to form a Cl^- ion; Cl_2 is reduced.

$$Cl_2 + \mathbf{2e^-} \longrightarrow 2Cl^-$$

b. Yes. A Mg atom loses electrons to form a Mg^{2+} ion; Mg is oxidized.

$$Mg \longrightarrow Mg^{2+} + \mathbf{2e^-}$$

Each O atom in O_2 gains 2 electrons to form an O^{2-} ion; O_2 is reduced.

$$O_2 + \mathbf{4e^-} \longrightarrow 2O^{2-}$$

c. No. $Ba^{2+}(aq) + 2Cl^-(aq) + 2Na^+(aq) + CO_3{}^{2-}(aq) \longrightarrow$
$$BaCO_3(s) + 2Na^+(aq) + 2Cl^-(aq)$$

There are no changes in ionic charges of the ions in the reactants and products. Barium ion has the same charge (Ba^{2+}) in $BaCl_2$ as in solid $BaCO_3$. Sodium and chlorine also have the same charge in both the reactants and products, Na^+ and Cl^-. There is no transfer of electrons in this double replacement reaction.

Study Check

Which of the following is an oxidation–reduction reaction?
a. $2Fe(s) + 3Br_2(l) \longrightarrow 2FeBr_3(s)$
b. $NaCl(aq) + AgNO_3(aq) \longrightarrow AgCl(s) + NaNO_3(aq)$

Sample Problem 15.2 **Oxidation and Reduction**

In the following reactions, what is oxidized and what is reduced?
a. $2Na(s) + Cl_2(g) \longrightarrow 2NaCl(s)$
b. $Ca(s) + 2H^+(aq) \longrightarrow Ca^{2+}(aq) + H_2(g)$

Solution

a. In this reaction, Na metal atoms are oxidized to Na^+ by losing electrons.

$$2Na \longrightarrow 2Na^+ + \mathbf{2e^-} \quad \text{Na is oxidized.}$$

Cl_2 gains electrons, which reduce Cl_2 to $2Cl^-$.

$$Cl_2 + \mathbf{2e^-} \longrightarrow 2Cl^- \quad \text{Cl in } Cl_2 \text{ is reduced.}$$

b. $Ca(s) + 2H^+(aq) \longrightarrow Ca^{2+}(aq) + H_2(g)$
By losing 2 electrons, Ca metal atoms are oxidized to Ca^{2+}.

$$Ca(s) \longrightarrow Ca^{2+}(aq) + \mathbf{2e^-} \quad \text{Ca is oxidized.}$$

To form the nonmetal element H_2, each H^+ must gain an electron.

$$2H^+(aq) + \mathbf{2e^-} \longrightarrow H_2(g) \quad H^+ \text{ is reduced.}$$

Study Check

In the following reaction, what is oxidized and what is reduced?

$$2Al(s) + 3Sn^{2+}(aq) \longrightarrow 2Al^{3+}(aq) + 3Sn(s)$$

Oxidizing and Reducing Agents

We have seen that an oxidation reaction must always be accompanied by a reduction reaction. The substance that is oxidized loses electrons, and the substance that is reduced gains those electrons. For example, Zn is oxidized to Zn^{2+} by losing 2 electrons that are transferred to Cl_2 to reduce it to $2Cl^-$.

$$Zn(s) + Cl_2(g) \longrightarrow ZnCl_2(s)$$

In an oxidation–reduction reaction, a **reducing agent** provides electrons and an **oxidizing agent** accepts electrons. In this reaction, Zn is a *reducing agent* because it provides electrons used to reduce Cl_2. At the same time, Cl_2 is an *oxidizing agent* because it accepts electrons provided by the oxidation of Zn. In any oxidization–reduction reaction, the reducing agent is oxidized, and the oxidizing agent is reduced.

$$Zn \longrightarrow Zn^{2+} + 2e^- \qquad \text{Zn is oxidized; Zn is the } \textit{reducing agent.}$$
$$Cl_2 + 2e^- \longrightarrow 2Cl^- \qquad \text{Cl in Cl}_2 \text{ is reduced; Cl}_2 \text{ is the } \textit{oxidizing agent.}$$

Sample Problem 15.3 ▸ Oxidizing and Reducing Agents

Identify the oxidizing agent and the reducing agent in each of the following reactions:

a. $2Na(s) + Cl_2(g) \longrightarrow 2NaCl(s)$
b. $Ca(s) + 2H^+(aq) \longrightarrow Ca^{2+}(aq) + H_2(g)$

Solution

a. In this reaction, Na metal atoms are oxidized to Na^+ by losing electrons. The Cl atoms in Cl_2 gain electrons, which reduces Cl_2 to $2Cl^-$. Thus Na, which is oxidized, is the *reducing agent*, and Cl_2, which is reduced, is the *oxidizing agent*.

$$2Na \longrightarrow 2Na^+ + 2e^- \qquad \text{Na is oxidized; Na is the } \textit{reducing agent.}$$
$$Cl_2 + 2e^- \longrightarrow 2Cl^- \qquad \text{Cl in Cl}_2 \text{ is reduced; Cl}_2 \text{ is the } \textit{oxidizing agent.}$$

b. $Ca(s) + 2H^+(aq) \longrightarrow Ca^{2+}(aq) + H_2(g)$
When Ca is oxidized to Ca^{2+}, 2 electrons are provided for the reduction of $2H^+$. Thus, Ca is a *reducing agent*. Because $2H^+$ gains electrons from the oxidation of Ca, H^+ is an *oxidizing agent*.

$$Ca(s) \longrightarrow Ca^{2+}(aq) + 2e^- \qquad \text{Ca is oxidized; Ca is the } \textit{reducing agent.}$$
$$2H^+(aq) + 2e^- \longrightarrow H_2(g) \qquad \text{H}^+ \text{ is reduced; H}^+ \text{ is the } \textit{oxidizing agent.}$$

Study Check

In the following reaction, what is the oxidizing agent and what is the reducing agent?

$$2Al(s) + 3Sn^{2+}(aq) \longrightarrow 2Al^{3+}(aq) + 3Sn(s)$$

Questions and Problems   **Oxidation and Reduction**

15.1 Indicate whether each of the following describes an oxidation or a reduction in a reaction:
a. $Na^+(aq) + e^- \longrightarrow Na(s)$
b. $Ni(s) \longrightarrow Ni^{2+}(aq) + 2e^-$
c. $Cr^{3+}(aq) + 3e^- \longrightarrow Cr(s)$
d. $2H^+(aq) + 2e^- \longrightarrow H_2(g)$

15.2 Indicate whether each of the following describes an oxidation or a reduction in a reaction:
a. $O_2(g) + 4e^- \longrightarrow 2O^{2-}(aq)$
b. $Al(s) \longrightarrow Al^{3+}(aq) + 3e^-$
c. $Fe^{3+}(aq) + e^- \longrightarrow Fe^{2+}(aq)$
d. $2Br^-(aq) \longrightarrow Br_2(l) + 2e^-$

15.3 In the following reactions, identify the substance that is oxidized and the substance that is reduced.
a. $Zn(s) + Cl_2(g) \longrightarrow ZnCl_2(s)$
b. $Cl_2(g) + 2NaBr(aq) \longrightarrow 2NaCl(aq) + Br_2(l)$
c. $2Pb(s) + O_2(g) \longrightarrow 2PbO(s)$
d. $2Fe^{3+}(aq) + Sn^{2+}(aq) \longrightarrow$
$2Fe^{2+}(aq) + Sn^{4+}(aq)$

15.4 In the following reactions, identify the substance that is oxidized and the substance that is reduced:
a. $2Li(s) + F_2(g) \longrightarrow 2LiF(s)$
b. $Cl_2(g) + 2KI(aq) \longrightarrow 2KCl(aq) + I_2(s)$
c. $Zn(s) + Cu^{2+}(aq) \longrightarrow Zn^{2+}(aq) + Cu(s)$
d. $Fe(s) + CuSO_4(aq) \longrightarrow FeSO_4(aq) + Cu(s)$

15.5 Indicate whether each of the following describes the oxidizing agent or the reducing agent in an oxidation–reduction reaction:
a. the substance that is oxidized
b. the substance that gains electrons

15.6 Indicate whether each of the following describes the oxidizing agent or the reducing agent in an oxidation–reduction reaction:
a. the substance that is reduced
b. the substance that loses electrons

15.7 In Problem 15.3, identify the oxidizing agent and the reducing agent in each reaction.

15.8 In Problem 15.4, identify the oxidizing agent and the reducing agent in each reaction.

Learning **Goal**

Assign an oxidation number to all the atoms in a compound; use oxidation numbers to identify what is oxidized, what is reduced, the reducing agent, and the oxidizing agent.

15.2 Oxidation Numbers

Up until now, the substances that are oxidized and reduced in an oxidation–reduction reaction have not been difficult to pick out. However, in more complex oxidation–reduction reactions, the changes in electrons are not so obvious. To identify the atoms that lose or gain electrons, values called **oxidation numbers** are assigned. While helpful in following the electron change for atoms and ions in oxidation and reduction reactions, it is important to recognize that oxidation numbers are not intended to represent actual charges.

Rules for Assigning Oxidation Numbers

The rules for assigning oxidation numbers to the atoms or ions in the reactants and products are given in Table 15.1.

Table 15.1   Rules for Assigning Oxidation Numbers

Rule	Examples	Oxidation Numbers	
1. An atom in the elemental state has an oxidation number of zero (0).	Cl_2 Cu Fe	$Cl = 0$ $Cu = 0$ $Fe = 0$	
2. The oxidation number of a monatomic ion is equal to its ionic charge.	$NaCl$ MgO	$Na^+ = +1$ $Mg^{2+} = +2$	$Cl^- = -1$ $O^{2-} = -2$
3. The sum of the oxidation numbers of the atoms in a compound equals zero (0).	$CuCl_2$	$Cu = +2$	$Cl^- = -1$
The sum of the oxidation numbers of the atoms in a polyatomic ion equals the charge of the ion.	OH^-	$H = +1$	$O = -2$

Table 15.1 (*Continued*)

Rule	Examples	Oxidation Numbers		
4. Oxidation numbers are assigned to elements in compounds, according to the following rules, listed in order of priority.				
Group 1A (1) $= +1$	Na_2S	Na $= +1$	S $= -2$	
Group 2A (2) $= +2$	$MgBr_2$	Mg $= +2$	Br $= -1$	
Hydrogen $= +1$	HCl	H $= +1$	Cl $= -1$	
	NaH	Na $= +1$	H $= -1$	
Fluorine $= -1$	SnF_2	Sn $= +2$	F $= -1$	
	ClF	Cl $= +1$	F $= -1$	
Oxygen $= -2$	N_2O	N $= +1$	O $= -2$	
	CO_3^{2-}	C $= +4$	O $= -2$	
Group 7A (17) $=$ usually -1 (This oxidation number can be adjusted if necessary.)	$BaCl_2$	Ba $= +2$	Cl $= -1$	
	$HBrO_2$	H $= +1$	Br $= +3$	O $= -2$

We can now look at how these rules are used to assign oxidation numbers to the atoms in a variety of chemical species. For each formula, the oxidation numbers are written *below* the symbols of the elements.

Examples of Assigning Oxidation Numbers

Formula	Explanation	Oxidation Numbers
Br_2	Each Br in the elemental state of bromine has an oxidation number of 0 (Rule 1).	Br_2 0
Ba^{2+}	Ba^{2+} is a monoatomic ion with an oxidation number of $+2$ (Rule 2).	Ba^{2+} $+2$
CO_2	O is assigned an oxidation number of -2 (Rule 4). Because CO_2 is neutral overall, the oxidation number of C is calculated as $+4$ (Rule 3). $C + 2O \quad = 0$ $C + 2(-2) = 0$ $C - 4 \quad\; = 0$ $C \qquad\;\; = +4$	CO_2 $+4 -2$ Check: $1C = 1(+4) = +4$ $\underline{2O = 2(-2) = -4}$ Sum $= \quad 0$ *Oxidation numbers for elements in covalent compounds are not actual charges.*

continued

Examples of Assigning Oxidation Numbers

Formula	Explanation	Oxidation Numbers
Al_2O_3	In compounds, the oxidation number of O is -2 (Rule 4). To give the compound a neutral charge, Al is calculated as $+3$ (Rule 3).	Al_2O_3 $+3-2$

For Al_2O_3:

$$2Al + 3\,O\ = 0$$
$$2Al + 3(-2) = 0$$
$$2Al - 6\ \ \ = 0$$
$$2Al\ \ \ \ \ \ \ \ = +6$$
$$Al\ \ \ \ \ \ \ \ \ = +3$$

Check:
$$2Al = 2(+3) = +6$$
$$3O\ = 3(-2) = -6$$
$$\text{Sum} = \ \ \ 0$$

| $HClO_3$ | From Rule 4, H is $+1$ and O is -2. The oxidation number of Cl is calculated as $+5$ for the neutral compound (Rule 3). | $HClO_3$
 $+1+5-2$ |

$$H + Cl + 3O\ \ \ = 0$$
$$(+1) + Cl + 3(-2) = 0$$
$$(+1) + Cl + (-6)\ \ = 0$$
$$Cl - 5\ \ \ \ \ \ = 0$$
$$Cl\ \ \ \ \ \ \ \ \ \ = +5$$

Check:
$$1H\ = 1(+1) = +1$$
$$1Cl = 1(+5) = +5$$
$$3O\ = 3(-2) = -6$$
$$\text{Sum} = \ \ \ 0$$

| SO_4^{2-} | From Rule 4, the oxidation number of O is -2. The oxidation number of S is calculated as $+6$ to give a -2 charge to the polyatomic ion (Rule 3). | SO_4^{2-}
 $+6-2$ |

$$S + 4O\ \ \ \ = -2$$
$$S + 4(-2) = -2$$
$$S - 8\ \ \ \ \ = -2$$
$$S\ \ \ \ \ \ \ \ \ = +6$$

Check:
$$1S\ = 1(+6) = +6$$
$$4O = 4(-2) = -8$$
$$\text{Sum} = -2$$
(charge of polyatomic ion $= -2$)

Oxidation numbers assigned to elements in polyatomic ions are not actual charges.

Sample Problem 15.4   Assigning Oxidation Numbers

Assign oxidation numbers to the elements in each of the following:
a. N_2 **b.** NCl_3 **c.** ClO_3^- **d.** SF_6

Solution

a. N_2: Because N_2 is in the elemental state, Rule 1 is used to assign an oxidation number of 0 to each N atom.

$$N_2$$
$$0$$

b. NCl_3: From the list in Rule 4, Cl has an oxidation number of -1. Because the sum of the oxidation numbers of N and Cl must be zero, the oxidation number of N is calculated as $+3$.

$$N + 3Cl\ \ \ = 0$$
$$N + 3(-1) = 0$$
$$N - 3\ \ \ \ \ = 0 \quad N\,Cl_3$$
$$N\ \ \ \ \ \ \ \ = +3 \quad +3-1$$

Check:
$$1N = 1(+3) = +3$$
$$3Cl = 3(-1) = -3$$
$$\text{Sum} = \ \ \ 0$$

c. ClO_3^-: Because in Rule 4, O is above Cl, O is assigned an oxidation number of -2. The oxidation number of Cl must be calculated so that the sum of oxidation numbers equals -1, the charge on the polyatomic ion.

$$Cl + 3O \quad = -1$$
$$Cl + 3(-2) = -1$$
$$Cl - 6 \qquad = -1 \quad ClO_3^-$$
$$Cl \qquad\qquad = +5 \quad {\scriptstyle +5-2}$$

$$\text{Check: } 1Cl = 1(+5) = +5$$
$$\underline{3O = 3(-2) = -6}$$
$$\text{Sum} = -1$$

d. SF_6: From Rule 4, F is assigned an oxidation number of -1. For the neutral compound, the oxidation number of S is calculated as $+6$.

$$S + 6F \quad = 0$$
$$S + 6(-1) = 0$$
$$S - 6 \qquad = 0 \quad SF_6$$
$$S \qquad\qquad = +6 \quad {\scriptstyle +6-1}$$

$$\text{Check: } S = 1(+6) = +6$$
$$\underline{6F = 6(-1) = -6}$$
$$\text{Sum} = \quad 0$$

Study Check

Assign oxidation numbers to the atoms in each of the following:
a. Cl_2O **b.** H_3PO_4 **c.** MnO_4^-

Identifying Oxidation–Reduction Using Oxidation Numbers

Oxidation numbers can be used to identify what is oxidized and what is reduced in a reaction. In oxidation, the loss of electrons increases the oxidation number so that it is higher (more positive) in the product than in the reactant. In reduction, the gain of electrons decreases the oxidation number so that it is lower (more negative) in the product than in the reactant.

Oxidation–Reduction Terminology

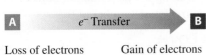

| A | e^- Transfer | B |

Loss of electrons	Gain of electrons
Oxidized	Reduced
Reducing agent	Oxidizing agent
Oxidation number increases	Oxidation number decreases

Reduction: oxidation number decreases ⟵

$$-7\ -6\ -5\ -4\ -3\ -2\ -1\quad 0\ +1\ +2\ +3\ +4\ +5\ +6\ +7$$

Oxidation: oxidation number increases ⟶

For example, the oxidation numbers can be used to identify the element that is oxidized, the element that is reduced, the oxidizing agent, and the reducing agent in the following reaction:

$$CO_2(g) + H_2(g) \longrightarrow CO(g) + H_2O(g)$$

In H_2, the oxidation number of H is 0. In H_2O, the oxidation number of H is $+1$. In CO_2, CO, and H_2O, the oxidation number of O is -2.

$$CO_2(g) + H_2(g) \longrightarrow CO(g) + H_2O(g)$$
$$\underset{?\ -2}{\quad}\qquad \underset{0}{\quad}\qquad\qquad \underset{?\ -2}{\quad}\qquad \underset{+1\ -2}{\quad} \qquad \text{Oxidation numbers}$$

Using the assigned values, the oxidation number of C is calculated as $+4$ in CO_2 and $+2$ in CO.

$$CO_2(g) + H_2(g) \longrightarrow CO(g) + H_2O(g)$$

| +4 −2 | 0 | +2 −2 | +1 −2 | Oxidation numbers |

In H_2, H is oxidized because its oxidation number increases from 0 in the reactant to $+1$ in the product. In CO_2, C is reduced because its oxidation number decreases from $+4$ to $+2$.

H is oxidized

$$CO_2(g) + H_2(g) \longrightarrow CO(g) + H_2O(g)$$

| +4 −2 | 0 | +2 −2 | +1 −2 | Oxidation numbers |

C is reduced

By accepting electrons, CO_2 is the oxidizing agent, and H_2 is the reducing agent because it provides electrons.

Element Oxidized	Element Reduced	Oxidizing Agent	Reducing Agent
H in H_2	C in CO_2	CO_2	H_2

Sample Problem 15.5 **Identifying Oxidized and Reduced Substances**

In each of the following, identify the substance that is oxidized, the substance that is reduced, the oxidizing agent, and the reducing agent:
a. $PbO(s) + CO(g) \longrightarrow Pb(s) + CO_2(g)$
b. $2Fe(s) + 3Cl_2(g) \longrightarrow 2FeCl_3(s)$

Solution

a. The O atom in PbO, CO, and CO_2 is assigned an oxidation number of -2. Pb, in the elemental state, has an oxidation number of 0.

$$PbO(s) + CO(g) \longrightarrow Pb(s) + CO_2(g)$$

| ? −2 | ? −2 | 0 | ? −2 | Oxidation numbers |

In PbO, the oxidation number of Pb is calculated as $+2$; the oxidation number of C is calculated as $+2$ in CO and $+4$ in CO_2. The C is oxidized because its oxidation number increases. In PbO, Pb is reduced because its oxidation number decreases.

C is oxidized

$$PbO(s) + CO(g) \longrightarrow Pb(s) + CO_2(g)$$

| +2 −2 | +2 −2 | 0 | +4 −2 | Oxidation numbers |

Pb^{2+} is reduced

Because Pb^{2+} accepts electrons, PbO is the oxidizing agent. The reducing agent is CO because it provides electrons.

b. An oxidation number of 0 is assigned to Fe and Cl atoms in their elemental states. For $FeCl_3$, the oxidation number of -1 is assigned to Cl, which gives Fe an oxidation number of $+3$.

$$2Fe(s) + 3Cl_2(g) \longrightarrow 2FeCl_3(s)$$
$$\mathbf{0} \qquad\qquad \mathbf{0} \qquad\qquad \mathbf{+3\,-1}$$

Fe is oxidized because its oxidation number increases from 0 to $+3$. Cl in Cl_2 is reduced because the oxidation number of each Cl atom decreases from 0 to -1.

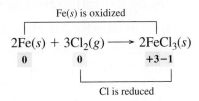

Because Cl_2 accepts the electrons, the Cl_2 is the oxidizing agent. Fe is the reducing agent because it provides electrons.

Study Check

Use oxidation numbers to identify the substances that are oxidized and reduced and to identify which are the oxidizing and reducing agents.

$$Zn(s) + CuCl_2(aq) \longrightarrow ZnCl_2(aq) + Cu(s)$$

Questions and Problems   Oxidation Numbers

15.9 Assign oxidation numbers to each of the following:
    a. Cu   **b.** F_2 **c.** Fe^{2+} **d.** O_2

15.10 Assign oxidation numbers to each of the following:
    a. Al   **b.** Al^{3+} **c.** F^- **d.** N_2

15.11 Assign oxidation numbers to all the elements in each of the following:
    a. KCl   **b.** MnO_2 **c.** CO   **d.** Mn_2O_3

15.12 Assign oxidation numbers to all the elements in each of the following:
    a. H_2S **b.** NO_2 **c.** SF_6 **d.** PCl_3

15.13 Assign oxidation numbers to all the elements in each of the following compounds or polyatomic ions:
    a. $AlPO_4$ **b.** SO_3^{2-} **c.** Cr_2O_3 **d.** NO_3^-

15.14 Assign oxidation numbers to all the elements in each of the following compounds or polyatomic ions:
    a. MnO_4^- **b.** $AlCl_3$ **c.** NH_4^+ **d.** $HBrO_4$

15.15 Assign oxidation numbers to all the elements in each of the following compounds or polyatomic ions:
    a. HSO_4^- **b.** H_3PO_3 **c.** $Cr_2O_7^{2-}$ **d.** Na_2CO_3

15.16 Assign oxidation numbers to all the atoms in each of the following compounds or polyatomic ions:
    a. N_2O **b.** LiOH   **c.** ClO_4^- **d.** IO_4^-

15.17 What is the oxidation number of the specified element in each compound or polyatomic ion?
    a. N in HNO_3 **b.** C in C_3H_6
    c. P in K_3PO_4 **d.** Cr in CrO_4^{2-}

15.18 What is the oxidation number of the specified element in each compound or polyatomic ion?
    a. C in $ZnCO_3$ **b.** Fe in $Fe(NO_3)_2$
    c. Cl in ClF_4^- **d.** S in $S_2O_3^{2-}$

15.19 In each of the following reactions, identify the substance that is oxidized and the substance that is reduced and identify the oxidizing and reducing agents:
    a. $2NiS(s) + 3O_3(g) \longrightarrow 2NiO(s) + 2SO_2(g)$
    b. $Sn^{2+}(aq) + 2Fe^{3+}(aq) \longrightarrow$
                           $Sn^{4+}(aq) + 2Fe^{2+}(aq)$
    c. $CH_4(g) + 2O_2(g) \longrightarrow CO_2(g) + 2H_2O(l)$
    d. $2Cr_2O_3(s) + 3Si(s) \longrightarrow 4Cr(s) + 3SiO_2(s)$

15.20 In each of the following reactions, identify the substance that is oxidized and the substance that is reduced and identify the oxidizing and reducing agents:
    a. $2HgO(s) \longrightarrow 2Hg(l) + O_2(g)$
    b. $Zn(s) + 2HCl(aq) \longrightarrow ZnCl_2(aq) + H_2(g)$
    c. $2Na(s) + 2H_2O(l) \longrightarrow$
                     $2Na^+(aq) + 2OH^-(aq) + H_2(g)$
    d. $6Fe^{2+}(aq) + Cr_2O_7^{2-}(aq) + 14H^+(aq) \longrightarrow$
                    $6Fe^{3+}(aq) + 2Cr^{3+}(aq) + 7H_2O(l)$

Learning Goal

Balance oxidation–reduction equations using the oxidation number or the half-reaction method.

15.3 Balancing Oxidation–Reduction Equations

Many oxidation–reduction equations are too complex to be balanced by inspection. Since the processes of oxidation and reduction always occur simultaneously, the same number of electrons are lost during oxidation and gained during reduction. This requirement of an equal loss and gain of electrons can be used to balance oxidation–reduction equations. The two methods of balancing are the oxidation number method and the half-reaction method.

Using Oxidation Numbers to Balance Oxidation–Reduction Equations

The oxidation number method consists of four steps. Oxidation numbers are assigned to all the substances in order to identify the atoms that are oxidized and reduced. Then the atoms involved in oxidation and reduction are multiplied by small integers to equalize the increase and decrease in oxidation numbers. The oxidation number method is typically used to balance equations written in the molecular form.

Guide to Balancing Equations Using Oxidation Numbers

STEP 1
Assign oxidation numbers to all elements.

STEP 2
Identify the oxidized and reduced elements from the changes in oxidation numbers.

STEP 3
Multiply the changes in oxidation numbers by small integers to equalize increase and decrease.

STEP 4
Balance the remaining elements by inspection.

Sample Problem 15.6 — Using Oxidation Numbers to Balance Equations

Use oxidation numbers to balance the following equation:

$$FeO(s) + C(s) \longrightarrow Fe(s) + CO_2(g)$$

Solution

Using the guide for balancing with oxidation numbers, we can proceed as follows:

STEP 1 Assign oxidation numbers to all the elements.

$$\underset{+2\ -2}{FeO(s)} + \underset{0}{C(s)} \longrightarrow \underset{0}{Fe(s)} + \underset{+4\ -2}{CO_2(g)}$$

STEP 2 Identify the oxidized and reduced elements from the changes in oxidation numbers. The oxidation number of C increases from 0 to +4; C is oxidized. The oxidation number of Fe decreases from +2 to +0; Fe is reduced.

Increases by 4

$$\underset{+2\ -2}{FeO(s)} + \underset{0}{C(s)} \longrightarrow \underset{0}{Fe(s)} + \underset{+4\ -2}{CO_2(g)}$$

Decreases by 2

STEP 3 Multiply the changes in oxidation numbers by small integers to equalize the increase and decrease. Iron needs a multiplying factor of 2 to equalize the increase and decrease of oxidation numbers.

Equalizing Changes in Oxidation Numbers

	Oxidation Numbers	Change	Multiplying Factor	Total
Oxidation:	$C(0) \longrightarrow C(+4)$	4 (increase)	× 1	= 4
Reduction:	$Fe(+2) \longrightarrow Fe(0)$	2 (decrease)	× 2	= 4

The multiplying factor of 2 is used as a coefficient for FeO and Fe. The coefficient of 1 from the multiplying factor for C and CO_2 is understood.

$$\overset{\text{1 × (Increase by 4)}}{2FeO(s) + C(s) \longrightarrow 2Fe(s) + CO_2(g)}$$
$$\underset{+2-2 \qquad\quad 0 \qquad\qquad 0 \qquad +4-2}{}$$
$$\text{2 × (Decrease by 2)}$$

STEP 4 **Balance the remaining elements by inspection.** All the atoms including the O atoms are balanced. The completely balanced equation is

$$2FeO(s) + C(s) \longrightarrow 2Fe(s) + CO_2(g)$$

Study Check

Use oxidation numbers to balance the equation for the oxidation–reduction reaction of

$$Al(s) + HCl(aq) \longrightarrow AlCl_3(aq) + H_2(g)$$

Sample Problem 15.7   **Balancing Equations with Oxidation Numbers**

Use oxidation numbers to balance the equation for the oxidation–reduction reaction of tin and nitric acid.

$$Sn(s) + HNO_3(aq) \longrightarrow SnO_2(s) + NO_2(g) + H_2O(g)$$

Solution

Using the guide for balancing with oxidation numbers, we can proceed as follows:

STEP 1 **Assign oxidation numbers to all the elements.**

$$\underset{0 \qquad\quad +1+5-2 \qquad\quad +4-2 \qquad +4-2 \qquad +1-2}{Sn(s) + HNO_3(aq) \longrightarrow SnO_2(s) + NO_2(g) + H_2O(g)}$$

STEP 2 **Identify the oxidized and reduced elements from the changes in oxidation numbers.** The oxidation number of Sn increases from 0 to +4; Sn is oxidized. The oxidation number of N decreases from +5 to +4; N is reduced.

$$\overset{\text{Increase by 4}}{Sn(s) + HNO_3(aq) \longrightarrow SnO_2(s) + NO_2(g) + H_2O(g)}$$
$$\underset{0 \qquad\quad +1+5-2 \qquad\quad +4\,-2 \qquad +4\,-2 \qquad +1\,-2}{}$$
$$\text{Decrease by 1}$$

STEP 3 **Multiply the changes in oxidation numbers by small integers to equalize the increase and decrease.** Nitrogen needs a multiplying factor of 4 to equalize the increase and decrease of oxidation numbers.

Equalizing Changes in Oxidation Number

	Oxidation Numbers	Change	Multiplying Factor	Total
Oxidation:	$Sn(0) \longrightarrow Sn(+4)$	4 (increase)	× **1**	= 4
Reduction:	$N(+5) \longrightarrow N(+4)$	1 (decrease)	× **4**	= 4

$$
\overset{\text{1 × (Increase by 4)}}{\underset{\substack{0 \qquad +1+5-2 \qquad\qquad +4-2 \qquad +4-2 \qquad +1-2}}{Sn(s) + \textbf{4}HNO_3(aq) \longrightarrow SnO_2(s) + \textbf{4}NO_2(g) + H_2O(g)}}
$$

4 × (Decrease by 1)

The multiplying factor of 4 is used as a coefficient for HNO_3 and NO_2. The coefficient of 1 from the multiplying factor for Sn and SnO_2 is understood.

STEP 4 **Balance the remaining elements by inspection.** The atoms H and O are balanced by using a coefficient of 2 for H_2O. The completely balanced equation is

$$Sn(s) + \textbf{4}HNO_3(aq) \longrightarrow SnO_2(s) + \textbf{4}NO_2(g) + \textbf{2}H_2O(g)$$

Study Check

Use oxidation numbers to balance the equation for the oxidation–reduction reaction of iron(III) oxide and carbon to form iron and carbon dioxide.

$$Fe_2O_3(s) + C(s) \longrightarrow Fe(s) + CO_2(g)$$

Using Half-Reactions to Balance Oxidation–Reduction Equations

In the **half-reaction method,** an oxidation–reduction reaction is written as two *half-reactions.* As each half-reaction is balanced for atoms and charge, it becomes apparent which one is oxidation and which one is reduction. Once the loss and gain of electrons are equalized for the half-reactions, they are combined to obtain the overall balanced equation. The half-reaction method is typically used to balance equations that are written as ionic equations. Let us consider the reaction between aluminum metal and a solution of Cu^{2+} as shown in Sample Problem 15.8.

Sample Problem 15.8 **Using Half-Reactions to Balance Equations**

Use half-reactions to balance the following reaction:

$$Al(s) + Cu^{2+}(aq) \longrightarrow Cu(s) + Al^{3+}(aq)$$

Solution

STEP 1 **Write two half-reactions for the equation.** The equation is divided into two half-reactions, one containing Al and the other containing Cu.

$$Al(s) \longrightarrow Al^{3+}(aq)$$
$$Cu^{2+}(aq) \longrightarrow Cu(s)$$

Guide to Balancing Redox Equations Using Half-Reactions

STEP 1
Write two half-reactions for the equation.

STEP 2
Balance elements other than H and O in each half-reaction. Add H_2O to the side that needs O and add H^+ to the side that needs H.

STEP 3
Balance each half-reaction for charge by adding electrons to the side with more positive charge.

STEP 4
Multiply each half-reaction by factors that equalize the loss and gain of electrons.

STEP 5
Add half-reactions, cancel electrons, and combine H_2O and H^+. Check balance of atoms and charge.

STEP 2 **Balance the elements other than H and O in each half-reaction.** In both half-reactions, the atoms Al and Cu are already balanced.

$$Al(s) \longrightarrow Al^{3+}(aq)$$
$$Cu^{2+}(aq) \longrightarrow Cu(s)$$

STEP 3 **Balance the charge in each half-reaction by adding electrons to the side with more positive charge.**

$$Al(s) \longrightarrow \underbrace{Al^{3+}(aq) + 3e^-}$$ Oxidation
0 charge = 0 charge

$$\underbrace{Cu^{2+}(aq) + 2e^-} \longrightarrow Cu(s)$$ Reduction
0 charge = 0 charge

STEP 4 **Multiply each half-reaction by factors that equalize the loss and gain of electrons.** Three electrons are lost in the oxidation of Al, and 2 electrons are gained in the reduction of Cu^{2+}. To equalize the loss and gain of electrons, we multiply the oxidation half-reaction by 2 and the reduction half-reaction by 3.

$$2 \times [Al(s) \longrightarrow Al^{3+}(aq) + 3e^-]$$
$$2Al(s) \longrightarrow 2Al^{3+}(aq) + 6e^- \quad 6e^- \text{ lost}$$
$$3 \times [Cu^{2+}(aq) + 2e^- \longrightarrow Cu(s)]$$
$$3Cu^{2+}(aq) + 6e^- \longrightarrow 3Cu(s) \quad 6e^- \text{ gained}$$

STEP 5 **Add balanced half-reactions and cancel electrons. Check balance of atoms and charge.**

$$2Al(s) \longrightarrow 2Al^{3+}(aq) + 6e^-$$
$$3Cu^{2+}(aq) + 6e^- \longrightarrow 3Cu(s)$$
$$\overline{2Al(s) + 3Cu^{2+}(aq) + \cancel{6e^-} \longrightarrow 2Al^{3+}(aq) + \cancel{6e^-} + 3Cu(s)}$$

Final balanced equation:

$$2Al(s) + 3Cu^{2+}(aq) \longrightarrow 2Al^{3+}(aq) + 3Cu(s)$$

Check balance of atoms and charge.

Atoms:	2Al	=	2Al
	3Cu	=	3Cu
Charge:	6+	=	6+

Study Check

Use the half-reaction method to balance the reaction

$$Zn(s) + Fe^{3+}(aq) \longrightarrow Zn^{2+}(aq) + Fe^{2+}(aq)$$

Sample Problem 15.9 **Using Half-Reactions to Balance Equations in Acidic Solutions**

Use half-reactions to balance the following reaction that takes place in acidic solution:

$$Fe^{2+}(aq) + MnO_4^-(aq) \longrightarrow Mn^{2+}(aq) + Fe^{3+}(aq)$$

Solution

STEP 1 **Write two half-reactions for the equation.** We separate the equation into two half-reactions by writing one half-reaction for Fe and one for Mn.

$$Fe^{2+}(aq) \longrightarrow Fe^{3+}(aq)$$
$$MnO_4^-(aq) \longrightarrow Mn^{2+}(aq)$$

STEP 2 **Balance the elements other than H and O in each half-reaction. In acidic solutions, balance O by adding H_2O and then balance H by adding H^+.**

$$Fe^{2+}(aq) \longrightarrow Fe^{3+}(aq)$$
$$MnO_4^-(aq) \longrightarrow Mn^{2+}(aq) + \textbf{4H}_2\textbf{O}(l) \qquad \text{H_2O balances O.}$$
$$\textbf{8H}^+(aq) + MnO_4^-(aq) \longrightarrow Mn^{2+}(aq) + 4H_2O(l) \qquad \text{H^+ balances H.}$$

STEP 3 **Balance the charge in each half-reaction by adding electrons to the side with more positive charge.**

$$Fe^{2+}(aq) \longrightarrow \underbrace{Fe^{3+}(aq) + \textbf{1}e^-}$$
$$\underset{\text{2+ charge}}{\phantom{Fe^{2+}(aq)}} \quad = \quad \underset{\text{2+ charge}}{\phantom{Fe^{3+}}}$$

$$\underbrace{\textbf{8H}^+(aq) + MnO_4^-(aq) + \textbf{5}e^-} \longrightarrow Mn^{2+}(aq) + 4H_2O(l)$$
$$\underset{\text{2+ charge}}{} \quad = \quad \underset{\text{2+ charge}}{}$$

STEP 4 **Multiply each half-reaction by factors that equalize the loss and gain of electrons.** The half-reaction with Fe is multiplied by 5 to equal the gain of $5e^-$ by Mn.

$$\textbf{5} \times [Fe^{2+}(aq) \longrightarrow Fe^{3+}(aq) + 1e^-]$$
$$\textbf{5}\,Fe^{2+}(aq) \longrightarrow \textbf{5}Fe^{3+}(aq) + \textbf{5}e^- \qquad \text{$5e^-$ lost}$$
$$8H^+(aq) + MnO_4^-(aq) + \textbf{5}e^- \longrightarrow Mn^{2+}(aq) + 4H_2O(l) \qquad \text{$5e^-$ gained}$$

STEP 5 **Add balanced half-reactions, cancel electrons, and combine any H_2O and H^+. Check balance of atoms and charge.**

$$5Fe^{2+}(aq) \longrightarrow 5Fe^{3+}(aq) + 5e^-$$
$$\underline{8H^+(aq) + MnO_4^-(aq) + \textbf{5}e^- \longrightarrow Mn^{2+}(aq) + 4H_2O(l)}$$
$$5Fe^{2+}(aq) + 8H^+(aq) + MnO_4^-(aq) + \cancel{5e^-} \longrightarrow 5Fe^{3+}(aq) + \cancel{5e^-} + Mn^{2+}(aq) + 4H_2O(l)$$

Final balanced equation:

$$5Fe^{2+}(aq) + 8H^+(aq) + MnO_4^-(aq) \longrightarrow$$
$$5Fe^{3+}(aq) + Mn^{2+}(aq) + 4H_2O(l)$$

Check balance of atoms and charge.

Atoms:	5Fe	=	5Fe
	1Mn	=	1Mn
	8H	=	8H
	4O	=	4O
Charge:	17+	=	17+

Study Check

Use half-reactions to balance the following equation in acidic solution:

$$NO(g) + MnO_4^-(aq) \longrightarrow NO_3^-(aq) + MnO_2(s)$$

Sample Problem 15.10   **Using Half-Reactions to Balance Equations in Acidic Solutions**

Use half-reactions to balance the following reaction that takes place in acidic solution:

$$I^-(aq) + Cr_2O_7^{2-}(aq) \longrightarrow I_2(s) + Cr^{3+}(aq)$$

Solution

STEP 1 **Write two half-reactions for the equation.** One of the half-reactions is written for I and the other for Cr.

$$I^-(aq) \longrightarrow I_2(s)$$
$$Cr_2O_7^{2-}(aq) \longrightarrow Cr^{3+}(aq)$$

STEP 2 **Balance the elements other than H and O in each half-reaction. In acidic solutions, balance O by adding H_2O and then balance H by adding H^+.** The two I atoms in I_2 are balanced with a coefficient of 2 for I^-.

$$2I^-(aq) \longrightarrow I_2(s)$$

The two Cr atoms are balanced with a coefficient of 2 for Cr^{3+}.

$$Cr_2O_7^{2-}(aq) \longrightarrow 2Cr^{3+}(aq)$$
$$\mathbf{14H^+}(aq) + Cr_2O_7^{2-}(aq) \longrightarrow 2Cr^{3+}(aq) + \mathbf{7H_2O}(l)$$

STEP 3 **Balance the charge in each half-reaction by adding electrons to the side with more positive charge.**

$$2I^-(aq) \qquad\qquad \longrightarrow I_2(aq) + \mathbf{2e^-}$$
$$\text{2− charge} \qquad\qquad = \qquad \text{2− charge}$$

$$\mathbf{6e^-} + 14H^+(aq) + Cr_2O_7^{2-}(aq) \longrightarrow 2Cr^{3+}(aq) + 7H_2O(l)$$
$$\text{6+ charge} \qquad = \qquad \text{6+ charge}$$

STEP 4 **Multiply each half-reaction by factors that equalize the loss and gain of electrons.** The half-reaction with I is multiplied by 3 to equal the gain of $6e^-$ by Cr.

$$\mathbf{3} \times [2I^-(aq) \longrightarrow I_2(s) + 2e^-]$$
$$6I^-(aq) \longrightarrow 3I_2(s) + \mathbf{6e^-} \qquad\qquad 6e^-\ \text{lost; oxidation}$$
$$\mathbf{6e^-} + 14H^+(aq) + Cr_2O_7^{2-}(aq) \longrightarrow 2Cr^{3+}(aq) + \mathbf{7H_2O}(l) \qquad 6e^-\ \text{gained; reduction}$$

STEP 5 **Add balanced half-reactions, cancel electrons, and combine any H_2O and H^+. Check balance of atoms and charge.**

$$6I^-(aq) \longrightarrow 3I_2(aq) + \mathbf{6e^-}$$
$$\mathbf{6e^-} + 14H^+(aq) + Cr_2O_7^{2-}(aq) \longrightarrow 2Cr^{3+}(aq) + 7H_2O(l)$$
$$\overline{\mathbf{6e^-} + 14H^+(aq) + Cr_2O_7^{2-}(aq) + 6I^-(aq) \longrightarrow 2Cr^{3+}(aq) + 3I_2(aq) + 7H_2O(l) + \mathbf{6e^-}}$$

Final balanced equation:

$$14H^+(aq) + Cr_2O_7{}^{2-}(aq) + 6I^-(aq) \longrightarrow 2Cr^{3+}(aq) + 3I_2(s) + 7H_2O(l)$$

Check balance of atoms and charge.

Atoms:	6I	=	6I
	2Cr	=	2Cr
	14H	=	14H
	7O	=	7O
Charge:	6+	=	6+

Study Check

Use half-reactions to balance the following equation in acidic solution:

$$Cu^{2+}(aq) + SO_2(g) \longrightarrow Cu(s) + SO_4{}^{2-}(aq)$$

Questions and Problems **Balancing Oxidation–Reduction Equations**

15.21 Use oxidation numbers to balance the following reactions:
  a. $PbS(s) + O_2(g) \longrightarrow PbO(s) + SO_2(g)$
  b. $Fe(s) + Cl_2(g) \longrightarrow FeCl_3(s)$
  c. $Al(s) + H_2SO_4(aq) \longrightarrow Al_2(SO_4)_3(aq) + H_2(g)$

15.22 Use oxidation numbers to balance the following reactions:
  a. $KClO_3(aq) + HBr(aq) \longrightarrow$
      $Br_2(l) + KCl(aq) + H_2O(l)$
  b. $Cu(s) + HNO_3(aq) \longrightarrow$
      $Cu(NO_3)_2(aq) + NO_2(g) + H_2O(l)$
  c. $C_2H_6(g) + O_2(g) \longrightarrow CO_2(g) + H_2O(g)$

15.23 Balance each of the following half-reactions in acidic solution:
  a. $Sn^{2+}(aq) \longrightarrow Sn^{4+}(aq)$
  b. $Mn^{2+}(aq) \longrightarrow MnO_4{}^-(aq)$
  c. $Cr_2O_7{}^{2-}(aq) \longrightarrow Cr^{3+}(aq)$
  d. $ClO_3{}^-(aq) \longrightarrow ClO_2(aq)$

15.24 Balance each of the following half-reactions in acidic solution:
  a. $Cu(s) \longrightarrow Cu^{2+}(aq)$
  b. $ClO^-(aq) \longrightarrow Cl^-(aq)$

  c. $BrO_3{}^-(aq) \longrightarrow Br^-(aq)$
  d. $MnO_4{}^-(aq) \longrightarrow MnO_2(s)$

15.25 Use the half-reaction method to balance each of the following in acidic solution:
  a. $Ag(s) + NO_3{}^-(aq) \longrightarrow Ag^+(aq) + NO_2(g)$
  b. $Fe^{2+}(aq) + ClO_3{}^-(aq) \longrightarrow Fe^{3+}(aq) + Cl^-(aq)$
  c. $NO_3{}^-(aq) + S(s) \longrightarrow NO(g) + SO_2(g)$
  d. $S_2O_3{}^{2-}(aq) + Cu^{2+}(aq) \longrightarrow$
      $S_4O_6{}^{2-}(aq) + Cu(s)$
  e. $PbO_2(s) + Mn^{2+}(aq) \longrightarrow$
      $Pb^{2+}(aq) + MnO_4{}^-(aq)$

15.26 Use the half-reaction method to balance each of the following in acidic solution:
  a. $Sn^{2+}(aq) + IO_4{}^-(aq) \longrightarrow Sn^{4+}(aq) + I^-(aq)$
  b. $Al(s) + ClO^-(aq) \longrightarrow Al^{3+}(aq) + Cl^-(aq)$
  c. $Mn(s) + NO_3{}^-(aq) \longrightarrow Mn^{2+}(aq) + NO_2(g)$
  d. $C_2O_4{}^{2-}(aq) + MnO_4{}^-(aq) \longrightarrow$
      $CO_2(g) + Mn^{2+}(aq)$
  e. $ClO_3{}^-(aq) + SO_3{}^{2-}(aq) \longrightarrow$
      $Cl^-(aq) + SO_4{}^{2-}(aq)$

15.4 Electrical Energy from Oxidation–Reduction Reactions

Learning **Goal**

Write the half-reactions that occur at the anode and cathode of a voltaic cell; write the shorthand cell notation.

We have seen that oxidation–reduction reactions involve a transfer of electrons. If we physically separate one half-reaction from the other in an apparatus called an **electrochemical cell,** the two half-reactions can still occur, but now electrons must flow through an external circuit. When the oxidation–reduction reaction generates electrical energy, the cell is called a **voltaic cell.** In section 15.5, we will describe electrochemical cells called *electrolytic cells,* which require electrical energy to make an oxidation–reduction reaction take place.

Voltaic Cells

When a piece of zinc metal is placed in a Cu^{2+} solution, the silvery zinc becomes coated with a rusty-brown coating of Cu while the blue color (Cu^{2+}) of the solution fades. (See Figure 15.1.) The oxidation of the zinc metal provides electrons for the reduction of the Cu^{2+} ions. We can write the two half-reactions as

$$Zn(s) \longrightarrow Zn^{2+}(aq) + 2e^- \qquad \text{Oxidation}$$
$$Cu^{2+}(aq) + 2e^- \longrightarrow Cu(s) \qquad \text{Reduction}$$

The overall reaction is

$$Zn(s) + Cu^{2+}(aq) \longrightarrow Cu(s) + Zn^{2+}(aq)$$

As long as the Zn metal and Cu^{2+} ions are in the same container, the electrons are transferred directly from Zn to Cu^{2+}. However, the components of the two half-reactions can be placed in separate containers, called *half-cells*, connected by an external circuit. When the electrons flow from one half-cell to the other, an electrical current is produced. In each half-cell, there is a strip of metal, called an *electrode*, in contact with the ionic solution. The electrode where oxidation takes place is called the **anode**; the **cathode** is where reduction takes place. In this example, the anode is a zinc metal strip placed in a Zn^{2+}(ZnSO₄) solution. The cathode is a copper metal strip placed in a Cu^{2+}(CuSO₄) solution. In this voltaic cell, the Zn anode and Cu cathode are connected by a wire that allows electrons to move from the oxidation half-cell to the reduction half-cell. (See Figure 15.2.)

The circuit is completed by a *salt bridge* containing positive and negative ions that is placed in the half-cell solutions. The purpose of the salt bridge is to provide ions, such as Na^+ and SO_4^{2-} ions, to maintain an electrical balance in each half-cell solution. As oxidization occurs at the Zn anode, there is an increase in Zn^{2+} ions, which is balanced by SO_4^{2-} anions from the salt bridge. At the cathode, there is a loss of positive charge as Cu^{2+} is reduced to Cu, which is balanced by SO_4^{2-} in the solution moving into the salt bridge and Na^+ moving out into solution. The complete circuit involves the flow of electrons from the anode to the cathode and the flow of anions from the cathode solution to the anode solution.

Figure 15.2 In this voltaic cell, the Zn anode is in a Zn^{2+} solution, and the Cu cathode is in a Cu^{2+} solution. Electrons produced by the oxidation of Zn move out of the Zn anode through the wire and into the Cu cathode where they reduce Cu^{2+} to Cu. As electrons flow through the wire, the circuit is completed by the flow of SO_4^{2-} through the salt bridge.

Q *Which electrode will be heavier when the reaction ends?*

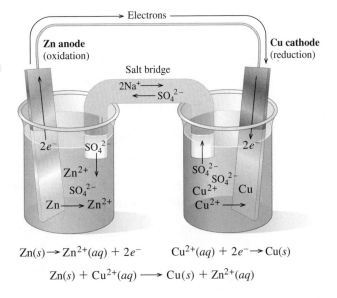

$$Zn(s) \longrightarrow Zn^{2+}(aq) + 2e^- \qquad Cu^{2+}(aq) + 2e^- \longrightarrow Cu(s)$$
$$Zn(s) + Cu^{2+}(aq) \longrightarrow Cu(s) + Zn^{2+}(aq)$$

Anode is where

    oxidation takes place

    electrons are produced

$$Zn(s) \longrightarrow Zn^{2+}(aq) + 2e^-$$

Cathode is where

    reduction takes place

    electrons are used up

$$Cu^{2+}(aq) + 2e^- \longrightarrow Cu(s)$$

An electrical current is produced as electrons flow from the anode through the wire to the cathode. Eventually, the loss of Zn decreases the mass of the Zn anode, while the formation of Cu increases the mass of the Cu cathode. We can diagram the cell using a shorthand notation as follows:

$$Zn(s) \,|\, Zn^{2+}(aq) \,\|\, Cu^{2+}(aq) \,|\, Cu(s)$$

The components of the oxidation half-cell (anode) are written on the left side in this shorthand notation, and the components of the reduction half-cell (cathode) are written on the right. A single vertical line separates the solid Zn anode from the ionic Zn^{2+} solution and the Cu^{2+} solution from the Cu cathode. A double vertical line separates the two half-cells. Reading the notation from left to right indicates that Zn is oxidized to Zn^{2+}, and Cu^{2+} is reduced to Cu as electrons move through the wire from left to right.

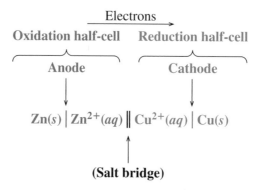

(Salt bridge)

In some voltaic cells, there is no component in the half-reactions that can be used as an electrode. When this is the case, electrodes made of graphite or platinum are used for the transfer of electrons. If there are two ionic components in a cell, their symbols are separated by a comma. For example, suppose a voltaic cell consists of a platinum anode placed in a Sn^{2+} solution such as $Sn(NO_3)_2$ and a silver cathode placed in a Ag^+ solution such as $AgNO_3$. The notation for the cell would be written as

$$Pt(s) \,|\, Sn^{2+}(aq),\, Sn^{4+}(aq) \,\|\, Ag^+(aq) \,|\, Ag(s)$$

The oxidation reaction at the anode is

$$Sn^{2+}(aq) \longrightarrow Sn^{4+}(aq) + 2e^-$$

The reduction reaction at the cathode is

$$Ag^+(aq) + e^- \longrightarrow Ag(s)$$

To balance the overall cell reaction, we multiply the cathode reduction by 2 and combine the two half-reactions.

$$
\begin{aligned}
2Ag^+(aq) + \mathbf{2}e^- &\longrightarrow 2Ag(s) \\
Sn^{2+}(aq) &\longrightarrow Sn^{4+}(aq) + \mathbf{2}e^- \\
\hline
Sn^{2+}(aq) + 2Ag^+(aq) &\longrightarrow Sn^{4+}(aq) + 2Ag(s)
\end{aligned}
$$

To operate the cell, a wire connects the Pt anode and the Ag cathode and a salt bridge is placed in the Sn^{2+} and Ag^+ solutions.

Sample Problem 15.11 — Diagramming a Voltaic Cell

A voltaic cell consists of an iron (Fe) anode in a Fe^{2+} solution [$Fe(NO_3)_2$] and a tin (Sn) cathode is placed in a Sn^{2+} solution [$Sn(NO_3)_2$]. Write the cell notation, the oxidation and reduction half-reactions, and the overall cell reaction.

Solution

The notation for the cell would be written as

$$Fe(s) \,|\, Fe^{2+}(aq) \,\|\, Sn^{2+}(aq) \,|\, Sn(s)$$

The oxidation reaction at the anode is

$$Fe(s) \longrightarrow Fe^{2+}(aq) + 2e^-$$

The reduction reaction at the cathode is

$$Sn^{2+}(aq) + 2e^- \longrightarrow Sn(s)$$

To write the overall cell reaction, we combine the two half-reactions.

$$
\begin{array}{l}
Fe(s) \longrightarrow Fe^{2+}(aq) + 2e^- \\
\underline{Sn^{2+}(aq) + 2e^- \longrightarrow Sn(s)} \\
Fe(s) + Sn^{2+}(aq) \longrightarrow Fe^{2+}(aq) + Sn(s)
\end{array}
$$

To operate the cell, a wire connects the Fe anode and the Sn cathode and a salt bridge is placed in the Fe^{2+} and Sn^{2+} solutions.

Study Check

Write the half-reactions and the overall cell reaction for the following notation of a voltaic cell:

$$Co(s) \,|\, Co^{2+}(aq) \,\|\, Cu^{2+}(aq) \,|\, Cu(s)$$

Prevention of Corrosion

Corrosion of metals has been a major problem for centuries. Because so many building materials involve iron or iron and carbon (steel), the corrosion of iron is detrimental to the strength of girders, cars, and ships. Buildings can collapse, the hulls of ships get holes, and pipes laid underground crumble. Many billions of dollars are spent each year to prevent corrosion and repair building materials made of iron. One way to prevent corrosion is to paint the bridges, cars, and ships with paints containing materials that seal the iron surface from H_2O and O_2. But it is necessary to repaint often and a scratch in the paint exposes the iron, which then begins to rust.

A more effective way to prevent corrosion is to place the iron in contact with a metal that substitutes for the anode region of iron. Metals such as Zn, Mg, or Al lose electrons more easily than iron. When one of these metals is in contact with iron, the metal acts as the anode instead of iron. For example, in a process called *galvanization,* an object made of iron is coated with zinc. The zinc becomes the anode because zinc loses electrons more easily than Fe. As long as Fe does not act as an anode, rust cannot form.

CHEM NOTE

CORROSION: OXIDATION OF METALS

Metals used in building materials, such as iron, eventually oxidize, which causes deterioration of the metal. Known as *corrosion,* this process results in rust and other corrosion on cars, bridges, ships, and underground pipes.

$$4Fe(s) + 3O_2(g) \longrightarrow 2Fe_2O_3(s)$$
Rust

The formation of rust requires both oxygen and water. The process of rusting requires an anode and cathode in different places on the surface of a piece of iron. In one area of the iron surface, called the anode region, the oxidation half-reaction takes place. (See Figure 15.3.)

Anode (oxidation): $Fe(s) \longrightarrow Fe^{2+}(aq) + 2e^-$

or $2Fe(s) \longrightarrow 2Fe^{2+}(aq) + 4e^-$

The electrons move through the iron metal from the anode to an area called the cathode region where oxygen dissolved in water is reduced to water.

Cathode (reduction):
$$O_2(g) + 4H^+(aq) + 4e^- \longrightarrow 2H_2O(l)$$

By combining the half-reactions that occur in the anode and cathode regions, we can write the overall oxidation–reduction process.

$$2Fe(s) + O_2(g) + 4H^+(aq) \longrightarrow 2Fe^{2+}(aq) + 2H_2O(l)$$

The formation of rust occurs as Fe^{2+} ions move out of the anode region and come in contact with dissolved oxygen (O_2).

Figure 15.3 Rust forms when regions on the surface of iron metal establish an electrochemical cell. Electrons from the oxidation of Fe flow from the anode to the cathode region where oxygen is reduced. As Fe^{2+} ions come in contact with O_2 and H_2O, rust forms.

Q *Why must both O_2 and H_2O be present for the corrosion of iron?*

The Fe^{2+} oxidizes to give Fe^{3+}, which reacts with oxygen to form rust.

$$4Fe^{2+}(aq) + O_2(g) + 4H_2O(l) \longrightarrow 2Fe_2O_3(s) + 8H^+(aq)$$
Rust

We can write the formation of rust starting with solid Fe reacting with O_2 as follows. There is no H^+ in the overall equation because H^+ is used and produced in equal quantities.

Corrosion of iron
$$4Fe(s) + 3O_2(g) \longrightarrow 2Fe_2O_3(s)$$
Rust

Other metals such as aluminum, copper, and silver also undergo corrosion, but at a slower rate than iron. The oxidation of Al on the surface of an aluminum object produces Al^{3+}, which reacts with oxygen in the air to form a protective coating of Al_2O_3. This Al_2O_3 coating can prevent any further oxidation of the aluminum underneath it.

$$Al(s) \longrightarrow Al^{3+}(aq) + 3e^-$$

When copper is used on a dome or a steeple, it oxidizes to Cu^{2+}, which is converted to a green patina of $Cu_2(OH)_2CO_3$.

$$Cu(s) \longrightarrow Cu^{2+}(aq) + 2e^-$$

When we use silver dishes and utensils, the Ag^+ ion from oxidation reacts with sulfides in food to form Ag_2S, which we call "tarnish."

$$Ag(s) \longrightarrow Ag^+(aq) + e^-$$

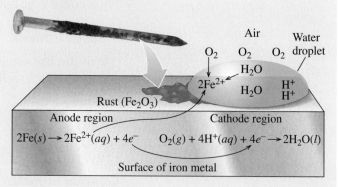

In a method called *cathodic protection,* structures such as iron pipes and underground storage containers are placed in contact with a piece of metal such as Mg, Al, or Zn. Again, because these metals lose electrons more easily than Fe, they become the anode, thereby preventing the rusting of the iron. As long as iron acts only as a cathode for the reduction of $O_2(g)$, the iron will not corrode. For example, a magnesium plate welded or bolted to a ship's hull loses electrons more easily than iron or steel and protects the hull from rusting. Occasionally, a new magnesium plate is added to replace the magnesium as it is used up. Magnesium stakes placed in the ground are connected to underground pipelines and storage containers to prevent corrosion damage.

$$Mg(s) \longrightarrow Mg^{2+}(aq) + 2e^-$$

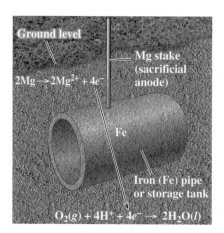

Batteries

Batteries are needed to power a cell phone, a watch, and a calculator. Batteries also are needed to make cars start and flashlights produce light. Within each of these batteries are voltaic cells that produce electrical energy. Let's look at some examples of commonly used batteries.

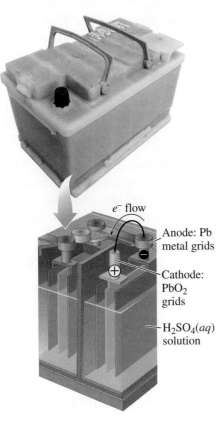

1. A *lead storage battery* is the type of battery used to operate the electrical system in a car. We need a car battery to start the engine, turn on the lights, or operate the radio. If the battery runs down, the car won't start and the lights won't turn on. A car battery or a lead storage battery is a type of voltaic cell. In a typical 12-V battery, there are six voltaic cells linked together. Each of the cells consists of a lead (Pb) plate that acts as the anode and a lead(IV) oxide (PbO_2) plate that acts as the cathode. Both half-cells contain a sulfuric acid (H_2SO_4) solution. When the car battery is producing electrical energy (discharging), the following half-reactions take place:

Anode (oxidation):
$$Pb(s) + SO_4^{2-}(aq) \longrightarrow PbSO_4(s) + 2e^-$$
Cathode (reduction):
$$PbO_2(s) + 4H^+(aq) + SO_4^{2-}(aq) + 2e^- \longrightarrow PbSO_4(s) + 2H_2O(l)$$

Overall cell reaction:
$$Pb(s) + PbO_2(s) + 4H^+(aq) + 2SO_4^{2-}(aq) \longrightarrow 2PbSO_4(s) + 2H_2O(l)$$

In both half-reactions, Pb^{2+} is produced and combines with SO_4^{2-} to form an insoluble salt $PbSO_4(s)$. As a car battery is used, there is a buildup of $PbSO_4$ on the electrodes. At the same time, there is a decrease in the concentrations of the sulfuric acid components, H^+ and SO_4^{2-}. As a car runs, the battery is continuously recharged by an alternator, which is powered by the engine. The recharging reactions restore the Pb and PbO_2 electrodes as well as the H_2SO_4. Without recharging, the car battery cannot continue to produce electrical energy.

2. *Dry-cell batteries* are used in calculators, watches, flashlights, and battery-operated toys. The term *dry cell* describes a battery that uses a paste rather than an aqueous solution. Dry cells can be *acidic* or *alkaline*. In an acidic dry cell, the anode is a zinc metal case that contains a paste of MnO_2, NH_4Cl, $ZnCl_2$, H_2O, and starch. Within this MnO_2 electrolyte mixture is a graphite cathode.

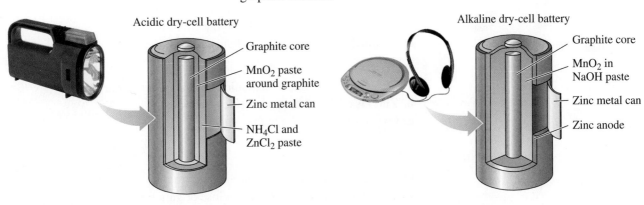

Anode (oxidation): $Zn(s) \longrightarrow Zn^{2+}(aq) + 2e^-$

Cathode (reduction): $2MnO_2(s) + 2NH_4^+(aq) + 2e^- \longrightarrow Mn_2O_3(s) + 2NH_3(aq) + H_2O(l)$

Overall cell reaction: $Zn(s) + 2MnO_2(s) + 2NH_4^+(aq) \longrightarrow Zn^{2+}(aq) + Mn_2O_3(s) + 2NH_3(aq) + H_2O(l)$

An alkaline battery has similar components except that NaOH or KOH replaces the NH_4Cl electrolyte. Under basic conditions, the product of oxidation is zinc oxide (ZnO). Alkaline batteries tend to be more expensive but they last longer and produce more power than acidic dry-cell batteries.

Anode (oxidation): $Zn(s) + 2OH^-(aq) \longrightarrow ZnO(s) + H_2O(l) + \boldsymbol{2e^-}$
Cathode (reduction): $2MnO_2(s) + H_2O(l) + \boldsymbol{2e^-} \longrightarrow Mn_2O_3(s) + 2OH^-(aq)$
___
Overall cell reaction: $Zn(s) + 2MnO_2(s) \longrightarrow ZnO(s) + Mn_2O_3(s)$

3. *Mercury and lithium batteries* are similar to alkaline dry-cell batteries. For example, a mercury battery has a zinc anode, but the cathode is steel in a mixture of HgO, KOH, and $Zn(OH)_2$. The reduced product Hg is toxic and an environmental hazard. Mercury batteries come with warnings on the label and should be disposed of properly.

Anode (oxidation): $Zn(s) + 2OH^-(aq) \longrightarrow ZnO(s) + H_2O(l) + \boldsymbol{2e^-}$
Cathode (reduction): $HgO(s) + H_2O(l) + \boldsymbol{2e^-} \longrightarrow Hg(l) + 2OH^-(aq)$
___
Overall cell reaction: $Zn(s) + HgO(s) \longrightarrow ZnO(s) + Hg(l)$

In a lithium battery, the anode is lithium, not zinc. Lithium is much less dense than zinc and a lithium battery can be made very small.

4. *Nickel–cadmium (NiCad) batteries* can be recharged. They use a cadmium anode and a cathode of solid nickel oxide $NiO(OH)(s)$.

Anode (oxidation): $Cd(s) + 2OH^-(aq) \longrightarrow Cd(OH)_2(s) + \boldsymbol{2e^-}$
Cathode (reduction): $2NiO(OH)(s) + 2H_2O(l) + \boldsymbol{2e^-} \longrightarrow 2Ni(OH)_2(s) + 2OH^-(aq)$
___
Overall cell reaction: $Cd(s) + 2NiO(OH)(s) + 2H_2O(l) \longrightarrow Cd(OH)_2(s) + 2Ni(OH)_2(s)$

NiCad batteries are expensive, but they can be recharged many times. A charger provides an electrical current that converts the solid $Cd(OH)_2$ and $Ni(OH)_2$ products in the NiCad battery back to the reactants needed for oxidation and reduction.

Sample Problem 15.12 ▸ Batteries

The following half-reaction takes place in a dry-cell battery used in portable radios and flashlights:

$$Zn(s) \longrightarrow Zn^{2+}(aq) + 2e^-$$

a. Is this half-reaction an oxidation or a reduction? Why?
b. What substance is oxidized or reduced? Why?
c. At which electrode does this half-reaction occur?

Solution

a. This half-reaction is an oxidation because electrons are lost.
b. $Zn(s)$ is oxidized because it loses electrons.
c. Oxidation takes place at the anode.

Study Check

In the half-reaction described in Sample Problem 15.12, does the electrode where the half-reaction takes place lose or gain mass? Why?

CHEM NOTE

FUEL CELLS: CLEAN ENERGY FOR THE FUTURE

Fuel cells are of interest to scientists because they provide an alternative source of electrical energy that is more efficient, does not use up oil reserves, and generates products that do not pollute the atmosphere. Fuels cells are considered to be a clean way to produce energy.

Like other cells, a fuel cell consists of an anode and a cathode connected by a wire. But unlike other cells, the reactants continuously enter the fuel cell to produce energy, and the fuel cells provide an electrical current only as long as the fuels are supplied. One type of hydrogen–oxygen fuel cell has been used in automobile prototypes. In this hydrogen cell, gas enters the fuel cell and comes in contact with a platinum catalyst embedded in a plastic membrane. The catalyst assists in the oxidation of hydrogen atoms to protons and electrons. (See Figure 15.4.)

$$2H_2(g) \longrightarrow 4H^+(aq) + 4e^- \qquad \text{Oxidation}$$

The electrons produce an electric current as they travel through the wire from the anode to the cathode. The protons flow through the plastic membrane to the cathode. At the cathode, oxygen molecules are reduced to oxide ions that combine with the protons to form water.

$$O_2(g) + 4H^+(aq) + 4e^- \longrightarrow 2H_2O(l) \qquad \text{Reduction}$$

The overall hydrogen–oxygen fuel cell reaction can be written as

$$2H_2(g) + O_2(g) \longrightarrow 2H_2O(l)$$

Fuel cells have already been used to power the space shuttle and may soon be available to produce energy for cars and buses. A major drawback to the practical use of fuel cells is the economic impact of converting cars to fuel cell operation. The storage and cost of producing hydrogen are also problems. Some manufacturers are experimenting with systems that convert gasoline or methanol to hydrogen for immediate use in fuel cells.

In homes, fuel cells may one day replace the batteries currently used to provide electrical power for cell phones, CD and DVD players, and laptop computers. Fuel cell design is still in the prototype phase, although there is much interest in their development. We already know they can work, but modifications must still be made before they become reasonably priced and part of our everyday lives.

Oxidation
$$H_2(g) \longrightarrow 4H^+(aq) + 4e^-$$

Reduction
$$O_2(g) + 4H^+(aq) + 4e^- \longrightarrow 2H_2O(l)$$

Figure 15.4 With a supply of hydrogen and oxygen, a fuel cell can generate electricity continuously.

Q *In most electrochemical cells, the electrodes are eventually used up. Is this true for a fuel cell? Why or why not?*

Questions and Problems ▷ **Electrical Energy from Oxidation–Reduction Reactions**

15.27 Write the half-reactions and the overall cell reaction from the following voltaic-cell diagrams:
  a. $Pb(s) | Pb^{2+}(aq) \| Cu^{2+}(aq) | Cu(s)$
  b. $Cr(s) | Cr^{2+}(aq) \| Ag^{+}(aq) | Ag(s)$

15.28 Write the half-reactions and the overall cell reaction from the following voltaic-cell diagrams:
  a. $Al(s) | Al^{3+}(aq) \| Cd^{2+}(aq) | Cd(s)$
  b. $Sn(s) | Sn^{2+}(aq) \| Fe^{3+}(aq), Fe^{2+}(aq) | C \text{ (graphite)}$

15.29 Describe the voltaic cell and half-cell components and write the shorthand notation for the following oxidation–reduction reactions:
  a. $Cd(s) + Sn^{2+}(aq) \longrightarrow Cd^{2+}(aq) + Sn(s)$
  b. $Zn(s) + Cl_2(g) \longrightarrow Zn^{2+}(aq) + 2Cl^{-}(aq)$
     (C graphite cathode)

15.30 Describe the voltaic cell and half-cell components and write the shorthand notation for the following oxidation–reduction reactions:
  a. $Mn(s) + Sn^{2+}(aq) \longrightarrow Mn^{2+}(aq) + Sn(s)$
  b. $Ni(s) + 2Ag^{+}(aq) \longrightarrow Ni^{2+}(aq) + 2Ag(s)$

15.31 The following half-reaction takes place in a nickel–cadmium battery used in a cordless drill:

$$Cd(s) + 2OH^{-}(aq) \longrightarrow Cd(OH)_2(s) + 2e^{-}$$

  a. Is the half-reaction an oxidation or a reduction?
  b. What substance is oxidized or reduced?
  c. At which electrode would this half-reaction occur?

15.32 The following half-reaction takes place in a mercury battery used in hearing aids:

$$HgO(s) + H_2O(l) + 2e^{-} \longrightarrow Hg(l) + 2OH^{-}(aq)$$

  a. Is the half-reaction an oxidation or a reduction?
  b. What substance is oxidized or reduced?
  c. At which electrode would this half-reaction occur?

15.33 The following half-reaction takes place in a mercury battery used in pacemakers and watches:

$$Zn(s) + 2OH^{-}(aq) \longrightarrow ZnO(s) + H_2O(l) + 2e^{-}$$

  a. Is the half-reaction an oxidation or a reduction?
  b. What substance is oxidized or reduced?
  c. At which electrode would this half-reaction occur?

15.34 The following half-reaction takes place in a lead storage battery used in automobiles:

$$Pb(s) + SO_4^{2-}(aq) \longrightarrow PbSO_4(s) + 2e^{-}$$

  a. Is the half-reaction an oxidation or a reduction?
  b. What substance is oxidized or reduced?
  c. At which electrode would this half-reaction occur?

15.5 Oxidation–Reduction Reactions That Require Electrical Energy

Learning **Goal**

> **Describe the half-cell reactions and the overall reactions that occur in electrolytic cells.**

Copper strip

Zn^{2+} solution

In the previous section we looked at oxidation–reduction reactions that were spontaneous. In each example, the reactant that lost electrons more easily was oxidized. For example, in the cell $Zn(s) | Zn^{2+}(aq) \| Cu^{2+}(aq) | Cu(s)$, the reaction was spontaneous because Zn is oxidized more easily than Cu. When we placed a zinc metal strip in a solution of Cu^{2+}, reddish-brown Cu metal accumulated on the Zn strip according to the following spontaneous reaction:

$$Zn(s) + Cu^{2+}(aq) \longrightarrow Cu(s) + Zn^{2+}(aq) \quad \text{Spontaneous}$$

Suppose that we wanted the reverse reaction to occur. If we place a Cu metal strip in a Zn^{2+} solution, nothing will happen. The reaction does not run spontaneously in the reverse direction because Cu does not lose electrons as easily as Zn. We can determine the direction of a spontaneous reaction from the activity series for metals and $H_2(g)$, which ranks the metals and H_2 in terms of how easily they lose electrons. The metals that lose electrons most easily are placed at the top, and the metals that do not lose electrons easily are at the bottom. We would also find that the metals whose ions gain electrons easily are at the bottom. Thus the metals that are most easily oxidized are above the metals whose ions are most easily reduced. The metal that loses electrons most easily is called the *most active* metal; the metal that loses electrons only with difficulty is considered *least active*. (See Table 15.2.) Metals listed below $H_2(g)$ will not react with H^{+} from acids.

Table 15.2   Activity Series for Some Metals

	Metal		Ion
Most active	$Li(s)$	$\longrightarrow$	$Li^+(aq) + e^-$
	$K(s)$	$\longrightarrow$	$K^+(aq) + e^-$
	$Ca(s)$	$\longrightarrow$	$Ca^{2+}(aq) + 2e^-$
	$Na(s)$	$\longrightarrow$	$Na^+(aq) + e^-$
	$Mg(s)$	$\longrightarrow$	$Mg^{2+}(aq) + 2e^-$
	$Al(s)$	$\longrightarrow$	$Al^{3+}(aq) + 3e^-$
	$Zn(s)$	$\longrightarrow$	$Zn^{2+}(aq) + 2e^-$
	$Cr(s)$	$\longrightarrow$	$Cr^{3+}(aq) + 3e^-$
	$Fe(s)$	$\longrightarrow$	$Fe^{2+}(aq) + 2e^-$
	$Ni(s)$	$\longrightarrow$	$Ni^{2+}(aq) + 2e^-$
	$Sn(s)$	$\longrightarrow$	$Sn^{2+}(aq) + 2e^-$
	$Pb(s)$	$\longrightarrow$	$Pb^{2+}(aq) + 2e^-$
	$H_2(g)$	$\longrightarrow$	$2H^+(aq) + 2e^-$
	$Cu(s)$	$\longrightarrow$	$Cu^{2+}(aq) + 2e^-$
	$Ag(s)$	$\longrightarrow$	$Ag^+(aq) + e^-$
Least active	$Au(s)$	$\longrightarrow$	$Au^{3+}(aq) + 3e^-$

According to the activity series, a metal will oxidize spontaneously when it is combined with the ions of any metal below it on the list. For the voltaic cell of $Zn(s)\,|\,Zn^{2+}(aq)\,||\,Cu^{2+}(aq)\,|\,Cu(s)$, Zn is above Cu in the activity series. This means that the spontaneous direction of the oxidation–reduction reaction is for the more active Zn to lose electrons and for Cu^{2+} to gain electrons to form the less active Cu metal. Nothing happens when a Cu strip is placed in Zn^{2+} because Cu is not as active as Zn.

$$Zn(s) + Cu^{2+}(aq) \longrightarrow Zn^{2+}(aq) + Cu(s) \quad \text{Spontaneous}$$

More active $\Longrightarrow$ Less active

We can use the activity series to help us predict the direction of the spontaneous reaction. Suppose we have two beakers. In one we place a Zn strip in a solution with Al^{3+} ions. In the other, we place an Al strip in a solution with Zn^{2+} ions. How can we predict whether or not a reaction will occur? Looking at the activity series we see that Al is listed above Zn, which means that Al is the more active metal and loses electrons more easily than Zn. Therefore, we predict the following half-reactions and overall reaction will occur:

More active metal: $Al(s) \longrightarrow Al^{3+}(aq) + 3e^-$ Spontaneous
Less active metal: $Zn^{2+}(aq) + 2e^- \longrightarrow Zn(s)$ Spontaneous

$$2Al(s) + 3Zn^{2+}(aq) \longrightarrow 2Al^{3+}(aq) + 3Zn(s) \quad \text{Spontaneous}$$

More active $\Longrightarrow$ Less active

Thus, there will be a coating of Zn on the Al strip as the oxidation–reduction reaction takes place. No reaction will occur spontaneously in the beaker containing the Zn strip and Al^{3+} ions.

We can see the activity of metals when various kinds of metals are placed in hydrochloric acid (HCl). Suppose we placed a Zn strip, a Mg strip, and a Cu strip in three beakers, each containing HCl. In the activity series, Zn and

Mg are above H_2, and Cu is below H_2. The strips of Mg and Zn disappear as they are oxidized, while the reduction of H^+ produces lots of H_2 bubbles. The Cu strip does not react with HCl, which means the Cu metal remains intact in the HCl solution and no H_2 bubbles form.

Sample Problem 15.13 ▶ **Predicting Spontaneous Reactions**

A Cr strip is placed in a beaker containing a solution of Ag^+ ions. In another beaker, an Ag strip is placed in a solution with Cr^{3+} ions. Write the half-reactions and the overall equation for the spontaneous reaction that takes place.

Solution

Looking at the activity series in Table 15.2, we see that Cr is listed above Ag, which means that Cr is the more active metal and loses electrons more easily than Ag. Therefore, we predict the following half-reactions and overall reaction will occur:

More active metal:	$Cr(s) \longrightarrow Cr^{3+}(aq) + 3e^-$	Spontaneous
Less active metal:	$Ag^+(aq) + e^- \longrightarrow Ag(s)$	Spontaneous
	$Cr(s) + 3Ag^+(aq) \longrightarrow Cr^{3+}(aq) + 3Ag(s)$	Spontaneous

More active ⟹ Less active

Study Check

Predict whether each of the following reactions will occur spontaneously:
a. $Sn(s) + Fe^{2+}(aq) \longrightarrow Fe(s) + Sn^{2+}(aq)$
b. $3Mg(s) + 2Cr^{3+}(aq) \longrightarrow 2Cr(s) + 3Mg^{2+}(aq)$

Electrolytic Cells

Suppose that we try to reduce Zn^{2+} to Zn in the presence of Cu and Cu^{2+}. When we look at the activity series, we see that Cu is below Zn. This means that the oxidation–reduction reaction in this direction is not spontaneous.

$$Cu(s) + Zn^{2+}(aq) \longrightarrow Zn(s) + Cu^{2+}(aq) \qquad \text{Not spontaneous}$$

Less active — More active

To make a nonspontaneous reaction take place, we need to use an electrical current, which is a process known as **electrolysis.** An **electrolytic cell** is an electrochemical cell in which electrical energy is used to drive a nonspontaneous oxidation–reduction reaction. We can think of the reactants in spontaneous oxidation–reduction reactions (voltaic cells) as rolling down a hill from higher to lower energy, which produces electrical energy. In electrolytic cells, energy must be provided by an outside energy source to push the reactants up a hill from lower energy to higher energy. (See Figure 15.5.)

Electrical power ⟹

$$Cu(s) + Zn^{2+}(aq) \longrightarrow Zn(s) + Cu^{2+}(aq) \qquad \text{Not spontaneous as written}$$

Figure 15.5 In an electrolytic cell, the Cu anode is in a Cu^{2+} solution, and the Zn cathode is in a Zn^{2+} solution. Electrons provided by a battery reduce Zn^{2+} to Zn and drive the oxidation of Cu to Cu^{2+} at the Cu anode.

Q *Why is an electrical current needed to make the reaction of Cu(s) and Zn^{2+} happen?*

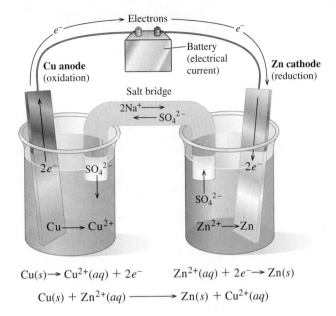

$$Cu(s) \rightarrow Cu^{2+}(aq) + 2e^- \qquad Zn^{2+}(aq) + 2e^- \rightarrow Zn(s)$$

$$Cu(s) + Zn^{2+}(aq) \longrightarrow Zn(s) + Cu^{2+}(aq)$$

Electrolysis of Sodium Chloride

When molten sodium chloride is electrolyzed, the products are sodium metal and chlorine gas. In this electrolytic cell, electrodes are placed in the mixture of Na^+ and Cl^- and connected to a battery. In this cell, the products are separated to prevent them from reacting spontaneously with each other. As electrons flow to the cathode, Na^+ is reduced to sodium metal. At the same time, electrons leave the anode as Cl^- is oxidized to Cl_2. The half-reactions and the overall reactions are

$$
\begin{array}{lll}
\text{Anode (oxidation):} & 2Cl^-(l) & \longrightarrow Cl_2(g) + 2e^- \\
\text{Cathode (reduction):} & 2Na^+(l) + 2e^- & \longrightarrow 2Na(l) \\
\hline
& 2Na^+(l) + 2Cl^-(l) & \longrightarrow Cl_2(g) + 2Na(l)
\end{array}
$$

$\Longrightarrow$ Electrical power

Electroplating

In industry, the process of electroplating uses electrolytic cells to coat a metal with a thin layer of silver, platinum, or gold. Car bumpers are electroplated with chromium to prevent rusting. Silver-plated utensils, bowls, and platters are made by electroplating objects with a layer of silver.

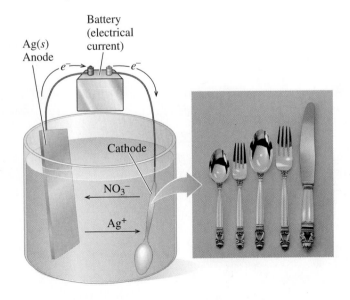

Sample Problem 15.14 **Electrolytic Cells**

Electrolysis is used to chrome-plate an iron hubcap by placing the hubcap in a Cr^{3+} solution.

a. What half-reaction takes place to plate the hubcap with metallic chromium?

b. Is the iron hubcap the anode or the cathode?

Solution

a. The Cr^{3+} ion in solution would gain electrons (reduction).

$$Cr^{3+}(aq) + 3e^- \longrightarrow Cr(s)$$

b. The iron hubcap is the cathode where reduction takes place.

Study Check

Why is electrolysis needed to chrome-plate the iron in Sample Problem 15.14?

Questions and Problems **Oxidation–Reduction Reactions That Require Electrical Energy**

15.35 What we call "tin cans" are really made of an iron can coated with a thin layer of tin.
 a. What half-reaction takes place to plate an iron can with tin?
 b. Is the iron the anode or the cathode?
 c. Why is electrolysis needed to tin-plate the iron?

15.36 Electrolysis is used to gold-plate jewelry made of stainless steel.
 a. What half-reaction takes place to plate Au^{3+} on a stainless steel earring?

b. Is the earring the anode or the cathode?
 c. Why is electrolysis needed to gold-plate the earring?

15.37 When the tin coating on an iron can is scratched, rust will form. Use the activity series to explain why this happens.

15.38 When the zinc coating on an iron can is scratched, rust does not form. Use the activity series to explain why this happens.

Concept Map **Oxidation–Reduction: Transfer of Electrons**

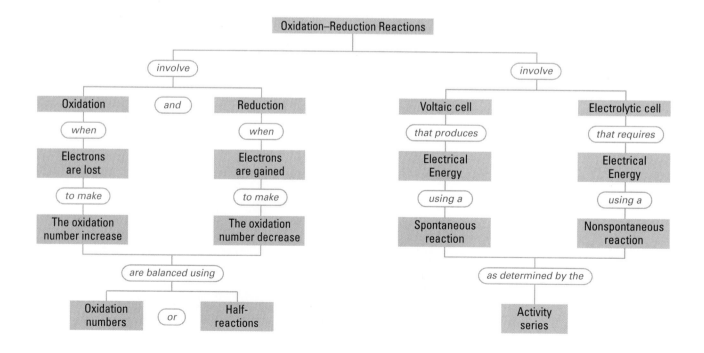

Chapter Review

15.1 Oxidation and Reduction

In oxidation–reduction reactions, electrons are transferred from one reactant to another. The reactant that loses electrons is oxidized, and the reactant that gains electrons is reduced. Oxidation must always occur with reduction. The reducing agent is the substance that provides electrons for reduction. The oxidizing agent is the substance that accepts the electrons from oxidation.

15.2 Oxidation Numbers

Oxidation numbers assigned to atoms keep track of the changes in oxidation numbers and the loss and gain of electrons. Oxidation is an increase in oxidation number; reduction is a decrease in oxidation number. In covalent compounds and polyatomic ions, oxidation numbers are assigned using a set of rules. The oxidation number of an element is zero, and the oxidation number of a monatomic ion is the same as the ionic charge of the ion. The sum of the oxidation numbers for a compound is equal to zero and for a polyatomic ion is equal to the overall charge.

15.3 Balancing Oxidation–Reduction Equations

Balancing oxidation–reduction equations using oxidation numbers involves the following: (1) assigning oxidation numbers; (2) determining the loss and gain of electrons; (3) equalizing the loss and gain of electrons; and (4) balancing the remaining substances by inspection. Balancing oxidation–reduction equations using half-reactions involves the following: (1) separating the equation into half-reactions; (2) balancing non-H, non-O elements, then O with H_2O and H with H^+; (3) balancing charge with electrons; (4) multiplying half-reactions by factors that equalize electrons; (5) combining half-reactions, canceling electrons, and combining H_2O and H^+.

15.4 Electrical Energy from Oxidation–Reduction Reactions

In a voltaic cell, the components of the two half-reactions of a spontaneous oxidation–reduction reaction are placed in separate containers called half-cells. With a wire connecting the half-cells, an electrical current is generated as electrons move from the anode where oxidation takes place to the cathode where reduction takes place.

15.5 Oxidation–Reduction Reactions That Require Electrical Energy

In an electrolytic cell, energy from an external source of electricity is used to make reactions take place that are not spontaneous. A method called electrolysis is used to plate chrome on hubcaps, tin or zinc on iron, or gold on stainless steel jewelry. The activity series, which lists metals with the most easily oxidized metal at the top, can be used to predict the direction of a spontaneous reaction.

Key Terms

anode The electrode where oxidation takes place.

cathode The electrode where reduction takes place.

electrochemical cell An apparatus that produces electrical energy from a spontaneous oxidation–reduction reaction or uses electrical energy to cause a nonspontaneous oxidation–reduction reaction to take place.

electrolysis The use of electrical energy to run a nonspontaneous oxidation–reduction reaction in an electrolytic cell.

electrolytic cell A cell in which electrical energy is used to make a nonspontaneous oxidation–reduction reaction happen.

half-reaction method A method of balancing oxidation–reduction reactions in which the half-reactions are balanced separately and then combined to give the complete reaction.

oxidation The loss of electrons by a substance.

oxidation number A number equal to zero in an element or the charge of a monatomic ion; in covalent compounds and polyatomic ions, oxidation numbers are assigned using a set of rules.

oxidation–reduction reaction A reaction in which electrons are transferred from one reactant to another.

oxidizing agent The reactant that gains electrons and is reduced.

reducing agent The reactant that loses electrons and is oxidized.

reduction The gain of electrons by a substance.

voltaic cell A type of electrochemical cell that uses spontaneous oxidation–reduction reactions to produce electrical energy.

Understanding the Concepts

15.39 Classify each of the following as oxidation or reduction:
   a. Electrons are lost.
   b. Requires an oxidizing agent.
   c. $O_2(g) \longrightarrow OH^-(aq)$
   d. $2Br^-(aq) \longrightarrow Br_2(l)$
   e. $Sn^{2+}(aq) \longrightarrow Sn^{4+}(aq)$

15.40 Classify each of the following as oxidation or reduction:
   a. Electrons are gained.
   b. Requires a reducing agent.
   c. $Al(s) \longrightarrow Al^{3+}(aq)$
   d. $MnO_4^-(aq) \longrightarrow MnO_2(s)$
   e. $Fe^{3+}(aq) \longrightarrow Fe^{2+}(aq)$

15.41 Assign oxidation numbers to the atoms in each of the following:
   a. VO_2
   b. Ag_2CrO_4
   c. $S_2O_8^{2-}$
   d. $FeSO_4$

15.42 Assign oxidation numbers to the atoms in each of the following:
   a. $NbCl_3$
   b. NbO
   c. NbO_2
   d. Nb_2O_5

15.43 Consider the following reaction:

$$Cr_2O_3(s) + Si(s) \longrightarrow Cr(s) + SiO_2(s)$$

   a. Identify the substance reduced.
   b. Identify the substance oxidized.
   c. Identify the oxidizing agent.
   d. Identify the reducing agent.
   e. Write the balanced equation for the overall reaction.

15.44 Consider the following reaction in acid:

$$MnO_4^-(aq) + Cl^-(aq) \longrightarrow Mn^{2+}(aq) + Cl_2(g)$$

   a. Identify the substance reduced.
   b. Identify the substance oxidized.
   c. Identify the oxidizing agent.
   d. Identify the reducing agent.
   e. Write the balanced equation for the overall reaction in acid.

15.45 Consider the following voltaic cell:

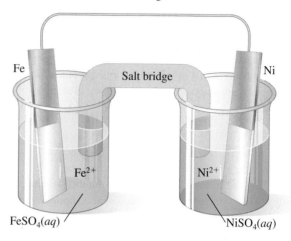

   a. What is the oxidation half-reaction?
   b. What is the reduction half-reaction?
   c. What metal is the anode?
   d. What metal is the cathode?
   e. What is the direction of electron flow?
   f. What is the overall reaction that takes place?
   g. Write the shorthand cell notation.

15.46 Consider the following voltaic cell:

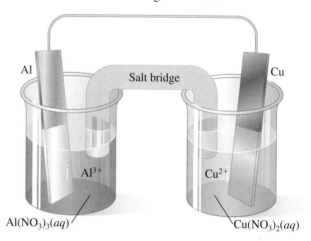

   a. What is the oxidation half-reaction?
   b. What is the reduction half-reaction?
   c. What metal is the anode?
   d. What metal is the cathode?
   e. What is the direction of electron flow?
   f. What is the overall reaction that takes place?
   g. Write the shorthand cell notation.

Additional Questions and Problems

15.47 Which of the following are oxidation–reduction reactions?
  a. $AgNO_3(aq) + NaCl(aq) \longrightarrow$
  $$AgCl(s) + NaNO_3(aq)$$
  b. $6Li(s) + N_2(g) \longrightarrow 2Li_3N(s)$
  c. $Ni(s) + Pb(NO_3)_2(aq) \longrightarrow$
  $$Ni(NO_3)_2(aq) + Pb(s)$$
  d. $2K(s) + 2H_2O(l) \longrightarrow 2KOH(aq) + H_2(g)$

15.48 Which of the following are oxidation–reduction reactions?
  a. $Ca(s) + F_2(g) \longrightarrow CaF_2(s)$
  b. $Fe(s) + 2HCl(aq) \longrightarrow FeCl_2(aq) + H_2(g)$
  c. $2NaCl(aq) + Pb(NO_3)_2(aq) \longrightarrow$
  $$PbCl_2(s) + 2NaNO_3(aq)$$
  d. $2CuCl(aq) \longrightarrow Cu(s) + CuCl_2(aq)$

15.49 In the mitochondria of human cells, energy is provided by the oxidation and reduction of the iron ions in the cytochromes. Identify each of the following reactions as an oxidation or reduction:
  a. $Fe^{3+} + e^- \longrightarrow Fe^{2+}$
  b. $Fe^{2+} \longrightarrow Fe^{3+} + e^-$

15.50 Chlorine (Cl_2) is used as a germicide to kill microbes in swimming pools. If the product is Cl^-, was the elemental chlorine oxidized or reduced?

15.51 State the oxidation numbers for each element in each of the following:
  a. Co_2O_3 **b.** $KMnO_4$ **c.** SF_6
  d. ClO_3^- **e.** PO_4^{3-}

15.52 Assign oxidation numbers to each element in each of the following:
  a. CO_3^{2-} **b.** $SbCl_5$ **c.** $Fe(OH)_2$
  d. HNO_3 **e.** $Cr_2O_7^{2-}$

15.53 Assign oxidation numbers to all the elements in the following reactions and identify the substance oxidized and the substance reduced. Balance the equation.
  a. $FeCl_2(aq) + Cl_2(g) \longrightarrow FeCl_3(aq)$
  b. $Si(s) + Cr_2O_3(s) \longrightarrow Cr(s) + SiO_2(s)$
  c. $Cr_2O_3(s) + Al(s) \longrightarrow Cr(s) + Al_2O_3(s)$
  d. $P_2O_5(s) + C(s) \longrightarrow P(s) + CO(g)$

15.54 Assign oxidation numbers to all the atoms in the following reactions and identify the substance oxidized and the substance reduced. Balance the equation.
  a. $Al(s) + O_2(g) \longrightarrow Al_2O_3(s)$
  b. $I_2O_5(s) + CO(g) \longrightarrow I_2(g) + CO_2(g)$
  c. $Fe_2O_3(s) + C(s) \longrightarrow Fe(l) + CO_2(g)$
  d. $H_2S(g) + HNO_3(aq) \longrightarrow$
  $$S(s) + NO(g) + H_2O(l)$$

15.55 Balance the following half-reactions in acidic solution:
  a. $Zn(s) \longrightarrow Zn^{2+}(aq)$
  b. $SnO_2^{2-}(aq) \longrightarrow SnO_3^{2-}(aq)$
  c. $Cr_2O_7^{2-}(aq) \longrightarrow Cr^{3+}(aq)$
  d. $NO_3^-(aq) \longrightarrow NO(g)$

15.56 Balance the following half-reactions in acidic solution:
  a. $I_2(s) \longrightarrow I^-(aq)$
  b. $MnO_4^-(aq) \longrightarrow Mn^{2+}(aq)$
  c. $Br_2(l) \longrightarrow BrO_3^-(aq)$
  d. $ClO_3^-(aq) \longrightarrow ClO_4^-(aq)$

15.57 Write a balanced ionic equation for the following reactions in acidic solution:
  a. $Zn(s) + NO_3^-(aq) \longrightarrow Zn^{2+}(aq) + NO_2(g)$
  b. $MnO_4^-(aq) + SO_3^{2-}(aq) \longrightarrow$
  $$Mn^{2+}(aq) + SO_4^{2-}(aq)$$
  c. $ClO_3^-(aq) + I^-(aq) \longrightarrow I_2(s) + Cl^-(aq)$
  d. $Cr_2O_7^{2-}(aq) + C_2O_4^{2-}(aq) \longrightarrow$
  $$Cr^{3+}(aq) + CO_2(g)$$

15.58 Write a balanced ionic equation for the following reactions in acidic solution:
  a. $Sn^{2+}(aq) + IO_4^-(aq) \longrightarrow Sn^{4+}(aq) + I^-(aq)$
  b. $S_2O_3^{2-}(aq) + I_2(s) \longrightarrow I^-(aq) + S_4O_6^{2-}(aq)$
  c. $Mg(s) + VO_4^{3-}(aq) \longrightarrow Mg^{2+}(aq) + V^{2+}(aq)$
  d. $Al(s) + Cr_2O_7^{2-}(aq) \longrightarrow Al^{3+}(aq) + Cr^{3+}(aq)$

15.59 Use the activity series to predict whether each of the following reactions will occur spontaneously:
  a. $Cu(s) + 2H^+(aq) \longrightarrow Cu^{2+}(aq) + H_2(g)$
  b. $Ni^{2+}(aq) + Fe(s) \longrightarrow Fe^{2+}(aq) + Ni(s)$
  c. $2Ag(s) + Cu^{2+}(aq) \longrightarrow 2Ag^+(aq) + Cu(s)$
  d. $3Ni^{2+}(aq) + 2Cr(s) \longrightarrow 3Ni(s) + 2Cr^{3+}(aq)$
  e. $Zn(s) + Cu^{2+}(aq) \longrightarrow Zn^{2+}(aq) + Cu(s)$
  f. $Pb^{2+}(aq) + Zn(s) \longrightarrow Pb(s) + Zn^{2+}(aq)$

15.60 Use the activity series to predict whether each of the following reactions will occur spontaneously:
  a. $2Ag(s) + 2H^+(aq) \longrightarrow 2Ag^+(aq) + H_2(g)$
  b. $Mg(s) + Cu^{2+}(aq) \longrightarrow Mg^{2+}(aq) + Cu(s)$
  c. $2Al(s) + 3Cu^{2+}(aq) \longrightarrow 2Al^{3+}(aq) + 3Cu(s)$
  d. $Mg^{2+}(aq) + Zn(s) \longrightarrow Mg(s) + Zn^{2+}(aq)$
  e. $Al^{3+}(aq) + 3Na(s) \longrightarrow Al(s) + 3Na^+(aq)$
  f. $Ni^{2+}(aq) + Mg(s) \longrightarrow Mg^{2+}(aq) + Ni(s)$

15.61 In a voltaic cell, one half-cell consists of nickel metal in a Ni^{2+} solution of $Ni(NO_3)_2$ and the other half-cell consists of iron metal in a Fe^{2+} solution of $Fe(NO_3)_2$. Indicate each of the following:
  a. the anode
  b. the cathode
  c. the half-reaction at the anode
  d. the half-reaction at the cathode
  e. the overall reaction
  f. the cell notation

15.62 In a voltaic cell, one half-cell consists of a zinc metal anode in a Zn^{2+} solution of $Zn(NO_3)_2$ and the other half-cell consists of a copper metal cathode in a Cu^{2+} solution of $Cu(NO_3)_2$. Indicate each of the following.
  a. the anode
  b. the cathode
  c. the half-reaction at the anode
  d. the half-reaction at the cathode
  e. the overall reaction
  f. the cell notation

15.63 Use the activity series to determine which of the following ions will be reduced when an iron strip is placed in an aqueous solution of that ion:
a. $Ca^{2+}(aq)$ **b.** $Ag^+(aq)$ **c.** $Ni^{2+}(aq)$
d. $Al^{3+}(aq)$ **e.** $Pb^{2+}(aq)$

15.64 Use the activity series to determine which of the following ions will be reduced when an aluminum strip is placed in an aqueous solution of that ion:
a. $Fe^{2+}(aq)$ **b.** $Au^{3+}(aq)$ **c.** $Zn^{2+}(aq)$
d. $H^+(aq)$ **e.** $Pb^{2+}(aq)$

15.65 Steel bolts made for sailboats are coated with zinc. Add the necessary components (electrodes, wires, batteries) to this diagram of a zinc-nitrate solution electrolytic cell to show how it could be used to zinc-plate a steel bolt.

$$Zn^{2+}(aq)$$
$$2NO_3^-(aq)$$

a. What is the anode?
b. What is the cathode?
c. What is the half-reaction that takes place at the anode?
d. What is the half-reaction that takes place at the cathode?
e. If steel is mostly iron, what is the purpose of the zinc coating?

15.66 Copper cooking pans are stainless steel pans plated with a layer of copper. Add the necessary components (electrodes, wires, batteries) to this diagram of a copper-nitrate solution electrolytic cell to show how it could be used to copper-plate a stainless steel (iron) pan.

$$Cu^{2+}(aq)$$
$$2NO_3^-(aq)$$

a. What is the anode?
b. What is the cathode?
c. What is the half-reaction that takes place at the anode?
d. What is the half-reaction that takes place at the cathode?

15.67 In the lead storage battery, the following unbalanced half-reaction takes place:

$$Pb(s) + SO_4^{2-}(aq) \longrightarrow PbSO_4(s)$$

a. Balance the half-reaction.
b. What reactant is oxidized or reduced?
c. Indicate whether the half-reaction takes place at the anode or cathode.

15.68 In an acidic dry-cell battery, the following unbalanced half-reaction takes place in acidic solution:

$$MnO_2(s) \longrightarrow Mn_2O_3(s)$$

a. Balance the half-reaction.
b. Is $MnO_2(s)$ oxidized or reduced?
c. Indicate whether the half-reaction takes place at the anode or cathode.

15.69 The following unbalanced reaction takes place in acidic solution:

$$Ag(s) + NO_3^-(aq) \longrightarrow Ag^+(aq) + NO(g)$$

a. Write the balanced equation.
b. How many liters of $NO(g)$ are produced at STP when 15.0 g silver reacts with excess nitric acid?

15.70 The following unbalanced reaction takes place in acidic solution:

$$MnO_4^-(aq) + Fe^{2+}(aq) \longrightarrow$$
$$Mn^{2+}(aq) + Fe^{3+}(aq)$$

a. Write the balanced equation.
b. How many milliliters of a 0.150 M $KMnO_4$ solution are needed to react with 25.0 mL of a 0.400 M $FeSO_4$ solution?

15.71 A concentrated nitric acid solution is used to dissolve copper(II) sulfide.

$$CuS(s) + HNO_3(aq) \longrightarrow$$
$$CuSO_4(aq) + NO(g) + H_2O(l)$$

a. Write the balanced equation.
b. How many milliliters of a 16.0 M HNO_3 solution are needed to dissolve 24.8 g CuS?

15.72 The following unbalanced reaction takes place in acidic solution:

$$Cr_2O_7^{2-}(aq) + Fe^{2+}(aq) \longrightarrow$$
$$Cr^{3+}(aq) + Fe^{3+}(aq)$$

a. Write the balanced equation.
b. How many milliliters of a 0.211 M $K_2Cr_2O_7$ solution are needed to react with 5.00 g $FeSO_4$?

Challenge Questions

15.73 Which of the following are oxidation–reduction reactions?
- **a.** $Ca(s) + 2H_2O(l) \longrightarrow Ca(OH)_2(aq) + H_2(g)$
- **b.** $CaCO_3(s) \longrightarrow CaO(s) + CO_2(g)$
- **c.** $Cl_2(g) + 2NaBr(aq) \longrightarrow Br_2(l) + 2NaCl(aq)$
- **d.** $BaCl_2(aq) + Na_2SO_4(aq) \longrightarrow$
  $$BaSO_4(s) + 2NaCl(aq)$$

15.74 Assign oxidation numbers to all the atoms in the following equation, and balance:

$$Fe_2O_3(s) + CO(g) \longrightarrow Fe(s) + CO_2(g)$$

- **a.** Which substance is oxidized?
- **b.** Which substance is reduced?
- **c.** Which substance is the oxidizing agent?
- **d.** Which substance is the reducing agent?

15.75 Determine the oxidation number of Br in each of the following:
- **a.** Br_2
- **b.** $HBrO_2$
- **c.** BrO_3^-
- **d.** $NaBrO_4$

15.76 Use half-reactions to balance the following reaction in acidic solution:

$$Cr_2O_7^{2-}(aq) + NO_2^-(aq) \longrightarrow$$
$$Cr^{3+}(aq) + NO_3^-(aq)$$

15.77 Draw a picture of a voltaic cell for $Ni(s) \mid Ni^{2+}(aq) \parallel Ag^+(aq) \mid Ag(s)$.
- **a.** What is the anode?
- **b.** What is the cathode?
- **c.** What is the half-reaction that takes place at the anode?
- **d.** What is the half-reaction that takes place at the cathode?
- **e.** What is the overall reaction for the cell?

15.78 Using the activity series for metals, indicate whether each of the following can be used to generate an electrical current or will require a battery:
- **a.** $Ca^{2+}(aq) + Zn(s) \longrightarrow Ca(s) + Zn^{2+}(aq)$
- **b.** $2Al(s) + 3Sn^{2+}(aq) \longrightarrow 2Al^{3+}(aq) + 3Sn(s)$
- **c.** $Mg(s) + 2H^+(aq) \longrightarrow Mg^{2+}(aq) + H_2(g)$
- **d.** $Cu(s) + Ni^{2+}(aq) \longrightarrow Cu^{2+}(aq) + Ni(s)$
- **e.** $2Cr(s) + 3Fe^{2+}(aq) \longrightarrow 2Cr^{3+}(aq) + 3Fe(s)$

15.79 In a process called anodizing, tumblers of silvery-colored aluminum are coated in bright colors. The aluminum anode is oxidized to form an oxide coating. When dyes are added, they attach to the coating to produce tumblers with bright colors. At the cathode, hydrogen gas is produced.

- **a.** Balance the half-reaction in acidic solution at the anode:
  $$Al(s) \longrightarrow Al_2O_3(s)$$
- **b.** Balance the half-reaction in acidic solution at the cathode:
  $$H^+(aq) \longrightarrow H_2(g)$$
- **c.** Write the overall reaction for the formation of the aluminum-oxide coating.

Answers

Answers to Study Checks

15.1 **a.** This is an oxidation–reduction reaction. Fe loses electrons; it is oxidized.
$$Fe \longrightarrow Fe^{3+} + 3e^-$$
Br_2 gains electrons; it is reduced.
$$Br_2 + 2e^- \longrightarrow 2Br^-$$

b. This is not an oxidation–reduction reaction because there is no change in the charges of the ions in the reactants and products.

15.2 Al loses 3 electrons and is oxidized. Sn^{2+} gains 2 electrons and is reduced.

15.3 Al loses electrons and is oxidized. It is the reducing agent. Sn^{2+} gains electrons and is reduced. It is the oxidizing agent.

15.4 **a.** O has an oxidation number of -2, which gives each Cl in Cl_2 an oxidation number of $+1$.

$$Cl_2O$$
$$+1 \; -2$$

Check: $2(+1) + 1(-2) = +2 - 2 = 0$

b. H is assigned an oxidation number of $+1$, and O a -2. To maintain a neutral overall charge, P will have an oxidation number of $+5$.

$$H_3PO_4$$
$$+1 \; +5 \; -2$$

Check: $3(+1) + 1(+5) + 4(-2) =$
$$+3 + 5 - 8 = 0$$

c. With O assigned an oxidation number of -2, Mn must be $+7$ to give an overall charge of -1.

$$MnO_4^-$$
$$+7 \; -2$$

Check: $1(+7) + 4(-2) = +7 - 8 = -1$

15.5 $Zn + CuCl_2 \longrightarrow ZnCl_2 + Cu$

     0    $+2 -1$ $+2 -1$ 0   Oxidation numbers

Zn is oxidized; Zn is the reducing agent. Cu^{2+} (in $CuCl_2$) is reduced; $CuCl_2$ is the oxidizing agent.

15.6 $2Al(s) + 6HCl(aq) \longrightarrow 2AlCl_3(aq) + 3H_2(g)$

     0      $+1 -1$ $+3 -1$ 0

                      Oxidation numbers

15.7 $2Fe_2O_3(s) + 3C(s) \longrightarrow 4Fe(s) + 3CO_2(g)$

   $+3 -2$ 0   0 $+4 -2$

                      Oxidation numbers

15.8 Balanced half-reactions:

$$Zn(s) \longrightarrow Zn^{2+}(aq) + 2e^-$$
$$2Fe^{3+}(aq) + 2e^- \longrightarrow 2Fe^{2+}(aq)$$

Overall equation:

$$Zn(s) + 2Fe^{3+}(aq) \longrightarrow Zn^{2+}(aq) + 2Fe^{2+}(aq)$$

15.9 Balanced half-reactions:

$$NO(g) + 2H_2O(l) \longrightarrow NO_3^-(aq) + 4H^+ + 3e^-$$
$$MnO_4^-(aq) + 4H^+(aq) + 3e^- \longrightarrow MnO_2(s) + 2H_2O(l)$$

Overall equation:

$$NO(g) + MnO_4^-(aq) \longrightarrow MnO_2(s) + NO_3^-(aq)$$

15.10 Balanced half-reactions:

$$Cu^{2+}(aq) + 2e^- \longrightarrow Cu(s)$$
$$SO_2(g) + 2H_2O(l) \longrightarrow SO_4^{2-}(aq) + 4H^+(aq) + 2e^-$$

Overall equation:

$$Cu^{2+}(aq) + SO_2(g) + 2H_2O(l) \longrightarrow Cu(s) + SO_4^{2-}(aq) + 4H^+(aq)$$

15.11 Anode reaction: $\quad Co(s) \longrightarrow Co^{2+}(aq) + 2e^-$
Cathode reaction: $\quad Cu^{2+}(aq) + 2e^- \longrightarrow Cu(s)$

Overall reaction: $\quad Co(s) + Cu^{2+}(aq) \longrightarrow Co^{2+}(aq) + Cu(s)$

15.12 The Zn anode will lose mass as the Zn metal is oxidized to Zn^{+2} ions.

15.13 a. Sn is below Fe in the activity series; the reaction is not spontaneous.
b. Mg is above Cr in the activity series; the reaction is spontaneous.

15.14 Since Fe is below Cr in the activity series, the plating of Cr^{3+} onto Fe is not spontaneous. Energy from electrolysis is needed to make the reaction proceed.

Answers to Selected Problems and Questions

15.1 a. Na^+ gains electrons; this is a reduction.
b. Ni loses electrons; this is an oxidation.
c. Cr^{3+} gains electrons; this is a reduction.
d. H^+ gains electrons; this is a reduction.

15.3 a. Zn loses electrons and is oxidized. Cl_2 gains electrons and is reduced.
b. Br^- (in NaBr) loses electrons and is oxidized. Cl_2 gains electrons and is reduced.
c. Pb loses electrons and is oxidized. O_2 gains electrons and is reduced.
d. Sn^{2+} loses electrons and is oxidized. Fe^{3+} gains electrons and is reduced.

15.5 a. The substance that is oxidized is the reducing agent.
b. The substance that gains electrons is reduced and is the oxidizing agent.

15.7 a. Zn is the reducing agent. Cl_2 is the oxidizing agent.
b. Br^- (in NaBr) is the reducing agent. Cl_2 is the oxidizing agent.
c. Pb is the reducing agent. O_2 is the oxidizing agent.
d. Sn^{2+} is the reducing agent. Fe^{3+} is the oxidizing agent.

15.9 a. 0   **b.** 0
     c. $+2$ **d.** 0

15.11 a. K is $+1$, Cl is -1.   **b.** Mn is $+4$, O is -2.
     c. C is $+2$, O is -2.   **d.** Mn is $+3$, O is -2.

15.13 a. Al is $+3$, P is $+5$, and O is -2.
b. S is $+4$, O is -2.
c. Cr is $+3$, O is -2.
d. N is $+5$, O is -2.

15.15 a. H is $+1$, S is $+6$, and O is -2.
b. H is $+1$, P is $+3$, and O is -2.
c. Cr is $+6$, O is -2.
d. Na is $+1$, C is $+4$, and O is -2.

15.17 a. +5   **b.** −2
c. +5   **d.** +6

15.19 a. S^{2-} (in NiS) is oxidized; O_2 is reduced. NiS is the reducing agent and O_2 is the oxidizing agent.
b. Sn^{2+} is oxidized; Fe^{3+} is reduced. Sn^{2+} is the reducing agent and Fe^{3+} is the oxidizing agent.
c. C (in CH_4) is oxidized; O_2 is reduced. CH_4 is the reducing agent and O_2 is the oxidizing agent.
d. Si is oxidized; Cr^{3+} (in Cr_2O_3) is reduced. Si is the reducing agent and Cr_2O_3 is the oxidizing agent.

15.21 a. $2PbS(s) + 3O_2(g) \longrightarrow 2PbO(s) + 2SO_2(g)$
b. $2Fe(s) + 3Cl_2(g) \longrightarrow 2FeCl_3(s)$
c. $2Al(s) + 3H_2SO_4(aq) \longrightarrow$
$Al_2(SO_4)_3(aq) + 3H_2(g)$

15.23 a. $Sn^{2+}(aq) \longrightarrow Sn^{4+}(aq) + 2e^-$
b. $Mn^{2+}(aq) + 4H_2O(l) \longrightarrow$
$MnO_4^-(aq) + 8H^+(aq) + 5e^-$
c. $14H^+(aq) + Cr_2O_7^{2-}(aq) + 6e^- \longrightarrow$
$2Cr^{3+}(aq) + 7H_2O(l)$
d. $ClO_3^-(aq) + 2H^+(aq) + e^- \longrightarrow$
$ClO_2(aq) + H_2O(l)$

15.25 a. $Ag(s) \longrightarrow Ag^+(aq) + e^-$
$2H^+(aq) + NO_3^-(aq) + e^- \longrightarrow$
$NO_2(g) + H_2O(l)$
Overall: $2H^+(aq) + Ag(s) + NO_3^-(aq) \longrightarrow$
$NO_2(g) + Ag^+(aq) + H_2O(l)$
b. $Fe^{2+}(aq) \longrightarrow Fe^{3+}(aq) + e^-$
$6H^+(aq) + ClO_3^-(aq) + 6e^- \longrightarrow$
$Cl^-(aq) + 3H_2O(l)$
Overall:
$6H^+(aq) + 6Fe^{2+}(aq) + ClO_3^-(aq) \longrightarrow$
$6Fe^{3+}(aq) + Cl^-(aq) + 3H_2O(l)$
c. $4H^+(aq) + NO_3^-(aq) + 3e^- \longrightarrow$
$NO(g) + 2H_2O(l)$
$2H_2O(l) + S(s) \longrightarrow SO_2(g) + 4H^+(aq) + 4e^-$
Overall: $4H^+(aq) + 4NO_3^-(aq) + 3S(s) \longrightarrow$
$4NO(g) + 2H_2O(l) + 3SO_2(g)$
d. $2S_2O_3^{2-}(aq) \longrightarrow S_4O_6^{2-}(aq) + 2e^-$;
$Cu^{2+}(aq) + 2e^- \longrightarrow Cu(s)$
Overall: $2S_2O_3^{2-}(aq) + Cu^{2+}(aq) \longrightarrow$
$S_4O_6^{2-}(aq) + Cu(s)$
e. $PbO_2(s) + 4H^+ + 2e^- \longrightarrow Pb^{2+}(aq) + 2H_2O(l)$
$Mn^{2+}(aq) + 4H_2O(l) \longrightarrow$
$MnO_4^-(aq) + 8H^+(aq) + 5e^-$
Overall: $4H^+(aq) + 5PbO_2(s) + 2Mn^{2+}(aq) \longrightarrow$
$5Pb^{2+}(aq) + 2MnO_4^-(aq) + 2H_2O(l)$

15.27 a. Anode reaction: $Pb(s) \longrightarrow Pb^{2+}(aq) + 2e^-$
Cathode reaction: $Cu^{2+}(aq) + 2e^- \longrightarrow Cu(s)$
Overall reaction: $Cu^{2+}(aq) + Pb(s) \longrightarrow$
$Cu(s) + Pb^{2+}(aq)$
b. Anode reaction: $Cr(s) \longrightarrow Cr^{2+}(aq) + 2e^-$
Cathode reaction: $Ag^+(aq) + e^- \longrightarrow Ag(s)$
Overall reaction: $2Ag^+(aq) + Cr(s) \longrightarrow$
$2Ag(s) + Cr^{2+}(aq)$

15.29 a. The anode is a Cd metal electrode in a Cd^{2+} solution. The anode reaction is
$$Cd(s) \longrightarrow Cd^{2+}(aq) + 2e^-$$
The cathode is a Sn metal electrode in a Sn^{2+} solution. The cathode reaction is
$$Sn^{2+}(aq) + 2e^- \longrightarrow Sn(s)$$
The shorthand notation for this cell is
$$Cd(s)\,|\,Cd^{2+}(aq)\,||\,Sn^{2+}(aq)\,|\,Sn(s)$$
b. The anode is a Zn metal electrode in a Zn^{2+} solution. The anode reaction is
$$Zn(s) \longrightarrow Zn^{2+}(aq) + 2e^-$$
The cathode is a C (graphite) electrode, where Cl_2 gas is reduced to Cl^-. The cathode reaction is
$$Cl_2(g) + 2e^- \longrightarrow 2Cl^-(aq)$$
The cell notation is
$$Zn(s)\,|\,Zn^{2+}(aq)\,||\,Cl_2(g), Cl^-(aq)\,|\,C\ (graphite)$$

15.31 a. The half-reaction is an oxidation.
b. Cd metal is oxidized.
c. Oxidation takes place at the anode.

15.33 a. The half-reaction is an oxidation.
b. Zn metal is oxidized.
c. Oxidation takes place at the anode.

15.35 a. The half-reaction to plate tin is
$$Sn^{2+}(aq) + 2e^- \longrightarrow Sn(s)$$
b. The iron is the cathode.
c. Since Fe is above Sn in the activity series, the plating of Sn^{2+} onto Fe is not spontaneous. Energy from electrolysis is needed to make the reaction proceed.

15.37 Since Fe is above Sn in the activity series, if the Fe is exposed to air and water, Fe will be oxidized and rust will form. To protect iron, Sn would have to be *more* active than Fe and it is not.

15.39 a. oxidation   **b.** reduction
c. reduction   **d.** oxidation
e. oxidation

15.41 a. VO_2 $V = +4, O = -2$
b. Ag_2CrO_4 $Ag = +1, Cr = +6, O = -2$
c. $S_2O_8^{2-}$ $S = +7, O = -2$
d. $FeSO_4$ $Fe = +2, S = +6, O = -2$

15.43 a. Cr^{3+} in Cr_2O_3 is reduced.
b. Si is oxidized.
c. Cr_2O_3 is the oxidizing agent.
d. Si is the reducing agent.
e. $2Cr_2O_3(s) + 3Si(s) \longrightarrow 4Cr(s) + 3SiO_2(s)$

15.45 **a.** $Fe(s) \longrightarrow Fe^{2+}(aq) + 2e^-$
b. $Ni^{2+}(aq) + 2e^- \longrightarrow Ni(s)$
c. Fe is the anode.
d. Ni is the cathode.
e. The electrons flow from Fe to Ni.
f. $Fe(s) + Ni^{2+}(aq) \longrightarrow Fe^{2+}(aq) + Ni(s)$
g. $Fe(s)\,|\,Fe^{2+}(aq)\,\|\,Ni^{2+}(aq)\,|\,Ni(s)$

15.47 Reactions b, c, and d all involve loss and gain of electrons; b, c, and d are oxidation–reduction reactions.

15.49 **a.** Fe^{3+} is gaining electrons; this is a reduction.
b. Fe^{2+} is losing electrons; this is an oxidation.

15.51 **a.** Co_2O_3 Co +3; O −2
b. $KMnO_4$ K +1, Mn +7, O −2
c. SF_6 S +6, F −1
d. ClO_3^- Cl +5, O −2
e. PO_4^{3-} P +5, O −2

15.53 **a.** $FeCl_2(aq) + Cl_2(g) \longrightarrow FeCl_3(aq)$
   +2 −1   0   +3 −1
  Fe^{2+} is oxidized and Cl_2 is reduced.
  $2FeCl_2(aq) + Cl_2(g) \longrightarrow 2FeCl_3(aq)$
b. $Si(s) + Cr_2O_3(s) \longrightarrow Cr(s) + SiO_2(s)$
   0   +3 −2   0   +4 −2
  Si is oxidized and Cr^{3+} is reduced.
  $3Si(s) + 2Cr_2O_3(s) \longrightarrow 4Cr(s) + 3SiO_2(s)$
c. $Cr_2O_3(s) + Al(s) \longrightarrow Cr(s) + Al_2O_3(s)$
   +3 −2   0   0   +3 −2
  Al is oxidized and Cr^{3+} is reduced.
  $Cr_2O_3(s) + 2Al(s) \longrightarrow 2Cr(s) + Al_2O_3(s)$
d. $P_2O_5(s) + C(s) \longrightarrow P(s) + CO(g)$
   +5 −2   0   0   +2 −2
  C is oxidized and P^{5+} is reduced.
  $P_2O_5(s) + 5C(s) \longrightarrow 2P(s) + 5CO(g)$

15.55 **a.** $Zn(s) \longrightarrow Zn^{2+}(aq) + 2e^-$
b. $SnO_2^{2-}(aq) + H_2O(l) \longrightarrow$
  $\qquad\qquad SnO_3^{2-}(aq) + 2H^+(aq) + 2e^-$
c. $Cr_2O_7^{2-}(aq) + 14H^+(aq) + 6e^- \longrightarrow$
  $\qquad\qquad 2Cr^{3+}(aq) + 7H_2O(l)$
d. $NO_3^-(aq) + 4H^+(aq) + 3e^- \longrightarrow$
  $\qquad\qquad NO(g) + 2H_2O(l)$

15.57 **a.** $Zn(s) \longrightarrow Zn^{2+}(aq) + 2e^-$;
  $NO_3^-(aq) + 2H^+(aq) + e^- \longrightarrow$
  $\qquad\qquad NO_2(g) + H_2O(l)$
  Overall:
  $Zn(s) + 2NO_3^-(aq) + 4H^+(aq) \longrightarrow$
  $\qquad\qquad Zn^{2+}(aq) + 2NO_2(g) + 2H_2O(l)$
b. $MnO_4^-(aq) + 8H^+(aq) + 5e^- \longrightarrow$
  $\qquad\qquad Mn^{2+}(aq) + 4H_2O(l)$
  $SO_3^{2-}(aq) + H_2O(l) \longrightarrow$
  $\qquad\qquad SO_4^{2-}(aq) + 2H^+(aq) + 2e^-$
  Overall:
  $2MnO_4^-(aq) + 5SO_3^{2-}(aq) + 6H^+(aq) \longrightarrow$
  $\qquad 2Mn^{2+}(aq) + 5SO_4^{2-}(aq) + 3H_2O(l)$

c. $2I^-(aq) \longrightarrow I_2(s) + 2e^-$;
  $ClO_3^-(aq) + 6H^+(aq) + 6e^- \longrightarrow$
  $\qquad\qquad Cl^-(aq) + 3H_2O(l)$
  Overall:
  $ClO_3^-(aq) + 6I^-(aq) + 6H^+(aq) \longrightarrow$
  $\qquad\qquad Cl^-(aq) + 3I_2(s) + 3H_2O(l)$
d. $C_2O_4^{2-}(aq) \longrightarrow 2CO_2(g) + 2e^-$
  $Cr_2O_7^{2-}(aq) + 14H^+(aq) + 6e^- \longrightarrow$
  $\qquad\qquad 2Cr^{3+}(aq) + 7H_2O(l)$
  Overall:
  $Cr_2O_7^{2-}(aq) + 3C_2O_4^{2-}(aq) + 14H^+(aq) \longrightarrow$
  $\qquad\qquad 2Cr^{3+}(aq) + 6CO_2(g) + 7H_2O(l)$

15.59 **a.** Since Cu is below H_2 in the activity series, the reaction will not be spontaneous.
b. Since Fe is above Ni in the activity series, the reaction will be spontaneous.
c. Since Ag is below Cu in the activity series, the reaction will not be spontaneous.
d. Since Cr is above Ni in the activity series, the reaction will be spontaneous.
e. Since Zn is above Cu in the activity series, the reaction will be spontaneous.
f. Since Zn is above Pb in the activity series, the reaction will be spontaneous.

15.61 **a.** The anode is Fe.
b. The cathode is Ni.
c. The half-reaction at the anode is
  $Fe(s) \longrightarrow Fe^{2+}(aq) + 2e^-$
d. The half-reaction at the cathode is
  $Ni^{2+}(aq) + 2e^- \longrightarrow Ni(s)$
e. The overall reaction is
  $Fe(s) + Ni^{2+}(aq) \longrightarrow Fe^{2+}(aq) + Ni(s)$
f. The cell notation is
  $Fe(s)\,|\,Fe^{2+}(aq)\,\|\,Ni^{2+}(aq)\,|\,Ni(s)$

15.63 **a.** $Ca^{2+}(aq)$ will not be reduced by an iron strip.
b. $Ag^+(aq)$ will be reduced by an iron strip.
c. $Ni^{2+}(aq)$ will be reduced by an iron strip.
d. $Al^{3+}(aq)$ will not be reduced by an iron strip.
e. $Pb^{2+}(aq)$ will be reduced by an iron strip.

15.65

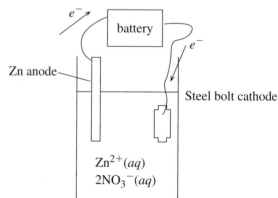

a. The anode is a bar of zinc.
b. The cathode is the steel bolt.
c. $Zn(s) \longrightarrow Zn^{2+}(aq) + 2e^-$
d. $Zn^{2+}(aq) + 2e^- \longrightarrow Zn(s)$
e. The purpose of the zinc coating is to prevent rusting of the bolt by H_2O and O_2.

15.67 a. $Pb(s) + SO_4^{2-}(aq) \longrightarrow PbSO_4(s) + 2e^-$
b. $Pb(s)$ is oxidized.
c. The half-reaction takes place at the anode.

15.69 a. $3Ag(s) + NO_3(aq) + 4H^+(aq) \longrightarrow$
$$3Ag^+(aq) + NO(g) + 2H_2O(l)$$
b. 1.04 L $NO(g)$ are produced at STP.

15.71 a. $3CuS(s) + 8HNO_3(aq) \longrightarrow$
$$3CuSO_4(aq) + 8NO(g) + 4H_2O(l)$$
b. $24.8 \text{ g CuS} \times \dfrac{1 \text{ mol CuS}}{95.62 \text{ g CuS}} \times \dfrac{8 \text{ mol HNO}_3}{3 \text{ mol CuS}}$

$\times \dfrac{1000 \text{ mL}}{16.0 \text{ mol HNO}_3} = 43.2 \text{ mL}$

15.73 a. Yes. $\underset{0}{Ca} \longrightarrow \underset{+2}{Ca(aq)}$ Oxidation

$\underset{+1}{H\,(H_2O)} \longrightarrow \underset{0}{H\,(H_2)}$ Reduction

b. No.   No change occurs in oxidation numbers of
$\underset{+2}{Ca}, \underset{+4}{C}, \underset{-2}{O}.$

c. Yes. $\underset{}{Cl\ 0} \longrightarrow Cl\ -1$ reduction
$Br\ -1 \longrightarrow Br\ 0$ oxidation
d. No.   No change occurs in oxidation numbers of
$\underset{+2}{Ba}, \underset{-1}{Cl}, \underset{+1}{Na}, \text{ or } \underset{+6}{S}.$

15.75 a. Br in $Br_2 = 0$
b. $HBrO_2$: $H = +1$. $O = -2$. Then $+1 + Br +2(-2) = 0$ $Br = -1 + 4 = +3$
c. BrO_3^-: $O = -2$.   Then $Br + 3(-2) = -1$ $Br = -1 + 6 = +5$
d. $NaBrO_4$: $Na^+ = +1$ $O = -2$.   Then $+1 + Br +4(-2) = 0$ $Br = -1 + 8 = +7$

15.77

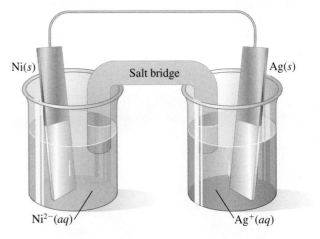

a. $Ni(s)$ is the anode.
b. $Ag(s)$ is the cathode.
c. $Ni(s) \longrightarrow Ni^{2+}(aq) + 2e^-$
d. $Ag^+(aq) + e^- \longrightarrow Ag(s)$
e. $Ni(s) + 2Ag^+(aq) \longrightarrow Ni^{2+}(aq) + 2Ag(s)$

15.79 a. $2Al(s) + 3H_2O(l) \longrightarrow$
$$Al_2O_3(s) + 6H^+(aq) + 6e^-$$
b. $2H^+(aq) + 2e^- \longrightarrow H_2(g)$
c. $2Al(s) + 3H_2O(l) \longrightarrow Al_2O_3(s) + 3H_2(g)$

Nuclear Radiation

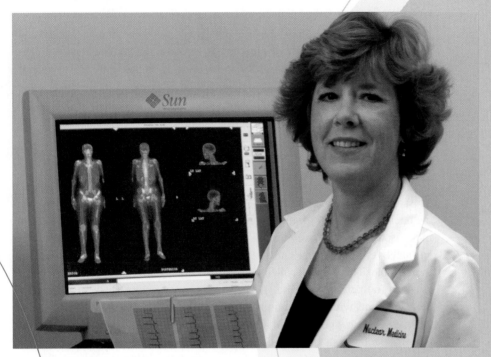

"Everything we do in this department involves radioactive materials," says Julie Goudak, nuclear medicine technologist at Kaiser Hospital. "The radioisotopes are given in several ways. The patient may ingest an isotope, breathe it in, or receive it by an IV injection. We do many diagnostic tests, particularly of the heart function, to determine if a patient needs a cardiac CAT scan."

A nuclear medicine technologist administers isotopes that emit radiation to determine the level of function of an organ such as the thyroid or heart, to detect the presence and size of a tumor, or to treat disease. A radioisotope locates in a specific organ and its radiation is used by a computer to create an image of that organ. From this data, a physician can make a diagnosis and design a treatment program.

the
Chemistry
place

Visit **www.aw-bc.com/chemplace** for extra quizzes, interactive tutorials, career resources, PowerPoint slides for chapter review, math help, and case studies.

With the production of artificial radioactive substances in 1934, the field of nuclear medicine was established. In 1937, the first radioactive isotope was used to treat a patient with leukemia at the University of California at Berkeley. Major strides in the use of radioactivity in medicine occurred in 1946, when a radioactive iodine isotope was successfully used to diagnose thyroid function and to treat hyperthyroidism and thyroid cancer. In the 1970s and 1980s, a variety of radioactive substances were used to produce an image of an organ, such as liver, spleen, thyroid gland, kidneys, and the brain, and to detect heart disease. Today, procedures in nuclear medicine provide information about the function and structure of every organ in the body, which allows the nuclear physician to diagnose and treat diseases early.

16.1 Natural Radioactivity

Learning **Goal**

Describe alpha, beta, and gamma radiation.

Most naturally occurring isotopes of elements up to atomic number 19 have stable nuclei. Elements with higher atomic numbers (20 to 83) usually have one or more isotopes, that have unstable nuclei. When a nucleus is unstable, it is **radioactive,** which means that it will spontaneously emit energy to become more stable. This energy, called **radiation,** may take the form of particles such as alpha (α) particles or beta (β) particles or pure energy such as gamma (γ) rays. Elements with atomic numbers of 84 and higher consist only of radioactive isotopes. So many protons and neutrons are crowded together in their nuclei that the strong repulsions between the protons makes those nuclei unstable.

In Chapter 4, we wrote symbols to distinguish the different isotopes of an element. Recall that an atom's mass number is equal to the sum of the protons and neutrons in the nucleus, and its atomic number is equal to the number of protons. In the symbol for an isotope, the mass number is written in the upper left corner of the symbol, and the atomic number is written in the lower left corner. For example, a radioactive isotope of iodine used in the diagnosis and treatment of thyroid conditions has a mass number of 131 and an atomic number of 53.

$$\text{Mass number (protons and neutrons)} \longrightarrow \quad {}^{131}_{53}\text{I}$$
$$\text{Element} \longrightarrow$$
$$\text{Atomic number (protons)} \longrightarrow$$

This isotope is also called iodine-131 or I-131. Radioactive isotopes are named by writing the mass number after the element's name or symbol. When necessary, we can obtain the atomic number from the periodic table. Table 16.1 compares some stable, nonradioactive isotopes with some radioactive isotopes.

Table 16.1 Stable and Radioactive Isotopes of Some Elements

Magnesium	Iodine	Uranium
Stable Isotopes		
${}^{24}_{12}\text{Mg}$	${}^{127}_{53}\text{I}$	None
magnesium-24	iodine-127	
Radioactive Isotopes		
${}^{23}_{12}\text{Mg}$	${}^{125}_{53}\text{I}$	${}^{235}_{92}\text{U}$
magnesium-23	iodine-125	uranium-235
${}^{27}_{12}\text{Mg}$	${}^{131}_{53}\text{I}$	${}^{238}_{92}\text{U}$
magnesium-27	iodine-131	uranium-238

Types of Radiation

Different forms of radiation are emitted from an unstable nucleus when a change takes place among its protons and neutrons. Symbols for different forms of radiation can include the charge and mass number, written in the lower and upper left corner. The release of radiation of any type produces a more stable, lower-energy nucleus. One type of radiation consists of alpha particles. An **alpha particle** contains 2 protons and 2 neutrons, which gives it a mass number of 4 and an atomic number of 2. Because it has 2 protons, an alpha particle has a charge of 2+. That makes it identical to a helium nucleus. In equations, it is written as the Greek letter alpha (α) or as the symbol for helium.

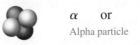

$$\alpha \quad \text{or} \quad {}^{4}_{2}\text{He}$$

Alpha particle

Another type of radiation occurs when a radioisotope emits **beta particles.** A beta particle, which is a high-energy electron, has a charge of $1-$ and, because its mass is so much less than the mass of a proton, it is given a mass number of 0. It is represented by the Greek letter beta (β) or by the symbol for the electron, with the charge, -1, written on the lower left.

$$\beta \quad \text{or} \quad {}^{0}_{-1}e$$

Beta particle

Beta particles are produced from an unstable nucleus when a neutron is transformed into a proton and an electron. The high-energy electron is emitted from the nucleus as beta radiation. This electron or beta particle did not exist in the nucleus until the following change occurred:

$${}^{1}_{0}\text{n} \longrightarrow {}^{1}_{1}\text{p} \quad + \quad {}^{0}_{-1}e$$

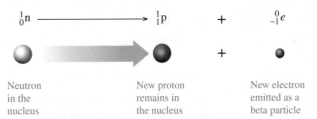

| Neutron in the nucleus | New proton remains in the nucleus | New electron emitted as a beta particle |

A **positron** is similar to an electron except that a positron has a positive ($+1$) charge. The electron and positron each have a mass number of 0. Instead of the atomic number, the charge is shown when we write the symbols of an electron and a positron as follows:

	Electron	**Positron**
Mass number Charge	${}^{0}_{-1}e$	${}^{0}_{+1}e$

A positron is produced by an ustable nucleus when a proton is transformed into a neutron and a positron.

$${}^{1}_{1}\text{p} \longrightarrow {}^{1}_{0}\text{n} \quad + \quad {}^{0}_{+1}e \text{ (or } \beta^{+})$$

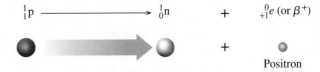

Positron

When high-energy radiation strikes molecules in its path, electrons may be knocked away. The result of this *ionizing radiation* is the formation of unstable ions or radicals. A free radical is a particle that has an unpaired electron. For example, when radiation passes through the human body, it may interact with water molecules, removing electrons and producing H_2O^+ ions or free radicals.

When ionizing radiation strikes the cells of the body, the unstable ions or free radicals that form can cause undesirable chemical reactions. The cells in the body most sensitive to radiation are the ones undergoing rapid division—those of the bone marrow, skin, reproductive organs, and intestinal lining, as well as all cells of growing children. Damaged cells may lose their ability to produce necessary materials. For example, if radiation damages cells of the bone marrow, red blood cells may no longer be produced. If sperm cells or ova or the cells of a fetus are damaged, birth defects may result. In contrast, cells of the nerves, muscles, liver, and adult bones are much less sensitive to radiation because they undergo little or no cellular division.

Cancer cells are another example of rapidly dividing cells. Because cancer cells are highly sensitive to radiation, large doses of radiation are used to destroy them. The surrounding normal tissue, dividing at a slower rate, shows a greater resistance to radiation and suffers less damage. In addition, normal tissue is able to repair itself more readily than cancerous tissue. However, this repair is not always complete. Possible long-term effects of ionizing radiation include a shortened life span, malignant tumors, leukemia, anemia, and genetic mutations.

WEB TUTORIAL
Radiation and Its Biological Effects

Table 16.2 Some Common Forms of Radiation

Type of Radiation	Symbol		Mass Number	Charge
Alpha particle	α	^4_2He	4	2+
Beta particle	β	$^{\ 0}_{-1}e$	0	1−
Positron	β^+	$^{\ 0}_{+1}e$	0	1+
Gamma ray	γ	$^0_0\gamma$	0	0
Proton	^1_1H	^1_1p	1	1+
Neutron	^1_0n	n	1	0

A positron is an example of *antimatter,* a term physicists use to describe a particle that is the exact opposite of a particle, in this case, an electron. When an electron and a positron collide, their minute masses are completely converted to energy in the form of gamma rays.

$$^{\ 0}_{-1}e + {}^{\ 0}_{+1}e \longrightarrow 2\,^0_0\gamma$$

When the symbol β is used with no charge, it is taken to mean a beta particle rather than a positron.

Gamma rays are high-energy radiation, released as an unstable nucleus undergoes a rearrangement of its particles to give a more stable, lower-energy nucleus. A gamma ray is shown as the Greek letter gamma (γ). Because gamma rays are energy only, there is no mass or charge associated with their symbol.

γ
Gamma ray

Table 16.2 summarizes the types of radiation we will use in nuclear equations.

Radiation Protection

Nuclear radiation is harmful because the particles are emitted with tremendous energy. Therefore, it is important that the radiologist, doctor, and nurse working with radioactive isotopes use proper radiation protection. Proper *shielding* is necessary to prevent exposure. Alpha particles are the heaviest of the radiation particles; they travel only a few centimeters in the air before they collide with air molecules, acquire electrons, and become helium atoms. A piece of paper, clothing, and our skin are protection against alpha particles. Lab coats and gloves will also provide sufficient shielding. However, if ingested or inhaled, alpha emitters can bring about serious internal damage because the mass and high charge of alpha particles causes much ionization in a short distance.

Beta particles have a very small mass and move much faster and farther than alpha particles, traveling as much as several meters through air. They can pass through paper and penetrate as far as 4–5 mm into body tissue. External exposure to beta particles can burn the surface of the skin, but they are stopped before they can reach the internal organs. Heavy clothing such as lab coats and gloves are needed to protect the skin from beta particles.

Gamma rays travel great distances through the air and pass through many materials, including body tissues. Only very dense shielding, such as lead or

Figure 16.1 A person working with radioisototopes wears protective clothing and gloves and stands behind a lead shield.

Q *What types of radiation does the lead shield block?*

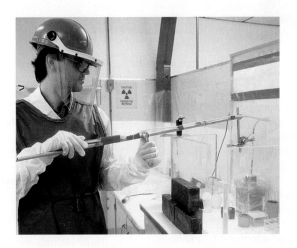

concrete, will stop them. Because they can penetrate so deeply, exposure to gamma rays can be extremely hazardous. Even the syringe used to give an injection of a gamma-emitting radioactive isotope is placed inside a special lead-glass cover. (See Figure 16.1.) Table 16.3 summarizes the shielding materials required for the various types of radiation.

Try to keep the time you must spend in a radioactive area to a minimum. A certain amount of radiation is emitted every minute. Remaining in a radioactive area twice as long exposes a person to twice as much radiation.

Keep your distance! The greater the distance from the radioactive source, the lower the intensity of radiation received. If you double your distance from the radiation source, the intensity of radiation drops to $\left(\frac{1}{2}\right)^2$, or one-fourth of its previous value.

Sample Problem 16.1 ▶ Radiation Protection

How does the type of shielding for alpha radiation differ from that used for gamma radiation?

Solution

Alpha radiation is stopped by paper and clothing. However, lead or concrete is needed for protection from gamma radiation.

Study Check

Besides shielding, what other methods help reduce exposure to radiation?

Table 16.3   Properties of Ionizing Radiation and Shielding Required

Property	Alpha	Beta	Gamma
Characteristics	Helium nucleus	Electron	High-energy rays
Symbols	$\alpha, {}_2^4\text{He}$	$\beta, {}_{-1}^0 e$	$\gamma, {}_0^0 \gamma$
Travel distance in air	2–4 cm	200–300 cm	500 m
Tissue depth	0.05 mm	4–5 mm	50 cm or more
Shielding	Paper, clothing	Heavy clothing, lab coats, gloves	Lead, thick concrete
Typical source	radium-226	carbon-14	technetium-99m

Questions and Problems   Natural Radioactivity

16.1 a. How are an alpha particle and a helium nucleus similar? different?
  b. What symbols are used for alpha particles?

16.2 a. How are a beta particle and an electron similar? different?
  b. What symbols are used for beta particles?

16.3 Naturally occurring potassium consists of three isotopes, potassium-39, potassium-40, and radioactive potassium-41.
  a. Write the atomic symbols for each isotope.
  b. In what ways are the isotopes similar, and in what ways do they differ?

16.4 Naturally occurring iodine is iodine-127. Iodine-125 and iodine-130 are radioactive isotopes of iodine.
  a. Write the atomic symbols for each isotope.
  b. In what ways are the isotopes similar, and in what ways do they differ?

16.5 Supply the missing information in the following table:

Medical Use	Atomic Symbol	Mass Number	Number of Protons	Number of Neutrons
Heart imaging	$^{201}_{81}\text{Tl}$			
Radiation therapy		60	27	
Abdominal scan			31	36
Hyperthyroidism	$^{131}_{53}\text{I}$			
Leukemia treatment		32		17

16.6 Supply the missing information in the following table:

Medical Use	Atomic Symbol	Mass Number	Number of Protons	Number of Neutrons
Cancer treatment	$^{60}_{27}\text{Co}$			
Brain scan		99	43	
Blood flow		141	58	
Bone scan		85		47
Lung function	$^{133}_{54}\text{Xe}$			

16.7 Write a symbol for each of the following:
  a. alpha particle   **b.** neutron
  c. beta particle   **d.** nitrogen-15
  e. iodine-125

16.8 Write a symbol for each of the following:
  a. proton   **b.** gamma ray
  c. positron   **d.** barium-131
  e. cobalt-60

16.9 Identify the symbol for X in each of the following:
  a. $^{0}_{-1}\text{X}$ **b.** $^{4}_{2}\text{X}$
  c. $^{1}_{0}\text{X}$ **d.** $^{24}_{11}\text{X}$
  e. $^{14}_{6}\text{X}$

16.10 Identify the symbol for X in each of the following:
  a. $^{1}_{1}\text{X}$ **b.** $^{32}_{15}\text{X}$
  c. $^{0}_{0}\text{X}$ **d.** $^{59}_{26}\text{X}$
  e. $^{85}_{38}\text{X}$

16.11 a. Why does beta radiation penetrate further in solid material than alpha radiation?
  b. How does ionizing radiation cause damage to cells of the body?
  c. Why does the x-ray technician leave the room when you receive an X ray?
  d. What is the purpose of wearing gloves when handling radioisotopes?

16.12 a. As a scientist you sometimes work with radioisotopes. What are three ways you can minimize your exposure to radiation?
  b. Why are cancer cells more sensitive to radiation than nerve cells?
  c. What is the purpose of placing a lead apron on a person who is receiving routine dental X rays?
  d. Why are the walls in a radiation lab built of thick concrete blocks?

16.2 Nuclear Equations

Learning Goal

Write an equation showing mass numbers and atomic numbers for radioactive decay.

the **Chemistry place**

WEB TUTORIAL
Radiation and Its Biological Effects

When a nucleus spontaneously breaks down by emitting radiation, the process is called *radioactive decay*. It can be shown as a *nuclear equation* using the symbols for the original radioactive nucleus, the new nucleus, and the radiation emitted.

$$\text{Radioactive nucleus} \longrightarrow \text{New nucleus} + \text{radiation } (\alpha, \beta, \gamma, \beta^+)$$

A nuclear equation is balanced when the sum of the mass numbers and the sum of the atomic numbers of the particles and atoms on one side of the equation are equal to their counterparts on the other side. The changes in mass number and atomic number of an atom that emits a radioactive particle are shown in Table 16.4.

Table 16.4 **Mass Number and Atomic Changes Due to Radiation**

Decay Process	Radiation Symbol	Change in Mass Number	Change in Atomic Number	Change in Neutron Number
Alpha emission	$^{4}_{2}\text{He}$ or α	-4	-2	-2
Beta emission	$^{0}_{-1}e$ or β	0	$+1$	-1
Positron emission	$^{0}_{+1}e$ or β^{+}	0	-1	$+1$
Gamma emission	$^{0}_{0}\gamma$	0	0	0

Alpha Emitters

Alpha emitters are radioisotopes that decay by emitting alpha particles. For example, uranium-238 decays to thorium-234 by emitting an alpha particle. The alpha particle emitted contains 2 protons, which means that the new nucleus has 2 fewer protons, or 90 protons. That means that the new nucleus has an atomic number of 90 and is therefore thorium (Th). Because the alpha particle has a mass number of 4, the mass number of the thorium isotope is 234, 4 less than that of the original uranium nucleus.

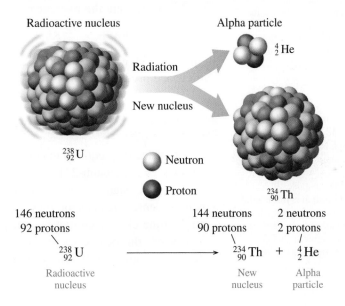

Radioactive nucleus   Alpha particle

Radiation $^{4}_{2}\text{He}$

New nucleus

$^{238}_{92}\text{U}$

◯ Neutron

● Proton $^{234}_{90}\text{Th}$

146 neutrons   144 neutrons   2 neutrons
92 protons   90 protons   2 protons

$^{238}_{92}\text{U} \longrightarrow \ ^{234}_{90}\text{Th} \ + \ ^{4}_{2}\text{He}$

Radioactive   New   Alpha
nucleus   nucleus   particle

Guide to Completing a Nuclear Equation

In another example of radioactive decay, radium-226 emits an alpha particle to form a new isotope whose mass number, atomic number, and identity we must determine.

STEP 1 **Write the incomplete nuclear equation.**

$$^{226}_{88}\text{Ra} \longrightarrow \ ? \ + \ ^{4}_{2}\text{He}$$

STEP 2 **Determine the missing mass number.**   In the equation, the mass number, 226, of the radium is equal to the combined mass numbers of the alpha particle and the new nucleus.

$$226 \quad = \ ? + 4$$
$$226 - 4 = \ ?$$
$$222 \quad = \ ? \ \text{(mass number of new nucleus)}$$

Guide to Completing a Nuclear Equation

STEP 1
Write the incomplete nuclear equation.

STEP 2
Determine the missing mass number.

STEP 3
Determine the missing atomic number.

STEP 4
Determine the symbol of the new nucleus.

STEP 5
Complete the nuclear equation.

The presence of radon has become a much publicized environmental and health issue because of radiation danger. Radioactive isotopes that produce radon, such as radium-226 and uranium-238, are naturally present in many types of rocks and soils. Radium-226 emits an alpha particle and is converted into radon gas, which diffuses out of the rocks and soil.

$$^{226}_{88}\text{Ra} \longrightarrow \, ^{222}_{86}\text{Rn} + \, ^{4}_{2}\text{He}$$

As uranium-238 decays, it also forms radium-226, which in turn produces radon. Uranium-238 has been found in particularly high levels in an area between Pennsylvania and New England.

Outdoors, radon gas poses little danger because it dissipates in the air. However, if the source of radon is under a house or building, the gas can enter the house through cracks in the foundation or other openings, where the radon can be inhaled by those living or working there. Inside the lungs, radon emits alpha particles to form polonium-218, which is known to cause cancer when present in the lungs.

$$^{222}_{86}\text{Rn} \longrightarrow \, ^{218}_{84}\text{Po} + \, ^{4}_{2}\text{He}$$

Some researchers have estimated that 10% of all lung cancer deaths in the United States are due to radon. The Environmental Protection Agency (EPA) recommends that the maximum level of radon not exceed 4 picocuries (pCi) per liter of air in a home. One (1) picocurie (pCi) is equal to 10^{-12} curies (Ci); curies are described in Section 16.3. In California, 1% of all the houses surveyed exceeded the EPA's recommended maximum radon level.

STEP 3 **Determine the missing atomic number.** The atomic number of radium, 88, must equal the sum of the atomic number of the alpha particle and the new nucleus.

$$88 \quad = ? + 2$$
$$88 - 2 = ?$$
$$86 \quad = ? \text{ (atomic number of new nucleus)}$$

STEP 4 **Determine the symbol of the new nucleus.** On the periodic table, the element that has atomic number 86 is radon, Rn. The nucleus of this isotope of Rn is written as

$$^{222}_{86}\text{Rn}$$

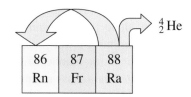

86	87	88
Rn	Fr	Ra

STEP 5 **Complete the nuclear equation.**

$$^{226}_{88}\text{Ra} \longrightarrow \, ^{222}_{86}\text{Rn} + \, ^{4}_{2}\text{He}$$

In this nuclear reaction, a radium-226 nucleus decays by releasing an alpha particle and produces a radon-222 nucleus.

Sample Problem 16.2 > **Writing an Equation for Alpha Decay**

Smoke detectors, required in homes and apartments, contain an alpha emitter such as americium-241. The alpha particles ionize air molecules, producing a constant stream of electrical current. However, when smoke particles enter the detector, they interfere with the formation of ions in the air, and the electric current is interrupted. This causes the alarm to sound and warns the occupants of the danger of fire. Complete the following nuclear equation for the decay of americium-241:

$$^{241}_{95}\text{Am} \longrightarrow ? + \, ^{4}_{2}\text{He}$$

Solution

STEP 1 **Write the incomplete nuclear equation.**

$$^{241}_{95}\text{Am} \longrightarrow ? + \, ^{4}_{2}\text{He}$$

STEP 2 **Determine the missing mass number.** In the equation, the mass number, 241, of the americium is equal to the sum of the mass numbers of the alpha particle and the new nucleus.

$$241 \quad = ? + 4$$
$$241 - 4 = ?$$
$$237 \quad = ? \text{ (mass number of new nucleus)}$$

STEP 3 **Determine the missing atomic number.** The atomic number of americium-95 must equal the sum of the atomic number of the alpha particle and the new nucleus.

$$95 \quad = ? + 2$$
$$95 - 2 = ?$$
$$93 \quad = ? \text{ (atomic number of new nucleus)}$$

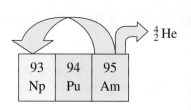

STEP 4 **Determine the symbol of the new nucleus.** On the periodic table, the element that has atomic number 93 is neptunium, Np. The symbol of this isotope of Np is written as

$$^{237}_{93}\text{Np}$$

STEP 5 **Complete the nuclear equation.**

$$^{241}_{95}\text{Am} \longrightarrow ^{237}_{93}\text{Np} + ^{4}_{2}\text{He}$$

In this nuclear reaction, an americium-241 nucleus decays by releasing an alpha particle and produces a neptunium-237 nucleus.

Study Check

Write a balanced nuclear equation for the alpha emitter polonium-214.

Beta Emitters

In beta emission the unstable nucleus converts a neutron into a proton and a beta particle, which is emitted from the nucleus. The newly formed proton adds to the number of protons already in the nucleus and increases the atomic number by 1. However, the mass number of the newly formed nucleus stays the same. For example, carbon-14 decays by emitting a beta particle and forming a nitrogen nucleus.

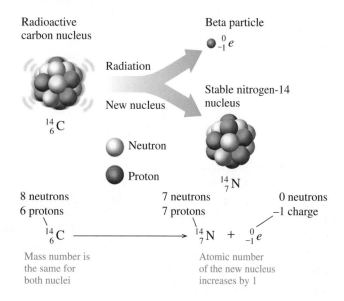

In the nuclear equation of a beta emitter, the mass number of the radioactive nucleus and the mass number of the new nucleus are the same, and the atomic number of the new nucleus increases by 1, indicating a change of one element into another.

Sample Problem 16.3 ▸ **Writing an Equation for Beta Decay**

Cobalt-60 decays by emitting a beta particle. Write the nuclear equation for its decay.

Solution

STEP 1 **Write the incomplete nuclear equation.**

$$^{60}_{27}Co \longrightarrow ? + ^{0}_{-1}e$$

STEP 2 **Determine the missing mass number.** In the equation, the mass number of cobalt-60 is equal to the sum of the mass numbers of the beta particle and the new nucleus.

$$60 = ? + 0$$
$$60 - 0 = ?$$
$$60 = ? \text{ (mass number of new nucleus)}$$

STEP 3 **Determine the missing atomic number.** The atomic number of cobalt-60 must equal the sum of the atomic number of the beta particle and the new nucleus.

$$27 = ? - 1$$
$$27 + 1 = ?$$
$$28 = ? \text{ (atomic number of new nucleus)}$$

STEP 4 **Determine the symbol of the new nucleus.** On the periodic table, the element that has atomic number 28 is nickel (Ni). The nucleus of this isotope of Ni is written as

$$^{60}_{28}Ni$$

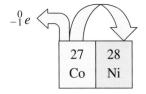

STEP 5 **Complete the nuclear equation.**

$$^{60}_{27}Co \longrightarrow ^{60}_{28}Ni + ^{0}_{-1}e$$

In this nuclear reaction, a cobalt-60 nucleus decays by releasing a beta particle and produces a nickel-60 nucleus.

Study Check

Write the nuclear equation for iodine-131, a beta emitter.

Positron Emitters

In positron emission, a proton in the nucleus is converted to a neutron and a positron, which leaves the nucleus.

Sample Problem 16.4 ▶   **Writing an Equation for Positron Emission**

Write the nuclear equation for manganese-49, which decays by emitting a positron.

Solution

STEP 1   **Write the incomplete nuclear equation.**

$$^{49}_{25}\text{Mn} \longrightarrow ? + ^{0}_{+1}e$$

STEP 2   **Determine the missing mass number.** In the equation, the mass number 49 of manganese is equal to the combined mass numbers of a positron and the new nucleus.

$$49 \quad = ? + 0$$
$$49 - 0 = ?$$
$$49 \quad = ? \text{ (mass number of new nucleus)}$$

STEP 3   **Determine the missing atomic number.** The atomic number of manganese-49 must equal the sum of the atomic number of the beta particle and the new nucleus.

$$25 \quad = ? + 1$$
$$25 - 1 = ?$$
$$24 \quad = ? \quad \text{(atomic number of new nucleus)}$$

STEP 4   **Determine the symbol of the new nucleus.** On the periodic table, the element that has atomic number 24 is chromium, Cr. The nucleus of this isotope of Cr is written as

$$^{49}_{24}\text{Cr}$$

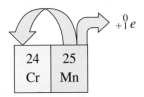

STEP 5   **Complete the nuclear equation.**

$$^{49}_{25}\text{Mn} \longrightarrow ^{49}_{24}\text{Cr} + ^{0}_{+1}e$$

In this nuclear reaction, a manganese-49 nucleus decays by releasing a positron and produces a chromium-49 nucleus.

Study Check

Write the nuclear equation for xenon-118, which decays by emitting a positron.

Figure 16.2 When the nuclei of alpha, beta, positron, and gamma emitters emit radiation, new, more stable nuclei are produced.

Q *What changes occur in the number of protons and neutrons when an alpha emitter gives off radiation?*

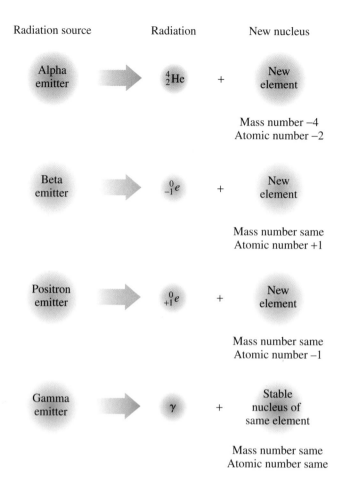

Radiation source   Radiation   New nucleus

Alpha emitter → $_2^4\text{He}$ + New element

Mass number −4
Atomic number −2

Beta emitter → $_{-1}^{0}e$ + New element

Mass number same
Atomic number +1

Positron emitter → $_{+1}^{0}e$ + New element

Mass number same
Atomic number −1

Gamma emitter → γ + Stable nucleus of same element

Mass number same
Atomic number same

Gamma Emitters

There are very few pure gamma emitters, although gamma radiation accompanies most alpha and beta radiation. In radiology, one of the most commonly used gamma emitters is technetium (Tc). The excited state called metastable technetium is written as technetium-99m, Tc-99m, or ^{99m}Tc. By emitting energy in the form of gamma rays, the excited nucleus becomes more stable.

$$_{43}^{99m}\text{Tc} \longrightarrow {}_{43}^{99}\text{Tc} + \gamma$$

Figure 16.2 summarizes the changes in the nucleus for alpha, beta, positron, and gamma radiation.

Producing Radioactive Isotopes

Today, more than 1500 radioisotopes are produced by converting stable, nonradioactive isotopes into radioactive ones. To do this, a stable atom is bombarded by fast-moving alpha particles, protons, or neutrons. When one of these particles is absorbed by a stable nucleus, the nucleus becomes unstable and the atom is now a radioactive isotope, or **radioisotope.** The process of changing one element into another is called **transmutation.**

When boron-10, a nonradioactive isotope, is bombarded by an alpha particle, it is converted to nitrogen-13, a radioisotope. In this bombardment reaction, a neutron is emitted.

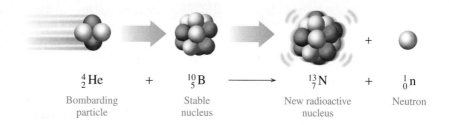

$$\underset{2}{^4}He \quad + \quad \underset{5}{^{10}}B \longrightarrow \underset{7}{^{13}}N \quad + \quad \underset{0}{^1}n$$

Bombarding   Stable   New radioactive   Neutron
particle   nucleus   nucleus

All of the known elements that have atomic numbers greater than 92 have been produced by bombardment; none of these elements occurs naturally. Most have been produced in only small amounts and exist for such a short time that it is difficult to study their properties. An example is element 105, dubnium, which is produced when californium-249 is bombarded with nitrogen-15.

$$\underset{98}{^{249}}Cf + \underset{7}{^{15}}N \longrightarrow \underset{105}{^{260}}Db + 4\,\underset{0}{^1}n$$

Technetium-99m is a radioisotope used in nuclear medicine for several diagnostic procedures, including the detection of brain tumors and examinations of the liver and spleen. The source of technetium-99m is molybdenum-99, which is produced in a nuclear reactor by neutron bombardment of molybdenum-98.

$$\underset{42}{^{98}}Mo + \underset{0}{^1}n \longrightarrow \underset{42}{^{99}}Mo$$

Many radiology laboratories have a small generator containing the radioactive molybdenum-99, which decays to give the technetium-99m radioisotope.

$$\underset{42}{^{99}}Mo \longrightarrow \underset{43}{^{99m}}Tc + \underset{-1}{^0}e$$

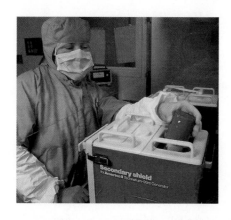

The technetium-99m radioisotope decays by emitting gamma rays. Gamma emission is desirable for diagnostic work because the gamma rays pass through the body to the detection equipment.

$$\underset{43}{^{99m}}Tc \longrightarrow \underset{43}{^{99}}Tc + \gamma$$

Sample Problem 16.5   **Completing a Nuclear Equation
for a Bombardment Reaction**

When a proton bombards nickel-58, the products are a new nucleus and an alpha particle. Write the balanced nuclear equation for the bombardment reaction.

Solution

STEP 1   **Write the incomplete nuclear equation.**

$$\underset{28}{^{58}}Ni + \underset{1}{^1}H \longrightarrow ? + \underset{2}{^4}He$$

STEP 2   **Determine the missing mass number.** In the equation, the sum of the mass numbers of nickel (58) and hydrogen (1) must be equal to the sum of the mass numbers of the alpha particle (4) and the new nucleus.

$$58 + 1 = ? + 4$$
$$59 - 4 = ?$$
$$55 \quad\quad = ? \text{ (mass number of new nucleus)}$$

STEP 3 **Determine the missing atomic number.** The atomic numbers of nickel (28) and hydrogen (1) must equal the sum of the atomic numbers of the alpha particle (2) and the new nucleus.

$$28 + 1 = ? + 2$$
$$29 - 2 = ?$$
$$27 \quad = ? \text{ (atomic number of new nucleus)}$$

STEP 4 **Determine the symbol of the new nucleus.** On the periodic table, the element that has atomic number 27 is cobalt, Co. The symbol of this isotope of Co is written as $^{55}_{27}\text{Co}$.

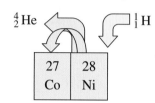

STEP 5 **Complete the nuclear equation.**

$$^{58}_{28}\text{Ni} + ^{1}_{1}\text{H} \longrightarrow ^{55}_{27}\text{Co} + ^{4}_{2}\text{He}$$

Proton   New isotope

Study Check

Complete the following bombardment equation:

$$? + ^{4}_{2}\text{He} \longrightarrow ^{17}_{8}\text{O} + ^{1}_{1}\text{H}$$

Questions and Problems   Nuclear Equations

16.13 Write a balanced nuclear equation for the alpha decay of each of the following radioactive isotopes:
 a. $^{208}_{84}\text{Po}$ **b.** $^{232}_{90}\text{Th}$
 c. $^{251}_{102}\text{No}$ **d.** $^{220}_{86}\text{Rn}$

16.14 Write a balanced nuclear equation for the alpha decay of each of the following radioactive isotopes:
 a. $^{243}_{96}\text{Cm}$ **b.** $^{252}_{99}\text{Es}$
 c. $^{251}_{98}\text{Cf}$ **d.** $^{261}_{107}\text{Bh}$

16.15 Write a balanced nuclear equation for the beta decay of each of the following radioactive isotopes:
 a. $^{25}_{11}\text{Na}$ **b.** $^{20}_{8}\text{O}$
 c. $^{92}_{38}\text{Sr}$ **d.** $^{42}_{19}\text{K}$

16.16 Write a balanced nuclear equation for the beta decay of each of the following radioactive isotopes:
 a. potassium-42 **b.** iron-59
 c. iron-60 **d.** barium-141

16.17 Write a balanced nuclear equation for the positron decay of each of the following radioactive isotopes:
 a. $^{26}_{14}\text{Si}$ **b.** $^{54}_{27}\text{Co}$
 c. $^{77}_{37}\text{Rb}$ **d.** $^{93}_{45}\text{Rh}$

16.18 Write a balanced nuclear equation for the positron decay of each of the following radioactive isotopes:
 a. $^{8}_{5}\text{B}$ **b.** $^{13}_{7}\text{N}$
 c. $^{40}_{19}\text{K}$ **d.** $^{118}_{54}\text{Xe}$

16.19 Complete each of the following nuclear equations:
 a. $^{28}_{13}\text{Al} \longrightarrow ? + ^{0}_{-1}e$
 b. $? \longrightarrow ^{86}_{36}\text{Kr} + ^{1}_{0}n$
 c. $^{66}_{29}\text{Cu} \longrightarrow ^{66}_{30}\text{Zn} + ?$
 d. $? \longrightarrow ^{4}_{2}\text{He} + ^{234}_{90}\text{Th}$
 e. $^{188}_{80}\text{Hg} \longrightarrow ? + ^{0}_{+1}e$

16.20 Complete each of the following nuclear equations:
 a. $^{11}_{6}\text{C} \longrightarrow ^{7}_{4}\text{Be} + ?$
 b. $^{35}_{16}\text{S} \longrightarrow ? + ^{0}_{-1}e$
 c. $? \longrightarrow ^{90}_{39}\text{Y} + ^{0}_{-1}e$
 d. $^{210}_{83}\text{Bi} \longrightarrow ? + ^{4}_{2}\text{He}$
 e. $? \longrightarrow ^{135}_{59}\text{Pr} + ^{0}_{+1}e$

16.21 Complete each of the following bombardment reactions:
 a. $^{9}_{4}\text{Be} + ^{1}_{0}n \longrightarrow ?$
 b. $^{32}_{16}\text{S} + ? \longrightarrow ^{32}_{15}\text{P}$
 c. $? + ^{1}_{0}n \longrightarrow ^{24}_{11}\text{Na} + ^{4}_{2}\text{He}$
 d. When Al-27 is bombarded by an alpha particle, it forms Si-30. What other particle is produced?

16.22 Complete each of the following bombardment reactions:
 a. $^{40}_{18}\text{Ar} + ? \longrightarrow ^{43}_{19}\text{K} + ^{1}_{1}\text{H}$
 b. $^{238}_{92}\text{U} + ^{1}_{0}n \longrightarrow ?$
 c. $? + ^{1}_{0}n \longrightarrow ^{14}_{6}\text{C} + ^{1}_{1}\text{H}$
 d. When an alpha particle bombards N-14, it forms a radioisotope and a proton. What is the radioisotope that is produced?

Learning Goal

Describe the detection and measurement of radiation.

16.3 Radiation Measurement

One of the most common instruments for detecting beta and gamma radiation is the Geiger counter. It consists of a metal tube filled with a gas such as argon. When radiation enters a window on the end of the tube, it produces ions in the gas; the ions produce an electrical current. Each burst of current is amplified to give a click and a readout on a meter.

$$Ar + radiation \longrightarrow Ar^+ + e^-$$

Radiation is measured in several different ways. We can measure the activity of a radioactive sample or determine the impact of radiation on biological tissue.

Activity

When a radiology laboratory obtains a radioisotope, the activity of the sample is measured in terms of the number of nuclear disintegrations per second. The **curie (Ci),** the original unit of activity, was defined as the number of disintegrations that occur in 1 second for 1 g radium, which is equal to 3.7×10^{10} disintegrations per second. The curie was named for Marie Curie, a Polish scientist, who along with her husband Pierre discovered the radioactive elements radium and polonium. The SI unit of radiation activity is the **becquerel (Bq),** which is one disintegration per second.

Biological Effect

After we measure the activity of a radioisotope, we often want to know how much radiation the tissues in the body absorb. The **rad (radiation absorbed**

Table 16.5   Some Units of Radiation Measurement

Measurement	Common Unit	SI Unit	Relationship
Activity	curie (Ci) = 3.7×10^{10} disintegrations/s	becquerel (Bq) = 1 disintegration/s	1 Ci = 3.7×10^{10} Bq
Absorbed Dose	rad	gray (Gy) = 1 J/kg tissue	1 Gy = 100 rad
Biological Damage	rem = rad $\times$ factor	1 sievert (Sv)	1 Sv = 100 rem

dose) is a unit that measures the amount of radiation absorbed by a gram of material such as body tissue. The SI unit for absorbed dose is the **gray (Gy),** which is defined as the joules of energy absorbed by 1 kg of tissue. The gray is equal to 100 rads.

The **rem (radiation equivalent in humans)** measures the biological effects of different kinds of radiation. Alpha particles don't penetrate the skin. But if they should enter the body by some other route, they cause a lot of damage even though the particles travel a short distance in tissue. High-energy radiation such as beta particles and high-energy protons and neutrons that penetrate the skin and travel into tissue cause more damage. Gamma rays are damaging because they travel a long way through tissue and create a great deal of ionization.

To determine the **equivalent dose** or rem dose, the absorbed dose (rads) is multiplied by a factor that adjusts for biological damage caused by a particular form of radiation. For beta and gamma radiation the factor is 1, so the biological damage in rems is the same as the absorbed radiation (rads). For high-energy protons and neutrons, the factor is about 10, and for alpha particles it is 20.

Biological damage (rem) = Absorbed dose (rad) $\times$ Factor

Often, the measurement for an equivalent dose will be in units of millirems (mrem). One rem is equal to 1000 mrem. The SI unit is the **sievert (Sv).** One sievert is equal to 100 rems. Table 16.5 summarizes the units used to measure radiation.

Sample Problem 16.6 ▶   Measuring Activity

A sample of phosphorus-32, a beta emitter, has an activity of 2 millicuries (mCi). How many beta particles are emitted by this sample of P-32 in 1 s?

Solution

STEP 1   **Given**   activity = 2 mCi; time = 1 s
            Need   number of beta particles

STEP 2   **Plan**

mCi   Metric factor   Ci   Curie factor β particles/s   Time β particles

STEP 3   **Equalities/Conversion Factors**

$$1 \text{ Ci} = 1000 \text{ mCi} \qquad \frac{3.7 \times 10^{10} \; \beta \text{ particles}}{1 \text{ s}} = 1 \text{ Ci}$$

$1 \text{ Ci} = 1000 \text{ mCi}$

$$\frac{1000 \text{ mCi}}{1 \text{ Ci}} \quad \text{and} \quad \frac{1 \text{ Ci}}{1000 \text{ mCi}}$$

$3.7 \times 10^{10} \; \beta \text{ particles/s} = 1 \text{ Ci}$

$$\frac{3.7 \times 10^{10} \; \beta \text{ particles/s}}{1 \text{ Ci}} \quad \text{and} \quad \frac{1 \text{ Ci}}{3.7 \times 10^{10} \; \beta \text{ particles/s}}$$

STEP 4   **Set Up Problem**   We calculate the number of beta particles emitted in 1 second from the activity of the radioisotope.

$$2 \text{ mCi} \times \frac{1 \text{ Ci}}{1000 \text{ mCi}} \times \frac{3.7 \times 10^{10} \; \beta \text{ particles/s}}{\text{Ci}} \times 1 \text{ s} = 7 \times 10^{7} \; \beta \text{ particles}$$

Study Check

An iodine-131 source has an activity of 0.25 Ci. How many radioactive atoms will disintegrate in 1.0 min?

CHEM NOTE

RADIATION AND FOOD

Foodborne illnesses caused by pathogenic bacteria such as *Salmonella, Listeria,* and *Escherichia coli* have become a major health concern in the United States. The Centers for Disease Control and Prevention estimates that each year *E. coli* in contaminated foods infects 20,000 people in the United States and that 500 people die. *E. coli* has been responsible for outbreaks of illness from contaminated ground beef, fruit juices, lettuce, and alfalfa sprouts.

The Food and Drug Administration (FDA) has approved the use of 0.3 kilogray (0.3 kGy) to 1 kGy of ionizing radiation produced by cobalt-60 or cesium-137 for the treatment of foods. The irradiation technology is much like that used to sterilize medical supplies. Cobalt pellets are placed in stainless steel tubes, which are arranged in racks. When food moves through the series of racks, the gamma rays pass through the food and kill the bacteria.

It is important for consumers to understand that when food is irradiated, it never comes in contact with the radioactive source. The gamma rays pass through the food to kill bacteria, but that does not make the food radioactive. The radiation kills bacteria because it stops their ability to divide and grow. We cook or heat food thoroughly for the same purpose. Radiation, as well as heat, has little effect on the food itself because its cells are no longer dividing or growing. Thus irradiated food is not harmed, although a small amount of vitamin B_1 and C may be lost.

Currently, tomatoes, blueberries, strawberries, and mushrooms are being irradiated to allow them to be harvested when completely ripe and extend their shelf life (see Figure 16.3). The FDA has also approved the irradiation of pork, poultry, and beef in order to decrease potential infections and to extend shelf life. Currently, irradiated vegetable and meat products are available in retail markets in South Africa. Apollo 17 astronauts

(a)

(b)

Figure 16.3 **(a)** The FDA requires this symbol to appear on irradiated retail foods. **(b)** After 2 weeks, the irradiated strawberries on the right show no spoilage. Mold is starting to grow on the nonirradiated ones on the left.

Q Why are irradiated foods used on spaceships and in nursing homes?

ate irradiated foods on the moon, and some U.S. hospitals and nursing homes now use irradiated poultry to reduce the possibility of infections among patients. The extended shelf life of irradiated food also makes it useful for campers and military personnel. Soon consumers concerned about food safety will have a choice of irradiated meats, fruits, and vegetables at the market.

Table 16.6	Average Annual Radiation Received by a Person in the United States	
Source	**Dose (mrem/yr)**	
Natural		
The ground	15	
Air, water, food	30	
Cosmic rays	40	
Wood, concrete, brick	50	
Medical		
Chest X ray	50	
Dental X ray	20	
Upper gastrointestinal tract X ray	200	
Other		
Television	2	
Air travel	1	
Radon	200[a]	
Cigarette smoking	35	

[a] Varies widely.

the Chemistry place

CASE STUDY
Food Irradiation

Table 16.7	Lethal Doses of Radiation for Some Life-Forms	
Life-Form	**LD_{50} (rem)**	
Insect	100,000	
Bacterium	50,000	
Rat	800	
Human	500	
Dog	300	

Background Radiation

We are all exposed to low levels of radiation every day. Naturally occurring radioactive isotopes are part of the atoms of wood, brick, and concrete in our homes and the buildings where we work and go to school. This radioactivity, called background radiation, is present in the soil, in the food we eat, in the water we drink, and in the air we breathe. For example, one of the naturally occurring isotopes of potassium, potassium-40, is radioactive. It is found in the body because it is always present in any potassium-containing food. Other naturally occurring radioisotopes in air and food are carbon-14, radon-222, strontium-90, and iodine-131. Table 16.6 lists some common sources of radiation.

We are also exposed to radiation (cosmic rays) produced in space by the sun. At higher elevations, the amount of cosmic radiation is greater because there are fewer air molecules to absorb the radiation. People living at high altitudes or flying in an airplane receive more radiation from cosmic rays than those who live at sea level. For example, a person living in Denver receives about twice the cosmic radiation as a person living in Los Angeles.

A person living close to a nuclear power plant normally does not receive much additional radiation, perhaps 0.1 millirem (mrem) in 1 year (1 rem equals 1000 mrem). However, in the accident at the Chernobyl nuclear power plant in 1986, it is estimated that people in a nearby town received as much as 1 rem/hr.

In addition to naturally occurring radiation from materials in our homes, we receive radiation from television. In the medical clinic, dental and chest X rays also add to our radiation exposure. The average person in the United States receives about 0.17 rem or 170 mrem of radiation annually.

Radiation Sickness

The larger the dose of radiation received at one time, the greater the effect on the body. Exposure to radiation under 25 rem usually cannot be detected. Whole-body exposure of 100 rem produces a temporary decrease in the number of white blood cells. If the exposure to radiation is 100 rem or higher, the person suffers the symptoms of radiation sickness: nausea, vomiting, fatigue, and a reduction in white-cell count. A whole-body dosage greater than 300 rem can lower the white-cell count to zero. The victim suffers diarrhea, hair loss, and infection. Exposure to radiation of about 500 rem is expected to cause death in 50% of the people receiving that dose. This amount of radiation is called the lethal dose for one-half the population, or the LD_{50}. The LD_{50} varies for different life-forms, as Table 16.7 shows. Radiation dosages of about 600 rem would be fatal to all humans within a few weeks.

16.23 a. How does a Geiger counter detect radiation?
   b. What are the SI unit and the older unit that describe the activity of a radioactive sample?
   c. What are the SI unit and the older unit that describe the radiation dose absorbed by tissue?
   d. What is meant by the term *kilogray*?

16.24 a. What is background radiation?
   b. What are the SI unit and the older unit that describe the biological effect of radiation?
   c. What is meant by the terms *mCi* and *mrem*?
   d. Why is a factor used to determine the dose equivalent?

16.25 a. A sample of iodine-131 has an activity of 3.0 Ci. How many disintegrations occur in the iodine-131 sample in 20. s?
   b. The recommended dosage of iodine-131 is 4.20 μCi/kg of body weight. How many microcuries of iodine-131 are needed for a 70.0-kg person with hyperthyroidism (1 mCi = 1000 μCi)?

16.26 a. The dosage of technetium-99m for a lung scan is 20 μCi/kg of body weight. How many millicuries should be given to a 50.0-kg person (1 mCi = 1000 μCi)?
   b. A person receives 50 mrads in a chest X ray. What is that amount in grays? What would be the dose equivalent in mrems?
   c. Suppose a person absorbed 50 mrads of alpha radiation. What would be the dose equivalent in mrems? How does it compare with the mrems in part b?

16.27 Why would an airline pilot be exposed to more background radiation than the person who works at the ticket counter?

16.28 In radiation therapy, a person receives high doses of radiation. What symptoms of radiation sickness might the patient exhibit?

16.4 Half-Life of a Radioisotope

Learning Goal

Given the half-life of a radioisotope, calculate the amount of radioisotope remaining after one or more half-lives.

The **half-life** of a radioisotope is the amount of time it takes for one-half of a sample to decay. For example, ^{131}I has a half-life of 8.0 days. As ^{131}I decays it produces a beta particle and the nonradioactive isotope ^{131}Xe.

$$^{131}_{53}\text{I} \longrightarrow ^{131}_{54}\text{Xe} + ^{0}_{-1}\beta$$

Suppose we have a sample that contains 20.0 grams of ^{131}I. We don't know which specific nucleus will emit radiation, but we do know that in 8.0 days, one-half of all the nuclei in the sample will decay to give ^{131}Xe. That means that after 8.0 days, there are one-half of the number of ^{131}I atoms in the sample, or 10.0 g of ^{131}I, remaining. The decay process has also produced 10.0 g of the product ^{131}Xe. After another half-life, or 8.0 days, passes, 5.00 g of the 10.0 g of ^{131}I will decay to ^{131}Xe. Now there are 5.00 g of ^{131}I left, while there is a total of 15.0 g of ^{131}Xe.

$$20 \text{ g }^{131}\text{I} \xrightarrow{\text{1 half-life}} 10 \text{ g }^{131}\text{I} \xrightarrow{\text{2 half-lives}} 5 \text{ g }^{131}\text{I}$$

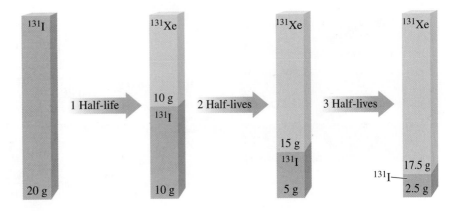

Figure 16.4 The decay curve for iodine-131 shows that one-half of the radioactive sample decays and one-half remains radioactive after each half-life of 8 days.

Q How many grams of the 20-g sample remain radioactive after 2 half-lives?

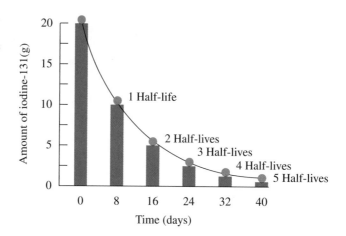

Table 16.8   Activity of an ^{131}I Sample with Time

Time elapsed	0 days	8.0 days	16.0 days	24.0 days
Half-lives	0	1	2	3
^{131}I remaining	20.0 g	10.0 g	5.00 g	2.50 g
^{131}Xe produced	0 g	10.0 g	15.0 g	17.5 g

A third half-life, or another 8.0 days, results in 2.50 g of the ^{131}I decaying to give ^{131}Xe, which leaves 2.50 g of ^{131}I still capable of producing radiation. This information is summarized in Table 16.8.

A **decay curve** is a diagram of the decay of a radioactive isotope. Figure 16.4 shows such a curve for the ^{131}I we have discussed.

Sample Problem 16.7   **Using Half-Lives of a Radioisotope**

Phosphorus-32, a radioisotope used in the treatment of leukemia, has a half-life of 14 days. If a sample contains 8.0 g of phosphorus-32, how many grams of phosphorus-32 remain after 42 days?

Solution

STEP 1   **Given**   8.0 g ^{32}P; 42 days; 14 days/half-life
           Need   g ^{32}P remaining

STEP 2   **Plan**

42 days   Half-life   Number of half-lives

8.0 g ^{32}P   Number of half-lives   g ^{32}P remaining

STEP 3   **Equalities/Conversion Factors**

1 half-life = 14 d

$$\frac{14 \text{ d}}{1 \text{ half-life}} \quad \text{and} \quad \frac{1 \text{ half-life}}{14 \text{ d}}$$

Guide to Using Half-Lives

STEP 1
State the given and needed amounts of radioisotope.

STEP 2
Write a plan to calculate amount of active radioisotope.

STEP 3
Write the half-life equality and conversion factors.

STEP 4
Set up problem to calculate amount of active radioisotope.

STEP 4 **Set Up Problem**   We can do this problem with two calculations. First, we determine the number of half-lives in the amount of time that has elapsed.

$$\text{Number of half-lives} = 42 \ \cancel{d} \ \times \ \frac{1 \ \text{half-life}}{14 \ \cancel{d}} \ = \ 3 \ \text{half-lives}$$

Now, we determine how much of the sample decays in 3 half-lives and how many grams of the phosphorus remain.

$$8.0 \text{ g }^{32}\text{P} \xrightarrow{\ 14 \text{ d}\ } 4.0 \text{ g }^{32}\text{P} \xrightarrow{\ 14 \text{ d}\ } 2.0 \text{ g }^{32}\text{P} \xrightarrow{\ 14 \text{ d}\ } 1.0 \text{ g }^{32}\text{P}$$

Study Check

Iron-59 has a half-life of 46 days. If the laboratory received a sample of 8.0 g iron-59, how many grams are still active after 184 days?

Naturally occurring isotopes of the elements usually have long half-lives, as shown in Table 16.9. They disintegrate slowly and produce radiation over a long period of time, even hundreds or millions of years. In contrast, many of the radioisotopes used in nuclear medicine have much shorter half-lives. They disintegrate rapidly and produce almost all their radiation in a short period of time. For example, technetium-99m emits half of its radiation in the first 6 hr. This means that a small amount of the radioisotope given to a patient is essentially gone within 2 days. The decay products of technetium-99m are totally eliminated by the body.

Table 16.9 · Half-Lives of Some Radioisotopes

Element	Radioisotope	Half-Life
Naturally Occurring Radioisotopes		
carbon	^{14}C	5730 yr
potassium	^{40}K	1.3×10^9 yr
radium	^{226}Ra	1600 yr
uranium	^{238}U	4.5×10^9 yr
Some Medical Radioisotopes		
chromium	^{51}Cr	28 days
iodine	^{131}I	8 days
iron	^{59}Fe	46 days
technetium	^{99m}Tc	6.0 hr

the Chemistry place

WEB TUTORIAL
Nuclear Chemistry

Sample Problem 16.8 ▸ Dating Using Half-Lives

The remains of ancient animals have been unearthed at La Brea tar pits in Los Angeles. Suppose a bone sample from the tar pits is subjected to the carbon-14 dating method. If the sample shows that two half-lives have passed, when did the animal live?

Solution

We can calculate the age of the bone sample by using the half-life of carbon-14 (5730 years).

$$2 \ \cancel{\text{half-lives}} \times \frac{5730 \text{ years}}{1 \ \cancel{\text{half-life}}} = 11\,500 \text{ years}$$

We would estimate that the animal lived 11 500 years ago, or about 9500 B.C.E.

Study Check

Suppose that a piece of wood found in a tomb had $\frac{1}{8}$ (3 half-lives) of its original C-14 activity. About how many years ago was the wood part of a living tree?

CHEM NOTE

DATING ANCIENT OBJECTS

A technique known as radiological dating is used by geologists, archaeologists, and historians as a way to determine the age of ancient objects. The age of an object derived from plants or animals (such as wood, fiber, natural pigments, bone, and cotton or woolen clothing) is determined by measuring the amount of carbon-14, a naturally occurring radioactive form of carbon. In 1960, Willard Libby received the Nobel Prize for the work he did developing carbon-14 dating techniques during the 1940s.

Carbon-14 is produced in the upper atmosphere by the bombardment of $^{14}_{7}N$ by high-energy neutrons from cosmic rays.

$$^{1}_{0}n \quad + \quad ^{14}_{7}N \quad \longrightarrow \quad ^{14}_{6}C \quad + \quad ^{1}_{1}H$$

Neutron from cosmic rays — Nitrogen in atmosphere — Radioactive carbon-14 — Proton

The carbon-14 reacts with oxygen to form radioactive carbon dioxide, $^{14}CO_2$. Because carbon dioxide is continuously absorbed by living plants during the process of photosynthesis, some carbon-14 will be taken into the plant. After the plant dies, no more carbon-14 is taken up, and the amount of carbon-14 contained in the plant steadily decreases as it undergoes radioactive β decay.

$$^{14}_{6}C \longrightarrow ^{14}_{7}N + ^{0}_{-1}e$$

Scientists use the half-life of carbon-14 (5730 years) to calculate the amount of time that has passed since the plant died, a process called *carbon dating*. The smaller the amount of carbon-14 remaining in the sample, the greater the number of half-lives that have passed. Thus, the approximate age of the sample can be determined. For example, a wooden beam found in an ancient Indian dwelling might have one-half of the carbon-14 found in living plants today. Thus, the dwelling was probably constructed about 5730 years ago, one half-life of

carbon-14. Carbon-14 dating was used to determine that the Dead Sea Scrolls are about 2000 years old.

A radiological dating method used for determining the age of rocks is based on the radioisotope uranium-238, which decays through a series of reactions to lead-206. The uranium-238 isotope has a very long half-life, about 4×10^9 (4 billion) years. Measurements of the amounts of uranium-238 and lead-206 enable geologists to determine the age of rock samples. The older rocks will have a higher percentage of lead-206 because more of the uranium-238 has decayed. The age of rocks brought back from the moon by the *Apollo* missions was determined using uranium-238. They were found to be about 4×10^9 years old, approximately the same age calculated for Earth.

Questions and Problems   Half-Life of a Radioisotope

16.29 What is meant by the term *half-life*?

16.30 Why are radioisotopes with short half-lives used for diagnosis in nuclear medicine?

16.31 Technetium-99m is an ideal radioisotope for scanning organs because it has a half-life of 6.0 hr and is a pure gamma emitter. Suppose that 80.0 mg were prepared in the technetium generator this morning. How many milligrams would remain after the following intervals?
   a. 1 half-life   **b.** 2 half-lives
   c. 18 hr   **d.** 24 hr

16.32 A sample of sodium-24 has an activity of 12 mCi. If sodium-24 has a half-life of 15 hr, what is the activity of the sodium after $2\frac{1}{2}$ days?

16.33 Strontium-85, used for bone scans, has a half-life of 64 days. How long will it take for the radiation level of strontium-85 to drop to one-fourth of its original level? to one-eighth?

16.34 Fluorine-18, which has a half-life of 110 min, is used in PET scans. If 100 mg of fluorine-18 is shipped at 8:00 A.M., how many milligrams of the radioisotope are still active if the sample arrives at the radiology laboratory at 1:30 P.M.?

Table 16.10 **Medical Applications of Radioisotopes**

Isotope	Half-Life	Medical Application
Ce-141	32.5 days	Gastrointestinal tract diagnosis; measuring blood flow to the heart
Ga-67	78 hr	Abdominal imaging; tumor detection
Ga-68	68 min	Detect pancreatic cancer
P-32	4.3 days	Treatment of leukemia, excess red blood cells, pancreatic cancer
I-125	60 days	Treatment of brain cancer; osteoporosis detection
I-131	8 days	Imaging thyroid; treatment of Graves' disease, goiter, and hyperthyroidism; treatment of thyroid and prostate cancer
Sr-85	65 days	Detection of bone lesions; brain scans
Tc-99m	6 hr	Imaging of skeleton and heart muscle, brain, liver, heart, lungs, bone, spleen, kidney, and thyroid; *most widely used radioisotope in nuclear medicine*

Learning **Goal**

Describe the use of radioisotopes in medicine.

16.5 Medical Applications Using Radioactivity

When a radiologist wants to determine the condition of an organ in the body, the patient is given a radioisotope that is known to concentrate in that organ. The cells in the body cannot differentiate between a nonradioactive atom and a radioactive one. However, radioactive atoms can be detected because they emit radiation and the nonradioactive atoms do not. Some radioisotopes used in nuclear medicine are listed in Table 16.10.

After a patient receives a radioisotope, an apparatus called a scanner produces an image of the organ. The scanner moves slowly across the region of the body where the organ containing the radioisotope is located. The gamma rays emitted from the radioisotope in the organ are used to expose a photographic plate, producing a *scan* of the organ.

A common method of determining thyroid function is the use of radioactive iodine uptake (RAIU). Taken orally, the radioisotope iodine-131 mixes with the iodine already present in the body. Twenty-four hours later, the amount of iodine taken up by the thyroid is determined. A detection tube detects the radiation coming from the iodine-131 that has located in the thyroid. (See Figure 16.5.) The iodine uptake is directly proportional to the activity of the thyroid. A patient with a hyperactive thyroid will have a higher-than-normal level of radioactive iodine, whereas a patient with a hypoactive thyroid will record lower values.

If the patient has hyperthyroidism, treatment is begun to lower the activity of the thyroid. One treatment involves giving the patient a therapeutic dosage of radioactive iodine, which has a higher radiation level than the diagnostic dose. The radioactive iodine goes to the thyroid where its radiation destroys some of the thyroid cells. The thyroid produces less thyroid hormone, bringing the hyperthyroid condition under control.

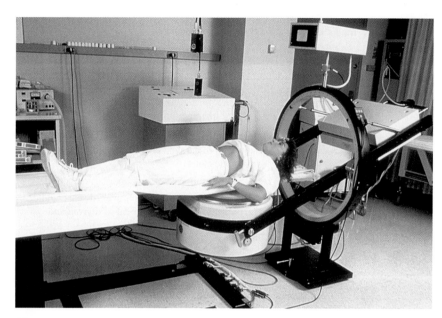

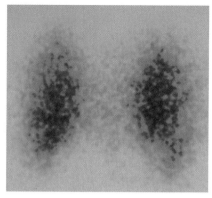

(b)

(a)

Figure 16.5 **(a)** A scanner is used to detect radiation from a radioisotope that has accumulated in an organ. **(b)** A scan of the thyroid shows the accumulation of radioactive iodine-131 in the thyroid.

Q What type of radiation would move through body tissues to create a scan?

Positron Emission Tomography (PET)

Radioisotopes that emit a positron are used in an imaging method called positron emission tomography (PET). Medically, positron emitters such as carbon-11, oxygen-15, and nitrogen-13 are used to diagnose conditions involving blood flow, metabolism, and particularly brain function. Glucose containing carbon-11 is used to detect damage in the brain from epilepsy, stroke, and Parkinson's disease.

Carbon-11 emits a positron when it decays

$$^{11}_{6}\text{C} \longrightarrow \ ^{11}_{5}\text{B} \ + \ ^{0}_{+1}e$$
$$\text{Positron}$$

The positron exists for only a moment before it collides with an electron, which produces gamma rays that can be detected.

$$^{0}_{+1}e \ + \ ^{0}_{-1}e \longrightarrow 2\gamma$$

Positron   Electron in   Gamma rays
        an atom   produced

The gamma rays from the positrons emitted are detected by computerized equipment to create a three-dimensional image of the organ. (See Figure 16.6.)

Figure 16.6 These PET scans of the brain show a normal brain on the left and a brain affected by Alzheimer's disease on the right.

Q When positrons collide with electrons, what type of radiation is produced that gives an image of an organ?

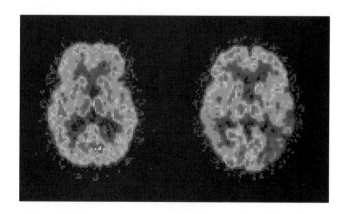

Sample Problem 16.9   Medical Application of Radioactivity

In hyperthyroidism, the thyroid produces an overabundance of thyroid hormone. Why would a radioisotope such as iodine-131 with an activity of 10 μCi be used to image the thyroid?

Solution

The thyroid takes up the iodine-131 because the thyroid uses most of the iodine in the body. An activity level of 10 μCi is high enough to determine the level of thyroid function, but low enough to keep cell damage to a minimum.

Study Check

The evaluation of the thyroid image determined that the patient had an enlarged thyroid. A few weeks later, that same patient was given 100 mCi of iodine-131 intravenously. Why was the activity of the iodine-131 sample used for treatment so much greater than the activity of the iodine-131 used in the diagnostic procedure?

Questions and Problems   Medical Applications Using Radioactivity

16.35 Bone and bony structures contain calcium and phosphorus.

  a. Why would the radioisotopes of calcium-47 and phosphorus-32 be used in the diagnosis and treatment of bone diseases?

  b. During nuclear tests, scientists were concerned that strontium-85, a radioactive product, would be harmful to the growth of bone in children. Explain.

16.36 a. Technetium-99m emits only gamma radiation. Why would this type of radiation be used in diagnostic imaging rather than an isotope that also emits beta or alpha radiation?

  b. A patient with an excess production of red blood cells receives radioactive phosphorus-32. Why would this treatment reduce the production of red blood cells in the bone marrow of the patient?

16.37 In a diagnostic test for leukemia, a person receives 4.0 mL of a solution containing selenium-75. If the activity of the selenium-75 is 45 μCi/mL, what is the dose received by the person?

16.38 A vial contains radioactive iodine-131 with an activity of 2.0 mCi per milliliter. If the thyroid test requires 3.0 mCi, how many milliliters are used to prepare the iodine-131 solution?

16.6 Nuclear Fission and Fusion

Learning Goal

Describe the processes of nuclear fission and fusion.

In the 1930s, scientists bombarding uranium-235 with neutrons discovered that the U-235 nucleus splits into two medium-weight nuclei and produces a great amount of energy. This was the discovery of a new kind of nuclear reaction called nuclear **fission.** The energy generated by nuclear fission, splitting the atom, was called atomic energy. When uranium-235 absorbs a neutron, it breaks apart to form two smaller nuclei, several neutrons, and a great amount of energy. A typical equation for nuclear fission is

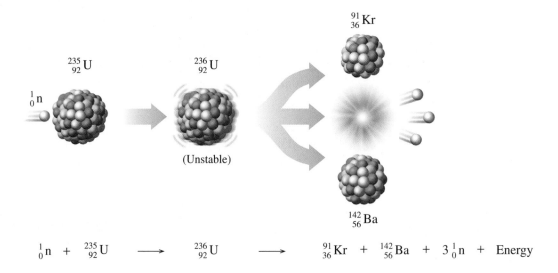

$$\ce{^{1}_{0}n + ^{235}_{92}U -> ^{236}_{92}U -> ^{91}_{36}Kr + ^{142}_{56}Ba + 3 ^{1}_{0}n + Energy}$$

If we could weigh the starting materials and products with great accuracy, we would find that the total mass of the products is slightly less than the mass of the starting materials. The missing mass has been converted into energy, consistent with the famous equation derived by Albert Einstein:

$$E = mc^2$$

E is the energy released, m is the mass lost, and c is the speed of light, 3×10^8 m/s. Even though the mass loss is very small, when it is multiplied by the speed of light squared the result is a large value for the energy released. The fission of 1 g of uranium-235 produces about as much energy as the burning of 3 tons of coal.

Chain Reaction

Fission begins when a neutron collides with the nucleus of a uranium atom. The resulting nucleus is unstable and splits into smaller nuclei. This fission process also releases several neutrons and large amounts of gamma radiation and energy. The neutrons emitted have high energies and bombard more uranium-235 nuclei. As fission continues, there is a rapid increase in the number of high-energy neutrons capable of splitting more uranium atoms, a process called a **chain reaction.** To sustain a nuclear chain reaction, sufficient quantities of uranium-235 must be brought together to provide a critical mass in which almost all the neutrons immediately collide with more uranium-235 nuclei. So much heat and energy build up that an atomic explosion occurs. (See Figure 16.7.)

Figure 16.7 In a nuclear chain reaction, the fission of each ^{235}U atom produces three neutrons that cause the nuclear fission of more and more ^{235}U atoms.

Q *Why is the fission of ^{235}U called a chain reaction?*

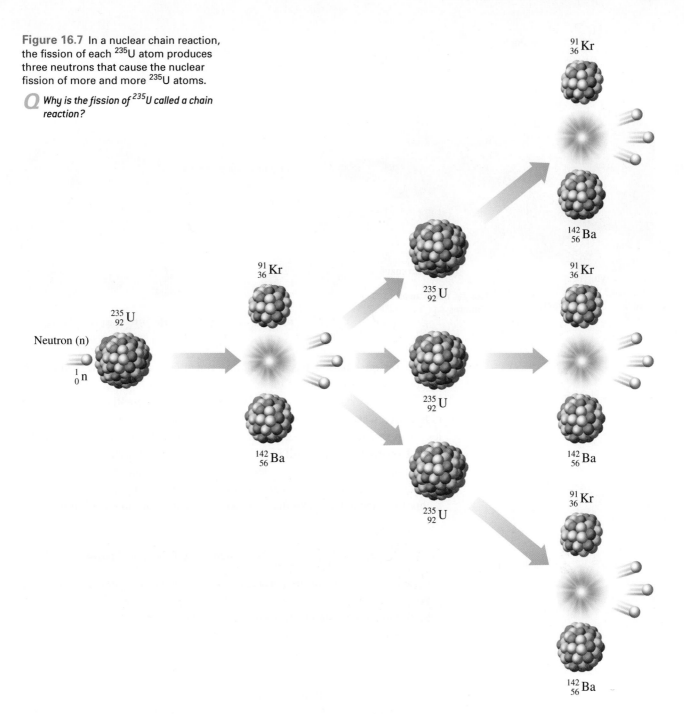

Nuclear Fusion

In **fusion,** two small nuclei, such as those in hydrogen, combine to form a larger nucleus. Mass is lost and a tremendous amount of energy is released, even more than the energy released from nuclear fission. However, a very high temperature (100 000 000°C) is required to overcome the repulsion of the hydrogen nuclei and cause them to undergo fusion. Fusion reactions occur continuously in the sun and other stars, providing us with heat and light. The huge amounts of energy produced by our sun come from the fusion of 6×10^{11} kilograms of hydrogen every second. The following fusion reaction involves the combination of two isotopes of hydrogen.

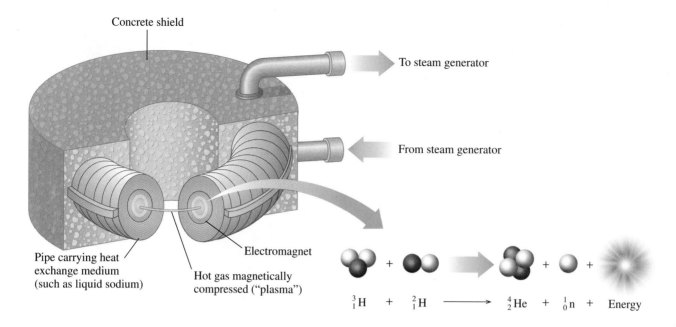

Concrete shield

To steam generator

From steam generator

Pipe carrying heat exchange medium (such as liquid sodium)

Electromagnet

Hot gas magnetically compressed ("plasma")

$$_{1}^{3}\text{H} \quad + \quad _{1}^{2}\text{H} \longrightarrow _{2}^{4}\text{He} \quad + \quad _{0}^{1}\text{n} \quad + \quad \text{Energy}$$

The fusion reaction has tremendous potential as a possible source for future energy needs. One of the advantages of fusion as an energy source is that hydrogen is plentiful in the oceans. Although scientists expect some radioactive waste from fusion reactors, the amount is expected to be much less than that from fission, and the waste products should have shorter half-lives. However, fusion is still in the experimental stage because the extremely high temperatures needed have been difficult to reach and even more difficult to maintain. Research groups around the world are attempting to develop the technology needed to make fusion power plants a reality in our lifetime.

Sample Problem 16.10 **Identifying Fission and Fusion**

Classify the following as pertaining to nuclear fission, nuclear fusion, or both:

a. Small nuclei combine to form larger nuclei.
b. Large amounts of energy are released.
c. Very high temperatures are needed for reaction.

Solution

a. fusion   **b.** both fusion and fission   **c.** fusion

Study Check

Would the following reaction be an example of a fission or fusion reaction?

$$_{1}^{2}\text{H} + _{1}^{1}\text{H} \longrightarrow _{2}^{3}\text{He}$$

NUCLEAR POWER PLANTS

In a nuclear power plant, the quantity of uranium-235 is held below a critical mass, so it cannot sustain a chain reaction. The fission reactions are slowed by placing control rods, which absorb some of the fast-moving neutrons, among the uranium samples. In this way, less fission occurs, and there is a slower, controlled production of energy. The heat from the controlled fission is used to produce steam. The steam drives a generator, which produces electricity. Approximately 10% of the electrical energy produced in the United States is generated in nuclear power plants.

Although nuclear power plants help meet some of our energy needs, there are some problems. One of the most serious problems is the production of radioactive by-products that have very long half-lives. It is essential that these waste products be stored safely for a very long time in a place where they do not contaminate the environment. Early in 1990, the Environmental Protection Agency gave its approval for the storage of radioactive hazardous wastes in chambers 2150 ft underground. In 1998, the Waste Isolation Pilot Plant (WIPP) repository site in New Mexico was ready to receive plutonium waste from former U.S. bomb factories. Although authorities have determined the caverns are safe, some people are concerned with the safe transport of the radioactive waste by trucks on the highways.

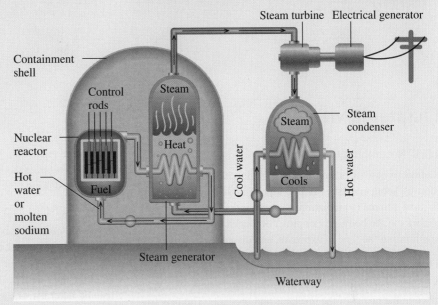

Questions and Problems   Nuclear Fission and Fusion

16.39 What is nuclear fission?

16.40 How does a chain reaction occur in nuclear fission?

16.41 Complete the following fission reaction:

$$^{235}_{92}U + ^{1}_{0}n \longrightarrow ^{131}_{50}Sn + ? + 2 \, ^{1}_{0}n + energy$$

16.42 In another fission reaction, U-235 bombarded with a neutron produces Sr-94, another small nucleus, and 3 neutrons. Write the complete equation for the fission reaction.

16.43 Indicate whether each of the following are characteristic of the fission or fusion process or both:

a. Neutrons bombard a nucleus.

b. The nuclear process occurring in the sun.

c. A large nucleus splits into smaller nuclei.

d. Small nuclei combine to form larger nuclei.

16.44 Indicate whether each of the following are characteristic of the fission or fusion process or both:

a. Very high temperatures are required to initiate the reaction.

b. Less radioactive waste is produced.

c. Hydrogen nuclei are the reactants.

d. Large amounts of energy are released when the nuclear reaction occurs.

Concept Map **Nuclear Chemistry**

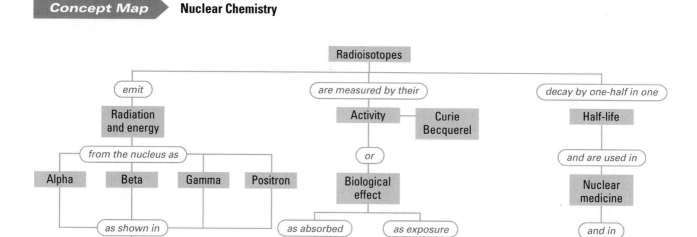

Chapter Review

16.1 Natural Radioactivity

Radioactive isotopes have unstable nuclei that break down (decay), spontaneously emitting alpha (α), beta (β), positron β^+, and gamma (γ) radiation. Because radiation can damage the cells in the body, proper protection must be used: shielding, limiting the time of exposure, and distance.

16.2 Nuclear Equations

A balanced equation is used to represent the changes that take place in the nuclei of the reactants and products. The new isotopes and the type of radiation emitted can be determined from the symbols that show the mass numbers and atomic numbers of the isotopes in the nuclear reaction. A radioisotope is produced artificially when a nonradioactive isotope is bombarded by a small particle such as a proton or an alpha or beta particle. Many radioactive isotopes used in nuclear medicine are produced in this way.

16.3 Radiation Measurement

In a Geiger counter, radiation ionizes gas in a metal tube, which produces an electrical current. The curie (Ci) measures the number of nuclear transformations of a radioactive sample. Activity is also measured in becquerel (Bq) units. The amount of radiation absorbed by a substance is measured in

rads or the gray (Gy). The rem and the sievert (Sv) are units used to determine the biological damage from the different types of radiation.

16.4 Half-Life of a Radioisotope

Every radioisotope has its own rate of emitting radiation. The time it takes for one-half of a radioactive sample to decay is called its half-life. For many medical radioisotopes, such as Tc-99m and I-131, half-lives are short. For other isotopes, usually naturally occurring ones such as C-14, Ra-226, and U-238, half-lives are extremely long.

16.5 Medical Applications Using Radioactivity

In nuclear medicine, radioisotopes are administered that go to specific sites in the body. By detecting the radiation they emit, an evaluation can be made about the location and extent of an injury, disease, tumor, or the level of function of a particular organ. Higher levels of radiation are used to treat or destroy tumors.

16.6 Nuclear Fission and Fusion

In fission, a large nucleus breaks apart into smaller pieces, releasing one or more types of radiation and a great amount of energy. In fusion, small nuclei combine to form a larger nucleus while great amounts of energy are released.

Key Terms

alpha particle A nuclear particle identical to a helium ($_2^4$He, or α) nucleus (2 protons and 2 neutrons).

becquerel (Bq) A unit of activity of a radioactive sample equal to one disintegration per second.

beta particle A particle identical to an electron ($_{-1}^{0}e$, or β) that forms in the nucleus when a neutron changes to a proton and an electron.

chain reaction A fission reaction that will continue once it has been initiated by a high-energy neutron bombarding a heavy nucleus such as U-235.

curie (Ci) A unit of radiation equal to 3.7×10^{10} disintegrations/s.

decay curve A diagram of the decay of a radioactive element.

equivalent dose The measure of biological damage from an absorbed dose that has been adjusted for the type of radiation.

fission A process in which large nuclei are split into smaller pieces, releasing large amounts of energy.

fusion A reaction in which large amounts of energy are released when small nuclei combine to form larger nuclei.

gamma ray High-energy radiation (γ) emitted to make a nucleus more stable.

gray (Gy) A unit of absorbed dose equal to 100 rads.

half-life The length of time it takes for one-half of a radioactive sample to decay.

positron A particle of radiation with no mass and a positive charge produced by an unstable nucleus when a proton is transformed into a neutron and a positron.

rad (radiation absorbed dose) A measure of an amount of radiation absorbed by the body.

radiation Energy or particles released by radioactive atoms.

radioactive The process by which an unstable nucleus breaks down with the release of high-energy radiation.

radioisotope A radioactive atom of an element.

rem (radiation equivalent in humans) A measure of the biological damage caused by the various kinds of radiation (rad $\times$ radiation biological factor).

sievert (Sv) A unit of biological damage (equivalent dose) equal to 100 rems.

transmutation The formation of a radioactive nucleus by bombarding a stable nucleus with fast-moving particles.

 # Understanding the Concepts

16.45 Consider the following nucleus of a radioactive isotope:

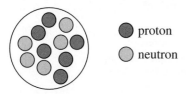

- proton
- neutron

    a. What is the nuclear symbol for this isotope?
    b. If this isotope decays by emitting a positron, what does the resulting nucleus look like?

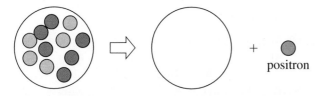

+ positron

16.46 Draw in the radioactive nucleus that emits a beta particle to form the following nucleus:

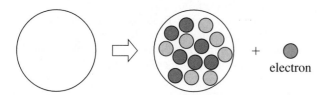

+ electron

16.47 Draw the nucleus of the atom to complete the following nuclear reaction:

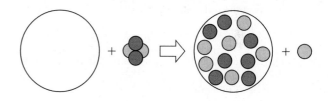

16.48 Draw the nucleus of the atom produced in the following nuclear reaction:

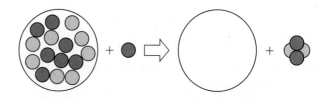

16.49 Carbon dating of small bits of charcoal used in cave paintings has determined that some of the paintings are from 10 000 to 30 000 years old. Carbon-14 has a half-life of 5730 years. In a 1 μg sample of carbon from a live tree, the activity of ^{14}C is 6.4 μCi. If researchers determine that 1 μg charcoal from a prehistoric cave painting in France has an activity of 0.80 μCi, what is the age of the painting?

16.50 a. Complete the values for the mass of radioactive ^{131}I on the vertical axis.
b. Complete the number of days on the horizontal axis.
c. What is the half-life in days of ^{131}I?

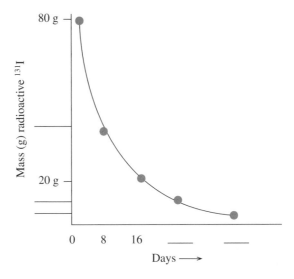

Additional Questions and Problems

16.51 Carbon-12 is a nonradioactive isotope of carbon, and carbon-14 is a radioactive isotope. What is similar and different about the two carbon isotopes?

16.52 Give the number of protons, neutrons, and electrons in atoms of the following isotopes:
a. boron-10   **b.** zinc-72
c. iron-59   **d.** gold-198

16.53 Describe alpha, beta, and gamma radiation in terms of the following:
a. type of radiation
b. symbols
c. depth of tissue penetration
d. type of shielding needed for protection

16.54 When you have dental X rays at the dentist's office, the technician places a heavy lead apron over you and then leaves the room to take your X rays. What is the purpose of these activities?

16.55 Write the balanced nuclear equations for each of the following emitters:
a. thorium-225 (α)
b. bismuth-210 (α)
c. cesium-137 (β)
d. tin-126 (β)

16.56 Write the balanced nuclear equations for each of the following emitters:
a. potassium-40 (β)
b. sulfur-35 (β)
c. platinum-190 (α)
d. radium-210 (α)

16.57 Write a balanced equation for each of the following radioactive emissions:
a. an alpha particle from Hg-180
b. a beta particle from Sn-126
c. a positron from Mn-49

16.58 Write a balanced equation for each of the following radioactive emissions:
a. an alpha particle from Gd-148
b. a beta particle from Sr-90
c. a positron from Al-25

16.59 Complete each of the following nuclear equations:
a. $^{14}_{7}\text{N} + ^{4}_{2}\text{He} \longrightarrow ? + ^{1}_{1}\text{H}$
b. $^{27}_{13}\text{Al} + ^{4}_{2}\text{He} \longrightarrow ^{30}_{14}\text{Si} + ?$
c. $^{235}_{92}\text{U} + ^{1}_{0}\text{n} \longrightarrow ^{90}_{38}\text{Sr} + 3^{1}_{0}\text{n} + ?$
d. $? \longrightarrow ^{127}_{54}\text{Xe} + ^{0}_{+1}e$

16.60 Complete each of the following nuclear equations:
a. $^{59}_{27}\text{Co} + ? \longrightarrow ^{56}_{25}\text{Mn} + ^{4}_{2}\text{He}$
b. $? \longrightarrow ^{14}_{7}\text{N} + ^{0}_{-1}e$
c. $^{76}_{36}\text{Kr} + ^{0}_{-1}e \longrightarrow ?$
d. $^{106}_{47}\text{Ag} \longrightarrow ^{106}_{46}\text{Pd} + ?$

16.61 Write the atomic symbols and a complete nuclear equation for the following:
a. When two oxygen-16 atoms collide, one of the products is an alpha particle.
b. When californium-249 is bombarded by oxygen-18, a new element, seaborgium-263, and 4 neutrons are produced.

c. Radon-222 emits an alpha particle, and the product emits another alpha particle. Write the two nuclear reactions.

d. An atom of strontium-80 emits a positron.

16.62 Write the symbols and complete the following nuclear equations:

a. Polonium-210 emits an alpha particle.

b. Bismuth-211 decays by emitting an alpha particle. The product is another radioisotope that emits a beta particle. Write equations for the two nuclear changes.

c. A radioisotope emits a positron to form titanium-48.

d. An atom of germanium-69 emits a positron.

16.63 If the amount of radioactive phosphorus-32 in a sample decreases from 1.2 g to 0.30 g in 28 days, what is the half-life of phosphorus-32?

16.64 If the amount of radioactive iodine-123 in a sample decreases from 0.4 g to 0.1 g in 26.2 hr, what is the half-life of iodine-123?

16.65 Iodine-131, a beta emitter, has a half-life of 8.0 days.

a. Write the nuclear equation for the beta-decay of iodine-131.

b. How many grams of a 12.0-g sample of iodine-131 would remain after 40 days?

c. How many days have passed if 48 g of iodine-131 decayed to 3.0 g iodine-131?

16.66 Cesium-137, a beta emitter, has a half-life of 30 years.

a. Write the nuclear equation for cesium-137.

b. How many grams of a 16-g sample of cesium-137 would remain after 90 yr?

c. How many years will be needed for 28 g cesium-137 to decay to 3.5 g cesium-137?

16.67 A 120-mg sample of technetium-99m is used for a diagnostic test. If technetium-99m has a half-life of 6.0 hr, how much of the technetium-99m sample remains 24 hr after the test?

16.68 The half-life of oxygen-15 is 124 s. If a sample of oxygen-15 has an activity of 4000 becquerels, how many minutes will elapse before it reaches an activity of 500 becquerels?

16.69 What is the purpose of irradiating meats, fruits, and vegetables?

16.70 The irradiation of foods was approved in the United States in the 1980s.

a. Why have we not seen many irradiated products in our markets?

b. Would you buy foods that have been irradiated? Why or why not?

16.71 What is the difference between fission and fusion?

16.72 **a.** What are the products in the fission of uranium-235 that make possible a nuclear chain reaction?

b. What is the purpose of placing control rods among uranium samples in a nuclear reactor?

16.73 Where does fusion occur naturally?

16.74 Why are scientists continuing to try to build a fusion reactor even though very high temperatures have been difficult to reach and maintain?

▶ Challenge Questions

16.75 Identify each of the following nuclear reactions as alpha decay, beta decay, positron emission, or gamma radiation.

a. $^{27m}_{13}\text{Al} \longrightarrow\ ^{27}_{13}\text{Al} +\ ^{0}_{0}\gamma$

b. $^{8}_{5}\text{B} \longrightarrow\ ^{8}_{4}\text{Be} +\ ^{0}_{+1}e$

c. $^{90}_{38}\text{Sr} \longrightarrow\ ^{90}_{39}\text{Y} +\ ^{0}_{-1}e$

d. $^{218}_{85}\text{At} \longrightarrow\ ^{214}_{83}\text{Bi} +\ ^{4}_{2}\text{He}$

16.76 Complete and balance each of the following nuclear equations:

a. $^{23m}_{12}\text{Mg} \longrightarrow\ \underline{\hspace{1cm}} +\ ^{0}_{0}\gamma$

b. $^{61}_{30}\text{Zn} \longrightarrow\ ^{61}_{29}\text{Cu} +\ \underline{\hspace{1cm}}$

c. $^{241}_{95}\text{Am} +\ ^{4}_{2}\text{He} \longrightarrow\ \underline{\hspace{1cm}} +\ 2^{0}_{1}n$

d. $^{126}_{50}\text{Sn} \longrightarrow\ \underline{\hspace{1cm}} +\ ^{0}_{-1}e$

16.77 The half-life for the radioactive decay of calcium-47 is 4.5 days. If a sample has an activity of 4.0 μCi after 18 days, what was the initial activity of the sample?

16.78 A 16-μg sample of sodium-24 decays to 2.0 μg in 45 hr. What is the half-life of ^{24}Na?

16.79 A nurse was accidentally exposed to potassium-42 while doing some brain scans for possible tumors. The error was not discovered until 36 hours later when the activity of the potassium-42 sample was 2.0 μCi. If potassium-42 has a half-life of 12 hr, what was the activity of the sample at the time the nurse was exposed?

16.80 A wooden object from the site of an ancient temple has a carbon-14 activity of 10 counts per minute compared with a reference piece of wood cut today that has an activity of 40 counts per minute. If the half-life for carbon-14 is 5730 years, what is the age of the ancient wood object?

Answers

Answers to Study Checks

16.1 Limiting the time one spends near a radioactive source and staying as far away as possible will reduce exposure to radiation.

16.2 $^{214}_{84}\text{Po} \longrightarrow \ ^{210}_{82}\text{Pb} + \ ^{4}_{2}\text{He}$

16.3 $^{131}_{53}\text{I} \longrightarrow \ ^{131}_{54}\text{Xe} + \ ^{0}_{-1}e$

16.4 $^{118}_{54}\text{Xe} \longrightarrow \ ^{118}_{53}\text{I} + \ ^{0}_{+1}e$

16.5 $^{14}_{7}\text{N}$

16.6 5.6×10^{11} iodine-131 atoms

16.7 0.50 g

16.8 17,200 years

16.9 An iodine-131 sample with a higher activity is used in radiation therapy when radiation is needed to destroy some of the cells in the thyroid gland.

16.10 fusion

Answers to Selected Questions and Problems

16.1 a. Both an alpha particle and a helium nucleus have 2 protons and 2 neutrons. However, an α particle is emitted from a nucleus during radioactive decay.
b. α, $^{4}_{2}\text{He}$

16.3 a. $^{39}_{19}\text{K}$, $^{40}_{19}\text{K}$, $^{41}_{19}\text{K}$
b. They all have 19 protons and 19 electrons, but they differ in the number of neutrons.

16.5

Medical Use	Isotope Symbol	Mass Number	Number of Protons	Number of Neutrons
Heart imaging	$^{201}_{81}\text{Tl}$	201	81	120
Radiation therapy	$^{60}_{27}\text{Co}$	60	27	33
Abdominal scan	$^{67}_{31}\text{Ga}$	67	31	36
Hyperthyroidism	$^{131}_{53}\text{I}$	131	53	78
Leukemia treatment	$^{32}_{15}\text{P}$	32	15	17

16.7 a. α, $^{4}_{2}\text{He}$ **b.** $^{1}_{0}\text{n}$ **c.** β, $^{0}_{-1}e$
d. $^{15}_{7}\text{N}$ **e.** $^{125}_{53}\text{I}$

16.9 a. β or e^- **b.** α or He   **c.** n
d. Na   **e.** C

16.11 a. Because β particles are so much less massive and move faster than α particles, they can penetrate further into tissue.
b. Ionizing radiation breaks bonds and forms reactive species that cause undesirable reactions in the cells.

c. X-ray technicians leave the room to increase the distance between them and the radiation. Also a wall that contains lead shields them.
d. Wearing gloves shields the skin from α and β radiation.

16.13 a. $^{208}_{84}\text{Po} \longrightarrow \ ^{204}_{82}\text{Pb} + \ ^{4}_{2}\text{He}$
b. $^{232}_{90}\text{Th} \longrightarrow \ ^{228}_{88}\text{Ra} + \ ^{4}_{2}\text{He}$
c. $^{251}_{102}\text{No} \longrightarrow \ ^{247}_{100}\text{Fm} + \ ^{4}_{2}\text{He}$
d. $^{220}_{86}\text{Rn} \longrightarrow \ ^{216}_{84}\text{Po} + \ ^{4}_{2}\text{He}$

16.15 a. $^{25}_{11}\text{Na} \longrightarrow \ ^{25}_{12}\text{Mg} + \ ^{0}_{-1}e$
b. $^{20}_{8}\text{O} \longrightarrow \ ^{20}_{9}\text{F} + \ ^{0}_{-1}e$
c. $^{92}_{38}\text{Sr} \longrightarrow \ ^{92}_{39}\text{Y} + \ ^{0}_{-1}e$
d. $^{42}_{19}\text{K} \longrightarrow \ ^{42}_{20}\text{Ca} + \ ^{0}_{-1}e$

16.17 a. $^{26}_{14}\text{Si} \longrightarrow \ ^{26}_{13}\text{Al} + \ ^{0}_{+1}e$
b. $^{54}_{27}\text{Co} \longrightarrow \ ^{54}_{26}\text{Fe} + \ ^{0}_{+1}e$
c. $^{77}_{37}\text{Rb} \longrightarrow \ ^{77}_{36}\text{Kr} + \ ^{0}_{+1}e$
d. $^{93}_{45}\text{Rh} \longrightarrow \ ^{93}_{44}\text{Ru} + \ ^{0}_{+1}e$

16.19 a. $^{28}_{14}\text{Si}$ **b.** $^{87}_{36}\text{Kr}$ **c.** $^{0}_{-1}e$
d. $^{238}_{92}\text{U}$ **e.** $^{188}_{79}\text{Au}$

16.21 a. $^{10}_{4}\text{Be}$ **b.** $^{0}_{-1}e$
c. $^{27}_{13}\text{Al}$ **d.** $^{1}_{1}\text{H}$

16.23 a. When radiation enters the Geiger counter, it ionizes a gas in the detection tube, which produces a burst of current that is detected by the instrument.
b. becquerel (Bq), curie (Ci)
c. gray (Gy), rad
d. 1000 Gy

16.25 a. 2.2×10^{12} disintegrations
b. 294 μCi

16.27 When pilots fly at high altitudes, there is less atmosphere to protect them from cosmic radiation.

16.29 A half-life is the time it takes for one-half of a radioactive sample to decay.

16.31 a. 40.0 mg   **b.** 20.0 mg
c. 10.0 mg   **d.** 5.00 mg

16.33 128 days, 192 days

16.35 a. Since the elements Ca and P are part of bone, their radioactive isotopes will also become part of the bony structures of the body, where their radiation can be used to diagnose or treat bone diseases.
b. Strontium (Sr) acts much like calcium (Ca) because both are Group 2A (2) elements. The body will accumulate radioactive strontium in bones in the same way that it incorporates calcium. Radioactive strontium is harmful to children because the radiation it produces causes more damage in cells that are dividing rapidly.

16.37 180 μCi

16.39 Nuclear fission is the splitting of a large atom into smaller fragments with the release of large amounts of energy.

16.41 $^{103}_{42}$Mo

16.43 **a.** fission **b.** fusion
c. fission **d.** fusion

16.45 **a.** $^{11}_{6}$C
b.

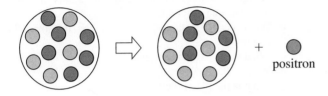

positron

16.47

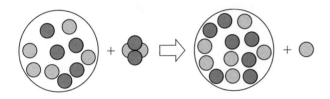

16.49 17 200 years old

16.51 Each has 6 protons and 6 electrons, but carbon-12 has 6 neutrons and carbon-14 has 8 neutrons. Carbon-12 is a stable isotope, but carbon-14 is radioactive and will emit radiation.

16.53 **a.** α radiation consists of a helium nucleus emitted from the nucleus of a radioisotope. β radiation is an electron, and γ radiation is high-energy radiation emitted from the nucleus of a radioisotope.
b. α particle: $^{4}_{2}$He
β particle: $^{0}_{-1}e$
γ radiation: γ
c. α particles penetrate 0.05 mm into tissue, β particles 4–5 mm, and γ rays 50 cm or more.
d. Paper or clothing will shield you from α particles; heavy clothing, lab coats, and gloves will shield you from β particles; and lead and concrete are needed for shielding from γ radiation.

16.55 **a.** $^{225}_{90}$Th $\longrightarrow$ $^{221}_{88}$Ra + $^{4}_{2}$He
b. $^{210}_{83}$Bi $\longrightarrow$ $^{206}_{81}$Tl + $^{4}_{2}$He
c. $^{137}_{55}$Cs $\longrightarrow$ $^{137}_{56}$Ba + $^{0}_{-1}e$
d. $^{126}_{50}$Sn $\longrightarrow$ $^{126}_{51}$Sb + $^{0}_{-1}e$

16.57 **a.** $^{180}_{80}$Hg $\longrightarrow$ $^{176}_{78}$Pt + $^{4}_{2}$He
b. $^{126}_{50}$Sn $\longrightarrow$ $^{126}_{51}$Sb + $^{0}_{-1}e$
c. $^{49}_{25}$Mn $\longrightarrow$ $^{49}_{24}$Cr + $^{0}_{+1}e$

16.59 **a.** $^{17}_{8}$O **b.** $^{1}_{1}$H
c. $^{143}_{54}$Xe **d.** $^{127}_{55}$Cs

16.61 **a.** $^{16}_{8}$O + $^{16}_{8}$O $\longrightarrow$ $^{4}_{2}$He + $^{28}_{14}$Si
b. $^{249}_{98}$Cf + $^{18}_{8}$O $\longrightarrow$ $^{263}_{106}$Sg + 4 $^{1}_{0}$n
c. $^{222}_{86}$Rn $\longrightarrow$ $^{218}_{84}$Po + $^{4}_{2}$He
$^{218}_{84}$Po $\longrightarrow$ $^{214}_{82}$Pb + $^{4}_{2}$He
d. $^{80}_{38}$Sr $\longrightarrow$ $^{0}_{+1}e$ + $^{80}_{37}$Rb

16.63 14 days

16.65 **a.** $^{131}_{53}$I $\longrightarrow$ $^{0}_{-1}e$ + $^{131}_{54}$Xe
b. 0.375 g **c.** 32 days

16.67 7.5 mg

16.69 The irradiation of meats, fruits, and vegetables kills bacteria such as *E. coli* that can cause foodborne illnesses. In addition, spoilage is deterred, and shelf life is extended.

16.71 In the fission process, an atom splits into smaller nuclei. In fusion, small nuclei combine (fuse) to form a larger nucleus.

16.73 Fusion occurs naturally in the sun and other stars.

16.75 **a.** gamma radiation
b. positron emission
c. beta decay
d. alpha decay

16.77 $\frac{1}{2}$ life = 4.5 days ^{47}Ca 4.0 μCi after 18 days

$$18 \text{ days} \times \frac{1 \text{ half-life}}{4.5 \text{ days}} = 4 \text{ half-lives}$$

$\quad\;$ (1) $\qquad$ (2) $\qquad$ (3) $\qquad$ (4)
64 μCi $\longrightarrow$ 32 μCi $\longrightarrow$ 16 μCi $\longrightarrow$ 8.0 μCi $\longrightarrow$ 4.0 μCi

16.79 16 μCi

Combining Ideas from Chapters 15 and 16

CI 27 Consider the reaction of sodium oxalate ($Na_2C_2O_4$) and potassium permanganate ($KMnO_4$) in acidic solution. The reactants and products are the following:

$$MnO_4^-(aq) + C_2O_4^{2-}(aq) \longrightarrow$$
$$Mn^{2+}(aq) + CO_2(g) \quad \text{(unbalanced)}$$

a. What is the oxidation half-reaction?
b. What is the reduction half-reaction?
c. What is the net ionic equation for the overall reaction?
d. If 28.25 mL $KMnO_4$ solution are needed to titrate 0.758 g sodium oxalate ($Na_2C_2O_4$), what is the molarity of the $KMnO_4$ solution?

CI 28 A strip of magnesium metal dissolves rapidly in 6.00 mL of 0.150 M hydrochloric acid producing magnesium chloride and hydrogen gas.

$$Mg(s) + HCl(aq) \longrightarrow MgCl_2(aq) + H_2(g)$$
$$\text{(unbalanced)}$$

a. Assign oxidation numbers to the atoms in the reactants and products.
b. What is the balanced equation for the overall reaction?
c. What is the oxidizing agent?
d. What is the reducing agent?
e. What is the pH of the HCl solution?
f. How many grams of magnesium can dissolve in the HCl solution?

CI 29 Using the reaction in CI 28, consider the following experiment. A piece of magnesium with a mass of 0.121 g is added to 50.0 mL of 1.00 M HCl at a temperature of 22.0°C. When the magnesium dissolves, the solution reaches a temperature of 33.0°C.
a. What is the limiting reactant?
b. What volume of hydrogen gas would be produced if the pressure is 750. mm Hg?
c. How many joules were released by the reaction of the magnesium? Assume the density of HCl solution is 1.00 g/mL and the specific heat of HCl solution is the same as for water.
d. What is the heat of reaction for Mg in J/g? in kJ/mol?

CI 30 The iceman known as "Otzi" was discovered in a high mountain pass on the Austrian-Italian border. Samples of his hair and bones had carbon-14 activity that was about 50% of that present in new hair or bone. Carbon-14 is a beta emitter.

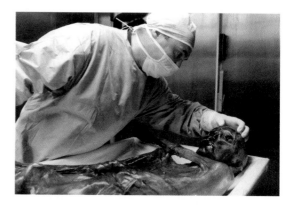

a. How long ago did "Otzi" live if the half-life for C-14 is 5730 years?
b. Write a nuclear equation for the decay of ^{14}C.

CI 31 A sample of silicon has the following isotopes: ^{27}Si, ^{28}Si, ^{29}Si, ^{30}Si, ^{31}Si.

Isotope	% Natural Abundance	Atomic Mass	Half-Life (radioactive)	Radiation
^{27}Si		26.995	4.9 s	Positron
^{28}Si	92.18	27.987		
^{29}Si	4.71	28.987		
^{30}Si	3.12	29.983		
^{31}Si		30.986	2.6 h	Beta

a. Indicate the number of protons, neutrons, and electrons for each isotope.

Isotope	Number of Protons	Number of Neutrons	Number of Electrons
^{27}Si			
^{28}Si			
^{29}Si			
^{30}Si			
^{31}Si			

b. What is the electron configuration and the abbreviated electron configuration of silicon?
c. Calculate the atomic mass for silicon using the isotopes that have a natural abundance.
d. Write the nuclear equation for the radioactive isotopes.
e. Write the electron-dot formula and predict the shape of $SiCl_4$.

CI 32 K^+ is an electrolyte required by the human body and found in many foods as well as salt substitutes. One of the isotopes of potassium is ^{40}K, which has a natural abundance of 0.012% and a half-life of 1.30×10^9 years. The isotope ^{40}K decays to ^{40}Ca or to ^{40}Ar. A typical activity for ^{40}K is 7.0 μCi per gram.

a. Write a nuclear equation for each type of decay.
b. Identify the particle emitted for each type of decay.
c. How many K^+ ions are in 3.5 oz KCl?
d. What is the activity of 25 g KCl in bequerels?

CI 33 Uranium-238 decays in a series of nuclear changes until stable ^{206}Pb is produced. Complete the following nuclear equations that are part of the ^{238}U decay series:
a. $^{238}U \longrightarrow ^{234}Th + ?$
b. $^{234}Th \longrightarrow ? + _{-1}^{0}\beta$
c. $? \longrightarrow _{86}^{222}Rn + _{2}^{4}\alpha$

CI 34

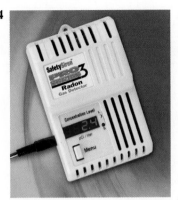

Of much concern to environmentalists is radon-222, which is a radioactive gas that can accumulate in basements of homes and buildings that are not well ventilated. In radon decay, alpha particles are emitted at a high rate because radon has a short half-life. The presence of alpha particles in the lungs is associated with the increased incidence of lung cancer. In a room that is 4.0 m wide, 6.0 m long, and 3.0 m high with a radon level of 4 pCi/L,
a. how many atoms of radon decay in 1.0 day?
b. how many moles of radon decay in 1.0 day?

Once emitted, alpha particles acquire 2 electrons to become He atoms. The last step in the decay series is Po-210, which is an alpha emitter.

c. Write the equation for the decay of Po-210.
d. If a 0.500-g sample of Po-210 is placed in a 0.230-L container at 25°C, what is the pressure of He gas after 276 days? The half-life of Po-210 is 138 days.

Answers to CIs

CI 27 a. $C_2O_4^{2-}(aq) \longrightarrow 2CO_2(g) + 2e^-$
b. $5e^- + MnO_4^-(aq) + 8H^+(aq) \longrightarrow Mn^{2+}(aq) + 4H_2O(l)$
c. $2MnO_4^-(aq) + 16H^+(aq) + 5C_2O_4^{2-}(aq) \longrightarrow 10CO_2(g) + 2Mn^{2+}(aq) + 8H_2O(l)$
d. 0.0800 M $KMnO_4$

CI 29 a. Mg is the limiting reactant.
b. $0.127 L H_2(g)$
c. $2.30 \times 10^3 J$
d. $1.90 \times 10^4 J/g$; 462 kJ/mol

CI 31 a.

Isotope	Number of Protons	Number of Neutrons	Number of Electrons
^{27}Si	14	13	14
^{28}Si	14	14	14
^{29}Si	14	15	14
^{30}Si	14	16	14
^{31}Si	14	17	14

b. $1s^2 2s^2 2p^6 3s^2 3p^2$; $[Ne]3s^2 3p^2$

c. Atomic mass is 28.10.

d. $^{27}_{14}Si \longrightarrow {}^{27}_{13}Al + {}^{0}_{+1}e$

$^{31}_{14}Si \longrightarrow {}^{31}_{15}P + {}^{0}_{-1}e$

e.

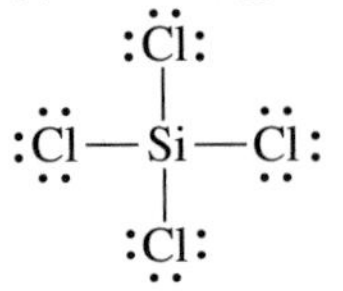

$$\overset{\displaystyle :\ddot{Cl}:}{\underset{\displaystyle :\ddot{Cl}:}{:\ddot{Cl} - Si - \ddot{Cl}:}} \qquad \text{tetrahedral}$$

CI 33 a. $^{238}_{92}U \longrightarrow {}^{234}_{90}Th + {}^{4}_{2}\alpha$

b. $^{234}_{90}Th \longrightarrow {}^{234}_{91}Pa + {}^{0}_{-1}\beta$

c. $^{226}_{88}Ra \longrightarrow {}^{222}_{86}Rn + {}^{4}_{2}\alpha$

Organic Chemistry

"*There are many carboxylic acids, including the alpha hydroxy acids, that are found today in skin products,*" says Dr. Ken Peterson, pharmacist and cosmetic chemist, Oakland. "*When you take a carboxylic acid called a fatty acid and react it with a strong base, you get a salt called soap. Soap has a high pH because the weak fatty acid and the strong base won't have a neutral pH of 7. If you take soap and drop its pH down to 7, you will convert the soap to the fatty acid. When I create fragrances, I use my nose and my chemistry background to identify and break down the reactions that produce good scents. Many fragrances are esters, which form when an alcohol reacts with a carboxylic acid. For example, the ester that smells like pineapple is made from ethanol and butyric acid.*"

the Chemistry place

Visit **www.aw-bc.com/chemplace** for extra quizzes, interactive tutorials, career resources, PowerPoint slides for chapter review, math help, and case studies.

rganic chemistry is the chemistry of carbon compounds that contain, primarily, carbon and hydrogen. The element carbon has a special role in chemistry because it bonds with other carbon atoms to give a vast array of molecules. The variety of molecules is so great that we find organic compounds in many common products we use such as gasoline, medicine, shampoos, plastic bottles, and perfumes. The food we eat is composed of different organic compounds that supply us with fuel for energy and the carbon atoms needed to build and repair the cells of our bodies.

Although many organic compounds occur in nature, chemists have synthesized even more. The cotton, wool, or silk in your clothes contain naturally occurring organic compounds, whereas materials such as polyester, nylon, or plastic have been synthesized through organic reactions. Sometimes it is convenient to synthesize a molecule in the lab even though that molecule is also found in nature. For example, vitamin C synthesized in a laboratory has the same structure as the vitamin C in oranges or lemons. In the next chapter, you will learn about the structures and reactions of organic molecules, which will provide a foundation for understanding the more complex molecules of biochemistry.

Learning Goal

Identify properties characteristic of organic or inorganic compounds.

WEB TUTORIAL
Introduction to Organic Molecules

17.1 Organic Compounds

At the beginning of the nineteenth century, scientists classified chemical compounds as inorganic and organic. An inorganic compound was a substance that was composed of minerals, and an organic compound was a substance that came from an organism, thus the use of the word *organic*. It was thought that some type of "vital force," which could only be found in living cells, was required to synthesize an organic compound. This perception was shown to be incorrect in 1828, when the German chemist Friedrick Wöhler synthesized urea, a product of protein metabolism, by heating an inorganic compound, ammonium cyanate.

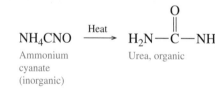

We now define organic chemistry as the study of carbon compounds. **Organic compounds** are usually nonpolar molecules, and the attractions between molecules are weak, which accounts for the low melting and boiling points of most organic compounds. Typically, organic compounds are not soluble in water. For example, vegetable oil, which is a mixture of organic compounds, does not dissolve in water, but floats on top. Many organic compounds undergo combustion and burn vigorously in air. Organic compounds contain C and H and may also have other nonmetallic elements like O, S, N, or Cl.

In contrast, many of the inorganic compounds are ionic, which leads to high melting and boiling points. Inorganic compounds that are ionic or polar covalent are usually soluble in water. Most inorganic substances do not burn in air. Table 17.1 contrasts some of the properties associated with organic and inorganic compounds, such as propane, C_3H_8, and sodium chloride, NaCl. (See Figure 17.1.)

Table 17.1 Some Properties of Organic and Inorganic Compounds

Property	Organic	Example: C_3H_8	Inorganic	Example: NaCl
Bonding	Mostly covalent	Covalent	Many are ionic, some covalent	Ionic
Polarity of bonds	Nonpolar, unless a more electronegative atom is present	Nonpolar	Most are ionic or polar covalent, a few are nonpolar covalent	Ionic
Melting point	Usually low	$-188°C$	Usually high	$801°C$
Boiling point	Usually low	$-42°C$	Usually high	$1413°C$
Flammability	High	Burns in air	Low	Does not burn
Solubility in water	Not soluble, unless a polar group is present	No	Most are soluble, unless nonpolar	Yes

Sample Problem 17.1   **Properties of Organic Compounds**

Indicate whether the following properties are characteristic of organic or inorganic compounds:

a. not soluble in water   **b.** high melting point   **c.** burns in air

Solution

a. Many organic compounds are not soluble in water.

b. Inorganic compounds are most likely to have high melting points.

c. Organic compounds are most likely to be flammable.

Study Check

Octane is not soluble in water. What type of compound is octane?

Figure 17.1 Propane, C_3H_8, is an organic compound, whereas sodium chloride, NaCl, is an inorganic compound.

Q Why is propane used as a fuel?

Bonding in Organic Compounds

The **hydrocarbons** are organic compounds that consist of only carbon and hydrogen. In the simplest hydrocarbon, methane (CH_4), the four valence electrons of carbon are shared with four hydrogen atoms to form an octet. In the electron-dot formula, each shared pair of electrons represents a single bond. In organic molecules, every carbon atom always has four bonds. An **expanded structural formula** is written when we show the bonds between all of the atoms.

$$\cdot \overset{\displaystyle .}{\underset{\displaystyle .}{C}} \cdot \ + \ 4H\cdot \ \longrightarrow \ H\overset{\displaystyle \overset{..}{H}}{\underset{\displaystyle \underset{..}{H}}{:\!C\!:}}H \ = \ H\!-\!\overset{\displaystyle H}{\underset{\displaystyle H}{\overset{|}{\underset{|}{C}}}}\!-\!H$$

Methane

The Tetrahedral Structure of Carbon

The VSEPR theory (Chapter 10) predicts that when four bonds are arranged as far apart as possible, they have a tetrahedral shape. In CH_4 the bonds to hydrogen are directed to the corners of a tetrahedron with bond angles of 109.5°. The three-dimensional structure of methane can be illustrated as a ball-and-stick model or a space-filling model. (See Figure 17.2.)

An organic compound with two atoms of carbon is ethane, C_2H_6. Each carbon atom is bonded to another carbon and three hydrogen atoms. When there are two or more carbon atoms in a molecule, each carbon retains the tetrahedral shape if bonded to four other atoms. For example, the ball-and-stick model of C_2H_6 is based on two tetrahedra attached to each other. (See Figure 17.3.) All the bond angles are close to 109.5°.

Figure 17.2 Three-dimensional representations of methane, CH_4: **(a)** tetrahedron, **(b)** ball-and-stick model, **(c)** space-filling model, **(d)** expanded structural formula.

Q *Why does methane have a tetrahedral shape and not a flat shape?*

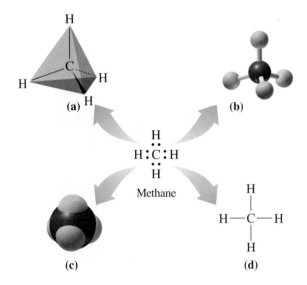

Figure 17.3 Three-dimensional representations of ethane, C_2H_6: **(a)** tetrahedral shape of each carbon, **(b)** ball-and-stick model, **(c)** space-filling model, **(d)** expanded structural formula.

Q How is the tetrahedral shape maintained in a molecule with two carbon atoms?

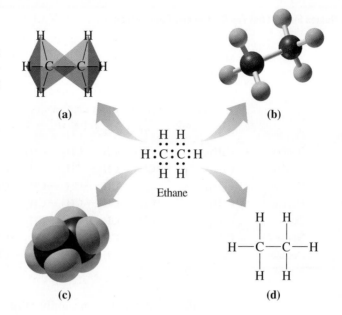

Ethane

(a)

(b)

(c)

(d)

Questions and Problems   **Organic Compounds**

17.1 Identify the following as formulas of organic or inorganic compounds:
- **a.** KCl
- **b.** C_4H_{10}
- **c.** CH_3CH_2OH
- **d.** H_2SO_4
- **e.** $CaCl_2$
- **f.** CH_3CH_2Cl

17.2 Identify the following as formulas of organic or inorganic compounds:
- **a.** $C_6H_{12}O_6$
- **b.** Na_2SO_4
- **c.** I_2
- **d.** C_2H_5Cl
- **e.** $C_{10}H_{22}$
- **f.** CH_4

17.3 Identify the following properties as most typical of organic or inorganic compounds:
- **a.** soluble in water
- **b.** low boiling point
- **c.** burns in air
- **d.** high melting point

17.4 Identify the following properties as most typical of organic or inorganic compounds:
- **a.** contains Na
- **b.** gas at room temperature
- **c.** covalent bonds
- **d.** produces ions in water

17.5 Match the following physical and chemical properties with the compounds ethane, C_2H_6, or sodium bromide, NaBr:
- **a.** boils at $-89°C$
- **b.** burns vigorously
- **c.** solid at $250°C$
- **d.** dissolves in water

17.6 Match the following physical and chemical properties with the compounds cyclohexane, C_6H_{12}, or calcium nitrate, $Ca(NO_3)_2$:
- **a.** melts at $500°C$
- **b.** insoluble in water
- **c.** produces ions in water
- **d.** is a liquid at room temperature

17.7 Why is the structure of the CH_4 molecule three-dimensional rather than two-dimensional?

17.8 In a propane molecule with three carbon atoms, what is the geometry around each carbon atom?

Propane

17.2 Alkanes

More than 90% of the known compounds in the world are organic compounds. Yet the predominant elements in organic compounds are carbon and hydrogen. The large number of carbon compounds is possible because the covalent bond between carbon atoms (C—C) is very strong, allowing carbon atoms to form long, stable chains.

In order to study organic compounds, the molecules are organized into classes of compounds that have similar structures and chemical properties.

Table 17.2   IUPAC Names for the First Ten Continuous-Chain Alkanes

Number of Carbon Atoms	Prefix	Name	Molecular Formula	Condensed Structural Formula
1	Meth	Methane	CH_4	CH_4
2	Eth	Ethane	C_2H_6	CH_3-CH_3
3	Prop	Propane	C_3H_8	$CH_3-CH_2-CH_3$
4	But	Butane	C_4H_{10}	$CH_3-CH_2-CH_2-CH_3$
5	Pent	Pentane	C_5H_{12}	$CH_3-CH_2-CH_2-CH_2-CH_3$
6	Hex	Hexane	C_6H_{14}	$CH_3-CH_2-CH_2-CH_2-CH_2-CH_3$
7	Hept	Heptane	C_7H_{16}	$CH_3-CH_2-CH_2-CH_2-CH_2-CH_2-CH_3$
8	Oct	Octane	C_8H_{18}	$CH_3-CH_2-CH_2-CH_2-CH_2-CH_2-CH_2-CH_3$
9	Non	Nonane	C_9H_{20}	$CH_3-CH_2-CH_2-CH_2-CH_2-CH_2-CH_2-CH_2-CH_3$
10	Dec	Decane	$C_{10}H_{22}$	$CH_3-CH_2-CH_2-CH_2-CH_2-CH_2-CH_2-CH_2-CH_2-CH_3$

The **alkanes** are a class or family of organic compounds that contain carbon and hydrogen atoms connected by single bonds. One of the most common uses of alkanes is as fuels. Methane, used in gas heaters and gas cooktops, is an alkane with one carbon atom. The alkanes ethane, propane, and butane contain two, three, and four carbon atoms connected in a row or a *continuous* chain. These names are part of the **IUPAC** (International Union of Pure and Applied Chemistry) **system,** which chemists use to name organic compounds. Alkanes with five or more carbon atoms in a chain are named using Greek prefixes: *pent* (5), *hex* (6), *hept* (7), *oct* (8), *non* (9), and *dec* (10). (See Table 17.2.)

Condensed Structural Formulas

In a **condensed structural formula,** we write each carbon atom and its attached hydrogen atoms as a group. A subscript indicates the number of hydrogen atoms bonded to each carbon atom.

$$
\begin{array}{ccccccc}
& H & & & & H & \\
& | & & & & | & \\
H-&C&- & = & CH_3- & \quad -&C&- & = & -CH_2- \\
& | & & & & | & \\
& H & & & & H &
\end{array}
$$

Expanded   Condensed   Expanded   Condensed

The molecular formula gives the total number of each kind of atom, but does not indicate the arrangement of the atoms in the molecule. Table 17.3 shows the molecular, expanded, and condensed structural formulas for alkanes with one, two, and three carbon atoms.

However, in chains of three carbon atoms or more, the carbon atoms do not actually lie in a straight line. The tetrahedral shape of carbon arranges the carbon bonds in a zigzag pattern, which is seen in the ball-and-stick model of hexane. (See Figure 17.4.) Butane can be depicted by a variety of two-dimensional structural formulas, as shown in Table 17.4. All of these structural formulas represent the same continuous chain of four carbon atoms.

Figure 17.4 A ball-and-stick model of hexane.

Q Why do the carbon atoms in hexane appear to be arranged in a zigzag chain?

Table 17.3   Writing Structural Formulas for Some Alkanes

Alkane	Methane	Ethane	Propane
Molecular formula	CH_4	C_2H_6	C_3H_8
Structural formulas			

Expanded, **Condensed** structural formulas for methane, ethane, and propane

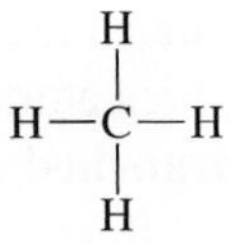

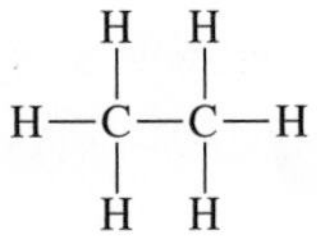

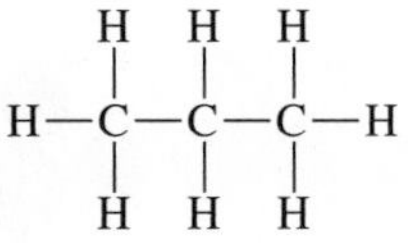

$$CH_4 \qquad CH_3-CH_3 \qquad CH_3-CH_2-CH_3$$

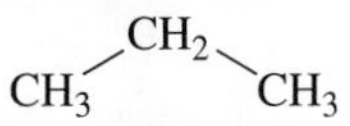

Sample Problem 17.2   **Drawing Expanded and Condensed Structural Formulas for Alkanes**

A molecule of butane, C_4H_{10}, has four carbon atoms in a row. What are its expanded and condensed structural formulas?

Solution

In the expanded structural formula, four carbon atoms are connected to each other and to hydrogen atoms using single bonds to give each carbon atom a total of four bonds. In the condensed structural formula, each carbon atom and its attached hydrogen atoms are written as CH_3-, or $-CH_2-$.

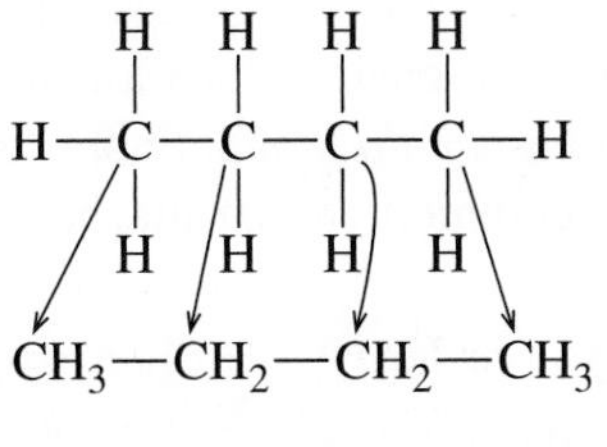

Expanded structural formula

$$CH_3-CH_2-CH_2-CH_3$$ Condensed structural formula

Study Check

Write the expanded and condensed structural formulas of pentane, C_5H_{12}.

Table 17.4   Some Structural Formulas for Butane, C_4H_{10}

Expanded structural formula for butane

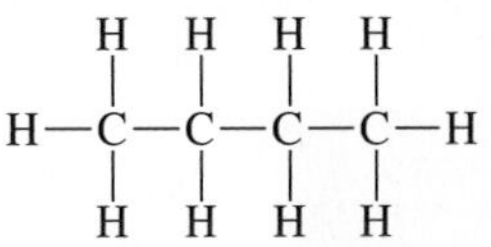

Condensed structural formulas that represent butane

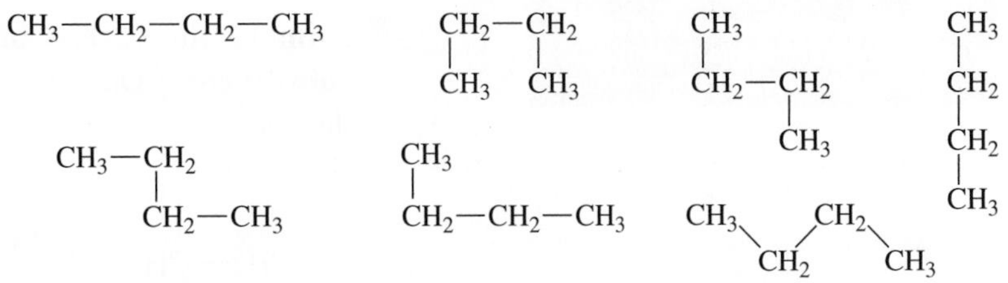

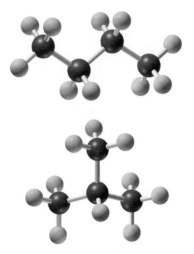

Figure 17.5 The isomers of C_4H_{10} have the same number and type of atoms, but bonded in a different order.

Q *What makes these molecules isomers?*

Naming Alkanes with Substituents

When an alkane has four or more carbon atoms, the atoms can be arranged so that a side group called a **branch** or **substituent** is attached to a carbon chain. For example, we can write two different structural formulas for the molecular formula C_4H_{10}. One formula contains the four carbon atoms in a continuous chain. In the other formula, a carbon atom in a *branch* is attached to a chain of three atoms. (See Figure 17.5.) An alkane with at least one branch is called a **branched alkane.** When two compounds have the same molecular formula but different arrangements of atoms, they are called **isomers.**

In another example, we can write the structural formulas of three different isomers with the molecular formula C_5H_{12} as follows:

Isomers of C_5H_{12}

Continuous Chain	Branched Chains

$$CH_3-CH_2-CH_2-CH_2-CH_3$$

$$\begin{array}{c} CH_3 \\ | \\ CH_3-CH-CH_2-CH_3 \end{array} \qquad \begin{array}{c} CH_3 \\ | \\ CH_3-C-CH_3 \\ | \\ CH_3 \end{array}$$

Table 17.5 Names and Formulas of Some Common Substituents

Substituent	Name
CH_3-	methyl
CH_3-CH_2-	ethyl
$CH_3-CH_2-CH_2-$	propyl
$CH_3-\overset{\textstyle |}{CH}-CH_3$	isopropyl
$F-, Cl-, Br-, I-$	fluoro, chloro, bromo, iodo

Substituents in Alkanes

In the IUPAC names for alkanes, a carbon branch is named as an **alkyl group,** which is an alkane that is missing one hydrogen atom. The alkyl group is named by replacing the *ane* ending of the corresponding alkane name with *yl.* Alkyl groups cannot exist on their own: they must be attached to a carbon chain. When a halogen atom is attached to a carbon chain, it is named as a *halo* group: fluoro (F), chloro (Cl), bromo (Br), or iodo (I). Some of the common groups attached to carbon chains are illustrated in Table 17.5.

Rules for Naming Alkanes with Substituents

In the IUPAC system of naming, the longest carbon chain is numbered to give the location of one or more substituents attached to it. Let's take a look at how we use the IUPAC system to name the following alkane:

$$\begin{array}{c} CH_3 \\ | \\ CH_3-CH-CH_2-CH_2-CH_3 \end{array}$$

Guide to Naming Alkanes

STEP 1
Write the alkane name of the longest continuous chain of carbon atoms.

STEP 2
Number the carbon atoms starting from the end nearest a substituent.

STEP 3
Give the location and name of each substituent (alphabetical order) as a prefix to the name of the main chain.

STEP 1 **Write the alkane name of the longest continuous chain of carbon atoms.** In this alkane, the longest chain has five carbon atoms, which is *pentane.*

$$\begin{array}{c} CH_3 \\ | \\ CH_3-CH-CH_2-CH_2-CH_3 \end{array} \qquad \text{pentane}$$

STEP 2 **Number the carbon atoms starting from the end nearest a substituent.** Once you start numbering, continue in that same direction.

$$\begin{array}{c} CH_3 \\ | \\ CH_3-CH-CH_2-CH_2-CH_3 \\ 1 \quad 2 \quad\; 3 \quad\;\; 4 \quad\;\; 5 \end{array} \qquad \text{pentane}$$

STEP 3 **Give the location and name of each substituent as a prefix to the alkane name.** Place a hyphen between the number and the substituent name.

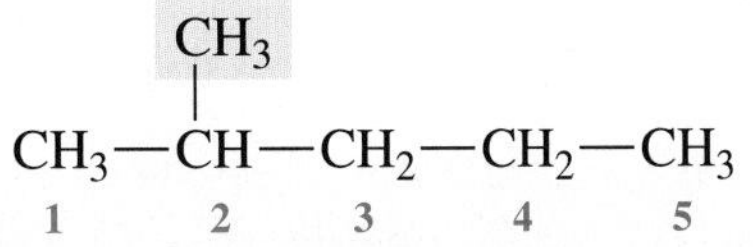

$$CH_3 \!-\! \underset{2}{\overset{\overset{\displaystyle CH_3}{|}}{CH}} \!-\! \underset{3}{CH_2} \!-\! \underset{4}{CH_2} \!-\! \underset{5}{CH_3}$$

2-methyl pentane

List the substituents in alphabetical order.

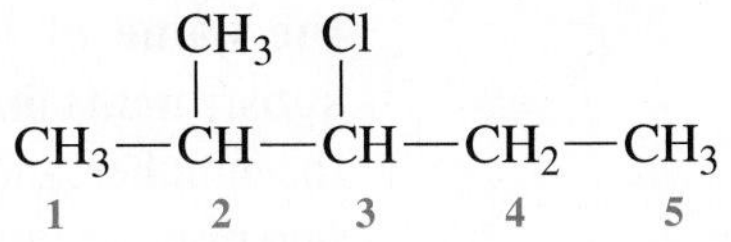

$$\underset{1}{CH_3} \!-\! \underset{2}{\overset{\overset{\displaystyle CH_3}{|}}{CH}} \!-\! \underset{3}{\overset{\overset{\displaystyle Cl}{|}}{CH}} \!-\! \underset{4}{CH_2} \!-\! \underset{5}{CH_3}$$

3-chloro-2-methylpentane

Use a prefix (di, tri, tetra) to indicate a group that appears more than once. Use commas to separate two or more numbers.

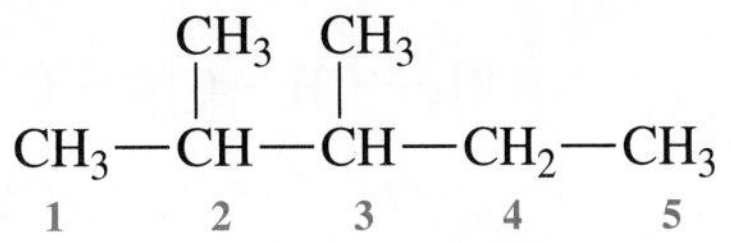

$$\underset{1}{CH_3} \!-\! \underset{2}{\overset{\overset{\displaystyle CH_3}{|}}{CH}} \!-\! \underset{3}{\overset{\overset{\displaystyle CH_3}{|}}{CH}} \!-\! \underset{4}{CH_2} \!-\! \underset{5}{CH_3}$$

2,3-dimethylpentane

When there are two or more substituents, the main chain is numbered in the direction that gives the lowest set of numbers.

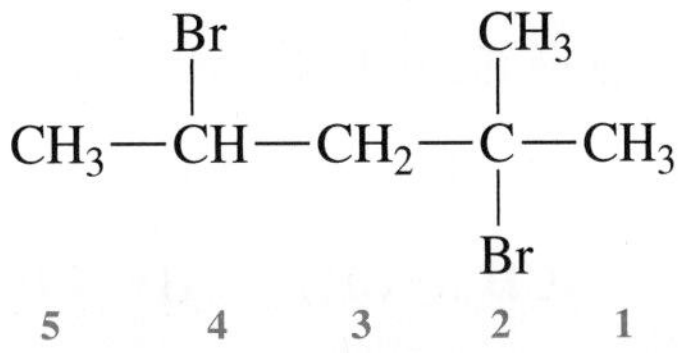

$$\underset{5}{CH_3} \!-\! \underset{4}{\overset{\overset{\displaystyle Br}{|}}{CH}} \!-\! \underset{3}{CH_2} \!-\! \underset{2}{\overset{\overset{\displaystyle CH_3}{|}}{\underset{\underset{\displaystyle Br}{|}}{C}}} \!-\! \underset{1}{CH_3}$$

2,4-dibromo-2-methylpentane

Sample Problem 17.3 **Writing IUPAC Names**

Give the IUPAC name for the following alkane:

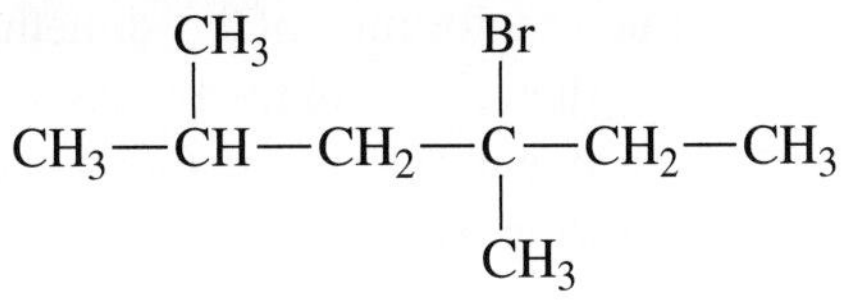

$$CH_3 \!-\! \overset{\overset{\displaystyle CH_3}{|}}{CH} \!-\! CH_2 \!-\! \overset{\overset{\displaystyle Br}{|}}{\underset{\underset{\displaystyle CH_3}{|}}{C}} \!-\! CH_2 \!-\! CH_3$$

Solution

STEP 1 **Write the alkane name of the longest continuous chain of carbon atoms.** In this alkane, the longest chain has six carbon atoms, which is *hexane*.

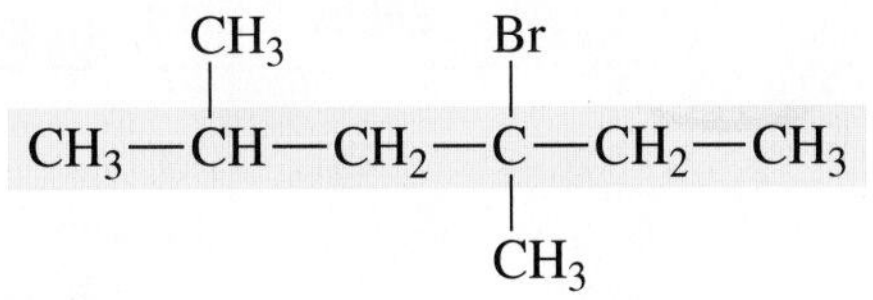

$$CH_3 \!-\! \overset{\overset{\displaystyle CH_3}{|}}{CH} \!-\! CH_2 \!-\! \overset{\overset{\displaystyle Br}{|}}{\underset{\underset{\displaystyle CH_3}{|}}{C}} \!-\! CH_2 \!-\! CH_3$$

hexane

STEP 2 **Number the carbon atoms starting from the end nearest a substituent.**

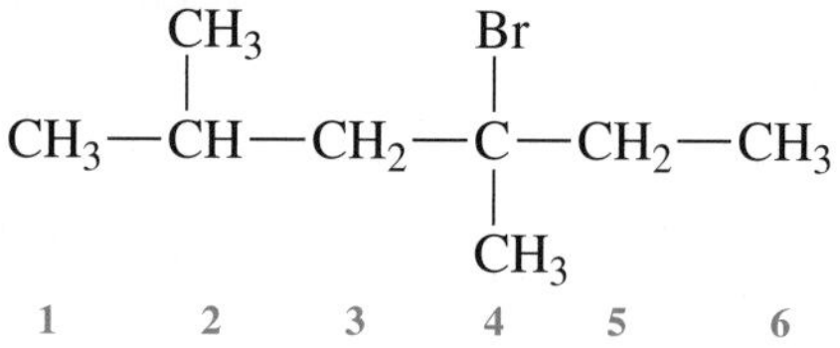

$$\underset{1}{CH_3}-\underset{2}{\underset{\underset{CH_3}{|}}{CH}}-\underset{3}{CH_2}-\underset{4}{\underset{\underset{CH_3}{|}}{\overset{\overset{Br}{|}}{C}}}-\underset{5}{CH_2}-\underset{6}{CH_3}\qquad \text{hexane}$$

STEP 3 **Give the location and name of each substituent in front of the name of the longest chain. List the names of different substituents in alphabetical order.** Place a hyphen between the number and the substituent names and commas to separate two or more numbers. A prefix (di, tri, tetra) indicates a group that appears more than once.

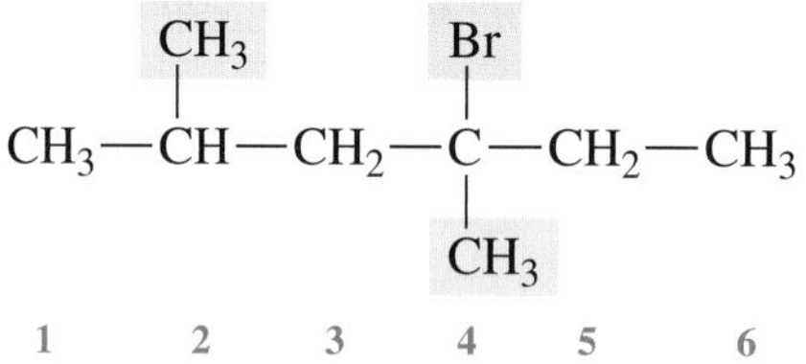

$$\underset{1}{CH_3}-\underset{2}{\underset{\underset{CH_3}{|}}{CH}}-\underset{3}{CH_2}-\underset{4}{\underset{\underset{CH_3}{|}}{\overset{\overset{Br}{|}}{C}}}-\underset{5}{CH_2}-\underset{6}{CH_3}\qquad \text{4-bromo- 2,4-dimethyl hexane}$$

Study Check

Give the IUPAC name for the following compound:

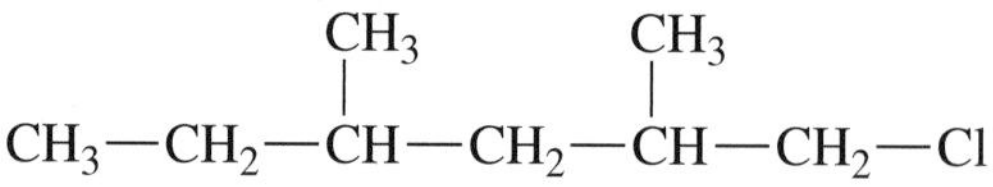

$$CH_3-CH_2-\underset{\underset{CH_3}{|}}{CH}-CH_2-\underset{\underset{CH_3}{|}}{CH}-CH_2-Cl$$

Drawing Structural Formulas for Alkanes

The IUPAC name gives all the information needed to draw the condensed structural formula of an alkane. Suppose you are asked to draw the condensed structural formula of 2,3-dimethylbutane. The alkane name gives the number of carbon atoms in the longest chain. The names in the beginning indicate the substituents and where they are attached. We can break down the name in the following way.

2,3-dimethylbutane

2,3-	di	methyl	but	ane
substituents on carbons 2 and 3	two identical groups	CH_3 — alkyl groups	4 C atoms in the main chain	single C — C bonds

Guide to Drawing Alkane Formulas

STEP 1
Draw the main chain of carbon atoms.

STEP 2
Number the chain and place the substituents on the carbons indicated by the numbers.

STEP 3
Add the correct number of hydrogen atoms to give four bonds to each C atom.

CASE STUDY
Hazardous Materials

Figure 17.6 In oil spills, large quantities of oil spread over the water.

Q What physical properties cause oil to remain on the surface of water?

Sample Problem 17.4   **Drawing Structures from IUPAC Names**

Write the condensed structural formula for 2,3-dimethylbutane.

Solution

We can use the following guide to draw the condensed structural formula.

STEP 1 **Draw the main chain of carbon atoms.**   For butane, we draw a chain of four carbon atoms and number it.

$$C—C—C—C$$
$$1 \quad 2 \quad 3 \quad 4$$

STEP 2 **Number the chain and place the substituents on the carbons indicated by the numbers.**   The first part of the name indicates two methyl groups ($CH_3—$), one on carbon 2 and one on carbon 3.

Methyl   Methyl
$$CH_3 \quad CH_3$$
$$| \qquad |$$
$$C—C—C—C$$
$$1 \quad 2 \quad 3 \quad 4$$

STEP 3 **Add the correct number of hydrogen atoms to give four bonds to each C atom.**

$$CH_3 \quad CH_3$$
$$| \qquad |$$
$$CH_3—CH—CH—CH_3$$
2,3-Dimethylbutane

Study Check

What is the structural formula for 2,4-dibromopentane?

Solubility and Density of Alkanes

Alkanes are nonpolar, which makes them insoluble in water. However, they are soluble in nonpolar solvents such as other alkanes. Alkanes have densities from 0.65g/mL to about 0.70g/mL, which is less dense than water (1.0 g/mL). If there is an oil spill in the ocean, the alkanes in the crude oil remain on the surface and spread over a large area. In the *Exxon Valdez* oil spill in 1989, 40 million liters of oil covered over 25 000 square kilometers of water in Prince William Sound, Alaska. (See Figure 17.6.) If the crude oil reaches the beaches and inlets, there can be considerable damage to shellfish, fish, birds, and wildlife habitats. Even today there is oil on the surface, or just beneath the surface, in some areas of Prince William Sound. Cleanup includes both mechanical and chemical methods. In one method, a nonpolar compound that is "oil-attracting" is used to pick up oil, which is then scraped off into recovery tanks.

The first four alkanes—methane, ethane, propane, and butane—are gases at room temperature and are widely used as heating fuels. Tanks of liquid propane and butane are used to provide fuels for heating homes and cooking on barbecues.

CHEM NOTE

CRUDE OIL

Crude oil or petroleum contains a wide variety of hydrocarbons. At an oil refinery, the components in crude oil are separated by fractional distillation, a process that removes groups or fractions of hydrocarbons by continually heating the mixture to higher temperatures. (See Table 17.6.) Fractions containing alkanes with longer carbon chains require higher temperatures before they reach their boiling temperature and form gases. The gases are removed and passed through a distillation column where they cool and condense back to liquids. The major use of crude oil is to obtain gasoline, but a barrel of crude oil is only about 35% gasoline. To increase the production of gasoline, heating oils are broken down to give the lower-weight alkanes.

Table 17.6 **Typical Alkane Mixtures Obtained by Distillation of Crude Oil**

Distillation Temperatures (°C)	Number of Carbon Atoms	Product
Below 30	1–4	Natural gas
30–200	5–12	Gasoline
200–250	12–16	Kerosene, jet fuel
250–350	15–18	Diesel fuel, heating oil
350–450	18–25	Lubricating oil
Nonvolatile residue	Over 25	Asphalt, tar

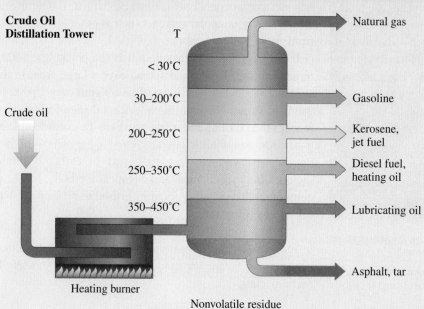

Figure 17.7 The solid alkanes that make up waxy coatings on fruits and vegetables help retain moisture, inhibit mold, and enhance appearance.

Q Why does the waxy coating help the fruits and vegetables retain moisture?

Alkanes having 5–8 carbon atoms (pentane, hexane, heptane, and octane) are liquids at room temperature. They are highly volatile, which makes them useful in fuels such as gasoline. Liquid alkanes with 9–17 carbon atoms have higher boiling points and are found in kerosene, diesel, and jet fuels. Motor oil is a mixture of liquid hydrocarbons and is used to lubricate the internal components of engines. Mineral oil is a mixture of liquid hydrocarbons and is used as a laxative and a lubricant. Alkanes with 18 or more carbon atoms are waxy solids at room temperature. Known as paraffins, these compounds are used in waxy coatings for fruits and vegetables to retain moisture, inhibit mold growth, and enhance appearance. (See Figure 17.7.) Petrolatum, or Vaseline, is a mixture of liquid hydrocarbons with low boiling points that are encapsulated in solid hydrocarbons. It is used in ointments and cosmetics and as a lubricant and a solvent.

Combustion of Alkanes

An alkane undergoes **combustion** when it reacts with oxygen to produce carbon dioxide, water, and energy.

$$Alkane + O_2 \longrightarrow CO_2 + H_2O + energy$$

Methane is used to cook our foods and heat our homes. Propane is the gas used in portable heaters and gas barbecues. (See Figure 17.8.) Gasoline, a mixture of liquid hydrocarbons, is the fuel that powers our cars, lawn mowers, and snow blowers. As alkanes, they all undergo combustion. The equations for the combustion of methane (CH_4) and propane (C_3H_8) follow:

$$CH_4 + 2O_2 \longrightarrow CO_2 + 2H_2O + energy$$
$$C_3H_8 + 5O_2 \longrightarrow 3CO_2 + 4H_2O + energy$$

Figure 17.8 The propane fuel in the tank undergoes combustion, which provides energy.

Q What is the balanced equation for the combustion of propane?

CHEM NOTE
INCOMPLETE COMBUSTION

You may already know that it is dangerous to burn natural gas, oil, or wood in a closed room where ventilation and fresh air are not adequate. A gas heater, fireplace, or wood stove must have proper ventilation. If the supply of oxygen is limited, incomplete combustion produces carbon monoxide. The incomplete combustion of methane in natural gas is written as

$$2CH_4(g) + 3O_2(g) \longrightarrow$$
$$2CO(g) + 4H_2O(g) + heat$$

Carbon monoxide (CO) is a colorless, odorless, poisonous gas. When inhaled, CO passes into the bloodstream, where it attaches to hemoglobin, which reduces the amount of oxygen (O_2) reaching the organs and cells. As a result, a healthy person can experience a reduction in exercise capability, visual perception, and manual dexterity.

When the amount of hemoglobin bound to CO (COHb) is 10% or less, a person may experience shortness of breath, mild headache, and drowsiness, which are symptoms that may be mistaken for the flu. Heavy smokers can have as high as 9% COHb in their blood. When as much as 30% of the hemoglobin is bound to CO, a person may experience more severe symptoms, including dizziness, mental confusion, severe headache, and nausea. If 50% or more of the hemoglobin is bound to CO, a person could become unconscious and die if not treated immediately with oxygen.

Questions and Problems ▸ Alkanes

17.9 Give the IUPAC name for each of the following alkanes:

a. CH₃
│
CH₂—CH₂—CH₂
│
CH₃
CH₂—CH₃
│
CH₂
│
c. CH₃—CH₂—CH₂

b. CH₃—CH₃

17.10 Give the IUPAC name for each of the following alkanes:

a. CH₄

b. CH₃—CH₂—CH₂—CH₃

CH₃
│
CH₂
│
c. CH₃

17.11 Write the condensed structural formulas for each of the following:
a. methane   **b.** ethane   **c.** pentane

17.12 Write the condensed structural formulas for each of the following:
a. propane   **b.** hexane   **c.** heptane

17.13 Indicate whether each of the following pairs of structural formulas represent isomers or the same molecule:

a. CH₃—CH—CH₃ and CH—CH₃
│ (CH₃)   │ (CH₃)
                               │
                               CH₃

b. CH₃—CH—CH₂—CH₃ and
│ (CH₃)
                    CH₃   CH₃
                    │   │
                    CH₂—CH₂—CH₂

c. CH₂—CH—CH₂—CH₃ and
│   │
CH₃   CH₃
            CH₃   CH₃
            │   │
            CH₃—CH—CH—CH₃

17.14 Indicate whether each of the following pairs of structural formulas represent isomers or the same molecule:

a. CH₃—C—CH₃ and CH—CH₂—CH₃
│ (CH₃)   │ (CH₃)
│   │
CH₃   CH₃

b. CH₃—CH—CH—CH₂ and
│   │   │
CH₃ CH₃ CH₃
                CH₃   CH₃
                │   │
                CH₃—CH—CH₂—CH—CH₃

c. CH₃—CH—CH₂—CH₃   and
│ (CH₃)
                    CH₃
                    │
                    CH₃—CH₂—CH—CH₃

17.15 Give the IUPAC name for each of the following alkanes:

a. CH₃—CH—CH₂—CH₃
│
F

CH₃
│
b. CH₃—C—CH₃
│
CH₃

CH₃ Cl
│   │
c. CH₃—CH₂—CH—CH—CH₃

17.16 Give the IUPAC name for each of the following alkanes:

CH₃
│
a. CH₃—CH—CH₂—CH₂—CH₂—Br

CH₃ CH₃
│   │
b. CH₃—CH—CH—CH₃

Br   Cl
│   │
c. CH₃—CH₂—CH—CH₂—CH—CH₃

17.17 Draw a condensed structural formula for each of the following alkanes:
a. 2-methylbutane
b. 3,3-dichloropentane
c. 2,3,5-trimethylhexane

17.18 Draw a condensed structural formula for each of the following alkanes:
a. 3-iodopentane
b. 3-ethyl-2-methylpentane
c. 2,2,3,5-tetrabromohexane

17.19 Heptane, C_7H_{16}, has a density of 0.68 g/mL.
a. What is the structural formula of heptane?
b. Is it a solid, liquid, or gas at room temperature?
c. Is it soluble in water?
d. Will it float or sink in water?

17.20 Nonane, C_9H_{20}, has a density of 0.79 g/mL.
a. What is the structural formula of nonane?
b. Is it a solid, liquid, or gas at room temperature?
c. Is it soluble in water?
d. Will it float or sink in water?

17.21 Write a balanced equation for the complete combustion of each of the following compounds:
a. ethane   **b.** propane   **c.** octane

17.22 Write a balanced equation for the complete combustion of each of the following compounds:
a. hexane   **b.** heptane   **c.** nonane

Table 17.7 Covalent Bonds for Elements in Organic Compounds

Element	Group	Covalent Bonds	Structure of Atoms
H	1A	1	H—
C	4A	4	$-\overset{\displaystyle |}{\underset{\displaystyle |}{C}}-$
N	5A	3	$-\overset{\displaystyle |}{\underset{\displaystyle \cdot\cdot}{N}}-$
O	6A	2	$-\overset{\displaystyle \cdot\cdot}{\underset{\displaystyle \cdot\cdot}{O}}-$
F, Cl, Br, I	7A	1	$-\overset{\displaystyle \cdot\cdot}{\underset{\displaystyle \cdot\cdot}{X}}\!:$ (X = F, Cl, Br, I)

Learning **Goal**

Classify organic molecules according to their functional groups.

WEB TUTORIAL
Functional Groups

17.3 Functional Groups

In organic compounds, carbon atoms are bonded to hydrogen, and some times oxygen, nitrogen, or a halogen such as chlorine. Hydrogen, with 1 valence electron, forms a single covalent bond. An octet is achieved by nitrogen forming three covalent bonds and oxygen forming two covalent bonds. The halogens, with 7 valence electrons, form one covalent bond. Table 17.7 lists the number of covalent bonds most often formed by elements found in organic compounds.

Organic compounds number in the millions and more are synthesized every day. It might seem that the task of learning organic chemistry would be overwhelming. However, within this vast number of compounds, there are characteristic structural features, which are specific groups of atoms called **functional groups** that cause compounds to undergo similar chemical reactions. Compounds with the same functional groups undergo similar chemical reactions.

Alkenes and Alkynes

Earlier in section 17.2, we learned that alkanes contain only carbon–carbon single bonds. The **alkenes** contain a functional group that is a double bond between two adjacent carbon atoms; **alkynes** contain a triple bond.

An alcohol

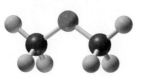

An ether

$$
\underset{\text{An alkane}}{H-\overset{\overset{\displaystyle H}{|}}{\underset{\underset{\displaystyle H}{|}}{C}}-\overset{\overset{\displaystyle H}{|}}{\underset{\underset{\displaystyle H}{|}}{C}}-H}
\qquad
\underset{\text{An alkene}}{H-\overset{\overset{\displaystyle H}{|}}{C}=\overset{\overset{\displaystyle H}{|}}{C}-H}
\qquad
\underset{\text{An alkyne}}{H-C\equiv C-H}
$$

Functional group:

$$
-\overset{|}{C}-\overset{|}{C}-
\qquad
-\overset{|}{C}=\overset{|}{C}-
\qquad
-C\equiv C-
$$

Condensed structural formula:

$$
CH_3-CH_3
\qquad
CH_2=CH_2
\qquad
HC\equiv CH
$$

Alcohols and Ethers

The characteristic functional group in **alcohols** is the **hydroxyl (—OH) group** bonded to a carbon atom. In **ethers,** the characteristic structural feature is an oxygen atom bonded to two carbon atoms. The oxygen atom also has two unshared pairs of electrons, but they will not be shown in the structural formulas.

$$
\underset{\text{An alcohol}}{CH_3-CH_2-OH}
\qquad
\underset{\text{An ether}}{CH_3-O-CH_3}
$$

An aldehyde

Aldehydes and Ketones

The aldehydes and ketones contain a **carbonyl group** ($C=O$), which is a carbon with a double bond to oxygen. In an **aldehyde,** the carbon atom of the carbonyl group is bonded to another carbon and one hydrogen atom. Only the simplest aldehyde, CH_2O, has a carbonyl group attached to two hydrogen atoms. In a **ketone,** the carbonyl group is bonded to two other carbon atoms.

A ketone

$$\underset{\text{An aldehyde}}{CH_3-\overset{\overset{\displaystyle O}{\|}}{C}-H} \qquad \underset{\text{A ketone}}{H_3C-\overset{\overset{\displaystyle O}{\|}}{C}-CH_3}$$

Carboxylic Acids and Esters

In **carboxylic acids,** the functional group is the **carboxyl group,** which is a combination of the *carbo*nyl and hydro*xyl* groups.

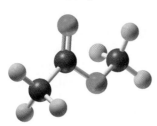

A carboxylic acid

$$\underset{\text{A carboxylic acid}}{CH_3-\overset{\overset{\displaystyle O}{\|}}{C}-O-H} \qquad \text{or} \qquad CH_3COOH \qquad \text{or} \qquad CH_3CO_2H$$

An **ester** is similar to a carboxylic acid, except the oxygen is attached to a carbon and not to hydrogen.

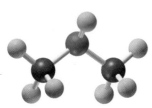

An ester

$$\underset{\text{An ester}}{CH_3-\overset{\overset{\displaystyle O}{\|}}{C}-O-CH_3} \qquad \text{or} \qquad CH_3COOCH_3 \qquad \text{or} \qquad CH_3CO_2CH_3$$

Amines and Amides

In **amines,** the central atom is a nitrogen atom. Amines are derivatives of ammonia, NH_3, in which carbon atoms replace one, two, or three of the hydrogen atoms.

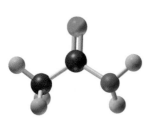

An amine

$$NH_3 \qquad CH_3-NH_2 \qquad CH_3-\underset{\underset{\displaystyle CH_3}{|}}{NH} \qquad CH_3-\underset{\underset{\displaystyle CH_3}{|}}{N}-CH_3$$

Ammonia   Examples of amines

In an **amide,** the hydroxyl group of a carboxylic acid is replaced by a nitrogen group.

$$\underset{\text{An amide}}{CH_3-\overset{\overset{\displaystyle O}{\|}}{C}-NH_2}$$

A list of the common functional groups in organic compounds is shown in Table 17.8.

An amide

Table 17.8   Classification of Organic Compounds

Class	Functional Group	Example
Alkene	$\diagdown$C=C$\diagup$	H_2C=CH_2
Alkyne	$-C\equiv C-$	$HC\equiv CH$
Haloalkane	$-F, -Cl, -Br, -I$	CH_3-Cl
Alcohol	$-OH$	CH_3-CH_2-OH
Ether	$-O-$	CH_3-O-CH_3
Aldehyde	$\overset{\displaystyle O}{\overset{\|}{-C-H}}$	$CH_3-\overset{\displaystyle O}{\overset{\|}{C}}-H$
Ketone	$\overset{\displaystyle O}{\overset{\|}{-C-}}$	$CH_3-\overset{\displaystyle O}{\overset{\|}{C}}-CH_3$
Carboxylic acid	$\overset{\displaystyle O}{\overset{\|}{-C-O-H}}$	$CH_3-\overset{\displaystyle O}{\overset{\|}{C}}-O-H$
Ester	$\overset{\displaystyle O}{\overset{\|}{-C-O-}}$	$CH_3-\overset{\displaystyle O}{\overset{\|}{C}}-O-CH_3$
Amine	$\overset{\|}{-N-}$	CH_3-NH_2
Amide	$\overset{\displaystyle O}{\overset{\|}{-C-}}\overset{\|}{N}-$	$CH_3-\overset{\displaystyle O}{\overset{\|}{C}}-NH_2$

Sample Problem 17.5   Identifying Functional Groups

Classify the following organic compounds according to their functional groups:

a. $CH_3-CH_2-NH-CH_3$ **b.** $CH_3-CH=CH-CH_3$

c. $CH_3-CH_2-\overset{\displaystyle O}{\overset{\|}{C}}-OH$ **d.** $CH_3-CH_2-CH_2-OH$

Solution

a. amine   **b.** alkene   **c.** carboxylic acid   **d.** alcohol

Study Check

How does a carboxylic acid differ from an ester?

Questions and Problems   Functional Groups

17.23 Identify the class of compounds that contains each of the following functional groups:
  a. hydroxyl group attached to a carbon chain
  b. carbon–carbon double bond
  c. carbonyl group attached to a hydrogen atom
  d. carboxyl group attached to two carbon atoms

17.24 Identify the class of compounds that contains each of the following functional groups:
  a. a nitrogen atom attached to one or more carbon atoms
  b. carboxyl group
  c. oxygen atom bonded to two carbon atoms
  d. a carbonyl group between two carbon atoms

17.25 Classify the following molecules according to their functional groups. The possibilities are alcohol, ether, ketone, carboxylic acid, or amine.

a. CH_3—CH_2—O—CH_2—CH_3

b. CH_3—$\overset{\overset{\displaystyle OH}{|}}{CH}$—$CH_3$

c. CH_3—$\overset{\overset{\displaystyle O}{\|}}{C}$—$CH_2$—$CH_3$

d. CH_3—CH_2—CH_2—COOH

e. CH_3—CH_2—NH_2

17.26 Classify the following molecules according to their functional groups. The possibilities are alkene, aldehyde, carboxylic acid, ester, or amine.

a. CH_3—$\overset{\overset{\displaystyle O}{\|}}{C}$—O—$CH_2$—$CH_3$

b. CH_3—$\overset{\overset{\displaystyle CH_3}{|}}{N}$—$CH_3$

c. CH_3—CH_2—CH_2—$\overset{\overset{\displaystyle O}{\|}}{C}$—H

d. CH_3—CH_2—CH_2—COOH

e. CH_3—CH=CH—CH_3

17.4 Alkenes and Alkynes

Learning Goal

Write the IUPAC names and formulas for alkenes and alkynes; draw formulas of products for the hydrogenation of alkenes.

Ethene

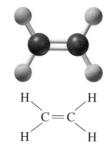

Ethyne

H—C≡C—H

Figure 17.9 Ball-and-stick models of ethene and ethyne show the functional groups of double or triple bonds.

Q   Why are these compounds called unsaturated hydrocarbons?

Alkenes and alkynes are **unsaturated hydrocarbons** because they contain double and triple bonds, respectively. An *alkene* contains at least one double bond between carbons. The double bond forms when two adjacent carbon atoms share two pairs of valence electrons. The simplest alkene is ethene, C_2H_4, which is more likely to be called by its common name, ethylene. In ethene, each carbon atom is attached to two H atoms and the other carbon atom in the double bond. The resulting molecule has a flat geometry because the carbon and hydrogen atoms all lie in the same plane. (See Figure 17.9.)

In an *alkyne,* a triple bond occurs when two carbon atoms share three pairs of valence electrons. The simplest alkyne is called ethyne, but is commonly known as acetylene, which is used in welding, where it burns at a very high temperature. In ethyne each carbon atom in the triple bond is attached to two other atoms.

Naming Alkenes and Alkynes

The IUPAC names for alkenes and alkynes are similar to those of alkanes. The simplest alkene, ethene (ethylene), is an important plant hormone involved in promoting the ripening of fruit. Commercially grown fruit, such as avocados, bananas, and tomatoes, are often picked before they are ripe. Before the fruit is brought to market, it is exposed to ethylene to accelerate the ripening process. Ethylene also accelerates the breakdown of cellulose in plants, which causes flowers to wilt and leaves to fall from trees.

The IUPAC name of the simplest alkyne is ethyne, although acetylene, its common name, is often used. See Table 17.9 for a comparison of the naming

Table 17.9   Comparison of Names for Alkanes, Alkenes, and Alkynes

Alkane	Alkene	Alkyne
H_3C-CH_3	$H_2C=CH_2$	$HC\equiv CH$
Ethane	Ethene (ethylene)	Ethyne (acetylene)
$CH_3-CH_2-CH_3$	$CH_3-CH=CH_2$	$CH_3-C\equiv CH$
Propane	Propene (propylene)	Propyne

for alkanes, alkenes, and alkynes. For alkenes and alkynes, the longest carbon chain must contain the double or triple bond.

STEP 1 **Name the longest carbon chain that contains the double or triple bond.** Replace the corresponding alkane ending with *ene* for an alkene and *yne* for an alkyne.

STEP 2 **Number the longest chain from the end nearest the double or triple bond.** Indicate the position of the double or triple bond with the number of the first unsaturated carbon.

$$CH_3-CH_2-CH=CH_2 \qquad CH_3-CH=CH-CH_3 \qquad CH_3-C\equiv C-CH_3$$
$$4321 \qquad\quad 1234 \qquad\quad 1234$$
$$\text{1-butene} \qquad\qquad\qquad \text{2-butene} \qquad\qquad\qquad \text{2-butyne}$$

STEP 3 **Give the location and name of each substituent (alphabetical order) as a prefix to the alkene or alkyne name.**

$$\overset{\displaystyle CH_3}{\underset{\displaystyle}{CH_2=CH-CH_2-CH-CH_3}}$$
$$12345$$
$$\text{4-methyl-1-pentene}$$

$$\overset{\displaystyle CH_3\ CH_3}{CH_3-C=C-CH_3}$$
$$1234$$
$$\text{2,3-dimethyl-2-butene}$$

$$\overset{\displaystyle Cl}{CH_3-CH-C\equiv CH}$$
$$4321$$
$$\text{3-chloro-1-butyne}$$

Sample Problem 17.6   **Naming Alkenes and Alkynes**

Write the IUPAC name for each of the following:

$$\textbf{a.}\quad \overset{\displaystyle CH_3}{\underset{\displaystyle}{CH_3-CH-CH=CH-CH_3}}$$

$$\textbf{b.}\quad CH_3-CH_2-C\equiv C-CH_2-CH_3$$

Solution

a.

STEP 1 **Name the longest carbon chain that contains the double or triple bond.** There are five carbon atoms in the longest carbon chain containing the double bond. Replacing the corresponding alkane ending with *ene* gives pentene.

STEP 2 **Number the longest chain from the end nearest the double or triple bond.** The number of the first carbon in the double bond is used to give the location of the double bond.

$$\overset{\displaystyle CH_3}{\underset{\displaystyle}{CH_3-CH-CH=CH-CH_3}} \qquad \text{2-pentene}$$
$$54321$$

Guide to Naming Alkenes and Alkynes

STEP 1
Name the longest carbon chain with a double or triple bond.

STEP 2
Number the carbon chain starting from the end nearest a double or triple bond.

STEP 3
Give the location and name of each substituent (alphabetical order) as a prefix to the name.

STEP 3   **Give the location and name of each substituent (alphabetical order) as a prefix to the alkene or alkyne name.**   The methyl group is located on carbon 4.

$$\underset{5}{CH_3}-\underset{4}{\overset{\overset{\displaystyle CH_3}{|}}{CH}}-\underset{3}{CH}=\underset{2}{CH}-\underset{1}{CH_3}\qquad\text{4-methyl-2-pentene}$$

b.

STEP 1   **Name the longest carbon chain that contains the double or triple bond.**   There are six carbon atoms in the longest chain containing the triple bond. Replacing the corresponding alkane ending with *yne* gives hexyne.

STEP 2   **Number the main chain from the end nearest the double or triple bond.**   The number of the first carbon in the triple bond is used to give the location of the double bond.

$$\underset{1}{CH_3}-\underset{2}{CH_2}-\underset{3}{C}\equiv\underset{4}{C}-\underset{5}{CH_2}-\underset{6}{CH_3}\qquad\text{3-hexyne}$$

STEP 3   **Give the location and name of each substituent (alphabetical order) as a prefix to the alkene or alkyne name.**   There are no substituents in this formula.

Study Check

Draw the structural formulas for each of the following:
a. 2-pentyne   **b.** 3-methyl-1-pentene

CHEM NOTE

PHEROMONES IN INSECT COMMUNICATION

Insects and many other organisms emit minute quantities of chemicals called pheromones. Insects use pheromones to send messages to individuals of the same species. Some pheromones warn of danger, others call for defense, mark a trail, or attract the opposite sex. In the last 40 years, the structures of many pheromones have been chemically determined. One of the most studied is bombykol, the sex pheromone produced by the female of the silkworm moth species. The bombykol molecule is a 16-carbon chain with two double bonds and an alcohol group. A few molecules of synthetic bombykol will attract male silkworm moths from distances of over 1 kilometer.

Scientists are interested in synthesizing pheromones for use as nontoxic alternatives to pesticides. When used in a trap, bombykol can be used to isolate male silkworm moths. When a synthetic pheromone is released in several areas of a field or crop, the males cannot locate the females, which disrupts the reproductive cycle. This technique has been successful with controlling the oriental fruit moth, the grapevine moth, and the pink bollworm.

Bombykol, sex attractant for the silkworm moth

Hydrogenation

In **hydrogenation,** atoms of hydrogen add to the carbons in a double or triple bond to form alkanes. A catalyst such as platinum (Pt), nickel (Ni), or palladium (Pd) is added to speed up the reaction. The general equation for hydrogenation can be written as follows:

$$\underset{\text{Double bond (unsaturated)}}{\diagdown \overset{\diagup}{C}=\overset{\diagdown}{C}\diagup} \quad + \quad H-H \quad \xrightarrow{\text{Catalyst}} \quad \underset{\text{Single bond (saturated)}}{-\overset{\overset{\displaystyle H}{|}}{C}-\overset{\overset{\displaystyle H}{|}}{C}-}$$

Some examples of the hydrogenation of alkenes and alkynes follow:

$$\underset{\text{2-Butene}}{CH_3-CH=CH-CH_3} \quad + \quad H-H \quad \xrightarrow{Pt} \quad \underset{\text{Butane}}{CH_3-\overset{\overset{\displaystyle H}{|}}{C}H-\overset{\overset{\displaystyle H}{|}}{C}H-CH_3}$$

The hydrogenation of alkynes requires two molecules of hydrogen to form the alkane product.

$$\underset{\text{2-Butyne}}{CH_3-C\equiv C-CH_3} \quad + \quad 2H-H \quad \xrightarrow{Pt} \quad \underset{\text{Butane}}{CH_3-\overset{\overset{\displaystyle H}{|}}{\underset{\underset{\displaystyle H}{|}}{C}}-\overset{\overset{\displaystyle H}{|}}{\underset{\underset{\displaystyle H}{|}}{C}}-CH_3}$$

Sample Problem 17.7   **Writing Equations for Hydrogenation**

Write the structural formula for the product of the following hydrogenation reactions:

a. $CH_3-CH=CH_2 + H_2 \xrightarrow{Pt}$

b. $HC\equiv CH + 2H_2 \xrightarrow{Ni}$

CHEM NOTE

HYDROGENATION OF UNSATURATED FATS

Vegetable oils such as corn oil or safflower oil are unsaturated fats composed of fatty acids that contain double bonds. The process of hydrogenation is used commercially to convert the double bonds in the unsaturated fats in vegetable oils to saturated fats such as margarine, which are more solid. Adjusting the amount of added hydrogen produces partially hydrogenated fats such as soft margarine, solid margarine in sticks, and shortenings, which are used in cooking. For example, oleic acid is a typical unsaturated fatty acid in olive oil and has a double bond at carbon 9. When oleic acid is hydrogenated, it is converted to stearic acid, a saturated fatty acid.

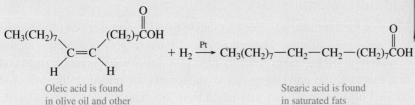

Oleic acid is found in olive oil and other unsaturated fats

Stearic acid is found in saturated fats

Solution

In an addition reaction, hydrogen adds to the double or triple bond to give an alkane.

a. $CH_3 - CH_2 - CH_3$　　　　**b.** $H_3C - CH_3$

Study Check

Draw the structural formula of the product of the hydrogenation of 2-methyl-1-butene using a platinum catalyst.

Questions and Problems　　　**Alkenes and Alkynes**

17.27 Identify the following as alkanes, alkenes, or alkynes:

a. 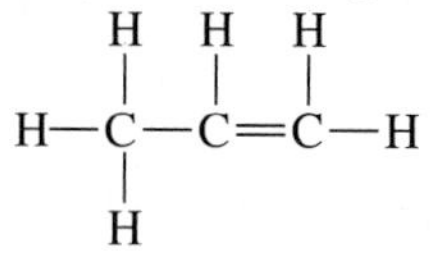
$$
\begin{array}{c}
\quad H \quad H \quad H \\
\quad | \quad\;\; | \quad\;\; | \\
H-C-C=C-H \\
\quad | \\
\quad H
\end{array}
$$

b. $CH_3 - CH_2 - C \equiv C - H$

17.28 Identify the following as alkanes, alkenes, or alkynes:

a. $CH_3 - CH_2 - C \equiv C - CH_3$

b. 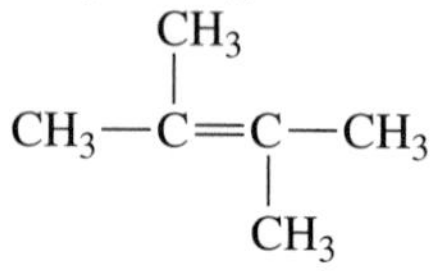
$$
\begin{array}{c}
\qquad\quad CH_3 \\
\qquad\quad | \\
CH_3 - C = C - CH_3 \\
\qquad\qquad | \\
\qquad\qquad CH_3
\end{array}
$$

17.29 Give the IUPAC name for each of the following:

a. $CH_2 = CH_2$

b. 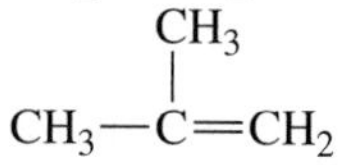
$$
\begin{array}{c}
\qquad CH_3 \\
\qquad | \\
CH_3 - C = CH_2
\end{array}
$$

c. $CH_3 - CH_2 - C \equiv C - CH_3$

17.30 Give the IUPAC name for each of the following:

a. $CH_2 = CH - CH_2 - CH_3$

b. 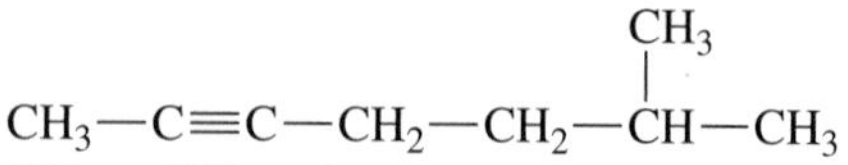
$$
\begin{array}{c}
\qquad\qquad\qquad\qquad CH_3 \\
\qquad\qquad\qquad\qquad | \\
CH_3 - C \equiv C - CH_2 - CH_2 - CH - CH_3
\end{array}
$$

c. $CH_3 - CH_2 - CH = CH - CH_3$

17.31 Draw the structural formula for each of the following compounds:

a. propene　　　　**b.** 1-pentene

c. 2-methyl-1-butene

17.32 Draw the structural formula for each of the following compounds:

a. 2-pentyne　　　　**b.** 3-methyl-1-butyne

c. 3,4-dimethyl-1-pentene

17.33 Give the condensed structural formulas and names of the products in each of the following reactions:

a. $CH_3 - CH_2 - CH_2 - CH = CH_2 + H_2 \xrightarrow{Pt}$

b. $CH_3 - CH = CH - CH_3 + H_2 \xrightarrow{Ni}$

c. 2-hexene $+ H_2 \xrightarrow{Pt}$

17.34 Give the condensed structural formulas and names of the products in each of the following reactions:

a. $CH_3 - CH_2 - CH = CH_2 + H_2 \xrightarrow{Pt}$

b. 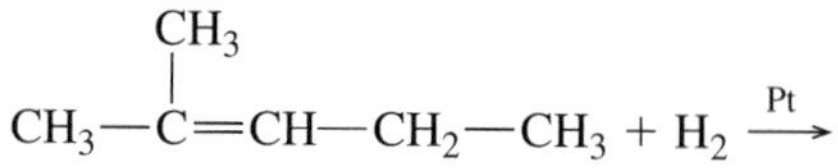
$$
\begin{array}{c}
\qquad\quad CH_3 \\
\qquad\quad | \\
CH_3 - C = CH - CH_2 - CH_3 + H_2 \xrightarrow{Pt}
\end{array}
$$

c.
$$
\begin{array}{c}
\qquad\quad CH_3 \\
\qquad\quad | \\
CH_3 - CH - C \equiv CH + 2H_2 \xrightarrow{Pt}
\end{array}
$$

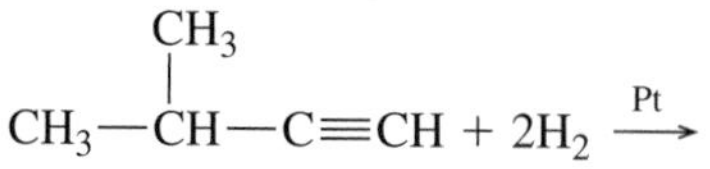

17.5 Polymers

Polymers are large molecules that consist of small repeating units called **monomers.** In the past hundred years, the plastics industry has made synthetic polymers that are in many of the materials we use every day, such as carpeting, plastic wrap, nonstick pans, plastic cups, and rain gear.

Addition Polymers

Many of the synthetic polymers are made by addition reactions of monomers that are small alkenes. The conditions for many polymerization reactions require high temperatures and very high pressure (over 1000 atm). In an addition reaction, the polymer grows as monomers are added to the end of the chain. Polyethylene, a polymer made from ethylene, $CH_2 = CH_2$, is used in

Learning **Goal**

Draw structural formulas of monomers that from a polymer or a three-monomer section of a polymer.

the
Chemistry place

WEB TUTORIAL

Polymers

Polyethylene

Polyvinyl chloride

Polypropylene

Polytetrafluoroethylene (Teflon)

Polydichloroethylene (Saran)

Polystyrene

Figure 17.10 Synthetic polymers provide a wide variety of items that we use every day.

Q What are some alkenes used to make the polymers in these plastic items?

plastic bottles, film, and plastic dinnerware. (See Figure 17.10.) In the polymerization, a series of addition reactions joins one monomer to the next until a long carbon chain forms that contains as many as 1000 monomers.

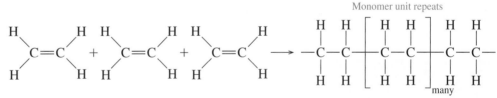

Ethene (ethylene) monomers   Polyethylene section

Table 17.10 lists several alkene monomers that are used to produce common synthetic polymers, and Figure 17.10 shows examples of each. The alkane-like nature of these plastic synthetic polymers makes them unreactive. Thus, they do not decompose easily (they are nonbiodegradable) and have become contributors to pollution. Efforts are being made to make them more degradable. It is becoming increasingly important to recycle plastic material, rather than add to our growing landfills.

Table 17.10 **Some Alkenes and Their Polymers**

Monomer	Polymer Section	Common Uses
$CH_2{=}CH_2$ Ethene (ethylene)	Polyethylene	Plastic bottles, film, insulation materials
$CH_2{=}CH$ $\quad\ \ \vert$ $\quad\ \ Cl$ Chloroethene (vinyl chloride)	Polyvinyl chloride (PVC)	Plastic pipes and tubing, garden hoses, garbage bags
$CH_2{=}CH$ $\quad\ \ \vert$ $\quad\ \ CH_3$ Propene (propylene)	Polypropylene	Ski and hiking clothing, carpets, artificial joints
$F{-}C{=}C{-}F$ Tetrafluoroethene	Polytetrafluoroethylene (Teflon)	Nonstick coatings
$CH_2{=}C{-}Cl$ $\quad\ \ \vert$ $\quad\ \ Cl$ 1,1-Dichloroethene	Polydichloroethylene (Saran)	Plastic film and wrap
$H_2C{=}CH$ Phenylethene (styrene)	Polystyrene	Plastic coffee cups and cartons, insulation

You can identify the type of polymer used to manufacture a plastic item by looking for the recycle symbol (arrows in a triangle) found on the label or on the bottom of the plastic container. For example, either the number 5 or the letters PP inside the triangle is a code for a polypropylene plastic.

1 PETE Polyethylene terephthalate	2 HDPE High-density polyethylene	3 PVC Polyvinyl chloride	4 LDPE Low-density polyethylene	5 PP Polypropylene	6 PS Polystyrene

Sample Problem 17.8 **Polymers**

What are the starting monomers for the following polymers?

a. polypropylene   **b.** Saran

Solution

a. propene (propylene) $CH_2{=}CH$
$$\overset{\displaystyle CH_3}{\underset{}{|}}$$

b. 1,1-dichloroethene, $CH_2{=}\overset{\displaystyle Cl}{\underset{}{|}}\!\!{C}{-}Cl$

Study Check

What is the monomer for PVC?

Questions and Problems ▷ **Polymers**

17.35 What is a polymer?

17.36 What is a monomer?

17.37 Write an equation that represents the formation of a part of the polyethylene polymer from three of the monomer units.

17.38 Write an equation that represents the formation of a part of the polystyrene polymer from three of the monomer units.

17.6 Aromatic Compounds

Learning **Goal**

Describe the bonding in benzene; name aromatic compounds, and write their structural formulas.

In 1825, Michael Faraday isolated a hydrocarbon called benzene, which had the molecular formula C_6H_6. Because many compounds containing benzene had fragrant odors, the family of benzene compounds became known as **aromatic compounds.** A molecule of **benzene** consists of a ring of six carbon atoms with one hydrogen atom attached to each carbon. Each carbon atom uses 3 valence electrons to bond to the hydrogen atom and two adjacent carbons. That leaves 1 valence electron to share in a double bond with an adjacent carbon. In 1865, August Kekulé proposed that the carbon atoms were arranged in a flat ring with alternating single and double bonds between the carbon atoms. This idea led to two ways of writing the benzene structure, as follows:

Structures for benzene

However, there is only one structure of benzene. Today we know that all the bonds in benzene are identical. In the benzene ring, scientists determined that the electrons are shared equally, a unique feature that makes aromatic compounds especially stable. To show this stability, the benzene structure is also represented as a hexagon with a circle in the center.

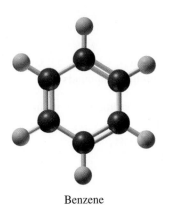

Benzene

Naming Aromatic Compounds

Aromatic compounds that contain a benzene ring with a single substituent are usually named as benzene derivatives. However, many of these compounds

CHEM NOTE

SOME COMMON AROMATIC COMPOUNDS

Aromatic compounds are common in nature and in medicine. Toluene is used as a reactant to make drugs, dyes, and explosives such as TNT (trinitrotoluene). The benzene ring is found in some pain relievers such as aspirin, acetaminophen, and ibuprofen, and in flavorings such as vanillin.

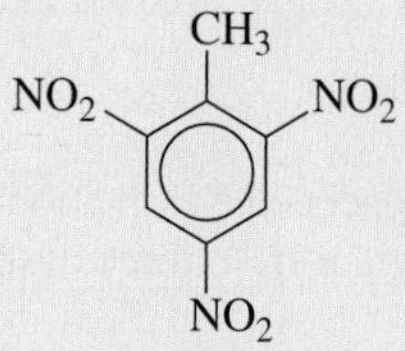

TNT (2,4,6-trinitrotoluene)

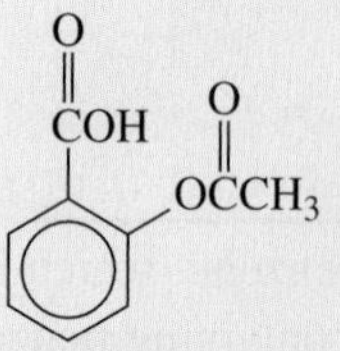

Aspirin

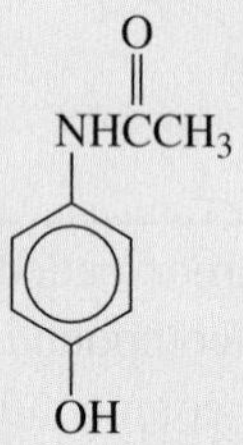

Acetaminophen

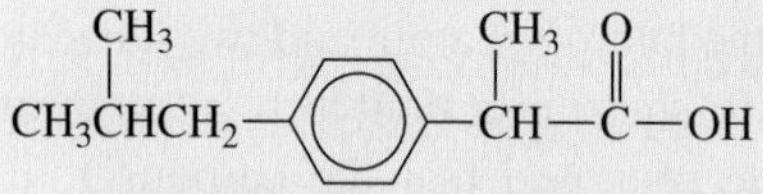

Ibuprofen

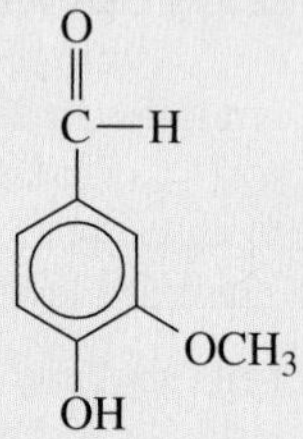

Vanillin

have been important in chemistry for many years and still use their common names. Some widely used names such as toluene, aniline, and phenol are allowed by IUPAC rules.

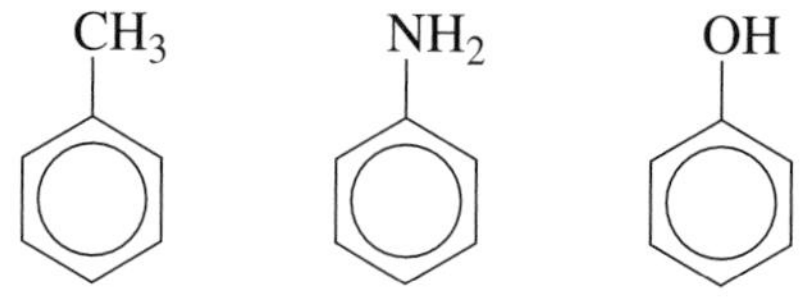

Toluene (methylbenzene) Aniline (benzenamine) Phenol (hydroxybenzene)

When there are two or more substituents on a benzene ring, the ring is numbered to give the lowest numbers to the substituents.

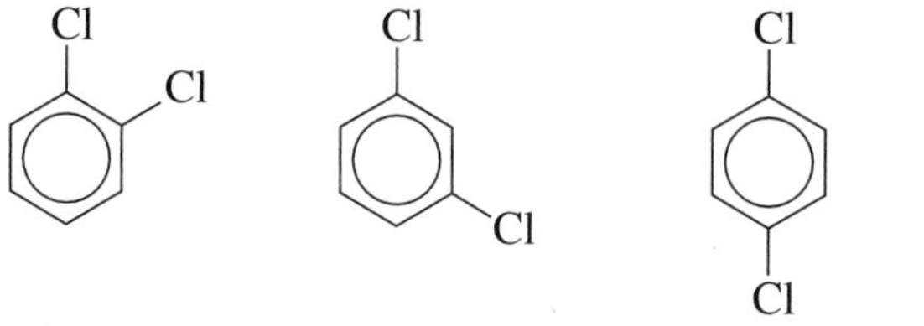

1,2-Dichlorobenzene 1,3-Dichlorobenzene 1,4-Dichlorobenzene

The substituents are named alphabetically.

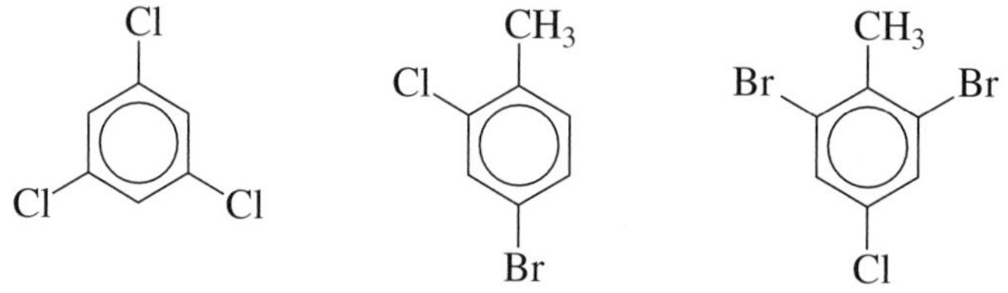

1,3,5-Trichlorobenzene 4-Bromo-2-chlorotoluene 2,6-Dibromo-4-chlorotoluene

When a benzene ring is a substituent, C_6H_5—, it is named as a phenyl group.

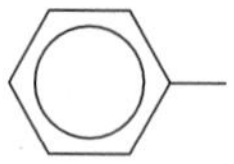

Phenyl group

Sample Problem 17.9 **Naming Aromatic Compounds**

Give the IUPAC name for each of the following aromatic compounds:

a. **b.** **c.**

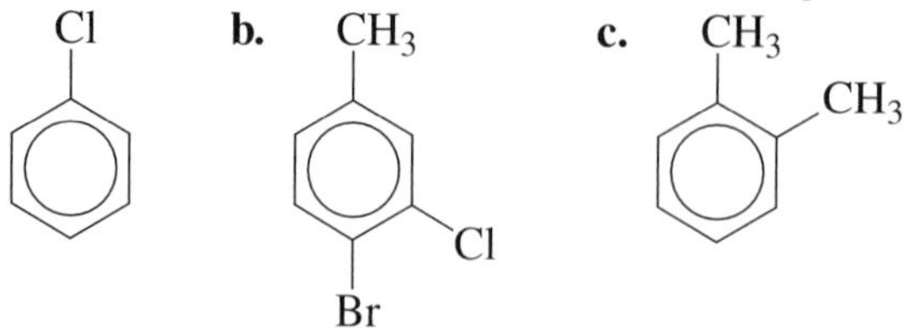

Solution

a. chlorobenzene

b. 4-bromo-3-chlorotoluene

c. 1,2-dimethylbenzene

Study Check

Name the following compound:

CH₂—CH₃

CH₂—CH₃

Aromatic Compounds

17.39 Give the IUPAC name for each of the following:
a. CH₃ Cl **b.** CH₂—CH₃

c. Cl Cl Cl

17.40 Give the IUPAC name for each of the following:
a. **b.** CH₃ **c.** Cl Cl

17.41 Draw the structural formulas for each of the following compounds:
a. toluene
b. 1,3-dichlorobenzene
c. 4-ethyltoluene

17.42 a. benzene
b. 2-chlorotoluene
c. propylbenzene

17.7 Alcohols and Ethers

As we learned in section 17.3, alcohols and ethers are two classes of organic compounds that contain an oxygen (O) atom. In an alcohol, the oxygen atom is part of a *hydroxyl* group (—OH) that is attached to a carbon atom in an alkane. In an ether, the oxygen atom is attached to two carbon atoms. Both alcohols and ethers have bent structures similar to water. One hydrogen atom of water is replaced by an alkyl group in an alcohol and by a benzene ring in a **phenol**. (See Figure 17.11.)

Figure 17.11 A hydrogen atom in water is replaced by an alkyl group in methanol and by an aromatic ring in phenol.

Q Why are the structures of alcohols and phenols similar to water?

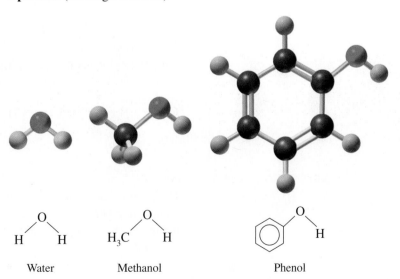

Water   Methanol   Phenol

Naming Alcohols

In the IUPAC system, the alcohol family is indicated by the *ol* ending.

STEP 1 **Name the longest carbon chain containing the —OH group.** Replace the *e* in the alkane name with *ol*. Consider the following alcohol:

$$CH_3—CH_2—CH_2—OH \qquad \text{propanol}$$

STEP 2 **Number the longest chain to give the —OH group the lowest number.** For simple alcohols, the common name (shown in parentheses) gives the name of the carbon chain as an alkyl group followed by *alcohol*.

$$CH_3—CH_2—CH_2—OH$$
$$321$$

1-propanol
(propyl alcohol)

STEP 3 **Name and number other substituents relative to the —OH group.**

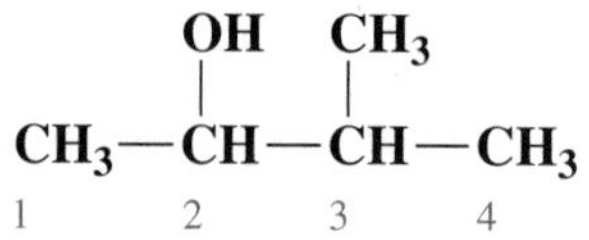

$$\overset{\displaystyle OH}{\underset{}{|}} \quad \overset{\displaystyle CH_3}{\underset{}{|}}$$
$$CH_3—\overset{}{CH}—\overset{}{CH}—CH_3$$
$$1 \quad\ 2 \quad\ 3 \quad 4$$
3-Methyl-2-butanol

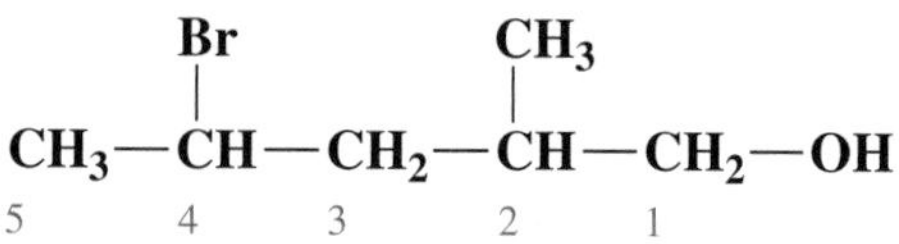

$$\overset{\displaystyle Br}{\underset{}{|}} \qquad\qquad \overset{\displaystyle CH_3}{\underset{}{|}}$$
$$CH_3—\overset{}{CH}—CH_2—\overset{}{CH}—CH_2—OH$$
$$5 \quad\ 4 \quad\ 3 \quad\ 2 \quad\ 1$$
4-Bromo-2-methyl-1-pentanol

STEP 4 **When the —OH group is attached to a benzene ring, it is named *phenol*.** When there is a second substituent on the benzene ring, the ring is numbered from carbon 1, which is attached to the —OH group, to give the lower possible number to the substituent.

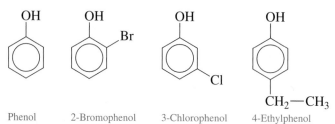

Phenol   2-Bromophenol   3-Chlorophenol   4-Ethylphenol

Sample Problem 17.10 **Naming Alcohols**

Give the IUPAC name for each of the following:

$$\overset{\displaystyle CH_3}{\underset{}{|}} \qquad \overset{\displaystyle OH}{\underset{}{|}}$$
a. 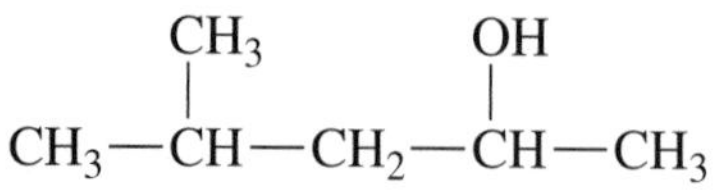 $CH_3—CH—CH_2—CH—CH_3$ **b.**

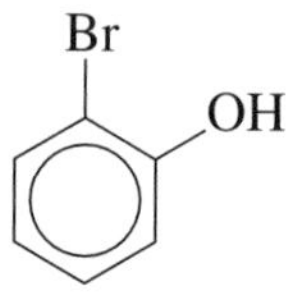

Solution

a. The parent chain is pentane; the alcohol is named *pentanol*. The carbon chain is numbered to give the position of the —OH group on carbon 2 and the methyl group on carbon 4. The compound is named *4-methyl-2-pentanol.*

b. The compound is a *phenol* because the —OH is attached to a benzene ring. The ring is numbered with carbon 1 attached to the —OH in the direction that gives the bromine the lower number. The compound is named *2-bromophenol*.

Study Check

Give the IUPAC name for the following:

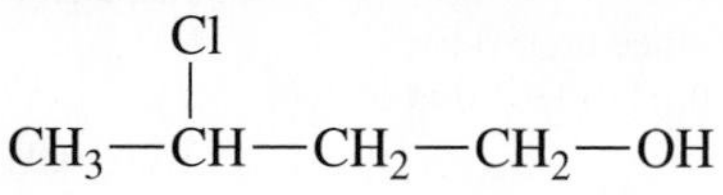

$$CH_3-\underset{\underset{\displaystyle Cl}{|}}{CH}-CH_2-CH_2-OH$$

Classification of Alcohols

Alcohols are classified by the number of carbon groups attached to the carbon atom bonded to the hydroxyl (—OH) group. A **primary (1°) alcohol** has one alkyl group attached to the carbon atom bonded to the —OH, a **secondary (2°) alcohol** has two alkyl groups, and a **tertiary (3°) alcohol** has three alkyl groups.

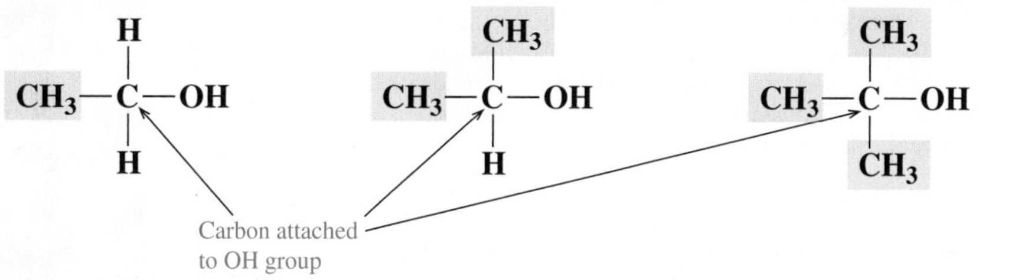

Primary (1°) alcohol	Secondary (2°) alcohol	Tertiary (3°) alcohol

$$CH_3-\underset{\underset{\displaystyle H}{|}}{\overset{\overset{\displaystyle H}{|}}{C}}-OH \qquad CH_3-\underset{\underset{\displaystyle H}{|}}{\overset{\overset{\displaystyle CH_3}{|}}{C}}-OH \qquad CH_3-\underset{\underset{\displaystyle CH_3}{|}}{\overset{\overset{\displaystyle CH_3}{|}}{C}}-OH$$

Carbon attached to OH group

Sample Problem 17.11 | **Classifying Alcohols**

Classify each of the following alcohols as primary, secondary, or tertiary:

a. $CH_3-CH_2-CH_2-OH$

b. 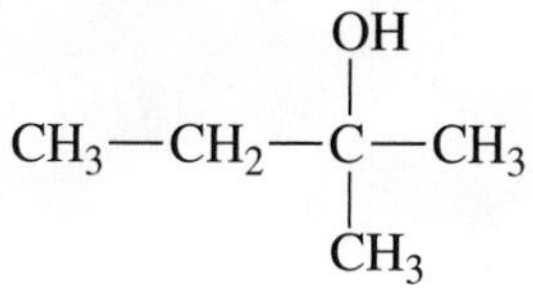 $CH_3-CH_2-\underset{\underset{\displaystyle CH_3}{|}}{\overset{\overset{\displaystyle OH}{|}}{C}}-CH_3$

Solution

a. One alkyl group attached to the carbon atom bonded to the —OH makes this a primary alcohol.

b. Three alkyl groups attached to the carbon atom bonded to the —OH makes this a tertiary alcohol.

Study Check

Classify the following as primary, secondary, or tertiary:

$$CH_3-\underset{\underset{\displaystyle OH}{|}}{CH}-CH_3$$

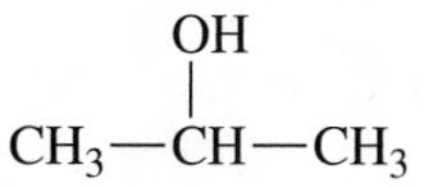

SOME IMPORTANT ALCOHOLS AND PHENOLS

Methanol (methyl alcohol), the simplest alcohol, is found in many solvents and paint removers. If ingested, methanol is oxidized to formaldehyde, which can cause headaches, blindness, and death. Methanol is used to make plastics, medicines, and fuels. In car racing, it is used as a fuel because it is less flammable and has a higher octane rating than gasoline.

Ethanol (ethyl alcohol) has been known since prehistoric times as an intoxicating product formed by the fermentation of grains and starches.

$$C_6H_{12}O_6 \xrightarrow{\text{fermentation}} 2CH_3-CH_2-OH + 2CO_2$$

Today, ethanol for commercial uses is produced from ethene and water allowed to react at high temperatures and pressures. It is used as a solvent for perfumes, varnishes, and some medicines, such as tincture of iodine. "Gasohol" is a mixture of ethanol and gasoline used as a fuel.

$$H_2C=CH_2 + H_2O \xrightarrow{\text{300°C, 200 atm, catalyst}} CH_3-CH_2-OH$$

1,2-Ethanediol (ethylene glycol) is used as antifreeze in heating and cooling systems. It is also a solvent for paints, inks, and plastics, and is used in the production of synthetic fibers such as Dacron. If ingested, it is extremely toxic. In the body, it is oxidized to oxalic acid, which forms insoluble salts in the kidneys that cause renal damage, convulsions, and death. Because its sweet taste is attractive to pets and children, ethylene glycol solutions must be carefully stored.

$$HO-CH_2-CH_2-OH \xrightarrow{[O]} HO-\overset{\overset{\displaystyle O}{\|}}{C}-\overset{\overset{\displaystyle O}{\|}}{C}-OH$$
1,2-Ethanediol (ethylene glycol)   Oxalic acid

1,2,3-Propanetriol (glycerol or glycerin), a trihydroxy alcohol, is a viscous liquid obtained from oils and fats during the production of soaps. The presence of several polar —OH groups makes it strongly attracted to water, a feature that makes glycerin useful as a skin softener in products such as skin lotions, cosmetics, shaving creams, and liquid soaps.

$$HO-CH_2-\overset{\overset{\displaystyle OH}{|}}{CH}-CH_2-OH$$
1,2,3-Propanetriol (glycerol)

Several of the essential oils of plants, which produce the odor or flavor of the plant, are derivatives of phenol. Eugenol is found in cloves, vanillin in vanilla bean, isoeugenol in nutmeg, and thymol in thyme and mint. Thymol has a pleasant, minty taste and is used in mouthwashes and by dentists to disinfect a cavity before adding a filling compound. (See Figure 17.12.)

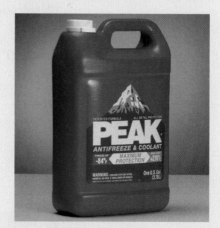

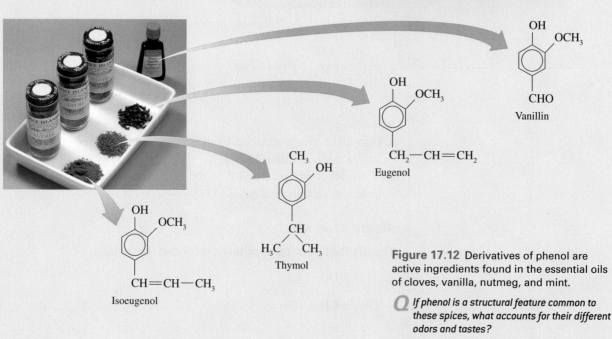

Figure 17.12 Derivatives of phenol are active ingredients found in the essential oils of cloves, vanilla, nutmeg, and mint.

Q *If phenol is a structural feature common to these spices, what accounts for their different odors and tastes?*

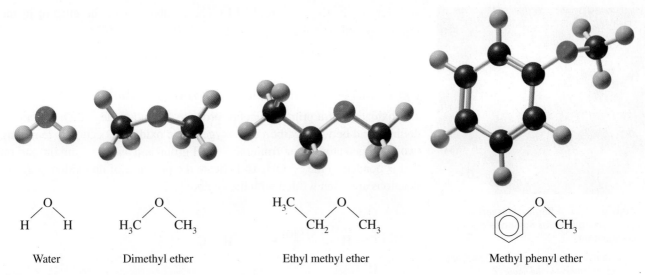

| Water | Dimethyl ether | Ethyl methyl ether | Methyl phenyl ether |

Figure 17.13 The structures of ethers are similar to that of water.

Q *What part of the structure of water is found in ethers?*

Ethers

As we saw in section 17.3, an *ether* contains an oxygen atom that is attached by single bonds to two carbon groups that are alkyls or aromatic rings. Ethers have a bent structure like water and alcohols except both hydrogen atoms are replaced by alkyl groups. (See Figure 17.13.)

Naming Ethers

Most ethers are named by their common names. Write the name of each alkyl or aryl (aromatic) group attached to the oxygen atom in alphabetical order followed by the word *ether*.

Methyl group

$$CH_3—O—CH_2—CH_2—CH_3 \quad \text{Propyl group}$$

Common name: methyl propyl ether

Sample Problem 17.12   **Ethers**

Give the common name for the following ether:

$$CH_3—CH_2—O—CH_2—CH_2—CH_3$$

Solution

The groups attached to the oxygen are an ethyl group and a propyl group. The common name is *ethyl propyl ether*.

Study Check

Draw the structure of methyl phenyl ether.

Reactions of Alcohols

Earlier in this chapter, we learned that hydrocarbons undergo combustion in the presence of oxygen. Alcohols burn with oxygen too. For example, in a restaurant, a dessert may be prepared by pouring a liquor on fruit or ice cream

Figure 17.14 A flaming dessert is prepared using a liquor that undergoes combustion.

Q What is the equation for the combustion of the ethanol in the liquor?

and lighting it. (See Figure 17.14.) The combustion of the ethanol in the liquor proceeds as follows:

$$CH_3-CH_2-OH + 3O_2 \longrightarrow 2CO_2 + 3H_2O + energy$$

Oxidation of Primary and Secondary Alcohols

The oxidation of a primary alcohol produces an aldehyde, which contains a double bond between carbon and oxygen. The oxidation occurs by removing two hydrogen atoms, one from the —OH group and another from the carbon that is bonded to the —OH. To indicate the presence of an oxidizing agent, reactions are often written with the symbol [O].

OH
|
H—C—H $\xrightarrow{[O]}$ H—C—H
| ‖
H   O

methyl alcohol   formaldehyde

OH   O
| ‖
CH₃—CH₂ $\xrightarrow{[O]}$ CH₃—C—H

ethyl alcohol   acetaldehyde

In the oxidation of secondary alcohols, the products are ketones. One hydrogen is removed from the —OH and another from the carbon bonded to the —OH group. The result is a ketone that has the carbon–oxygen double bond attached to alkyl groups on both sides.

OH   O
| ‖
CH₃—C—CH₃ $\xrightarrow{[O]}$ CH₃—C—CH₃
|
H

isopropyl alcohol   dimethyl ketone: acetone

Tertiary alcohols do not oxidize readily because there are no hydrogen atoms on the carbon bonded to the —OH group. Because C—C bonds are usually too strong to oxidize, tertiary alcohols resist oxidation.

No   No hydrogen on
double   this carbon
bond   O—H
forms |
CH₃—C—CH₃ $\xrightarrow{[O]}$ No oxidation product readily formed
             |
             CH₃

3° Alcohol

Sample Problem 17.13   **Oxidation of Alcohols**

Draw the structural formula of the aldehyde or ketone formed by the oxidation of each of the following:

OH
|
a. CH₃—CH₂—CH—CH₃   **b.** CH₃—CH₂—CH₂—OH

Solution

a. The oxidation of a secondary alcohol produces a ketone.

$$CH_3-CH_2-\overset{\displaystyle O}{\overset{\|}{C}}-CH_3$$

b. The oxidation of a primary alcohol produces an aldehyde.

$$CH_3-CH_2-\overset{\displaystyle O}{\overset{\|}{C}}-H$$

the Chemistry place

CASE STUDY
Alcohol Toxicity

Study Check

Draw the structural formula of the aldehyde or ketone formed by the oxidation of 3-pentanol.

CHEM NOTE

OXIDATION OF ALCOHOL IN THE BODY

Ethanol is the most commonly abused drug in the United States. When ingested in small amounts, ethanol may produce a feeling of euphoria in the body although it is a depressant. In the liver, enzymes such as alcohol dehydrogenases oxidize ethanol to acetaldehyde, a substance that impairs mental and physical coordination. If the blood alcohol concentration exceeds 0.4%, coma or death may occur. Table 17.11 gives some of the typical behaviors exhibited at various levels of blood alcohol.

$$CH_3CH_2OH \xrightarrow{[O]} CH_3\overset{\displaystyle O}{\overset{\|}{C}}H \xrightarrow{[O]} 2CO_2 + H_2O$$

Ethanol   Acetaldehyde

The acetaldehyde produced from ethanol in the liver is further oxidized to acetic acid, which is converted to carbon dioxide and water in the citric acid (Krebs) cycle. Thus, the enzymes in the liver can eventually break down ethanol, but the aldehyde and carboxylic acid intermediates can cause considerable damage while they are present within the cells of the liver.

A person weighing 150 lb requires about 1 hour to completely metabolize 10 ounces of beer. However, the rate of metabolism of ethanol varies between nondrinkers and drinkers. Typically, nondrinkers and social drinkers can metabolize 12–15 mg of ethanol/dL of blood in 1 hour, but an alcoholic can metabolize as much as 30 mg of ethanol/dL in 1 hour. Some effects of alcohol metabolism include an increase in liver lipids (fatty liver), an increase in serum triglycerides, gastritis, pancreatitis, ketoacidosis, alcoholic hepatitis, and psychological disturbances.

Table 17.11 **Typical Behaviors Exhibited by a 150-lb Person Consuming Alcohol**

Number of Beers (12 oz) or Glasses of Wine (5 oz)	Blood Alcohol Level (%)	Typical Behavior
1	0.025	Slightly dizzy, talkative
2	0.05	Euphoria, loud talking, and laughing
4	0.10	Loss of inhibition, loss of coordination, drowsiness, legally drunk in most states
8	0.20	Intoxicated, quick to anger, exaggerated emotions
12	0.30	Unconscious
16–20	0.40–0.50	Coma and death

When the Breathalyzer test is used for suspected drunk drivers, the driver exhales a volume of breath into a solution containing the orange Cr^{6+} ion. If there is ethyl alcohol present in the exhaled air, the alcohol is oxidized, and the Cr^{6+} is reduced to give a green solution of Cr^{3+}.

Questions and Problems ▸ Alcohols and Ethers

17.43 Give the IUPAC name for each of the following:

   a. CH_3—CH_2—OH

   b. CH_3—CH_2—$\overset{\displaystyle OH}{\underset{|}{CH}}$—$CH_3$

   c. CH_3—$\overset{\displaystyle OH}{\underset{|}{CH}}$—$CH_2$—$CH_2$—$CH_3$

   d.

17.44 Give the IUPAC name for each of the following alcohols:

   a. CH_3—CH_2—$\overset{\displaystyle CH_3}{\underset{|}{CH}}$—$CH_2$—$OH$

   b. CH_3—CH_2—$\overset{\displaystyle CH_3}{\underset{|}{CH}}$—$\overset{\displaystyle CH_3}{\underset{|}{CH}}$—$CH_2$—$OH$

   c. CH_3—CH_2—CH_2—$\overset{\displaystyle OH}{\underset{|}{CH}}$—$CH_3$

   d.

17.45 Write the condensed structural formula of each of the following alcohols:

   a. 1-propanol   **b.** methyl alcohol

   c. 3-pentanol   **d.** 2-methyl-2-butanol

17.46 Write the condensed structural formula of each of the following alcohols:

   a. ethyl alcohol

   b. 3-methyl-1-butanol

   c. 2,4-dichloro-3-hexanol

   d. propyl alcohol

17.47 Classify each of the following as a primary, secondary, or tertiary alcohol:

   a. CH_3—$\overset{\displaystyle CH_3}{\underset{|}{CH}}$—$CH_2$—$CH_2$—$OH$

   b. CH_3—CH_2—CH_2—CH_2—OH

   c. CH_3—$\overset{\displaystyle OH}{\underset{\displaystyle CH_3}{\overset{|}{\underset{|}{C}}}}$—$CH_2$—$CH_3$

17.48 Classify each of the following as a primary, secondary, or tertiary alcohol:

   a. CH_3—$\overset{\displaystyle CH_3}{\underset{|}{CH}}$—$CH_2$—$OH$

   b.

   c. CH_3—CH_2—CH_2—$\overset{\displaystyle CH_3}{\underset{\displaystyle CH_3}{\overset{|}{\underset{|}{C}}}}$—$OH$

17.49 Give a common name for each of the following ethers:

   a. CH_3—O—CH_2—CH_3

   b. CH_3—CH_2—CH_2—O—CH_2—CH_2—CH_3

   c. CH_3—O—CH_2—CH_2—CH_3

17.50 Give the common name for each of the following ethers:

   a. CH_3—CH_2—O—CH_2—CH_2—CH_3

   b.

   c. CH_3—O—CH_3

17.51 Draw the condensed structural formula of the aldehyde or ketone when each of the following alcohols is oxidized [O] (if no reaction, write *none*):

   a. CH_3—CH_2—CH_2—CH_2—CH_2—OH

   b. CH_3—CH_2—$\overset{\displaystyle OH}{\underset{|}{CH}}$—$CH_3$

   c. CH_3—$\overset{\displaystyle OH}{\underset{|}{CH}}$—$CH_2$—$\overset{\displaystyle CH_3}{\underset{|}{CH}}$—$CH_3$

17.52 Draw the condensed structural formula of the organic product when each of the following alcohols is oxidized [O] (if no reaction, write *none*):

   a. CH_3—$\overset{\displaystyle CH_3}{\underset{|}{CH}}$—$CH_2$—$CH_2$—$OH$

   b. CH_3—CH_2—$\overset{\displaystyle OH}{\underset{\displaystyle CH_3}{\overset{|}{\underset{|}{C}}}}$—$CH_3$

   c. CH_3—CH_2—$\overset{\displaystyle OH}{\underset{|}{CH}}$—$CH_2$—$CH_3$

17.8 Aldehydes and Ketones

The *carbonyl group* consists of a carbon–oxygen double bond. The double bond in the carbonyl group is similar to that of alkenes, except the carbonyl group has a dipole. The oxygen atom is much more electronegative than the carbon atom. Therefore, the carbonyl group has a strong dipole with a partial negative charge (δ^-) on the oxygen and a partial positive charge (δ^+) on the carbon. The polarity of the carbonyl group strongly influences the physical and chemical properties of aldehydes and ketones.

$$\begin{array}{c} O^{\delta-} \\ \| \\ C^{\delta+} \end{array}$$

In an *aldehyde,* the carbon of the carbonyl group is bonded to at least one hydrogen atom. That carbon may also be bonded to another hydrogen, a carbon of an alkyl group, or an aromatic ring. (See Figure 17.15.) In a *ketone,* the carbon of the carbonyl group is bonded to two alkyl groups or aromatic rings.

Naming Aldehydes

In the IUPAC names of aldehydes, the *e* of the alkane name is replaced with *al.*

STEP 1 **Name the longest carbon chain containing the carbonyl group by replacing the *e* in the corresponding alkane name by *al.*** No number is needed for the aldehyde group because it always appears at the end of the chain.

Figure 17.15 The carbonyl group in aldehydes and ketones.

Q If aldehydes and ketones both contain a carbonyl group, how can you differentiate between compounds from each family?

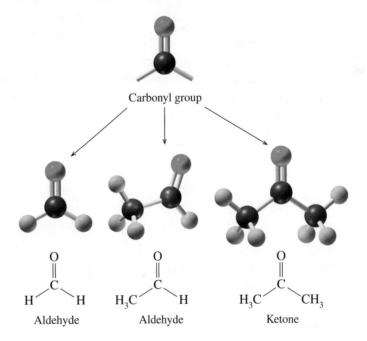

Carbonyl group

$$\begin{array}{ccc} O & O & O \\ \| & \| & \| \\ H-C-H & H_3C-C-H & H_3C-C-CH_3 \\ \text{Aldehyde} & \text{Aldehyde} & \text{Ketone} \end{array}$$

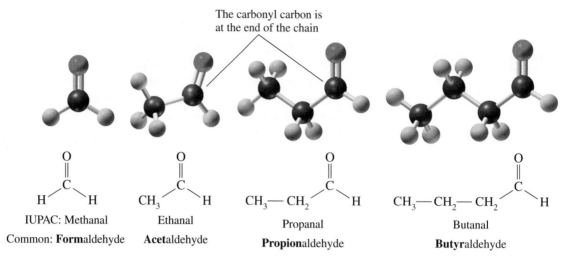

Figure 17.16 In the structures of aldehydes, the carbonyl group is always the end carbon.

Q *Why is the carbon in the carbonyl group in aldehydes always at the end of the chain?*

The IUPAC system names the aldehyde of benzene as benzaldehyde.

Benzaldehyde

The first four unbranched aldehydes are often referred to by their common names, which end in *aldehyde*. (See Figure 17.16.) The roots of these common names are derived from Latin or Greek words that indicate the source of the corresponding carboxylic acid. We will study carboxylic acids in the next section.

STEP 2 **Name and number any substituents on the carbon chain by counting the carbonyl carbon as carbon 1.**

2-Methylpropanal

4-Methylpentanal

Sample Problem 17.14 **Naming Aldehydes**

Give the IUPAC names for the following aldehydes:

a. $CH_3-CH_2-CH_2-CH_2-\overset{\overset{\displaystyle O}{\|}}{C}-H$

b.

c.

Solution

a. pentanal

b. 4-chlorobenzaldehyde

c. The longest unbranched chain has four atoms with a methyl group on the third carbon. The IUPAC name is *3-methylbutanal.*

Study Check

What are the IUPAC and common names of the aldehyde with three carbon atoms?

Naming Ketones

Aldehydes and ketones are some of the most important classes of organic compounds. Because they have played a major role in organic chemistry for more than a century, the common names for unbranched ketones are still in use. In the common names the alkyl groups bonded to the carbonyl group are named as substituents and listed alphabetically followed by *ketone.* Acetone, which is another name for propanone, has been retained by the IUPAC system.

In the IUPAC system, the name of a ketone is obtained by replacing the *e* in the corresponding alkane name with *one.*

> **STEP 1** **Name the longest carbon chain containing the carbonyl group by replacing the *e* in the corresponding alkane name by *one*.**

> **STEP 2** **Number the main chain starting from the end nearest the carbonyl group.** Place the number of the carbonyl carbon in front of the ketone name. (Propanone and butanone do not require numbers.)

$$CH_3-\overset{\overset{\displaystyle O}{\|}}{C}-CH_3 \qquad CH_3-CH_2-\overset{\overset{\displaystyle O}{\|}}{C}-CH_3 \qquad CH_3-CH_2-\overset{\overset{\displaystyle O}{\|}}{C}-CH_2-CH_3$$

Propanone   Butanone   3-Pentanone
(dimethyl ketone: acetone)   (ethyl methyl ketone)   (diethyl ketone)

> **STEP 3** **Name and number any substituents on the carbon chain.**

$$CH_3-\overset{\overset{\displaystyle O}{\|}}{C}-\overset{\overset{\displaystyle CH_3}{|}}{CH}-CH_3 \qquad CH_3-\overset{\overset{\displaystyle Br}{|}}{CH}-\overset{\overset{\displaystyle O}{\|}}{C}-CH_2-CH_3$$

3-Methylbutanone   2-Bromo-3-pentanone

Sample Problem 17.15   **Names of Ketones**

Give the IUPAC name for the following ketone:

$$CH_3-\overset{\overset{\displaystyle CH_3}{|}}{CH}-CH_2-\overset{\overset{\displaystyle O}{\|}}{C}-CH_3$$

Solution

The longest chain is five carbon atoms. Counting from the right, the carbonyl group is on carbon 2 and a methyl group is on carbon 4. The IUPAC name is *4-methyl-2-pentanone.*

Study Check

What is the common name of 3-hexanone?

CHEM NOTE

SOME IMPORTANT ALDEHYDES AND KETONES

Formaldehyde, the simplest aldehyde, is a colorless gas with a pungent odor. Industrially, it is a reactant in the synthesis of polymers used to make fabrics, insulation materials, carpeting, pressed wood products such as plywood, and plastics for kitchen counters. An aqueous solution called formalin, which contains 40% formaldehyde, is used as a germicide and to preserve biological specimens. Exposure to formaldehyde fumes can irritate the eyes, nose, and upper respiratory tract and cause skin rashes, headaches, dizziness, and general fatigue.

Acetone, or propanone (dimethyl ketone), which is the simplest ketone, is a colorless liquid with a mild odor that has wide use as a solvent in cleaning fluids, paint, nail-polish removers, and rubber cement. (See Figure 17.17.) It is extremely flammable and care must be taken when using acetone. In the body, acetone may be produced in uncontrolled diabetes, fasting, and high-protein diets when large amounts of fats are metabolized for energy.

Several naturally occurring aromatic aldehydes are used to flavor food and as fragrances in perfumes. Benzaldehyde is found in almonds, vanillin in vanilla beans, and cinnamaldehyde in cinnamon.

Benzaldehyde
(almond)

Vanillin
(vanilla)

Cinnamaldehyde
(cinnamon)

$$CH_3-\overset{\overset{\displaystyle O}{\|}}{C}-\overset{\overset{\displaystyle O}{\|}}{C}-CH_3$$

Butanedione

The flavor of butter or margarine comes from butanedione, muscone is used to make musk perfumes, and oil of spearmint contains carvone.

Figure 17.17 Acetone is used as a solvent in paint and nail-polish removers.

Q What is the IUPAC name of acetone?

Butanedione
(butter flavor)

Muscone
(musk)

Carvone
(spearmint oil)

17.53 Give a common name for each of the following compounds:

a. CH₃—C(=O)—H

b. CH₃—C(=O)—CH₂—CH₂—CH₃

c. H—C(=O)—H

17.54 Give the common name for each of the following compounds:

a. CH₃—C(=O)—CH₂—CH₃

b. CH₃—CH₂—C(=O)—CH₂—CH₃

c. CH₃—CH₂—C(=O)—H

17.55 Give the IUPAC name for each of the following compounds:

a. CH₃—CH₂—C(=O)—H

b. CH₃—CH₂—C(=O)—CH(CH₃)—CH₃

c.

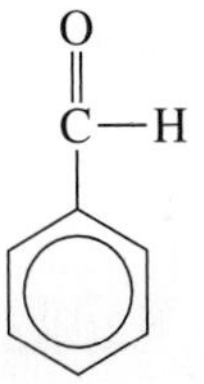

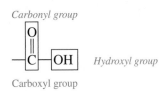

17.56 Give the IUPAC name for each of the following compounds:

a. CH₃—CH₂—CH₂—C(=O)—H

b. CH₃—CH₂—CH₂—C(=O)—CH₃

c.

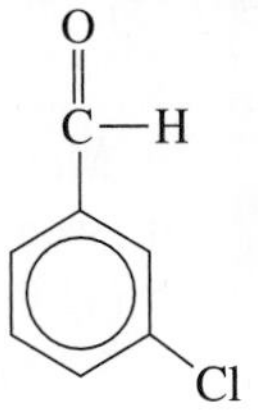

17.57 Write the condensed structural formula for each of the following compounds:
a. acetaldehyde **b.** 2-pentanone
c. butyl methyl ketone

17.58 Write the condensed structural formula for each of the following compounds:
a. propionaldehyde **b.** butanal
c. 4-bromobutanone

17.9 Carboxylic Acids and Esters

Learning Goal

Give the common names, IUPAC names, and condensed structural formulas of carboxylic acids and esters.

Carboxylic acids are more examples of weak acids we studied in Chapter 14. They have a sour or tart taste, produce hydronium ions in water, and neutralize bases. You encounter carboxylic acids when you use a vinegar salad dressing, which is a solution of acetic acid and water, or experience the sour taste of citric acid in a grapefruit or lemon.

Earlier in this chapter, we described the carbonyl group (C=O) as the functional group in aldehydes and ketones. In a *carboxylic acid,* a hydroxyl group is attached to the carbonyl group, forming a *carboxyl group*. The carboxyl functional group may be attached to an alkyl group or an aromatic group.

the Chemistry place
WEB TUTORIAL
Carboxylic Acids

CH₃—C(=O)—OH
Acetic acid

CH₃(CH₂)₁₆—C(=O)—OH
Stearic acid (a fatty acid in fats and oils)

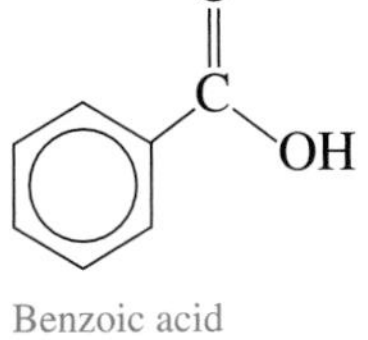

Benzoic acid (aromatic acid)

Naming Carboxylic Acids

The IUPAC names of carboxylic acids use the alkane names of the corresponding carbon chains.

STEP 1 Identify the longest carbon chain containing the carboxyl group and replace the *e* of the alkane name by *oic acid.*

STEP 2 Number the carbon chain beginning with the carboxyl carbon as carbon 1.

STEP 3 Give the location and names of substituents on the main chain. The carboxyl function group takes priority over all the functional groups we have discussed.

$$
\begin{array}{ccc}
\underset{\text{Methanoic acid}}{\text{H}-\overset{\displaystyle O}{\overset{\|}{\text{C}}}-\text{OH}} &
\underset{\text{2-Methylpropanoic acid}}{\text{CH}_3-\overset{\displaystyle \text{CH}_3}{\underset{}{\text{CH}}}-\overset{\displaystyle O}{\overset{\|}{\text{C}}}-\text{OH}} &
\underset{\text{3-Hydroxybutanoic acid}}{\text{CH}_3-\overset{\displaystyle \text{OH}}{\underset{}{\text{CH}}}-\text{CH}_2-\overset{\displaystyle O}{\overset{\|}{\text{C}}}-\text{OH}}
\end{array}
$$

Many carboxylic acids are still named by their common names, which are derived from their natural sources. In the last section, we named aldehydes using the prefixes that represent the typical sources of carboxylic acids.

Formic acid is injected under the skin from bee or red ant stings and other insect bites. (See Figure 17.18.) Acetic acid is the oxidation product of the ethanol in wines and apple cider. The resulting solution of acetic acid and water is known as vinegar. (See Figure 17.19.) Butyric acid gives the foul odor to rancid butter. (See Table 17.12.) Some ball-and-stick models of carboxylic acids are shown in Figure 17.20.

Figure 17.18 Red ants inject formic acid under the skin, which causes burning and irritation.

Q What is the IUPAC name of formic acid?

Figure 17.19 Vinegar is a 5% solution of acetic acid and water.

Q What is the IUPAC name for acetic acid?

Table 17.12 Names and Natural Sources of Carboxylic Acids

Condensed Structural Formulas	IUPAC Name	Common Name	Occurs In
$\text{H}-\overset{\displaystyle O}{\overset{\|}{\text{C}}}-\text{OH}$	Methanoic acid	Formic acid	Ant and bee stings (Latin *formica*, "ant")
$\text{CH}_3-\overset{\displaystyle O}{\overset{\|}{\text{C}}}-\text{OH}$	Ethanoic acid	Acetic acid	Vinegar (Latin *acetum*, "vinegar")
$\text{CH}_3-\text{CH}_2-\overset{\displaystyle O}{\overset{\|}{\text{C}}}-\text{OH}$	Propanoic acid	Propionic acid	Dairy products (Greek *pro*, "first," *pion*, "fat")
$\text{CH}_3-\text{CH}_2-\text{CH}_2-\overset{\displaystyle O}{\overset{\|}{\text{C}}}-\text{OH}$	Butanoic acid	Butyric acid	Rancid butter (Latin *butyrum*, "butter")

Figure 17.20 In carboxylic acids, a carbonyl group and a hydroxyl group are bonded to the same carbon atom.

What is the IUPAC and common name of a carboxylic acid with a chain of four carbons?

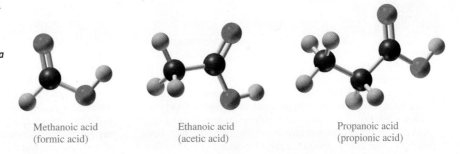

Methanoic acid
(formic acid)

Ethanoic acid
(acetic acid)

Propanoic acid
(propionic acid)

The aromatic carboxylic acid is called benzoic acid. With the carboxyl carbon bonded to carbon 1, the ring is numbered in the direction that gives substituents the smallest possible numbers.

Benzoic acid   4-Aminobenzoic acid   3,4-Dichlorobenzoic acid

---

Sample Problem 17.16 **Naming Carboxylic Acids**

Give the IUPAC and common name, if any, for each of the following carboxylic acids:

a. $CH_3-CH_2-\overset{\displaystyle O}{\overset{\displaystyle \|}{C}}-OH$ b. $CH_3-CH_2-\overset{\displaystyle CH_3}{\overset{\displaystyle |}{C}}H-\overset{\displaystyle O}{\overset{\displaystyle \|}{C}}-OH$

c. $\overset{\displaystyle O}{\overset{\displaystyle \|}{C}}-OH$

Solution

a. This carboxylic acid has three carbon atoms. In the IUPAC system, the *e* in propane is replaced by *oic acid,* to give the name, *propanoic acid.* Its common name is *propionic acid.*
b. This carboxylic acid has a methyl group on the second carbon. It has the IUPAC name *2-methylbutanoic acid.*
c. An aromatic carboxylic acid is named as benzoic acid.

Study Check

Write the condensed structural formula of pentanoic acid.

ALPHA HYDROXY ACIDS

Alpha hydroxy acids (AHAs) are naturally occurring carboxylic acids found in fruits, milk, and sugarcane. Cleopatra reportedly bathed in sour milk to smooth her skin. Dermatologists have been using products with a high concentration of AHAs to remove acne scars and reduce irregular pigmentation and age spots. Now lower concentrations (8–10%) of AHAs have been added to skin care products for the purpose of smoothing fine lines, improving skin texture, and cleansing pores. Several alpha hydroxy acids may be found in skin care products singly or in combination. Glycolic acid and lactic acid are most frequently used.

Recent studies indicate that products with AHAs increase sensitivity of the skin to sun and UV radiation. It is recommended that a sunscreen with a sun protection factor (SPF) of at least 15 be used when treating the skin with products that include AHAs. Products containing AHAs at concen-

trations under 10% and pH values greater than 3.5 are generally considered safe. However, the Food and Drug Administration has reports of AHAs causing skin irritation, including blisters, rashes, and discoloration of the skin. The FDA does not require product safety reports from cosmetic manufacturers, although they are responsible for marketing safe products. The FDA advises that you test any product containing AHAs on a small area of skin before you use it on a large area.

Alpha Hydroxy Acid (Source)	Structure
Glycolic acid (Sugarcane, sugar beet)	$HO-CH_2-\overset{\overset{\textstyle O}{\|}}{C}-OH$
Lactic acid (Sour milk)	$CH_3-\overset{\overset{\textstyle OH}{\|}}{C}H-\overset{\overset{\textstyle O}{\|}}{C}-OH$
Tartaric acid (Grapes)	$HO-\overset{\overset{\textstyle O}{\|}}{C}-\overset{\overset{\textstyle OH}{\|}}{C}H-\overset{\overset{\textstyle OH}{\|}}{C}H-\overset{\overset{\textstyle O}{\|}}{C}-OH$
Malic acid (Apples, grapes)	$HO-\overset{\overset{\textstyle O}{\|}}{C}-CH_2-\overset{\overset{\textstyle OH}{\|}}{C}H-\overset{\overset{\textstyle O}{\|}}{C}-OH$

Esters

A carboxylic acid reacts with an alcohol to form an **ester.** In an ester, the —H of the carboxylic acid is replaced by an alkyl group. Aspirin is an ester as well as a carboxylic acid. Fats and oils in our diets contain esters of glycerol and fatty acids, which are long-chain carboxylic acids. The aromas and flavors of many fruits including bananas, oranges, and strawberries are due to esters.

Carboxylic acid

$CH_3-\overset{\overset{\textstyle O}{\|}}{C}-O-H$

Ethanoic acid
(acetic acid)

Ester

$CH_3-\overset{\overset{\textstyle O}{\|}}{C}-O-CH_3$

Methyl ethanoate
(methyl acetate)

Esterification

In a reaction called **esterification,** a carboxylic acid reacts with an alcohol when heated in the presence of an acid catalyst (usually H_2SO_4). In the reaction, water is produced from the —OH removed from the carboxylic acid and an —H lost by the alcohol.

$$CH_3-\overset{\overset{\textstyle O}{\|}}{C}-O-H + H-O-CH_3 \xrightarrow{H^+,\ heat} CH_3-\overset{\overset{\textstyle O}{\|}}{C}-O-CH_3 + H-O-H$$

Acetic acid   Methyl alcohol   Methyl acetate

Sample Problem 17.17 ▸ Writing Esterification Equations

The ester that gives the flavor and odor of apples can be synthesized from butyric acid and methyl alcohol. What is the equation for the formation of the ester in apples?

Solution

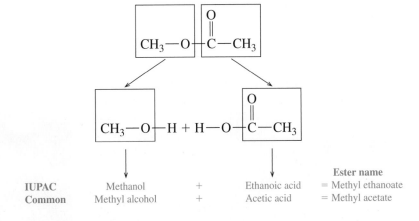

$$CH_3-CH_2-CH_2-\overset{\displaystyle O}{\overset{\|}{C}}-OH + H-O-CH_3 \underset{\xrightarrow{\hspace{1cm}}}{\overset{H^+, \text{ heat}}{\xleftarrow{\hspace{1cm}}}}$$

Butyric acid   Methyl alcohol

$$CH_3-CH_2-CH_2-\overset{\displaystyle O}{\overset{\|}{C}}-O-CH_3 + H_2O$$

Methyl butyrate

Study Check

What carboxylic acid and alcohol are needed to form the following ester, which gives the flavor and odor to apricots?

$$CH_3-CH_2-\overset{\displaystyle O}{\overset{\|}{C}}-O-CH_2-CH_2-CH_2-CH_2-CH_3$$

Figure 17.21 The ester methyl ethanoate (methyl acetate) is made from methyl alcohol and ethanoic acid (acetic acid).

Q *What change is made in the name of the carboxylic acid used to make the ester?*

Naming Esters

The name of an ester consists of two words taken from the names of the alcohol and the acid from which it was formed. The first word indicates the *alkyl* part of the alcohol. The second word is the *carboxylate* name of the carboxylic acid. The IUPAC names of esters use the IUPAC names for the alkyl group and the carboxylate ion, while the common names of esters use the common names of each.

Let's take a look at the following ester and break it into two parts, one from the alcohol and one from the acid. By writing and naming the alcohol and carboxylic acid that produced the ester, we can determine the name of the ester (Figure 17.21).

$$CH_3-O\,|\,\overset{\displaystyle O}{\overset{\|}{C}}-CH_3$$

$$CH_3-O\,|\,H + H-O\,|\,\overset{\displaystyle O}{\overset{\|}{C}}-CH_3$$

				Ester name
IUPAC	Methanol	+	Ethanoic acid	= Methyl ethanoate
Common	Methyl alcohol	+	Acetic acid	= Methyl acetate

Figure 17.22 Esters are responsible for part of the odor and flavor of oranges, bananas, pears, pineapples, and strawberries.

Q *What is the ester found in pineapple?*

Esters in Plants

Many of the fragrances of perfumes and flowers and the flavors of fruits are due to esters. Small esters are volatile, so we can smell them, and soluble in water, so we can taste them. (See Figure 17.22.) Several of these are listed in Table 17.13.

Table 17.13 Some Esters in Fruits and Flavorings

Condensed Structural Formula and Name	Flavor/Odor

$$CH_3-\overset{\overset{\textstyle O}{\|}}{C}-O-CH_2-CH_2-CH_3$$

Propyl ethanoate
(propyl acetate) — Pears

$$CH_3-\overset{\overset{\textstyle O}{\|}}{C}-O-CH_2-CH_2-CH_2-CH_2-CH_3$$

Pentyl ethanoate
(pentyl acetate) — Bananas

$$CH_3-\overset{\overset{\textstyle O}{\|}}{C}-O-CH_2-CH_2-CH_2-CH_2-CH_2-CH_2-CH_2-CH_3$$

Octyl ethanoate
(octyl acetate) — Oranges

$$CH_3-CH_2-CH_2-\overset{\overset{\textstyle O}{\|}}{C}-O-CH_2-CH_3$$

Ethyl butanoate
(ethyl butyrate) — Pineapples

$$CH_3-CH_2-CH_2-\overset{\overset{\textstyle O}{\|}}{C}-O-CH_2-CH_2-CH_2-CH_2-CH_3$$

Pentyl butanoate
(pentyl butyrate) — Apricots

CHEM NOTE

SALICYLIC ACID AND ASPIRIN

Chewing on a piece of willow bark was used as a way of relieving pain for many centuries. By the 1800s, chemists discovered that salicylic acid was the agent in the bark responsible for the relief of pain. However, salicylic acid, which has both a carboxylic group and a hydroxyl group, irritates the stomach lining. A less irritating ester of salicylic acid and acetic acid, called acetylsalicylic acid or "aspirin," was prepared in 1899 by the Bayer chemical company in Germany. In some aspirin preparations, a buffer is added to neutralize the carboxylic acid group and lessen its irritation of the stomach. Aspirin is used as an analgesic (pain reliever), antipyretic (fever reducer), and antiinflammatory agent.

Salicylic acid + Acetic acid →

Acetylsalicylic acid, "aspirin" + H₂O

Salicylic acid + Methyl alcohol →

Methyl salicylate (oil of wintergreen) + H₂O

Oil of wintergreen, or methyl salicylate, has a spearmint odor and flavor. Because it can pass through the skin, methyl salicylate is used in skin ointments where it acts as a counter-irritant, producing heat to soothe sore muscles.

Questions and Problems ▶ Carboxylic Acids and Esters

17.59 Give the IUPAC and common names (if any) for the following carboxylic acids:

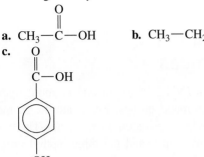

17.60 Give the IUPAC and common names (if any) for the following carboxylic acids:

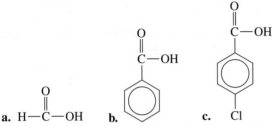

17.61 Draw the condensed structural formulas of each of the following carboxylic acids:
  a. propionic acid   **b.** benzoic acid
  c. 2-chloroethanoic acid   **d.** 3-hydroxypropanoic acid

17.62 Draw the condensed structural formulas of each of the following carboxylic acids:
  a. pentanoic acids   **b.** 3-ethylbenzoic acid
  c. 2-hydroxyacetic acid   **d.** 2,4-dibromobutanoic acid

17.63 Identify each of the following as an aldehyde, a ketone, a carboxylic acid, or an ester:

  a. CH$_3$—C(=O)—H   **b.** CH$_3$—C(=O)—O—CH$_3$

  c. CH$_3$—CH$_2$—C(=O)—CH$_3$

  d. CH$_3$—CH$_2$—C(=O)—O—H

17.64 Identify each of the following as an aldehyde, a ketone, a carboxylic acid, or an ester:

  a. CH$_3$—C(=O)—OH

  b. CH$_3$—C(=O)—O—CH$_2$—CH$_3$

  c. CH$_3$—CH$_2$—C(=O)—H

  d. CH$_3$—CH(CH$_3$)—C(=O)—O—CH$_2$—CH$_3$

17.65 Write the condensed structural formula of the ester formed when each of the following reacts with methyl alcohol:
  a. acetic acid   **b.** butyric acid

17.66 Write the condensed structural formula of the ester formed when each of the following reacts with methyl alcohol:
  a. formic acid   **b.** propionic acid

17.67 Draw the condensed structural formulas of the ester formed when each of the following carboxylic acids and alcohols react:
  a.
  CH$_3$—CH$_2$—C(=O)—OH + OH—CH$_2$—CH$_2$—CH$_3$ $\overset{H^+}{\rightleftharpoons}$
  b.
  CH$_3$—CH$_2$—CH$_2$—CH$_2$—C(=O)—OH + HO—CH(CH$_3$)—CH$_3$ $\overset{H^+}{\rightleftharpoons}$

17.68 Draw the condensed structural formula of the ester formed when each of the following carboxylic acids and alcohols react:

  a. CH$_3$—CH$_2$—C(=O)—OH + HO—CH$_3$ $\overset{H^+}{\rightleftharpoons}$

  b. C$_6$H$_5$—C(=O)—OH + HO—CH$_2$—CH$_2$—CH$_2$—CH$_3$ $\overset{H^+}{\rightleftharpoons}$

17.69 Name each of the following esters:

  a. CH$_3$—O—C(=O)—H

  b. CH$_3$—O—C(=O)—CH$_3$

  c. CH$_3$—O—C(=O)—CH$_2$—CH$_2$—CH$_3$

17.70 Name each of the following esters:

  a. CH$_3$—CH$_2$—O—C(=O)—CH$_2$—CH$_2$—CH$_3$

  b. CH$_3$—O—C(=O)—CH$_2$—CH$_2$—CH$_2$—CH$_2$—CH$_3$

  c. CH$_3$—O—C(=O)—CH$_2$—CH$_2$—CH$_3$

17.71 Draw the condensed structural formulas of each of the following esters:
a. methyl acetate
b. butyl formate
c. ethyl pentanoate
d. propyl propanoate

17.72 Draw the condensed structural formulas of each of the following esters:
a. hexyl acetate
b. propyl propionate
c. ethyl butanoate
d. methyl benzoate

Ammonia

Methylamine

Dimethylamine

Trimethylamine

Figure 17.23 Amines have one or more carbon atoms bonded to the N atom.

Q How many carbon atoms are bonded to the nitrogen atom in dimethylamine?

17.10 Amines and Amides

Amines are derivatives of ammonia (NH_3) in which one or more hydrogen atoms is replaced with alkyl or aromatic groups. For example, in methylamine, a methyl group replaces one hydrogen atom in ammonia. The bonding of two methyl groups gives dimethylamine, and the three methyl groups in trimethylamine replace all the hydrogen atoms in ammonia. (See Figure 17.23.)

Naming Amines

There are several systems in use for naming amines. For simple amines, the common names are often used. In the common name, the alkyl groups bonded to the nitrogen atom are listed in alphabetical order. The prefixes *di* and *tri* are used to indicate two and three identical substituents.

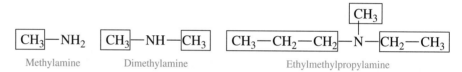

Methylamine   Dimethylamine   Ethylmethylpropylamine

Aromatic Amines

The aromatic amines use the name *aniline,* which is approved by IUPAC.

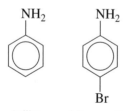

Aniline   4-Bromoaniline

Sample Problem 17.18 ▶ **Naming Amines**

Give the common name for each of the following amines:
a. $CH_3—CH_2—NH_2$

b. $CH_3—\overset{\overset{\displaystyle CH_3}{|}}{N}—CH_3$

c.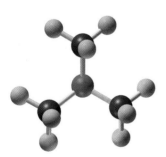

Solution

a. This amine has one ethyl group attached to the nitrogen atom; its name is *ethylamine.*

b. This amine has three methyl groups attached to the nitrogen atom; its name is *trimethylamine.*

c. This aromatic amine is *aniline.*

Study Check

Draw the structure of ethylpropylamine.

CHEM NOTE

AMINES IN HEALTH AND MEDICINE

In response to allergic reactions or injury to cells, the body increases the production of histamine, which causes blood vessels to dilate and increases the permeability of the cells.

Redness and swelling occur in the area. Administering an antihistamine such as diphenylhydramine helps block the effects of histamine.

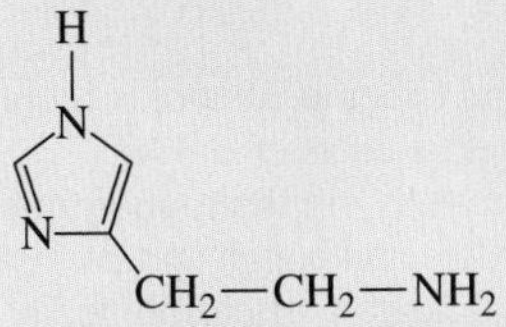

Histamine

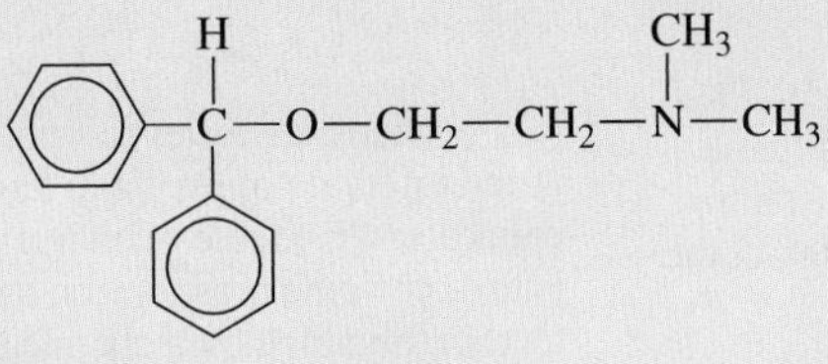

Diphenylhydramine

In the body, hormones called biogenic amines carry messages between the central nervous system and nerve cells. Epinephrine (adrenaline) and norepinephrine (noradrenaline) are released by the adrenal medulla in "fight or flight" situations to raise the blood glucose level and move the blood to the muscles. Used in remedies for colds, hay fever, and asthma, the

norepinephrine contracts the capillaries in the mucous membranes of the respiratory passages. The prefix *nor* in a drug name means there is one less CH_3— group on the nitrogen atom. Parkinson's disease is a result of a deficiency in another biogenic amine called dopamine.

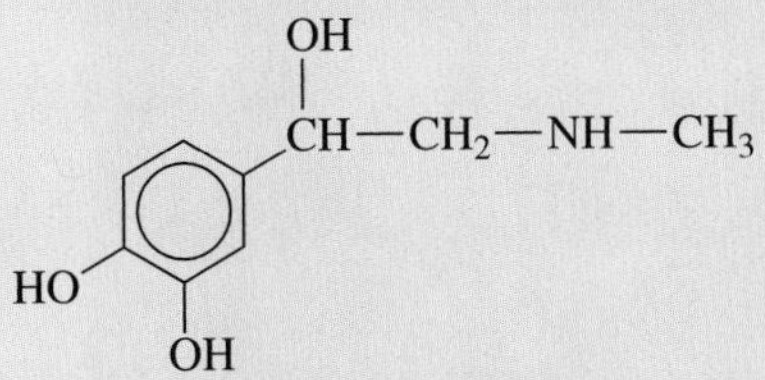

Epinephrine (adrenaline)

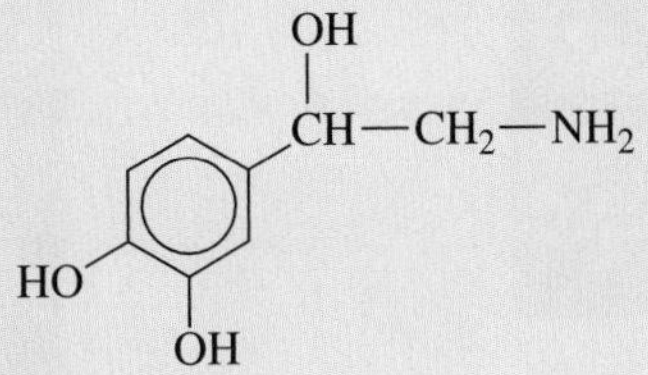

Norepinephrine (noradrenaline)

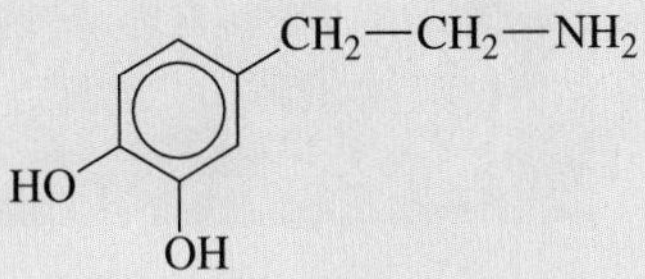

Dopamine

Produced synthetically, amphetamines (known as "uppers") are stimulants of the central nervous system much like epinephrine, but they also increase cardiovascular activity and depress the appetite. They are sometimes used to bring about weight loss, but they can cause chemical dependency. Benzedrine and Neo-Synephrine (phenylephrine) are used in medications to

reduce respiratory congestion from colds, hay fever, and asthma. Sometimes, Benzedrine is taken internally to combat the desire to sleep, but it has side effects. Methedrine is used to treat depression and in the illegal form is known as "speed" or "crank." The prefix *meth* means that there is one more methyl group on the nitrogen atom.

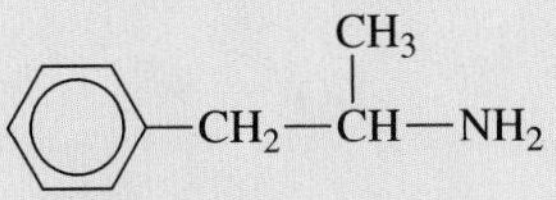

Benzedrine (amphetamine)

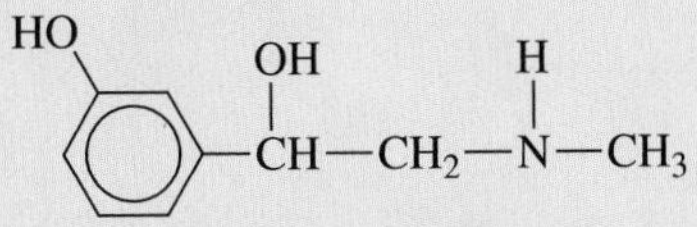

Neo-Synephrine (phenylephrine)

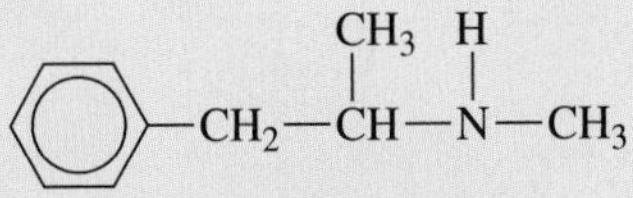

Methamphetamine (methedrine)

ALKALOIDS: AMINES IN PLANTS

Alkaloids are physiologically active nitrogen-containing compounds produced by plants. The term *alkaloid* refers to the "alkali-like" or basic characteristics of amines. Certain alkaloids are used in anesthetics, in antidepressants, and as stimulants, although many are habit forming.

As a stimulant, nicotine increases the level of adrenaline in the blood, which increases the heart rate and blood pressure. It is well known that smoking cigarettes can damage the lungs and that exposure to tars and other carcinogens in cigarette smoke can lead to lung cancer. However, nicotine is responsible for the addiction of smoking. Coniine, which is obtained from hemlock, is an extremely toxic alkaloid.

Nicotine

Coniine

Caffeine is a stimulant of the central nervous system. Present in coffee, tea, soft drinks, chocolate, and cocoa, caffeine increases alertness, but may cause nervousness and insomnia. Caffeine is also used in certain pain relievers to counteract the drowsiness caused by an antihistamine.

Caffeine

Several alkaloids are used in medicine. Quinine obtained from the bark of the cinchona tree has been used in the treatment of malaria since the 1600s. Atropine from belladonna is used in low concentrations to accelerate slow heart rates and as an anesthetic for eye examinations.

Quinine

Atropine

For many centuries morphine and codeine, alkaloids found in the oriental poppy plant, have been used as effective painkillers. Codeine, which is structurally similar to morphine, is used in some prescription painkillers and cough syrups. Heroin, obtained by a chemical modification of morphine, is strongly addicting and is not used medically.

Morphine

Codeine

Amides

CASE STUDY
Death by Chocolate

The *amides* are derivatives of carboxylic acids in which a nitrogen group replaces the hydroxyl group. (See Figure 17.24.) An amide is produced when a carboxylic acid reacts with ammonia or an amine. A molecule of water is eliminated, and the fragments of the carboxylic acid and amine molecules join to form the amide, much like the formation of esters.

$$CH_3-CH_2-\overset{\overset{\displaystyle O}{\|}}{C}-OH + H-\overset{\overset{\displaystyle H}{|}}{N}-H \xrightarrow{\text{Heat}} CH_3-CH_2-\boxed{\overset{\overset{\displaystyle O}{\|}}{C}-\overset{\overset{\displaystyle H}{|}}{N}}-H + H_2O$$

Propanoic acid   Ammonia   Propanamide
(propionic acid)   (propionamide)

$$CH_3-CH_2-\overset{\overset{\displaystyle O}{\|}}{C}-OH + H-\overset{\overset{\displaystyle H}{|}}{N}-CH_3 \xrightarrow{\text{Heat}} CH_3-CH_2-\boxed{\overset{\overset{\displaystyle O}{\|}}{C}-\overset{\overset{\displaystyle H}{|}}{N}}-CH_3 + H_2O$$

Propanoic acid   Methylamine   *N*-Methylpropanamide
(propionic acid)   (*N*-methylpropionamide)

Carboxylic acid

Ethanoic acid
(Acetic acid)

Amide

Ethanamide
(Acetamide)

Figure 17.24 Amides are derivatives of carboxylic acids in which an amino group replaces the hydroxyl group (—OH).

Q What is the amide of pentanoic acid?

Sample Problem 17.19   **Formation of Amides**

Give the structural formula of the amide product in each of the following reactions:

a.
$$\text{(benzene ring)}-\overset{\overset{\displaystyle O}{\|}}{C}-OH + NH_3 \xrightarrow{\text{Heat}}$$

b. $CH_3-\overset{\overset{\displaystyle O}{\|}}{C}-OH + NH_2-CH_2-CH_3 \xrightarrow{\text{Heat}}$

Solution

a. The structural formula of the amide product can be written by attaching the carbonyl group from the acid to the nitrogen atom of the amine. —OH is removed from the acid and —H from the amine to form water.

$$\text{(benzene ring)}-\overset{\overset{\displaystyle O}{\|}}{C}-NH_2$$

b. $CH_3-\overset{\overset{\displaystyle O}{\|}}{C}-\overset{\overset{\displaystyle H}{|}}{N}-CH_2-CH_3$

Study Check

What are the condensed structural formulas of the carboxylic acid and amine needed to prepare the following amide?

$$H-\overset{\overset{\displaystyle O}{\|}}{C}-\overset{\overset{\displaystyle CH_3}{|}}{N}-CH_3$$

Naming Simple Amides

In both the common and IUPAC names, simple amides are named by dropping the *ic acid* or *oic acid* from the carboxylic acid names and adding the suffix *amide*.

$$H-\underset{\underset{\displaystyle }{\|}}{\overset{\overset{\displaystyle O}{\|}}{C}}-NH_2$$

Methanamide
(formamide)

$$CH_3-\underset{\underset{\displaystyle }{\|}}{\overset{\overset{\displaystyle O}{\|}}{C}}-NH_2$$

Ethanamide
(acetamide)

$$CH_3-CH_2-CH_2-\underset{\underset{\displaystyle }{\|}}{\overset{\overset{\displaystyle O}{\|}}{C}}-NH_2$$

Butanamide
(butyramide)

Benzamide

Sample Problem 17.20 | Naming Amides

Give the common and IUPAC names for

$$CH_3-CH_2-\underset{\underset{\displaystyle }{\|}}{\overset{\overset{\displaystyle O}{\|}}{C}}-NH_2$$

Solution

The IUPAC name of the carboxylic acid is propanoic acid; the common name is propionic acid. Replacing the *oic acid* or *ic acid* ending with *amide* gives the IUPAC name of *propanamide* and common name of *propionamide*.

Study Check

Draw the condensed structural formula of benzamide.

Questions and Problems | Amines and Amides

17.73 Write the common names for each of the following:
a. $CH_3-CH_2-NH_2$
b. $CH_3-NH-CH_2-CH_2-CH_3$
c. $CH_3-CH_2-\underset{\underset{\displaystyle CH_3}{|}}{N}-CH_2-CH_3$

17.74 Write the common names for each of the following:
a. $CH_3-CH_2-CH_2-NH_2$
b. $CH_3-NH-CH_2-CH_3$
c. $CH_3-CH_2-CH_2-CH_2-NH_2$

17.75 Give the IUPAC and common names (if any) for each of the following amides:
a. $CH_3-\underset{\underset{\displaystyle }{\|}}{\overset{\overset{\displaystyle O}{\|}}{C}}-NH_2$

b. $CH_3-CH_2-CH_2-\underset{\underset{\displaystyle }{\|}}{\overset{\overset{\displaystyle O}{\|}}{C}}-NH_2$

c. $H-\underset{\underset{\displaystyle }{\|}}{\overset{\overset{\displaystyle O}{\|}}{C}}-NH_2$

17.76 Give the IUPAC and common names (if any) for each of the following amides:
a. $CH_3-CH_2-\underset{\underset{\displaystyle }{\|}}{\overset{\overset{\displaystyle O}{\|}}{C}}-NH_2$

b. $CH_3-CH_2-CH_2-CH_2-CH_2-\underset{\underset{\displaystyle }{\|}}{\overset{\overset{\displaystyle O}{\|}}{C}}-NH_2$

c. $\underset{\underset{\displaystyle }{\|}}{\overset{\overset{\displaystyle O}{\|}}{C}}-NH_2$

17.77 Draw the condensed structural formulas for each of the following amides:
a. propionamide
b. 2-methylpentanamide
c. methanamide

17.78 Draw the condensed structural formulas for each of the following amides:
a. formamide
b. benzamide
c. 3-methylbutyramide

Concept Map ▶ **Organic Chemistry**

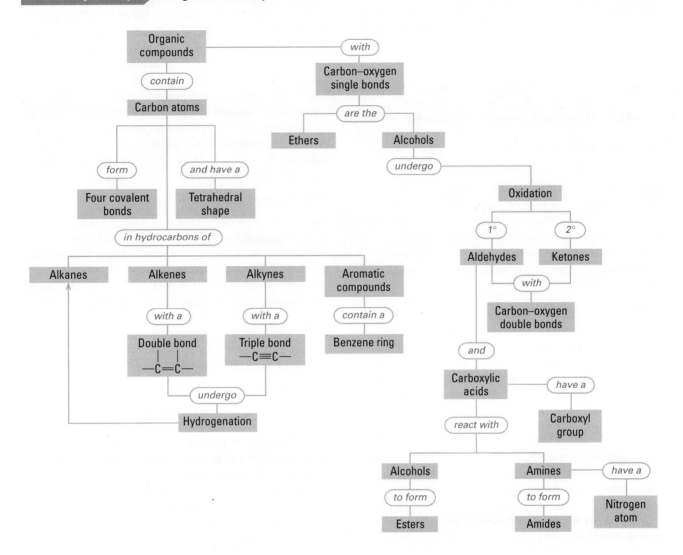

Chapter Review

17.1 Organic Compounds

Most organic compounds have covalent bonds and form nonpolar molecules. Often they have low melting points and low boiling points, are not very soluble in water, and burn vigorously in air. In contrast, many inorganic compounds are ionic or contain polar covalent bonds, have high melting and boiling points, are usually soluble in water, produce ions in water, and do not burn in air.

Carbon atoms share 4 valence electrons to form four covalent bonds. In the simplest organic molecule, methane, CH_4, the four bonds that bond hydrogen to the carbon atom are directed out to the corners of a tetrahedron.

17.2 Alkanes

Alkanes are hydrocarbons that have only C—C single bonds. In the expanded structural formula, a separate line is drawn for every bonded atom. A condensed structural formula depicts groups composed of each carbon atom and its attached hydrogen atoms. The IUPAC name indicates the number of carbon atoms. For a continuous alkane, the carbon atoms are connected in a chain and bonded to hydrogen atoms. Substituents such as alkyl groups and halogen atoms (named as fluoro, chloro, bromo, or iodo) can replace hydrogen atoms on the main chain. As nonpolar molecules, alkanes are not soluble in water. They are less dense than water. In combustion, alkanes react with oxygen to produce carbon dioxide and water.

17.3 Functional Groups

An organic molecule contains a characteristic group of atoms called a functional group that determines the molecule's family name and chemical reactivity. Functional groups are used to classify organic compounds, act as reactive sites in the molecule, and provide a system of naming for organic compounds. Some common functional groups include the hydroxyl group ($-OH$) in alcohols, the carbonyl group ($C=O$) in aldehydes and ketones, and a nitrogen atom $\left(\underset{|}{-N-} \right)$ in amines.

17.4 Alkenes and Alkynes

Alkenes are unsaturated hydrocarbons that contain carbon–carbon double bonds ($C=C$). Alkynes contain a triple bond ($C\equiv C$). The IUPAC names of alkenes end with *ene*, while alkyne names end with *yne*. The main chain is numbered from the end nearest the double or triple bond. Hydrogenation adds hydrogen atoms to double or triple bonds to yield an alkane.

17.5 Polymers

Polymers are long-chain molecules that consist of many repeating units of smaller carbon molecules called monomers. Many materials that we use every day, including carpeting, plastic wrap, nonstick pans, and nylon, are made by addition reactions in which a catalyst links the carbon atoms from various kinds of alkene molecules.

17.6 Aromatic Compounds

Most aromatic compounds contain benzene, C_6H_6, a cyclic structure represented as a hexagon with a circle in the center. Many aromatic compounds use the parent name benzene, although common names such as toluene are retained. The benzene ring is numbered and the branches are listed in alphabetical order.

17.7 Alcohols and Ethers

The functional group of an alcohol is the hydroxyl group $-OH$ bonded to a carbon chain. In a phenol, the hydroxyl group is bonded to an aromatic ring. In the IUPAC system, the names of alcohols have *ol* endings, and the location of the $-OH$ group is given by numbering the carbon chain. Simple alcohols are generally named by their common names with the alkyl name preceding the term *alcohol.* Alcohols are classified according to the number of alkyl or aromatic groups bonded to the carbon that holds the $-OH$ group. In a primary (1°) alcohol one alkyl group is attached to the hydroxyl carbon. In a secondary (2°) alcohol two alkyl groups are attached, and in a tertiary (3°) alcohol there are three alkyl groups bonded to the hydroxyl carbon.

In an ether, an oxygen atom is connected by single bonds to two alkyl or aromatic groups, $C-O-C$. In the common names of ethers, the alkyl groups are listed alphabetically followed by the name *ether.* Primary alcohols are oxidized to aldehydes. Secondary alcohols are oxidized to ketones. Tertiary alcohols do not oxidize.

17.8 Aldehydes and Ketones

Aldehydes and ketones contain a carbonyl group ($C=O$), which consists of a double bond between a carbon and an oxygen atom. In aldehydes, the carbonyl group appears at the end of carbon chains. In ketones, the carbonyl group occurs between two carbon groups. In the IUPAC system, the *e* in the corresponding alkane is replaced with *al* for aldehydes and *one* for ketones. For ketones with more than four carbon atoms in the main chain, the carbonyl group is numbered to show its location. Many of the simple aldehydes and ketones use common names.

17.9 Carboxylic Acids and Esters

A carboxylic acid contains the carboxyl functional group, which is a hydroxyl group connected to the carbonyl group. In the presence of a strong acid, a carboxylic acid reacts with an alcohol to produce an ester. A molecule of water is removed: $-OH$ from the carboxylic acid and $-H$ from the alcohol molecule. The names of esters consist of two words, one from the alcohol and the other from the carboxylic acid with the *ic* ending replaced by *ate.*

17.10 Amines and Amides

A nitrogen atom attached to alkyl or aromatic groups forms an amine. In the common names of simple amines, the alkyl groups are listed alphabetically followed by the suffix *amine.* Amides are derivatives of carboxylic acids in which the hydroxyl group is replaced by $-NH_2$. Amides are named by replacing the *ic acid* or *oic acid* with *amide.*

Summary of Naming

Structure	Family	IUPAC Name	Common Name
$CH_3-CH_2-CH_3$	Alkane	Propane	
$CH_3-\overset{\displaystyle CH_3}{\underset{\displaystyle \vert}{CH}}-CH_3$		Methylpropane	
$CH_3-CH_2-CH_2-Cl$	Haloalkane	1-Chloropropane	Propyl chloride
$CH_3-CH=CH_2$	Alkene	Propene	Propylene
$CH_3C\equiv CH$	Alkyne	Propyne	
(benzene ring)	Aromatic	Benzene	
(toluene ring, CH_3)		Methylbenzene or toluene	
CH_3-OH	Alcohol	Methanol	Methyl alcohol
(benzene ring)$-OH$	Phenol	Phenol	Phenol
CH_3-O-CH_3	Ether		Dimethyl ether
$H-\overset{\displaystyle O}{\overset{\displaystyle \|}{C}}-H$	Aldehyde	Methanal	Formaldehyde
$CH_3-\overset{\displaystyle O}{\overset{\displaystyle \|}{C}}-CH_3$	Ketone	Propanone	Acetone; Dimethyl ketone
$CH_3-\overset{\displaystyle O}{\overset{\displaystyle \|}{C}}-OH$	Carboxylic acid	Ethanoic acid	Acetic acid
$CH_3-\overset{\displaystyle O}{\overset{\displaystyle \|}{C}}-OCH_3$	Ester	Methyl ethanoate	Methyl acetate
$CH_3-CH_2-NH_2$	Amine		Ethylamine
$CH_3-\overset{\displaystyle O}{\overset{\displaystyle \|}{C}}-NH_2$	Amide	Ethanamide	Acetamide

Summary of Reactions

Combustion

$$CH_4 + 2O_2 \longrightarrow CO_2 + 2H_2O$$

Hydrogenation

$$CH_2{=}CH{-}CH_3 + H_2 \xrightarrow{Pt} CH_3{-}CH_2{-}CH_3$$

$$CH_3{-}C{\equiv}CH + 2H_2 \xrightarrow{Pt} CH_3{-}CH_2{-}CH_3$$

Combustion of Alcohols

$$CH_3{-}CH_2{-}OH + 3OH_2 \longrightarrow 2CO_2 + 3H_2O$$

Ethanol   Oxygen   Carbon dioxide Water

Oxidation of Primary Alcohols to Form Aldehydes

$$\underset{\text{Ethanol}}{CH_3{-}\overset{\displaystyle OH}{\underset{\displaystyle |}{CH_2}}} \xrightarrow{[O]} \underset{\text{Acetaldehyde}}{CH_3{-}\overset{\displaystyle O}{\overset{\displaystyle \|}{C}}{-}H} + H_2O$$

Oxidation of Secondary Alcohols to Form Ketones

$$\underset{\text{2-Propanol}}{CH_3{-}\overset{\displaystyle OH}{\underset{\displaystyle |}{CH}}{-}CH_3} \xrightarrow{[O]} \underset{\text{Propanone}}{CH_3{-}\overset{\displaystyle O}{\overset{\displaystyle \|}{C}}{-}CH_3} + H_2O$$

Esterification: Carboxylic Acid and an Alcohol

$$\underset{\substack{\text{Ethanoic acid} \\ \text{(acetic acid)}}}{CH_3{-}\overset{\displaystyle O}{\overset{\displaystyle \|}{C}}{-}OH} + \underset{\substack{\text{Methanol} \\ \text{(methyl alcohol)}}}{HO{-}CH_3} \underset{\longleftarrow}{\overset{H^+}{\longrightarrow}}$$

$$\underset{\substack{\text{Methyl ethanoate} \\ \text{(methyl acetate)}}}{CH_3{-}\overset{\displaystyle O}{\overset{\displaystyle \|}{C}}{-}O{-}CH_3} + H_2O$$

Formation of Amides

$$\underset{\substack{\text{Propanoic acid} \\ \text{(propionic acid)}}}{CH_3{-}CH_2{-}\overset{\displaystyle O}{\overset{\displaystyle \|}{C}}{-}OH} + \underset{\text{Ammonia}}{H{-}\overset{\displaystyle H}{\underset{\displaystyle |}{N}}{-}H} \xrightarrow{\text{Heat}}$$

$$\underset{\substack{\text{Propanamide} \\ \text{(propionamide)}}}{CH_3{-}CH_2{-}\overset{\displaystyle O}{\overset{\displaystyle \|}{C}}{-}\overset{\displaystyle H}{\underset{\displaystyle |}{N}}{-}H} + H_2O$$

Key Terms

alcohols A class of organic compounds that contains the hydroxyl ($-$OH) group bonded to a carbon atom.

aldehydes A class of organic compounds that contains a carbonyl group (C$=$O) bonded to at least one hydrogen atom.

alkanes Hydrocarbons containing only single bonds between carbon atoms.

alkenes Hydrocarbons that contain carbon–carbon double bonds (C$=$C).

alkyl group An alkane minus one hydrogen atom. Alkyl groups are named like the alkanes except a *yl* ending replaces *ane*.

alkynes Hydrocarbons that contain carbon–carbon triple bonds (C$\equiv$C).

amides Organic compounds in which a carbon containing the carbonyl group is attached to a nitrogen atom.

amines A class of organic compounds that contains a nitrogen atom bonded to one or more carbon atoms.

aromatic compounds Compounds that contain the ring structure of benzene.

benzene A ring of six carbon atoms each of which is attached to one hydrogen atom, C_6H_6.

branch A carbon group or halogen bonded to the main carbon chain.

branched alkane A hydrocarbon containing a hydrocarbon substituent bonded to the main chain.

carbonyl group A functional group that contains a double bond between a carbon atom and an oxygen atom (C$=$O).

carboxyl group A functional group found in carboxylic acids composed of carbonyl and hydroxyl groups.

$$\overset{\displaystyle O}{\overset{\displaystyle \|}{-}C{-}OH} \quad \text{Carboxyl group}$$

carboxylic acids A class of organic compounds that contains the functional group $-$COOH.

combustion A chemical reaction in which an alkane or alcohol reacts with oxygen to produce CO_2, H_2O, and energy.

condensed structural formula A structural formula that shows the arrangement of the carbon atoms in a molecule but groups each carbon atom with its bonded hydrogen atoms (CH_3, CH_2, or CH).

esterification The formation of an ester from a carboxylic acid and an alcohol with the elimination of a molecule of water in the presence of an acid catalyst.

esters A class of organic compounds that contains a $-$COO$-$ group with an oxygen atom bonded to carbon.

ethers A class of organic compounds that contains an oxygen atom bonded to two carbon atoms.

expanded structural formula A type of structural formula that shows the arrangement of the atoms by drawing each bond in the hydrocarbon as C—H or C—C.

functional group A group of atoms that determine the physical and chemical properties and naming of a class of organic compounds.

hydrocarbons Organic compounds consisting of only carbon and hydrogen.

hydrogenation The addition of hydrogen (H_2) to the double bond of alkenes or alkynes to yield alkanes.

hydroxyl group The group of atoms (—OH) characteristic of alcohols.

isomers Organic compounds in which identical molecular formulas have different arrangements of atoms.

IUPAC system A system for naming organic compounds determined by the International Union of Pure and Applied Chemistry.

ketones A class of organic compounds in which a carbonyl group is bonded to two carbon atoms.

monomer The small organic molecule that is repeated many times in a polymer.

organic compounds Compounds made of carbon that typically have covalent bonds, nonpolar molecules, low melting and boiling points, are insoluble in water, and flammable.

phenol An organic compound that has an —OH group attached to a benzene ring.

polymer A very large molecule that is composed of many small, repeating structural units that are identical.

primary (1°) alcohol An alcohol that has one alkyl group bonded to the alcohol carbon atom.

secondary (2°) alcohol An alcohol that has two alkyl groups bonded to the carbon atom with the —OH group.

substituent Groups of atoms such as an alkyl group or a halogen bonded to the main chain or ring of carbon atoms.

tertiary (3°) alcohol An alcohol that has three alkyl groups bonded to the carbon atom with the —OH.

unsaturated hydrocarbons A compound of carbon and hydrogen in which the carbon chain contains at least one double (alkene) or triple carbon–carbon bond (alkyne). An unsaturated compound is capable of an addition reaction with hydrogen, which converts the double or triple bonds to single carbon–carbon bonds.

▶ Understanding the Concepts

17.79 Match the following physical and chemical properties with the compounds butane, C_4H_{10}, used in lighters, or potassium chloride, KCl, in salt substitutes:

- **a.** melts at $-138°C$
- **b.** burns vigorously in air
- **c.** melts at $770°C$
- **d.** produces ions in water
- **e.** is a gas at room temperature

17.80 Match the following physical and chemical properties with the compounds cyclohexane, C_6H_{12}, or calcium nitrate, $Ca(NO_3)_2$:
- **a.** contains only covalent bonds
- **b.** melts above $500°C$
- **c.** insoluble in water
- **d.** liquid at room temperature
- **e.** produces ions in water

17.81 Classify the following as alkenes, alkynes, alcohols, ethers, aldehydes, ketones, carboxylic acids, esters, or amines:

$$\textbf{a. } CH_3-CH_2-\overset{\overset{\displaystyle O}{\|}}{C}-OH$$

b. $CH_3-CH=CH-CH_3$

$$\textbf{c. } CH_3-\overset{\overset{\displaystyle O}{\|}}{C}-O-CH_3$$

d. $CH_3-CH_2-NH-CH_3$

$$\textbf{e. } CH_3-CH_2-\overset{\overset{\displaystyle O}{\|}}{C}-H$$

17.82 Classify the following as alkenes, alkynes, alcohols, ethers, aldehydes, ketones, carboxylic acids, esters, or amines:
- **a.** CH_3-NH_2

$$\textbf{b. } CH_3-\overset{\overset{\displaystyle O}{\|}}{C}-CH_3$$

$$\textbf{c. } CH_3-\overset{\overset{\displaystyle OH}{|}}{CH}-CH_3$$

- **d.** $CH_3-C\equiv CH$
- **e.** $CH_3-O-CH_2-CH_3$

17.83 Match each of the following terms with the corresponding description:
alkane, alkene, alkyne, alcohol, ether, aldehyde, ketone, carboxylic acid, ester, amine, functional group, tetrahedral

a. an organic compound that contains a hydroxyl group bonded to a carbon

b. a hydrocarbon that contains one or more carbon–carbon double bonds

c. an organic compound in which the carbon of a carbonyl group is bonded to a hydrogen

d. a hydrocarbon that contains only carbon–carbon single bonds

e. an organic compound in which the carbon of a carbonyl group is bonded to a hydroxyl group

f. an organic compound that contains a nitrogen atom bonded to one or more carbon atoms

17.84 Match each of the following terms with the corresponding description:
alkane, alkene, alkyne, alcohol, ether, aldehyde, ketone, carboxylic acid, ester, amine, functional group, tetrahedral

a. the three-dimensional shape of a carbon bonded to four hydrogen atoms

b. an organic compound in which the hydrogen atom of a carboxyl group is replaced by a carbon atom

c. an organic compound that contains an oxygen atom bonded to two carbon atoms

d. a hydrocarbon that contains a carbon–carbon triple bond

e. a charactistic group of atoms that makes compounds behave and react in a particular way

f. an organic compound in which the carbonyl group is bonded to two carbon atoms

17.85 Draw a part of the polymer (use four monomers) of Teflon made from 1,1,2,2-tetrafluoroethene.

17.86 A garden hose is made of polyvinylchloride (PVC) from chloroethene (vinyl chloride). Draw a part of the polymer (use four monomers) for PVC.

17.87 Identify the functional groups in each of the following:

a.

b. CH=CH—CH

c. CH₃—C—C—CH₃

17.88 Identify the functional groups in each of the following:

a. BHA, an antioxidant used as a preservative in foods such as baked goods, butter, meats, and snack foods

b. vanillin, a flavoring, obtained from the seeds of the vanilla bean

17.89 The sweetener aspartame is made from two amino acids: aspartic acid and phenylalanine. Identify the functional groups in aspartame.

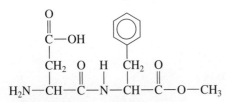

17.90 Some aspirin substitutes contain phenacetin to reduce fever. Identify the functional groups in phenacetin.

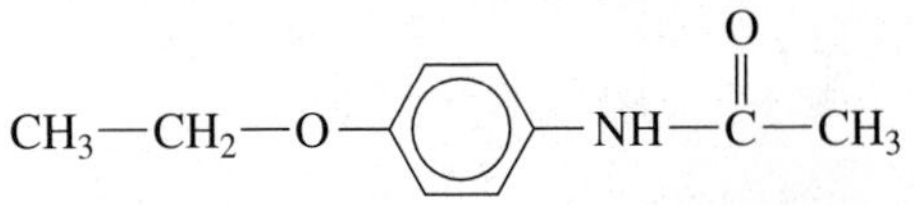

Additional Questions and Problems

17.91 Write the condensed structural formula for each of the following molecules:
a. 3-ethylhexane
b. 2, 3-dimethylpentane
c. 1, 3-dichloro-3-methylheptane

17.92 Give the IUPAC names for each of the following molecules:

a. 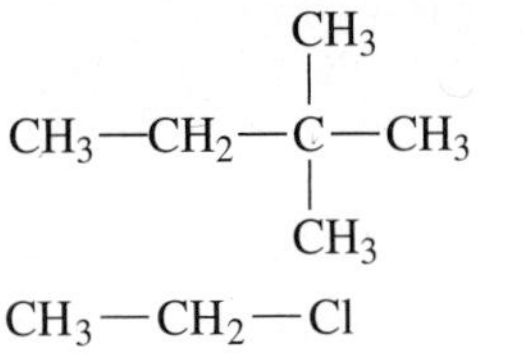 $CH_3-CH_2-\underset{\underset{CH_3}{|}}{\overset{\overset{CH_3}{|}}{C}}-CH_3$

b. CH_3-CH_2-Cl

c. 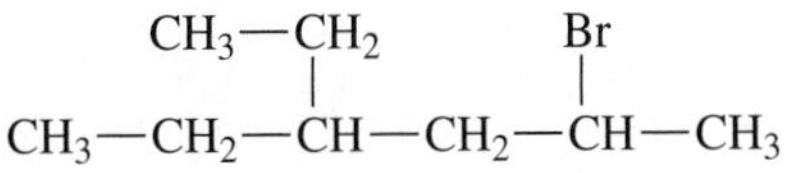 $CH_3-CH_2-\underset{\underset{CH_3}{|}}{\overset{\overset{CH_3-CH_2}{|}}{CH}}-CH_2-\overset{\overset{Br}{|}}{CH}-CH_3$

17.93 Give the IUPAC name for each of the following compounds:

a. 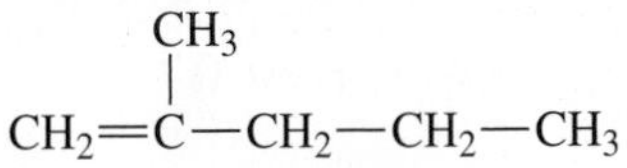 $CH_2{=}\overset{\overset{CH_3}{|}}{C}-CH_2-CH_2-CH_3$

b. $CH_3-CH_2-C{\equiv}CH$
c. $CH_3-CH{=}CH-CH_2-CH_3$

17.94 Write the condensed structures of each of the following compounds:
a. 2-pentyne   **b.** 2-heptene
c. 3-hexene

17.95 Name each of the following aromatic compounds:

a. 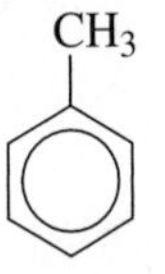   CH₃ (benzene ring)

b.   CH₃, Cl (benzene ring)

c. 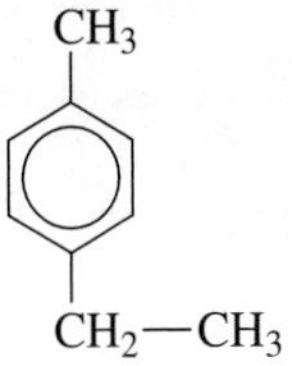   CH₃ (benzene ring), CH₂—CH₃

17.96 Write the structural formulas for each of the following:
a. ethylbenzene
b. 1, 3-dichlorobenzene
c. 1, 2, 4-trimethylbenzene

17.97 Classify each of the following as primary, secondary, or tertiary alcohols:

a. 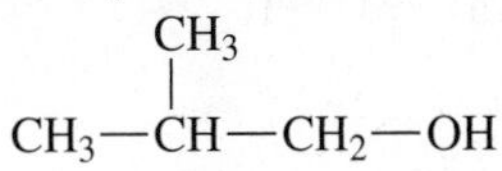 $CH_3-\overset{\overset{CH_3}{|}}{CH}-CH_2-OH$

b. 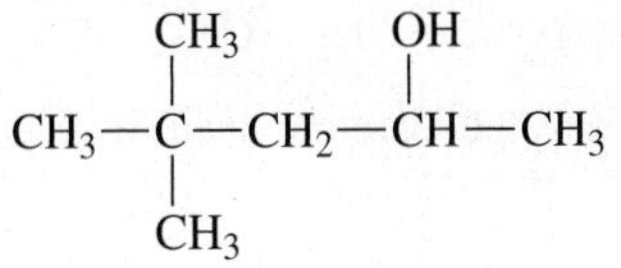 $CH_3-\underset{\underset{CH_3}{|}}{\overset{\overset{CH_3}{|}}{C}}-CH_2-\overset{\overset{OH}{|}}{CH}-CH_3$

c. $HO-CH_2-CH_2-CH_3$

17.98 Classify each of the following as primary, secondary, or tertiary alcohols:

a. 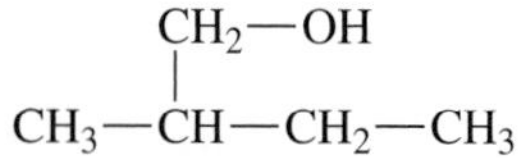 $CH_3-\overset{\overset{\displaystyle CH_2-OH}{|}}{CH}-CH_2-CH_3$

b. 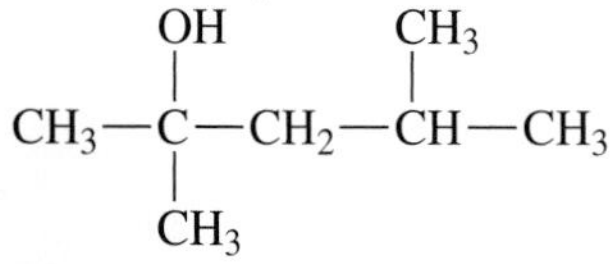 $CH_3-\overset{\overset{\displaystyle OH}{|}}{\underset{\underset{\displaystyle CH_3}{|}}{C}}-CH_2-\overset{\overset{\displaystyle CH_3}{|}}{CH}-CH_3$

c. $CH_3-CH_2-CH_2-CH_2-OH$

17.99 Draw the condensed structural formula of each of the following compounds:
a. 4-chlorophenol   b. 2-methyl-3-pentanol
c. 3-pentanone

17.100 Draw the condensed structural formula of each of the following compounds:
a. 3-pentanol   b. 2-pentanol
c. methyl propyl ether

17.101 Draw the condensed structural formula for the product of each of the following reactions:

a. $CH_3-CH_2-CH_2-OH \xrightarrow{[O]}$

b. 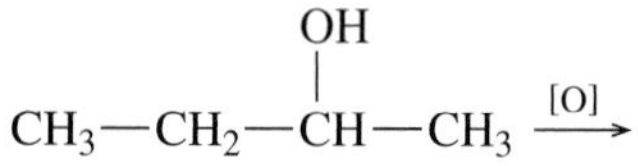 $CH_3-CH_2-\overset{\overset{\displaystyle OH}{|}}{CH}-CH_3 \xrightarrow{[O]}$

17.102 Draw the condensed structural formula for the product of each of the following reactions:

a. 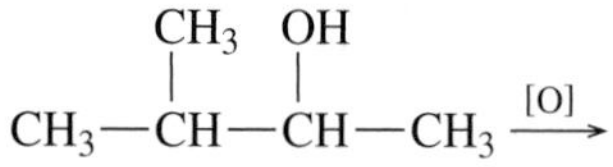 $CH_3-\overset{\overset{\displaystyle CH_3}{|}}{CH}-\overset{\overset{\displaystyle OH}{|}}{CH}-CH_3 \xrightarrow{[O]}$

b. 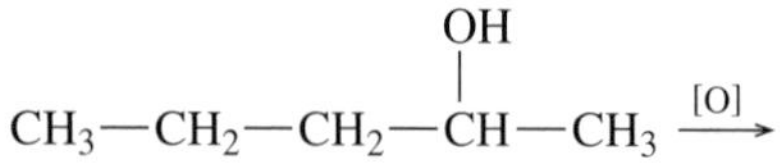 $CH_3-CH_2-CH_2-\overset{\overset{\displaystyle OH}{|}}{CH}-CH_3 \xrightarrow{[O]}$

17.103 Give the IUPAC and common names (if any) for each of the following compounds:

a.

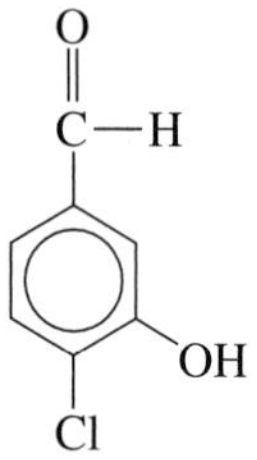

b. 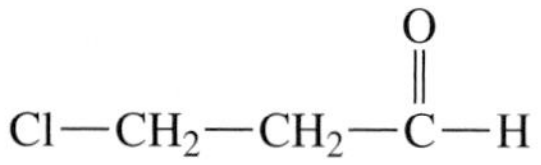 $Cl-CH_2-CH_2-\overset{\overset{\displaystyle O}{||}}{C}-H$

c. $CH_3-\overset{\overset{\displaystyle Cl}{|}}{CH}-\overset{\overset{\displaystyle O}{||}}{C}-CH_2-CH_3$

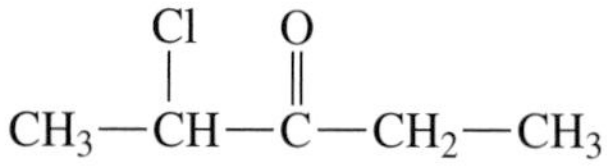

17.104 Give the IUPAC and common names (if any) for each of the following compounds:

a. 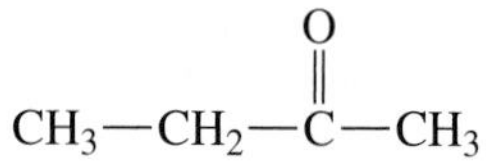 $CH_3-CH_2-\overset{\overset{\displaystyle O}{||}}{C}-CH_3$

b.

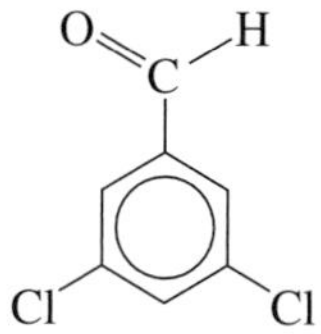

c. 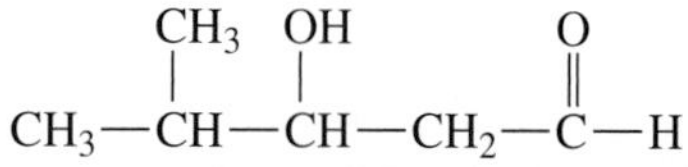 $CH_3-\overset{\overset{\displaystyle CH_3}{|}}{CH}-\overset{\overset{\displaystyle OH}{|}}{CH}-CH_2-\overset{\overset{\displaystyle O}{||}}{C}-H$

17.105 Draw the condensed structural formulas of each of the following:
a. 4-chlorobenzaldehyde
b. 3-chloropropionaldehyde
c. ethyl methyl ketone

17.106 Draw the condensed structural formulas of each of the following:
a. propionaldehyde   b. 2-chlorobutanal
c. 3,5-dimethylhexanal

17.107 Draw the structural formula of the aldehyde or ketone formed when each of the following is oxidized:
a. $CH_3-CH_2-CH_2-OH$

b. 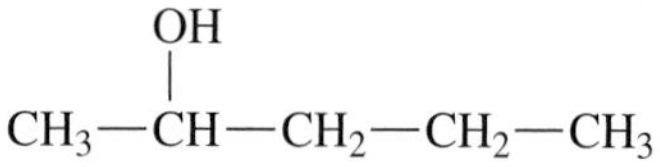 $CH_3-\overset{\overset{\displaystyle OH}{|}}{CH}-CH_2-CH_2-CH_3$

c. $CH_3-CH_2-CH_2-CH_2-OH$

17.108 Draw the structural formula of the aldehyde or ketone formed when each of the following is oxidized:

a. 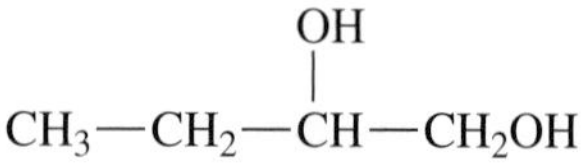 $CH_3-CH_2-\overset{\overset{\displaystyle OH}{|}}{CH}-CH_2OH$

b. 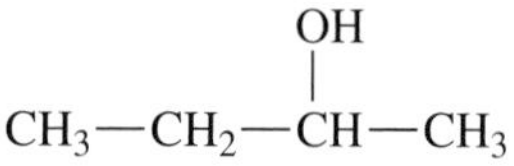 $CH_3-CH_2-\overset{\overset{\displaystyle OH}{|}}{CH}-CH_3$

c. 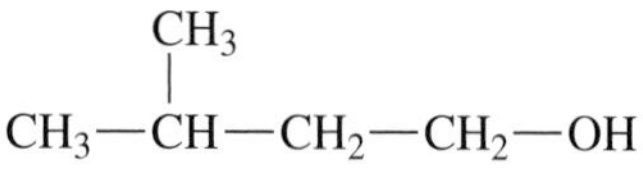 $CH_3-\overset{\overset{\displaystyle CH_3}{|}}{CH}-CH_2-CH_2-OH$

17.109 Give the IUPAC and common names (if any) for each of the following compounds:

a. 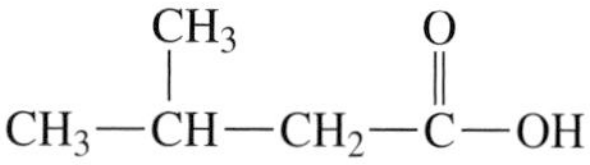 $CH_3-\overset{\overset{\displaystyle CH_3}{|}}{CH}-CH_2-\overset{\overset{\displaystyle O}{||}}{C}-OH$

b.

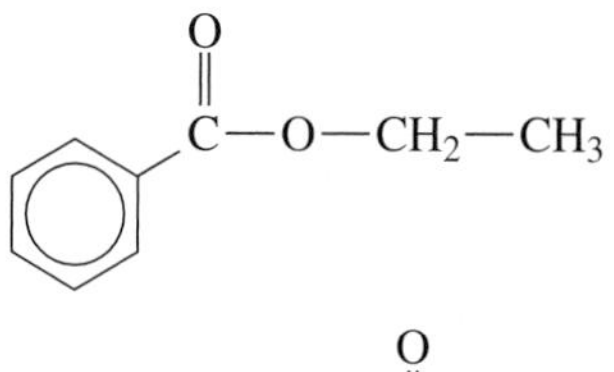

c. 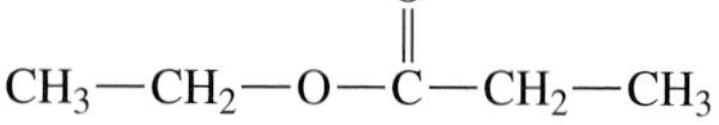 $CH_3-CH_2-O-\overset{\overset{\displaystyle O}{||}}{C}-CH_2-CH_3$

17.110 Give the IUPAC and common names (if any) for each of the following compounds:

a. 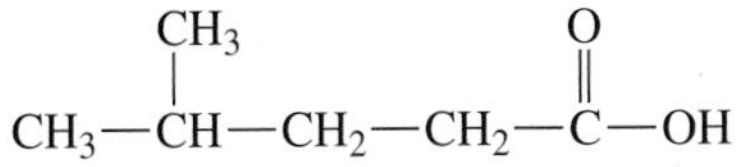
$$CH_3-\underset{\overset{\displaystyle CH_3}{|}}{CH}-CH_2-CH_2-\underset{\overset{\displaystyle O}{||}}{C}-OH$$

b.

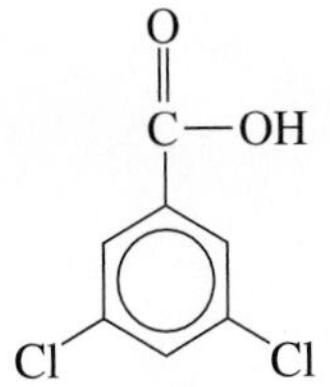

c.

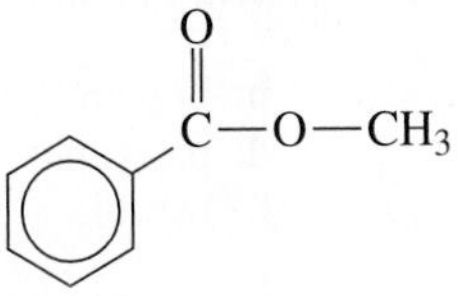

17.111 Draw the structure of each of the following compounds:
a. ethylamine
b. dimethylamine
c. triethylamine

17.112 Give the IUPAC name for each of the following amides:

a. 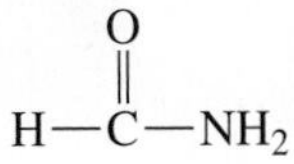
$$H-\underset{\overset{\displaystyle O}{||}}{C}-NH_2$$

b.
$$CH_3-CH_2-\underset{\overset{\displaystyle O}{||}}{C}-NH_2$$

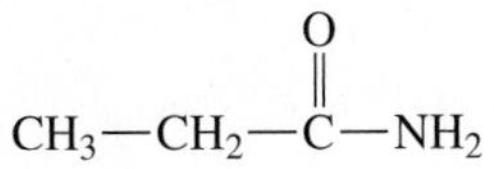

c.
$$CH_3-\underset{\overset{\displaystyle O}{||}}{C}-NH_2$$

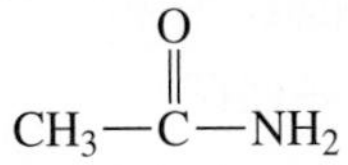

Challenge Questions

17.113 Toradol is used in dentistry to relieve pain. Name the functional groups in this molecule.

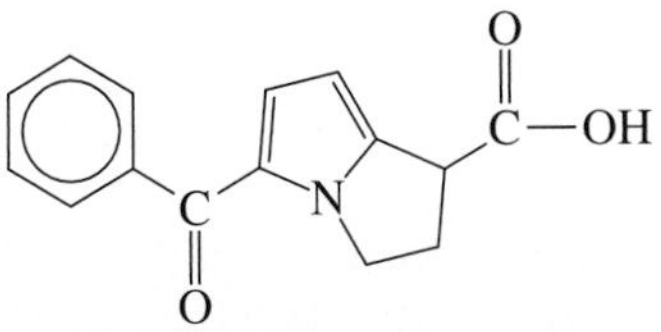

17.114 Voltaren is indicated for acute and chronic treatment of the symptoms of rheumatoid arthritis. Name the functional groups in this molecule.

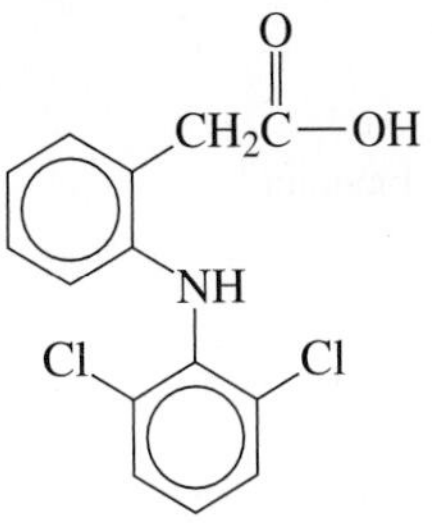

17.115 Identify the functional group and name each of the following:
a. $CH_3-CH_2-CH_2-OH$
b. $CH_3-CH_2-CH_2-NH-CH_3$

c. 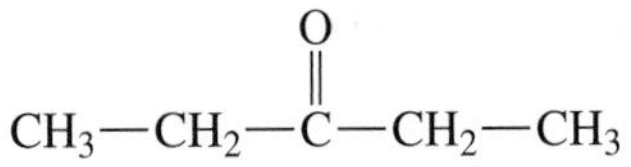
$$CH_3-CH_2-\underset{\overset{\displaystyle O}{||}}{C}-CH_2-CH_3$$

d. 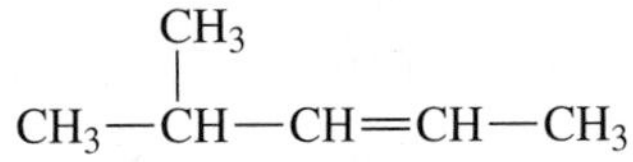
$$CH_3-\underset{\overset{\displaystyle CH_3}{|}}{CH}-CH=CH-CH_3$$

e. 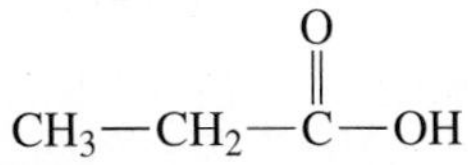
$$CH_3-CH_2-\underset{\overset{\displaystyle O}{||}}{C}-OH$$

17.116 Write the condensed structural formula for each of the following:
a. 2,4-dimethylpentane
b. 1,2-dichlorobenzene
c. ethylmethylamine
d. dimethylketone

17.117 Write the condensed structural formulas for all the compounds with the formula $C_4H_{10}O$.

17.118 Complete and balance each of the following reactions:

a. $C_5H_{12} + O_2 \xrightarrow{\text{Heat}}$

b. $CH_3-CH=CH-CH_3 + H_2 \xrightarrow{\text{Ni}}$

c. 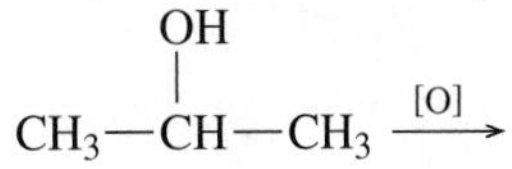 $CH_3-\underset{\overset{\displaystyle OH}{|}}{CH}-CH_3 \xrightarrow{[O]}$

Answers

Answers to Study Checks

17.1 Octane is not soluble in water; it is an organic compound.

17.2

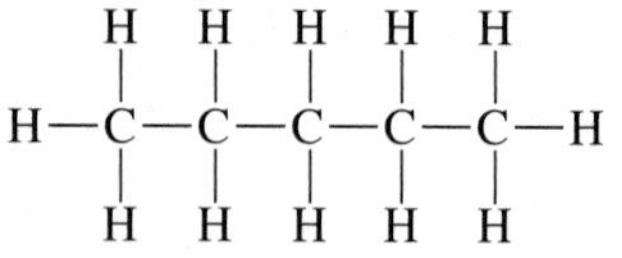

$$CH_3-CH_2-CH_2-CH_2-CH_3$$

17.3 1-chloro-2,4-dimethylhexane

17.4

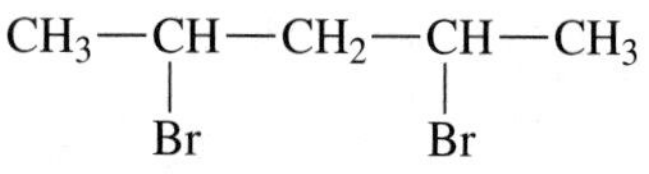

17.5 A carboxylic acid has a carboxyl group COOH. In an ester, the oxygen atom of the hydroxyl group is attached to a carbon atom, not hydrogen.

17.6 **a.** $CH_3-C\equiv C-CH_2-CH_3$

b.

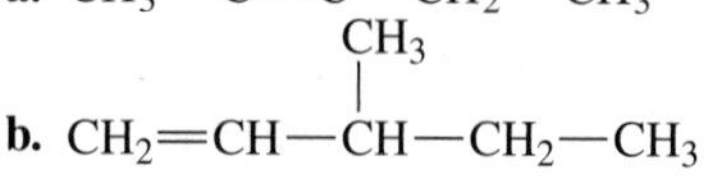

17.7

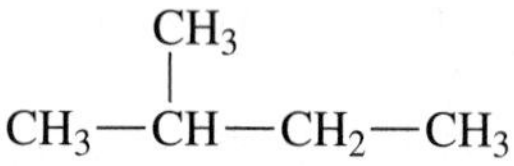

17.8 The monomer of PVC, polyvinyl chloride, is chloroethene:

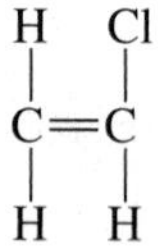

17.9 1,3-diethylbenzene

17.10 3-chloro-1-butanol

17.11 secondary

17.12

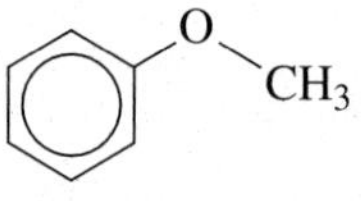

17.13

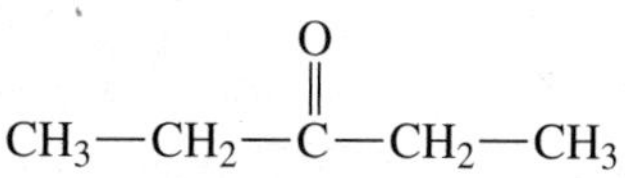

17.14 propanal (IUPAC), propionaldehyde (common)

17.15 ethyl propyl ketone

17.16

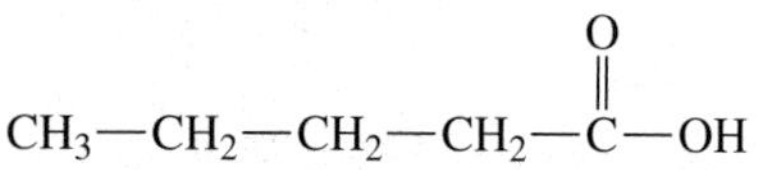

17.17 propanoic (propionic) acid and 1-pentanol

17.18

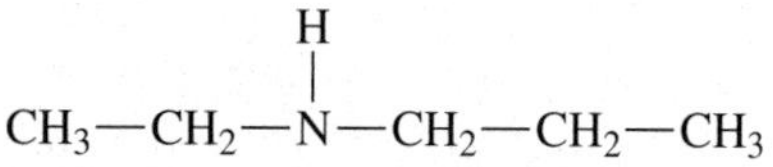

17.19

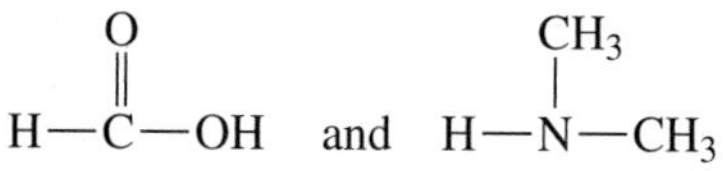

17.20

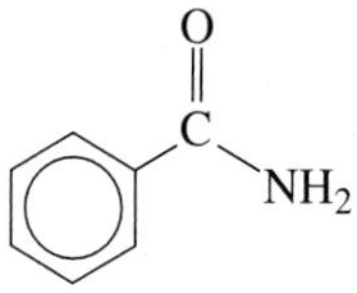

Answers to Selected Questions and Problems

17.1 **a.** inorganic   **b.** organic
        c. organic   **d.** inorganic
        e. inorganic   **f.** organic

17.3 **a.** inorganic   **b.** organic
        c. organic   **d.** inorganic

17.5 **a.** ethane   **b.** ethane
        c. NaBr   **d.** NaBr

17.7 VSEPR theory predicts that the four bonds in CH_4 will be as far apart as possible, which means that the hydrogen atoms are at the corners of a tetrahedron.

17.9 **a.** pentane
        b. ethane
        c. hexane

17.11 **a.** CH_4
        b. CH_3-CH_3
        c. $CH_3-CH_2-CH_2-CH_2-CH_3$

17.13 **a.** same molecule
        b. isomers of C_5H_{12}
        c. isomers of C_6H_{14}

17.15 **a.** 2-fluorobutane
        b. 2,2-dimethylpropane
        c. 2-chloro-3-methylpentane

17.17

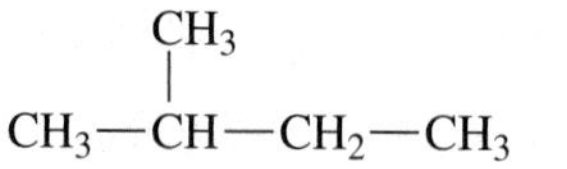

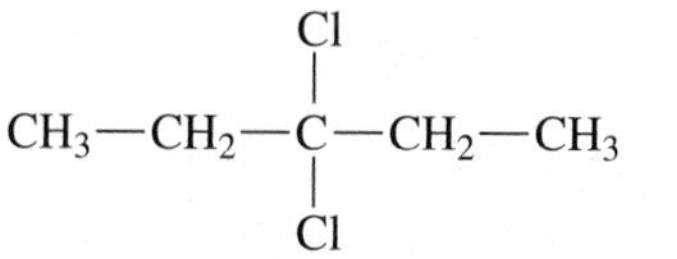

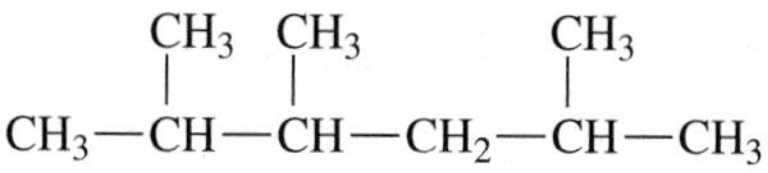

17.19 **a.** $CH_3-CH_2-CH_2-CH_2-CH_2-CH_2-CH_3$
        b. liquid
        c. insoluble in water
        d. float

17.21 **a.** $2C_2H_6 + 7O_2 \longrightarrow 4CO_2 + 6H_2O$
        b. $C_3H_8 + 5O_2 \longrightarrow 3CO_2 + 4H_2O$
        c. $2C_8H_{18} + 25O_2 \longrightarrow 16CO_2 + 18H_2O$

17.23 a. alcohol **b.** alkene
  c. aldehyde **d.** ester

17.25 a. ether **b.** alcohol
  c. ketone **d.** carboxylic acid
  e. amine

17.27 a. An alkene has a double bond.
  b. An alkyne has a triple bond.

17.29 a. ethane **b.** 2-methylpropene
  c. 2-pentyne

17.31 a. $CH_3-CH=CH_2$
  b. $CH_2=CH-CH_2-CH_2-CH_3$
  c. $CH_2=\overset{\overset{\displaystyle CH_3}{|}}{C}-CH_2-CH_3$

17.33 a. $CH_3-CH_2-CH_2-CH_2-CH_3$ Pentane
  b. $CH_3-CH_2-CH_2-CH_3$ Butane
  c. $CH_3-CH_2-CH_2-CH_2-CH_2-CH_3$
            Hexane

17.35 A polymer is a very large molecule composed of small units that are repeated many times.

17.37

$3H-\overset{\overset{\displaystyle H\ H}{|\ |}}{\underset{\underset{\displaystyle}{}}{C}}=C-H \longrightarrow$

17.39 a. 2-chlorotoluene **b.** ethylbenzene
  c. 1,3,5-trichlorobenzene

17.41 a. CH_3 (on benzene ring) **b.** Cl, Cl (on benzene ring)
  c. CH_2-CH_3 and CH_3 (on benzene ring)

17.43 a. ethanol **b.** 2-butanol
  c. 2-pentanol **d.** phenol

17.45 a. $CH_3-CH_2-CH_2-OH$
  b. CH_3-OH
  c. $CH_3-CH_2-\overset{\overset{\displaystyle OH}{|}}{CH}-CH_2-CH_3$
  d. $CH_3-\overset{\overset{\displaystyle OH}{|}}{\underset{\underset{\displaystyle CH_3}{|}}{C}}-CH_2-CH_3$

17.47 a. 1° **b.** 1° **c.** 3°

17.49 a. ethyl methyl ether
  b. dipropyl ether
  c. methyl propyl ether

17.51 a. $CH_3-CH_2-CH_2-CH_2-\overset{\overset{\displaystyle O}{||}}{C}-H$
  b. $CH_3-CH_2-\overset{\overset{\displaystyle O}{||}}{C}-CH_3$
  c. $CH_3-\overset{\overset{\displaystyle O}{||}}{C}-CH_2-\overset{\overset{\displaystyle CH_3}{|}}{CH}-CH_3$

17.53 a. acetaldehyde **b.** methyl propyl ketone
  c. formaldehyde

17.55 a. propanal **b.** 2-methyl-3-pentanone
  c. benzaldehyde

17.57 a. $CH_3-\overset{\overset{\displaystyle O}{||}}{C}-H$
  b. $CH_3-\overset{\overset{\displaystyle O}{||}}{C}-CH_2-CH_2-CH_3$
  c. $CH_3-\overset{\overset{\displaystyle O}{||}}{C}-CH_2-CH_2-CH_2-CH_3$

17.59 a. ethanoic acid (acetic acid)
  b. propanoic acid (propionic acid)
  c. 4-hydroxybenzoic acid

17.61 a. $CH_3-CH_2-\overset{\overset{\displaystyle O}{||}}{C}-OH$
  b. benzene ring $-\overset{\overset{\displaystyle O}{||}}{C}-OH$
  c. $Cl-CH_2-\overset{\overset{\displaystyle O}{||}}{C}-OH$
  d. $HO-CH_2-CH_2-\overset{\overset{\displaystyle O}{||}}{C}-OH$

17.63 a. aldehyde
  b. ester
  c. ketone
  d. carboxylic acid

17.65 a. $CH_3-\overset{\overset{\displaystyle O}{||}}{C}-O-CH_3$
  b. $CH_3-CH_2-CH_2-\overset{\overset{\displaystyle O}{||}}{C}-O-CH_3$

17.67 a. $CH_3-CH_2-\overset{\overset{\displaystyle O}{||}}{C}-O-CH_2-CH_2-CH_3$
  b. $CH_3-CH_2-CH_2-CH_2-\overset{\overset{\displaystyle O}{||}}{C}-O-\overset{\overset{\displaystyle CH_3}{|}}{CH}-CH_3$

17.69 a. methyl formate (methyl methanoate)
  b. methyl acetate (methyl ethanoate)
  c. methyl butyrate (methyl butanoate)

17.71 a. $CH_3-\overset{\displaystyle O}{\overset{\|}{C}}-O-CH_3$

  b. $H-\overset{\displaystyle O}{\overset{\|}{C}}-O-CH_2-CH_2-CH_2-CH_3$

  c. $CH_3-CH_2-CH_2-CH_2-\overset{\displaystyle O}{\overset{\|}{C}}-O-CH_2-CH_3$

  d. $CH_3-CH_2-\overset{\displaystyle O}{\overset{\|}{C}}-O-CH_2-CH_2-CH_3$

17.73 a. ethylamine
  b. methylpropylamine
  c. diethylmethylamine

17.75 a. ethanamide (acetamide)
  b. butanamide (butyramide)
  c. methanamide (formamide)

17.77 a. $CH_3-CH_2-\overset{\displaystyle O}{\overset{\|}{C}}-NH_2$

  b. $CH_3-CH_2-CH_2-\overset{\displaystyle CH_3}{\underset{}{\overset{|}{C}H}}-\overset{\displaystyle O}{\overset{\|}{C}}-NH_2$

  c. $H-\overset{\displaystyle O}{\overset{\|}{C}}-NH_2$

17.79 a. butane
  b. butane
  c. potassium chloride
  d. potassium chloride
  e. butane

17.81 a. carboxylic acid
  b. alkene
  c. ester
  d. amine
  e. aldehyde

17.83 a. alcohol   **b.** alkene
  c. aldehyde   **d.** alkane
  e. carboxylic acid **f.** amine

17.85

$-\overset{F}{\underset{F}{\overset{|}{\underset{|}{C}}}}-\overset{F}{\underset{F}{\overset{|}{\underset{|}{C}}}}-\overset{F}{\underset{F}{\overset{|}{\underset{|}{C}}}}-\overset{F}{\underset{F}{\overset{|}{\underset{|}{C}}}}-\overset{F}{\underset{F}{\overset{|}{\underset{|}{C}}}}-\overset{F}{\underset{F}{\overset{|}{\underset{|}{C}}}}-\overset{F}{\underset{F}{\overset{|}{\underset{|}{C}}}}-\overset{F}{\underset{F}{\overset{|}{\underset{|}{C}}}}-$

17.87 a. aromatic, aldehyde
  b. aromatic, aldehyde, alkene
  c. ketone

17.89 carboxylic acid, aromatic, amine, amide, ester

17.91 a.

$CH_3-CH_2-\overset{\displaystyle CH_2-CH_3}{\underset{}{\overset{|}{C}H}}-CH_2-CH_2-CH_3$

b.

$CH_3-\overset{\displaystyle CH_3}{\underset{}{\overset{|}{C}H}}-\overset{\displaystyle CH_3}{\underset{Cl}{\overset{|}{C}H}}-CH_2-CH_3$

$Cl-CH_2-CH_2-\overset{\displaystyle Cl}{\underset{CH_3}{\overset{|}{\underset{|}{C}}}}-CH_2-CH_2-CH_2-CH_3$

17.93 a. 2-methyl-1-pentene
  b. 1-butyne
  c. 2-pentene

17.95 a. toluene
  b. 2-chlorotoluene
  c. 4-ethyltoluene

17.97 a. 1°
  b. 2°
  c. 1°

17.99 a.

b. $CH_3-\overset{\displaystyle CH_3}{\underset{}{\overset{|}{C}H}}-\overset{\displaystyle OH}{\underset{}{\overset{|}{C}H}}-CH_2-CH_3$

c. $CH_3-CH_2-\overset{\displaystyle O}{\overset{\|}{C}}-CH_2-CH_3$

17.101 a. $CH_3-CH_2-\overset{\displaystyle O}{\overset{\|}{C}}-H$

  b. $CH_3-CH_2-\overset{\displaystyle O}{\overset{\|}{C}}-CH_3$

17.103 a. 4-chloro-3-hydroxybenzaldehyde
  b. 3-chloropropanal
  c. 2-chloro-3-pentanone

17.105 a.

b. $Cl-CH_2-CH_2-\overset{\displaystyle O}{\overset{\|}{C}}-H$

c. $CH_3-CH_2-\overset{\displaystyle O}{\overset{\|}{C}}-CH_3$

17.107 a. $CH_3-CH_2-\overset{\displaystyle O}{\overset{\|}{C}}-H$

b.
$$CH_3-\overset{\overset{\displaystyle O}{\|}}{C}-CH_2-CH_2-CH_3$$

c.
$$CH_3-CH_2-CH_2-\overset{\overset{\displaystyle O}{\|}}{C}-H$$

17.109 **a.** 3-methylbutanoic acid
b. ethyl benzoate
c. ethyl propanoate; ethyl propionate

17.111 **a.** $CH_3-CH_2-NH_2$
b. $CH_3-NH-CH_3$
c.
$$CH_3-CH_2-\overset{\overset{\displaystyle CH_2-CH_3}{|}}{N}-CH_2-CH_3$$

17.113 aromatic, ketone, amine, carboxylic acid

17.115 **a.** alcohol; 1-propanol
b. amine; methylpropylamine
c. ketone; 3-pentanone
d. alkene; 4-methyl-2-pentene
e. carboxylic acid; propanoic acid

17.117 $CH_3-CH_2-CH_2-CH_2-OH$

$$CH_3-CH_2-\overset{\overset{\displaystyle OH}{|}}{CH}-CH_3$$

$$CH_3-\overset{\overset{\displaystyle CH_3}{|}}{CH}-CH_2-OH$$

$$CH_3-\overset{\overset{\displaystyle CH_3}{|}}{\underset{\underset{\displaystyle CH_3}{|}}{C}}-OH$$

$$CH_3-CH_2-O-CH_2-CH_3$$

$$CH_3-\overset{\overset{\displaystyle CH_3}{|}}{CH}-O-CH_3$$

$$CH_3-CH_2-CH_2-O-CH_3$$

Biochemistry

"*The purpose of our research was to create a way to make Taxol,*" *says Paul Wender, Francis W. Bergstrom Professor of Chemistry and head of the Wender research group at Stanford University.* "*Taxol is a chemotherapy drug originally derived from the bark of the Pacific yew tree. However, removing the bark from yew trees destroys them, so we need a renewable resource. We worked out a synthesis that began with turpentine, which is both renewable and inexpensive. Initially, Taxol was used with patients who did not respond to chemotherapy. The first person to be treated was a woman who was diagnosed with terminal ovarian cancer and given three to six months to live. After a few treatments with Taxol, she was declared 98% disease—free. A drug like Taxol can save many lives, which is one reason that a study of biochemistry is so important.*"

the Chemistry place

Visit **www.aw-bc.com/chemplace** for extra quizzes, interactive tutorials, career resources, PowerPoint slides for chapter review, math help, and case studies.

In **biochemistry,** we study the structures and reactions of chemicals that occur in living systems. In this chapter, we will focus on four important types of biomolecules: carbohydrates, lipids, proteins, and nucleic acids. Each of these consists of small molecules that link together to form large molecules.

Carbohydrates are the most abundant organic compounds in nature. In plants, energy from the sun converts carbon dioxide and water into the carbohydrate glucose. Many glucose molecules link to form long-chain polymers of energy-storing starch or into cellulose to build the structure of the plant. Each day you may enjoy the polysaccharides called starches in bread and pasta. The table sugar used to sweeten cereal, tea, or coffee is sucrose, a disaccharide that consists of two simple sugars, glucose and fructose.

Lipids and proteins are also important nutrients that we obtain from food. Lipids include the fats and oils in our diets, steroids, and cholesterol. In the body, lipids store energy, insulate organs, and build cell membranes. Proteins have many functions in the body including building muscle and cartilage, transporting oxygen in blood, and directing biological reactions. All proteins are composed of building blocks called amino acids.

Nucleic acids are molecules in our cells that store information for cellular growth and reproduction and direct the use of this information. Deoxyribonucleic acid (DNA) contains the directions for making proteins, whereas ribonucleic acids (RNA) are used to decode this information for the production of proteins.

18.1 Carbohydrates

Learning Goal

Classify carbohydrates as aldose or ketose; draw the open-chain and cyclic structures for glucose, galactose, and fructose.

Carbohydrates such as table sugar, lactose in milk, and cellulose are all made of carbon, hydrogen, and oxygen. Simple sugars, which have formulas of $C_n(H_2O)_n$, were once thought to be hydrates of carbon, thus the name *carbohydrate*. In a series of reactions called photosynthesis, energy from the sun is used to combine the carbon atoms from carbon dioxide (CO_2) and the hydrogen and oxygen atoms of water into the carbohydrate glucose.

$$6CO_2 + 6H_2O + \text{energy} \underset{\text{Respiration}}{\overset{\text{Photosynthesis}}{\rightleftharpoons}} \underset{\text{Glucose}}{C_2H_{12}O_6} + 6O_2$$

In our body, glucose is oxidized in a series of metabolic reactions known as respiration, which releases chemical energy to do work in the cells. Carbon dioxide and water are produced and returned to the atmosphere. The combination of photosynthesis and respiration is called the carbon cycle, in which energy from the sun is stored in plants by photosynthesis and made available to us when the carbohydrates in our diets are metabolized. (See Figure 18.1.)

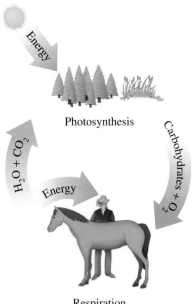

Photosynthesis

Respiration

Figure 18.1 During photosynthesis, energy from the sun combines CO_2 and H_2O to form glucose, $C_6H_{12}O_6$, and O_2. During respiration in the body, carbohydrates are oxidized to CO_2 and H_2O, while energy is produced.

Q What are the reactants and products of respiration?

Monosaccharides

Monosaccharides are simple sugars that have an unbranched chain of three to six carbon atoms, with one carbon in a carbonyl group and the rest attached to hydroxyl groups. In an **aldose,** the carbonyl group is on the first carbon as an aldehyde (—CHO); a **ketose** contains the carbonyl group on the second carbon atom as a ketone (C=O).

the
Chemistry
place

WEB TUTORIAL
Carbohydrates

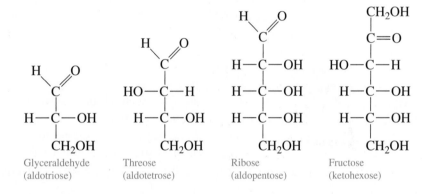

A monosaccharide with three carbon atoms is a *triose,* one with four carbon atoms is a *tetrose,* a *pentose* has five carbons, and a *hexose* contains six carbons. Thus, an aldopentose is a five-carbon monosaccharide that is an aldehyde; a ketohexose would be a six-carbon monosaccharide that is a ketone. Some examples are

Glyceraldehyde (aldotriose)　Threose (aldotetrose)　Ribose (aldopentose)　Fructose (ketohexose)

Sample Problem 18.1 ▶ **Monosaccharides**

Classify each of the following monosaccharides to indicate their carbonyl group and number of carbon atoms:

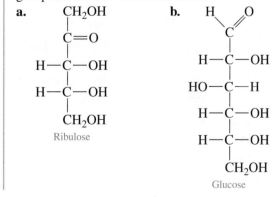

HYPERGLYCEMIA AND HYPOGLYCEMIA

A doctor may order a glucose tolerance test to evaluate the body's ability to return to normal glucose concentrations (70–90 mg/dL plasma) in response to the ingestion of a specified amount of glucose. The patient fasts for 12 hours and then drinks a solution containing glucose. If the blood glucose exceeds 140 mg/dL in plasma and remains high, hyperglycemia may be indicated. The term *glyc* or *gluco* refers to "sugar." The prefix *hyper* means above or over, and *hypo* is below or under. Thus the blood sugar level in *hyperglycemia* is above normal and below normal in *hypoglycemia*.

An example of a disease that can cause hyperglycemia is diabetes mellitus, which occurs when the pancreas is unable to produce sufficient quantities of insulin. As a result, glucose levels in the body fluids can rise as high as 350 mg/dL plasma. Symptoms of diabetes in people under the age of 40 include thirst, excessive urination, increased appetite, and weight loss. In older persons, diabetes is sometimes a consequence of excessive weight gain.

When a person is hypoglycemic, the blood glucose level rises and then decreases rapidly to levels as low as 40 mg/dL plasma. In some cases, hypoglycemia is caused by overproduction of insulin by the pancreas. Low blood glucose can cause dizziness, general weakness, and muscle tremors. A diet may be prescribed that consists of several small meals high in protein and low in carbohydrate. Some hypoglycemic patients are finding success with diets that include more complex carbohydrates rather than simple sugars.

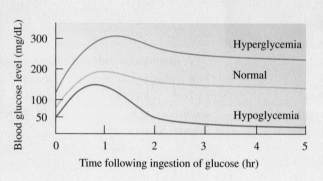

CASE STUDY
Diabetes and Blood Glucose

Solution

a. The structural formula has a ketone group; ribulose is a ketose. Because there are five carbon atoms, it is a pentose. Combining these classifications makes it a ketopentose.

b. The structural formula has an aldehyde group; glucose is an aldose. Because there are six carbon atoms, it is an aldohexose.

Study Check

The simplest ketose is a triose named dihydroxyacetone. Draw its structural formula.

Structures of Some Important Monosaccharides

The most common hexose, **glucose,** $C_6H_{12}O_6$, is found in fruits, vegetables, corn syrup, and honey. (See Figure 18.2.) It is a building block of the disaccharides sucrose, lactose, and maltose and polysaccharides such as starch, cellulose, and glycogen.

Galactose is an aldohexose that does not occur in the free form in nature. It is obtained from the disaccharide lactose, a sugar found in milk and milk products. Galactose is important in the cellular membranes of the brain and nervous system. The only difference in the structures of glucose and galactose is the arrangement of the —OH group on carbon 4.

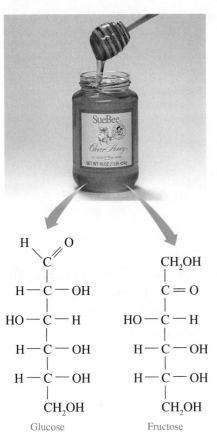

Glucose   Galactose

In contrast to glucose and galactose, **fructose** is a ketohexose. The structure of fructose differs from glucose at carbons 1 and 2 by the location of the carbonyl group.

Glucose   Fructose

Figure 18.2 The sweet taste of honey is due to the monosaccharides glucose and fructose.

What are some differences in the structures of glucose and fructose?

Fructose is the sweetest of the carbohydrates, twice as sweet as sucrose (table sugar). This makes fructose popular with dieters because less fructose, and therefore fewer calories, are needed to provide a pleasant taste. Fructose is found in fruit juices and honey.

Cyclic Structures of Monosaccharides

Up until now we have drawn the structures of monosaccharides such as glucose as open chains; however, these molecules normally exist in a cyclic structure. For example, in glucose, a ring of six atoms forms when the —OH group on carbon 5 (C5) reacts with the carbonyl group (C=O) on carbon 1 (C1). Let's look at how the cyclic structure of glucose is drawn, starting with the open chain.

STEP 1 Think of turning the open chain of glucose clockwise to the right. Then the —OH groups written on the right, other than the one

the **C**hemistry place

WEB TUTORIAL
Forms of Carbohydrates

on carbon 6, are drawn down and the —OH group on the left is up.

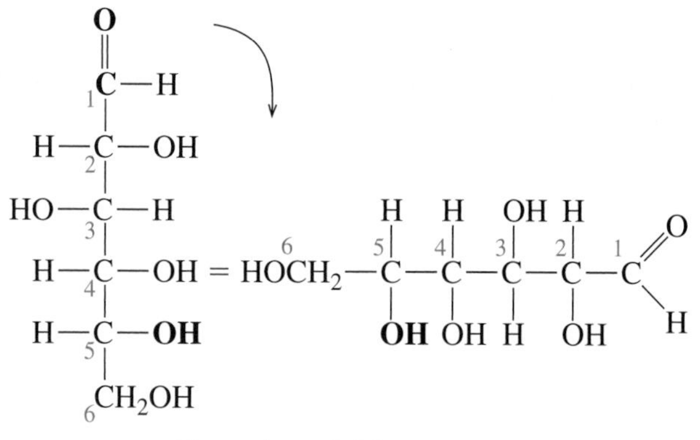

Glucose (open chain)

Guide to Drawing Cyclic Structures

STEP 1
Turn the open chain clockwise 90°.

STEP 2
Fold the chain into a hexagon and bond the O on carbon 5 to carbon 1 of the carbonyl group.

STEP 3
Write the new —OH group on carbon 1 down to give the α anomer or up to give the β anomer.

STEP 2 Rotate the groups around carbon 5 placing the —CH$_2$OH up and the —OH group close to the carbonyl carbon 1. Form the cyclic structure by bonding the oxygen in the —OH group to the carbonyl carbon.

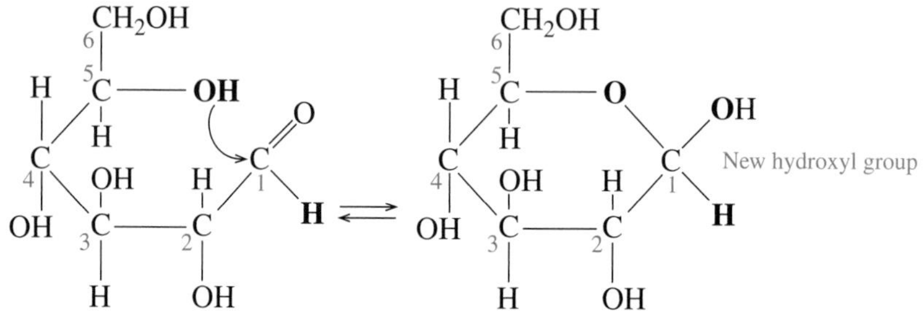

Carbon-5 oxygen bonds to carbonyl   Cyclic structure

STEP 3 In the cyclic structure, carbon 1 is now bonded to a new —OH group. There are two ways to place the —OH, either up or down. The new —OH group is down in the α (alpha) form and up in the β (beta) form.

For convenience, the cyclic structure is simplified by indicating the carbon atoms as the corners of a hexagon and showing only the attached groups other than hydrogen.

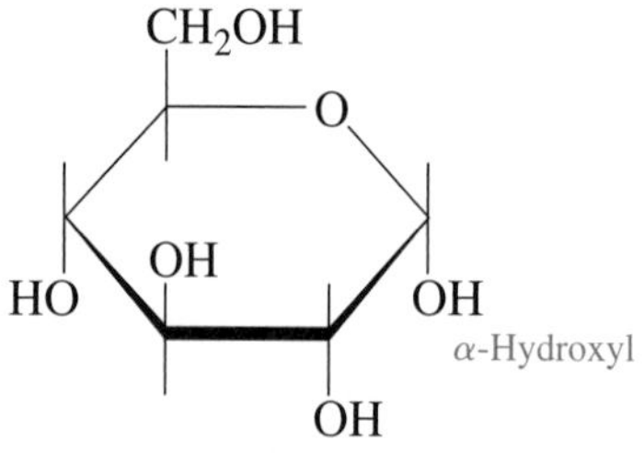

α-Galactose

α-Fructose

α-Glucose (simplified structure)

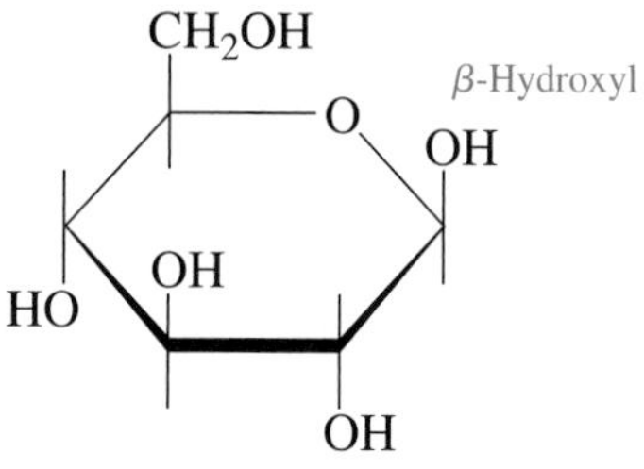

β-Glucose

Galactose is an aldohexose like glucose, differing only in the arrangement of the —OH group on carbon 4. Thus, its cyclic structure is also similar to glucose, except that in galactose the —OH on carbon 4 is up. Galactose also exists in α and β forms.

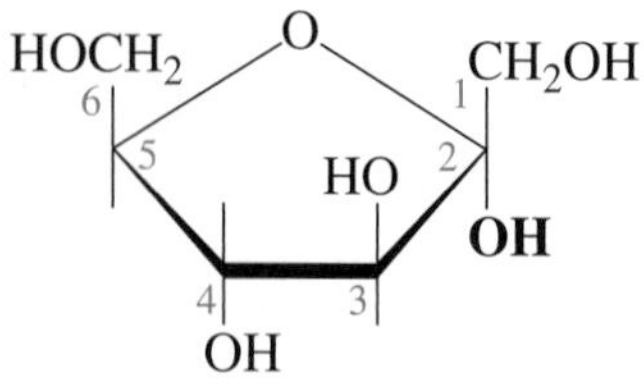

In contrast to glucose and galactose, fructose is a ketohexose. It forms a five-atom ring when a hydroxyl group on carbon 5 reacts with the carbon of the ketone group. The new hydroxyl group is on carbon 2.

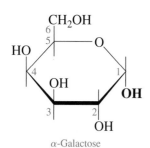

18.1 What functional groups are found in all monosaccharides?

18.2 What is the difference between an aldose and a ketose?

18.3 What are the functional groups and number of carbons in a ketopentose?

18.4 What are the functional groups and number of carbons in an aldohexose?

18.5 Classify each of the following monosaccharides as an aldose or ketose:

a.

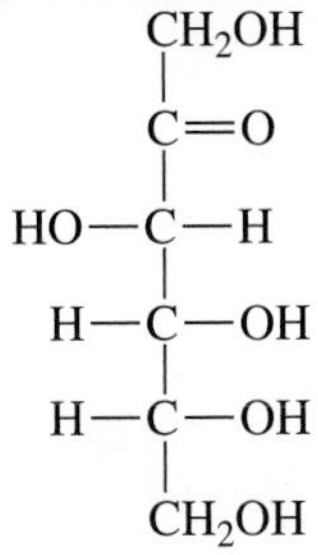

$$
\begin{array}{c}
CH_2OH \\
| \\
C=O \\
| \\
HO-C-H \\
| \\
H-C-OH \\
| \\
H-C-OH \\
| \\
CH_2OH
\end{array}
$$

Fructose

b.

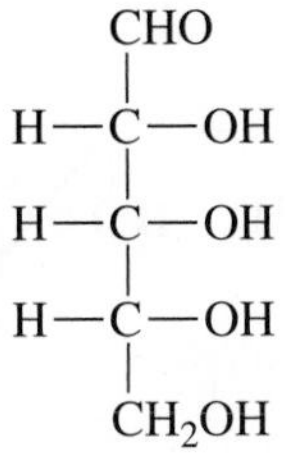

$$
\begin{array}{c}
CHO \\
| \\
H-C-OH \\
| \\
H-C-OH \\
| \\
H-C-OH \\
| \\
CH_2OH
\end{array}
$$

Ribose

c.

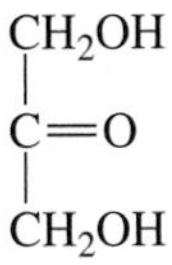

$$
\begin{array}{c}
CH_2OH \\
| \\
C=O \\
| \\
CH_2OH
\end{array}
$$

Dihydroxyacetone

d.

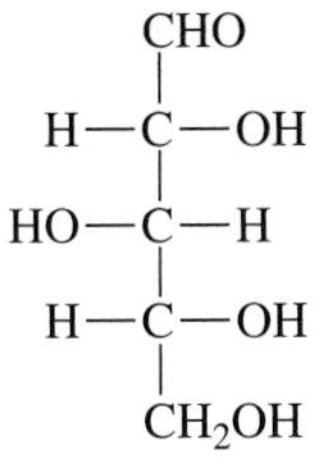

$$
\begin{array}{c}
CHO \\
| \\
H-C-OH \\
| \\
HO-C-H \\
| \\
H-C-OH \\
| \\
CH_2OH
\end{array}
$$

Xylose

e.

$$
\begin{array}{c}
CHO \\
| \\
H-C-OH \\
| \\
HO-C-H \\
| \\
HO-C-H \\
| \\
H-C-OH \\
| \\
CH_2OH
\end{array}
$$

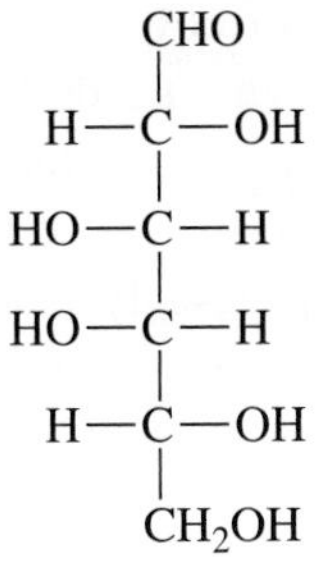

Galactose

18.6 Classify each of the monosaccharides in problem 18.5 according to the number of carbon atoms in the chain.

18.7 How does the open-chain structure of galactose differ from glucose?

18.8 How does the open-chain structure of fructose differ from glucose?

18.9 What are the kind and number of atoms in the ring portion of the cyclic structure of glucose?

18.10 What are the kind and number of atoms in the ring portion of the cyclic structure of fructose?

18.11 Identify each of the following cyclic structures as the α or β form:

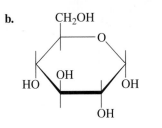

18.12 Identify each of the following cyclic structures as the α or β form:

a.

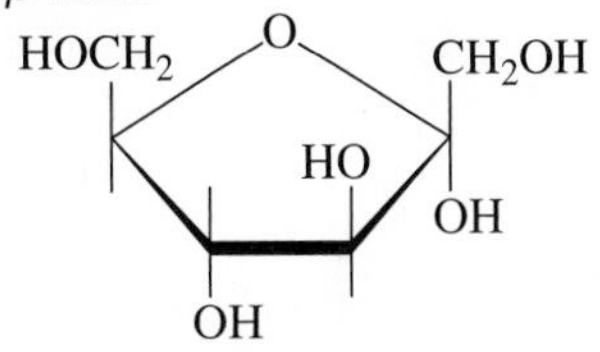

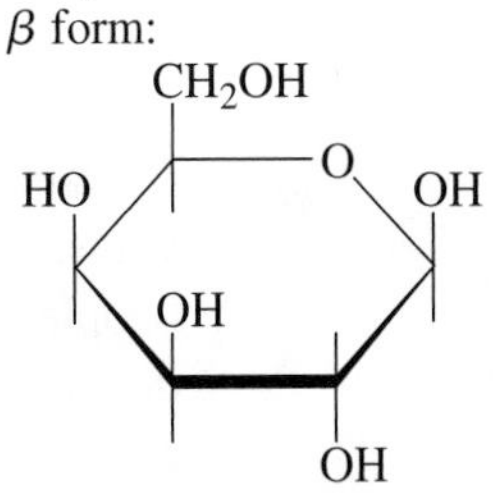

b.

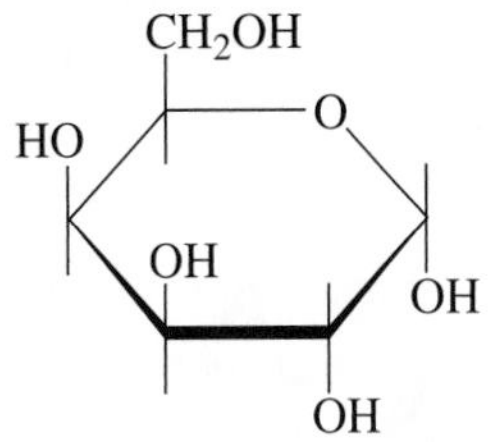

18.2 Disaccharides and Polysaccharides

Learning **Goal**

Describe the monosaccharide units and linkages in disaccharides.

A **disaccharide** is composed of two monosaccharides linked together. The most common disaccharides are maltose, lactose, and sucrose, which consist of the following monosaccharides:

$$\text{Maltose} + H_2O \xrightarrow{\ H^+\ } \text{glucose} + \text{glucose}$$

$$\text{Lactose} + H_2O \xrightarrow{\ H^+\ } \text{glucose} + \text{galactose}$$

$$\text{Sucrose} + H_2O \xrightarrow{\ H^+\ } \text{glucose} + \text{fructose}$$

Maltose, or malt sugar, is obtained from starch. Maltose is used in cereals, candies, and the brewing of beverages.

In maltose, a *glycosidic bond* joins the two glucose molecules with the loss of a molecule of water. The glycosidic bond is designated as an α-1,4 linkage to show that the α —OH on carbon 1 is joined to carbon 4 of the second glucose. Because the second glucose molecule has a free —OH on carbon 1, there are α and β forms of maltose.

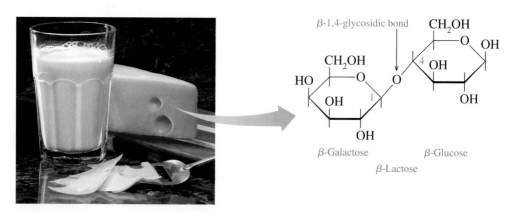

Lactose, milk sugar, is found in milk and milk products. (See Figure 18.3.) It makes up 6–8% of human milk and about 4–5% of cow's milk. Some people do not produce sufficient quantities of the enzyme needed to break down lactose, and the sugar remains undigested, causing abdominal cramps and diarrhea. In some commercial milk products, an enzyme called lactase is added to break down lactose. The bond in lactose is a β-1,4-glycosidic bond because the β form of galactose links to a hydroxyl group on carbon 4 of glucose.

Figure 18.3 Lactose is a disaccharide found in milk and milk products.

Q What types of glycosidic bond links galactose and glucose in lactose?

Figure 18.4 Sucrose is a disaccharide obtained from sugar beets and sugar cane.

Q *What monosaccharides form sucrose?*

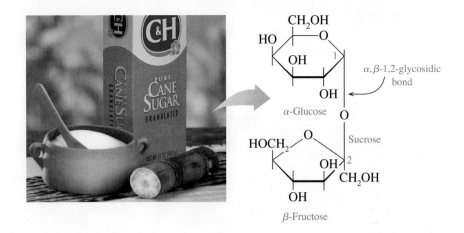

Sucrose, ordinary table sugar, is a disaccharide that is the most abundant carbohydrate in the world. Most of the sucrose for table sugar comes from sugar cane (20% by mass) or sugar beets (15% by mass). (See Figure 18.4.) Both the raw and refined forms of sugar are sucrose. Some estimates indicate that each person in the United States consumes an average of 45 kg (100 lb) of sucrose every year either by itself or in a variety of food products. Sucrose consists of glucose and fructose molecules joined by an α, β-1,2-glycosidic bond.

Sample Problem 18.2 **Glycosidic Bonds in Disaccharides**

Melibiose is a disaccharide that has a sweetness of about 30 compared with sucrose (= 100).

CH$_2$OH

HO

OH

OH

O—CH$_2$ *melibiose*

O

OH

HO   OH

OH

a. What are the monosaccharide units in melibiose?
b. What type of glycosidic bond links the monosaccharides?

Solution
a. The monosaccharide on the left side is α-galactose; on the right is α-glucose.
b. The monosaccharide units are linked by an α-1,6-glycosidic bond.

Study Check

Cellobiose is a disaccharide composed of two β-glucose molecules linked by a β-1,4-glycosidic linkage. Draw a structural formula for β-cellobiose.

CHEM NOTE

HOW SWEET IS MY SWEETENER?

Although many of the monosaccharides and disaccharides taste sweet, they differ considerably in their degree of sweetness. Dietetic foods contain sweeteners that are noncarbohydrate or carbohydrates that are sweeter. Some examples of sweeteners compared with sucrose are shown in Table 18.1.

Sucralose is made from sucrose by replacing some of the hydroxyl groups with chlorine atoms.

sucralose

Aspartame, which is marketed as NutraSweet, is used in a large number of sugar-free products. It is a noncarbohydrate sweetener made of aspartic acid and a methyl ester of phenylalanine. It does have some caloric value, but it is so sweet that a very small quantity is needed. However, one of the breakdown products, phenylalanine, poses a danger to anyone who cannot metabolize it properly, a condition called phenylketonuria (PKU).

From aspartic acid   From phenylalanine

Aspartame (Nutra-Sweet)

Saccharin has been used as a noncarbohydrate artificial sweetener for the past 25 years. The use of saccharin has been

Table 18.1 **Relative Sweetness of Sugars and Artificial Sweeteners**

	Sweetness Relative to Sucrose (=100)
Monosaccharides	
galactose	30
sorbitol	36
glucose	75
fructose	175
Disaccharides	
lactose	16
maltose	33
sucrose	100 ← reference standard
Artificial Sweeteners (Noncarbohydrate)	
sucralose	60 000
aspartame	18 000
saccharin	45 000

banned in Canada because studies indicate that it may cause bladder tumors. However, it is still approved for use by the FDA in the United States.

saccharin

Polysaccharides

WEB TUTORIAL

Polymers

A **polysaccharide** is a polymer of many monosaccharides joined together. Three biologically important polysaccharides—starch, cellulose, and glycogen—are all polymers of glucose, which differ only in the type of glycosidic bonds and the amount of branching in the molecule. Starch, a storage form of glucose in plants, is composed of two kinds of polysaccharides. **Amylose,** which makes up about 20% of starch, consists of α-glucose molecules connected by

α-1,4-glycosidic bonds in a continuous chain. A typical polymer of amylose may contain from 250 to 4000 glucose units. Sometimes called a straight-chain polymer, polymers of amylose are actually coiled in helical fashion.

Amylopectin, which makes up as much as 80% of plant starch, is a branched-chain polysaccharide. Like amylose, the glucose molecules are connected by α-1,4-glycosidic bonds. However, at about every 25 glucose units, there is a branch of glucose molecules attached by an α-1,6-glycosidic bond between carbon 1 of the branch and carbon 6 in the main chain. (See Figure 18.5.)

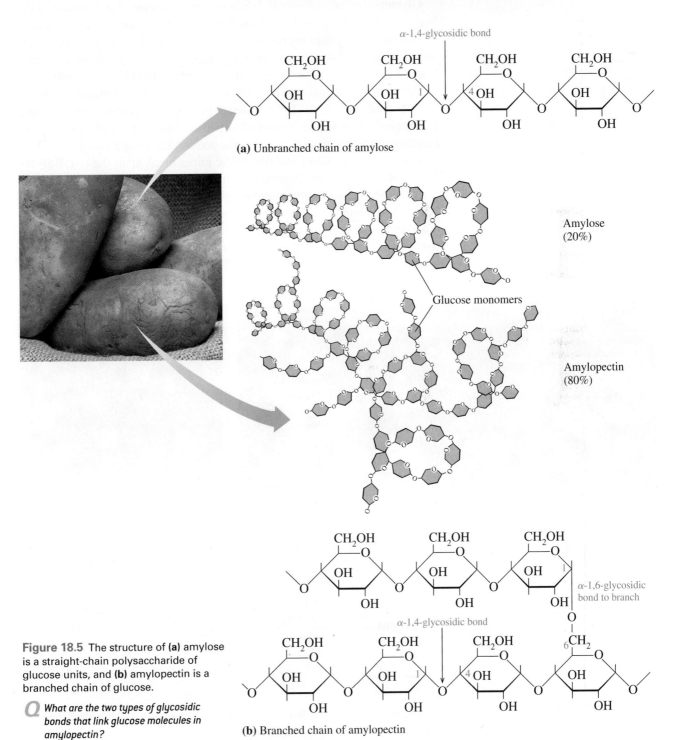

(a) Unbranched chain of amylose

Amylose (20%)

Glucose monomers

Amylopectin (80%)

(b) Branched chain of amylopectin

Figure 18.5 The structure of **(a)** amylose is a straight-chain polysaccharide of glucose units, and **(b)** amylopectin is a branched chain of glucose.

Q *What are the two types of glycosidic bonds that link glucose molecules in amylopectin?*

Starches hydrolyze easily in water and acid to give smaller saccharides called dextrins, which then hydrolyze to maltose and finally glucose. In our bodies, these complex carbohydrates are digested by the enzymes amylase (in saliva) and maltase. The glucose obtained provides about 50% of our nutritional calories.

$$\text{Amylose, amylopectin} \xrightarrow{\text{H}^+ \text{ or amylase}} \text{dextrins} \xrightarrow{\text{H}^+ \text{ or amylase}} \text{maltose} \xrightarrow{\text{H}^+ \text{ or maltase}} \text{many glucose units}$$

Glycogen, or animal starch, is a polymer of glucose that is stored in the liver and muscle of animals. It is used in our cells to maintain the blood level of glucose and provide energy between meals. The structure of glycogen is very similar to that of amylopectin except that glycogen is more highly branched. In glycogen, branches occur about every 10–15 glucose units.

Cellulose is the major structural material of wood and plants. Cotton is almost pure cellulose. In cellulose, glucose molecules form a long unbranched chain similar to that of amylose. However, the glucose units in cellulose are linked by β-1,4-glycosidic bonds. (See Figure 18.6.)

Enzymes in our saliva and pancreatic juices break apart the α-1,4-glycosidic bonds of starches. However, there are no enzymes in humans that can break down the β-1,4-glycosidic bonds of cellulose; we cannot digest cellulose. Some animals such as goats and cows and insects such as termites are able to obtain glucose from cellulose. Their digestive systems contain bacteria and protozoa with enzymes that can break apart β-1,4-glycosidic bonds.

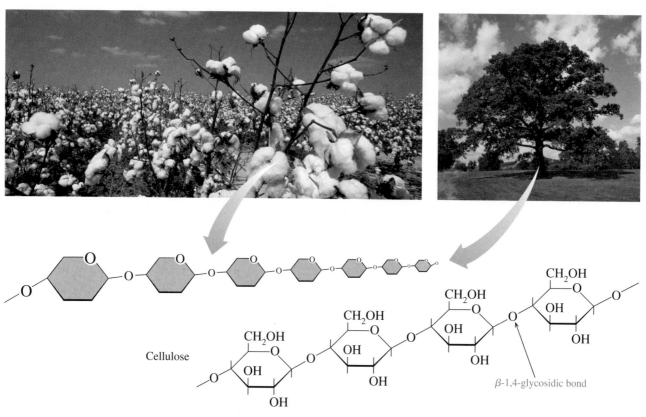

Figure 18.6 The polysaccharide cellulose is composed of β-1,4-glycosidic bonds.

Q Why are humans unable to digest cellulose?

Sample Problem 18.3 **Structures of Polysaccharides**

Identify the polysaccharide described by each of the following:
a. an unbranched polysaccharide with α-1,4-glycosidic bonds
b. an unbranched polysaccharide containing β-1,4-glycosidic bonds
c. a starch containing α-1,4- and α-1,6-glycosidic bonds

Solution
a. amylose   **b.** cellulose   **c.** amylopectin, glycogen

Study Check

Cellulose and amylose are both unbranched glucose polymers. How do they differ?

Questions and Problems   **Disaccharides and Polysaccharides**

18.13 For each of the following disaccharides, give the mono-saccharide units, the type of glycosidic bond, and the identity of the disaccharide:
a.

b.

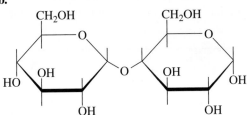

b.

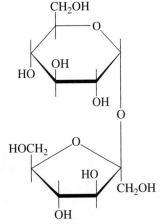

18.14 For each of the following disaccharides, give the mono-saccharide units, the type of glycosidic bond, and the identity of the disaccharide:
a.

18.15 Identify disaccharides that fit each of the following descriptions:
a. ordinary table sugar
b. found in milk and milk products
c. also called *malt* sugar
d. contains galactose and glucose

18.16 Identify disaccharides that fit each of the following descriptions:
a. used in brewing
b. composed of two glucose units
c. also called *milk* sugar
d. contains glucose and fructose

18.17 Give the name of one or more polysaccharides that matches each of the following descriptions:
a. not digestible by humans
b. the storage form of carbohydrates in plants
c. contains only α-1,4-glycosidic bonds
d. the most highly branched polysaccharide

18.18 Give the name of one or more polysaccharides that matches each of the following descriptions:
a. the storage form of carbohydrates in animals
b. contains only β-1,4-glycosidic bonds
c. contains both α-1,4- and α-1,6-glycosidic bonds
d. produces maltose during digestion

18.3 Lipids

Lipids are a family of biomolecules that have the common property of being soluble in organic solvents but not very soluble in water. The word *lipid* comes from the Greek word *lipos,* meaning "fat." Within the lipid family, there are certain structures that distinguish the different types of lipids. Lipids such as fats and oils are esters of glycerol and fatty acids. Steroids are characterized by the steroid nucleus of four fused carbon rings.

Fatty Acids

A **fatty acid** contains a long carbon chain with a carboxylic acid group at one end. An example is lauric acid, a 12-carbon acid found in coconut oil, which has a structure that can be written in several forms.

Writing Formulas for Lauric Acid

$$CH_3-(CH_2)_{10}-\overset{\overset{\displaystyle O}{\|}}{C}-OH \qquad CH_3-(CH_2)_{10}-COOH$$

$$CH_3-CH_2-CH_2-CH_2-CH_2-CH_2-CH_2-CH_2-CH_2-CH_2-CH_2-\overset{\overset{\displaystyle O}{\diagup\!\!\diagup}}{C}\diagdown_{OH}$$
Condensed structural formula

Line-bond structural formula

Saturated fatty acids such as lauric acid contain only single bonds between carbons. *Monounsaturated fatty acids* have one double bond in the carbon chain, and *polyunsaturated fatty acids* have two or more double bonds. Table 18.2 lists some of the typical fatty acids in lipids.

Compounds with double bonds can have two structures known as *cis* and *trans* isomers. A cis isomer has groups attached on the same side of the

Table 18.2 Structures and Melting Points of Common Fatty Acids

Name	Carbon Atoms	Double Bonds	Melting Structure	Point (°C)	Source
Saturated					
Lauric acid	12	0	$CH_3-(CH_2)_{10}-COOH$	43	Coconut
Myristic acid	14	0	$CH_3-(CH_2)_{12}-COOH$	54	Nutmeg
Palmitic acid	16	0	$CH_3-(CH_2)_{14}-COOH$	62	Palm
Stearic acid	18	0	$CH_3-(CH_2)_{16}-COOH$	69	Animal fat
Unsaturated					
Palmitoleic acid	16	1	$CH_3-(CH_2)_5-CH=CH-(CH_2)_7-COOH$	0	Butter
Oleic acid	18	1	$CH_3-(CH_2)_7-CH=CH-(CH_2)_7-COOH$	13	Olives, corn
Linoleic acid	18	2	$CH_3-(CH_2)_4-CH=CH-CH_2-CH=CH-(CH_2)_7-COOH$	−9	Soybean, safflower, sunflower
Linolenic acid	18	3	$CH_3-CH_2-CH=CH-CH_2-CH=CH-CH_2-CH=CH-(CH_2)_7-COOH$	−17	Corn

double bond, and the trans isomer has groups attached on the opposite sides of the double bond.

CH₃— groups on same side of the double bond

CH₃— groups on opposite side of the double bond

CH₃   CH₃
  \\C=C//
  H   H

cis-2-butene

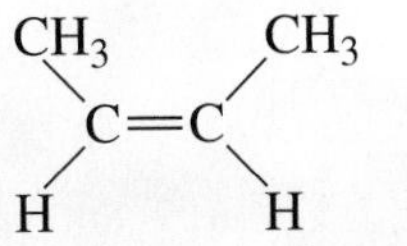

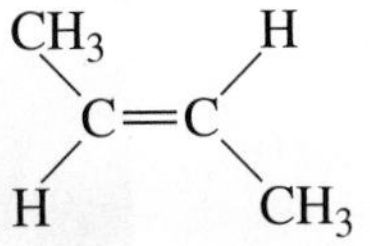

CH₃   H
  \\C=C//
  H   CH₃

trans-2-butene

As seen in the following, *cis*-oleic acid is not linear, but has a "kink" at the double bond. In contrast, *trans*-oleic acid has a regular, linear order of atoms. In naturally occurring unsaturated fatty acid, the double bonds are predominantly the cis isomers. As we will see, the cis bonds have a major impact on the physical properties of unsaturated fatty acids.

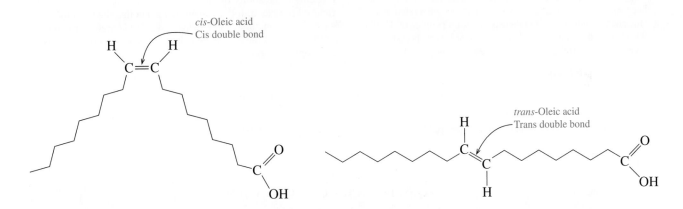

cis-Oleic acid
Cis double bond

trans-Oleic acid
Trans double bond

Sample Problem 18.4 **Structures and Properties of Fatty Acids**

Consider the structural formula of oleic acid:

$$CH_3(CH_2)_7CH=CH(CH_2)_7\overset{\displaystyle O}{\overset{\|}{C}}OH$$

a. Why is the substance called an acid?
b. How many carbon atoms are in oleic acid?
c. Is it a saturated or unsaturated fatty acid?

Solution
a. Oleic acid contains a carboxylic acid group.
b. It contains 18 carbon atoms.
c. It is an unsaturated fatty acid.

Study Check

Palmitoleic acid is a fatty acid with the following formula:

$$CH_3(CH_2)_5CH=CH(CH_2)_7\overset{\displaystyle O}{\overset{\|}{C}}OH$$

a. How many carbon atoms are in palmitoleic acid?
b. Is it a saturated or unsaturated fatty acid?

OMEGA-3 FATTY ACIDS IN FISH OILS

Because unsaturated fats are now recognized as being more beneficial to health than saturated fats, Americans have been changing their diets to include more unsaturated fats. Unsaturated fats contain two types of unsaturated fatty acids, omega-3 and omega-6. Counting from the CH_3— end, the first double bond occurs at carbon 6 in an omega-6 fatty acid, whereas in the omega-3 type, the first double bond occurs at the third carbon.

The omega-6 fatty acids are mostly found in grains, oils from plants, and eggs, and the omega-3 fatty acids are mostly found in cold-water fish such as tuna and salmon. The benefits of omega-3 fatty acids were first recognized when a study was conducted of the diets of the Inuit peoples of Alaska, who have a high-fat diet and high levels of blood cholesterol but a very low occurrence of coronary heart disease. The fats in the

Inuit diet are omega-3 fatty acids obtained primarily from fish.

In coronary heart disease, cholesterol forms plaque that adheres to the walls of the blood vessels. Blood pressure rises as blood has to squeeze through a smaller opening in the blood vessel. As more plaque forms, there is also a possibility of blood clots blocking the blood vessels and causing a heart attack. Omega-3 fatty acids lower the tendency of blood platelets to stick together, thereby reducing the possibility of blood clots. However, high levels of omega-3 fatty acids can increase bleeding if the ability of the platelets to form blood clots is reduced too much. It does seem that a diet that includes fish such as salmon, tuna, and herring can provide higher amounts of the omega-3 fatty acids, which help lessen the possibility of developing heart disease.

Omega-6 Fatty Acids

Linoleic acid
$$CH_3\underset{1}{\text{—}}(CH_2)_4\text{—}\underset{6}{CH}\text{=}CH\text{—}CH_2\text{—}CH\text{=}CH\text{—}(CH_2)_7\text{—}COOH$$

Arachidonic acid
$$CH_3\underset{1}{\text{—}}(CH_2)_4\text{—}(\underset{6}{CH}\text{=}CH\text{—}CH_2)_4\text{—}(CH_2)_2\text{—}COOH$$

Omega-3 Fatty Acids

Linolenic acid
$$CH_3\text{—}CH_2\text{—}(\underset{1}{CH}\text{=}\underset{2}{CH}\text{—}\underset{3}{CH_2})_3\text{—}(CH_2)_6\text{—}COOH$$

Eicosapentaenoic acid (EPA)
$$CH_3\text{—}CH_2\text{—}(\underset{1}{CH}\text{=}\underset{2}{CH}\text{—}\underset{3}{CH_2})_5\text{—}(CH_2)_2\text{—}COOH$$

Docosahexaenoic acid (DHA)
$$CH_3\text{—}CH_2\text{—}(\underset{1}{CH}\text{=}\underset{2}{CH}\text{—}\underset{3}{CH_2})_6\text{—}CH_2\text{—}COOH$$

WEB TUTORIAL
Triacylglycerols

Fats and Oils: Triacylglycerols

Fats and oils are known as **triacylglycerols.** These substances, also called *triglycerides,* are triesters of glycerol (a trihydroxy alcohol) and fatty acids. A triacylglycerol is produced by *esterification,* a reaction in which the hydroxyl groups of glycerol form ester bonds with the carboxyl groups of fatty acids. For example, glycerol and three molecules of stearic acid form tristearin (glyceryl tristearate).

Ester bonds

g ← Fatty acid
l
y
c — Fatty acid
e
r
o
l — Fatty acid

$$
\begin{array}{l}
CH_2O\text{—}H + HO\text{—}\overset{\displaystyle O}{\overset{\|}{C}}\text{—}(CH_2)_{16}CH_3 \\[6pt]
CHO\text{—}H + HO\text{—}\overset{\displaystyle O}{\overset{\|}{C}}\text{—}(CH_2)_{16}CH_3 \\[6pt]
CH_2O\text{—}H + HO\text{—}\overset{\displaystyle O}{\overset{\|}{C}}\text{—}(CH_2)_{16}CH_3
\end{array}
\longrightarrow
\begin{array}{l}
CH_2\text{—}O\text{—}\overset{\displaystyle O}{\overset{\|}{C}}\text{—}(CH_2)_{16}CH_3 \\[6pt]
CH\text{—}O\text{—}\overset{\displaystyle O}{\overset{\|}{C}}\text{—}(CH_2)_{16}CH_3 + \mathbf{3H_2O} \\[6pt]
CH_2\text{—}O\text{—}\overset{\displaystyle O}{\overset{\|}{C}}\text{—}(CH_2)_{16}CH_3
\end{array}
$$

glycerol   3 stearic acid molecules   glyceryl tristearate, (tristearin, a fat)

Triacylglycerols are the major form of energy storage for animals. Animals that hibernate eat large quantities of plants, seeds, and nuts that contain large amounts of fats and oils. They gain as much as 14 kilograms a week. As the external temperature drops, the animal goes into hibernation. The body temperature drops to nearly freezing, and there is a dramatic reduction in cellular activity, respiration, and heart rate. Animals who live in extremely cold climates will hibernate for 4–7 months. During this time, stored fat is the only source of energy.

| **Sample Problem 18.5** | **Writing Structures for a Triacylglycerol** |

Draw the structural formula of triolein, a triacylglycerol that uses oleic acid.

Solution

Triolein is the triacylglycerol of glycerol and three oleic acid molecules. Each fatty acid is attached by an ester bond to one of the hydroxyl groups in glycerol.

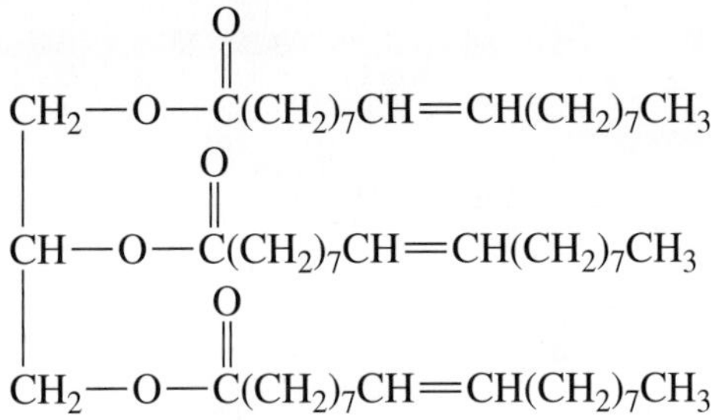

$$
\begin{array}{l}
\text{CH}_2-\text{O}-\overset{\displaystyle \overset{\text{O}}{\|}}{\text{C}}(\text{CH}_2)_7\text{CH}=\text{CH}(\text{CH}_2)_7\text{CH}_3 \\
\text{CH}-\text{O}-\overset{\displaystyle \overset{\text{O}}{\|}}{\text{C}}(\text{CH}_2)_7\text{CH}=\text{CH}(\text{CH}_2)_7\text{CH}_3 \\
\text{CH}_2-\text{O}-\overset{\displaystyle \overset{\text{O}}{\|}}{\text{C}}(\text{CH}_2)_7\text{CH}=\text{CH}(\text{CH}_2)_7\text{CH}_3
\end{array}
$$
glyceryl trioleate (triolein)

Study Check

Write the structure of the triacylglycerol containing three molecules of myristic acid.

Melting Points of Fats and Oils

A **fat** is a triacylglycerol that is solid at room temperature, such as fats in meat, whole milk, butter, and cheese. An **oil** is a triacylglycerol that is usually liquid at room temperature. The most common oils, such as olive oil, peanut oil, and corn oil, come from plant sources. (See Figure 18.7.)

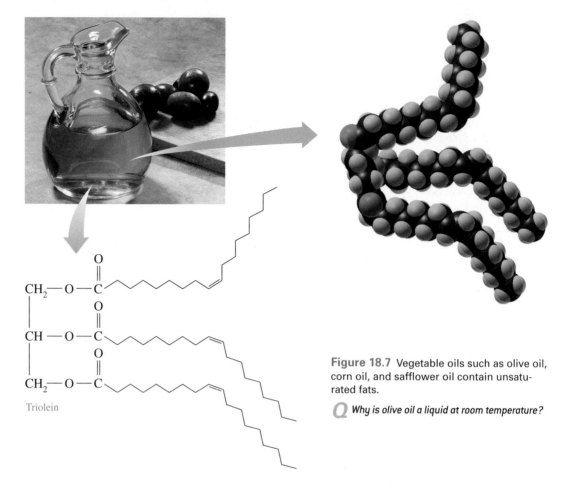

Triolein

Figure 18.7 Vegetable oils such as olive oil, corn oil, and safflower oil contain unsaturated fats.

Q Why is olive oil a liquid at room temperature?

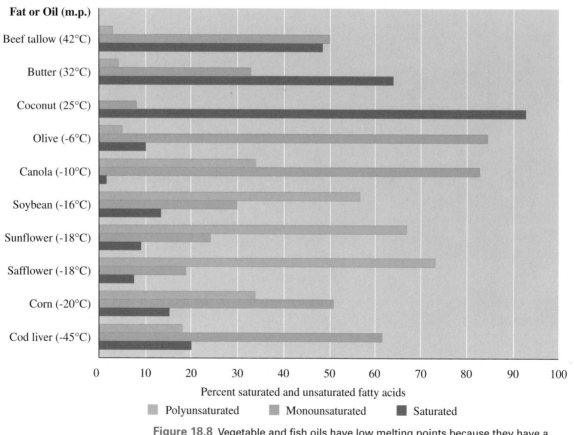

Fat or Oil (m.p.)
Beef tallow (42°C)
Butter (32°C)
Coconut (25°C)
Olive (-6°C)
Canola (-10°C)
Soybean (-16°C)
Sunflower (-18°C)
Safflower (-18°C)
Corn (-20°C)
Cod liver (-45°C)

Percent saturated and unsaturated fatty acids

■ Polyunsaturated   ■ Monounsaturated   ■ Saturated

Figure 18.8 Vegetable and fish oils have low melting points because they have a higher percentage of unsaturated fatty acids than do animal fats.

Q Why is the melting point of butter higher than that of olive or canola oil?

OLESTRA: A FAT SUBSTITUTE

In 1968, food scientists designed an artificial fat called *olestra* as a source of nutrition for premature babies. However, olestra could not be digested and was never used for that purpose. Then scientists realized that olestra had the flavor and texture of a fat without the calories.

Olestra is manufactured by obtaining the fatty acids from the fats in cottonseed or soybean oils and bonding the fatty acids with the hydroxyl groups on sucrose. Chemically, olestra is composed of six to eight long-chain fatty acids attached by ester links to a sugar (sucrose) rather than to a glycerol molecule found in fats. This makes olestra a very large molecule, which cannot be absorbed through the intestinal walls. The enzymes and bacteria in the intestinal tract are unable to break down the olestra molecule and it travels through the intestinal tract undigested.

In 1996, the Food and Drug Administration approved olestra for use in potato chips, tortilla chips, crackers, and fried snacks. Olestra snack products were test-marketed in 1996 in parts of Iowa, Wisconsin, Indiana, and Ohio. By 1997, there were reports of some adverse reactions, including diarrhea, abdominal cramps, and anal leakage, indicating that olestra may act as a laxative in some people. However, the manufacturers contend there is no direct proof that olestra is the cause of those effects.

The large molecule of olestra also combines with fat-soluble vitamins (A, D, E, and K) as well as the carotenoids from the foods we eat before they can be absorbed through the intestinal wall. Carotenoids are plant pigments in fruits and vegetables that protect against cancer, heart disease, and macular degeneration, a form of blindness in the elderly. The FDA now requires manufacturers to add the four vitamins, but not the carotenoids, to olestra products. The label on an olestra product must state the following: "This product contains olestra. Olestra may cause abdominal cramping and loose stools. Olestra inhibits the absorption of some vitamins and other nutrients. Vitamins A, D, E, and K have been added." Some snack foods made with olestra are now in supermarkets nationwide. Since there are already low-fat snacks on the market, it remains to be seen whether olestra will have any significant effect on reducing the problem of obesity.

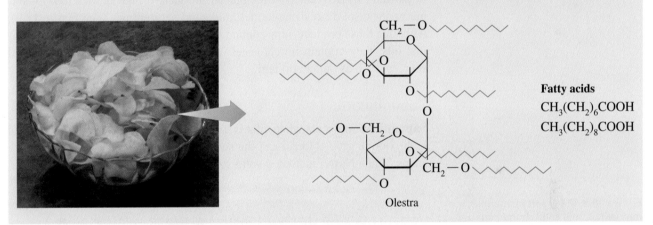

Fatty acids

$CH_3(CH_2)_6COOH$

$CH_3(CH_2)_8COOH$

Olestra

The amounts of saturated, monounsaturated, and polyunsaturated fatty acids in some typical fats and oils are shown in Figure 18.8. Saturated fatty acids have higher melting points than unsaturated fatty acids because they pack together more tightly. Animal fats usually contain more saturated fatty acids than do vegetable oils. Therefore, the melting points of animal fats are higher than those of vegetable oils.

Reactions of Fats and Oils

The **hydrogenation** of unsaturated fats converts carbon–carbon double bonds to single bonds. The hydrogen gas is bubbled through the heated oil in the presence of a nickel catalyst.

$$-CH=CH- + H_2 \xrightarrow{Ni} -\overset{\overset{\displaystyle H}{|}}{\underset{\underset{\displaystyle H}{|}}{C}}-\overset{\overset{\displaystyle H}{|}}{\underset{\underset{\displaystyle H}{|}}{C}}-$$

For example, when hydrogen adds to all of the double bonds of triolein using a nickel catalyst, the product is the saturated fat tristearin.

$$CH_2-O-\overset{\displaystyle O}{\overset{\|}{C}}-(CH_2)_7CH=CH(CH_2)_7CH_3$$
$$CH-O-\overset{\displaystyle O}{\overset{\|}{C}}-(CH_2)_7CH=CH(CH_2)_7CH_3 + 3H_2 \xrightarrow{Ni}$$
$$CH_2-O-\overset{\displaystyle O}{\overset{\|}{C}}-(CH_2)_7CH=CH(CH_2)_7CH_3$$

glyceryl trioleate
(triolein)

$$CH_2-O-\overset{\displaystyle O}{\overset{\|}{C}}-(CH_2)_{16}CH_3$$
$$CH-O-\overset{\displaystyle O}{\overset{\|}{C}}-(CH_2)_{16}CH_3$$
$$CH_2-O-\overset{\displaystyle O}{\overset{\|}{C}}-(CH_2)_{16}CH_3$$

glyceryl tristearate
(tristearin)

In commercial hydrogenation, the addition of hydrogen is stopped before all the double bonds in an oil are completely saturated. Complete hydrogenation gives a very brittle product, whereas the partial hydrogenation of a liquid vegetable oil changes it to a soft, semisolid fat. As the oil becomes more saturated, the melting point increases and the fat becomes more solid at room temperature. Control of the degree of hydrogenation gives the various types of partially hydrogenated vegetable oil products on the market today—soft margarines, solid stick margarines, and solid shortenings. (See Figure 18.9). Although these products now contain more saturated fatty acids than the original oils, they contain no cholesterol, unlike similar products from animal sources, such as butter and lard.

Saponification

Saponification occurs when a fat is heated with a strong base such as sodium hydroxide to give glycerol and the sodium salts of the fatty acids, which are soaps. When NaOH is used, a solid soap is produced that can be molded into

Figure 18.9 Many soft margarines, stick margarines, and solid shortenings are produced by the partial hydrogenation of vegetable oils.

Q How does hydrogenation change the structure of the fatty acids in the vegetable oils?

Vegetable oils
(liquids)

Shortening
(solid)

Tub (soft)
margarine

Stick margarine
(soft and solid)

TRANS FATTY ACIDS AND HYDROGENATION

Margarine is produced by partially hydrogenating the unsaturated fats in vegetable oils such as safflower oil, corn oil, canola oil, cottonseed oil, and sunflower oil, which usually contain cis double bonds. As hydrogenation occurs, some of the cis double bonds are converted to trans double bonds. If the label on a product states that the oils have been "partially" or "fully hydrogenated," that product will also contain trans fatty acids.

The concern about trans fatty acids is that their altered structure may make them behave like saturated fatty acids in the body. Several studies reported that trans fatty acids raise the levels of LDL-cholesterol, low-density lipoproteins containing cholesterol that can accumulate in the arteries. Some studies also report that trans fatty acids lower HDL-cholesterol, high-density lipoproteins that carry cholesterol to the liver to be excreted. Current evidence does not yet indicate that the intake of trans fatty acids is a significant risk factor for heart disease.

The trans fatty acids controversy will continue to be debated as more research is done.

Foods containing trans fatty acids include milk, bread, fried foods, ground beef, baked goods, stick margarine, butter, soft margarine, cookies, crackers, and vegetable shortening. The American Heart Association recommends that margarine should have no more than 2 grams of saturated fat per tablespoon. They also recommend the use of soft margarine, which is lower in trans fatty acids because soft margarine is only slightly hydrogenated, and diet margarine because it has less fat and therefore fewer trans fatty acids.

There are several products including peanut butter and butter-like spreads on the market that have 0% trans fatty acids. On the labels, they state that their products are nonhydrogenated, which avoids the production of the undesirable trans fatty acid. By 2006, food labels had to show the grams of trans fat per serving.

Cis-oleic acid

H₂/Ni

Double bond opens

Ni catalyst

Double bond reforms as trans

Addition of H₂

Undesired side product (*trans*-oleic acid)

Desired saturated product (stearic acid)

a desired shape; KOH produces a softer, liquid soap. Oils that are polyunsaturated produce softer soaps. Names like "coconut" or "avocado shampoo" tell you the sources of the oil used in the hydrolysis reaction.

Fat or oil + strong base → glycerol + salts of fatty acids (soaps)

glyceryl tripalmitate (tripalmitin)

glycerol

3 sodium palmitate (soap)

Steroids: Cholesterol and Steroid Hormones

Steroids are compounds containing the steroid nucleus, which consists of four carbon rings fused together. Although they are large molecules, steroids do not contain fatty acids.

steroid

Attaching other atoms and groups of atoms to the steroid structure forms a wide variety of steroid compounds. **Cholesterol,** which is one of the most important and abundant steroids in the body, is a *sterol* because it contains an oxygen atom as a hydroxyl (—OH) group. Like many steroids, cholesterol has methyl groups and a carbon chain with a double bond. In other steroids, the hydroxyl group is replaced by a carbonyl (C=O) group. Cholesterol in the body is obtained from eating meats, milk, and eggs, and it is also synthesized by the liver from fats, carbohydrates, and proteins. There is no cholesterol in vegetable and plant products. High levels of cholesterol are also associated with the accumulation of lipid deposits (plaque) that line and narrow the coronary arteries. (See Figure 18.10.)

cholesterol

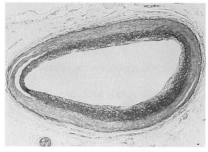

(a)

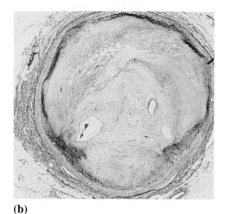

(b)

Figure 18.10 Excess cholesterol forms plaque that can block an artery, resulting in a heart attack. **(a)** A normal, open artery shows no buildup of plaque. **(b)** An artery that is almost completely clogged by atherosclerotic plaque.

Q What property of cholesterol would cause it to form deposits along the coronary arteries?

Steroid Hormones

The word *hormone* comes from the Greek "to arouse" or "to excite." Hormones are chemical messengers that serve as a kind of communication system from one part of the body to another. The *steroid* hormones, which include the sex hormones and the adrenocortical hormones, are closely related in structure to cholesterol and depend on cholesterol for their synthesis.

Two important male sex hormones, *testosterone* and *androsterone,* promote the growth of muscle and of facial hair and the maturation of the male sex organs and of sperm.

The *estrogens,* a group of female sex hormones, direct the development of female sexual characteristics. *Progesterone* prepares the uterus for the implantation of a fertilized egg. The structures of some steroid hormones follow:

testosterone (androgen)

estradiol (estrogen)

progesterone

18.19 Which of the following fatty acids are saturated and which are unsaturated? (See Table 18.2.)
 a. lauric acid   **b.** linolenic acid
 c. palmitoleic acid   **d.** stearic acid

18.20 Which of the following fatty acids are saturated and which are unsaturated? (See Table 18.2.)
 a. linoleic acid   **b.** palmitic acid
 c. myristic acid   **d.** oleic acid

18.21 Draw the structure of tripalmitin.

18.22 Draw the structure of triolein.

18.23 Safflower oil is called a polyunsaturated oil, whereas olive oil is a monounsaturated oil. Explain.

18.24 Why does olive oil have a lower melting point than butter fat?

18.25 A label on a container of margarine states that it contains partially hydrogenated corn oil.
 a. How has the liquid corn oil been changed?
 b. Why is the margarine product solid?

18.26 Write the product of the hydrogenation of the following triacylglycerol:

$$CH_2-O-\overset{\displaystyle O}{\overset{\|}{C}}-(CH_2)_{16}CH_3$$
$$CH-O-\overset{\displaystyle O}{\overset{\|}{C}}-(CH_2)_7CH=CH(CH_2)_7CH_3$$
$$CH_2-O-\overset{\displaystyle O}{\overset{\|}{C}}-(CH_2)_{16}CH_3$$

18.27 Draw the structure for the steroid nucleus.

18.28 What are the functional groups on the steroid nucleus in the sex hormones estradiol and testosterone?

18.4 Proteins

Learning Goal

Describe protein functions, and draw structures for amino acids and dipeptides.

Proteins perform many different functions in the body. There are proteins that form structural components such as cartilage, muscles, hair, and nails. Wool, silk, feathers, and horns are proteins made by animals. Proteins called enzymes regulate biological reactions such as digestion and cellular metabolism. Still other proteins, hemoglobin and myoglobin, carry oxygen in the blood and muscle. Table 18.3 gives examples of proteins that are classified by their functions in biological systems.

the Chemistry place

WEB TUTORIAL
Functions of Proteins

Table 18.3   Classification of Some Proteins and Their Functions

Class of Protein	Function in the Body	Examples
Structural	Provide structural components	*Collagen* is in tendons and cartilage. *Keratin* is in hair, skin, wool, and nails.
Contractile	Movement of muscles	*Myosin* and *actin* contract muscle fibers.
Transport	Carry essential substances throughout the body	*Hemoglobin* transports oxygen. *Lipoproteins* transport lipids.
Storage	Store nutrients	*Casein* stores protein in milk. *Ferritin* stores iron in the spleen and liver.
Hormone	Regulate body metabolism and nervous system	*Insulin* regulates blood glucose level. *Growth hormone* regulates body growth.
Enzyme	Catalyze biochemical reactions in the cells	*Sucrase* catalyzes the hydrolysis of sucrose. *Trypsin* catalyzes the hydrolysis of proteins.
Protection	Recognize and destroy foreign substances	*Immunoglobulins* stimulate immune responses.

Amino Acids

Proteins are composed of molecular building blocks called **amino acids,** which contain an amino group ($-NH_2$) and a carboxylic acid group ($-COOH$) bonded to a central carbon atom. Amino acids with this structure are called α (alpha) amino acids. Although there are many amino acids, only 20 different amino acids are usually present in the proteins in humans. The unique characteristics of the 20 amino acids are due to a side chain (R), which can include alkyl, hydroxyl, thiol ($-SH$), amino, sulfur, or aromatic groups or rings containing carbon and nitrogen. In biological systems, amino acids ionize.

General Structure of an α-Amino Acid

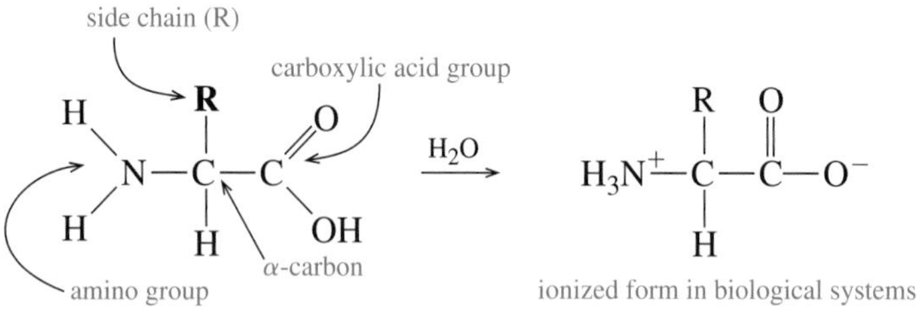

Nonpolar amino acids, which have hydrocarbon side chains, are **hydrophobic** ("water-fearing"). **Polar amino acids,** which have polar or ionic side chains, are **hydrophilic** ("water-attracting"). The side chain of an acidic amino acid contains a carboxylic acid group that donates a proton (H^+). The amino group in the side chain of a basic amino acid accepts a proton. The structures of 20 amino acids found in proteins, their common names, and their three-letter abbreviations are listed in Table 18.4.

Sample Problem 18.6   **Structural Formulas of Amino Acids**

Write the structural formulas and abbreviations for the following amino acids:

a. alanine (R = $-CH_3$)   **b.** serine (R = $-CH_2OH$)

Solution

a. The structure of the amino acids is written by attaching the side group (R) to the central carbon atom of the general structure of an amino acid.

alanine (Ala)

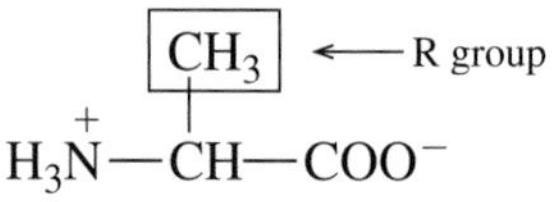

b. serine (Ser)

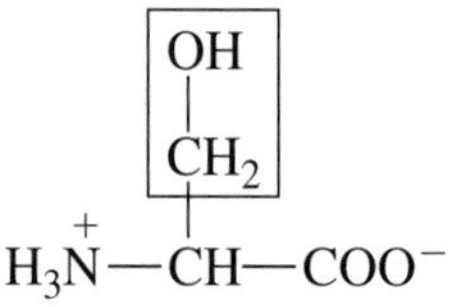

Study Check

Classify the amino acids in the sample problem as polar or nonpolar.

Table 18.4   The 20 Amino Acids in Proteins

Nonpolar Amino Acids

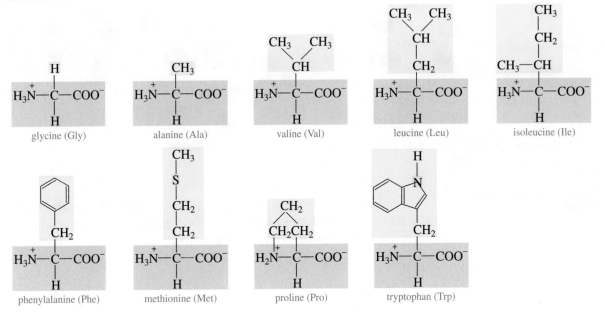

glycine (Gly)   alanine (Ala)   valine (Val)   leucine (Leu)   isoleucine (Ile)

phenylalanine (Phe)   methionine (Met)   proline (Pro)   tryptophan (Trp)

Polar Amino Acids (Neutral)

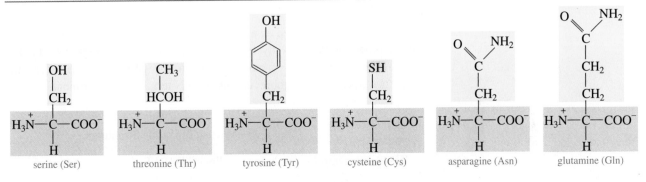

serine (Ser)   threonine (Thr)   tyrosine (Tyr)   cysteine (Cys)   asparagine (Asn)   glutamine (Gln)

Acidic Amino Acids   Basic Amino Acids

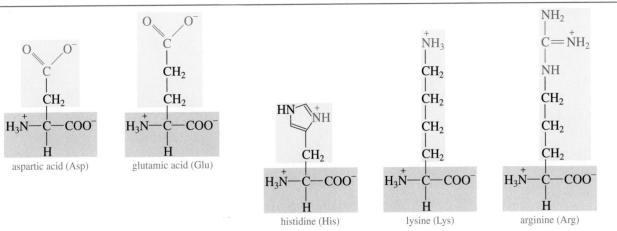

aspartic acid (Asp)   glutamic acid (Glu)   histidine (His)   lysine (Lys)   arginine (Arg)

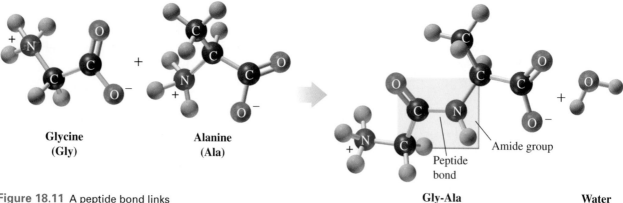

Glycine
(Gly)

Alanine
(Ala)

Amide group

Peptide
bond

Gly-Ala

Water

Figure 18.11 A peptide bond links glycine and alanine to form the dipeptide Gly-Ala.

Q *What functional groups in glycine and alanine form the peptide bond?*

Peptides

The linking of two or more amino acids forms a **peptide.** A **peptide bond** is an amide bond that forms when the —COO⁻ group of one amino acid reacts with the H₃N⁺— group of the next amino acid.

$$H_3\overset{+}{N}-CH-\boxed{\underset{R}{\overset{\overset{O}{\parallel}}{C}}-O^-} + H_3\overset{+}{N}-\underset{R}{\overset{\overset{O}{\parallel}}{CH-C}}-O^- \longrightarrow H_3\overset{+}{N}-\underset{R}{CH}-\boxed{\underset{}{\overset{\overset{O\;\;\;H}{\parallel\;\;|}}{C-N}}}-\underset{R}{CH}-\overset{\overset{O}{\parallel}}{C}-O^- + H_2O$$

amino acid 1   amino acid 2   dipeptide

N terminal   C terminal

Two amino acids linked by a peptide bond form a *dipeptide*. The formation of the dipeptide between glycine and alanine is shown in Figure 18.11. For convenience, the order of amino acids in the peptide is written as the sequence of three-letter abbreviations from N terminal to C terminal.

Sample Problem 18.7 **Writing Dipeptide Structures**

Write a structural formula for the dipeptide Val-Ser.

Solution

Valine is joined to serine by a peptide bond; valine is the N terminal and serine is the C terminal.

$$H_3\overset{+}{N}-CH-\overset{\overset{O}{\parallel}}{C}-NH-CH-\overset{\overset{O}{\parallel}}{C}-O^-$$

CH—CH₃   CH₂OH

CH₃

From valine   From serine

Val-Ser

Study Check

Aspartame, an artificial sweetener 200 times sweeter than sucrose, contains the dipeptide Asp-Phe. Give the structure of the dipeptide in aspartame.

18.29 Describe the functional groups found in all α-amino acids.

18.30 How does the polarity of the side chain in leucine compare to that of the side chain in serine?

18.31 Draw the structural formula for each of the following amino acids:

    a. alanine   **b.** threonine   **c.** phenylalanine

18.32 Draw the structural formula for each of the following amino acids:

    a. serine   **b.** leucine   **c.** tyrosine

18.33 Classify the amino acids in problem 18.31 as hydrophobic (nonpolar) or hydrophilic (polar, neutral).

18.34 Classify the amino acids in problem 18.32 as hydrophobic (nonpolar) or hydrophilic (polar, neutral).

18.35 Give the name of the amino acid represented by each of the following three-letter abbreviations:

    a. Ala   **b.** Val   **c.** Lys   **d.** Cys

18.36 Give the name of the amino acid represented by each of the following three-letter abbreviations:

    a. Trp   **b.** Met   **c.** Pro   **d.** Gly

18.37 Draw the structural formula of each of the following peptides:

    a. Ala-Cys   **b.** Ser-Phe   **c.** Gly-Ala-Val

18.38 Draw the structural formula of each of the following peptides:

    a. Met-Asp   **b.** Ala-Trp   **c.** Met-Glu-Lys

18.5 Protein Structure

Primary Structure

Learning Goal

Identify the levels of structure of a protein.

When there are more than 50 amino acids in a chain, a polypeptide is called a **protein.** The **primary structure** is the order of the amino acids held together by peptide bonds.

    The first protein to have its primary structure determined was insulin, which is a hormone that regulates the glucose level in the blood. The primary structure of human insulin contains two polypeptide chains. Chain A has 21 amino acids, and chain B has 30 amino acids. The polypeptide chains are held together by disulfide bonds formed by the side chains of the cysteine amino acids in each of the chains. (See Figure 18.12.)

Secondary Structure

The **secondary structure** of a protein describes the way the amino acids next to or near each other along the polypeptide are arranged in space. The three most common types of secondary structure are the *alpha helix,* the *beta-pleated sheet,* and the *triple helix.* The corkscrew shape of an **alpha helix (α helix)** is held in place by hydrogen bonds between each N—H group and the oxygen of a C=O group in the next turn of the helix, four amino acids down the chain. (See Figure 18.13.) Because many hydrogen bonds form along the peptide backbone, this portion of the protein takes the shape of a strong, tight coil that looks like a telephone cord or a Slinky toy.

    Another type of secondary structure is known as the **beta-pleated sheet (β-pleated sheet).** In a β-pleated sheet, polypeptide chains are held together side by side by hydrogen bonds between the peptide chains. The hydrogen bonds holding the sheets tightly in place account for the strength and durability of proteins such as silk. (See Figure 18.14.)

the Chemistry place

WEB TUTORIAL

Structure of Proteins

the Chemistry place

WEB TUTORIAL

Primary and Secondary Structure

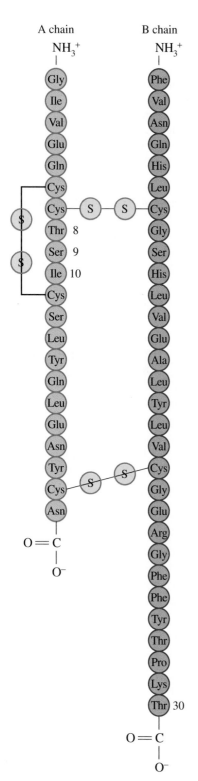

A chain   B chain

Figure 18.12 The sequence of amino acids in human insulin is its primary structure.

Q *What kinds of bonds occur in the primary structure of a protein?*

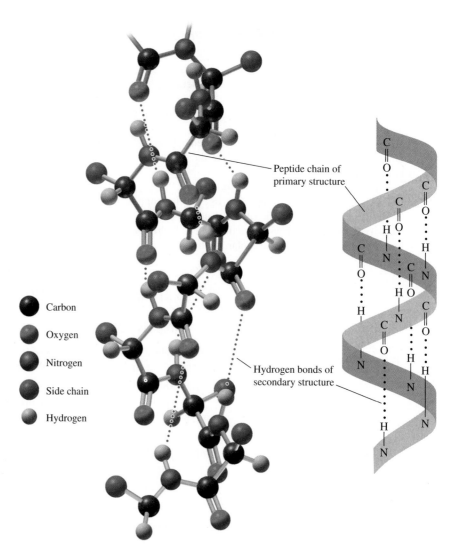

Figure 18.13 The α (alpha) helix acquires a coiled shape from hydrogen bonds between the N—H of the peptide bond in one loop and the C═O of the peptide bond in the next loop.

Q *What are partial charges of the H in N—H and the O in C═O that permit hydrogen bonds to form?*

Collagen, the most abundant protein, makes up as much as one-third of all the protein in vertebrates. It is found in connective tissue, blood vessels, skin, tendons, ligaments, the cornea of the eye, and cartilage. The strong structure of collagen is a result of three polypeptides woven together like a braid to form a **triple helix,** as seen in Figure 18.15. When several triple helixes wrap together, they form the fibrils that make up connective tissues and tendons. In a young person, collagen is elastic. As a person ages, additional cross-links form between the fibrils, which make collagen less elastic. Cartilage and tendons become more brittle, and wrinkles are seen in the skin.

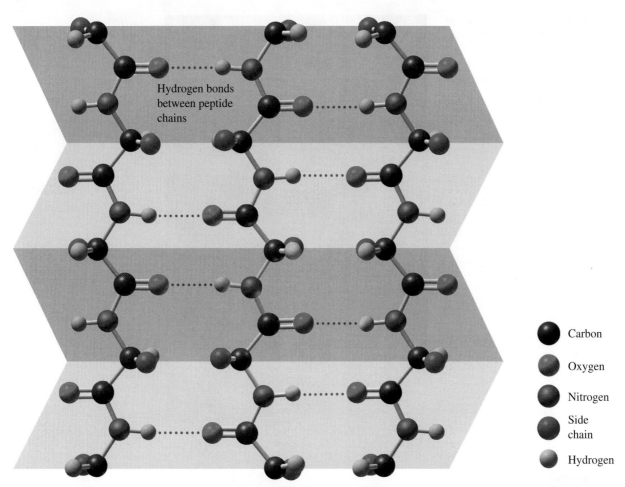

Figure 18.14 In a β (beta)-pleated sheet secondary structure, hydrogen bonds form between the peptide chains.

Q How do the hydrogen bonds differ in a β-pleated sheet from those in the alpha helix?

Carbon

Oxygen

Nitrogen

Side chain

Hydrogen

CHEM NOTE

PROTEIN STRUCTURE AND MAD COW DISEASE

Up until recently, researchers thought that only viruses or bacteria were responsible for transmitting diseases. Now a group of diseases have been found in which the infectious agents are proteins called *prions.* Bovine spongiform encephalopathy (BSE), or "mad cow disease," is a fatal brain disease of cattle in which the brain fills with cavities resembling a sponge. In the noninfectious form of the prion PrP^c, the N-terminal portion is a random coil. Although the noninfectious form may be ingested from meat products, its structure can change to what is known as PrP^{sc} or *prion-related protein scrapie.* In this infectious form, the end of the peptide chain folds into a β-pleated sheet, which has disastrous effects on the brain and spinal cord. The conditions that cause this structural change are not yet known.

The human variant is called Creutzfeldt–Jakob (CJD) disease. Around 1955, Dr. Carleton Gajdusek was studying a disease known as kuru, a neurological disease that was killing members of a tribe in Papua New Guinea. Because their diets were low in protein, it was a ritual to eat members of the tribe who died. As a result, the infectious agent kuru was transmitted from one member to another. After Gajdusek identified the infectious agent in kuru as similar to the prions that cause BSE, he received the Nobel prize.

BSE was diagnosed in Great Britain in 1986. The protein is present in nerve tissue but is not found in meat. Control measures that exclude brain and spinal cord from animal feed are now in place to reduce the incidence of BSE. In 2003, the first case of BSE was identified in the United States.

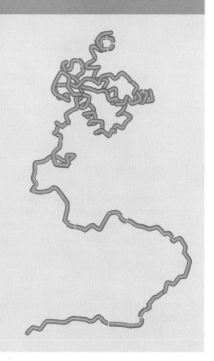

Figure 18.15 Hydrogen bonds between polar R groups in three polypeptide chains form the triple helixes that combine to make fibers of collagen.

Q How does the structure of collagen differ from a β-pleated sheet?

Triple helix   3 α-Helix peptide chains

CHEM NOTE

ESSENTIAL AMINO ACIDS

Of the 20 amino acids used to build the proteins in the body, only 10 can be synthesized in adequate amounts by the body. The other 10 amino acids, listed in Table 18.5, are *essential amino acids* that cannot be synthesized and must be obtained from the proteins in the diet.

Table 18.5 Essential Amino Acids

arginine (Arg)*	methionine (Met)
histidine (His)*	phenylalanine (Phe)
isoleucine (Ile)	threonine (Thr)
leucine (Leu)	tryptophan (Trp)
lysine (Lys)	valine (Val)

* Required in diets of children, not adults.

Complete proteins, which contain all of the essential amino acids are found in most animal products such as eggs, milk, meat, fish, and poultry. However, gelatin and plant proteins such as grains, beans, and nuts are *incomplete proteins* because they are deficient in one or more of the essential amino acids. Diets that rely on plant foods for protein must contain a variety of protein sources to obtain all the essential amino acids. For example, a diet of rice and beans contains all the essential amino acids because they are complementary proteins. Rice contains the methionine and tryptophan deficient in beans, while beans contain the lysine that is lacking in rice. (See Table 18.6.)

Table 18.6 Amino Acid Deficiency in Selected Vegetables and Grains

Food Source	Amino Acids Missing
Eggs, milk, meat, fish, poultry	none
Wheat, rice, oats	lysine
Corn	lysine, tryptophan
Beans	methionine, tryptophan
Peas	methionine
Almonds, walnuts	lysine, tryptophan
Soy	low in methionine

Sample Problem 18.8 ▸ Identifying Secondary Structures

Indicate the secondary structure (α helix, β-pleated sheet, or triple helix) described in each of the following:

a. a coiled peptide chain held in place by hydrogen bonding between peptide bonds in the same chain

b. a structure that has hydrogen bonds between polypeptide chains arranged side by side

Solution

a. α helix   **b.** β-pleated sheet

Study Check

What is the secondary structure in collagen?

WEB TUTORIAL
Tertiary and Quaternary Structure

Tertiary Structure

The **tertiary structure** of a protein involves attractions and repulsions between the side chain groups of the amino acids in the polypeptide chain. As interactions occur between different parts of the peptide chain, segments of the chain twist and bend until the protein acquires a specific three-dimensional shape. It is this unique molecular shape that determines the biological function of the molecule. The tertiary structure of a protein is stabilized by interactions between the R groups of the amino acids in one region of the polypeptide chain with R groups of amino acids in other regions of the protein. (See Figure 18.16.) Table 18.7 lists the stabilizing interactions of tertiary structures.

Figure 18.16 Interactions between amino acid side chains fold a protein into a specific three-dimensional shape called its tertiary structure.

Q *Why would one section of the protein chain move to the center while another section remains on the surface of the tertiary structure?*

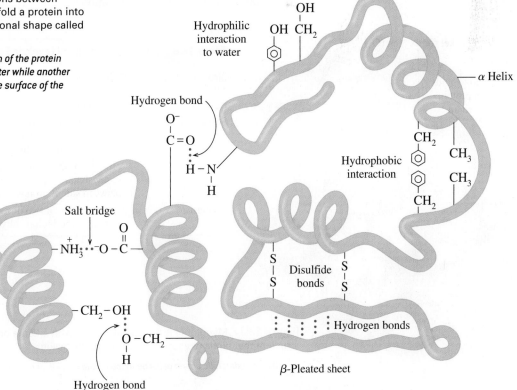

Table 18.7   Some Cross-Links in Tertiary Structures

	Nature of Bonding
Hydrophobic interactions	Attractions between nonpolar groups
Hydrophilic interactions	Attractions between polar groups and water
Salt bridges (ionic bonds)	Ionic interactions between acidic and basic amino acids
Hydrogen bonds	Occur between H and O or N
Disulfide bonds	Strong covalent links between sulfur atoms of two cysteine amino acids

Sample Problem 18.9   Cross-Links in Tertiary Structures

What type of interaction would you expect between the side chains of the following amino acids?
a. cysteine and cysteine
b. glutamic acid and lysine

Solution

a. Because the cysteine contains —SH, a disulfide bond will form.
b. An ionic bond (salt bridge) can form by the interaction of the —COO$^-$ of glutamic acid and the —NH$_3^+$ of lysine.

Study Check

What type of interaction would you expect between valine and leucine in a globular protein?

Myoglobin, a protein that stores oxygen in skeletal muscle, contains 153 amino acids in a single polypeptide chain. It forms a compact tertiary structure that contains a pocket of amino acids and a heme group that binds and stores oxygen (O_2). (See Figure 18.17.) When a biologically active protein consists of two or more polypeptide subunits, the structural level is referred to as a **quaternary structure.** Hemoglobin, a protein that transports oxygen in blood, consists of four polypeptide chains or subunits. The subunits are held together in the quaternary structure by the same interactions that stabilize the tertiary structure. Each subunit of the hemoglobin contains a heme group that binds oxygen. In the adult hemoglobin molecule, the quaternary structure of hemoglobin can bind and transport four molecules of oxygen. Table 18.8 and Figure 18.18 summarize the structural levels of proteins.

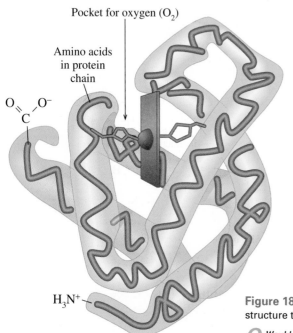

Pocket for oxygen (O_2)

Amino acids in protein chain

$O\!=\!\overset{}{C}\!-\!O^-$

H_3N^+—

Table 18.8   Summary of Structural Levels in Proteins

Structural Level	Characteristics
Primary	The sequence of amino acids
Secondary	The coiled α helix, β-pleated sheet, or a triple helix formed by hydrogen bonding between peptide bonds along the chain
Tertiary	A folding of the protein into a compact, three-dimensional shape stabilized by interactions between side R groups of amino acids
Quaternary	A combination of two or more protein subunits to form a larger, biologically active protein

Figure 18.17 Myoglobin is a globular protein with a heme pocket in its tertiary structure that binds oxygen to be carried to the tissues.

Q Would hydrophilic amino acids be found on the outside or inside of the myoglobin structure?

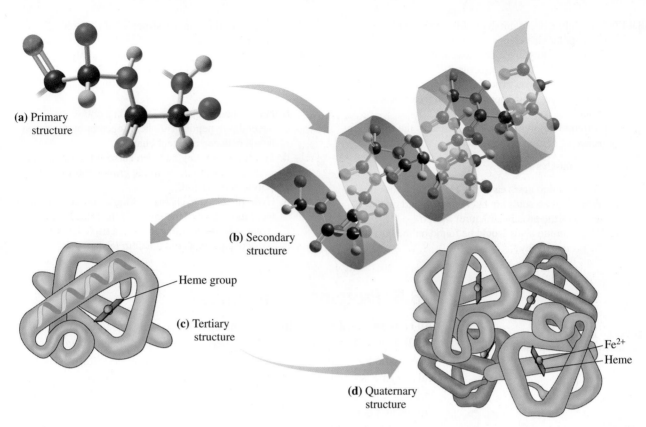

(a) Primary structure

(b) Secondary structure

Heme group

(c) Tertiary structure

Fe²⁺
Heme

(d) Quaternary structure

Figure 18.18 The quaternary structure of hemoglobin consists of four polypeptide subunits, each containing a heme group that binds an oxygen molecule.

Q *What is the difference between a tertiary structure and a quaternary structure?*

Sample Problem 18.10   **Identifying Protein Structure**

Indicate whether the following conditions are responsible for primary, secondary, tertiary, or quaternary protein structures:

a. Disulfide bonds form between portions of a protein chain.

b. Peptide bonds form a chain of amino acids.

Solution

a. Disulfide bonds help to stabilize the tertiary structure of a protein.

b. The sequence of amino acids in a polypeptide is a primary structure.

Study Check

What structural level is represented by the grouping of two subunits in insulin?

Questions and Problems   **Protein Structure**

18.39 Two peptides each contain one molecule of valine and two molecules of serine. What are their possible primary structures?

18.40 What are three different types of secondary protein structure?

18.41 What is the difference in bonding between an α helix and a β-pleated sheet?

18.42 How is the secondary structure of a β-pleated sheet different from that of a triple helix?

18.43 What type of interaction would you expect from the following amino acids in a tertiary structure?
a. two cysteine residues
b. serine and aspartic acid
c. two leucine residues

18.44 In myoglobin, about one-half of the 153 amino acids have nonpolar side chains.
  a. Where would you expect those amino acids to be located in the tertiary structure?
  b. Where would you expect the polar side chains to be?
  c. Why is myoglobin more soluble in water than silk or wool?

18.45 A portion of a polypeptide chain contains the following sequence of amino acid residues:

  -Leu-Val-Cys-Asp-

  a. Which amino acids can form a disulfide cross-link?
  b. Which amino acids are likely to be found on the inside of the protein structure? Why?
  c. Which amino acids would be found on the outside of the protein? Why?

  d. How does the primary structure of a protein affect its tertiary structure?

18.46 State whether the following statements apply to primary, secondary, tertiary, or quaternary protein structure:
  a. Side groups interact to form disulfide bonds or ionic bonds.
  b. Peptide bonds join amino acids in a polypeptide chain.
  c. Several polypeptides are held together by hydrogen bonds between adjacent chains.
  d. Hydrogen bonding between carbonyl oxygen atoms and nitrogen atoms of amide groups causes a polypeptide to coil.
  e. Hydrophobic side chains seeking a nonpolar environment move toward the inside of the folded protein.
  f. Protein chains of collagen form a triple helix.
  g. An active protein contains four tertiary subunits.

18.6 Proteins as Enzymes

Describe the role of an enzyme in an enzyme-catalyzed reaction.

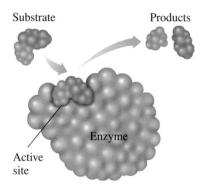

Figure 18.19 On the surface of an enzyme, a small region called an active site binds a substrate and catalyzes a reaction of that substrate.

Q Why does an enzyme catalyze a reaction of only certain substrates?

An **enzyme** has a unique three-dimensional shape that recognizes and binds a small group of reacting molecules called *substrates.* In a catalyzed reaction, an enzyme must first bind to a substrate in a way that favors catalysis. A typical enzyme is much larger than its substrate. However, within its large tertiary structure, there is a region called the **active site** that binds a substrate or substrates and catalyzes the reaction. This active site is often a small pocket that closely fits the structure of the substrate. (See Figure 18.19.) In an early theory of enzyme action called the **lock-and-key model,** the active site is described as having a rigid, nonflexible shape. Thus only those substrates with shapes that fit exactly into the active site are able to bind with that enzyme. The shape of the active site is analogous to a lock, and the proper substrate is the key that fits into the lock. (See Figure 18.20a.)

While the lock-and-key model explains the binding of substrates for many enzymes, certain enzymes have a broader range of specificity than the lock-and-key model allows. In the **induced-fit model,** there is an interaction between both the enzyme and substrate. (See Figure 18.20b.) The active site adjusts to fit the shape of the substrate more closely. At the same time the substrate adjusts its shape to better adapt to the geometry of the active site. In the induced-fit model, substrate and enzyme work together to acquire a geometrical arrangement that lowers the activation energy, thereby increasing the rate of reaction. A different substrate could not induce these structural changes and no catalysis would occur. (See Figure 18.20c.)

Sample Problem 18.11 **Active Site and Enzyme Activity**

What is the function of the active site in an enzyme?

Solution

The active site in an enzyme binds the substrate and contains the amino acid side chains that bind the substrate and catalyze the reaction.

Study Check

How do the lock-and-key and the induced-fit models differ in their description of the active site in an enzyme?

Figure 18.20 **(a)** In the lock-and-key model, a substrate fits the shape of the active site and forms an enzyme–substrate complex. **(b)** In the induced-fit model, a flexible active site and substrate adapt to provide a close fit to a substrate and proper orientation for reaction. **(c)** A substrate that does not fit or induce a fit in the active site cannot undergo catalysis by the enzyme.

Q *How does the induced-fit model differ from the lock-and-key model?*

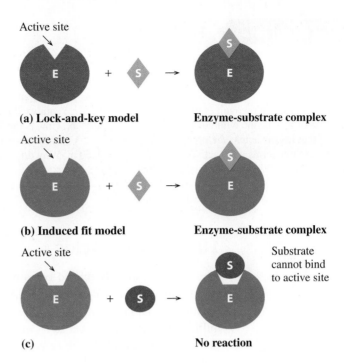

Active site

(a) Lock-and-key model   **Enzyme-substrate complex**

Active site

(b) Induced fit model   **Enzyme-substrate complex**

Active site

Substrate cannot bind to active site

(c)   **No reaction**

Enzyme Catalyzed Reaction

The proper alignment of a substrate within the active site forms an **enzyme–substrate (ES) complex.** This combination of enzyme and substrate provides an alternative pathway for the reaction that has a lower activation energy. Within the active site, the amino acid side chains take part in catalyzing the chemical reaction. For example, acidic and basic side chains remove protons from or provide protons for the substrate. As soon as the catalyzed reaction is complete, the products are quickly released from the enzyme so it can bind to a new substrate molecule. We can write the catalyzed reaction of an enzyme (E) with a substrate (S) to form product (P) as follows:

Step 1 $E + S \rightleftharpoons ES$

Step 2 $ES \longrightarrow E + P$

$$E + S \rightleftharpoons ES \longrightarrow E + P$$
Enzyme + substrate   ES complex   Enzyme + product

the
Chemistry
place

WEB TUTORIAL

How Enzymes Work

Let's consider the hydrolysis of sucrose by sucrase. When sucrose binds to the active site of sucrase, the glycosidic bond of sucrose is placed into a geometry favorable for reaction. The amino acid side chains catalyze the cleavage of the sucrose to give the products glucose and fructose.

$$\text{Sucrase} + \text{sucrose} \rightleftharpoons \text{sucrase-sucrose complex} \longrightarrow \text{sucrase} + \text{glucose} + \text{fructose}$$
E   +   S   ES complex   E   + P_1 + P_2

Because the structures of the products are no longer attracted to the active site, they are released and the sucrase binds another sucrose substrate.

Questions and Problems ▸ Proteins as Enzymes

18.47 Match the following three terms—(1) enzyme–substrate complex, (2) enzyme, and (3) substrate—with these phrases:
 a. has a tertiary structure that recognizes the substrate
 b. the combination of an enzyme with the substrate
 c. has a structure that fits the active site of an enzyme
18.48 Match the following three terms—(1) active site, (2) lock-and-key model, and (3) induced-fit model—with these phrases:
 a. the portion of an enzyme where catalytic activity occurs
 b. an active site that adapts to the shape of a substrate
 c. an active site that has a rigid shape
18.49 a. Write an equation that represents an enzyme-catalyzed reaction.
 b. How is the active site different from the whole enzyme structure?
18.50 a. Why does an enzyme speed up the reaction of a substrate?
 b. After the products have formed, what happens to the enzyme?

18.7 Nucleic Acids

Learning Goal

Describe the structure of the nucleic acids in DNA and RNA.

There are two closely related types of nucleic acids: *deoxyribonucleic acid* (**DNA**) and *ribonucleic acid* (**RNA**). Both are polymers of repeating monomer units known as *nucleotides*. A DNA molecule may contain several million nucleotides; smaller RNA molecules may contain up to several thousand. Each nucleotide has three components: a base, a five-carbon sugar, and a phosphate group. (See Figure 18.21.)

The *nitrogen-containing bases* in nucleic acids are derivatives of *pyrimidine* or *purine*.

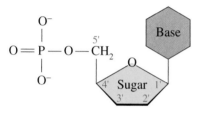

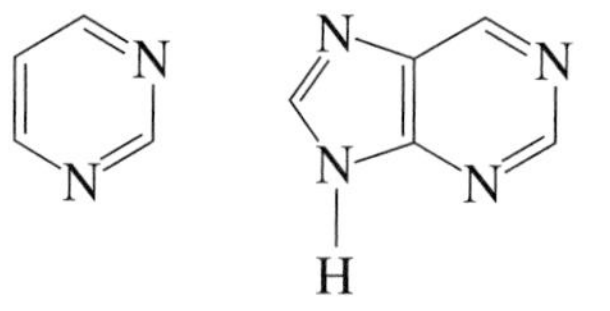

Figure 18.21 A diagram of the general structure of a nucleotide found in nucleic acids.

In a nucleotide, what types of groups are bonded to a five-carbon sugar?

Pyrimidine   Purine

In DNA, there are two purines, adenine (A) and guanine (G), and two pyrimidines, cytosine (C) and thymine (T). RNA contains the same bases, except thymine (5-methyluracil) is replaced by uracil (U). (See Figure 18.22.)

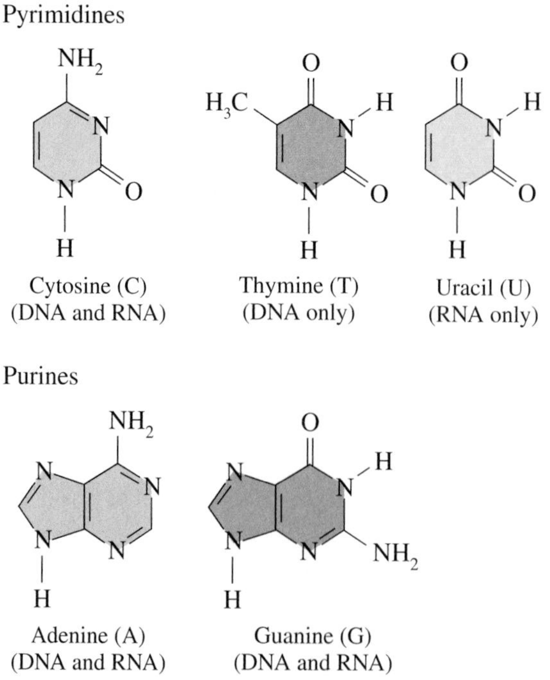

Figure 18.22 DNA contains the bases A, G, C, and T; RNA contains A, G, C, and U.

Which bases are found in DNA?

Pyrimidines

Cytosine (C)
(DNA and RNA)

Thymine (T)
(DNA only)

Uracil (U)
(RNA only)

Purines

Adenine (A)
(DNA and RNA)

Guanine (G)
(DNA and RNA)

Figure 18.23 The five-carbon pentose sugar found in RNA is ribose and deoxyribose in DNA.

Q What is the difference between ribose and deoxyribose?

Pentose sugars in RNA and DNA

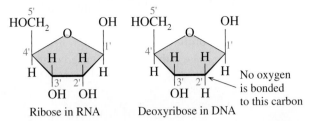

Ribose in RNA   Deoxyribose in DNA

No oxygen is bonded to this carbon

Ribose and Deoxyribose Sugars

The nucleotides of RNA and DNA contain five-carbon pentose sugars. In RNA, the five-carbon sugar is *ribose,* which gives the letter R in the abbreviation RNA. In DNA, the five-carbon sugar is *deoxyribose,* which is similar to ribose except that there is no hydroxyl group (—OH) on C2′ of ribose. The *deoxy* prefix means "without oxygen" and provides the D in DNA. The atoms in the pentose sugars are numbered with primes (1′, 2′, 3′, 4′, and 5′) to differentiate them from the atoms in the nitrogen bases. (See Figure 18.23.)

the Chemistry place

WEB TUTORIAL
DNA and RNA Structure

Nucleosides and Nucleotides

A **nucleoside** is produced when a base forms a glycosidic bond to C1′ of a sugar.

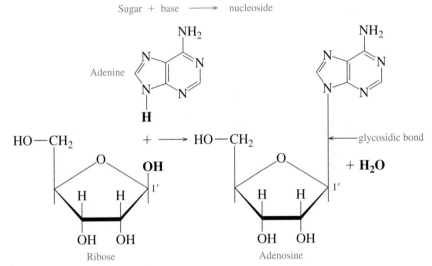

Nucleotides are formed when the C5′ —OH group of a sugar in a nucleoside bonds with phosphoric acid.

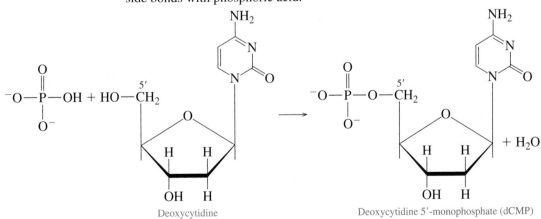

Free 5' end

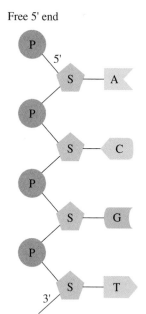

Free 3' end

The name of a nucleotide is obtained from the name of the nucleoside followed by 5'-monophosphate. A nucleotide of RNA is named as a *nucleoside 5'-monophosphate,* or NMP. Nucleotides of DNA have the prefix *deoxy.* Although the letters A, G, C, U, and T represent the bases, they are often used in the abbreviations of the respective nucleotides as well.

Structure of Nucleic Acids

The **nucleic acids** consist of polymers of many nucleotides in which the 3'—OH group of the sugar in one nucleotide bonds to the phosphate group on the 5'—carbon atom in the sugar of the next nucleotide to form a **phosphodiester bond.** As more nucleotides are added, a backbone forms that consists of alternating sugar and phosphate groups.

Along a DNA or RNA chain, the bases attached to each of the sugars extend out from the nucleic acid backbone. A nucleic acid sequence is often written using only the letters of the bases. Thus, the nucleotides in a section of RNA shown in Figure 18.24 are read from 5' $\longrightarrow$ 3' as ACGU.

RNA (ribonucleic acid)

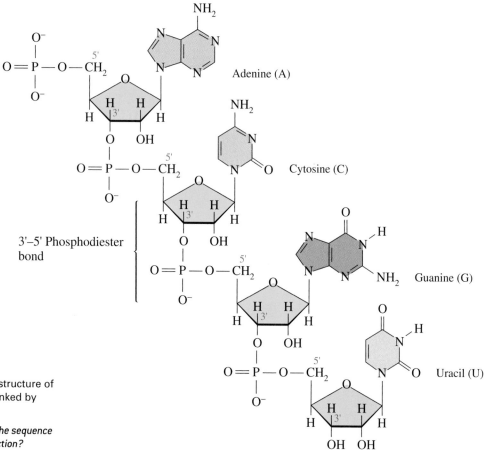

Figure 18.24 In the primary structure of an RNA, the nucleotides are linked by 3'–5' phosphodiester bonds.

Q What is the abbreviation for the sequence of nucleotides in this RNA section?

DNA Double Helix: A Secondary Structure

In 1953, James Watson and Francis Crick proposed that DNA was a **double helix** that consists of two polynucleotide strands winding about each other like a spiral staircase. (See Figure 18.25.) The hydrophilic sugar-phosphate backbones are analogous to the outside railings with the hydrophobic bases arranged like steps along the inside. Each of the bases along one polynucleotide strand forms hydrogen bonds to a specific base on the opposite DNA strand. Adenine only bonds to thymine, and guanine only bonds to cytosine. (See Figure 18.26.) The pairs A—T and G—C are called **complementary base pairs.** The specificity of the base pairing is due to the fact that adenine and thymine form two hydrogen bonds, while cytosine and guanine form three hydrogen bonds. Thus, DNA has equal amounts of A and T bases and equal amounts of G and C.

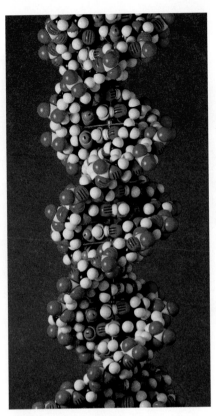

Figure 18.25 This space-filling model shows the double helix that is the characteristic shape of DNA molecules.

Q What is meant by the term double helix?

Sample Problem 18.12   **Complementary Base Pairs**

Write the base sequence of the complementary segment for the following segment of a strand of DNA:

    A—C—G—A—T—C—T

Solution

In the complementary strand of DNA, the base A pairs with T, and G pairs with C.

Given segment of DNA: A—C—G—A—T—C—T

                               :   :   :   :   :   :   :

Complementary segment: T—G—C—T—A—G—A

Study Check

What is the sequence of bases that is complementary to a portion of DNA with a base sequence of G—G—T—T—A—A—C—C?

DNA Replication

When cells divide, copies of DNA must be produced in order to transfer the genetic information to the new cells. This is the process of DNA replication. In DNA **replication,** the strands in the parent DNA separate, which allows each of the original strands to makes copies by synthesizing complementary strands. The replication process begins when an enzyme catalyzes the unwinding of a portion of the double helix by breaking the hydrogen bonds between the complementary bases. These single strands now act as templates for the synthesis of new complementary strands. (See Figure 18.27.) Within the nucleus, nucleoside triphosphates for each base are available so that each exposed base on the template strand can form hydrogen bonds with its complementary base in the nucleoside triphosphate. After the base pairs are formed, *DNA polymerase* catalyzes the formation of phosphodiester bonds between the nucleotides.

the Chemistry place

WEB TUTORIAL
DNA Replication

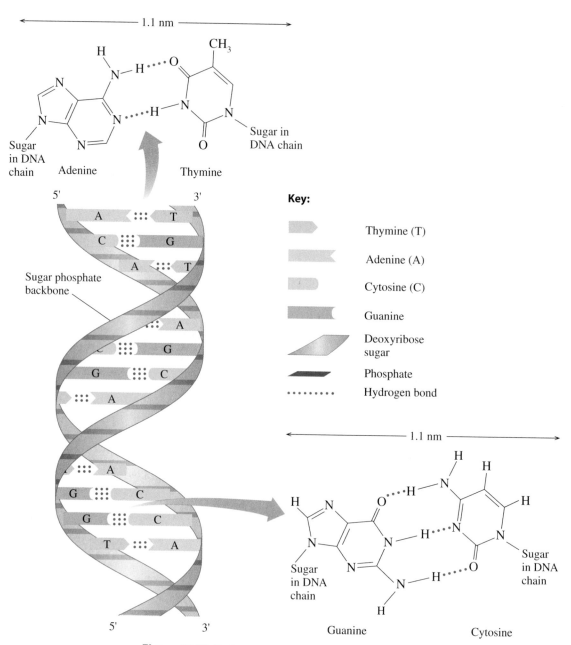

Figure 18.26 Hydrogen bonds between complementary base pairs hold the polynucleotide strands together in the double helix of DNA.

Q *Why are G — C base pairs more stable than A — T base pairs?*

Eventually, the entire double helix of the parent DNA is copied. In each new DNA molecule, one strand of the double helix is from the original DNA and one is a newly synthesized strand. This process produces two new DNAs called *daughter DNA* that are identical to each other and exact copies of the original parent DNA. In DNA replication, complementary base pairing ensures the correct placements of bases in the new DNA strands.

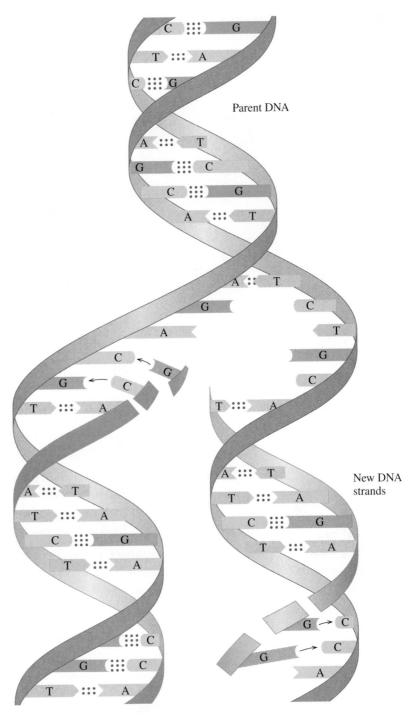

Figure 18.27 In DNA replication, the separate strands of the parent DNA are the templates for the synthesis of complementary strands, which produces two exact copies of DNA.

Q *How many strands of the parent DNA are in each of the new double-stranded copies of DNA?*

Questions and Problems   Nucleic Acids

18.51 Identify the following base as present in RNA, DNA, or both:
   a. thymine

   b.

18.52 Identify the following base as present in RNA, DNA, or both:
   a. guanine

   b.

18.53 How are the two strands of nucleic acid in DNA held together?

18.54 What is meant by complementary base pairing?

18.55 Complete the base sequence in a second DNA strand if a portion of one strand has the following base sequence:
   a. AAAAAA   b. GGGGGG
   c. AGTCCAGGT   d. CTGTATACGTTA

18.56 Complete the base sequence in a second DNA strand if a portion of one strand has the following base sequence:
   a. TTTTTT   b. CCCCCCCCC
   c. ATGGCA   d. ATATGCGCTAAA

18.57 What process ensures that the replication of DNA produces identical copies?

18.58 What is the function of the enzyme DNA polymerase in DNA replication?

18.8 Protein Synthesis

Learning **Goal**

Describe the synthesis of protein from mRNA.

WEB TUTORIAL
Overview of Protein Synthesis

Ribonucleic acid, RNA, the most prevalent nucleic acid in the cell, is involved with transmitting the genetic information needed to operate the cell. Similar to DNA, RNA molecules are polymers of nucleotides. However, there are several important differences:

1. The sugar in RNA is ribose rather than the deoxyribose found in DNA.
2. The base uracil replaces thymine.
3. RNA molecules are single, not double stranded.
4. RNA molecules are much smaller than DNA molecules.

There are three major types of RNA in the cells: *messenger RNA, ribosomal RNA,* and *transfer RNA,* which are classified according to their location and function in Table 18.9.

Table 18.9   **Types of RNA Molecules**

Type	Abbreviation	Function in the Cell
Ribosomal RNA	rRNA	Major component of the ribosomes
Messenger RNA	mRNA	Carries information for protein synthesis from the DNA in the nucleus to the ribosomes
Transfer RNA	tRNA	Brings amino acids to the ribosomes for protein synthesis

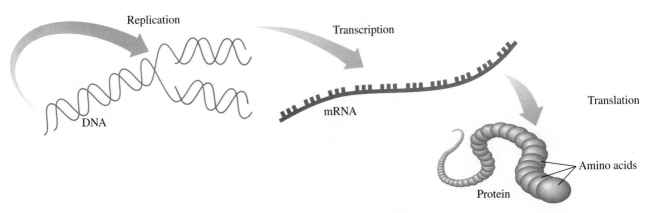

Figure 18.28 The genetic information in DNA is replicated in cell division and used to produce messenger RNAs. The mRNAs are converted into amino acids for protein synthesis.

Q *What is the difference between transcription and translation?*

Transcription: Synthesis of mRNA

In replication, the genetic information in DNA is reproduced by making identical copies of DNA. In **transcription,** the information contained in DNA is transferred to mRNA molecules. In **translation,** the genetic information now present in the mRNA is used to build the sequence of amino acids of the desired protein. (See Figure 18.28.)

Transcription begins when the section of a DNA that contains the gene to be copied unwinds. One strand of DNA acts as a template as bonds are formed to each complementary base: C is paired with G, T pairs with A, and A pairs with U (not T).

WEB TUTORIAL

Transcription

Section of bases on DNA template: —G—A—A—C—T—

Complementary base sequence in mRNA: —C—U—U—G—A—

Sample Problem 18.13 ▷ **RNA Synthesis**

The sequence of bases in a part of the DNA template for mRNA is CGATCA. What is the corresponding mRNA produced?

Solution

The nucleotides in DNA pair up with the ribonucleotides as follows: G ⟶ C, C ⟶ G, T ⟶ A, and A ⟶ U.

Portion of DNA template: C—G—A—T—C—A

Complementary base in mRNA: G—C—U—A—G—U

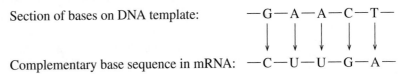

Study Check

What is the DNA template that codes for the mRNA having the ribonucleotide sequence GGGUUUAAA?

The Genetic Code

In the **genetic code,** a sequence of three bases in the mRNA, called a **codon,** specifies each amino acid in the protein. Early work on protein synthesis showed that repeating triplets of uracil (UUU) produced a polypeptide that contained only phenylalanine. Therefore, a sequence of —UUU—UUU—UUU— codes is for three phenylalanines.

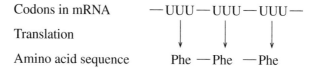

The codons have now been determined for all 20 amino acids. A total of 64 codons are possible from the triplet combinations of A, G, C, and U. Three of these, UGA, UAA, and UAG, are stop signals that code for the termination of protein synthesis.

Protein Synthesis: Translation

Once the mRNA is synthesized, it migrates out of the nucleus into the cytoplasm to the ribosomes. At the ribosomes, the *translation* process converts the codons on mRNA into amino acids to make a protein.

Protein synthesis begins when a mRNA combines with a ribosome. There, tRNA molecules, which carry amino acids, align with mRNA, and a peptide bond forms between the amino acids. After the first tRNA detaches from the ribosome, the ribosome shifts to the next codon on the mRNA. Each time the ribosome shifts and the next tRNA aligns with the mRNA, a peptide bond joins the new amino acid to the growing polypeptide chain. After all the amino acids for a particular protein have been linked together by peptide bonds, the ribosome encounters a stop codon. Because there are no tRNAs to complement the termination codon, protein synthesis ends and the completed polypeptide chain is released from the ribosome. Then interactions between the amino acids in the chain form the protein into the three-dimensional structure that makes the polypeptide into a biologically active protein. (See Fig. 18.29.)

WEB TUTORIAL

Translation

Questions and Problems ▶ **Protein Synthesis**

18.59 What are the three different types of RNA?

18.60 What are the functions of each type of RNA?

18.61 What is meant by the term *transcription*?

18.62 What bases in mRNA are used to complement the bases A, T, G, and C in DNA?

18.63 Write the corresponding section of mRNA produced from the following section of DNA template:
CCGAAGGTTCAC

18.64 Write the corresponding section of mRNA produced from the following section of DNA template:
TACGGCAAGCTA

18.65 What is a codon?

18.66 Where does protein synthesis take place?

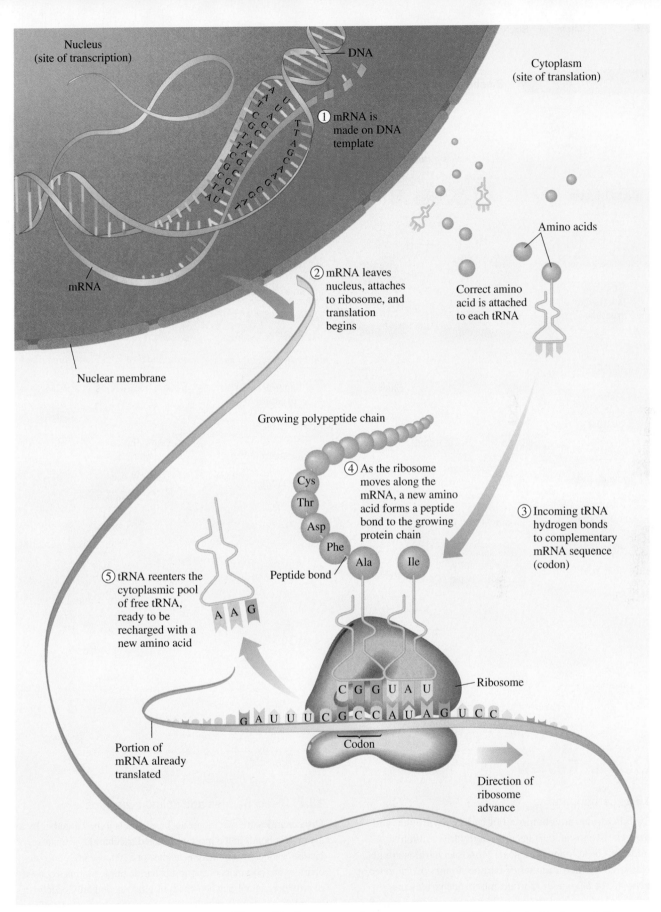

Figure 18.29 In the translation process, the mRNA synthesized by transcription attaches to a ribosome and tRNAs pick up their amino acids and place them in a growing peptide chain.

Q *How is the correct amino acid placed in the peptide chain?*

Concept Map ▶ **Biochemistry**

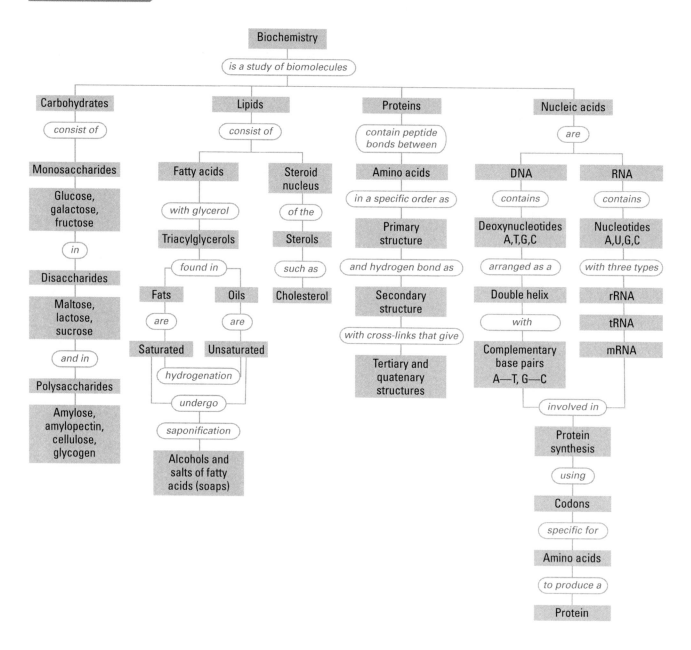

Chapter Review

18.1 Carbohydrates

Carbohydrates are composed of carbon, hydrogen, and oxygen. Monosaccharides are polyhydroxy aldehydes (aldoses) or ketones (ketoses). Monosaccharides are also classified by their number of carbon atoms: *triose, tetrose, pentose,* or *hexose*. Important monosaccharides are glucose, galactose, and fructose. The predominant form of monosaccharides is the cyclic form of five or six atoms. The cyclic structure forms by a reaction between an OH (usually the one on carbon 5 in hexoses) with the carbonyl group of the same molecule.

18.2 Disaccharides and Polysaccharides

Disaccharides are two monosaccharide units joined together by a glycosidic bond. In the most common disaccharides—maltose, lactose, and sucrose—there is at least one glucose unit. Polysaccharides are polymers of monosaccharide units. Starches consist of amylose, an unbranched chain of glucose, and amylopectin, a branched polymer of glucose. Glycogen, the storage form of glucose in animals, is similar to amylopectin with more branching. Cellulose is also a polymer of glucose, but in cellulose the glycosidic bonds are β bonds rather than α bonds as in the starches. Humans can digest starches, but not cellulose, to obtain energy. However, cellulose is important as a source of fiber in our diets.

18.3 Lipids

Lipids are nonpolar compounds that are not soluble in water. Classes of lipids include fats, oils, and steroids. Fatty acids are unbranched carboxylic acids that typically contain an even number (12–18) of carbon atoms. Fatty acids may be saturated or unsaturated. Triacylglycerols are esters of glycerol with three fatty acids. Fats contain more saturated fatty acids and have higher melting points than most vegetable oils. The hydrogenation of unsaturated fatty acids converts double bonds to single bonds. In saponification, a fat heated with a strong base produces glycerol and the salts of the fatty acids, or soaps. Steroids are lipids containing the steroid nucleus, which is a fused structure of four rings. The steroid hormones are closely related in structure to cholesterol. The sex hormones, such as estrogen and testosterone, are responsible for sexual characteristics and reproduction.

18.4 Proteins

A group of 20 amino acids provides the molecular building blocks of proteins. Attached to the central (alpha) carbon of each amino acid is an amino group, a carboxyl group, and a characteristic side group (R). Amino acids form peptides when a peptide bond joins the carboxyl group of one amino acid and the amino group of the second. Long chains of amino acids are called proteins.

18.5 Protein Structure

The primary structure of a protein is its sequence of amino acids. In the secondary structure, hydrogen bonds between peptide groups produce a characteristic shape such as an α helix, β-pleated sheet, or a triple helix. A tertiary structure is stabilized by interactions between R groups of amino acids in one region of the polypeptide chain with R groups in different regions of the protein. In a quaternary structure, two or more tertiary subunits combine for biological activity.

18.6 Proteins as Enzymes

Enzymes are proteins that act as biological catalysts by accelerating the rate of cellular reactions. Within the tertiary structure of an enzyme, a small pocket called the active site binds the substrates. In the lock-and-key model, a substrate precisely fits the shape of the active site. In the induced-fit model, substrates induce the active site to change structure to give an optimal fit by the substrate. In the enzyme–substrate complex, catalysis takes place when amino acid side chains react with a substrate. The products are released and the enzyme is available to bind another substrate molecule.

18.7 Nucleic Acids

Nucleic acids, deoxyribonucleic acid (DNA) and ribonucleic acid (RNA), are polymers of nucleotides. A nucleotide is composed of three parts: a base, a sugar, and a phosphate group. In DNA, the sugar is deoxyribose and the nitrogen-containing base can be adenine, thymine, guanine, or cytosine. In RNA, the sugar is ribose and uracil replaces thymine. Each nucleic acid has its own unique sequence of bases known as its primary structure. A DNA molecule consists of two strands of nucleotides. The two strands are held together by hydrogen bonds between complementary base pairs, A with T, and G with C. During DNA replication, complementary base pairing ensures that an identical copy of the original DNA is made.

18.8 Protein Synthesis

Transcription is the process by which mRNA is produced from one strand of DNA. The bases in the mRNA are complementary to the DNA, except A in DNA is paired with U in RNA.

The genetic code consists of a sequence of three bases (codon) in mRNA that specifies the order for the amino acids in a protein. Proteins are synthesized at the ribosomes. During translation, tRNAs bring the appropriate amino acids to the mRNA at the ribosome and peptide bonds form. When the polypeptide is released, it takes on its secondary and tertiary structures and becomes a functional protein in the cell.

Key Terms

active site A pocket in a part of the tertiary enzyme structure that binds substrate and catalyzes a reaction.

aldose Monosaccharides that contain an aldehyde group.

α (alpha) helix A secondary level of protein structure in which hydrogen bonds connect the NH of one peptide bond with the C=O of a peptide bond later in the chain to form a coiled or corkscrew structure.

amino acid The building block of proteins, consisting of an amino group, a carboxylic acid group, and a unique side group attached to the alpha carbon.

amylopectin A branched-chain polymer of starch composed of glucose units joined by α-1,4- and α-1,6-glycosidic bonds.

amylose An unbranched polymer of starch composed of glucose units joined by α-1,4-glycosidic bonds.

β (beta)-pleated sheet A secondary level of protein structure that consists of hydrogen bonds between peptide links in parallel polypeptide chains.

biochemistry The study of the structure and reactions of chemicals that occur in living systems.

carbohydrate A simple or complex sugar composed of carbon, hydrogen, and oxygen.

cellulose An unbranched polysaccharide composed of glucose units linked by β-1,4-glycosidic bonds that cannot be hydrolyzed by the human digestive system.

cholesterol The most prevalent of the steroid compounds found in cellular membranes.

codon A sequence of three bases in mRNA that specifies a certain amino acid to be placed in a protein. A few codons signal the start or stop of transcription.

complementary base pairs In DNA, adenine is always paired with thymine (A—T or T—A), and guanine is always paired with cytosine (G—C or C—G). In forming RNA, adenine is paired with uracil (A—U).

disaccharide Carbohydrate composed of two monosaccharides joined by a glycosidic bond.

DNA Deoxyribonucleic acid; the genetic material of all cells containing nucleotides with deoxyribose sugar, phosphate, and the four nitrogenous bases adenine, thymine, guanine, and cytosine.

double helix The helical shape of the double chain of DNA that is like a spiral staircase with a sugar-phosphate backbone on the outside and base pairs like stair steps on the inside.

enzyme A protein that catalyzes a biological reaction.

enzyme–substrate (ES) complex An intermediate consisting of an enzyme that binds to a substrate in an enzyme-catalyzed reaction.

fat Another term for solid triacylglycerols.

fatty acids Long-chain carboxylic acids found in fats.

fructose A monosaccharide found in honey and fruit juices; it is combined with glucose in sucrose.

galactose A monosaccharide that occurs combined with glucose in lactose.

genetic code The sequence of codons in mRNA that specifies the amino acid order for the synthesis of protein.

glucose The most prevalent monosaccharide in the diet. An aldohexose that is found in fruits, vegetables, corn syrup, and honey. Combines in glycosidic bonds to form most of the polysaccharides.

glycogen A polysaccharide formed in the liver and muscles for the storage of glucose as an energy reserve. It is composed of glucose in a highly branched polymer joined by α-1,4- and α-1,6-glycosidic bonds.

hydrogenation The addition of hydrogen to unsaturated fats.

hydrophilic amino acid An amino acid having polar, acidic, or basic R groups that are attracted to water; "water-loving."

hydrophobic amino acid A nonpolar amino acid with hydrocarbon R groups; "water-fearing."

induced-fit model A model of enzyme action in which a substrate induces an enzyme to modify its shape to give an optimal fit with the substrate structure.

ketose A monosaccharide that contains a ketone group.

lactose A disaccharide consisting of glucose and galactose found in milk and milk products.

lipids A family of compounds that is nonpolar in nature and not soluble in water; includes fats, waxes, and steroids.

lock-and-key model A model of an enzyme in which the substrate, like a key, exactly fits the shape of the lock, which is the specific shape of the active site.

maltose A disaccharide consisting of two glucose units; it is obtained from the hydrolysis of starch and in germinating grains.

monosaccharide A polyhydroxy compound that contains an aldehyde or ketone group.

nonpolar amino acids Amino acids that are not soluble in water because they contain a nonpolar side chain.

nucleic acids Large molecules composed of nucleotides, found as a double helix in DNA and as the single strands of RNA.

nucleoside The combination of a pentose sugar and a nitrogen-containing base.

nucleotides Building blocks of a nucleic acid consisting of a nitrogen-containing base, a pentose sugar (ribose or deoxyribose), and a phosphate group.

oil Another term for liquid triacylglycerols.

peptide The combination of two or more amino acids joined by peptide bonds.

peptide bond The amide bond that joins amino acids in polypeptides and proteins.

phosphodiester bond The phosphate link that joins the hydroxyl group in one nucleotide to the phosphate group on the next nucleotide.

polar amino acids Amino acids that are soluble in water because their R group is polar: hydroxyl (OH), thiol (SH), carbonyl (C=O), amino (NH_2), or carboxyl (COOH).

polysaccharides Polymers of many monosaccharide units, usually glucose. Polysaccharides differ in the types of glycosidic bonds and the amount of branching in the polymer.

primary structure The sequence of the amino acids in a protein.

protein A term used for biologically active polypeptides that have many amino acids linked together by peptide bonds.

quaternary structure A protein structure in which two or more protein subunits form an active protein.

replication The process of duplicating DNA by pairing the bases on each parent strand with their complementary base.

RNA Ribonucleic acid, a type of nucleic acid that is a single strand of nucleotides containing adenine, cytosine, guanine, and uracil.

saponification The reaction of a fat with a strong base to form glycerol and salts of fatty acids (soaps).

secondary structure The formation of an α helix, β-pleated sheet, or triple helix.

steroids Types of lipid composed of a multicyclic ring system.

sucrose A disaccharide composed of glucose and fructose; commonly called table sugar or "sugar."

tertiary structure The folding of the secondary structure of a protein into a compact structure that is stabilized by the interactions of R groups such as ionic and disulfide bonds.

triacylglycerols A family of lipids composed of three fatty acids bonded through ester bonds to glycerol, a trihydroxy alcohol.

transcription The transfer of genetic information from DNA by the formation of mRNA.

translation The interpretation of the codons in mRNA as amino acids in a peptide.

triple helix The protein structure found in collagen consisting of three polypeptide chains woven together like a braid.

Understanding the Concepts

18.67 Melezitose is a saccharide with the following structure:

a. Is melezitose a mono-, di-, tri-, or polysaccharide?
b. What are the monosaccharides in melezitose?

18.68 What are the disaccharides and polysaccharides present in each of the following?

(a)   (b)

(c)   (d)

18.69 Palmitic acid is obtained from palm oil as glyceryl tripalmitate. Draw the structure of glyceryl tripalmitate.

18.70 Identify each of the following as saturated, monounsaturated, polyunsaturated, omega-3, or omega-6 fatty acids:

a. $CH_3-(CH_2)_4-(CH=CH-CH_2)_2-(CH_2)_6-COOH$
b. linolenic acid
c. $CH_3-(CH_2)_{14}-COOH$
d. $CH_3-(CH_2)_7-CH=CH-(CH_2)_7-COOH$

18.71 Sunflower oil can be used to make margarine. A triacylglycerol in sunflower oil consists of two linoleic acids and one oleic acid.
a. Write two isomers for the triacylglycerol in sunflower oil.
b. Using one of the isomers, write the reaction that would be used when sunflower oil is used to make solid margarine.

18.72 Seeds and vegetables are often deficient in one or more essential amino acids. Using the following table, state whether the following combinations would provide all the essential amino acids:

Source	Lysine	Tryptophan	Methionine
Oatmeal	No	Yes	Yes
Rice	No	Yes	Yes
Garbanzo beans	Yes	No	Yes
Lima beans	Yes	No	No
Cornmeal	No	Yes	Yes

a. rice and garbanzo beans
b. lima beans and cornmeal
c. a salad of garbanzo beans and lima beans
d. rice and lima beans
e. rice and oatmeal
f. oatmeal and lima beans

18.73 For each of the following pairs of side chains, identify the amino acids and the type of cross-link that forms between them:

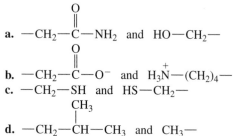

a. $-CH_2-C(=O)-NH_2$ and $HO-CH_2-$

b. $-CH_2-C(=O)-O^-$ and $H_3\overset{+}{N}-(CH_2)_4-$

c. $-CH_2-SH$ and $HS-CH_2-$

d. $-CH_2-CH(CH_3)-CH_3$ and CH_3-

18.74 Answer the following questions for the given section of DNA.

a. Complete the bases in the parent and template strands.

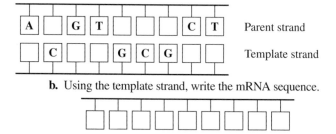

b. Using the template strand, write the mRNA sequence.

Additional Questions and Problems

18.75 What are the structural differences in glucose and galactose?

18.76 What are the structural differences in glucose and fructose?

18.77 Consider the sugar gulose.

Gulose

Draw the cyclic structure for α- and β-gulose.

18.78 From the compounds shown, select those that match with the following:

a. a ketopentose   **b.** an aldopentose
c. a ketohexose

18.79 Gentiobiose, which is found in saffron, contains two glucose molecules linked by a β-1,6-glycoside bond. Draw the structure of α-gentiobiose.

18.80 Why would an animal that lives in a cold climate have more unsaturated triacylglycerols in its body fat than an animal that lives in a warm climate?

18.81 Draw the structure of Ser-Lys-Asp.

18.82 Draw the structure of Val-Ala-Leu.

18.83 Identify the base and sugar in each of the following nucleosides:

a. deoxythymidine   **b.** adenosine
c. cytidine   **d.** deoxyguanosine

18.84 Identify the base and sugar in each of the following nucleotides:
a. CMP   b. dAMP
c. dGMP   d. UMP

18.85 Write the complementary base sequence for each of the following DNA segments:
a. GACTTAGGC
b. TGCAAACTAGCT
c. ATCGATCGATCG

18.86 Write the complementary base sequence for each of the following DNA segments:
a. TTACGGACCGC
b. ATAGCCCTTACTGG
c. GGCCTACCTTAACGACG

18.87 Match the following statements with rRNA, mRNA, or tRNA:
a. carries genetic information from the nucleus to the ribosomes
b. acts as a template for protein synthesis

18.88 Match the following statements with rRNA, mRNA, or tRNA:
a. found in ribosome
b. brings amino acids to the ribosomes for protein synthesis

Challenge Questions

18.89 Raffinose is a trisaccharide found in Australian manna and in cottonseed meal. It is composed of three different monosaccharides. Identify the monosaccharides in raffinose.

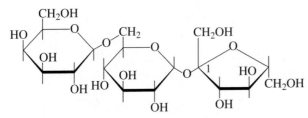

18.90 One mole of triolein is completely hydrogenated. What is the product? How many moles of hydrogen are required? How many grams of hydrogen? How many liters of hydrogen are needed if the reaction is run at STP?

18.91 What are some differences between the following pairs?
a. secondary and tertiary protein structures
b. essential and nonessential amino acids

c. polar and nonpolar amino acids
d. di- and tripeptides
e. an ionic bond (salt bridge) and a disulfide bond
f. α helix and β-pleated sheet
g. tertiary and quaternary structures of proteins

18.92 What type of interaction would you expect from the following amino acids in a tertiary structure?
a. threonine and asparagine
b. valine and alanine
c. arginine and aspartic acid

18.93 a. If the DNA double helix in salmon contains 28% adenine, what is the percent of thymine, guanine, and cytosine?
b. If the DNA double helix in humans contains 20% cytosine, what is the percent of guanine, adenine, and thymine?

18.94 Why are there no base pairs in DNA between adenine and guanine, or thymine and cytosine?

Answers

Answers to Study Checks

18.1   CH$_2$OH
|
C$=$O
|
CH$_2$OH

18.2

CH$_2$OH ... CH$_2$OH ... OH ... O ... O ... OH ... OH ... HO ... OH ... OH

18.3 Cellulose contains glucose units connected by β-1,4-glycosidic bonds, whereas the glucose units in amylose are connected by α-1,4-glycosidic bonds.

18.4 a. 16   b. unsaturated

18.5

CH$_2$—O—C—(CH$_2$)$_{12}$—CH$_3$
| ‖
CH—O—C—(CH$_2$)$_{12}$—CH$_3$
| ‖
CH$_2$—O—C—(CH$_2$)$_{12}$—CH$_3$

18.6 a. nonpolar   b. polar

18.7

18.8 a triple helix

18.9 Both are nonpolar and would have hydropholic interactions.

18.10 quaternary

18.11 In the lock-and-key model, the shape of a substrate fits the shape of the active site exactly. In the induced-fit model, the substrate induces the active site to adjust its shape to fit the substrate.

18.12 C—C—A—A—T—T—G—G

18.13 CCCAAATTT

Answers to Selected Questions and Problems

18.1 Hydroxyl groups are found in all monosaccharides along with a carbonyl on the first or second carbon.

18.3 A ketopentose contains hydroxyl and ketone functional groups and has five carbon atoms.

18.5 **a.** ketose **b.** aldose
  c. ketose **d.** aldose
  e. aldose

18.7 In galactose the hydroxyl on carbon 4 extends to the left. In glucose this hydroxyl goes to the right.

18.9 In the cyclic structure of glucose, there are five carbon atoms and an oxygen.

18.11 **a.** α-form **b.** α-form

18.13 **a.** galactose and glucose; β-1,4 bond; lactose
  b. glucose and glucose; α-1,4 bond; maltose

18.15 **a.** sucrose **b.** lactose
  c. maltose **d.** lactose

18.17 **a.** cellulose **b.** amylose, amylopectin
  c. amylose **d.** glycogen

18.19 **a.** saturated **b.** unsaturated
  c. unsaturated **d.** saturated

18.21

18.23 Safflower oil contains fatty acids with two or three double bonds; olive oil contains a large amount of oleic acid, which has only one (monounsaturated) double bond.

18.25 **a.** Some of the double bonds in the unsaturated fatty acids have been converted to single bonds by the addition of hydrogen.
  b. It is mostly saturated fatty acids.

18.27

18.29 All amino acids contain a carboxylic acid group and an amino group on the α carbon.

18.31 **a.**

  b.

  c.

18.33 **a.** hydrophobic nonpolar
  b. hydrophilic polar
  c. hydrophobic nonpolar

18.35 **a.** alanine **b.** valine
  c. lysine **d.** cysteine

18.37 **a.**

Ala-Cys

  b.

Ser-Phe

  c.

Gly-Ala-Val

18.39 Val-Ser-Ser, Ser-Val-Ser, or Ser-Ser-Val

18.41 In the α helix, hydrogen bonds form between the carbonyl oxygen atom and the amino hydrogen atom of the fourth amino acid in the sequence. In the β-pleated sheet, hydrogen bonds occur between parallel peptides or across sections of a long polypeptide chain.

18.43 **a.** a disulfide bond   **b.** hydrogen bond
c. hydrophobic interaction

18.45 **a.** cysteine
b. Leucine and valine will be found on the inside of the protein because they are hydrophobic.
c. The cysteine and aspartic acid would be on the outside of the protein because they are polar.
d. The order of the amino acids (the primary structure) provides the R groups, whose interactions determine the tertiary structure of the protein.

18.47 **a.** enzyme   **b.** enzyme–substrate complex
c. substrate

18.49 **a.** E + S $\rightleftharpoons$ ES $\longrightarrow$ E + P
b. The active site is a region or pocket within the tertiary structure of an enzyme that accepts the substrate, aligns the substrate for reaction, and catalyzes the reaction.

18.51 **a.** DNA   **b.** both DNA and RNA

18.53 The two DNA strands are held together by hydrogen bonds between the bases in each strand.

18.55 **a.** TTTTT
b. CCCCCC
c. TCAGGTCCA
d. GACATATGCAAT

18.57 The DNA strands separate and the DNA polymerase pairs each of the bases with its complementary base and produces two exact copies of the original DNA.

18.59 ribosomal RNA, messenger RNA, and transfer RNA

18.61 In transcription, the sequence of nucleotides on a DNA template (one strand) is used to produce the base sequences of a messenger RNA.

18.63 GGCUUCCAAGUG

18.65 a three-base sequence in mRNA that codes for a specific amino acid in a protein

18.67 **a.** Melezitose is a trisaccharide.
b. Melezitose contains two glucose molecules and a fructose molecule.

18.69

$$
\begin{array}{l}
CH_2-O-\overset{\displaystyle O}{\overset{\|}{C}}-(CH_2)_{14}-CH_3 \\
\overset{\displaystyle O}{} \\
H-\overset{}{C}-O-\overset{\displaystyle \|}{C}-(CH_2)_{14}-CH_3 \qquad \text{glyceryl tripalmitate}\\
\overset{\displaystyle O}{} \\
CH_2-O-\overset{\displaystyle \|}{C}-(CH_2)_{14}-CH_3
\end{array}
$$

18.71 **a.**

$$
\begin{array}{l}
CH_2-O-\overset{O}{\overset{\|}{C}}-(CH_2)_7-CH{=}CH-CH_2-CH{=}CH-(CH_2)_4-CH_3 \\
O \\
H-C-O-\overset{\|}{C}-(CH_2)_7-CH{=}CH-(CH_2)_7-CH_3 \\
O \\
CH_2-O-\overset{\|}{C}-(CH_2)_7-CH{=}CH-CH_2-CH{=}CH-(CH_2)_4-CH_3
\end{array}
$$

$$
\begin{array}{l}
CH_2-O-\overset{O}{\overset{\|}{C}}-(CH_2)_7-CH{=}CH-CH_2-CH{=}CH-(CH_2)_4-CH_3 \\
O \\
H-C-O-\overset{\|}{C}-(CH_2)_7-CH{=}CH-CH_2-CH{=}CH-(CH_2)_4-CH_3 \\
O \\
CH_2-O-\overset{\|}{C}-(CH_2)_7-CH{=}CH-(CH_2)_7-CH_3
\end{array}
$$

b.

$$
\begin{array}{l}
CH_2-O-\overset{O}{\overset{\|}{C}}-(CH_2)_7-CH{=}CH-CH_2-CH{=}CH-(CH_2)_4-CH_3 \\
O \\
HC-O-\overset{\|}{C}-(CH_2)_7-CH{=}CH-(CH_2)_7-CH_3 \\
O \\
CH_2-O-\overset{\|}{C}-(CH_2)_7-CH{=}CH-CH_2-CH{=}CH-(CH_2)_4-CH_3
\end{array}
\;+\;5H_2 \;\xrightarrow{\;Pt\;}\;
\begin{array}{l}
CH_2-O-\overset{O}{\overset{\|}{C}}-(CH_2)_{16}-CH_3 \\
O \\
CH-O-\overset{\|}{C}-(CH_2)_{16}-CH_3 \\
O \\
CH_2-O-\overset{\|}{C}-(CH_2)_{16}-CH_3
\end{array}
$$

18.73 **a.** asparagine and serine; hydrogen bond
b. aspartic acid and lysine; salt bridge
c. two cysteines; disulfide bond
d. valine and alanine; hydrophobic attraction

18.75 They differ only at carbon 4 where the —OH in glucose is on the right side and in galactose it is on the left side.

18.77

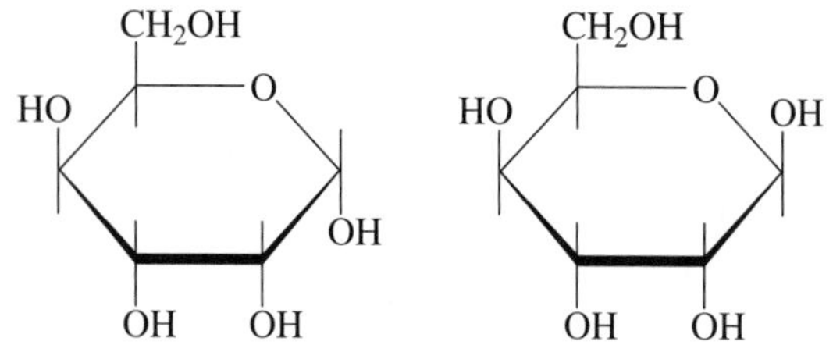

α-Gulose   β-Gulose

18.79

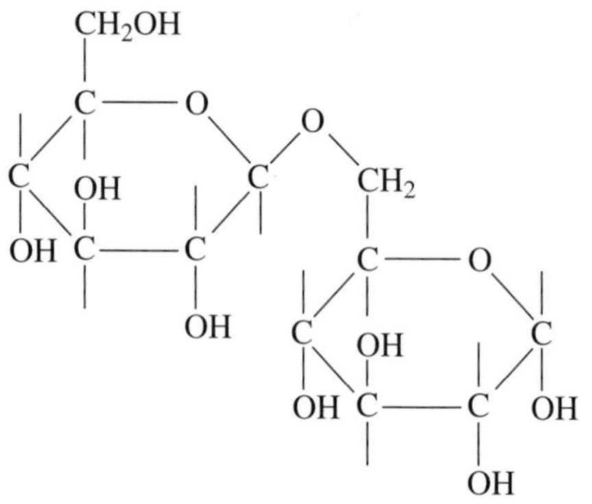

18.81

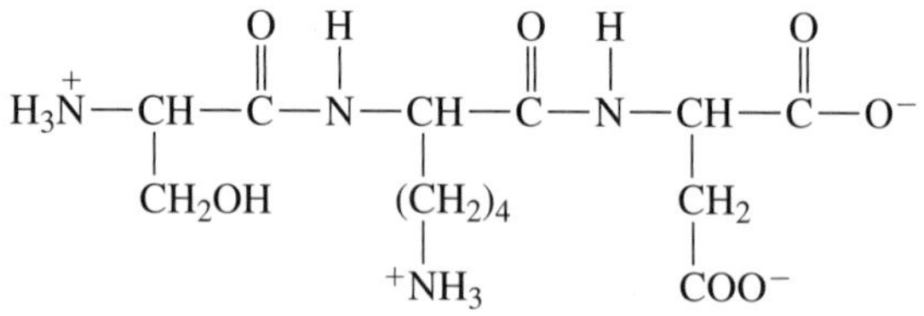

18.83 **a.** thymine and deoxyribose
b. adenine and ribose
c. cytosine and ribose
d. guanine and deoxyribose

18.85 **a.** CTGAATCCG
b. ACGTTTGATCGA
c. TAGCTAGCTAGC

18.87 **a.** mRNA   **b.** mRNA

18.89 galactose, glucose, and fructose

18.91 **a.** The secondary structure of a protein depends on hydrogen bonds to form a helix or a pleated sheet. The tertiary structure is determined by the interaction of side chains and determines the three-dimensional structure of the protein.
b. Nonessential amino acids are synthesized by the body, but essential amino acids must be supplied by the diet.
c. Polar amino acids have hydrophilic side groups; nonpolar amino acids have hydrophobic side groups.
d. Dipeptides contain two amino acids; tripeptides contain three.
e. An ionic bond is an interaction between a basic and acidic side group; a disulfide bond links the sulfides of two cysteines.
f. The alpha helix is the secondary shape like a spiral staircase or corkscrew. The beta-pleated sheet is a secondary structure that is formed by many proteins side by side.
g. The tertiary structure of a protein is its three-dimensional structure. In the quaternary structure, two or more peptide subunits are grouped.

18.93 Because A bonds with T, T is also 28%. Since A and T = 56%, there is 44% for the other nucleotides, or 22% G and 22% C.

Combining Ideas from Chapters 17 and 18

CI 35 Suncreens contain compounds that absorb UV light, such as oxybenzone and 2-ethylhexyl-4-methoxycinnamate.

Identify the functional groups in each of the following UV-absorbing compounds used in suncreens:

a. oxybenzone

CH₃—O— (benzophenone structure) —OH

b. 2-ethylhexyl-4-methoxycinnamate

CH₃—O— (aromatic ring) —CH=CH—C(=O)—O—CH₂—CH—(CH₂)₃—CH₃ with CH₂—CH₃ substituent

CI 36 Identify the functional groups in each of the following:

a. Oxymetazoline is a vasoconstrictor used in nasal decongestant sprays such as Afrin.

oxymetazoline

(CH₃)₃C— (aromatic ring with OH, CH₃, CH₂ substituents) —N=, NH (imidazoline ring), CH₃

b. Decimemide is used as an anticonvulsant.

decimemide

CH₃O, CH₃(CH₂)₉O— (aromatic ring) —C(=O)—NH₂, CH₃O

CI 37 Acetylene gas reacts with oxygen and burns at high temperature in an acetylene torch.

a. Write the balanced equation for the complete combustion of acetylene.

b. How many grams of oxygen are needed to react with 8.5 L acetylene at STP?

c. How many liters of CO_2 (at STP) are produced when 30.0 g acetylene undergoes combustion?

CI 38 A compound with the formula C_4H_8O is synthesized by oxidation of 2-methyl-1-propanol. What is the structure of the compound?

CI 39 Methyl *tert*-butyl ether (MTBE) or methyl 2-methyl-propyl ether ($C_5H_{12}O$) is a fuel additive for gasoline to boost the octane rating. It increases the oxygen content, which reduces CO emissions to an acceptable level determined by the Clean Air Act.

a. If fuel mixtures are required to contain 2.7% oxygen by mass, how many grams MTBE must be present in each 100 g gasoline?

b. How many liters of MTBE would be in a liter of fuel if the density of both gasoline and MTBE is 0.740 g/mL?

c. Write the equation for the complete combustion of MTBE.

d. How many liters of air containing 21% (v/v) O_2 are required at STP to completely react (combust) 1.00 L liquid MTBE?

CI 40 If a female silkworm moth secretes 50 ng Bombykol, a sex attractant, how many molecules did she secrete? (See Chem Note, "Pheromones in Insect Communication," on page 600.)

CI 41 A sink drain can become clogged with solid fat such as glyceryl tristearate.

a. How would adding lye (NaOH) to the sink drain remove the blockage?

b. Write an equation for the reaction that occurs.

CI 42 Olive oil consists of a high percentage of triolein.

a. Draw the structure for triolein.

b. How many liters of H_2 gas at STP are needed to completely saturate 100. g of triolein?

c. How many milliliters of 6.00 M NaOH are needed to completely saponify 100. g of triolein?

Answers to CIs

CI 35 a. aromatic, ether, alcohol, ketone
b. aromatic, ether, alkene, ester

CI 37 a. acetylene $H-C\equiv C-H$ or C_2H_2

$$2C_2H_2(g) + 5O_2(g) \longrightarrow 4CO_2(g) + 2H_2O(g)$$

b. C_2H_2 26.04 g/mol

$$8.5 \text{ L } C_2H_2 \times \frac{1 \text{ mol } C_2H_2}{22.4 \text{ L } C_2H_2} \times \frac{5 \text{ mol } O_2}{2 \text{ mol } C_2H_2}$$
$$\times \frac{32.00 \text{ g } O_2}{1 \text{ mol } O_2} = 30. \text{ g } O_2$$

c. $30.0 \text{ g } C_2H_2 \times \frac{1 \text{ mol } C_2H_2}{26.04 \text{ g } C_2H_2} \times \frac{4 \text{ mol } CO_2}{2 \text{ mol } C_2H_2}$
$$\times \frac{22.4 \text{ L } CO_2}{1 \text{ mol } CO_2} = 51.6 \text{ L } CO_2$$

CI 39 a. 15 g MTBE
b. 0.15 L MTBE
c. $2 C_5H_{12}O + 15 O_2 \longrightarrow 10 CO_2 + 12 H_2O$
MTBE
d. 6700 L air or 6.7×10^3 L air

CI 41 a. Adding NaOH will hydrolyze the tristearate lipid, breaking it up to wash down the rain.

b.

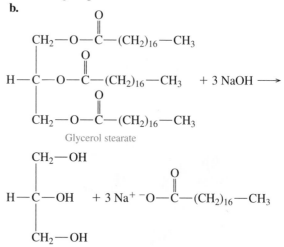

Glycerol stearate

Glycerol   Salts of stearic acid

Appendix

Periodic Table of Elements

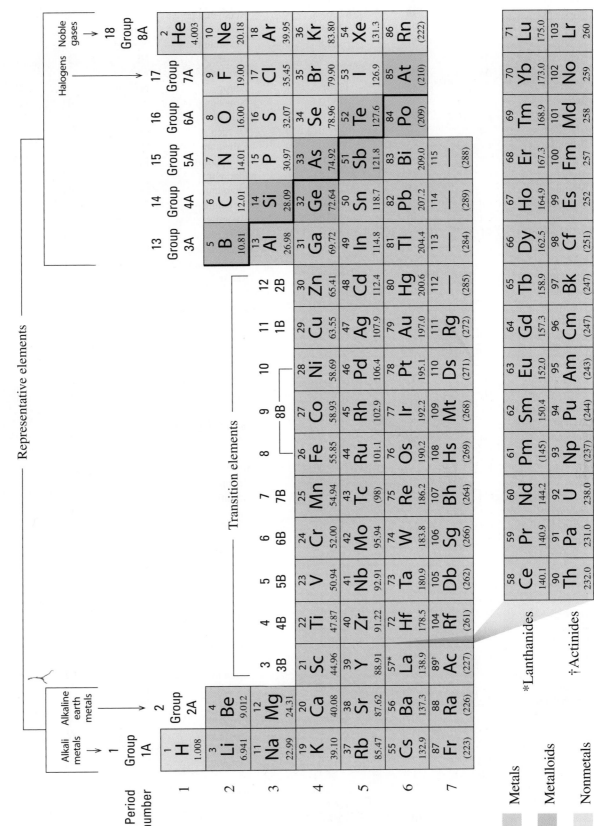

Atomic Masses of the Elements

Name	Symbol	Atomic Number	Atomic Mass[a]	Name	Symbol	Atomic Number	Atomic Mass[a]
Actinium	Ac	89	(227)	Molybdenum	Mo	42	95.94
Aluminum	Al	13	26.98	Neodymium	Nd	60	144.2
Americium	Am	95	(243)	Neon	Ne	10	20.18
Antimony	Sb	51	121.8	Neptunium	Np	93	(237)
Argon	Ar	18	39.95	Nickel	Ni	28	58.69
Arsenic	As	33	74.92	Niobium	Nb	41	92.91
Astatine	At	85	(210)	Nitrogen	N	7	14.01
Barium	Ba	56	137.3	Nobelium	No	102	(259)
Berkelium	Bk	97	(247)	Osmium	Os	76	190.2
Beryllium	Be	4	9.012	Oxygen	O	8	16.00
Bismuth	Bi	83	209.0	Palladium	Pd	46	106.4
Bohrium	Bh	107	(264)	Phosphorus	P	15	30.97
Boron	B	5	10.81	Platinum	Pt	78	195.1
Bromine	Br	35	79.90	Plutonium	Pu	94	(244)
Cadmium	Cd	48	112.4	Polonium	Po	84	(209)
Calcium	Ca	20	40.08	Potassium	K	19	39.10
Californium	Cf	98	(251)	Praseodymium	Pr	59	140.9
Carbon	C	6	12.01	Promethium	Pm	61	(145)
Cerium	Ce	58	140.1	Protactinium	Pa	91	231.0
Cesium	Cs	55	132.9	Radium	Ra	88	(226)
Chlorine	Cl	17	35.45	Radon	Rn	86	(222)
Chromium	Cr	24	52.00	Rhenium	Re	75	186.2
Cobalt	Co	27	58.93	Rhodium	Rh	45	102.9
Copper	Cu	29	63.55	Roentgenium	Rg	111	(272)
Curium	Cm	96	(247)	Rubidium	Rb	37	85.47
Darmstadtium	Ds	110	(271)	Ruthenium	Ru	44	101.1
Dubnium	Db	105	(262)	Rutherfordium	Rf	104	(261)
Dysprosium	Dy	66	162.5	Samarium	Sm	62	150.4
Einsteinium	Es	99	(252)	Scandium	Sc	21	44.96
Erbium	Er	68	167.3	Seaborgium	Sg	106	(266)
Europium	Eu	63	152.0	Selenium	Se	34	78.96
Fermium	Fm	100	(257)	Silicon	Si	14	28.09
Fluorine	F	9	19.00	Silver	Ag	47	107.9
Francium	Fr	87	(223)	Sodium	Na	11	22.99
Gadolinium	Gd	64	157.3	Strontium	Sr	38	87.62
Gallium	Ga	31	69.72	Sulfur	S	16	32.07
Germanium	Ge	32	72.64	Tantalum	Ta	73	180.9
Gold	Au	79	197.0	Technetium	Tc	43	(98)
Hafnium	Hf	72	178.5	Tellurium	Te	52	127.6
Hassium	Hs	108	(269)	Terbium	Tb	65	158.9
Helium	He	2	4.003	Thallium	Tl	81	204.4
Holmium	Ho	67	164.9	Thorium	Th	90	232.0
Hydrogen	H	1	1.008	Thulium	Tm	69	168.9
Indium	In	49	114.8	Tin	Sn	50	118.7
Iodine	I	53	126.9	Titanium	Ti	22	47.87
Iridium	Ir	77	192.2	Tungsten	W	74	183.8
Iron	Fe	26	55.85	Uranium	U	92	238.0
Krypton	Kr	36	83.80	Vanadium	V	23	50.94
Lanthanum	La	57	138.9	Xenon	Xe	54	131.3
Lawrencium	Lr	103	(260)	Ytterbium	Yb	70	173.0
Lead	Pb	82	207.2	Yttrium	Y	39	88.91
Lithium	Li	3	6.941	Zinc	Zn	30	65.41
Lutetium	Lu	71	175.0	Zirconium	Zr	40	91.22
Magnesium	Mg	12	24.31	—	—	112	(285)
Manganese	Mn	25	54.94	—	—	113	(284)
Meitnerium	Mt	109	(268)	—	—	114	(289)
Mendelevium	Md	101	(258)	—	—	115	(288)
Mercury	Hg	80	200.6				

[a] Values in parentheses are the mass number of the most stable isotope.

SI Units and Some Useful Conversion Factors

Length   SI unit meter (m)

1 meter (m) = 100 centimeters (cm)

1 meter (m) = 1000 millimeters (mm)

1 cm = 10 mm

1 kilometer (km) = 0.6214 mile (mi)

1 inch (in.) = 2.54 cm (exact)

Volume   SI unit cubic meter (m^3)

1 liter (L) = 1000 milliliters (mL)

1 mL = 1 cm^3

1 L = 1.057 quart (qt)

Mass   SI unit kilogram (kg)

1 kilogram (kg) = 1000 grams (g)

1 g = 1000 milligrams (mg)

1 kg = 2.205 lb

1 lb = 453.6 g

1 mol = 6.022×10^{23} particles

Water

density = 1.00 g/mL

Temperature   SI unit kelvin (K)

$°F = 1.8(°C) + 32$

$°C = \dfrac{(°F - 32)}{1.8}$

$K = °C + 273$

Pressure   SI unit pascal (Pa)

1 atm = 760 mm Hg

1 atm = 101.3 kPa

1 atm = 760 torr

1 mol (STP) = 22.4 L

$R = 0.0821$ L·atm/mol·K

$R = 62.4$ L·mm Hg/mol·K

Energy   SI unit joule (J)

1 calorie (cal) = 4.184 J

1 kcal = 1000 cal

Water

Heat of fusion = 334 J/g

Heat of vaporization = 2260 J/g

SH = 4.184 J/g°C

Prefixes for Metric (SI) Units

Prefix	Symbol	Power of Ten
Values greater than 1		
giga	G	10^9
mega	M	10^6
kilo	k	10^3
Values less than 1		
deci	d	10^{-1}
centi	c	10^{-2}
milli	m	10^{-3}
micro	μ	10^{-6}
nano	n	10^{-9}
pico	p	10^{-12}

Formulas and Molar Masses of Some Typical Compounds

Name	Formula	Molar Mass (g/mol)	Name	Formula	Molar Mass (g/mol)
ammonia	NH_3	17.03	hydrogen chloride	HCl	36.46
ammonium chloride	NH_4Cl	53.49	iron(III) oxide	Fe_2O_3	159.70
ammonium sulfate	$(NH_4)_2SO_4$	132.15	magnesium oxide	MgO	40.31
bromine	Br_2	159.80	methane	CH_4	16.04
butane	C_4H_{10}	58.12	nitrogen	N_2	28.02
calcium carbonate	$CaCO_3$	100.09	oxygen	O_2	32.00
calcium chloride	$CaCl_2$	110.98	potassium carbonate	K_2CO_3	138.21
calcium hydroxide	$Ca(OH)_2$	74.10	potassium nitrate	KNO_3	101.11
calcium oxide	CaO	56.08	propane	C_3H_8	44.09
carbon dioxide	CO_2	44.01	sodium chloride	NaCl	58.44
chlorine	Cl_2	79.90	sodium hydroxide	NaOH	40.00
copper(II) sulfide	CuS	95.62	sulfur trioxide	SO_3	80.07
glucose	$C_6H_{12}O_6$	180.16	water	H_2O	18.02
hydrogen	H_2	2.016			

Formulas and Charges of Some Common Cations

Cations (Fixed Charge)

1+	2+	3+
Li^+ lithium	Mg^{2+} magnesium	Al^{3+} aluminum
Na^+ sodium	Ca^{2+} calcium	
K^+ potassium	Sr^{2+} strontium	
NH_4^+ ammonium	Ba^{2+} barium	
H_3O^+ hydronium		

Cations with Variable Valence

1+ or 2+		1+ or 3+	
Cu^+ copper(I)	Cu^{2+} copper(II)	Au^+ gold(I)	Au^{3+} gold(III)

2+ or 3+		2+ or 4+	
Fe^{2+} iron(II)	Fe^{3+} iron(III)	Sn^{2+} tin(II)	Sn^{4+} tin(IV)
Co^{2+} cobalt(II)	Co^{3+} cobalt(III)	Pb^{2+} lead(II)	Pb^{4+} lead(IV)
Cr^{2+} chromium(II)	Cr^{3+} chromium(III)		
Mn^{2+} manganese(II)	Mn^{3+} manganese(III)		

Formulas and Charges of Some Common Anions

1−		2−	3−
F^- fluoride	Br^- bromide	O^{2-} oxide	N^{3-} nitride
Cl^- chloride	I^- iodide	S^{2-} sulfide	P^{3-} phosphide

Polyatomic Ions

HCO_3^- hydrogen carbonate (bicarbonate)

$C_2H_3O_2^-$ acetate CN^- cyanide

SCN^- thiocyanate

NO_3^- nitrate NO_2^- nitrite

$H_2PO_4^-$ dihydrogen phosphate HPO_4^{2-} hydrogen phosphate PO_4^{3-} phosphate

$H_2PO_3^-$ dihydrogen phosphite HPO_3^{2-} hydrogen phosphite PO_3^{3-} phosphite

HSO_4^- hydrogen sulfate (bisulfate) SO_4^{2-} sulfate

HSO_3^- hydrogen sulfite (bisulfite) SO_3^{2-} sulfite

ClO_4^- perchlorate ClO_3^- chlorate

ClO_2^- chlorite ClO^- hypochlorite

OH^- hydroxide CrO_4^{2-} chromate

MnO_4^- permanganate $Cr_2O_7^{2-}$ dichromate

Credits

Unless otherwise acknowledged, all photographs are the property of Prentice Hall.

Glossary/Index

A

Absolute zero, 69
Acetaldehyde, 613
Acetaminophen, 606
Acetic acid
  heat of vaporization and fusion, 317
  making vinegar, 620,
  structural formula for, 619
Acetone, 618
Acetylene
  molecular and empirical formula of, 181
  as simplest alkyne, 598
Acid A substance that dissolves in water and produces hydrogen ions (H$^+$), according to the Arrhenius theory. All acids are proton donors, according to the Brønsted–Lowry theory.
  adding to carbonate or bicarbonate, 479
  characteristics of, 458, 466
  naming, 454–455
  neutralization of, 479
  proton transfer, 458
  reaction with metal, 478
  strength of, 461–462
  titration of, 482
  typical reactions of, 478
  See also Conjugate acid–base pair
Acid–base conjugate pair, 488
Acid–base reaction, 463
Acid–base titration, 481
Acid dissociation constant (K_a) The product of the concentrations of the ions from the dissociation of a weak acid divided by the concentration of the weak acid. 465
Acidic amino acid, 669
Acidic blood pH, 163
Acidic dry cell battery, 526
Acidic salt solution, 486
Acidic solution
  cation and anion of salts in, 486
  examples of hydronium and hydroxide, 469
  measuring pH value of, 471
Acid ionization constant, 465
Acidity, 470
Acid rain, 477
Actinide, 97
Activation energy The energy that must be provided by a collision to break apart the bonds of the reacting molecules. 413, 414
Active learning, 11
Active metal, 478
Active site A pocket in a part of the tertiary enzyme structure that binds substrate and catalyzes a reaction.
  amino acid catalyzing chemical reaction, 679
  catalysis by enzyme, 679
  of enzyme, 678

Activity series
  for metal, 528, 529
  predicting direction of spontaneous reaction, 529
Actual yield The actual amount of product produced by a reaction. 242
Adding, 29–30
Addition reaction, 602
Age at death, 1
Aging
  collagen in person, 672
  process of, 412
Air
  as component of nature, 7
  compounds burning in, 582
  as gas mixture, 364
  as a mixture, 59
  typical composition of, 364
Alchemist, 7
Alcohol A class of organic compounds that contains the hydroxyl (—OH) group bonded to a carbon atom. 595
  classification of, 609
  functional group and example of, 597
  increasing levels of uric acid, 390
  naming, 608
  as organic compound containing oxygen, 607
  oxidation in body, 613
  reacting with carboxylic acid, 581
  reactions of, 611–612
  types of, 610
  typical behavior of person ingesting, 613
Aldehyde A class of organic compounds that contains a carbonyl group (C═O) bonded to at least one hydrogen atom.
  containing carbonyl group, 596
  functional group and example of, 597
  naming, 615–616
  role in organic chemistry, 617
  structural formula for, 616
  types of, 618
Aldopentose, 647
Aldose Monosaccharides that contain an aldehyde group. 647
Alkali metal Elements of Group 1A (1) except hydrogen; these are soft, shiny metals.
  elements as, 98, 99
  on periodic table, 97
  valence electrons in, 275
Alkaline battery, 526
Alkaline dry-cell battery, 526
Alkaline earth metal Group 2A (2) elements.
  elements as, 98
  on periodic table, 97
  valence electrons in, 115, 275
Alkaloid, 628
Alkalosis, 491

Alkane Hydrocarbons containing only single bonds between carbon atoms.
  carbon–carbon single bond, 595
  combustion of, 593
  comparing names with alkenes and alkynes, 599
  drawing structural formulas for, 590, 591
  as gas, 591
  mixtures from crude oil, 592
  rules for naming, 588–589
  solubility of, 591
  structural formula for, 586, 587
  substituents in, 588
  as waxy coating, 593
Alkene Hydrocarbons that contain carbon–carbon double bonds (C═C). 595
  comparing names with alkanes and alkynes, 599
  functional group and example of, 597
  polymers of, 604
  as unsaturated hydrocarbon, 598
Alkyl group An alkane minus one hydrogen atom. Alkyl groups are named like the alkanes except a *yl* ending replaces *ane*. 609
Alkyne Hydrocarbons that contain carbon–carbon triple bonds (C≡C). 595
  comparing names with alkanes and alkenes, 599
  functional group and example of, 597
  as unsaturated hydrocarbon, 598
Allergy testing, 6
Alpha emitter
  causing internal damage, 546
  as decaying radioisotope, 549
  producing stable nuclei, 554
Alpha helix (α helix) A secondary level of protein structure in which hydrogen bonds connect the NH of one peptide bond with the C═O of a peptide bond later in the chain to form a coiled or corkscrew structure.
  secondary structure of, 671–672
  shape of, 672
Alpha hydroxy acid
  as naturally occurring carboxylic acid, 622
  in skin products, 581
  sources and structures of, 622
Alpha (α) particle A nuclear particle identical to a helium ($_2^4$He, or α) nucleus (2 protons and 2 neutrons). 544
  damaging effects of, 558
  mass of, 546
  radioisotopes emitting, 549
  as type of radiation, 545
Alternator, 526
Altitude, 339
  atmospheric pressure and, 349
Altitude sickness, 434

Base A substance that dissolves in water and produces hydroxide ions (OH⁻) according to the Arrhenius theory. All bases are proton acceptors, according to the Brønsted–Lowry theory.
  characteristics of, 458
  as ionic compound, 455
  naming, 456
  proton transfer, 458
  strength of, 462–463
  typical reactions of, 478
  See also Conjugate acid–base pair

Base dissociation constant (Kb) The product of the concentrations of the ions from the dissociation of a weak base divided by the concentration of the weak base. 466

Basic solution
  cation and anion of salts in, 486
  examples of hydronium and hydroxide, 469
  measuring pH value of, 471

Battery
  connecting sodium and chloride to, 531
  types of, 525–526

Becquerel (Bq) A unit of activity of a radioactive sample equal to one disintegration per second. 557

Beef, ground, 83
Belladonna, 628
Bends, 340

Bent The shape of a molecule with two bonded atoms and two lone pairs. 299

Benzaldehyde, 618
Benzedrine (amphetamine), 627

Benzene A ring of six carbon atoms each of which is attached to one hydrogen atom, C₆H₆.
  heat of vaporization and fusion, 317
  molecular and empirical formula of, 181
  structures for, 605

Benzene ring, 606
Benzoic acid, 619
Beryllium (Be), 96
  as alkaline earth metal, 99
  electron level arrangement for, 114
  number of electrons in energy level, 115

Beta emitter
  converting neutron to proton, 551
  nuclear equation of, 552
  producing stable nuclei, 554

Beta (β) particle A particle identical to an electron (₋₁⁰e, or β) that forms in the nucleus when a neutron changes to a proton and an electron. 544
  damaging effects of, 558
  mass of, 546
  as type of radiation, 545

Beta-pleated sheet (β-pleated sheet) A secondary level of protein structure that consists of hydrogen bonds between peptide links in parallel polypeptide chains.
  formation of hydrogen bonds, 673
  secondary structure of, 671–672

Beta radiation, 557
Bicarbonate, 479

Biochemistry The study of the structure and reactions of chemicals that occur in living systems. 645

Biogeochemistry, 411
Blood
  acidity level of, 470
  buffers in, 491
  dissolving oxygen, 340

extracting drug from, 453
  normal values for buffer in, 491
  oxygen levels in, 333
  testing for mercury levels, 101
Blood clot, 660
Blood droplet, 1
Blood gas, 367
Blood gas analyzer, 163
Blood pressure, 17, 337
Body
  combustion reaction, 210
  injury of, 2
Body fat mass, 38
Body fluid, 390
Body temperature
  comparing temperature scales, 66
  regulation of, 440
  variation in, 71
Boiling The formation of bubbles of gas throughout a liquid. 315
Boiling point (bp) The temperature at which a substance exists as a liquid and gas; liquid changes to gas (boils), and gas changes to liquid (condenses). 315
  comparing temperature scales, 66
  of liquid alkane, 593
  molecule polarity and, 306
  pressure and, 349
  vapor pressure and, 348
Bond
  electronegativity difference and types of, 305
  predicting type from electronegativity differences, 305
Bonding electron, 303
Bonding pair A pair of electrons shared between two atoms. 149
Bonding pattern
  attractive force and, 310
  of nonmetals in covalent compound, 292
  variation in, 305
Bone analysis, 1
Bone mass, 144
Bone structure, 144
Boron (B), 96, 114
Bovine spongiform encephalopathy (BSE), 673
Boyle's law A gas law stating that the pressure of a gas is inversely related to the volume when temperature and moles of the gas do not change; that is, if volume decreases, pressure increases.
  mechanics of breathing and, 343
  using pressure–volume relationship, 341
Branch A carbon group or halogen bonded to the main carbon chain. 588
Branched alkane A hydrocarbon containing a hydrocarbon substituent bonded to the main chain. 588
Brass, 59
Breakfast cereal, 3, 4
Breathalyzer test, 613
Breathing
  gas laws and, 333
  pressure-volume relationship, 343
Bromide, 136
Bromine (Br)
  as halogen, 99, 100
  number of electrons in energy level, 115
Bromo, 588
Brønsted–Lowry acid An acid is a proton donor. 457
Brønsted–Lowry base A base is a proton acceptor. 457

Buffer
  adding to aspirin preparations, 624
  in blood, 491
  guide to calculating pH of, 490
  normal values in arterial blood, 491
  preventing pH change, 488
  types of, 489
Buffer solution A mixture of a weak acid or a weak base and its salt that resists changes in pH when small amounts of an acid or a base are added.
  acid–base conjugate pair as, 488
  maintaining hydronium ion concentration, 489
Butane
  isomers of, 588
  providing fuel for heat, 591
  structural formula for, 587
  two-dimensional structural formula of, 586
Butanedione, 618
Butter
  butanedione in, 618
  combustion of, 82
Butyric acid
  rancid butter and, 620
  reacting with ethanol, 581
  as source of carboxylic acid, 620

C

Cadmium (Cd), 137
Caffeine, 628
  chemical processes of, 4
  LD50 value of, 34
Calcium (Ca)
  as alkali earth metal, 99
  DDT reducing, 9
  dietary recommendations for deficiency, 93
  as element essential to health, 104
  function of ion in body, 132
  number of electrons in energy level, 115
Calcium carbonate, 422
Calcium oxide, 422
Calculator
  adding significant zeros, 28–29
  addition and subtraction, 29–30
  multiplication and division, 28–29
  scientific notation and, 23
Californium (Cf), 94
Calorie (cal) The amount of heat energy that raises the temperature of exactly 1 g water exactly 1°C.
  content in food, 83
  as form of energy, 74
Calorimeter
  determining energy value of food, 82
Cancer, 257
Cancer cell, 546
Canned soda. *See* Carbonated soft drink
Carbohydrate A simple or complex sugar composed of carbon, hydrogen, and oxygen.
  chemical processes of, 4
  elements in, 104
  energy value for, 82
  as organic compound, 646
  plants and, 3
Carbon (C), 96
  arranging atoms in flat ring, 605
  atomic mass of, 111
  attaining octet, 150
  electronegativity value of, 615
  electron level arrangement for, 114
  as element essential to health, 104
  forming long stable chains, 585

International System of Units, 18
Intoxication, 613, 613
Inverse relationship A relationship in which two properties change in opposite directions. 341
Investigation, 1
Iodide, 136
Iodine (I)
    activity of with time, 562
    decay curve for radioactive, 562
    determining thyroid function, 565
    dietary recommendations for deficiency, 93
    as element essential to health, 104
    half-life of, 561
    as halogen, 99, 100
    source of name, 94
Iodo, 588
Ion An atom or group of atoms having an electrical charge because of a loss or gain of electrons.
    in the body, 132
    in bones and teeth, 144
    as conductor of electricity, 380
    dissolving in water, 378
    electron configuration and, 289
    forming, 129, 130
    formulas and names of, 136
    octet rule and, 128
    polyatomic, 141
    shape of, 298
    size of, 278
    solubility of salt when present, 444
Ionic bond The attraction between oppositely charged ions. 128, 129, 310
Ionic charge The difference between the number of protons (positive) and the number of electrons (negative), written in the upper right corner of the symbol for the ion. 129
    from group number, 130
    writing ionic formula from, 135
Ionic compound A compound of positive and negative ions held together by ionic bonds. 128
    electron configuration and, 289
    flowchart for naming, 146, 153
    forming, 148
    guide to naming, 136
    guide to naming with polyatomic ions, 146
    guide to naming with variable charge metals, 138
    heat of vaporization and fusion, 317
    names and formulas for, 155
    property of, 133
    rules for naming, 147
    subscripts in formula of, 135
    writing formulas from name of, 139
Ionic equation An equation for a reaction in solution that gives all the individual ions, both reacting ions and spectator ions. 516
Ionic formula
    naming, 136
    writing from ionic charge, 135
Ionic solid
    melting, 309
    rules for solubility in water, 386
    structure of, 133
Ionization energy The energy needed to remove the least tightly bound electron from the outermost energy level of an atom. 117, 279
    increasing and decreasing, 280
    removing tightly bound electron, 279
Ionizing radiation

effects of, 546
properties of, 547
protection from, 547
Ion-product constant of water (K_w) The product of $[H_3O^+]$ and $[OH^-]$ in solution; $K_w = [H_3O^+][OH^-]$. 468
Iron (Fe)
    chemical reaction of, 199, 200
    composition of atoms in, 109
    corrosion of, 523
    dietary recommendations for deficiency, 93
    as element essential to health, 104
    as metal, 101
    specific heat of, 76
Isomer Organic compounds in which identical molecular formulas have different arrangements of atoms. 588
Isopropyl, 588
Isotope An atom that differs only in mass number from another atom of the same element. Isotopes have the same atomic number (number of protons) but different numbers of neutrons.
    atomic mass and, 110, 111
    patient ingesting, 543
    symbol for, 544
IUPAC (International Union of Pure and Applied Chemistry) system A system for naming organic compounds determined by the International Union of Pure and Applied Chemistry.
    naming alcohols, 608
    naming aldehydes, 615–616
    naming alkanes with substituents, 588–589
    naming alkenes and alkynes, 598
    naming amides, 630
    naming amines, 626
    naming aromatic compounds, 606
    naming carboxylic acids, 620
    naming continuous-chain alkanes, 586
    naming esters, 623
    naming ketones, 617
    using to name organic compounds, 586

J

Jewelry, 3
Joule (J) The SI unit of heat energy; 4.184 J = 1 cal. 73, 74

K

Kekulé, August, 605
Kelvin (K) scale A temperature scale on which the lowest possible temperature is 0 K, which makes the freezing point of water 273 K and the boiling point of water 373 K.
    comparing to Celsius and Fahrenheit scales, 66–67
    converting to Celsius (°C) scale, 70
    gas molecule and, 334
    as unit of measurement, 20, 21
Ketohexose, 647
Ketone A class of organic compounds in which a carbonyl group is bonded to two carbon atoms.
    carbonyl group and, 615
    containing carbonyl group, 596
    functional group and example of, 597
    naming, 617
    oxidation of secondary alcohol and, 612
    role in organic chemistry, 617
    types of, 618
Ketose A monosaccharide that contains a ketone group. 647

Kidney stone
    composition of, 441
    formation and prevention of, 445
    occurrence of, 390
Kilo (k), 31, 32
Kilocalorie (kcal), 74
Kilogram (kg) A metric mass of 1000 g, equal to 2.205 lb. The kilogram is the SI standard unit of mass. 20, 21
Kilojoule (kJ), 73, 74
Kilometer (km), 18
Kilopascals (kPa), 339
Kinetic energy A type of energy that is required for actively doing work; energy of motion.
    of gas molecule, 334
    losing as temperature drops, 312
    as motion, 72
    temperature of gas and, 336
Kinetic molecular theory of gas A model used to explain the behavior of gases. 334
Krypton (Kr), 99

L

Laboratory equipment, 7
Laboratory test, 18
Lactic acid, 622
Lactose A disaccharide consisting of glucose and galactose found in milk and milk products. 651, 652
Lambda (λ), 254
Lanthanide, 97
Laser, 8
Lauric acid, 658
Law of conservation of mass In a chemical reaction, the total mass of the reactants is equal to the total mass of the product; matter is neither lost nor gained. 226
LD50 value, 34
Lead storage battery, 525
Le Châtelier's Principle When a stress is placed on a system at equilibrium, the equilibrium shifts to relieve that stress. 430, 431, 432–433, 434
Length
    conversion factors with powers, 37
    equality of, 32, 33, 36
    as unit of measurement, 18–19, 21
    using ruler to measure, 24–25
Lethal dose, 34
Libby, Willard, 564
Light, 255
Light energy, 72, 258, 260
"Like dissolves like", 378, 379
Limestone, 477
Limiting reactant The reactant used up during a chemical reaction, which limits the amount of product that can form.
    calculating mass of product from, 239–241
    calculating moles of product from, 236–238
    examples of, 235–236
Linear The shape of a molecule that has two bonded atoms and no lone pair. 299
Lipid A family of compounds that is nonpolar in nature and not soluble in water; includes fats, waxes, and steroids.
    as necessary nutrient, 646
    solubility of, 658
Liquid A state of matter that has it own volume, but takes the shape of the container.
    adding heat to form gas, 318
    boiling point of, 314, 315, 348
    change of state, 311